New Spaces in Mathematics

After the development of manifolds and algebraic varieties in the previous century, mathematicians and physicists have continued to advance concepts of space. This book and its companion explore various new notions of space, including both formal and conceptual points of view, as presented by leading experts at the *New Spaces in Mathematics and Physics* workshop held at the Institut Henri Poincaré in 2015.

The chapters in this volume cover a broad range of topics in mathematics, including diffeologies, synthetic differential geometry, microlocal analysis, topos theory, infinity-groupoids, homotopy type theory, category-theoretic methods in geometry, stacks, derived geometry, and noncommutative geometry. It is addressed primarily to mathematicians and mathematical physicists, but also to historians and philosophers of these disciplines.

MATHIEU ANEL is a Visiting Assistant Professor at Carnegie Mellon University. His research interests include higher category theory, algebraic topology, and topos theory.

GABRIEL CATREN is Permanent Researcher in philosophy of physics at the French National Centre for Scientific Research (CNRS). His research interests include the foundations of classical and quantum mechanics, and the foundations of gauge theories.

New Spaces in Mathematics
Formal and Conceptual Reflections

Edited by

MATHIEU ANEL
Carnegie Mellon University

GABRIEL CATREN
CNRS - Université de Paris

CAMBRIDGE
UNIVERSITY PRESS

University Printing House, Cambridge CB2 8BS, United Kingdom

One Liberty Plaza, 20th Floor, New York, NY 10006, USA

477 Williamstown Road, Port Melbourne, VIC 3207, Australia

314–321, 3rd Floor, Plot 3, Splendor Forum, Jasola District Centre,
New Delhi – 110025, India

103 Penang Road, #05-06/07, Visioncrest Commercial, Singapore 238467

Cambridge University Press is part of the University of Cambridge.

It furthers the University's mission by disseminating knowledge in the pursuit of
education, learning, and research at the highest international levels of excellence.

www.cambridge.org
Information on this title: www.cambridge.org/9781108490634
DOI: 10.1017/9781108854429

© Cambridge University Press 2021

This publication is in copyright. Subject to statutory exception
and to the provisions of relevant collective licensing agreements,
no reproduction of any part may take place without the written
permission of Cambridge University Press.

First published 2021
Reprinted 2021

Printed in the United Kingdom by TJ Books Limited, Padstow Cornwall

A catalogue record for this publication is available from the British Library.

Library of Congress Cataloging-in-Publication Data
Names: Anel, Mathieu, editor. | Catren, Gabriel, 1973– editor.
Title: New spaces in mathematics : formal and conceptual reflections / edited
by Mathieu Anel, Carnegie Mellon University, Pennsylvania, Gabriel
Catren, Centre National de la Recherche Scientifique (CNRS), Paris.
Description: New York : Cambridge University Press, 2020. |
Includes bibliographical references and index.
Identifiers: LCCN 2020006667 | ISBN 9781108490634 (hardback) |
ISBN 9781108854399 (epub)
Subjects: LCSH: Mathematical physics–Research.
Classification: LCC QC20.8 .N495 2020 | DDC 539.7/258–dc23
LC record available at https://lccn.loc.gov/2020006667

ISBN - 2 Volume Set 978-1-108-85436-8 Hardback
ISBN - Volume I 978-1-108-49063-4 Hardback
ISBN - Volume II 978-1-108-49062-7 Hardback

Cambridge University Press has no responsibility for the persistence or accuracy of
URLs for external or third-party internet websites referred to in this publication
and does not guarantee that any content on such websites is, or will remain,
accurate or appropriate.

Contents

Contents for New Spaces in Physics   *page* vii

    Introduction   1

    Introduction   1
    Mathieu Anel and Gabriel Catren

    PART I   DIFFERENTIAL GEOMETRY   29

1   **An Introduction to Diffeology**   31
    Patrick Iglesias-Zemmour
2   **New Methods for Old Spaces: Synthetic Differential Geometry**   83
    Anders Kock
3   **Microlocal Analysis and Beyond**   117
    Pierre Schapira

    PART II   TOPOLOGY AND ALGEBRAIC TOPOLOGY   153

4   **Topo-logie**   155
    Mathieu Anel and André Joyal
5   **Spaces as Infinity-Groupoids**   258
    Timothy Porter
6   **Homotopy Type Theory: The Logic of Space**   322
    Michael Shulman

    PART III   ALGEBRAIC GEOMETRY   405

7   **Sheaves and Functors of Points**   407
    Michel Vaquié
8   **Stacks**   462
    Nicole Mestrano and Carlos Simpson

9	**The Geometry of Ambiguity: An Introduction to the Ideas of Derived Geometry** *Mathieu Anel*	505
10	**Geometry in dg-Categories** *Maxim Kontsevich*	554

Contents for *New Spaces in Physics*

Contents for New Spaces in Mathematics   *page* vii

Introduction   1
Mathieu Anel and Gabriel Catren

PART I   NONCOMMUTATIVE AND SUPERCOMMUTATIVE GEOMETRIES   21

1. **Noncommutative Geometry, the Spectral Standpoint**   23
   Alain Connes
2. **The Logic of Quantum Mechanics (Revisited)**   85
   Klaas Landsman
3. **Supergeometry in Mathematics and Physics**   114
   Mikhail Kapranov

PART II   SYMPLECTIC GEOMETRY   153

4. **Derived Stacks in Symplectic Geometry**   155
   Damien Calaque
5. **Higher Prequantum Geometry**   202
   Urs Schreiber

PART III   SPACETIME   279

6. **Struggles with the Continuum**   281
   John C. Baez

7	**Twistor Theory: A Geometric Perspective for Describing the Physical World**	327
	Roger Penrose	
8	**Quantum Geometry of Space**	373
	Muxin Han	
9	**Stringy Geometry and Emergent Space**	407
	Marcos Mariño	

Introduction

Mathieu Anel and Gabriel Catren

Contents

1	A Brief History of Space	1
2	Contemporary Mathematical Spaces	3
3	Summaries of the Chapters	12
Acknowledgments		22
References		22

1 A Brief History of Space

Space is a central notion in both mathematics and physics and has always been at the heart of their interactions. From Greek geometry to Galileo experiences, mathematics and physics have been rooted in constructions performed in the same ambient physical space. But both mathematics and physics have eventually left the safe experience of this common ground for more abstract notions of space.

The 17th century witnessed the development of projective geometry and the strange, yet effective, idea of points at infinity. This century witnessed also the advent of analytic geometry (with its use of coordinates) with Descartes and of differential calculus with Newton and Leibniz. Both have led to an approach to geometry fundamentally based on the manipulation of algebraic formulas. The capacity to manipulate spaces without relying on a spatial intuition has laid the foundations for one of the most important revolutions in geometry: the conception of spaces of arbitrary dimension.

During the same time, the successful geometrization of astronomy, optics, and mechanics anchored physics to the paradigm of a differential Euclidean space.

The 18th century essentially develops differential calculus for spaces of high dimensions. Analytic geometry and physics converge into analytical mechanics. This is a revolution that introduces abstract spaces in physics (like the six-dimensional spaces of trajectories) and reduces the actual physical space to be a mere starting point out of which other relevant spaces can be constructed. In mathematics, the invention of complex numbers laid the groundwork for the future algebraic geometry. In turn, the contradictions of logarithm theory and the study of polyhedra and graphs planted the seeds of algebraic topology. Moreover, analytical methods forecast a new notion of space: the infinite-dimensional spaces of functions.

In the 19th century, geometry exploded in diversity. The use of local coordinates in analytical mechanics gave rise to the intrinsic theory of manifolds and the fundamental local–global dialectic. The points at infinity, the points with complex coordinates, and the multiple points of intersection theory are all unified in the framework of algebraic geometry. The geometric study of linear equations led to the notion of vector space. The development of Lie group theory created a completely new branch of geometry centered on the characterization of the symmetries of spaces. The construction of models for non-Euclidean geometry revived the old synthetic/axiomatic geometry, and the development of analysis prepares the notions of metric spaces. In physics, thermodynamics and electromagnetism are successfully developed within the framework of the differential calculus of $\mathbb{R}^n$. The latter entangles space with time in an unusual way, but the geometric paradigm of classical mechanics remains well secured.

The mathematics of the 20th century started with the successful definition of topological spaces. At the heart of the notion of space are now the set of its points, the open subsets, and the continuous-discontinuous dialectic. From function spaces to manifolds and unseparated spaces, topological spaces are powerful enough to unify many kinds of spaces. Lie groups and the paradigm of symmetry are also everywhere, from differential equations to manifolds and linear algebra. Another major revolution was the discovery that the spaces of high dimensions have specific shapes and can be different from each other. This qualitative study of spaces gave birth to algebraic topology and its two branches of homotopy and homology theories. With higher-dimensional spaces and algebraic topology, figures have essentially disappeared from geometry books, and geometry has become the study of spaces that cannot be "seen" anymore.

Introduction

In physics, the contradictions raised by the constancy of the speed of light and the spectrum of the black body were the source of a schism on the role of space. Relativity grounded the geometry of physics in the new dynamical object that is spacetime and successfully formalized gravity in purely geometrico-differential terms. Yang–Mills theory extended this program to the other (electromagnetic and nuclear) fundamental interactions. On the other side, the formalism of quantum mechanics required abandoning geometric intuition and, rather, focusing on algebras of operators. Despite the fundamental role played by symmetries and Lie group theory in both theories, the geometric unity of physics was to a certain extent broken.

By the middle of the 20th century, mathematics and physics are much better structured than they were at the beginning. The notions of sets, topological spaces, manifolds (Riemannian or not), algebraic varieties, and vector spaces organize the geography of mathematical spaces. General relativity, classical mechanics, and quantum mechanics divide the physical space in three scales, each with its own geometric formalism. The situation is summarized in Tables 0.1 and 0.2. All things seems to fall into place, and, for our purposes in this book, we shall refer to this situation as the *classical paradigm of space*. The conceptual categories that organize this paradigm on the mathematical side are points, open and closed subsets, coordinates and functions, local/global, measure of distances, continuity/discontinuity, infinitesimal variations, and approximation. On the physical side, the classical paradigm relies on a differentiable spacetime, trajectories and fields, infinitesimal equations, and symmetries and covariance. The intuition of space has been pushed far away from the original intuition of the ambient physical space, but in a clear continuity.

The evolution of the notion of space in mathematics and physics has continued until now. However, the results of these developments are less universally known in the mathematical and physical communities where the common background stays, even nowadays, the classical paradigm. It is the purpose of this book and its companion to illustrate and explain some of these "postclassical" developments.

2 Contemporary Mathematical Spaces

2.1 Algebraic Topology

One of the most important geometric achievements of the postclassical period is the revisitation of *algebraic topology* (homotopy and homology theory) in terms of higher category theory. Homotopy theory evolved from the definition

Table 0.1. *The "classical" kinds of mathematical spaces*

Basic structures	Topology	Differential geometry	Linear spaces	Algebraic geometry
Sets	Topological space	Differential manifold, Lie groups	Vector spaces, function spaces, modules over rings	Algebraic variety, algebraic groups
Preorder, equivalence relations	Metric space	Riemannian manifold	Banach, Fréchet, Hilbert... spaces	

Table 0.2. *The "classical" kinds of physical spaces*

	General relativity (large scale)	Mechanics and thermodynamics (medium scale)	Quantum mechanics (small scale)
Ambient space and time	Lorentzian 4-manifold	Galilean spacetime $\mathbb{R} \times \mathbb{R}^3$	Galilean or Poincaré Lie group
Phase spaces	Spaces with action of the local Poincaré group	Manifolds with action of the Galilean group	Representations of various Lie groups in Hilbert spaces

of the fundamental group of a space (and its applications to classify covering spaces and to explain the multiple values of analytic continuations) to a general study of continuous maps and spaces up to continuous deformations (homotopies and homotopy equivalences) [25, 43]. The central object ended up to be that of the *homotopy type* of a space, that is, the equivalent classes of this space up to homotopy equivalence. Homological algebra evolved from a computation of numbers and groups to a calculus of resolutions of modules over a ring (or sheaves of such) [16, 63]. The notion of abelian category put some order in this calculus [13, 36], but it is only with triangulated categories that a central object emerged: *chain complexes* up to quasi-isomorphisms [23, 93]. In a separated approach, the axiomatization of homology theories in terms of functors had also led to a new kind of object: *spectra*, of which chain complexes are a particular instance [1, 57, 77]. Any space defines both a homotopy type and a spectrum (its *stable homotopy type*), but until the 1970s, the nature of these two objects was somehow elusive.

The development in the 1970s of *homotopical algebra* (i.e., model category theory) provided for the first time a unified framework for both homotopy and homology theories [31, 73]. But even with this unification, the theory was still highly technical and, many times, ad hoc. The concepts that revealed the meaning of these constructions were only found in the 1980s, when higher category theory emerged [11, 28, 38, 72]. The main progress was to understand that homotopy types of spaces were the same thing as ∞-groupoids, that is, a particular kind of higher category in which all morphisms are invertible (see Chapter 5). By viewing homotopy types as ∞-groupoids, it was possible to revisit homotopical algebra from the standpoint provided by the whole conceptual apparatus of higher category theory. This has provided conceptual simplification of many of the homotopical constructions, but this story lies beyond the scope of this book (see [18, 61]). We have limited our study to the utilization of ∞-groupoids in geometry, namely, in topos theory (Chapter 4), in stack theory (Chapter 8), and in the theory of derived schemes (Chapter 9). We have also included a chapter explaining how ∞-groupoids have permitted us to revisit the foundations of mathematics (Chapter 6). Moreover, Chapters 4 and 5 of the companion volume, *New Spaces in Physics*, show how ∞-groupoids are useful in symplectic geometry and physics.

2.2 Algebraic Geometry

The field of geometry that has undergone the deepest postclassical transformation is *algebraic geometry*. From the 1950s to the 1980s, Grothendieck's school brought many definitions and improvements for the objects of algebraic geometry. The definition of Zariski spectra and schemes as ringed spaces permitted for the first time the unification of all the notions of algebraic varieties. Moreover, the notion of affine scheme provided a perfect duality between some geometric objects and arbitrary commutative rings of coordinates [39]. An important difference that schemes have with manifolds is the fact that they can accommodate singular points. This singular structure is encoded algebraically by the existence of nilpotent elements in the ring of local coordinates, a feature that is possible only if arbitrary rings are considered. Nilpotent elements provide an efficient infinitesimal calculus, which is one of the nicest achievements of algebraic geometry (see [21] and volume IV of [39]).[1]

---

[1] This calculus is also at the core of *synthetic differential geometry*; see Chapter 2.

The theory of ringed spaces was efficient to define general schemes by pasting of affine schemes. However, motivated by the study of algebraic groups and the construction of moduli spaces, schemes were almost immediately redefined as functors, making the previous construction somehow superfluous (see Chapter 7 on the functor of points and [24, 37]). Later on, the definition of étale spectra of rings (which was needed to define cohomology theories with étale descent) came back to a definition in terms of ringed spaces with the difference that the base space was now a *topos* (see Chapter 4 and [5, 23, 64]). The functorial point of view continued to be used simultaneously.[2]

The definitive approach to constructing *moduli spaces* (e.g., the spaces of curves or bundles on a given space) was eventually found with *stacks*, which are a variation on the notion of sheaf (see Chapter 8 and [3, 22, 33, 38, 55, 81]). Essentially, stacks provide a notion of space where the set of points is enhanced into a groupoid of points. This feature makes them perfectly suited to classifying objects (such as curves or bundles) together with their symmetries. From a geometric point of view, stack theory is a formalism intended to deal properly with the possible singularities created by taking a quotient (see Chapter 9).

The most recent development has been *derived algebraic geometry*. This formalism enhances the theory of stacks in order also to tame the singularities created by nontransverse intersections (see Chapter 9 and [62, 86, 87]). At the end of the story, derived algebraic geometry provides by far the most sophisticated notion of space ever invented.[3] Derived stacks have become a powerful archetype for a new paradigm of geometric spaces (see Chapter 9 and [48, 58, 70, 84, 88]). However, so many turns in only 60 years have been hard to follow, and the community of algebraic geometers is largely spread out between different technologies and viewpoints on its objects.

An important field related to algebraic geometry is *complex geometry*. In comparison with their differential analogs, complex manifolds have the problem that they admit too few globally defined holomorphic functions. This has deeply grounded the field in sheaf theory and cohomological methods and kept it close to algebraic geometry, where the same methods were used for similar reasons [34, 44, 79, 94]. Nonetheless, complex manifolds have

---

[2] For example, the notion of a connection on a singular scheme X was successfully defined by means of the *de Rham shape* of X, which is the quotient of X by the equivalence relation identifying two infinitesimally closed points. The result of such a quotient is not a scheme, but it can be described nicely as a sheaf on schemes (see Chapters 4 and 5 of *New Spaces in Physics* and [21, 80]).

[3] Algebraic geometry was able to deal successfully with "multiple points with complex coordinates at infinity"; derived algebraic geometry added to these features the possibility to work with quotients by nonfree group actions and self-intersection of such points.

not really evolved into more sophisticated types of spaces (incorporating singularities and points with symmetries). The recent rise of *derived analytic geometry* might change this [59, 70, 71].

Algebraic geometry depends on the existence of a well-defined dictionary between the geometric features of affine schemes and the algebraic features of commutative rings. This successful translation has led to several attempts to generalize it for other kinds of algebraic structures. The most famous attempt is given by the geometry of noncommutative rings. The attempts to build an actual topological space (a spectrum) from noncommutative rings have not been entirely satisfactory [74, 91],[4] but the dual attempt to characterize geometric features in noncommutative terms has had more success (see Chapter 10 and Chapter 1 of the companion volume, and references therein). However, some important geometric notions are absent from both these approaches (e.g., open subsets, étale maps, the local/global dialectic), preventing a geometric intuition of noncommutative features in classical terms. Other offsprings of algebraic geometry have been *relative geometry*, which develops a geometry for various contexts of commutative monoids (see Chapter 7 and [88]), the geometry of *Berkovich spaces* dual to non-Archimedean fields [9, 12], the *tropical geometry* dual to tropical semirings [35, 66], and the conjectural *geometry over the field with one element* [19, 27, 82].

2.3 Topology

The notion of topological space has been robust enough to successfully deal with some of the new spaces invented in the second half of the 20th century, such as fractals, strange attractors, and nonseparated spaces (such as the Zariski spectrum of a commutative ring). Even the study of topological spaces by means of rings of continuous functions (motivated by Stone and Gelfand dualities) has not introduced new objects [32].

Nonetheless, new spaces have been invented for the needs of topology. For example, the close relationship between topology and intuitionist logic à la Heyting has led to *locale theory*, a variation on topological spaces well suited to define interpretations of logical theories (see Chapter 4 and [45, 92]). Also, in algebraic geometry, the remarkable analogy between Galois theory of fields and the theory of covering spaces [26] has motivated the search for a functor associating a topological space to a commutative ring (a spectrum) that could transport, so to speak, the former theory into the latter. The Zariski spectrum fails to satisfy this, and the proper answer was found with the *étale*

---

[4] Mostly by lack of functoriality of the spectra.

spectrum. However, étale spectra could no longer be defined as topological spaces anymore but rather were defined as topoi [5, 23]. Essentially, a *topos* is a new kind of space defined by its category of sheaves instead of its poset of open subspaces. This broader definition led to many new topological objects that are not topological spaces (see Chapter 4 and [46, 64]).

Another motivation for enhancing the notion of topological space was the study of badly separated spaces [2], for example, spaces that have many points but a trivial topology, such as the irrational torus $\mathbb{T}_\alpha := \mathbb{R}/(\mathbb{Z} \oplus \alpha\mathbb{Z})$ ($\alpha \notin \mathbb{Q}$), the leaf spaces of foliations with dense leaves, or even bizarre quotients like $\mathbb{R}/\mathbb{R}_{dis}$ (the continuous $\mathbb{R}$ quotiented by the discrete $\mathbb{R}$). The theory of topoi turned out to be well suited to studying these spaces.[5] But other methods have been developed, like topological sheaves and stacks (inspired by algebraic geometry) [10, 17, 20] or noncommutative geometry à la Connes (see Chapter 1 of the companion volume, and references therein), diffeologies (see Chapter 1, and references therein), or orbifolds and Lie groupoids [56, 67, 75, 90].

2.4 Differential Geometry

Differential geometry has not escaped the development of new types of spaces, but the size of the field has perhaps kept most of it within the classical paradigm. From Riemannian geometry to knot theory, the basic notion is still that of the manifold. Overall, the field does not seem to be in a hurry to incorporate the developments of algebraic geometry (e.g., duality algebra/geometry, singular spaces, functorial approach to moduli spaces and infinite dimensions, relativization with respect to a base space, tangent complexes). Many attempts have been made to improve manifolds, but none of them seems to have become central. An example is *diffeology* theory, which provides a nice framework to deal with infinite-dimensional spaces as well as quotients (see Chapter 1, and references therein). Another one is *synthetic differential geometry*, which enhances the notion of the manifold by authorizing singular points and nilpotent coordinates (see Chapter 2, and references therein).[6] Related approaches have tried to ground differential geometry in the algebraic notion of the C^∞-ring [48, 68, 69]. The most

---

[5] They are called *étendues* in topos theory; see [5, 46].
[6] Synthetic differential geometry, as its name suggests, also promulgates an axiomatic approach to geometry.

successful new notion of differentiable space is perhaps that of *orbifolds*, motivated among other things by Thurston's geometrization program [75, 85]. Orbifolds have brought to the field some tools from higher category theory like stacks [56, 67] and equivariant homotopy theory [76].

Another domain using such methods is *microlocal analysis*, where sheaves and their derived categories are of great help for dealing with the problem of extending local solutions of differential equations (see Chapter 3, and references therein).

The most impressive display of postclassical methods in differential geometry can be found in *symplectic geometry* (together with the related fields of *Poisson* and *contact geometries*). Symplectic geometry is a contemporary descendent of analytical mechanics. The notions of symplectic manifold and their Lagrangian submanifolds have given a new geometrical meaning to many constructions of mechanics (e.g., extremal principles and generating functions, covariant phase spaces, Nœther symmetries and reduction [40, 50, 83]). A central operation in the theory is *symplectic reduction*, which combines the restriction to a subspace of a symplectic manifold with a group quotient [65]. Since these two operations might create singularities, symplectic geometry has been forced to deal with both nontransverse intersections and quotients of non-free group actions. These issues have led to the use of new formalisms, such as cohomological methods [41, 54], Lie groupoids and stacks [95, 96], and, eventually, derived geometry (see Chapter 4 of the companion volume and [14, 89]). Also, the application of symplectic geometry to physics has imported many methods from higher category theory: cohomological methods in deformation quantization [15, 53], Fukaya categories in mirror symmetry [52, 78], and, more recently, a whole new interpretation of gauge theory in terms of stacks (see Chapter 5 of the companion volume, and references therein). In fact, more than a simpler user of higher categories and derived algebraic geometry, symplectic geometry has been an important catalyzer in the development of these theories.

Another important innovation with respect to the notion of manifold has been the interpretation of manifolds with boundaries and cobordisms in terms of *higher categories with duals*, a viewpoint that was inspired by topological field theories in physics [6, 7, 60]. In the same way that homotopy theory has transformed topological spaces into tools that can be used to work with ∞-groupoids, this view on cobordisms does not address manifolds as an object on its own but rather as a tool to encode the combinatorial structure of some higher categories.

2.5 Conclusion

We have referred to the understanding of the notion of space in the middle of the 20th century as the "classical paradigm." This raises the question of whether the many evolutions undergone by the notion of space since then qualify as a new paradigm. As the previous presentation and Table 0.3 illustrate, the classical geography of mathematical spaces is still pertinent today. The conceptual categories organizing the intuition of space have not fundamentally changed (points, functions, local/global dialectic, etc.). If something radical has changed, it will not be found there.

In our opinion, the most important postclassical change has in fact not concerned spaces directly – although it had a tremendous impact on them – but *sets*. If there has been a paradigm shift in mathematics, it has been the enhancement of *set theory* in *category theory* (in which we include higher categories). Category theory is responsible for most of the new spatial features:

1. The most important change has been that sets of points have been enhanced in categories of points (in particular, points can have symmetries).
2. The definition of a space by means of a poset of open subsets has been enhanced in a definition by means of categories of sheaves (topoi, dg-categories, stable categories, etc.).
3. Functions with values in set-based objects (numbers, manifolds, etc.) have been enhanced by functions with values in category-based objects (stacks, moduli spaces, etc.).
4. Many spaces are defined as functors (schemes, moduli spaces, stacks, diffeologies, etc.).
5. Homotopy types are now seen as ∞-groupoids.
6. Also, the relation with logic and axiomatization is made by means of categorical semantics for logical theories.

In the classical paradigm, sets can be thought of as the most primitive notion of space – collecting things together in a minimalist way – from which other notions of space are formally derived. In the new paradigm, categories, and particularly higher categories, are the new primitive spatial notion from which the others are derived. Nowadays, categories are everywhere in topology and geometry, from the definition of the basic objects to the problems and methods of study. The reader will realize that category theory is central in *all* the chapters of this volume.

Table 0.3. Classical and "new" kinds of mathematical spaces

	Basic structures	Topology	Differential geometry	Linear spaces	Algebraic geometry	New geometries
Classical notions (1-categorical)	Sets	Topological space	Differential manifold, Lie groups	Vector spaces, function spaces, modules over rings	Algebraic variety, algebraic groups	—
	Preorder, equivalence relations	Metric space	Riemannian manifold	Banach, Fréchet, Hilbert... spaces		
New 1-categorical notions	—	Locales	Diffeologies, SDG, C^∞-manifolds	—	Zariski spectra, schemes, algebraic spaces	Noncommutative spaces (Connes), super-manifolds, tropical geometry
New ∞-categorical notions	Groupoids, cellular complexes, ∞-groupoids (homotopy types)	Topoi, ∞-topoi	Lie groupoids, orbifolds, stacks, moduli spaces, derived manifolds	Chain complexes, spectra (homology theories)	Étale spectra, algebraic stacks, derived schemes	Noncommutative spaces (Kontsevich)

3 Summaries of the Chapters

3.1 Part I Differential Geometry

3.1.1 An Introduction to Diffeology (Patrick Iglesias-Zemmour)

The theory of diffeologies – started by the French mathematician J.-M. Souriau in the early 1980s – provides a formal setting in which the main tools of differential calculus can be extended to infinite-dimensional spaces (such as spaces of functions between manifolds or symmetry groups of manifolds). Recall that a manifold M of dimension n can be described by the data given by all differentiable maps $\mathbb{R}^n \to M$ that are open immersions. A diffeology X will be similarly described by the data of all differentiable maps $\mathbb{R}^n \to M$ (called *plots*), but without the assumption that the maps $\mathbb{R}^n \to X$ have to be open immersions, and without the restriction that n has to be fixed. For example, if E is an infinite-dimensional topological vector space, the corresponding diffeology is defined by the data given by all (nonlinear) differentiable maps $\mathbb{R}^n \to E$ for all $\mathbb{R}^n$. In analogy with the fact that an infinite set is always the union of its finite subsets, a diffeology can be understood as the "union" of all its finite-dimensional plots.[7]

This definition permits the definition of fiber bundles, differential forms, de Rham cohomology, and other classical notions of manifold theory. The methods to do so are very close to the sheaf theoretic methods of Chapter 7 but with a more classical flavor. Diffeologies provide an efficient setting extending the classical notion of manifolds at a rather low technical cost. Despite their original application to infinite-dimensional spaces, they have proved also to be well suited to defining the differentiable structure of some "bad quotients" (such as the irrational torus $\mathbb{T}_\alpha = \mathbb{R}/(\mathbb{Z} \oplus \alpha\mathbb{Z})$ for $\alpha \notin \mathbb{Q}$ or other leaf spaces of dense foliations) and of manifolds with boundaries.

3.1.2 New Methods for Old Spaces: Synthetic Differential Geometry (Anders Kock)

Synthetic differential geometry (SDG in what follows) started with Lawvere's work on continuum mechanics in the 1960s and relies on two main ideas. First, SDG provides a *synthetic* – or *axiomatic* – framework for differential geometry. In the same way that points and lines are just assumed and not constructed in Euclidean geometry, manifolds are just assumed collectively as primitive objects in SDG. This idea is opposed to the analytic description of manifolds

---

[7] This point of view becomes clearer when diffeologies are defined as sheaves over the category of manifolds [8]. In this framework, the notion of "union" is given by the categorical notion of *colimit*. By the Yoneda lemma, any sheaf on a category C is always a colimit of objects of C. A diffeology is then a colimit of $\mathbb{R}^n$s.

individually in terms of coordinates. The central object of SDG is a ring object R playing the role of the field $\mathbb{R}$ of real numbers. The axioms are chosen so that the theory recovers all classical constructions (tangent vectors, differential forms, connections) and more.

The second idea is to provide a setting encompassing manifolds with singularities (like the cusp $\{(x, y)|x^2 = y^3\}$). The definition of singular objects is inspired from algebraic geometry, where singular points can be defined by the property to have local coordinates that are *nilpotent*. Having nilpotent elements is the main difference between R and $\mathbb{R}$ in SDG. The entire differential calculus can be deployed from these elements. For example, the subspace $D_1 = \{x \in R | x^2 = 0\}$ of elements of square zero plays the role of the first-order infinitesimal neighborhood of 0 in R. Then a tangent vector of a manifold M is simply a map $D_1 \to M$. The space D_1 has a canonical point, which is 0, and the base point of the vector is simply the image of 0.[8]

The classical construction of $\mathbb{R}$ does not allow nilpotent elements, and the requirement of SDG may seem strange.[9] But the definition of tangent vectors shows that their introduction simplifies classical constructions. Other examples are given by the pleasant definition of differential forms and affine connections (see the chapter).

3.1.3 Microlocal Analysis and Beyond (Pierre Schapira)

An important problem with differential equations is to know if a solution over a domain U can be extended over a bigger domain V, and this problem can naturally be formulated in terms of sheaves. Recall that the notion of a sheaf on a space X encodes the data of local functions defined on X. Let U be an open domain, σ a section of a sheaf F on U, and x a point of the boundary of U. The problem at stake can be locally formulated as follows: *is it possible to find a neighborhood of x and an extension of σ in this neighborhood?*

By developing the work of Sato and Hormander from the 1970s, Kashiwara and Schapira's microlocal analysis tackles this question on manifolds. The differentiable structure of a manifold M provides tools to answer the problem. If U has a smooth boundary, the tangent hyperplane at x is always the kernel of a differential form p that is negative on U and positive outside (sometimes called the codirection of the hyperplane). The question about the propagation of sections can then be formulated in terms of p and no longer U. The

---

[8] Recall that in classical differential calculus, a vector is defined as an equivalence class of paths having the same 1-jet. But this definition cannot work at singular points, since singular points are precisely points with 1-jets not integrable into actual paths.

[9] It might help to look at the explicit model, closer to the classical analytical approach, given in terms of sheaves in [68].

microsupport of a sheaf F is the set of points (x, p) of T^*M through which the sections of F cannot be extended uniquely. In other terms, microlocal analysis introduces a notion of locality that refers not only to the points of the manifold (classical locality) but also to the codirections around that point (microlocality).

A remarkable result is that the microsupport of a sheaf F is within the zero section of T^*M if and only if F is a locally constant sheaf. This is analogous to the fact that the graph of the differential df of a function $f : M \to \mathbb{R}$ is in the zero section of T^*M if and only if f is a locally constant function. This fact suggests that the microsupport can be understood as a sort of "derivative" of the sheaf F. From this perspective, microlocal analysis may be the beginning of a differential calculus for sheaves on manifolds.

3.2 Part II Topology and Algebraic Topology

3.2.1 Topo-logie (Mathieu Anel and André Joyal)

This chapter is about the two evolutions of the notion of topological space, which are *locales* and *topoi*. The *theory of locales* – also known as *point-free topology* – is rooted in the close relationship between topology and intuitionist logic. The main idea in defining a locale is to forget the underlying set of points of a topological space (hence the name of the field) and to define locales directly by their *frames* of open subsets, which are commutative ring-like structures. The category of locales is then formally defined as the opposite of that of frames. The theory is then based on a dictionary between geometric features of locales and their translation in algebraic features of frames, very much as in algebraic geometry. Locales have some new features compared to topological spaces. For example, any intersection of dense subspaces is always dense, and there exist nontrivial locales with an empty set of points.[10] Overall, locales provide a nicer topological setting than topological spaces, but the two notions are too close, and the latter is too well established for the former to pretend to replace it.

The notion of *topos* is similarly defined as dual to an algebraic structure that the authors call a *logos*. A logos is intuitively a category of sheaves on a space, and it is equipped with operations that make it look like a commutative ring. As for locales, the theory of topoi provides a dictionary between geometric and algebraic features. Every topological space or locale defines a topos, but there are much more topoi than topological spaces. A big difference between topoi and classical spaces is that the former can have a category of points instead

---

[10] For this reason, it is a pun to refer to point-free topology as "pointless" topology [47].

of a mere set. In particular, there exists a topos $\mathbb{A}$ whose category of points is the category of sets. This topos plays a central role in the theory since one can describe the logos $Sh(X)$ of sheaves on a topos X as the category of morphisms of topoi $X \to \mathbb{A}$. In other words, a topos is the object dual to an algebra of functions with values in the "space of sets."

The theory of topoi has mostly been popular in logic, where it turned out to be well suited to providing interpretations of higher-order theories. The topological aspects of topos theory are less known. Nevertheless, the theory of topoi is also quite useful in topology, where it can encode badly separated spaces, such as foliation spaces[11] or some moduli spaces. It also provides a nice setting where the homotopy and homology theories of spaces can be defined.

3.2.2 Spaces as Infinity-Groupoids (Timothy Porter)

The homotopy theory of topological spaces grew from the study of the fundamental group $\pi_1(X)$ of a connected space X to the definition of a whole collection of homotopy groups $\pi_n(X)$ indexed by natural numbers. For this reason, homotopy types of spaces (i.e., topological spaces up to weak homotopy equivalence) were first understood as an algebraic structure akin to groups. As Porter explains in his chapter, the quest for this structure has led to many definitions, but they were never able to encompass all homotopy types. Significant progress was made when it was understood that the notion of algebraic structure based on functional relations (such as the composition law) and conditions written as equations (such as associativity) was too strict. Another kind of algebra was needed to capture the features of homotopy theory: functions had to be multivalued (correspondences), and equations had to be replaced by the existence of paths.

The problem starts when one tries to define the composition of paths in a space: not only are there many ways to define such a composition but none of them are associative. The classical solution is to look at paths up to homotopy for which the composition becomes uniquely defined and associative. However, this strategy truncates the higher homotopical structure and only captures $\pi_1(X)$. The insight was to recognize that the existence of multiple compositions for paths was not a problem but a feature of the theory. The composition does not exist uniquely, but between two choices of compositions, there always exists a homotopy; moreover, between any two such homotopies, there always exists a higher homotopy, and so on. In other words, the regular structure was found when all the possible compositions were

---

[11] See the notion of *étendue* in [5, 46].

considered together and not individually: the composition becomes "unique" because the space of compositions is *contractible*. The same idea can be used to deal with associativity.[12] The existence of homotopies is formally encoded by *lifting conditions*, for example, those defining *Kan complexes*, which are one of the best definitions of ∞-groupoids.

Another important step forward was made when it was understood that homotopy types of spaces were a particular kind of higher category called ∞-groupoids. Even if this equivalence does not entail a simplification of the definition of homotopy types/∞-groupoids, it has permitted us to understand many constructions of homotopy theory in light of concepts coming from higher category theory.

3.2.3 Homotopy Type Theory: The Logic of Space (Michael Shulman)

The most unexpected consequence of the idea that homotopy types are ∞-groupoids has been in foundations of mathematics. The development of ∞-groupoids in algebraic topology and algebraic geometry has produced algebraic objects such as groups or rings with an underlying ∞-groupoid instead of an underlying set. This has led topologists and geometers to the idea that ∞-groupoids are objects as fundamental as sets and could be used as a primitive notion (or "background structure" in Shulman's terms) to build other mathematical objects.

A similar idea was found in logic when Martin-Löf's theory of *dependent types with identity types* – which has been designed as a language for set theory – was given a successful interpretation in terms of ∞-groupoids by Awodey–Warren [4] and Voevodsky [42, 49]. More precisely, the homotopical idea that the paths between two elements a and b of a space X are the elements of the path space $\Omega_{a,b}X$ turned out to be perfectly suited to encoding the logical idea that proofs of equality between two terms a and b of a type X should be the terms of an identity type $a =_X b$. *Homotopy type theory* is the offspring of homotopy theory and type theory based on this idea.

From a logical perspective, this homotopical interpretation has provided a completely new understanding of types and their axioms. In particular, thinking of the proofs of an equality $a =_X b$ as paths has explained why such a proof need not be unique (not all paths are homotopic).[13] For the

---

[12] This idea has been promoted to an important working philosophy: whenever some choice has to be made, the good choice is to consider all choices together.

[13] When types are interpreted as sets, the type $a =_X b$ is either empty or a one-point set; however, this fact cannot be deduced from the axioms of type theory. This posited a long-standing puzzle. The homotopical interpretation solved it since it provided a model where this is not true.

mathematician, the homotopic semantic of type theory has offered the luxury of a formal language to work with homotopy types, which is an alternative to higher category theory, independent of set theory and implementable on proof assistants.

In addition to semantics in terms of sets, logical theories and type theories also have interpretations in terms of topological spaces. When these semantics are crossed with the homotopical semantics, these give rise to interpretations in higher topological stacks, that is, ∞-groupoids enriched over topological spaces. Moreover, variations can be defined where stacks are defined in the context of differential geometry or algebraic geometry. In this way, type theory provides a common language for a variety of geometric contexts. It is in this sense that type theory is indeed a "logic of spaces."

3.3 Part III Algebraic Geometry

3.3.1 Sheaves and Functors of Points (Michel Vaquié)

Locally, a differential manifold looks like $\mathbb{R}^n$, and any manifold can be obtained by pasting the elementary pieces $\mathbb{R}^n$. Moreover, every manifold can be embedded in some sufficiently large $\mathbb{R}^n$. The situation is not as simple in algebraic geometry. Only the affine schemes dual to commutative rings can be embedded in affine spaces $\mathbb{A}^n$. Also, because schemes can have singularities, it is false that every scheme is locally like $\mathbb{A}^n$.[14] A general scheme is defined as a pasting of affine schemes, and, unfortunately, there is no simpler and smaller class of objects (like the $\mathbb{R}^n$s or the handlebodies of Morse theory) that could also generate all schemes.

This is where the general methods of category theory are helpful. Given the category Aff of affine schemes, there is a simple description of all possible pastings of objects of Aff: they are the presheaves on Aff.[15] Presheaves are functors, and this is where the *functor of points* approach has its roots. Any presheaf is the pasting of a diagram of affine schemes, but not every presheaf is nice enough to be considered as a geometric object (for instance, they do not all have tangent spaces). This is why presheaves satisfying extra *geometric conditions* are considered. Among these conditions, there is always a *sheaf condition*, which guarantees that the embedding of affine schemes into the geometric presheaves preserves the pastings of affine schemes that are affine. The other conditions are conditions on the type of pasting producing geometric

---

[14] This is only true for smooth schemes and for the étale topology.
[15] Presheaves are functors Aff^{op} $\to$ Set. The category of presheaves on a category C has the universal property to be the free completion of C for colimits, i.e., for pasting.

objects (for example, the only pastings allowed for schemes are along open subsets).

The framework given by the "functors of points" is efficient enough to provide a definition of schemes different than the classical presentation by ringed spaces. This functorial framework turned out to be perfectly suited to study *moduli spaces* classifying some structure S (e.g., the Hilbert schemes that classify all the closed subschemes of a given scheme). By definition, moduli spaces define almost tautologically a presheaf on Aff: the value of the functor on an affine scheme X is the set of families of objects with the structure S parameterized by X (see the next chapter on stacks for an example). Once this presheaf is defined, the *moduli problem* is to know whether it is the functor of points of a scheme (or another kind of geometric object, such as an *algebraic space* [51]). This setting as been efficient to construct many moduli spaces: projective spaces, Grassmannians, flag manifolds, Hilbert schemes, Picard schemes, and so on.

3.3.2 Stacks (Nicole Mestrano and Carlos Simpson)

When a moduli space is intended to classify objects that have symmetries, the sheaves with values in *sets* have to be replaced by sheaves with values in groupoids (*stacks*), where the morphisms encode the corresponding symmetries. The paradigmatic example in Mestrano and Simpson's chapter is given by the *moduli space of curves*.[16] A family of curves parameterized by an affine scheme X[17] is a bundle $E \to X$ whose fibers are curves. Such a family can intuitively be thought of as a function on X with values in some "space of curves" M. The *moduli problem* of curves is to construct this space M. The corresponding functor of points sends an affine scheme X to the set $\mathcal{M}(X)$ of bundles of curves over X, and the question is whether there exists a scheme M such that $\mathcal{M}(X)$ is in bijection with the set of morphisms $X \to M$. Unfortunately, the answer is negative. A necessary condition for the functor $\mathcal{M}$ to be represented by a scheme M is that it be a sheaf, but this is not the case. The problem comes from the fact that the data of bundles on an open cover U_i of X is patched together into a bundle on the whole of X by using *isomorphisms* on $U_i \cap U_j$. But the sheaf condition only patches them up to *equality* on $U_i \cap U_j$, which is too strict. This problem can be bypassed by

---

[16] The problem at stake is really to study Riemann surfaces. They are called *curves* by algebraic geometers because such surfaces are in fact of dimension 1 relative to the field of complex numbers.

[17] The reader unfamiliar with scheme theory can assume that they are manifolds since the peculiarities of schemes will not play a role here. Although the chapter presents the moduli problem of curves in the framework of algebraic geometry, the problem can also be formulated in differential or complex geometries.

incorporating the data associated to the isomorphisms between bundles into the values of the functor $\mathcal{M}$. This means that $\mathcal{M}(X)$ – rather than being a mere set – is now the *groupoid* of curve bundles together with their isomorphisms. This new functor does satisfy a sheaf-like condition: the *stack condition* (see the chapter). When the identifications between the objects are unique, the functor is valued in groupoids that are just sets, and the stack is simply a sheaf.

An important class of stacks comprises the *geometric stacks* (e.g., Deligne–Mumford and Artin stacks). These are the stacks that admit *local coordinates* (*atlases*) and for which it is possible to define tangent spaces, local dimensions, and so on. The moduli stack of curves is an example of a geometric stack. Stacks have been defined first in algebraic geometry, but they are a general notion that also plays a role in differential geometry (orbifolds) and in topology.

3.3.3 The Geometry of Ambiguity: An Introduction to the Ideas of Derived Geometry (Mathieu Anel)

Two fundamental geometric operations are intersecting subspaces and taking quotients. Both operations can create singular points. Intersection singularities include the *multiple points* appearing in nontransverse intersections (like intersecting a circle with one of its tangent lines) and the singular points of a function (like the cusp of $x^3 = y^2$). Quotient singularities are typically created by fixed points of group actions (like the origin in the action of $\mathbb{Z}/2\mathbb{Z}$ on the affine line $\mathbb{A}^1$ or that of $SO(2)$ on the affine plane $\mathbb{A}^2$). The notion of a scheme is able to deal efficiently with multiple points and singular points of functions but finds a limit when self-intersections are involved.[18] Also, the quotients of group actions do not in general have local coordinates when computed in schemes, but a solution was found with stacks. Because of the use of groupoids, stacks are perfectly suited to keeping track of the possible multiple identifications between the points of a quotient (isotropy groups). This has the consequence that quotients do have local coordinates (atlas) when computed in stacks. This property has been another incentive for introducing stacks (the quotients $\mathbb{A}^1/(\mathbb{Z}/2\mathbb{Z})$ or $\mathbb{A}^2/SO(2)$ are geometric stacks).

Both types of singular points have in common that the tangent space at a singular point has a dimension that is strictly bigger than the local dimension of the space around it.[19] In practice, these tangent spaces are always computed as

---

[18] The "nonschematic" intersections are the ones for which the Serre intersection formula, involving the derived tensor product, has been invented.

[19] This is actually how singular points are defined in algebraic geometry. Intuitively, this is a way to say that not all 1-jets of paths can be integrated into actual paths.

the zeroth homology group of some chain complex whose Euler characteristic is always the good local dimension. A point is regular if and only if the homology of this "tangent complex" is concentrated in degree 0. When the point is singular, the positive homology of the tangent complex measures the "quotient complexity" of the singularity, and the negative homology measures the "intersection complexity."

Derived geometry can then be understood as an enhancement of stack theory (which already deals satisfactorily with quotient singularities) that incorporates tools permitting us to deal with intersection singularities and justify the computation of tangent complexes. As with stacks, the methods to do so are deeply rooted in homotopy theory (see the chapter for why and how).

3.3.4 Geometry in dg-Categories (Maxim Kontsevich)

The 20th century has witnessed the rise of many algebraic methods to study spaces. Beyond the classical correspondence between spaces and rings of functions/coordinates, other algebraic devices have been invented, notably *categories of sheaves*. Sheaves of sets are the core of topos theory, sheaves of ∞-groupoids are the core of higher topos theory and homotopy theory, sheaves of modules over a ring are widely used in algebraic and complex geometry, and sheaves of chain complexes are central in homology theory. Beyond sheaves, categories have also become central in the study of representations of groups and algebras.

Overall, these constructions of categories provide a unity of structure to topology, geometry, and algebra that has brought M. Kontsevich to an original geometric program of unified notions and methods for these three fields. The common structure is that of a category enriched over chain complexes (differential graded category or dg-category) that is *derived* from abelian categories (in the sense of derived functors and derived categories). The categories produced from topology or geometry are usually equipped with a monoidal structure (often commutative), but those from algebra may not be. This leads Kontsevich to make the leap to the noncommutative realm not only by forgetting the commutativity of the monoidal structure but by removing entirely the monoidal structure. The basic object of the resulting *derived noncommutative geometry* is, then, simply a bare dg-category.[20]

The first part of the chapter presents the main examples in algebraic topology, algebraic geometry, and algebra from which Kontsevich draws his inspiration. The second part continues with a list of geometric notions that

---

[20] Perhaps the name "nonmonoidal geometry" would be more accurate than "noncommutative geometry."

Table 0.4. *Motivations for the new mathematical spaces*

		Infinite dimension (function spaces)	Better local structure (singular points)	Better quotients	Better intersections	Better basic objects than sets	Moduli spaces	Axiomatic approach	Geometry/Algebra duality	Application to physics
Diff. geom.	Diffeology [Ch. 1]	good		a few						some
	SDG [Ch. 2]	some	some		some			good	partial	
	Micro-local analysis [Ch. 3]		some							
Top. and alg. top.	Locales [Ch. 4]			some	some		some		yes	
	Topos [Ch. 4]			some	some		some		yes	
	∞-Groupoids [Ch. 5]			good		good				
	HoTT [Ch. 6]			some		good		good		
Alg. geom.	Schemes [Ch. 7]	good	some		some		a few		partial	
	Sheaves [Ch. 7]	good	some	some			some			
	Stacks [Ch. 8]		some	good			good			good
	Derived geometry [Ch. 9]	good	good	good	good		good		partial	good
	NCG [Ch. 10]			some					partial	some

survive the removal of the monoidal structure (e.g., smoothness, properness, finiteness, deformations). The third part continues with new features and methods specifically provided by this new context. The resulting definitions permit the extension of the notions to nongeometric contexts, thereby enriching the comprehension of these contexts with a geometric intuition. Dually, the study of actual geometrical objects without the constraint to preserve the monoidal structure brings a freedom of operations that is similar to what complex numbers bring to the study of real numbers.

Acknowledgments

The material for this book and its companion (*New Spaces in Physics: Formal and Conceptual Reflections*) grew up from the conference New Spaces in Mathematics and Physics organized by the editors in 2015 at the Institut Henri Poincaré in Paris. Both the conference and the book project have been realized in the framework of the ERC project "Philosophy of Canonical Quantum Gravity" piloted by G. Catren. Information on the conference (including videos of the talks) is available at https://www.youtube.com/playlist?list=PLRxtDuSeiaXYy17D56Era8ns3V4vmWmqE.

The history of the evolution of space in mathematics and physics attempted here does not pretend to be comprehensive. For various reasons, important chapters are missing (orbifolds; tropical geometry; Berkovich spaces; $\mathbb{F}_1$-geometry; motives; quantales; constructive analysis and its real numbers; formal topology; or other approaches to quantum gravity, such as causal sets, group field theory, or dynamical triangulations). and we apologize to the reader who might have expected to find these topics in this work.

This project has received funding from the European Research Council under the European Community's Seventh Framework Programme (FP7/2007-2013 Grant Agreement 263523, project "Philosophy of Canonical Quantum Gravity").

References

[1] J. F. Adams, *Stable homotopy and generalized homology*, Chicago Lectures in Mathematics (1974)

[2] M. Anel, *What is a space?*, slides from a talk at the 6th Workshop on Formal Topology, Birmingham (2019), http://mathieu.anel.free.fr/mat/doc/Anel-2019-Birmingham.pdf

[3] M. Artin, *Versal deformations and algebraic stacks*, Inventiones Math. 27 (1974) 165–89

[4] S. Awodey and M. Warren, *Homotopy theoretic models of identity types*, Mathematical Proceedings of the Cambridge Philosophical Society, 2009. Available as arXiv:0709.0248

[5] M. Artin, A. Grothendieck, and J.-L. Verdier, *Théorie des topos et cohomologie étale des schémas*, Lecture Notes in Mathematics 269 (Tome 1) 270 (Tome 2) 305 (Tome 3), Springer (1972–73)

[6] D. Ayala, J. Francis, and N. Rozenblyum, *Factorization homology I: higher categories*, to appear in Advances of Mathematics

[7] J. C. Baez and J. Dolan, *Higher dimensional algebra and topological quantum field theory*, J. Math. Phys. 36 (1995) 6073–6105

[8] J. C. Baez and A. E. Hoffnung, *Convenient categories of smooth spaces*, Trans. Amer. Math. Soc. 363 (2011) 5789–5825

[9] O. Ben-Bassat and K. Kremnizer, *Non-Archimedean analytic geometry as relative algebraic geometry*, Ann. Fac. Sci. Toulouse Math. 26 (2017) 49–126

[10] K. Behrend, G. Ginot, B. Noohi, and P. Xu, *String topology for stacks*, Astérisque 343 (2012)

[11] J. Bergner, *A survey of $(\infty, 1)$-categories*, in J. Baez and J. P. May, eds., *Towards Higher Categories*, IMA Volumes in Mathematics and Its Applications, Springer (2010) 69–83

[12] V. Berkovich, *Spectral theory and analytic geometry over non-Archimedean fields*, Math. Surv. Monogr. 33 (1990) 169 pp.

[13] D. A. Buchsbaum, *Exact categories and duality*, Trans. Am. Math. Soc. 80 (1995) 1–34

[14] D. Calaque, T. Pantev, B. Toën, M. Vaquié, and G. Vezzosi, *Shifted Poisson structures*, J. Topol. 10 (2017) 483–584

[15] A. Cannas da Silva and A. Weinstein, *Geometric models for noncommutative algebras*, Berkeley Mathematics Lecture Notes 10 (1999)

[16] E. Cartan and S. Eilenberg, *Homological Algebra*, Princeton University Press (1956)

[17] D. Carchedi, *Compactly generated stacks: A Cartesian closed theory of topological stacks*, Adv. Math. 229 (2012) 3339–97

[18] D.-C. Cisinksi, *Higher Categories and Homotopical Algebra*, Cambridge University Press (2019)

[19] A. Connes, C. Consani, and M. Marcolli, *Fun with $\mathbb{F}_1$*, J. Number Theory 129 (2009) 1532–61

[20] T. Coyne and B. Noohi, *Singular chains on topological stacks, I*, Adv. Math. 303 (2016) 1190–1235

[21] P. Deligne, *Équations différentielles à points singuliers réguliers*, Lecture Notes in Mathematics 163, Springer (1970)

[22] P. Deligne and D. Mumford, *The irreducibility of the space of curves of given genus*, Publ. Math. l'IHÉS (Paris) 36 (1969) 75–109

[23] P. Deligne, *Cohomologie étale (SGA $4\frac{1}{2}$)*, Lecture Notes in Mathematics 569, Springer (1977)

[24] M. Demazure and P. Gabriel, *Groupes algébriques. Tome I: Géométrie algébrique, généralités, groupes commutatifs*, North-Holland (1970)

[25] J. Dieudonné, *A History of Algebraic and Differential Topology, 1900–1960*, Modern Birkhäuser Classics (2009)
[26] A. Douady and R. Douady, *Algèbre et théories galoisiennes*, Fernand Nathan (1977)
[27] N. Durov, *New Approach to Arakelov Geometry*, Preprint, arXiv:0704.2030
[28] W. Dwyer and D. Kan, *Simplicial localizations of categories*, J. Pure Appl. Algebra 17 (1980) 267–84
[29] S. Eilenberg and S. Mac Lane, *General theory of natural equivalences*, Trans. Am. Math. Soc. 58 (1945) 231–294
[30] S. Eilenberg and N. E. Steenrod, *Foundations of Algebraic Topology*, Princeton University Press (1952)
[31] P. Gabriel and M. Zisman, *Calculus of Fractions and Homotopy Theory*, Ergebnisse der Mathematik und ihrer Grenzgebiete, Band 35, Springer (1967)
[32] L. Gillman and M. Jerison *Rings of Continuous Functions*, Springer (1960)
[33] J. Giraud, *Cohomologie non abélienne*, Springer (1971)
[34] P. Griffiths and J. Harris, *Principles of Algebraic Geometry*, Wiley-Interscience (1994)
[35] M. Gross, *Tropical geometry and mirror symmetry*, CBMS Regional Conf. ser. 114 (2011)
[36] A. Grothendieck, *Sur quelques points d'algèbre homologique*, Tôhoku Math. J. 9 (1957)
[37] A. Grothendieck, *Technique de descente et théorèmes d'existence en géométrie algébrique. II. Le théorème d'existence en théorie formelle des modules*, Séminaire Bourbaki, Astérisque 5 (1958–60)
[38] A. Grothendieck, *À la poursuite des champs*, Manuscript (1983), https://webusers.imj-prg.fr/ georges.maltsiniotis/ps.html
[39] A. Grothendieck and A. Dieudonné, *Éléments de géométrie algébrique, I–IV*, Publ. Math. l'IHÉS (1960–67)
[40] V. Guillemin and S. Sternberg, *Symplectic Techniques in Physics*, Cambridge University Press (1990)
[41] M. Henneaux and C. Teitelboim, *Quantization of Gauge Systems*, Princeton University Press (1992)
[42] Univalent Foundations Program, *Homotopy Type Theory: Univalent Foundations of Mathematic*, 1st ed. (2013), http://homotopytypetheory.org/book/
[43] W. Hurewicz, *Beiträge zur Topologie der Deformationen*, Proc. Koninkl. Akad. Amsterdam. *I. Höherdimensionale Homotopiegruppen*, 38 (1935) 112–19. *II. Homotopie und Homologiegruppen*, 38 (1935) 521–28. *III. Klassen und Homologietypen von Abbidungen*, 39 (1936) 117–26. *IV. Asphärische Räumen*, 39 (1936) 215–24
[44] D. Huybrechts, *Complex Geometry: An Introduction*, Springer (2005)
[45] P. T. Johnstone, *Stone Spaces*, Cambridge Studies in Advanced Mathematics 3, Cambridge University Press (1982)
[46] P. T. Johnstone, *Sketches of an Elephant*, 2 vols., Oxford Logic Guide: 43 and 44
[47] P. T. Johnstone, *The point of pointless topology*, Bull. Amer. Math. Soc. 8 (1983) 41–53
[48] D. Joyce, *Algebraic geometry over C-infinity rings*, to appear in Memoirs of the A.M.S.

[49] C. Kapulkin and P. LeFanu Lumsdaine, *The simplicial model of univalent foundations (after Voevodsky)*, Preprint, arXiv:1211.2851.

[50] J. Kijowski and W. Tulczyjew, *A Symplectic Framework for Field Theories*, Lecture Notes in Physics 107, Springer (1979)

[51] D. Knutson, *Algebraic spaces*, Lecture Notes in Mathematics 203, Springer (1971)

[52] M. Kontsevich, *Homological algebra of mirror symmetry*, Proc. ICM Zürich (1994)

[53] M. Kontsevich, *Deformation quantization of Poisson manifolds*, Lett. Math. Phys. 66 (2003) 157–216

[54] B. Kostant and S. Sternberg, *Symplectic reduction, BRS cohomology and infinite-dimensional Clifford algebras*, Ann. of Phys. 176 (1987)

[55] G. Laumon and L. Moret-Bailly, *Champs algébrique*, Ergebn. Grenzgebiete 39 (2000)

[56] E. Lerman, *Orbifolds as stacks?*, L'Enseign. Math. 56 (2010) 315–63

[57] E. L. Lima, *The Spanier–Whitehead duality in new homotopy categories*, Summa Brasil. Math. 4 (1959) 91–148

[58] J. Lurie, *DAG V – Structured spaces*, Preprint, http://people.math.harvard.edu/~lurie/papers/DAG-V.pdf

[59] J. Lurie, *DAG IX – Closed immersions*, Preprint, http://people.math.harvard.edu/~lurie/papers/DAG-IX.pdf

[60] J. Lurie, *On the classification of topological field theories*, Curr, Dev. Math. 2008 (2009) 129–280

[61] J. Lurie, *Higher Algebra*, http://people.math.harvard.edu/~lurie/papers/HA.pdf

[62] J. Lurie, *Spectral Algebraic Geometry*, Book in preparation, http://people.math.harvard.edu/~lurie/papers/SAG-rootfile.pdf

[63] S. Mac Lane, *Homology*, Springer (1995)

[64] S. Mac Lane and I. Moerdijk, *Sheaves in Geometry and Logic: A First Introduction to Topos Theory*, Springer (1992)

[65] J. Marsden and A. Weinstein, *Reduction of symplectic manifolds with symmetry*, Rep. Math. Phys. 5 (1974) 121–30

[66] G. Mikhalkin, *Tropical geometry and its applications*, Eur. Math. Soc. Zürich 2 (2006) 827–52

[67] I. Moerdijk and D. Pronk, *Orbifolds, sheaves and groupoids*, K-Theory 12 (1997) 3–21

[68] I. Moerdijk and G. Reyes, *Models for Smooth Infinitesimal Analysis*, Springer (1991)

[69] J. A. Navarro González and J. B. Sancho de Salas, *C^∞-Differentiable Spaces*, Lecture Notes in Mathematics 1824, Springer (2003)

[70] M. Porta, *Derived analytic geometry*, PhD thesis, Université Paris Diderot (2016)

[71] M. Porta and T. Y. Yu, *Higher analytic stacks and GAGA theorems*, Adv. Math. 302 (2016) 351–409

[72] T. Porter, *S-categories, S-groupoids, Segal categories and quasicategories*, Notes, arXiv:math/0401274

[73] D. Quillen, *Homotopical Algebra*, Lecture Notes in Mathematics 43, Springer (1967)

[74] A. Rosenberg, *Noncommutative schemes*, Compos. Math. 112 (1998) 93–125
[75] I. Satake, *On a generalisation of the notion of manifold*, Proc. Natl. Acad. Sci. U.S.A. 42 (1956) 359–63
[76] S. Schwede, *Categories and orbispaces*, Algebraic & Geometric Topology (forthcoming)
[77] S. Schwede and B. Shipley, *Stable model categories are categories of modules*, Topology 42 (2003) 103–53
[78] P. Seidel, *Fukaya categories and Picard–Lefschetz theory*, Zurich Lectures in Advanced Mathematics, Zürich (2008)
[79] J.-P. Serre, *Géométrie algébrique et géométrie analytique*, Ann. Inst. Fourier 6 (1956) 1–42
[80] C. Simpson, *Homotopy over the complex numbers and generalized de Rham cohomology*, in M. Maruyama, ed., *Moduli of Vector Bundles*, Dekker (1996) 229–63
[81] C. Simpson, *Algebraic (geometric) n-stacks*, Preprint, arXiv:alg-geom/9609014
[82] C. Soulé, *Les variétés sur le corps à un élément*, Mosc. Math. J. 4 (2004) 217–44, 312
[83] J.-M. Souriau, *Structure of Dynamical Systems: A Symplectic View of Physics*, Birkhäuser (1997)
[84] D. Spivak, *Derived smooth manifolds*, Duke Math. J. 153 (2010) 55–128
[85] W. Thurston, *Three-Dimensional Geometry and Topology*, vol. 1, Princeton University Press (1997)
[86] B. Toën, *Higher and derived stacks: A global overview*, Algebraic geometry – Seattle 2005, Part 1, 435–87, Proc. Sympos. Pure Math. 80, Amer. Math. Soc. (2009)
[87] B. Toën, *Derived algebraic geometry*, EMS Surv. Math. Sci. 1 (2014), 153–240
[88] B. Toën and M. Vaquié, *Au-dessous de Spec Z*, J. K-Theory 3 (2009) 437–500
[89] T. Pantev, B. Toën, M. Vaquié, and G. Vezzosi, *Shifted symplectic structures*, Publ. Math. Inst. Hautes Études Sci. 117 (2013) 271–328
[90] J. Pradines, *In Ehresmann's footsteps: From group geometries to groupoid geometries*, in *Geometry and Topology of Manifolds*, Banach Center Publ. 76, Polish Academy of Sciences, Warsaw (2007) 87–157
[91] F. Van Oystaeyen and A. Verschoren, *Noncommutative algebraic geometry*, Lecture Notes in Mathematics 887, Springer (1981)
[92] S. Vickers, *Topology via logic*, Cambridge University Press (1989)
[93] J.-L. Verdier, *Des Catégories Dérivées des Catégories Abéliennes*, Asterisque, 239, Société Mathématique de France (1996)
[94] C. Voisin, *Hodge theory and complex algebraic geometry I-II*, Cambridge Stud. Adv. Math. 76, 77 (2002–3)
[95] A. Weinstein, *Symplectic groupoids and Poisson manifolds*, Bull. Amer. Math. Soc. 16 (1987) 101–4
[96] P. Xu, *Morita equivalent symplectic groupoids*, in *Symplectic Geometry, Groupoids, and Integrable Systems*, Math. Sci. Res. Inst. Publ. 20, Springer (1991) 291–311

Mathieu Anel
Department of Philosophy, Carnegie Mellon University
mathieu.anel@gmail.com

Gabriel Catren
Laboratoire SPHERE – Sciences, Philosophie, Histoire
(UMR 7219, CNRS, Université de Paris)
gabrielcatren@gmail.com

PART I

Differential Geometry

1
An Introduction to Diffeology

Patrick Iglesias-Zemmour

Contents

1	Introduction	31
2	The Unexpected Example: The Irrational Torus	34
3	What Is a Diffeology?	38
4	Fiber Bundles	43
5	Homotopy Theory	46
6	Modeling Diffeology	53
7	Cartan–de Rham Calculus	61
8	Symplectic Diffeology	68
9	In Conclusion	76
References		79

1 Introduction

Since its creation in the early 1980s, diffeology has become an alternative, or rather a natural extension of, traditional differential geometry. With its developments in higher homotopy theory, fiber bundles, modeling spaces, Cartan–de Rham calculus, moment maps, and symplectic[1] programs, diffeology now covers a large spectrum of traditional fields and deploys them from singular quotients to infinite-dimensional spaces – and mixes the two – treating mathematical objects that are or are not strictly speaking manifolds, and other constructions, on an equal footing in a common framework. We shall see some

---

[1] About what it means "to be symplectic" in diffeology, see [21, 24, 27].

of its achievements through a series of examples, chosen because they are not covered by the geometry of manifolds, because they involve either infinite-dimensional spaces or singular quotients, or both.

The growing interest in diffeology comes from the conjunction of two strong properties of the theory:

1. Mainly, the category {Diffeology} is stable under all set-theoretic constructions: sums, products, subsets, and quotients. It is then a complete and cocomplete category. It is also cartesian closed; the space of smooth maps in diffeology has itself a natural *functional diffeology*.
2. Just as importantly, quotient spaces, which are trivial under almost all generalizations of differential calculus,[2] get naturally a meaningful diffeology. That is in particular the case of *irrational tori*, quotients of the real line[3] by dense subgroups. They own, as we shall see, a nontrivial diffeology, capturing faithfully the intrication of the subgroup into its ambient space. This crucial property will be the raison d'être of many new constructions, or wide generalizations of classical constructions, that cannot exist in almost all the other extensions of differential geometry.[2]

The treatment of any kind of singularities, maybe more than the inclusion of infinite-dimensional spaces, reveals how diffeology changes the way we understand smoothness and discriminates this theory among the various alternatives; see, for example, the use of dimension in diffeology [20], which distinguishes between the different quotients $\mathbf{R}^n/O(n)$.

The diagram in Figure 1.1 shows the inclusivity of diffeology, with respect to differential constructions, in comparison with the traditional theory.

1.1 Connecting a Few Dots

The story began in the early 1980s, when Jean-Marie Souriau introduced the *difféologies* in a paper titled "Groupes Différentiels" [47]. It was defined as a formal but light structure on groups,[4] and it was designed for dealing easily with infinite-dimensional groups of diffeomorphisms, in particular the group of symplectomorphisms or quantomorphisms. He named the groups equipped with such a structure *groupes différentiels*,[5] as announced in the title

---

[2] I am referring here to the various generalizations of C^∞ differential geometry à la Sikorski or Frölicher [34]. I am not considering the various algebraic generalizations that do not play on the same level of intuition and generality and do not concern exactly the same kind of objects, lattices instead of quotients, etc.
[3] There exists a concept of higher dimensional irrational torus, we do not consider it here.
[4] Compared to functional analysis heavy structures.
[5] Which translates into English as "differential" or "differentiable groups."

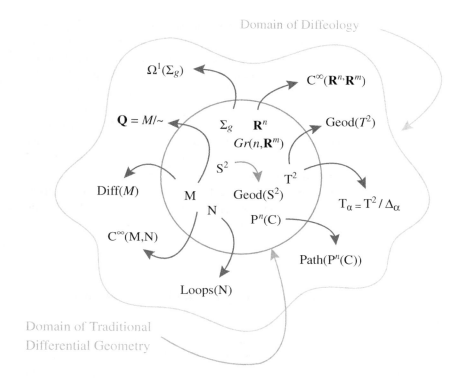

Figure 1.1 The scope of diffeology.

of his paper.[6] His definition was made of five axioms that we can decompose today into a first group of three that gave later the notion of diffeology on arbitrary sets, and the last two, for the compatibility with the internal group multiplication. But it took three years, from 1980 to 1983, to separate the first three general axioms from the last two specific ones and to extract the general structure of *espace différentiel* from the definition of *groupe différentiel*. That was the ongoing work of Paul Donato on the covering of *homogeneous spaces* of differential groups, for one part, and mostly our joint work on the *irrational torus*, which made urgent and unavoidable a formal separation between groups and spaces in the domain of Souriau's *differential structures*, as that gave a new spin to the theory. Actually, the first occurrence of the wording *espace*

---

[6] Actually, difféologies are built on the model of K.-T. Chen, *Differentiable Spaces* [3], for which the structure is defined over convex Euclidean subsets instead of open Euclidean domains. That makes diffeology more suitable to extending differential geometry than Chen's differentiable spaces, which focus more on homology and cohomology theories.

différentiel (for "differential space") appears in the paper "Exemple de groupes différentiels: flots irrationnels sur le tore," published in July 1983 [7]. The expression was used informally, for the purposes of the case, without giving the precise definition. The formal definition was published a couple of months later, in October 1983, in "Groupes différentiels et physique mathématique" [48]. It took then a couple of years to bring the theory of *espaces différentiel* to a new level: with Souriau on generating quantum structures [49]; with Donato's doctoral dissertation on covering of homogeneous spaces, defended in 1984 [8]; and with my doctoral dissertation, defended in 1985, in which I introduced higher homotopy theory and diffeological fiber bundles [16].

What follows is an attempt to introduce the main constructions and results in diffeology, past and recent, through several meaningful examples. The details on the theory can be found in the textbook *Diffeology* [22] and in related papers.

2 The Unexpected Example: The Irrational Torus

Let us begin with the *irrational torus* T_α. At this time, in the early 1980s, physicists were interested in quantizing one-dimensional systems with a quasi-periodic potential, that is, a function u from $\mathbf{R}$ to $\mathbf{R}$, which is the pullback of a smooth function U on a torus $T^2 = \mathbf{R}^2/\mathbf{Z}^2$ along a line of slope α, with $\alpha \in \mathbf{R} - \mathbf{Q}$. Explicitly, $u(x) = U(e^{2i\pi x}, e^{2i\pi\alpha x})$, where $U \in C^\infty(T^2, \mathbf{R})$. That problem has drawn the attention of physicists and some mathematicians to the question of the statute of the quotient space

$$T_\alpha = T^2/\Delta_\alpha,$$

where $\Delta_\alpha \subset T^2$ is a one-parameter subgroup, a projection from $\mathbf{R}^2$ of the line $y = \alpha x$, that is, $\Delta_\alpha = \{(e^{2i\pi x}, e^{2i\pi\alpha x})\}_{x \in \mathbf{R}}$.

As a topological space, T_α is trivial, because α is irrational and Δ_α "fills" the torus, that is, its closure is T^2. And a trivial topological space is of no help. The various differentiable approaches[7] lead also to dead ends; the only smooth maps from T_α to $\mathbf{R}$ are constant maps, because the composition with the projection $\pi: T^2 \to T_\alpha$ should be smooth. For these reasons, the irrational torus was regarded by everyone as an extremely *singular* space.

But when we think about it, the irrational torus T_α is a group, moreover, an abelian group, and there is nothing more regular and homogeneous than a group. And that was the point that made us eager to explore T_α through

---

[7] Sikorski, Frölicher, ringed spaces, etc.

the approach of *diffeologies*. But diffeologies were invented to study infinite-dimensional groups, such as groups of symplectomorphisms, and it was not clear that they could be of any help for the study of such "singular" spaces as T_α, except that, because a diffeology on a set is defined by its *smooth parameterizations*, and because smooth parameterizations on T_α are just (locally) the composites of smooth parameterizations into T^2 by the projection $\pi : T^2 \to T_\alpha$, it was already clear that the diffeology of T_α was not the *trivial diffeology* made of all parameterizations, and neither was it the *discrete diffeology* made only of locally constant parameterizations. Indeed, considering two smooth parameterizations[8] P and P' in T^2, there is a great chance that $\pi \circ P$ and $\pi \circ P'$ are different. And because P and P' are not any parameterizations but smooth ones, that makes T_α neither trivial nor discrete. That was already a big difference with the traditional topological or differential approches we mentioned above that make T_α coarse.

But how to measure this nontriviality? To what ends did this new approach lead? That was the true question. We gave some answers in the paper "Exemple de groupes différentiels: flots irrationnels sur le tore" [7]. Thanks to what Paul Donato had already developed at this time on the covering of *homogeneous differential spaces*, and which made the core of his doctoral dissertation [8], we could compute the fundamental group of T_α and its universal covering $\tilde{T}_\alpha$. We found that

$$\pi_1(T_\alpha) = \mathbf{Z} \times \mathbf{Z} \quad \text{and} \quad \tilde{T}_\alpha = \mathbf{R},$$

with $\pi_1(T_\alpha)$ included in $\mathbf{R}$ as $\mathbf{Z} + \alpha \mathbf{Z}$. That was a first insightful result showing the capability of diffeology concerning these spaces regarded ordinarily as (highly) singular. Needless to say, since then, they have become completely admissible.

In the second half of the 1970s, the quantum mechanics of quasi-periodic potentials hit the field of theoretical physics,[9] and the question of the structure of the space of leaves of the linear foliation of the 2-torus sparked a new level of interest. In particular, French theoretical physicists used the techniques of *noncommutative geometry* introduced by Alain Connes. So, the comparison between the two theories became a natural question. The notion of fundamental group or universal covering was missing at that time in noncommutative

---

[8] A *parameterization* in a set X is just any map P defined on some open subset U of some numerical space $\mathbf{R}^n$ into X.

[9] The most famous paper on the question was certainly from Dinaburg and Sinai, on "The One-Dimensional Schrödinger Equation with a Quasiperiodic Potential" [5].

geometry,[10] so it was not with these invariants that we could compare the two approaches. That came eventually from the following result:

Theorem 2.1 (Donato-Iglesias, 1983) *Two irrational tori T_α and T_β are diffeomorphic if and only if α and β are conjugate modulo $GL(2,\mathbf{Z})$, that is, if there exists a matrix*

$$M = \begin{pmatrix} a & b \\ c & d \end{pmatrix} \in GL(2,\mathbf{Z}) \quad \text{such that} \quad \beta = \frac{a\alpha + b}{c\alpha + d}.$$

That result had its correspondence in Connes's theory, due to Marc Rieffel [41]: the $\mathbf{C}^*$-algebras associated with α and β are Morita-equivalent if and only if α and β are conjugate modulo $GL(2,\mathbf{Z})$. At this moment, it was clear that diffeology could be a possible alternative to noncommutative geometry. The advantage of diffeology was to stay close to the special intuition and concepts developed by geometers over centuries, if not millennia.

Actually, because of irrationality of α and β, we showed that a diffeomorphism $\varphi \colon T_\alpha \to T_\beta$ could be fully lifted at the level of the covering $\mathbf{R}^2$ of T^2 into an affine diffeomorphism $\Phi(Z) = AZ + B$, where A preserves the lattice $\mathbf{Z}^2 \subset \mathbf{R}^2$, that is, $A \in GL(2,\mathbf{Z})$, and $B \in \mathbf{R}^2$. The fact that these natural isomorphisms are preserved *a minima* in diffeology was of course an encouragement to continue the exploration of this example and to push the test of diffeology even further.

And that is not all that could be said, and has been said, on irrational tori. Indeed, the category {Diffeology} has many nice properties; in particular, as we shall see in the following, it is cartesian closed. That means in particular that the set of smooth maps between diffeological spaces has a natural diffeology. We call it the *functional diffeology*. Thus any set of smooth maps between diffeological spaces has a fundamental group, in particular, the group $\text{Diff}(T_\alpha)$ of diffeomorphisms of T_α. And its computation gives us another surprise:

Theorem 2.2 (Donato-Iglesias, 1983) *The connected component of $\text{Diff}(T_\alpha)$ is T_α, acting by multiplication on itself. Its group of connected components is*

$$\pi_0(\text{Diff}(T_\alpha)) = \begin{cases} \{\pm 1\} \times \mathbf{Z} & \text{if } \alpha \text{ is quadratic,} \\ \{\pm 1\} & \text{otherwise.} \end{cases}$$

We recall that a number is quadratic if it is a solution of a quadratic polynomial with integer coefficients. That result was indeed discriminating,

---

[10] It is only recently that we have built formal links between the two theories [29].

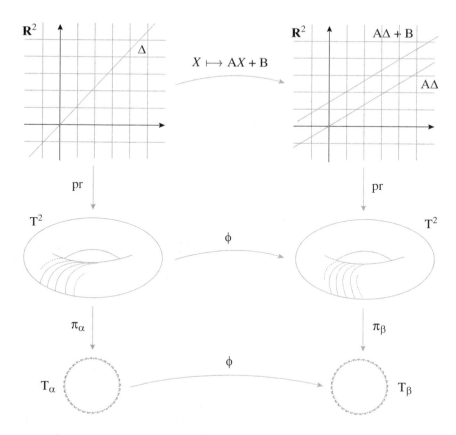

Figure 1.2 A diffeomorphism from T_α to T_β.

since it did not appear in any other theory extending differential geometry. That clearly showed that diffeology, even for such a twisted example, was subtle enough to distinguish between numbers, quadratic or not. This property was not without reminding us about the periodicity of the continued fraction of real numbers.

This computation has been generalized on the $\pi_0(\mathrm{Diff}(T_H))$ [18], where $H \subset \mathbf{R}^n$ is a totally irrational hyperplane and $T_H = T^n/H$.

Theorem 2.3 (Iglesias-Lachaud, 1990) *Let* $H \subset \mathbf{R}^n$ *be a totally irrational hyperplane, that is,* $H \cap \mathbf{Z}^n = \{0\}$. *Let* $T_H = T^n/\mathrm{pr}(H)$, *where* pr *is the projection from* $\mathbf{R}^n$ *to* T^n. *Let* $w = (1, w_2, \ldots, w_n)$ *be the normalized 1-form such that* $H = \mathrm{Ker}(w)$. *The coefficients* w_i *are independent on* $\mathbf{Q}$. *Let*

$E_w = \mathbf{Q} + w_2\mathbf{Q} + \cdots + w_n\mathbf{Q} \subset \mathbf{R}$ *be the* $\mathbf{Q}$*-vector subspace of* $\mathbf{R}$ *generated by the* w_i. *The subset*

$$K_w = \{\lambda \in \mathbf{R} \mid \lambda E_w \subset E_w\}$$

is an algebraic number field, a finite extension of $\mathbf{Q}$. *Then*, $\pi_0(\mathrm{Diff}(T_H))$ *is the group of the units of an order of* K_w. *Thanks to the Dirichlet theorem,*

$$\pi_0(\mathrm{Diff}(T_H)) \simeq \{\pm 1\} \times \mathbf{Z}^{r+s-1},$$

where r and s are the number of real and complex places of K_w.

Now that we have seen a couple of unexpected results from diffeology, we may have gotten your attention, and it is time to give some details on what exactly a *diffeology* on an arbitrary set is. Then we shall see other applications of diffeology and examples, some of them famous.

3 What Is a Diffeology?

It is maybe time to give a precise meaning to what we claimed on the irrational torus. As a preamble, let us say that, contrarily to many constructions in differential geometry, diffeologies are defined just on sets – dry sets – without any preexisting structure, neither topology nor anything else. That is important enough to be underlined, and that is also what makes the difference. A diffeology on a set X consists in declaring what parameterizations are smooth. Let us first introduce formally a fundamental word of this theory.

3.1 Parameterization

We call *parameterization* in a set X any map $P\colon U \to X$ such that U is some open subset of a Euclidean space. If we want to be specific, we say that P is an n-parameterization when U is an open subset of $\mathbf{R}^n$. The set of all parameterizations in X is denoted by Params (X).

Note that there is no condition of injectivity on P, neither is there any topology precondition on X a priori. And also, we shall say *Euclidean domain* for "open subset of a Euclidean space." Now,

3.2 Definition of a Diffeology

A *diffeology* on X is any subset $\mathcal{D}$ of Params (X) that satisfies the following axioms:

1. COVERING : $\mathcal{D}$ contains the constant parameterizations.
2. LOCALITY : Let P be a parameterization in X. If, for all $r \in \text{dom}(P)$, there is an open neighborhood V of r such that $P \restriction V \in \mathcal{D}$, then $P \in \mathcal{D}$.
3. SMOOTH COMPATIBILITY: For all $P \in \mathcal{D}$, for all $F \in C^\infty(V, \text{dom}(P))$, where V is a Euclidean domain, $P \circ F \in \mathcal{D}$.

A space equipped with a diffeology is called a *diffeological space*. The elements of the diffeology of a diffeological space are called the *plots* of (or in) the space.[11]

The first and foremost examples of diffeological spaces are the Euclidean domains, equipped with their *smooth diffeology*, that is, the ordinary smooth parameterizations. Pick, for example, the smooth $\mathbf{R}^2$; we have a diffeology on $\mathrm{T}^2 = \mathbf{R}^2/\mathbf{Z}^2$ by lifting locally the parameterizations in $\mathbf{R}^2$. That is, a plot of T^2 will be a parameterization $P: r \mapsto (z_r, z'_r)$ such that, for every point in the domain of P, there exist two smooth parameterizations θ and θ', in $\mathbf{R}$, defined in the neighborhood of this point, with $(z_r, z'_r) = (e^{2i\pi\theta(r)}, e^{2i\pi\theta'(r)})$, that is, the usual diffeology that makes T^2 the manifold we know. But that procedure can be extended naturally to T_α. Indeed, a parameterization $P: r \mapsto \tau_r$ in T_α is a plot if there exists locally, in the neighborhood of every point in the domain of P, a parameterization $\zeta: r \mapsto (z_r, z'_r)$, such that $\tau_r = \pi(\zeta(r))$. That construction is summarized by the sequence of arrows

$$\mathbf{R}^2 \xrightarrow{\text{pr}} \mathrm{T}^2 \xrightarrow{\pi} \mathrm{T}_\alpha.$$

That is exactly the diffeology we consider when we talk about the irrational torus. But this construction of diffeologies by *pushforward* is actually one of the fundamental constructions of the theory. However, to go there, we need first to define an important property of diffeologies.

3.3 Comparing Diffeologies

Inclusion defines a partial order in diffeology. If $\mathcal{D}$ and $\mathcal{D}'$ are two diffeologies on a set X, one says that $\mathcal{D}$ is finer than $\mathcal{D}'$ if $\mathcal{D} \subset \mathcal{D}'$, or $\mathcal{D}'$ is coarser than

---

[11] There is a discussion about diffeology as a sheaf theory in [17, Annex]. But we do not develop this formal point of view in general, because the purpose of diffeology is to minimize the technical tools in favor of a direct, more geometrical, intuition.

$\mathcal{D}$. Moreover, diffeologies are stable by intersection, which gives the following property:

Proposition 3.1 *This partial order, called fineness, makes the set of diffeologies on a set* X *a lattice. That is, every set of diffeologies has an infimum and a supremum. As usual, the infimum of a family is obtained by intersection of the elements of the family, and the supremum is obtained by intersecting the diffeologies coarser than any element of the family.*

The infimum of every diffeology, the finest diffeology, is called the *discrete diffeology*. It consists of locally constant parameterizations. The supremum of every diffeology, the coarsest diffeology, is called the *coarse* or *trivial diffeology*, and it is made of all the parameterizations. As we shall see, these bounds will be useful for defining diffeologies defined by properties (boolean functions).

Now, we can write the construction by pushforward:

3.4 Pushing Forward Diffeologies

Let $f: X \to X'$ be a map, and let X be a diffeological space with diffeology $\mathcal{D}$. Then, there exists a finest diffeology on X' such that f is smooth. It is called the *pushforward* of the diffeology of X. We denote it by $f_*(\mathcal{D})$. If f is surjective, its plots are the parameterizations P in X' that can be written $\text{Sup}_i \, f \circ P_i$, where the P_i are plots of X such that the $f \circ P_i$ are compatible, that is, coincide on the intersection of their domains, and Sup denotes the smallest common extension of the family $\{f \circ P_i\}_{i \in \mathcal{I}}$.

In particular, the diffeology of T^2 is the pushforward of the smooth diffeology of $\mathbf{R}^2$ by pr, and the diffeology on T_α is the pushforward of the diffeology of T^2 by π, or, equivalently, the pushforward of the smooth diffeology of $\mathbf{R}^2$ by the projection $\pi \circ \text{pr}$.

Note 1 Let $\pi: X \to X'$ be a map between diffeological spaces. We say that π is a *subduction* if it is surjective and if the pushforward of the diffeology of X coincides with the diffeology of X'. In particular pr: $\mathbf{R}^2 \to T^2$ and $\pi: T^2 \to T_\alpha$ are two subductions. Subductions make a subcategory, since the composite of two subductions is again a subduction.

Note 2 Let X be a diffeological space and $\sim$ be an equivalence relation on X. Let $Q = X/\sim$ be the quotient set,[12] that is,

$$Q = \{\text{class}(x) \mid x \in X\} \quad \text{and} \quad \text{class}(x) = \{x' \mid x' \sim x\}.$$

---

[12] I regard always a quotient set as a subset of the set of all subsets of a set.

The pushforward on Q of the diffeology of X by the projection class is called the *quotient diffeology*. Equipped with the quotient diffeology, Q is called the *quotient space* of X by $\sim$. This is the first important property of the category {Diffeology}, it is closed by quotient.

Then, after having equipped the irrational tori with a diffeology (actually the quotient diffeology), we would compare different irrational tori with respect to diffeomorphisms. For that, we need a precise definition.

3.5 Smooth Maps

Let X and X' be two diffeological spaces. A map $f : X \to X'$ is *smooth* if, for any plot P in X, $f \circ P$ is a plot in X'. The set of smooth maps from X to X' is denoted, as usual, by $C^\infty(X, X')$.

Note 3 The composition of smooth maps is smooth. Diffeological spaces, together with smooth maps, make a category that we denote by {Diffeology}.

Note 4 The isomorphisms of the category {Diffeology} are the bijective maps, smooth as well as their inverses. They are called *diffeomorphisms*.

3.6 What about Manifolds?

The time has come to make a comment on manifolds. Every manifold is naturally a diffeological space; its plots are the smooth parameterizations in the usual sense. That makes the category {Manifolds} a full and faithful subcategory of {Diffeology}. But we should insist that diffeology not be understood as a generalization of the theory of manifolds. It happens that, between many other things, diffeology extends the theory of manifolds, but its true nature is to extend the differential calculus on domains in Euclidean spaces, and this is the way it should be regarded.

Let us continue to explore our example of the irrational torus T_α. We still have to describe its fundamental group and its universal covering. Actually, the way it was treated in the founding paper [7] used special definitions adapted only to groups and homogeneous spaces, because at this time, diffeology was only about groups. It was clear at this moment that considering diffeology only on groups was insufficient and that we missed a real independent theory of fiber bundles and homotopy in diffeology. That was the content of my doctoral dissertation "Fibrations difféologiques et homotopie" [16].

Let us begin with the covering thing. A covering is a special kind of fiber bundle with a discrete fiber. But all these terms must be understood in the sense of diffeology, especially the word *discrete*. Let me give an example:

3.7 Proposition

The rational numbers **Q** are discrete in **R**.

This is completely natural for everyone, except that this is false as far as topology is concerned. But we are talking diffeology: if we equip **Q** with the diffeology *induced* by **R**, it is not difficult to prove[13] that **Q** is discrete, that is, its diffeology is discrete. And that is the meaning we want to give to be a *discrete subset* of a diffeological space. Well, we still have to elaborate a little bit about *induced diffeology*.

3.8 Pulling Back Diffeologies

Let $f : X \to X'$ be a map, and let X' be a diffeological space with diffeology $\mathcal{D}'$. Then, there exists a coarsest diffeology on X such that f is smooth. It is called the *pullback* of the diffeology of X'. We denote it by $f^*(\mathcal{D}')$. Its plots are the parameterizations P in X such that $f \circ P$ is a plot of X'.

In particular, that gives to any subset $A \subset X$, where X is a diffeological space, a *subset diffeology*, that is, $j^*(\mathcal{D})$, where $j : A \to X$ is the inclusion and $\mathcal{D}$ is the diffeology of X. A subset equipped with the subset diffeology is called a *diffeological subspace*. Now it is clear what is meant by a *discrete subset* of a diffeological space: it is a subset such that its induced diffeology is discrete.

Note 5 Let $j : X \to X'$ be a map between diffeological spaces. We say that j is an *induction* if j is injective and if the pullback of the diffeology of X' coincides with the diffeology of X. For example, in the case of the irrational torus, the injection $t \mapsto (e^{2i\pi t}, e^{2i\pi \alpha t})$ from **R** to T^2 is an induction. That means precisely that if $r \mapsto (z_r, z'_r)$ is smooth in T^2 but takes its values in Δ_α, then there exists a smooth parameterization $r \mapsto t_r$ in **R** such that $z_r = e^{2i\pi t_r}$ and $z'_r = e^{2i\pi \alpha t_r}$.

Then, with this construction, the category {Diffeology}, which was closed by quotient, is also closed by inclusion.

Now, continuing with our example, there are two reasons we need a good concept of fiber bundle in diffeology.

1. Coverings of diffeological spaces should be defined as fiber bundles with discrete fiber. (We shall see then that we are even able to build a universal covering, unique up to isomorphism, for every diffeological space.)

---

[13] A nice application of the intermediate value theorem.

2. If we look closely at the projection $\pi : T^2 \to T_\alpha$, we observe that this looks like a fiber bundle with fiber $\Delta_\alpha \simeq \mathbf{R}$. Thus, if the long homotopy sequence could apply to diffeological fiber bundles, we would get immediately $\pi_1(T_\alpha) = \pi_1(T^2) = \mathbf{Z}^2$, since the fiber $\mathbf{R}$ is contractible.

4 Fiber Bundles

Of course, the classical definition of a locally trivial bundle is useless here since T_α has a trivial topology. The situation is more subtle – we are looking for a definition[14] that satisfies the following two conditions:

1. The quotient of a diffeological group by any subgroup is a diffeological fibration, whatever the subgroup is.
2. For diffeological fibrations, the long exact homotopy sequence applies.

Since we refer to diffeological groups, we have to clarify their definition.

4.1 Diffeological Group

A diffeological group is a group G that is also a diffeological space such that the multiplication $m: (g, g') \mapsto gg'$ and the inversion inv: $g \mapsto g^{-1}$ are smooth.

That needs a comment on the product of diffeological spaces, since we refer to the multiplication in a diffeological group G, which is defined on the product $G \times G$.

4.2 Product of Diffeological Spaces

Let $\{X_i\}_{i \in \mathcal{I}}$ be any family of diffeological spaces. There exists on the product $X = \prod_{i \in \mathcal{I}} X_i$ a coarsest diffeology such that every projection $\pi_i : X \to X_i$ is smooth. It is called the *product diffeology*. A plot in X is just a parameterization $r \mapsto (x_{i,r})_{i \in \mathcal{I}}$ such that each $r \mapsto x_{i,r}$ is a plot of X_i.

Note 6 The category {Diffeology} is then closed by products.

There are two equivalent definitions of diffeological bundles; the following one is the pedestrian version.

---

[14] The definition of fiber bundles in diffeology – their two equivalent versions – has been introduced in "Fibrés difféologiques et homotopie" [16].

4.3 Diffeological Fiber Bundles

Let $\pi: Y \to X$ be a map with X and Y two diffeological spaces. We say that π is a fibration, with fiber F, if, for every plot P: $U \to X$, the pullback

$$\mathrm{pr}_1: \mathrm{P}^*(Y) \to U \quad \text{with} \quad \mathrm{P}^*(Y) = \{(r, y) \in U \times Y \mid P(r) = \pi(y)\}$$

is locally trivial with fiber F. That is, every point in U has an open neighborhood V such that there exists a diffeomorphism $\phi: V \times F \to \mathrm{pr}_1^{-1}(V) \subset \mathrm{P}^*(Y)$, with $\mathrm{pr}_1 \circ \phi = \mathrm{pr}_1$:

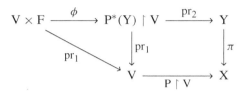

This definition extends the usual definition of smooth fiber bundles in differential geometry. But it contains more:

Proposition 4.1 *The projection $\pi: G \to G/H$, where G is a diffeological group and $H \subset G$ is any subgroup, is a diffeological fibration.*

We have now the formal framework where $\pi: T^2 \to T_\alpha$ is a legitimate fiber bundle. This definition of fiber bundle satisfies also the long sequence of homotopy; we shall come back to that subject later.

Since we have a definition of fiber bundles, we inherit naturally the notion of a diffeological covering:

Definition 4.2 (Definition of coverings) A covering of a diffeological space X is a diffeological fibration $\pi: \hat{X} \to X$ with a discrete fiber.

Note 7 The projection $\pi: \mathbf{R} \to \mathbf{R}/(\mathbf{Z} + \alpha \mathbf{Z})$ is a simply connected covering, since $\mathbf{R}$ is a diffeological group and $\mathbf{Z} + \alpha \mathbf{Z}$ is a subgroup. The fact that $\mathbf{R}/(\mathbf{Z} + \alpha \mathbf{Z})$ is diffeomorphic to T_α is an exercise left to the reader.

There is an alternative to the definition of fiber bundles involving a groupoid of diffeomorphisms. This alternate definition is maybe less intuitive, but it is more internal to diffeology. It is based on the existence of a natural diffeology on the set of smooth maps between diffeological spaces.

4.4 The Functional Diffeology

Let X and X' be two diffeological spaces. There exists on $C^\infty(X, X')$ a coarsest diffeology such that the *evaluation map*

$$\mathrm{ev}\colon C^\infty(X,X') \times X \to X' \quad \text{defined by} \quad \mathrm{ev}(f,x) = f(x)$$

is smooth. Thus diffeology is called the *functional diffeology*.

Note 8 A parameterization $r \mapsto f_r$ is a plot for the functional diffeology if the map $(r,x) \mapsto f_r(x)$ is smooth.

Note 9 There exists a natural diffeomorphism between $C^\infty(X, C^\infty(X',X''))$ and $C^\infty(X \times X', X')$. That makes the category {Diffeology} *cartesian closed*, which is a pretty nice property.

4.5 Fiber Bundles: The Groupoid Approach

Let $\pi\colon Y \to X$ be a map with X and Y two diffeological spaces. Consider the groupoid **K** whose objects are the points of X and the arrows from x to x' are the diffeomorphisms from Y_x to $Y_{x'}$, where the preimages $Y_x = \pi^{-1}(x)$ are equipped with the subset diffeology.

There is a *functional diffeology* on **K** that makes it a *diffeological groupoid*. The set Obj(**K**) = X is obviously equipped with its own diffeology. Next, a parameterization $r \mapsto \phi_r$ in Mor(**K**), defined on a domain U, will be a plot if

1. $r \mapsto (\mathrm{src}(\phi_r), \mathrm{trg}(\phi_r))$ is a plot of $X \times X$;
2. the maps $\mathrm{ev}\colon U_\mathrm{src} \to Y$ and $\overline{\mathrm{ev}}\colon U_\mathrm{trg} \to Y$, defined by $\mathrm{ev}(r,y) = \phi_r(y)$ and $\overline{\mathrm{ev}}(r,y) = \phi_r^{-1}(y)$, on $U_\mathrm{src} = \{(r,y) \in U \times Y \mid y \in \mathrm{def}(\phi_r)\}$ and $U_\mathrm{trg} = \{(r,y) \in U \times Y \mid y \in \mathrm{def}(\phi_r^{-1})\}$ are smooth, where these two sets are equipped with the subset diffeology of the product $U \times Y$.

We have, then, the following theorem [16]:

Theorem 4.3 *The map π is a fibration if and only if the characteristic map*

$$\chi\colon \mathrm{Mor}(\mathbf{K}) \to X \times X, \quad \text{defined by} \quad \chi(f) = (\mathrm{src}(f), \mathrm{trg}(f)),$$

is a subduction.

Thanks to this approach, we can construct, for every diffeological fiber bundle, a principal fiber bundle – by splitting the groupoid – with which the fiber bundle is associated. It is possible then to refine this construction and define fiber bundles with structures (e.g., linear).

The next step concerning fiber bundles will be to establish the long homotopy sequence, but that requires preliminary preparation, beginning with the definition of the homotopy groups.

5 Homotopy Theory

The basis of homotopy begins by understanding what it means to be *homotopic*, that is, in the same "place" or "component."

5.1 Homotopy and Connexity

Let X be a diffeological space; we denote by Paths (X) the space of (smooth) paths in X, that is, $C^\infty(\mathbf{R}, X)$. The ends of a path γ are denoted by

$$\hat{0}(\gamma) = \gamma(0), \quad \hat{1}(\gamma) = \gamma(1) \quad \text{and} \quad \text{ends}(\gamma) = (\gamma(0), \gamma(1)).$$

We say that two points x and x' are *connected* or *homotopic* if there exists a path γ such that $x = \hat{0}(\gamma)$ and $x' = \hat{1}(\gamma)$.

To be connected defines an equivalence relation whose equivalence classes are called *connected components*, or simply *components*. The set of components is denoted by $\pi_0(X)$.

Proposition 5.1 *The space* X *is the sum of its connected components, that is,*

$$X = \coprod_{X_i \in \pi_0(X)} X_i.$$

Moreover, the partition in connected components is the finest partition of X *that makes* X *the sum of its parts.*

It is time to give the precise definition of the sum of diffeological spaces that founds the previous proposition.

5.2 Sum of Diffeological Spaces

Let $\{X_i\}_{i \in \mathcal{I}}$ be a family of diffeological spaces. There exists a finest diffeology on the sum $X = \coprod_{X_i \in \mathcal{I}} X_i$ such that each injection $j_i = x \mapsto (i, x)$, from X_i to X, is smooth. We call this the *sum diffeology*. The plots of X are the parameterization $r \mapsto (i_r, x_r)$ such that $r \mapsto i_r$ is locally constant.

With that definition, we close, by the way, one of the most interesting aspects of the category {Diffeology}. This category is stable by all the set-theoretic constructions: sum, product, part, quotient. It is a complete and cocomplete category, every direct or inverse limit of diffeological spaces having their natural diffeology. Moreover, the category is cartesian closed.

That is by itself very interesting, as a generalization of the smooth category of Euclidean domains. But there are a few other generalizations that have these same properties (Frölicher spaces, for example). What really makes the difference is that these nice properties apply to a category that includes nontrivially extremely singular spaces, as we have seen with the irrational tori, and that is what makes {Diffeology} different. But let us now come back to the homotopy theory.

5.3 The Fundamental Group and Coverings

Let X be a connected diffeological space, that is, $\pi_0(X) = \{X\}$. Let $x \in X$ be some point. We denote by Loops $(X, x) \subset$ Paths (X) the subspaces of *loops* in X, based at x, that is, the subspace of paths ℓ such that $\ell(0) = \ell(1) = x$. As a diffeological space, with its functional diffeology, this space has a set of components. We define the *fundamental group*[15] of X, at the point x, as

$$\pi_1(X, x) = \pi_0(\text{Loops}(X, x), \hat{x}),$$

where $\hat{x}$ is the constant loop $t \mapsto x$. The group multiplication on $\pi_1(X, x)$ is defined as usual, by concatenation:

$$\tau \cdot \tau' = \text{class}\left[t \mapsto \begin{cases} \ell(2t) & \text{if } t \leq 1/2 \\ \ell'(2t - 1) & \text{if } t > 1/2 \end{cases}\right],$$

where $\tau = \text{class}(\ell)$ and $\tau' = \text{class}(\ell')$. The inverse is given by $\tau^{-1} = \text{class}[t \mapsto \ell(1 - t)]$. Actually, we work with stationary paths for the concatenation to be smooth. A stationary path is a path that is constant on a small interval around 0 and also around 1. We have now a few main results.

Proposition 5.2 *The groups $\pi_1(X, x)$ are conjugate to each other when x runs over X. We denote $\pi_1(X)$ their type.*

We say that X is *simply connected* if its fundamental group is trivial, $\pi_1(X) = \{0\}$. Now we have the following results.

Theorem 5.3 (universal covering) *Every connected diffeological space X has a unique – up to isomorphism – simply connected covering space $\pi : \tilde{X} \to X$. It is a principal fiber bundle with group $\pi_1(X)$. It is called the universal covering; every other connected covering is a quotient of $\tilde{X}$ by a subgroup of $\pi_1(X)$.*

Actually, the universal covering is *half* of the *Poincaré groupoid*, quotient of the space Paths (X) by fixed-end homotopy relation [22, Section 5.15].

---

[15] Formally speaking, the homotopy groups are objects of the category {Pointed Sets}. In particular, $\pi_0(X, x) = (\pi_0(X), x)$.

This is the second meaningful construction of groupoid in the development of diffeology.

Theorem 5.4 (Monodromy theorem) *Let $f: Y \to X$ be a smooth map, where Y is a simply connected diffeological space. With the notations above, there exists a unique lifting $\tilde{f}: Y \to \tilde{X}$ once we fix $\tilde{f}(y) = \tilde{x}$, with $x = f(y)$ and $\tilde{x} \in \pi^{-1}(x)$.*

These are the more relevant constructions and results, familiar to the differential geometer, concerning the fundamental group on diffeological spaces.

5.4 Two Examples of Coverings

Let us now come back to our irrational torus $T_\alpha = T^2/\Delta_\alpha = [R^2/Z^2]/\Delta_\alpha$, that is, $T_\alpha = R^2/[Z^2 \times \{(x, \alpha x)\}_{x \in R}] = [R^2/\{(x, \alpha x)\}_{x \in R}]/Z^2$. The quotient $R^2/\{(x, \alpha x)\}_{x \in R}$ can be realized by R with the projection $(x, y) \mapsto y - \alpha x$. Then, the action of Z^2 on R^2 induces the action on Z^2 on R by $(n, m)(x, y) = (x+n, y+m) \mapsto y + m - \alpha(x+n) = y - \alpha x + (m - \alpha n)$. That is, $(n, m): t \mapsto t + m - \alpha n$. Therefore, $T_\alpha \simeq R/Z + \alpha Z$. Since R is simply connected, thanks to the theorem on existence and unicity of the universal covering, $\tilde{T}_\alpha = R$, and the $\pi_1(T_\alpha)$ injects in R as $Z + \alpha Z$. Of course, there is no need for all these sophisticated tools to get this result, as we have seen in [7]. But this shows how the pedestrian computation integrates seamlessly the general theory.

Another example[16] of this theory is one in infinite dimensions: the universal covering of $\mathrm{Diff}(S^1)$. The group is equipped with its functional diffeology. Let f be a diffeomorphism of $S^1 = R/Z$, and assume that f fixes 1. Let $\pi: t \mapsto e^{2i\pi t}$, from R to S^1, be the universal covering. The composite $f \circ \pi$ is a plot. Thanks to the monodromy theorem, since R is simply connected, $f \circ \pi$ has a unique smooth lifting $\tilde{f}: R \to R$, such that $\tilde{f}(0) = 0$:

$$\begin{array}{ccc} R & \xrightarrow{\tilde{f}} & R \\ \pi \downarrow & & \downarrow \pi \\ S^1 & \xrightarrow{f} & S^1 \end{array}$$

Then, $\pi \circ \tilde{f} = f \circ \pi$ implies that $\tilde{f}(t+1) = \tilde{f}(t) + k$, $k \in Z$, but f cannot be injective unless $k = \pm 1$. Next, f being a diffeomorphism implies that $\tilde{f}$ is a diffeomorphism of R, that is, a strictly increasing or

---

[16] It was first elaborated by Paul Donato in his dissertation [8]. We just reinterpret it with our tools.

decreasing function. Assume that $\tilde{f}$ is increasing; then $\tilde{f}(t+1) = \tilde{f}(t) + 1$. Thus, the *positive diffeomorphisms* of S^1 are the quotient of the increasing diffeomorphisms of **R** satisfying that condition. Now, let $\tilde{f}_s(t) = f(t) + s(t - \tilde{f}(t))$, with $s \in [0, 1]$. We still have $\tilde{f}_s(t+1) = \tilde{f}_s(t) + 1$ and $\tilde{f}'_s(t) = s + (1-s)\tilde{f}'(t)$, which still is positive. Thus, since $\tilde{f}_0(t) = \tilde{f}(t)$ and $\tilde{f}_1(t) = t$, the group

$$\widetilde{\mathrm{Diff}}_+(S^1) = \{\tilde{f} \in \mathrm{Diff}_+(\mathbf{R}) \mid \tilde{f}(t+1) = \tilde{f}(t) + 1\}$$

is contractible, hence simply connected. It is the universal covering of $\mathrm{Diff}_+(S^1)$, the group of the positive diffeomorphisms of S^1. The monodromy theorem indicates that there are only **Z** different liftings $\tilde{f}$ of a given diffeomorphism f. Hence, $\pi_1(\mathrm{Diff}_+(S^1)) = \mathbf{Z}$.

5.5 Higher Homotopy Groups

Let X be a diffeological space and $x \in X$. Since Loops (X, x) is a diffeological space that contains $\hat{x}: t \mapsto x$, there is no obstruction to defining the *higher homotopy groups* by recursion:

$$\pi_n(X, x) = \pi_{n-1}(\mathrm{Loops}\,(X, x), \hat{x}), \quad n \geq 1.$$

One can also define the recursion of diffeological spaces:

$$X_0 = X, \ x_0 = x \in X_0 \ ; \ X_1 = \mathrm{Loops}\,(X_0, x_0), \ x_1 = [t \mapsto x_0] \in X_1 \ ; \ \ldots$$
$$\ldots \ ; \ X_n = \mathrm{Loops}\,(X_{n-1}, x_{n-1}), \ x_n = [t \mapsto x_{n-1}] \in X_n \ ; \ \ldots$$

Thus, $\pi_n(X, x) = \pi_{n-1}(\mathrm{Loops}\,(X, x), \hat{x})$, that is, $\pi_n(X, x) = \pi_{n-1}(X_1, x_1) = \cdots = \pi_1(X_{n-1}, x_{n-1}) = \pi_0(X_n, x_n)$. Since $\pi_n(X, x)$ is the fundamental group of a diffeological space, it is a group. And that is the formal definition of the *nth homotopy group*[17] of X at the point x.

We can feel in particular here the benefits of considering all these spaces — X, Paths (X), Loops (X), and so on — on an equal footing. Being all diffeological spaces, the recursion does not need any supplementary construction than the ones already defined.

5.6 The Homotopy Sequence of a Fiber Bundle

One of the most important properties of diffeological fiber bundles is their long homotopy sequence. Let $\pi : Y \to X$ be a fiber bundle with fiber F. Then, there is a long exact sequence of group homomorphisms [16],

---

[17] It is abelian for $n = 2$.

$$\cdots \to \pi_n(F) \to \pi_n(Y) \to \pi_n(X) \to \pi_{n-1}(F) \to \cdots$$
$$\cdots \to \pi_0(F) \to \pi_0(Y) \to \pi_0(X) \to 0.$$

As usual in these cases, if the fiber is homotopy trivial, then the base space has the homotopy of the total space. And that is what happens for the irrational torus $\pi_k(T_\alpha) = \pi_k(T^2)$, $k \in \mathbf{N}$.

Let us consider another example, in infinite dimensions this time. Let S^∞ be the *infinite-dimensional sphere* in the Hilbert space $\mathcal{H} = \ell^2(\mathbf{C})$. We equip first $\mathcal{H}$ with the fine diffeology of vector space [22, Section 3.7]. Then we prove that $S^\infty \subset \mathcal{H}$, equipped with the subset diffeology, is contractible [22, Section 4.10]. Then we consider the *infinite projective space* $\mathbf{CP}^\infty = S^\infty/S^1$, where S^1 acts on the ℓ^2 sequences by multiplication. The projection $\pi: S^\infty \to \mathbf{CP}^\infty$ is then a diffeological principal fibration with fiber S^1. The homotopy exact sequence gives then $\pi_2(\mathbf{CP}^\infty) = \mathbf{Z}$ and $\pi_k(\mathbf{CP}^\infty) = 0$ if $k \neq 2$, which is what we expected.

That shows how we can work on singular constructions or infinite-dimensional spaces using the same tools and the same intuition as when we deal with differential geometry of manifolds.

5.7 Connections on Fiber Bundles and Homotopy Invariance

Let $\pi: Y \to X$ be a principal fiber bundle with group G. That fabricates a new principal fibration $\pi_*:$ Paths $(Y) \to$ Paths (X) with structure group Paths (G). Roughly speaking, a *connection* on π is a reduction of this *paths fiber bundle* to the subgroup $G \subset$ Paths (G), consisting of constant paths [22, Section 8.32]. We require for this reduction to satisfy a few axioms: locality (sheaf condition on the interval of $\mathbf{R}$), compatibility with concatenation, and so on. The main point is that once we have a path γ in X and a point y over $x = \gamma(t)$, there exists a unique *lift* $\tilde{\gamma}$ of γ such that $\tilde{\gamma}(t) = y$; this is called the *horizontal lift*. Moreover, if $y' = g_Y(y)$, the lift γ' of γ passing at y' at the time t is the shifted $\gamma' = g_Y \circ \gamma$. That property is exactly what we call a reduction of π_* to G.

An important consequence of the existence of a connection on a principal fiber bundle is the homotopy invariance of pullbacks.

Proposition 5.5 *Let* $\pi: Y \to X$ *be a principal fiber bundle with group G, equipped with a connection. Let* $t \mapsto f_t$ *be a smooth path in* $C^\infty(X', X)$, *where X' is any diffeological space. Then, the pullbacks* $\mathrm{pr}_0: f_0^*(Y) \to X'$ *and* $\mathrm{pr}_1: f_1^*(Y) \to X'$ *are equivalent.*

Corollary 5.6 *Any diffeological fiber bundle equipped with a connection over a contractible space is trivial.*

We know that this is always true in traditional differential geometry, because every principal bundle over a manifold can be equipped with a connection.

5.8 The Group of Flows of a Space

Connections are usually defined in traditional differential geometry by a differential form with values in some Lie algebra. As we have seen, that is not the way chosen in diffeology, for a few good reasons. First, for such an important property as the homotopy invariance of pullbacks, a broad definition of connection is enough. Moreover, we have no indisputable concept of Lie algebra in diffeology,[18] and choosing one definition rather than another would link a universal concept, such as *parallel transport*, to an arbitrary choice.

And there is also a new diffeological construction where the difference between general connections and form-valued connections is meaningful enough to justify a posteriori our choice. That is the computation of the *group of flows* over the irrational torus [22, Section 8.39]. Let us begin with a definition:

Definition 5.7 We shall call *flow* over a diffeological space X any $(\mathbf{R},+)$-principal bundle over X.

There is an additive operation on the set Flows(X) of (equivalence classes of) flows. Let $a = \text{class}(\pi: Y \to X)$ and $a' = \text{class}(\pi': Y' \to X)$ be two classes of flows. Consider the pullback $\pi^*(Y') = \{(y,y') \in Y \times Y' \mid \pi(y) = \pi'(y')\}$. It is a $(\mathbf{R}^2, +)$ principal bundle over X by $(y, y') \mapsto \pi(y) = \pi'(y')$. Let Y'' be the quotient of $\pi^*(Y')$ by the antidiagonal action of $\mathbf{R}$, that is, $\underline{t}(y, y') = (t_Y(y), -t_{Y'}(y'))$, and let $\pi'': Y'' \to X$ be the projection $\pi''(\text{class}(y, y')) = \pi(y) = \pi'(y')$. We define then $a + a' = a''$ with $a'' = \text{class}(\pi'': Y'' \to X)$.

The set Flows(X), equipped with this addition, is an abelian group. The neutral element is the class of the trivial bundle, and the inverse of a flow is the same bundle but with the inverse action of $\mathbf{R}$. Note that this group is a kind of Picard group on a diffeological space, but with $\mathbf{R}$ instead of S^1 as structure group. And if this group does not appear in traditional differential geometry, it is because every principal bundle with fiber $\mathbf{R}$ over a manifold is trivial. But that is not the case in diffeology, and we know one such nontrivial bundle, the irrational torus $\pi: T^2 \to T_\alpha$.

---

[18] Even for the moment map, in symplectic diffeology, we do not need the definition of a Lie algebra, as we shall see later on.

Let $\pi : Y \to T_\alpha$ be a flow. Consider the pullback $\mathrm{pr}_1 : \mathrm{pr}^*(Y) \to \mathbf{R}$, where $\mathrm{pr} : \mathbf{R} \to T_\alpha$ is the universal covering. It is an $\mathbf{R}$-principal fiber bundle over $\mathbf{R}$, so it is trivial. Let $\phi : \mathbf{R} \times \mathbf{R} \to \mathrm{pr}^*(Y)$ be an isomorphism. Thus $Y \simeq \mathbf{R} \times \mathbf{R}/\mathrm{pr}_2 \circ \phi$. But $\mathrm{pr}_2 \circ \phi$ is any lifting on the second factor of $\mathbf{R} \times \mathbf{R}$, of the action of $\mathbf{Z} \oplus \alpha \mathbf{Z}$ on the first one:

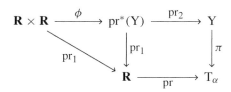

A lifting of a subgroup $\Gamma \subset \mathbf{R}$ on the second factor of $\mathbf{R} \times \mathbf{R}$, where Γ acts by translation on the first factor, is given, for all $k \in \Gamma$ and $(x, \tau) \in \mathbf{R} \times \mathbf{R}$, by

$$k : (x, t) \mapsto (x + k, t + \tau(k)(x)),$$

where $\tau : \Gamma \to C^\infty(\mathbf{R})$ is a cocycle satisfying

$$\tau(k + k')(x) = \tau(k)(x + k') + \tau(k')(x).$$

Two cocycles τ and τ' define the same flow if they differ from a coboundary $\delta \sigma$:

$$\tau'(k)(x) = \tau(k)(x) + \sigma(x + k) - \sigma(x), \quad \text{with} \quad \sigma \in C^\infty(\mathbf{R}).$$

In other words,

$$\mathrm{Flows}(\mathbf{R}/\Gamma) = H^1(\Gamma, C^\infty(\mathbf{R})).$$

Applied to $\Gamma = \mathbf{Z} + \alpha\, \mathbf{Z}$, that gives $\mathrm{Flows}(T_\alpha)$ equivalent to the group of real 1-periodic functions f, after some normalization, modulo the relation

$$f \sim f' \quad \text{if} \quad f'(x) = f(x) + g(x + \alpha) - g(x).$$

This relation is known as the *small divisors Arnold's cohomology relation*. The solution depends on the arithmetic of α: if α is a diophantine or a Liouville number, $\mathrm{Flows}(T_\alpha)$ is one-dimensional or ∞-dimensional.[19]

Moreover, every flow $\pi : Y \to T = \mathbf{R}/\Gamma$ defined by a cocycle τ can be naturally equipped with the connection associated with the covering $\mathrm{pr} : \mathbf{R} \to T$ [22, Section 8.36]. However, not all these bundles support a connection form [22, Section 8.37], only those whose cocycle τ defining π is equivalent to a homomorphism from Γ to $\mathbf{R}$; see [22, Exercise 139]. In other words, if the cocycle τ is not cohomologous to a homomorphism, then there is no connection that can be defined by a connection form. In particular, for T_α,

---

[19] As $H^1(\Gamma, C^\infty(\mathbf{R}))$, the group $\mathrm{Flows}(\mathbf{R}/\Gamma)$ is obviously a real vector space.

the only flow equipped with connections defined by a connection form is the Kronecker flow (with arbitrary speeds).

6 Modeling Diffeology

Now we have seen a few constructions in diffeology and applications to unusual situations: singular quotients and infinite-dimensional spaces. It will be interesting to revisit some constructions of differential geometry and see how diffeology views them.

6.1 Manifolds

Every manifold owns a natural diffeology, for which the plots are the smooth parameterizations. There is a definition internal to the category diffeology[20]:

Definition 6.1 An n-manifold is a diffeological space locally diffeomorphic to $\mathbf{R}^n$ at each point.

With this definition, as diffeological spaces, {Manifolds} form a full subcategory of the category {Diffeology}.

Of course, this definition needs a precise use of the wording *locally diffeomorphic*.

6.2 Local Smooth Maps, D-Topology, and So On

Very soon after the initial works in diffeology, it was clear that we needed to enrich the theory with local considerations, which were missing until then. To respect the spirit of diffeology, we defined directly the concept of local smoothness [16], as follows:

Definition 6.2 Let X and X' be two diffeological spaces. Let f be a map from a subset $A \subset X$ into X'. We say that f is *local smooth* if, for each plot P in X, $f \circ P$ is a plot of X'.

Note that $f \circ P$ is defined on $P^{-1}(A)$, and a first condition for $f \circ P$ to be a plot of X' is that $P^{-1}(A)$ be open. That leads to the second definition:

Definition 6.3 We say that a subset $A \subset X$ is *D-open* if $P^{-1}(A)$ is open for all plots P in X. The D-open subsets in X define a topology on X called the *D-topology*.

---

[20] We could use too the concept of generating families; see [22, Section 1.66].

We have, then, the following proposition linking these two definitions:

Proposition 6.4 *A map f defined on a subset $A \subset X$ to X' is local smooth if and only if A is D-open, and $f : A \to X'$ is smooth when A is equipped with the subset diffeology.*

To avoid misunderstanding and signify that f is local smooth — not just smooth for the subset diffeology – we note $f : X \supset A \to X$.

Next, since we have local smooth maps, we have *local diffeomorphisms* too.

Definition 6.5 We say that $f : X \supset A \to X'$ is a *local diffeomorphism* if f is injective, if it is local smooth, and if its inverse $f^{-1} : X' \supset f(A) \to X$ is local smooth. We say that f is a local diffeomorphism at $x \in X$ if there is a superset A of x such that $f \upharpoonright A : X \supset A \to X'$ is a local diffeomorphism.

In particular, these definitions give a precise meaning to the sentence "the space X is locally diffeomorphic to X' at each/some point."

6.3 Orbifold as Diffeologies

The word *orbifold* was coined by William P. Thurston [51] in 1978 as a replacement for *V-manifold*, a structure invented by Ichiro Satake in 1956 [42].

These new objects have been introduced to describe the smooth structure of spaces that look like manifolds, except around a few points, where they look like quotients of Euclidean domains by finite linear groups. Satake captured the smooth structure around the singularities by a family of compatible *local uniformizing systems* defining the orbifold.[21]

Figure 1.3 gives an idea about what would be a uniformizing system for the "teardrop" with one conic singularity.

The main problem with Satake's definition is that it does not lead to a satisfactory notion of smooth maps between orbifolds and therefore prevents the conception of a category of orbifolds. Indeed, in [43, p. 469], Satake writes this footnote:

> *The notion of C^∞-map thus defined is inconvenient in the point that a composite of two C^∞-maps defined in a different choice of defining families is not always a C^∞ map.*

For a mathematician, that is very annoying.

Considering orbifolds as diffeologies solved the problem. Indeed, in [28], we defined a *diffeological orbifold* by a modeling process in the same spirit as for smooth manifolds

---

[21] We will not discuss this construction here. The original description by Satake is found in [42], and a discussion of this definition is in [28].

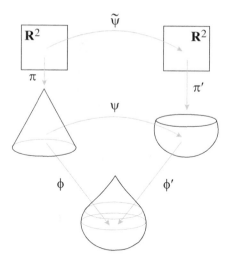

Figure 1.3 The teardrop as Satake's orbifold.

Definition 6.6 An orbifold is a diffeological space that is localy diffeomorphic, at each point, to some quotient space $\mathbf{R}^n / \Gamma$, for some finite subgroup Γ of the linear group $\mathrm{GL}(n, \mathbf{R})$, *depending possibly on the point.*

Figure 1.4 gives an idea about what is a diffeological orbifold: the teardrop as some diffeology on the sphere S^2.

Once this definition is given, we could prove op. cit. that, according to this definition, every Satake defining family of a local uniformizing system was associated with a diffeology of orbifold and, conversely, that every diffeological orbifold was associated with a Satake defining family of a local uniformizing system. And we proved that these constructions are inverse to each other, modulo equivalence.

Thus, the diffeology framework fulfilled Satake's program by embedding orbifolds into diffeological spaces and providing them naturally with good, workable, smooth mappings.

The difficulty met by Satake is subtle and can be explained as follows: he tried to define smooth maps between orbifolds as maps that have equivariant liftings on the level of Euclidean domains, before quotienting. But the embedding of orbifolds into {Diffeology} shows that, if that is indeed satisfied for local diffeomorphisms (see [28, Lemma 20, 21, 22]), it is not necessarily the case for ordinary smooth maps, as this counterexample shows.

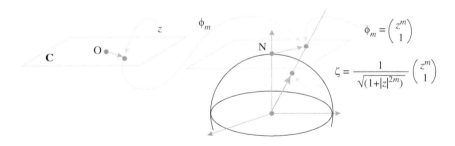

Figure 1.4 The teardrop as diffeology.

Consider the cone orbifolds $\mathcal{Q}_n = \mathbf{R}^2/\mathbf{Z}_n$, and let $f: \mathbf{R}^2 \to \mathbf{R}^2$:

$$f(x,y) = \begin{cases} 0 & \text{if } r > 1 \text{ or } r = 0, \\ e^{-1/r}\rho_n(r)(r,0) & \text{if } \frac{1}{n+1} < r \leq \frac{1}{n} \text{ and } n \text{ is even,} \\ e^{-1/r}\rho_n(r)(x,y) & \text{if } \frac{1}{n+1} < r \leq \frac{1}{n} \text{ and } n \text{ is odd,} \end{cases}$$

where $r = \sqrt{x^2 + y^2}$ and ρ_n is a smooth bump function that is zero outside the interval $[1/(n+1), 1/n]$. Then, for all integers m dividing n, f projects onto a smooth map $\phi \colon \mathcal{Q}_m \to \mathcal{Q}_n$ that cannot be lifted locally in an equivariant smooth map, over a neighborhood of 0.

Again, diffeology structured the problem in a way that almost solved it.

6.4 Noncommutative Geometry and Diffeology: The Case of Orbifolds

The question of a relation between diffeology and noncommutative geometry appeared immediately with the study of the irrational torus [7]. The condition of diffeomorphy between two irrational tori T_α and T_β, that is, α and β conjugate modulo $GL(2,\mathbf{Z})$, coincided clearly in noncommutative geometry, with the Morita-equivalent of the $\mathbf{C}^*$-algebras associated with the foliations [41]. That suggested a structural relationship between diffeology and noncommutative geometry deserving to be explored, but that question had been left aside since then.

We recently reopened the case of this relationship, considering orbifolds. And we exhibited a simple construction associating naturally with every diffeological orbifold a $\mathbf{C}^*$-algebra, such that two diffeomorphic orbifolds give two Morita-equivalent algebras [29].

Before continuing, we need to recall a general definition [22, Sections 1.66 and 1.76]:

Definition 6.7 Let X be a diffeological space. We define the *nebula* of a set $\mathcal{F}$ of plots in X as the diffeological sum

$$\mathcal{N} = \coprod_{P \in \mathcal{F}} \mathrm{dom}(P) = \{(P, r) \mid P \in \mathcal{F},\, r \in \mathrm{dom}(P)\},$$

where each domain is equipped with its standard smooth diffeology. Then, we define the *evaluation map*

$$\mathrm{ev}\colon \mathcal{N} \to X \quad \text{by} \quad \mathrm{ev}\colon (P, r) \mapsto P(r).$$

We say that $\mathcal{F}$ is a *generating family* of X if ev is a subduction.

Now, let $\mathcal{Q}$ be an orbifold. A local diffeomorphism F from a quotient $\mathbf{R}^n / \Gamma$ to $\mathcal{Q}$ will be called a *chart*. A set $\mathcal{A}$ of charts whose images cover $\mathcal{Q}$ will be called an *atlas*. With every atlas $\mathcal{A}$ is associated a special generating family $\mathcal{F}$ by considering the strict lifting of the charts $F \in \mathcal{A}$ to the corresponding $\mathbf{R}^n$. Precisely, let $\pi_\Gamma = \mathbf{R}^n \to \mathbf{R}^n / \Gamma$, $F\colon \mathbf{R}^n / \Gamma \supset \mathrm{dom}(F) \to \mathcal{Q}$; then $\mathcal{F} = \{F \circ \pi_\Gamma\}_{F \in \mathcal{A}}$. We call $\mathcal{F}$ the *strict generating family* associated with $\mathcal{A}$, and we denote by $\mathcal{N}$ its nebula.

Next, we consider the groupoid **G** whose objects are the points of the nebula $\mathcal{N}$ and the arrows the germs of local diffeomorphisms φ of $\mathcal{N}$ that project on the identity by ev, that is, $\mathrm{ev} \circ \varphi = \mathrm{ev}$. We call **G** the *structure groupoid* of $\mathcal{Q}$. Then, for a suitable but natural *functional diffeology* on **G**, we have the following [29]:

Theorem 6.8 *The groupoid **G** is Hausdorff and etale. The groupoids associated with different atlases are equivalent as categories.*

Hence, thanks to the etale property, we can associate a $\mathbf{C}^*$-algebra $\mathfrak{A}$ with **G** by the process described by Renaud in [40]. We have, then, the following:

Theorem 6.9 *The groupoids associated with different atlases of an orbifold are equivalent in the sense of Muhly–Renault–Williams [37]. Therefore, their $\mathbf{C}^*$-algebras are Morita-equivalent.*

This Morita-equivalence between the $\mathbf{C}^*$-algebras associated with different atlases of the orbifold is the condition required to make this construction meaningful and categorical. That construction is the first bridge between diffeology and noncommutative geometry; it gives an idea where and how diffeology and noncommutative geometry respond to each other, at least at the level of orbifolds.[22]

---

[22] *Quasifolds* are a more general subcategory of {Diffeology} for which the same construction leads to the same conclusion. It is the subject of a recent work, "Quasifolds, diffeology and noncommutative geometry."

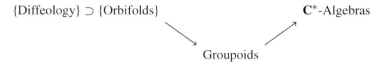

As an example, let us consider the simple orbifold $\Delta_1 = \mathbf{R}/\{\pm 1\}$. The structure of the orbifold is represented by the pushforward of the standard diffeology from $\mathbf{R}$ to $[0, \infty[$, by the square map sqr: $t \mapsto t^2$. The singleton $\mathcal{F} = \{\text{sqr}\}$ is a strict generating family, and the structure groupoid $\mathbf{G}$ is the groupoid of the action of $\Gamma = \{\pm 1\}$, that is,

$$\text{Obj}(\mathbf{G}) = \mathbf{R} \quad \text{and} \quad \text{Mor}(\mathbf{G}) = \{(t, \varepsilon, \varepsilon t) \mid \varepsilon = \pm 1\} \simeq \mathbf{R} \times \{\pm 1\}.$$

A continuous function f on $\text{Mor}(\mathbf{G})$ to $\mathbf{C}$ is a pair of functions $f = (a, b)$, where $a(t) = f(t, 1)$ and $b(t) = f(t, -1)$. With this convention, the algebra of the orbifold is then represented by a submodule of $M_2(\mathbf{C}) \otimes C^0(\mathbf{R}, \mathbf{C})$,

$$f = (a, b) \mapsto M = \left[t \mapsto \begin{pmatrix} a(t) & b(-t) \\ b(t) & a(-t) \end{pmatrix} \right],$$

with $M^*(t) = [{}^\tau M(t)]^*$. The superscript τ represents the transposition, and the asterisk represents the complex conjugation element by element.

Note that we still trace the orbifold in the characteristic polynomial $P_M(\lambda)$, which is invariant by the action of $\{\pm 1\}$, and is then defined on the orbifold Δ_1 itself, $P_M(\lambda): t \mapsto \lambda^2 - \lambda \text{tr}(M(t)) + \det(M(t))$, where $\text{tr}(M(t)) = a(t) + a(-t)$ and $\det(M(t)) = a(t)a(-t) - b(t)b(-t)$ are obviously $\{\pm 1\}$ invariant.

6.5 Manifolds with Boundary and Corners

One day, in 2007, I received an email from a colleague, wondering how diffeology behaves around the corners. Here is an excerpt:

> *I have just one worry about this theory. I found it very difficult to check that there's a diffeology on the closed interval* $[0, 1]$ *such that a smooth function* $f: [0, 1] \to \mathbf{R}$ *is smooth in the usual sense, even at the endpoints. . . . This problem would become very easy to solve using Chen's definition of smooth space, which allows for plots whose domain is any convex subset of* $\mathbf{R}^n$.

He ended his remark with this wish: that people interested in diffeology "are hoping you could either solve this problem in the context of diffeologies, or switch to Chen's definition."

The good news is that it is not necessary to give up diffeology. There is indeed a diffeology on $[0, 1]$ such that "a smooth function $f: [0, 1] \to \mathbf{R}$ is smooth in the usual sense." And that diffeology is simply the subset diffeology, precisely [22, Section 4.13],

Theorem 6.10 *Let $H_n \subset \mathbf{R}^n$ be the half-space defined by $x_1 \geq 0$, equipped with the subset diffeology. Let $f \in C^\infty(H_n, \mathbf{R})$; then there exists a smooth function F, defined on an open superset of H_n in $\mathbf{R}^n$, such that $f = F \restriction H^n$.*

This proposition is a consequence of a famous Whitney theorem on extension of smooth even functions [52]. By the way, it gives a solid basis to the vague concept of "smooth in the usual sense," as to be smooth in the usual sense means then to be smooth for the diffeology.

We can investigate further and characterize the local diffeomorphisms of half-spaces [22, Section 4.14]:

Theorem 6.11 *A map $f : A \to H_n$, with $A \subset H_n$, is a local diffeomorphism for the subset diffeology if and only if A is open in H_n, f is injective, $f(A \cap \partial H_n) \subset \partial H_n$, and for all $x \in A$ there exist an open ball $\mathcal{B} \subset \mathbf{R}^n$ centered at x and a local diffeomorphism $F: \mathcal{B} \to \mathbf{R}^n$ such that f and F coincide on $\mathcal{B} \cap H_n$.*

Thanks to this theorem, it is then easy to include the manifolds with boundary into the category {Diffeology}, in the same way we included the categories {Manifolds} and {Orbifolds}.

Definition 6.12 An n-dimensional *manifold with boundary* is a diffeological space X which is diffeomorphic, at each point, to the half-space H_n. We say that X is *modeled* on H_n.

Thanks to the previous theorems, it is clear that this definition covers completely, and not more, the usual definition of manifold with boundary one can find for example in [9] or [36]. In other words, the ordinary category {Manifolds with Boundary} is a natural full subcategory of {Diffeology}, the category {Manifolds} being itself a full subcategory of {Manifolds with Boundary}.

Moreover, we have a similar result for the subset diffeology of corners, thanks to a Schwartz theorem[23] [44]:

Theorem 6.13 *Let K^n be the positive n-corner in $\mathbf{R}^n$ defined by $x_i \geq 0$, with $i = 1, \ldots, n$, equipped with the subset diffeology. Let $f \in C^\infty(K^n, \mathbf{R})$; then there exists a smooth function F, defined on an open superset of K^n in $\mathbf{R}^n$, such that $f = F \restriction K^n$.*

Actually, this property extends to any differential k-form on K^n: it is the restriction of a smooth k-form on $\mathbf{R}^n$ [12].

---

[23] In this case, it is a simple corollary of the Whitney theorem [52, Remark p. 310].

As for manifolds with boundary, it is natural to define the n-dimensional *manifolds with corners* as diffeological spaces that are locally diffeomorphic to K^n at each point. We get naturally then the category {Manifolds with Corners} as a new subcategory of {Diffeology}.

Back to the alternative Chen versus Souriau: the three axioms of diffeologies are indeed identical to the preceding three axioms of Chen's differentiable spaces [3], except for the domains of plots that are open instead of being convex. Chen's spaces had been introduced with homology and cohomology in mind, and that is why he chose the convex subsets as domains for his plots. On another side, the choice of open subsets positions diffeology in the same conceptual category as differential geometry. And now, the fact that smooth maps for half-spaces or corners, equipped with the subset diffeology, coincide with what was guessed heuristically to describe manifolds with boundary or corners is another confirmation that there is no need to amend the theory in any way. For example, we could define smooth simplices in diffeological spaces as smooth maps from the standard simplices, equipped with its subset diffeology. And that would cover the usual situation in differential geometry.

6.6 Frölicher Spaces as Reflexive Diffeological Spaces

We recall that a Frölicher structure on a set X is defined by a pair of sets $\mathcal{F} \subset \mathrm{Maps}(X, \mathbf{R})$ and $\mathcal{C} \subset \mathrm{Maps}(\mathbf{R}, X)$ that satisfies the double condition

$$\mathcal{C} = \{c \in \mathrm{Maps}(\mathbf{R}, X) \mid \mathcal{F} \circ c \subset C^\infty(\mathbf{R}, \mathbf{R})\}$$
$$\mathcal{F} = \{f \in \mathrm{Maps}(X, \mathbf{R}) \mid f \circ \mathcal{C} \subset C^\infty(\mathbf{R}, \mathbf{R})\}.$$

A set X equipped with a Frölicher structure is called a Frölicher space [34]. Now, let X be a diffeological space, $\mathcal{D}$ be its diffeology, and $C^\infty(X, \mathbf{R})$ be its set of real smooth maps.

Definition 6.14 We say that X is *reflexive* if $\mathcal{D}$ coincides with the coarsest diffeology on X, for which the set of real smooth maps is exactly $C^\infty(X, \mathbf{R})$.

Then, thanks to Boman's theorem [1], one can show[24] that a Frölicher space, equipped with the coarsest diffeology such that the elements of $\mathcal{F}$ are smooth, is reflexive. And one can check conversely that a reflexive diffeological space satisfies the Frölicher conditions above; see [22, Exercises 79 and 80]. In other words,

---

[24] The concept of reflexive space has been suggested by Yael Karshon, and we established the equivalence with Frölicher spaces, together with Augustin Batubenge and Jordan Watts, at a seminar in Toronto in 2010.

Proposition 6.15 *The category of Frölicher spaces coincides with the subcategory of reflexive diffeological spaces.*

7 Cartan–de Rham Calculus

With fiber bundles and homotopy, differential calculus is one of the most developed domains in diffeology. We begin first with the definition of a differential form on a diffeological space.

7.1 Differential Forms

Let $U \subset \mathbf{R}^n$; we denote by $\Lambda^k(\mathbf{R}^n)$ the vector space of linear k-forms on $\mathbf{R}^n$, $k \in \mathbf{N}$. We call *smooth k-form* on U any smooth map $a \colon U \to \Lambda^k(\mathbf{R}^n)$.

Now, let X be a diffeological space.

Definition 7.1 We call *differential k-form* on X any map α that associates, with every plot $P \colon U \to X$, a smooth k-form $\alpha(P)$ on U that satisfies the chain-rule condition

$$\alpha(P \circ F) = F^*(\alpha(P))$$

for all smooth parameterizations F in U.

The set of k-forms on X is denoted by $\Omega^k(X)$. Note that $\Omega^0(X) = C^\infty(X, \mathbf{R})$.

Note also that one can consider an n-domain U as a diffeological space; in this case, a differential k-form α is immediately identified with its value $a = \alpha(\mathbf{1}_U)$ on the identity, $\alpha \in \Omega^k(U)$ and $a \in C^\infty(U, \Lambda^k(\mathbf{R}^n))$.

The two main operations on the differential forms on a diffeological space are as follows:

1. THE PULLBACK. Let $f \in C^\infty(X', X)$ be a smooth map between diffeological spaces, and let $\alpha \in \Omega^k(X)$. Then $f^*(\alpha) \in \Omega^k(X')$ is the k-form defined by

$$[f^*(\alpha)](P') = \alpha(f \circ P'),$$

   for all plots P' in X'.
2. THE EXTERIOR DERIVATIVE. Let $\alpha \in \Omega^k(X)$; its exterior derivative $d\alpha \in \Omega^{k+1}(X)$ is defined by

$$[d\alpha](P) = d[\alpha(P)],$$

   for all plots P in X.

Then, we have a de Rham complex $\Omega^*(X)$, with an endomorphism d that satisfies

$$d \circ d = 0, \quad \text{and} \quad \begin{cases} Z_{dR}^*(X) = \operatorname{Ker}(d: \Omega^*(X) \to \Omega^{*+1}(X)) \\ B_{dR}^*(X) = d(\Omega^{*-1}(X)) \subset Z_{dR}^*(X). \end{cases}$$

This defines a de Rham cohomology series of groups

$$H_{dR}^k(X) = Z_{dR}^k(X)/B_{dR}^k(X).$$

Note that this series begins with $k = 0$, for which $B_{dR}^0(X) = \{0\}$.

The first cohomology group $H_{dR}^0(X)$ is easy to compute. The differential df of a smooth function $f \in \Omega^0(X)$ vanishes if and only if f is constant on the connected components of X. Thus, $H_{dR}(X)$ is the real vector space generated by $\pi_0(X)$, that is, $\operatorname{Maps}(\pi_0(X), \mathbf{R})$.

7.2 Quotienting Differential Forms

One of the main procedures on differential forms is *quotienting forms*. That means the following: let X and X' be two diffeological spaces, and let $\pi: X \to X'$ be a subduction. The following criterion [49] identifies $f^*(\Omega^*(X'))$ into $\Omega^*(X)$:

Proposition 7.2 *Let $\alpha \in \Omega^*(X)$. There exists $\beta \in \Omega^*(X')$ such that $\alpha = \pi^*(\beta)$ if and only if, for all pairs of plots P and P' in X, if $\pi \circ P = \pi \circ P'$, then $\alpha(P) = \alpha(P')$.*

That proposition helps us to compute $H_{dR}(T_\alpha)$, for example. Actually, the criterion above has a simple declination for coverings.

Proposition 7.3 *Let X be a diffeological space and $\pi: \tilde{X} \to X$ its universal covering. Let $\tilde{\alpha} \in \Omega^*(\tilde{X})$. There exists $\alpha \in \Omega^*(X)$ such that $\tilde{\alpha} = \pi^*(\alpha)$ if and only if $\tilde{\alpha}$ is invariant by $\pi_1(X)$, that is, $k^*(\tilde{\alpha}) = \tilde{\alpha}$ for all $k \in \pi_1(X)$.*

Indeed, for the criterion above, $\pi \circ P = \pi \circ P'$ if and only if, locally on each ball in the domain of the plots, there exists an element $k \in \pi_1(X)$ such that $P' = k \circ P$ on the ball.

Now, let us apply this criterion to any irrational torus $T = \mathbf{R}/\Gamma$, where Γ is a strict dense subgroup of $\mathbf{R}$. A 1-form $\tilde{\alpha} \in \mathbf{R}$ writes $\tilde{a}(x)\, dx$. It is invariant by Γ if and only if $a(x) = a$ is constant. Let θ be the 1-form whose pullback is dx; then

$$\Omega^1(T) = \mathbf{R}\theta \quad \text{and} \quad H_{dR}^1(T) = \mathbf{R}.$$

Obviously, $H_{dR}^0(T) = \mathbf{R}$ and $H_{dR}^k(T) = \{0\}$ if $k > 1$.

7.3 Parasymplectic Form on the Space of Geodesics

It is well known that, if the space Geod(M) of (oriented) *geodesic trajectories* (aka unparameterized geodesics) of a Riemannian manifold (M, g) is a manifold, then this manifold is naturally symplectic for the quotient of the presymplectic form defining the geodesic flow. A famous example is the geodesics of the sphere S^2, for which the space of geodesics is also S^2, equipped with the standard surface element.[25] In this case, the mapping from the unit bundle US^2 to $Geod(S^2)$ is realized by the moment map of the rotations:

$$\ell \colon US^2 = \{(x, u) \in S^2 \times S^2 \mid u \cdot x = 0\} \to Geod(S^2) \text{ with } \ell(x, u) = x \wedge u.$$

Now, what about the space of geodesics of the 2-torus $T^2 = \mathbf{R}^2/\mathbf{Z}^2$, for example? It is certainly not a manifold because of the mix of closed and unclosed geodesics. And about the canonical symplectic structure, does it remain something from it? And what? That is exactly the kind of question diffeology is able to answer.

The geodesics of T^2 are the characteristics of the differential $d\lambda$ of the Liouville 1-form λ on UT^2, associated with the ordinary Euclidean product

$$\lambda(\delta y) = u \cdot \delta x, \quad \text{with} \quad y = (x, u) \in UT^2 \quad \text{and} \quad \delta y \in T_y(UT^2).$$

And then,

$$Geod(T^2) = \left\{ pr\{x + tu\}_{t \in \mathbf{R}} \times \{u\} \subset T^2 \times S^1 \mid (x, u) \in UT^2 \right\}.$$

The direction of the geodesic $pr_2 \colon (pr\{x + tu\}_{t \in \mathbf{R}}, u) \mapsto u$ is a natural projection on S^1:

$$\begin{array}{ccc} UT^2 & \xrightarrow{\pi} & Geod(T^2) \\ & \searrow pr_2 \quad pr_2 \swarrow & \\ & S^1 & \end{array}$$

The fiber $pr_2^{-1}(u) \subset Geod(T^2)$ is the torus T_u of all lines with slope u. We have seen that, depending on whether the slope is rational, we get a circle or an irrational torus. As we claimed, $Geod(T^2)$ equipped with the quotient diffeology of UT^2 is not a manifold. However, there exists on $Geod(T^2)$ a closed 2-form ω such that $d\lambda = \pi^*(\omega)$ [25]. We say that $Geod(T^2)$ is parasymplectic.[26]

---

[25] For a judicious choice of constant.
[26] The meaning of the word *symplectic* in diffeology is still under debate. That is why I use the wording *parasymplectic* to indicate a closed 2-form.

Actually, this example is just a special case of the general situation [26].

Theorem 7.4 *Let* M *be a Riemannian manifold. Let* Geod(M) *be the space of geodesics, defined as the characteristics of the canonical presymplectic 2-form* $d\lambda$ *on the unit bundle* UM. *Then, there exists a closed 2-form* ω *on* Geod(M) *such that* $d\lambda = \pi^*(\omega)$.

This result is a direct application of the criterion above on quotienting forms. It is strange that we had to wait so long to clarify this important point, which should have been one of the first results in diffeology.

7.4 Differential Forms on Manifolds with Corners

We claimed previously that the smooth maps $f : K_n \to \mathbf{R}$, where K_n is the n-dimensional corner, are the restrictions of smooth functions defined on some open neighborhood of K_n in $\mathbf{R}^n$. We have more [10]:

Proposition 7.5 *Let* $\omega \in \Omega^k(K^n)$ *be a differential k-form on* K^n. *Then, there exists a smooth k-form* $\bar\omega$ *defined on some open neighborhood of* $K^n \subset \mathbf{R}^n$ *such that* $\omega = \bar\omega \upharpoonright K^n$.

This proposition has, then, a corollary (op. cit.):

Theorem 7.6 *Let* M *be a smooth manifold. Let* $W \subset M$, *equipped with the subset diffeology, be a submanifold with boundary and corners. Any differential form on* W *is the restriction of a smooth form defined on an open neighborhood.*

That closes the discussion about the compatibility between diffeology and manifolds with boundary and corners for any question relative to the de Rham complex.

7.5 The Problem with the de Rham Homomorphism

Let us focus on 1-forms, precisely, the 1-forms on T_α, to take an example. The integration on paths defines the first de Rham homomorphism. Let ϵ be a closed 1-form on T_α (actually any 1-form, since they are all closed). Consider the map

$$\gamma \mapsto \int_\gamma \epsilon = \int_0^1 \epsilon(\gamma)_x(1)\, dx,$$

where $\gamma \in \text{Paths}(T_\alpha)$. Because ϵ is closed and because the integral depends only on the fixed end homotopy class of γ, restricted to $\text{Loops}(T_\alpha)$, this integral defines a homomorphism from $\pi_1(T_\alpha)$ to $\mathbf{R}$. And since the integral on a loop of a differential df vanishes, the integral depends only on the cohomology class of ϵ. Thus, we get the first de Rham homomorphism:[27]

$$h\colon H^1_{dR}(T_\alpha) \to \mathrm{Hom}(\pi_1(T_\alpha), \mathbf{R}) \quad \text{defined by} \quad \epsilon \mapsto [\ell \mapsto \int_\ell \epsilon].$$

Now, since $H^1_{dR}(T_\alpha) = \mathbf{R}$ and $\mathrm{Hom}(\pi_1(T_\alpha), \mathbf{R}) = \mathbf{R}^2$, the de Rham homomorphism cannot be an isomorphism, as it is the case for Euclidean domains, or more generally ordinary manifolds. Actually, it is precisely given by

$$h\colon a \mapsto [(n,m) \mapsto a(n + \alpha m)],$$

where $a \in H^1_{dR}(T_\alpha)$ is represented by $a\,dt$ on $\mathbf{R}$, and $\pi_1(T_\alpha) = \mathbf{Z} + \alpha \mathbf{Z} \subset \mathbf{R}$.

This hiatus is specific to diffeology versus differential geometry. It is, however, still true that, for any diffeological space, the first de Rham homomorphism is injective, and we can interpret geometrically its cokernel.[28]

Let $\tilde{T}_\alpha(=\mathbf{R})$ denote the universal covering of T_α. Consider then a homomorphism ρ from $\pi_1(T_\alpha)$ to $(\mathbf{R}, +)$. Then, build the associated bundle pr: $\tilde{T}_\alpha \times_\rho \mathbf{R} \to T_\alpha$, where $\pi_1(T_\alpha)$ acts diagonally on the product $\tilde{T}_\alpha \times \mathbf{R}$. That is, for all $(x,t) \in \tilde{T}_\alpha \times \mathbf{R}$ and all $k \in \pi_1(T_\alpha)$, $k\colon (x,t) \mapsto (x+k, t+\rho(k))$. Let class : $\tilde{T}_\alpha \times \mathbf{R} \to \tilde{T}_\alpha \times_\rho \mathbf{R}$ be the projection.

$$\begin{array}{ccc} \tilde{T}_\alpha \times \mathbf{R} & \xrightarrow{\text{class}} & \tilde{T}_\alpha \times_\rho \mathbf{R} \\ \mathrm{pr}_1 \downarrow & & \downarrow \mathrm{pr} \\ \tilde{T}_\alpha & \xrightarrow{\pi} & T_\alpha \end{array}$$

The right down arrow pr: class $(x,t) \mapsto \pi(x)$ is a principal $(\mathbf{R}, +)$ fiber bundle for the action $s\colon$ class $(x,t) \mapsto$ class $(x, t+s)$. This principal fiber bundle has a natural connection induced by the connection of the universal covering. Pick a path γ in T_α and a point $\tilde{x}$ over $x = \gamma(t)$; there exists a unique lifting $\tilde{\gamma}$ such that $\tilde{\gamma}(t) = \tilde{x}$. Then, we define the horizontal lifting $\bar{\gamma}$ of γ passing through class $(\tilde{x}, s)$ by $\bar{\gamma}(t') =$ class $(\tilde{\gamma}(t'), t' - t + s)$. This connection is, by construction, flat. Indeed, the subspace $\{$class $(\tilde{x}, 0) \mid \tilde{x} \in \tilde{T}_\alpha\}$ is a reduction of the principal fiber bundle pr: $\tilde{T}_\alpha \times_\rho \mathbf{R} \to T_\alpha$ to the group $\pi_1(T_\alpha)/\mathrm{Ker}(\rho)$. Now, in [22, Section 8.30], we prove that a homomorphism $\rho\colon \pi_1(T_\alpha) \to \mathbf{R}$ gives a trivial fiber bundle pr if and only if ρ is the de Rham homomorphism of a closed 1-form ϵ. And eventually, we prove the following:

---

[27] See [22, Section 6.74] for the general construction and for the justifications needed.
[28] For differential forms in higher degree, this is still a work in progress.

Proposition 7.7 *The cokernel of the de Rham homomorphism is equivalent to the set of equivalence classes of* $(\mathbf{R},+)$-*principal bundle over* T_α, *equipped with a flat connection. This result is actually general for any diffeological space* X.

Note that this analysis puts on an equal footing the surjectivity of the first de Rham homomorphism and the triviality of principal $(\mathbf{R},+)$-principal bundles over manifolds. That deserves to be noticed.

7.6 The Chain-Homotopy Operator

The *chain-homotopy operator* K is a fundamental construction in diffeology for differential calculus [22, Section 6.83]. It is related in particular to integration of closed differential forms, homotopic invariance of de Rham cohomology, and the moment map in symplectic geometry, as we shall see in the following.

Let X be a diffeological space; there exists a smooth linear operator

$$\mathrm{K}\colon \Omega^p(\mathrm{X}) \to \Omega^{p-1}(\mathrm{Paths}\,(\mathrm{X})) \quad \text{with} \quad p \geq 1$$

that satisfies

$$\mathrm{K} \circ d + d \circ \mathrm{K} = \hat{1}^* - \hat{0}^*,$$

where $\hat{0}, \hat{1} \colon \mathrm{Paths}\,(\mathrm{X}) \to \mathrm{X}$ are defined by $\hat{t}(\gamma) = \gamma(t)$.

Explicitly, let α be a p-form of X, with $p > 1$, and $\mathrm{P}\colon \mathrm{U} \to \mathrm{Paths}\,(\mathrm{X})$ be an n-plot. The value of $\mathrm{K}\alpha$ on the plot P, at the point $r \in \mathrm{U}$, evaluated on $(p-1)$ vectors $(v)_{i=2}^p = (v_2)\ldots(v_p)$ of $\mathbf{R}^n$, is given by

$$\mathrm{K}\alpha\,(\mathrm{P})_r(v)_{i=2}^p = \int_0^1 \alpha\left(\binom{t}{r} \mapsto \mathrm{P}(r)(t)\right)_{\binom{t}{r}} \binom{1}{0}\binom{0}{v_i}_{i=2}^p dt.$$

For $p = 1$,

$$\mathrm{K}\alpha\colon \gamma \mapsto \int_\gamma \alpha = \int_0^1 \alpha(\gamma)_t\, dt$$

is the usual integration along the paths.

Note that this operator is specific to diffeology since there is no concept of differential forms on the space of paths in traditional differential geometry. Of course, there are a few bypasses, but none as direct or efficient as the operator K, as we shall see now.

7.7 Homotopic Invariance of de Rham Cohomology

Consider an homotopy $t \mapsto f_t$ in Paths $(C^\infty(X, X'))$, where X and X' are two diffeological spaces.

Proposition 7.8 *Let $\alpha \in \Omega^k(X')$ and $d\alpha = 0$. Then, $f_1^*(\alpha) = f_0^*(\alpha) + d\beta$, with $\beta \in \Omega^{k-1}(X)$.*

In other words, the de Rham cohomology is homotopic invariant:

$$\text{class}(f_1^*(\alpha)) = \text{class}(f_0^*(\alpha)) \in H_{dR}^k(X).$$

Let us prove this rapidly. Consider the smooth map $\varphi \colon X \to$ Paths (X') defined by $\varphi(x) = [t \mapsto f_t(x)]$. Take the pullback of the chain-homotopy identity, $\varphi^*(K(d\alpha) + d(K\alpha)) = \varphi^*(\hat{1}^*(\alpha)) - \varphi^*(\hat{0}^*(\alpha))$. That is, $\varphi^*(d(K\alpha)) = d(\varphi^*(K\alpha)) = (\hat{1} \circ \varphi)^*(\alpha) - (\hat{0} \circ \varphi)^*(\alpha)$. But, $\hat{t} \circ \varphi = f_t$, thus $f_1^*(\alpha) - f_0^*(\alpha) = d\beta$ with $\beta = \varphi^*(K\alpha)$.

That is one of the most striking uses of this chain-homotopy operator, and it proves at the same time how one can take advantage of diffeology, even in a traditional course on differential geometry.

7.8 Integration of Closed 1-Forms

Consider a manifold M and a closed 1-form α on M. We know that if α is integral, that is, its integral on every loop is a multiple of some number called the *period*, then there exists a smooth function f from M to the circle S^1 such that $\alpha = f^*(\theta)$, where θ is the canonical length element. This specific construction has an ultimate generalization in diffeology that avoids the integral condition – and that is what diffeology is for, indeed.

Let α be a closed 1-form on a connected diffeological space X. Consider the equivalence relation on Paths (X) defined by

$$\gamma \sim \gamma' \quad \text{if} \quad \text{ends}(\gamma) = \text{ends}(\gamma') \quad \text{and} \quad \int_\gamma \alpha = \int_{\gamma'} \alpha.$$

The quotient $\mathfrak{X}_\alpha = $ Paths $(X)/\sim$ is a groupoid for the addition[29] class $(\gamma) + $ class $(\gamma') = $ class $(\gamma \vee \gamma')$, when $\gamma(1) = \gamma'(0)$. Because the integral of α on γ does not depend on its fixed-endpoints homotopy class, the groupoid $\mathfrak{X}_\alpha$ is a covering groupoid, a quotient of the Poincaré groupoid $\mathfrak{X}$. Let $F_\alpha \colon \mathfrak{X}_\alpha \to \mathbf{R}$ be $F_\alpha(\text{class}(\gamma)) = K\alpha(\gamma) = \int_\gamma \alpha$.

---

[29] Actually, this addition is defined on the stationary paths, but since the space of stationary paths is a deformation retract of the space of paths, that does not really matter.

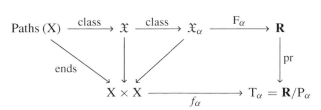

The function F integrates α on $\mathfrak{X}_\alpha$, in the sense that, thanks to the chain-homotopy identity, $dF_\alpha = \hat{1}^*(\alpha) - \hat{0}^*(\alpha)$. Let $o \in X$ be a point defined as an *origin* of X, and define X_α to be the subspace of $\mathfrak{X}_\alpha$ made of classes of paths with origin o. Let $\pi = \hat{1} \restriction X_\alpha$; then $dF = \pi^*(\alpha)$, with $F = F_\alpha \restriction X_\alpha$. The covering X_α of X is the smallest covering where the pullback of α is exact. We call it the integration covering of α. Its structure group P_α is the *group of periods* of α, that is,

$$P_\alpha = \left\{ \int_\ell \alpha \mid \ell \in \text{Loops}(X) \right\}.$$

Proposition 7.9 *If the group of periods P_α is a strict subgroup of $\mathbf{R}$, then* pr: $\mathbf{R} \to T_\alpha = \mathbf{R}/P_\alpha$ *is a covering, and there exists a smooth map* $f: X \to T_\alpha$ *such that* $f^*(\theta) = \alpha$, *with θ being the projection of dt on T_α by* pr.

This proposition [22, Section 8.29] is the best generalization of the integration of integral 1-forms in differential geometry, allowed by diffeology. We can notice that the usual condition of second countability for manifolds is in fact a sufficient precondition. The real obstruction, valid for manifolds as well as for general diffeological spaces, is that the group of periods P_α is discrete in $\mathbf{R}$, and that is equivalent to being a strict subgroup of $\mathbf{R}$.

8 Symplectic Diffeology

During the 1990s, a lot of symplectic-like geometry situations were explored, essentially in infinite-dimensional spaces, but not only, with more or less success. What these attempts at generalization missed was a uniform framework of concepts and vocabulary, precise definitions framing the context of their studies. Each example came with its own heuristic and context, for example, the nature of the moment maps was not clearly stated: for the case of the moment of imprimitivity [53], it was a function with values of Dirac delta functions (distributions); for another example involving the connections of a torus bundle [6], it was the curvature of the connection. And Elisa Prato's quasifolds [39], for which the moment map is defined on a space that is not a legitimate manifold but a singular quotient, adds up to these infinite-dimensional examples.

What we shall see now is how diffeology is a common framework, where all these examples find their places, are treated on an equal footing, and give what we are waiting for from them.

The general objects of interest will be arbitrary closed 2-forms on diffeological spaces, for which we introduce this new terminology:

Definition 8.1 We call *parasymplectic space*[30] any diffeological space X equipped with a closed 2-form ω.

Next, we consider the group of *symmetries* (or *automorphisms*) of ω, denoted by Diff(X, ω). The pseudogroup $\text{Diff}_{\text{loc}}(X, \omega)$ of local symmetries will play some role too. Then, to introduce the *moment map* for any group of symmetries G, we need to define some vocabulary and notations:[31]

Definition 8.2 We shall call *momentum*[32] of a diffeological group G any left-invariant 1-form. We denote by $\mathcal{G}^*$ its space of *momenta*, that is,

$$\mathcal{G}^* = \{\varepsilon \in \Omega^1(G) \mid L(g)^*(\varepsilon) = \varepsilon, \text{ for all } g \in G\}.$$

The set $\mathcal{G}^*$ is naturally a real vector space[33].

8.1 The Moment Map

Let (X, ω) be a parasymplectic space and G be a diffeological group. A *symmetric action* of G on (X, ω) is a smooth morphism $g \mapsto g_X$ from G to Diff(X, ω), where Diff(X, ω) is equipped with the functional diffeology. That is,

$$\text{for all } g \in G, \quad g_X^*(\omega) = \omega.$$

Now, to grab the essential nature of the moment map, which is a map from X to $\mathcal{G}^*$, we need to understand it in the simplest possible case. That is, when ω is exact, $\omega = d\alpha$, and when α is also invariant by G, $g_X^*(\alpha) = \alpha$. In these conditions, the moment map is given by

$$\mu : X \to \mathcal{G}^* \quad \text{with} \quad \mu(x) = \hat{x}^*(\alpha),$$

where $\hat{x} : G \to X$ is the *orbit map* $\hat{x}(g) = g_X(x)$. We check immediately that, since α is invariant by G, $\hat{x}^*(\alpha)$ is left invariant by G, and therefore $\mu(x) \in \mathcal{G}^*$.

---

[30] The quality of being *symplectic* or *presymplectic* will be discussed and get a precise meaning. The word *parasymplectic* seemed free and appropriate to denote a simple closed 2-form.
[31] Remember that a diffeological group is a group that is a diffeological space such that the multiplication and the inversion are smooth.
[32] Plural *momenta*.
[33] It is also a diffeological vector space for the functional diffeology, but we shall not discuss that point here.

But, as we know, not all closed 2-forms are exact, and even if they are exact, they do not necessarily have an invariant primitive. We shall see now how we can generally come to a situation, so close to the simple case above, that, modulo some minor subtleties, we can build a good moment map in all cases.

Let us consider now the general case, with X connected. Let K be the chain-homotopy operator, defined previously. Then, the differential 1-form $K\omega$, defined on Paths (X), satisfies $d[K\omega] = (\hat{1}^* - \hat{0}^*)(\omega)$, and $K\omega$ is invariant by G [22, Section 6.84]. Considering $\bar{\omega} = (\hat{1}^* - \hat{0}^*)(\omega)$ and $\bar{\alpha} = K\omega$, we are in the simple case $\bar{\omega} = d\bar{\alpha}$ and $\bar{\alpha}$ invariant by G. We can apply the construction above, and then we define the *paths moment map* by

$$\Psi: \text{Paths}(X) \to \mathcal{G}^* \quad \text{with} \quad \Psi(\gamma) = \hat{\gamma}^*(K\omega),$$

where $\hat{\gamma}: G \to \text{Paths}(X)$ is the orbit map $\hat{\gamma}(g) = g_X \circ \gamma$ of the path γ.

The paths moment map is additive with respect to the concatenation,

$$\Psi(\gamma \vee \gamma') = \Psi(\gamma) + \Psi(\gamma'),$$

and it is equivariant by G, which acts by composition on Paths (X) and by coadjoint action on $\mathcal{G}^*$. That is, for all $g, k \in G$ and $\epsilon \in \mathcal{G}^*$,

$$\text{Ad}(g): k \mapsto gkg^{-1} \quad \text{and} \quad \text{Ad}_*(g): \epsilon \mapsto \text{Ad}(g)_*(\epsilon) = \text{Ad}(g^{-1})^*(\epsilon).$$

Then, we define the *holonomy* of the action of G on X as the subgroup

$$\Gamma = \{\Psi(\ell) \mid \ell \in \text{Loops}(X)\} \subset \mathcal{G}^*.$$

The group Γ is made of (closed) Ad_*-invariant momenta. But $\Psi(\ell)$ depends only on the homotopy class of ℓ, so then Γ is a homomorphic image of $\pi_1(X)$, more precisely, it is abelianized.

Definition 8.3 If $\Gamma = \{0\}$, we say that the action of G on (X, ω) is *Hamiltonian*. The holonomy Γ is the obstruction for the action of the group G to be Hamiltonian.

Now, we can push forward the paths moment map on $\mathcal{G}^*/\Gamma$, as suggested by the commutative diagram

$$\begin{array}{ccc} \text{Paths}(X) & \xrightarrow{\Psi} & \mathcal{G}^* \\ {\scriptstyle \text{ends}}\downarrow & & \downarrow{\scriptstyle \text{class}} \\ X \times X & \xrightarrow{\psi} & \mathcal{G}^*/\Gamma \end{array}$$

We get then the *two-points moment map*:

$$\psi(x,x') = \text{class}(\Psi(\gamma)) \in \mathcal{G}^*/\Gamma, \text{ for any path } \gamma \text{ such that ends}(\gamma) = (x,x').$$

The additivity of Ψ becomes the Chasles's cocycle condition on ψ:

$$\psi(x,x') + \psi(x',x'') = \psi(x,x'').$$

The group Γ is invariant by the coadjoint action. Thus, the coadjoint action passes to the quotient group $\mathcal{G}^*/\Gamma$, and ψ is a natural group-valued moment map, equivariant for this quotient coadjoint action.

Because X is connected, there exists always a map

$$\mu: X \to \mathcal{G}^*/\Gamma \quad \text{such that} \quad \psi(x,x') = \mu(x') - \mu(x).$$

The solutions of this equation are given by

$$\mu(x) = \psi(x_0, x) + c,$$

where x_0 is a chosen point in X and c is a constant. These are the *one-point moment maps*. But these moment maps μ are a priori no longer equivariant. Their variance introduces a 1-cocycle θ of G with values in $\mathcal{G}^*/\Gamma$ such that

$$\mu(g(x)) = \text{Ad}_*(g)(\mu(x)) + \theta(g), \quad \text{with} \quad \theta(g) = \psi(x_0, g(x_0)) + \Delta c(g),$$

where Δc is the coboundary due to the constant c in the choice of μ. We say that the action of G on (X, ω) is *exact* when the cocycle θ is trivial. Defining

$$\text{Ad}^\theta_*(g): v \mapsto \text{Ad}_*(g)(v) + \theta(g), \quad \text{then} \quad \text{Ad}^\theta_*(gg') = \text{Ad}^\theta_*(g) \circ \text{Ad}^\theta_*(g').$$

The cocycle property of θ, that is, $\theta(gg') = \text{Ad}_*(g)(\theta(g')) + \theta(g)$, makes Ad^θ_* an action of G on the group $\mathcal{G}^*/\Gamma$. This action is called the *affine action*. For the affine action, the moment map μ is equivariant:

$$\mu(g(x)) = \text{Ad}^\theta_*(g)(\mu(x)).$$

This construction extends to the category {Diffeology}, the moment map for manifolds introduced by Souriau in [46]. When X is a manifold and the action of G is Hamiltonian, they are the standard moment maps he defined there. The remarkable point is that none of the constructions brought up above involves differential equations, and there is no need for considering a possible Lie algebra either. That is a very important point. The momenta appear as invariant 1-forms on the group, naturally, without intermediaries, and the moment map as a map in the space of momenta.

Note that the group of automorphisms $G_\omega = \text{Diff}(X, \omega)$ is a legitimate diffeological group. The above constructions apply and give rise to universal objects: *universal momenta* $\mathcal{G}^*_\omega$, *universal path moment map* Ψ_ω, *universal* *holonomy* Γ_ω, *universal two-points moment map* ψ_ω, *universal moment maps* μ_ω, and *universal Souriau's cocycles* θ_ω.

A *parasymplectic action* of a diffeological group G is a smooth morphism $h\colon G \to G_\omega$, and the objects, associated with G, introduced by the above moment map constructions are naturally subordinate to their universal counterparts.

We shall illustrate this construction by two examples in the next paragraphs. More examples can be found in [22, Sections 9.27–9.34].

8.2 Example 1: The Moment of Imprimitivity

Consider the cotangent space T*M of a manifold M, equipped with the standard symplectic form $\omega = d\lambda$, where λ is the Liouville form. Let G be the abelian group $C^\infty(M, \mathbf{R})$. Consider the action of G on T*M by

$$f\colon (x,a) \mapsto (x, a - df_x),$$

where $x \in M$, $a \in T_x^*M$, and df_x is the differential of f at the point x. Then, the moment map is given by

$$\mu\colon (x,a) \mapsto d[f \mapsto f(x)] = d[\delta_x],$$

where δ_x is the Dirac distribution $\delta_x(f) = f(x)$.

Since δ_x is a smooth function on $C^\infty(M, \mathbf{R})$, its differential is a 1-form. Let us check that this 1-form is invariant:

Let $h \in C^\infty(M, \mathbf{R})$, $L(h)^*(\mu(x)) = L(h)^*(d[\delta_x]) = d[L(h)^*(\delta_x)] = d[\delta_x \circ L(h)]$, but $\delta_x \circ L(h)\colon f \mapsto \delta_x(f + h) = f(x) + h(x)$. Then, $d[\delta_x \circ L(h)] = d[f \mapsto f(x) + h(x)] = d[f \mapsto f(x)] = \mu(x)$.

We see that in this case, the moment map identifies with a function with values distributions but still has the definite formal statute of a map into the space of momenta of the group of symmetries.

Moreover, this action is Hamiltonian and exact. This example, generalized to diffeological space, is developed in [22, Exercise 147].

8.3 Example 2: The 1-Forms on a Surface

Let Σ be a closed surface, oriented by a 2-form Surf. Consider $\Omega^1(\Sigma)$, the infinite-dimensional vector space of 1-forms on Σ, equipped with the functional diffeology. Let ω be the antisymmetric bilinear map defined on $\Omega^1(\Sigma)$ by

$$\omega(\alpha,\beta) = \int_\Sigma \alpha \wedge \beta,$$

for all α, β in $\Omega^1(\Sigma)$.[34] With this bilinear form is naturally associated a differential 2-form ω on $\Omega^1(\Sigma)$, defined by

$$\omega(P)_r(\delta r, \delta'r) = \int_\Sigma \frac{\partial P(r)}{\partial r}(\delta r) \wedge \frac{\partial P(r)}{\partial r}(\delta'r),$$

for all n-plots $P\colon U \to X$, for all $r \in U$, δr and $\delta'r$ in $\mathbf{R}^n$. Moreover, ω is the differential of the 1-form λ on $\Omega^1(\Sigma)$:

$$\omega = d\lambda \quad \text{with} \quad \lambda(P)_r(\delta r) = \frac{1}{2}\int_\Sigma P(r) \wedge \frac{\partial P(r)}{\partial r}(\delta r).$$

Define now the action of the additive group $C^\infty(\Sigma, \mathbf{R})$ on $\Omega^1(\Sigma)$ by

$$f\colon \alpha \mapsto \alpha + df.$$

Then, $C^\infty(\Sigma, \mathbf{R})$ acts by symmetry on $(\Omega^1(\Sigma), \omega)$, for all f in $C^\infty(\Sigma, \mathbf{R})$, $f^*(\omega) = \omega$.

The moment map of $C^\infty(\Sigma, \mathbf{R})$ on $\Omega^1(\Sigma)$ is given (up to a constant) by

$$\mu\colon \alpha \mapsto d\left[f \mapsto \int_\Sigma f \times d\alpha\right].$$

The moment map is invariant by the action of $C^\infty(\Sigma, \mathbf{R})$, that is, exact, and Hamiltonian. And here again, the moment map is a function with values distributions.

Now, $\mu(\alpha)$ is fully characterized by $d\alpha$. This is why we find in the literature on the subject that the moment map for this action is the exterior derivative (or curvature, depending on the authors) $\alpha \mapsto d\alpha$. As we see again in this example that diffeology gives a precise meaning by procuring a unifying context. One can find the complete conputation of this example in [22, Section 9.27].

8.4 Symplectic Manifolds Are Coadjoint Orbits

Because symplectic forms of manifolds have no local invariants, as we know thanks to Darboux's theorem, they have a huge group of automorphisms. This group is big enough to be transitive [2], so that the universal moment map identifies the symplectic manifold with its image, which, by equivariance, is a coadjoint orbit (affine or not) of its group of symmetries. In other words, coadjoint orbits are the universal models of symplectic manifolds.

---

[34] Since the exterior product $\alpha \wedge \beta$ is a 2-form of Σ, there exists $\varphi \in C^\infty(\Sigma, \mathbf{R})$ such that $\alpha \wedge \beta = \varphi \times \text{Surf}$. By definition, $\int_\Sigma \alpha \wedge \beta = \int_\Sigma \varphi \times \text{Surf}$.

Precisely, let M be a connected Hausdorff manifold, and let ω be a closed 2-form on M. Let $G_\omega = \mathrm{Diff}(M, \omega)$ be its group of symmetries and $\mathcal{G}_\omega^*$ its space of momenta. Let Γ_ω be the holonomy and μ_ω be a universal moment map with values in $\mathcal{G}_\omega^*/\Gamma_\omega$. We have, then, the following:

Theorem 8.4 *The form ω is symplectic, that is, nondegenerate, if and only if*

1. *the group G_ω is transitive on M;*
2. *the universal moment map $\mu_\omega \colon M \to \mathcal{G}_\omega^*/\Gamma_\omega$ is injective.*

This theorem is proved in [22, Section 9.23], but let us make some comments on the key elements. Consider the closed 2-form $\omega = (x^2+y^2)\, dx \wedge dy$; one can show that it has an injective universal moment map μ_ω. But its group G_ω is not transitive, since ω is degenerate in $(0,0)$, and only at that point. Thus, the transitivity of G_ω is necessary.

Let us give some hint about the sequel of the proof. Assume ω is symplectic. Let $m_0, m_1 \in M$ and p be a path connecting these points. For all $f \in C^\infty(M, \mathbf{R})$ with compact support, let

$$F \colon t \mapsto e^{t\,\mathrm{grad}_\omega(f)}$$

be the exponential of the symplectic gradient of the f. Then, F is a 1-parameter group of automorphisms, and its value on $\Psi_\omega(p)$ is

$$\Psi_\omega(p)(F) = [f(m_1) - f(m_0)] \times dt.$$

Now, if $\mu_\omega(m_0) = \mu_\omega(m_1)$, then there exists a loop ℓ in M such that $\Psi_\omega(p) = \Psi_\omega(\ell)$. Applied to the 1-plot F, we deduce $f(m_1) = f(m_0)$ for all f. Therefore $m_0 = m_1$, and μ_ω is injective.

Conversely, let us assume that G_ω is transitive and μ_ω is injective. By transitivity, the rank of ω is constant. Now, let us assume that ω is degenerate, that is, $\dim(\mathrm{Ker}(\omega)) > 0$. Since the distribution $\mathrm{Ker}(\omega)$ is integrable, given two different points m_0 and m_1 in a characteristic, there exists a path p connecting these two points and drawn entirely in the characteristic, that is, such that $dp(t)/dt \in \mathrm{Ker}(\omega)$ for all t. But that implies $\Psi_\omega(p) = 0$ [22, Section 9.20]. Hence, $\mu_\omega(m_0) = \mu_\omega(m_1)$. But we assumed μ_ω to be injective. Thus, ω is nondegenerate, that is, symplectic.

8.5 Hamiltonian Diffeomorphisms

Let (X, ω) be a parasymplectic diffeological space. We have seen above that the universal moment map μ_ω takes its values into the quotient $\mathcal{G}_\omega^*/\Gamma_\omega$, where the holonomy $\Gamma_\omega \subset \mathcal{G}_\omega^*$ is a subgroup made of closed Ad_*-invariant 1-forms

An Introduction to Diffeology

on G. This group Γ_ω is the very obstruction for the action of the group of symmetries $G_\omega = \text{Diff}(X, \omega)$ to be Hamiltonian.

Proposition 8.5 *There exists a largest connected subgroup* $\text{Ham}(X, \omega) \subset G_\omega$ *with vanishing holonomy. This group is called the group of* Hamiltonian diffeomorphisms. *Every Hamiltonian smooth action on* (X, ω) *factorizes through* $\text{Ham}(X, \omega)$.

Hence, the moment map $\bar{\mu}_\omega$ with respect to the action of $H_\omega = \text{Ham}(X, \omega)$ takes its values in the vector space of momenta $\mathcal{H}_\omega^*$. One can prove also that the Hamiltonian moment map $\bar{\mu}_\omega$ is still injective when X is a manifold. It is probably an embedding in the diffeological sense [22, Section 2.13], but that has still to be proved.

Perhaps most interesting is how this group is built. Let G_ω° be the neutral component of G_ω and $\pi : \tilde{G}_\omega^\circ \to G_\omega^\circ$ be its universal covering.[35] For all $\gamma \in \Gamma_\omega$, let $F_\gamma \in \text{Hom}^\infty(\tilde{G}_\omega^\circ, \mathbf{R})$ be the primitive of $\pi^*(\gamma)$, that is, $dF_\gamma = \pi^*(\gamma)$ and $F_\gamma(\mathbf{1}_{\tilde{G}_\omega^\circ}) = 0$. Next, let

$$F: \tilde{G}_\omega^\circ \to \mathbf{R}^{\Gamma_\omega} = \prod_{\gamma \in \Gamma_\omega} \mathbf{R} \quad \text{defined by} \quad F(\tilde{g}) = (F_\gamma(\tilde{g}))_{\gamma \in \Gamma_\omega}.$$

The map F is a smooth homomorphism, where $\mathbf{R}^{\Gamma_\omega}$ is equipped with the product diffeology. Then [22, Section 9.15],

$$\text{Ham}(X, \omega) = \pi(\text{Ker}(F)).$$

This definition gives the same group of Hamiltonian diffeomorphisms when X is a manifold [22, Section 9.16].

Now, let P_γ be the group of periods of $\gamma \in \Gamma_\omega$, that is,

$$P_\gamma = F(\pi_1(G_\omega^\circ, \mathbf{1}_{G_\omega^\circ})) = \left\{ \int_\ell \gamma \mid \ell \in \text{Loops}(G_\omega^\circ, \mathbf{1}) \right\}.$$

Then, let

$$\mathbf{P}_\omega = \prod_{\gamma \in \Gamma_\omega} P_\gamma \quad \text{and} \quad \mathbf{T}_\omega = \prod_{\gamma \in \Gamma_\omega} T_\gamma \quad \text{with} \quad T_\gamma = \mathbf{R}/P_\gamma.$$

Each homomorphism F_γ projects onto a smooth homomorphism $f_\gamma : G_\omega^\circ \to T_\gamma$, and the homomorphism F projects onto a smooth homomorphism $f : G_\omega^\circ \to \mathbf{T}_\omega$, according to the following commutative diagram of smooth homomorphisms:

---

[35] For universal covering of diffeological groups, see [22, Section 7.10].

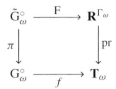

with $\mathrm{Ker}(\pi) = \pi_1(G_\omega^\circ, \mathbf{1}_{G_\omega^\circ})$, $\mathrm{Ker}(\mathrm{pr}) = \mathbf{P}_\omega = F(\mathrm{Ker}(\pi))$. We get, then,

$$\mathrm{Ham}(X,\omega) = \mathrm{Ker}(f),$$

an alternative, somewhat intrinsic, definition of the group of Hamiltonian diffeomorphisms as the kernel of the *holonomy homomorphism f*.

9 In Conclusion

We have proposed in this text to switch from the rigid category {Manifolds} to the flexible category {Diffeology}, which is now well developed. This is a category closed under every set-theoretic operation, complete, cocomplete, and cartesian closed, which includes on an equal footing manifolds, singular quotients, and infinite-dimensional spaces. It is an ideal situation already, from the pure point of view of categoricians, and mainly the reason for the interest in this theory in that discipline; see, for example, "Model Category Structures" [4], [31] or "Differentiable Homotopy Theory"[36] [30].[37]

But of course, the primary interest of diffeology lies first and foremost in its very strength in geometry. The geometer will find pleasant and useful the flexibility of diffeology, to extend in a unique formal and versatile framework different constructions in various fields, without inventing each time a heuristic framework that momentarily satisfies its needs. For example, the construction of the moment map and the integration of any closed 2-form on any diffeological space [22, Section 8.42] are prerequisite for an extension of symplectic geometry on spaces that are not manifolds but that have burst into mathematical physics these last decades with the problems of quantization and field theory.

Then, beyond all these circumstances and technicalities, what does diffeology have to offer on a more formal or conceptual level? The answer lies partly in Felix Klein's Erlangen program [33]:

---

[36] Different from geometric homotopy theory, which we described previously in this chapter.

[37] The Japanese school is very productive in these fields these days, using diffeology as a tool or general framework. One can consult other papers on the subject, already published or not, for examples [14, 15, 32, 35, 45].

The totality of all these transformations we designate as the principal group *of space-transformations;* geometric properties are not changed by the transformations of the principal group... *And, conversely,* geometric properties are characterized by their remaining invariant under the transformations of the principal group.

As a generalization of geometry arises then the following comprehensive problem:

Given a manifoldness and a group of transformations of the same; to investigate the configurations belonging to the manifoldness with regard to such properties as are not altered by the transformations of the group.

As we know, these considerations are regarded by mathematicians as the modern understanding of the word/concept of *geometry*. A geometry is given as soon as a space and a group of transformations of this space are given.[38]

Consider Euclidean geometry, defined by the group of Euclidean transformations, our *principal group*, on the Euclidean space. We can interpret, for example, the *Euclidean distance* as the invariant associated with the action of the Euclidean group on the set of pairs of points. We can superpose a pair of points onto another pair of points, by an Euclidean transformation, if and only if the distance between the points is the same for the two pairs.[39] Hence, *geometric properties* or *geometric invariants* can be regarded as the orbits of the *principal group* in some spaces built on top of the principal space and also as fixed/invariant points, since an orbit is a fixed point in the set of all the subsets of that space. In brief, what emerges from these considerations suggested by Felix Klein's principle is the following.

Claim A geometry is associated with/defined by a *principal group* of transformations of some space, according to Klein's statement. The various geometric properties/invariants are described by the various actions of the principal group on spaces built on top of the principal space: products, sets of subsets, and so on. Each one of these properties, embodied as orbits, stabilizers, quotients, and so on, captures a part of this geometry.

Now, how does diffeology fit into this context?

- One can regard a diffeological space as the collection of the plots that gives its structure. That is the *passive approach*.
- Or we can look at the space through the action of its group of diffeomorphisms:[40] on itself, but also on its powers or parts or maps. That is the *active approach*.

---

[38] Jean-Marie Souriau reduces the concept of geometry to the group itself [50]. But this is an extreme point of view I am not confident to share, for several reasons.

[39] We could continue with the case of triangles and other elementary constructions – circles, parallels, etc. – and a comparison between Euclidean and symplectic geometry, for example, from a strict Kleinian point of view. See the discussion in [19].

[40] Personally, I consider precisely the action of the pseudo-group of local diffeomorphisms.

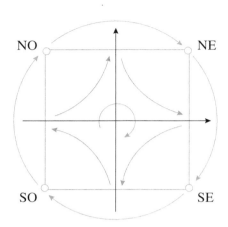

Figure 1.5 A diffeomorphism of the square.

This dichotomy appears already for manifolds, where the change of coordinates (transition functions of an atlas) is the passive approach. The active approach, as the action of the group of diffeomorphisms, is often neglected, and there are a few reasons for that. Among them, the group of diffeomorphisms is not a Lie group *stricto sensu* – it does not fit in the category {Manifolds} – and that creates a psychological issue. A second reason is that its action on the manifold itself is transitive[41] – there are no immediate invariants, one having first to consider some secondary/subordinate spaces to make the first invariants appear.

These obstacles, psychological or real, vanish in diffeology. First of all, the group of diffeomorphisms is naturally a diffeological group. And here is an example where the action of the group of diffeomorphisms, the principal group in the sense of Klein, captures in the space a good preliminary image of its geometry.[42]

Consider the square $Sq = [-1, +1]^2$, equipped with the subset diffeology. Its decomposition in orbits, by its group of diffeomorphisms, is the following:

1. the 4-corners-orbit;
2. the 4-edges-orbit; and
3. the interior-orbit.

Any diffeomorphism preserves separately the interior of the square and its border, which is a consequence of the D-topology.[43] But the fact that a

---

[41] Generally, manifolds are regarded as connected, Haussdorf, and second countable.
[42] And not just of its topology.
[43] A diffeomorphism is in particular a homomorphism for the D-topology. We can use that, or homotopy.

diffeomorphism of the square cannot map a corner into the interior of an edge is a typical smooth property (see [11, 12] and some comments on more general stratified spaces [13]).

On the basis of this simple example, we can experiment with Klein's principle for the group of diffeomorphisms of a nontransitive diffeological space. The square being naturally an object of the theory, there is no need for heuristic extension here.[44]

Claim Hence, considering the group of diffeomorphisms of a diffeological space as its principal group, we can look at diffeology as the formal framework that makes *differential geometry* the geometry – in the sense of Felix Klein – of the group of diffeomorphisms.[45]

That principle, in the framework of diffeology, can be regarded as the definitive expression of Souriau's point of view, developed in his paper "Les groupes comme universaux" [50].

Because diffeology is a such large and stable category that encompasses satisfactorily so many various situations, from singular quotients to infinite dimensions, mixing even these cases [24], one can believe that this interpretation of diffeology fulfills its claim and answers in some sense to the subtitle of this book.

References

[1] J. Boman, *Differentiability of a function and of its compositions with functions of one variable*, Math. Scand. 20 (1967) 249–68

[2] W. M. Boothby, *Transitivity of the automorphisms of certain geometric structures*, Trans. Amer. Math. Soc.137 (1969) 93–100

[3] K.-T. Chen, *Iterated path integrals*, Bull. Am. Math. Soc. 83 (1977) 831–79

[4] D. Christensen and E. Wu, *The homotopy theory of diffeological spaces*, N. Y. J. Math. 20 (2014) 1269–1303

[5] E. Dinaburg and Y. Sinai, *The one-dimensional Schrödinger equation with a quasiperiodic potential*, Funct. Anal. Appl. 9 (1975) 279–89

[6] S. K. Donaldson, *Moment maps and diffeomorphisms*, Asian J. Math. 3 (1999) 1–16

[7] P. Donato and P. Iglesias, *Exemple de groupes différentiels: flots irrationnels sur le tore*, Preprint, CPT-83/P.1524 http://math.huji.ac.il/~piz/documents/EDGDFISLT.pdf

---

[44] See Section 6.5.

[45] It is possible to weaken this claim by considering instead the (larger) pseudogroup of local diffeomorphisms, which I prefer. The introduction of the D-topology has mutated diffeology into *local diffeology*, which has enriched its geometry.

[8] P. Donato, *Revêtement et groupe fondamental des espaces différentiels homogènes*, Thèse de doctorat d'état, Université de Provence, Marseille (1984)

[9] V. Guillemin and A. Pollack, *Differential Topology*, Prentice Hall (1974)

[10] S. Gürer and P. Iglesias-Zemmour, *k-Forms on half-spaces*, Blog post (2016), http://math.huji.ac.il/~piz/documents/DBlog-Rmk-kFOHS.pdf

[11] ——- *Diffeomorphisms of the square*, Blog post (2016), http://math.huji.ac.il/~piz/documents/DBlog-Rmk-DOTS.pdf

[12] ——- *Differential forms on manifolds with boundary and corners*, ePrint (2017), http://math.huji.ac.il/~piz/documents/OMWBAC.pdf

[13] ——- *Differential forms on stratified spaces I & II*, Bull. Australian Math. Soc. 98 (2018) 319–30; 99 (2019) 311–18, http://math.huji.ac.il/~piz/documents/DFOSS.pdf, http://math.huji.ac.il/~piz/documents/DFOSS-A.pdf

[14] T. Haraguchi, *Homotopy structures of smooth CW complexes*, Preprint, arXiv:1811.06175

[15] T. Haraguchi and K. Shimakawa, *A model structure on the category of diffeological spaces*, Preprint, arXiv:1311.5668

[16] P. Iglesias, *Fibrations difféologique et homotopie*, Thèse d'état, Université de Provence, Marseille (1985)

[17] ——- *Connexions et difféologie*, in *Aspects dynamiques et topologiques des groupes infinis de transformation de la mécanique*, vol. 25 of *Travaux en cours*, Hermann (1987) 61–78, http://math.huji.ac.il/~piz/documents/CED.pdf

[18] P. Iglesias and G. Lachaud, *Espaces différentiables singuliers et corps de nombres algébriques*, Ann. Inst. Fourier, Grenoble 40 (1990) 723–37

[19] P. Iglesias-Zemmour, *Aperçu des Origines de la Mécanique Symplectique*, in *Actes du Colloque "Histoire des géométries,"* vol. I, Maison des Sciences de l'Homme (2004), http://math.huji.ac.il/~piz/documents/AOGS-MSH.pdf

[20] ——- *Dimension in diffeology*, Indagationes Math. 18 (2007)

[21] ——- *The moment maps in diffeology*, Mem. Amer. Math. Soc. 207 (2010)

[22] ——- *Diffeology*, Math. Surv. Monogr. 185 (2013), http://www.ams.org/bookstore-getitem/item=SURV-185

[23] ——- *Variations of integrals in diffeology*, Can. J. Math. 65 (2013) 1255–86

[24] ——- *Example of singular reduction in symplectic diffeology*, Proc. Amer. Math. Soc. 144 (2016) 1309–24

[25] ——- *The geodesics of the 2-torus*, Blog post (2016), http://math.huji.ac.il/~piz/documents/DBlog-Rmk-TGOT2T.pdf

[26] ——- *Diffeology of the space of geodesics*, Work in progress

[27] ——- *Every symplectic manifolds is a coadjoint orbit*, in Proceedings of Science, vol. 224 (2017), http://math.huji.ac.il/~piz/documents/ESMIACO.pdf

[28] P. Iglesias, Y. Karshon, and M. Zadka, *Orbifolds as diffeology*, Trans. Amer. Math. Soc. 362 (2010) 2811–31

[29] P. Iglesias-Zemmour and J.-P. Laffineur, *Noncommutative geometry and diffeologies: The case of orbifolds*, J. Noncommutative Geometry 12 (2018) 1551–72, http://math.huji.ac.il/~piz/documents/CSAADTCOO.pdf

[30] N. Iwase and N. Izumida, *Mayer-Vietoris sequence for differentiable/diffeological spaces*, Algebraic Topology Related Topics (2019) 123–51

[31] H. Kihara, *Model category of diffeological spaces*, J. Homotopy Relat. Struct. 14 (2019) 51–90

[32] H. Kihara, *Quillen equivalences between the model categories of smooth spaces, simplicial sets, and arc-gengerated spaces*, Preprint, arXiv:1702.04070

[33] F. Klein, *Vergleichende Betrachtungen über neuere geometrische Forschungen*, Math. Ann. 43 (1893) 63–100, http://www.deutschestextarchiv.de/book/view/klein_geometrische_1872

[34] A. Kriegl and P. Michor, *The convenient setting of global analysis*, Math. Surv. Monogr. 53 (1997)

[35] K. Kuribayashi, *Simplicial cochain algebras for diffeological spaces*, Preprint, arXiv:1902.10937

[36] J. M. Lee, *Introduction to Smooth Manifolds*, Graduate Texts in Mathematics, Springer (2006)

[37] P. Muhly, J. Renault, and D. Williams, *Equivalence and isomorphism for groupoid C^*-algebras*, J. Operator Theory 17 (1987) 3–22

[38] S. M. Omohundro, *Geometric Perturbation Theory in Physics*, World Scientific (1986)

[39] E. Prato, *Simple non-rational convex polytopes via symplectic geometry*, Topology 40 (2001) 961–75

[40] J. Renault, *A Groupoid Approach to C^*-Algebras*, Lecture Notes in Mathematics 793, Springer (1980)

[41] M. Rieffel, *C^*-algebras associated with irrationnal rotations*, Pac. J. Math. 93 (1981)

[42] I. Satake, *On a generalization of the notion of manifold*, Pro. Natl. Acad. Sci. 42 (1956) 359–63

[43] —— *The Gauss–Bonnet theorem for V-manifolds*, J. Math. Soc. Jpn. 9 (1957) 464–92

[44] G. Schwarz, *Smooth functions invariant under the action of a compact Lie group*, Topology 14 (1975) 63–68

[45] K. Shimakawa, K. Yoshida, and T. Haraguchi, *Homology and cohomology via enriched bifunctors*, Kyushu J. Math. 72 (2018) 239–52

[46] J.-M. Souriau, *Structure des systèmes dynamiques*, Dunod Editions (1970)

[47] —— *Groupes différentiels*, Lecture Notes in Mathematics 836, Springer (1980) 91–128

[48] —— *Groupes différentiels et physique mathématique*, Preprint, CPT-83/P.1547, http://math.huji.ac.il/~piz/documents-others/JMS-GDEPM-1983.pdf

[49] —— *Un algorithme générateur de structures quantiques*, Preprint, CPT-84/PE.1694

[50] —— *Les groupes comme universaux*, in D. Flament, ed., *Histoires de géométries*, Documents de travail (Équipe F2DS), Fondation Maison des Sciences de l'Homme (2003), http://semioweb.msh-paris.fr/f2ds/docs/geo_2002/Document02_Souriau1.pdf

[51] W. Thurston, *The Geometry and Topology of Three-Manifolds*, Princeton University Lecture Notes (1978–81), http://library.msri.org/books/gt3m/PDF/13.pdf

[52] H. Whitney, *Differentiable even functions*, Duke Math. J. 10 (1943 159–60

[53] F. Ziegler, *Théorie de Mackey symplectique*, in *Méthode des orbites et représentations quantiques*, Thèse de doctorat d'Université, Université de Provence, Marseille (1996), http://arxiv.org/pdf/1011.5056v1.pdf

Patrick Iglesias-Zemmour
Einstein Institute of Mathematics,
The Hebrew University of Jerusalem,
Campus Givat Ram, 9190401 Israel.
piz©math.huji.ac.il

2

New Methods for Old Spaces: Synthetic Differential Geometry

Anders Kock

Contents

1	Introduction	83
2	Some Differential Geometry in Terms of the Neighbour Relation	85
3	Neighbours in the Context of Euclidean Geometry	94
4	Coordinate Geometry and the Axiomatics	97
5	Models of the Axiomatics	104
6	New Spaces	106
7	The Role of Analysis	109
8	The Continuum and the Discrete	111
9	Looking Back	112
References		114

1 Introduction

The synthetic method consists in consideration of a class of objects in terms of their (often axiomatically assumed) mutual relationship, say, incidence relations – disregarding what the objects are "made up of." For geometry, this method goes back to the time of Euclid. With the advent of category theory, it became possible to make the notion of "relationship" more precise, in terms of the *maps* in some category $\mathcal{E}$ of "spaces," in some broad sense of this word.

A synthetic theory might be presented by giving axioms for some good category $\mathcal{E}$, possibly with some added structure. Thus, the list of axioms could begin "let $\mathcal{E}$ be a good category, and let R be a commutative ring object (to be thought of as the number line) in $\mathcal{E}$ ..." Even though this is the way most texts in synthetic differential geometry begin, some texts are purely synthetic/combinatorial, presupposing a category $\mathcal{E}$, but do not presuppose any ring object R; this applies, for example, to Sections 2.1, 2.2, and 2.4 below – and to a certain extent to Section 2.3.

When is a category $\mathcal{E}$ suitable for playing the role of a place, or scene, where a theory, say, axiomatic, of "spaces" and their geometry can be developed?

Experience since the 1950s has showed that many *topoi* $\mathcal{E}$ are suitable. Thus, for instance, the topos of simplicial sets was shown to have possibility for being an arena for homotopy theory, without recourse to the real numbers or to the notion of topological space.

A crucial point is that a topos is in many respects like the category of sets; in fact, the understanding of "the" category of sets is distilled out of our experience with categories of spaces, in a broad sense of the word. And many texts in synthetic differential geometry talk about the assumed $\mathcal{E}$ *as if* it were the category of sets, just making sure not to use the law of excluded middle; this law holds in the category of *abstract* sets (discrete spaces), but fails for most other categories of spaces. This is related to the contradiction between the discrete and the continuum; see Section 8.

So much for the question "why topos"? But why "ring"? Because we may hopefully use that ring for introducing *coordinates* and thereby supplement, or even replace, geometric reasoning with algebraic calculation. Such a ring (a "number line") is, however, not the central geometric notion in differential geometry.

I would like to advance the thesis that the notion of pairs of *neighbour* points in a space M is a more basic notion in differential geometry; that central differential geometric concepts can be formulated in terms of that relation; and that this allows one to present such a concept in terms of *pictures*; see, for example, (2.1). There is no notion of "limiting positions" involved. A "line type" ring will be introduced later, together with an account of some of the standard synthetic differential geometry, and this will also provide "coordinate" models for the axiomatics about a category of spaces with a neighbour relation, and will thus make this wishful thinking in Section 1 come true.

The guideline for this is the theory developed in algebraic geometry by Kähler, Grothendieck, and many others: the first neighbourhood of the diagonal of a scheme, or the (first) prolongation space of a smooth manifold ([27, p. 52]). From these sources, combined with standard synthetic differential geometry, models for the axiomatics are drawn (and they are briefly recalled in Section 5).

We use the abbreviation SDG for "synthetic differential geometry."

It is not a mathematical *field*, but a *method*. Not really a *new* one, it has, as initiated by Lawvere, over several decades by now, contributed by making the synthetic method more explicit; see also Section 9.

2 Some Differential Geometry in Terms of the Neighbour Relation

The spaces M considered in differential geometry come equipped with a reflexive symmetric relation $\sim_M$, the (first-order) neighbour relation, often mentioned in the heuristic part of classical texts but rarely made precise – neither in how it is defined nor in how one reasons with it.

One aim in SDG is to make precise how one reasons with the neighbour relation (and this is done axiomatically); it is not a main aim to describe how it is constructed in concrete contexts.

The relation $\sim_M$ is reflexive and symmetric; it is *not* transitive (see the Remark after Corollary 4.1 why transitivity is incompatible with the axiomatics to be presented).

The subobject $M_{(1)} \subseteq M \times M$ defining the relation $\sim_M$ is called the *(first) neighbourhood of the diagonal* of M. This terminology is borrowed from the theory of schemes in algebraic geometry or in differential geometry (see, e.g., [27], who call it the *(first) prolongation space* of M).

To state some notions of differential-geometric nature, we shall talk about the category $\mathcal{E}$ as if it were the category of sets. The objects of $\mathcal{E}$ we call "sets" or "spaces." If the space M is understood from the context, we write $\sim$ instead of $\sim_M$. There are also higher-order neighbour relations $\sim_2, \sim_3, \ldots$ on M; they satisfy $(x \sim y) \Rightarrow (x \sim_2 y) \Rightarrow (x \sim_3 y), \ldots$ The neighbour relations $\sim_k$ will not be transitive, but $x \sim_k y$ and $y \sim_l z$ will imply $x \sim_{k+l} z$. For $x \in M$, we call $\{y \in M \mid y \sim_k x\} \subseteq M$ the kth-order *monad* around x, and we denote it $\mathfrak{M}_k(x)$. In the axiomatics to be presented, it represents the notion of k-jet at x. The first-order monad $\mathfrak{M}_1(x)$ will also be denoted $\mathfrak{M}(x)$.

The higher neighbour relations will not be discussed in the present text, but see, for example, [22].

The spaces one considers live in some category $\mathcal{E}$ of spaces; maps in $\mathcal{E}$ preserve the neighbour relations (which is a "continuity" property). But the neighbour relation on a product space $M \times N$ will be more restrictive than the product relation: we will not in general have that $m \sim m'$ and $n \sim n'$ implies that $(m,n) \sim (m',n')$; see the Remark after Proposition 4.4.

For all $x' \in \mathfrak{M}(x)$, we thus have by definition $x' \sim x$, and for sufficiently good spaces, the monad $\mathfrak{M}(x)$ will have the property that x is the *only* point in $\mathfrak{M}(x)$ with this property.

We present some differential geometric notions that may be expressed in terms of the (first-order) neighbour relation $\sim$. The argument that they comprise the classical notions with the corresponding names, may, for most of them, be found in [22].

2.1 Touching

From the neighbour relation, one derives a fundamental geometric notion, namely, what does it mean to say that two subspaces S and T of a space M *touch* each other at a point $x \in S \cap T$? We take that to mean that $\mathfrak{M}_1(x) \cap S = \mathfrak{M}_1(x) \cap T$. (The intended interpretation is that S and T are subspaces "of the same dimension"; there is clearly also a notion of, say, when a curve touches a surface, which also can be expressed in terms of $\sim$.) To "touch each other at x" is clearly an equivalence relation on the set of subspaces of M containing x. (In the intended models, say, where $\mathcal{E}$ is a topos containing the category of smooth manifolds and M is a smooth manifold, this becomes the relation that S and T has *first-order contact* at x.)

Pictures can conveniently be drawn for the touching notion: in the picture below, M is the plane of the paper, the bullet indicates x, and the interior of the circle indicates $\mathfrak{M}_1(x)$. Ignore the fact that T looks like a line; the notion of *line* is an invention of the age of civilization, whereas the notion of *touching* is known already from precivilized stone ages and before. So the present section may be thought of as Stone Age geometry. The same applies to Sections 2.4 and 2.5.

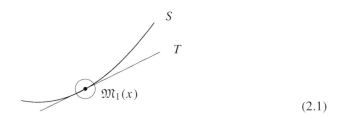

(2.1)

2.2 Characteristics and Envelopes

Assume that a space T parameterizes a family $\{S_t \mid t \in T\}$ of subspaces S_t of a space M. Then, for $t_0 \in T$, the *characteristic* set C_{t_0} (at the parameter value t_0) is the intersection of all the neighbouring sets of S_{t_0}, precisely,

$$C_{t_0} := \cap_{t \sim t_0} S_t,$$

and the *envelope* E of the family may be defined as

$$E := \cup_{t_0 \in T} C_{t_0} = \cup_{t_0 \in T} \cap_{t \sim t_0} S_t. \quad (2.2)$$

Under nonsingularity assumptions, E is the *disjoint* union of the characteristics, that is, there is a function $\tau : E \to T$ associating to a point Q of E the parameter value t such that $Q \in C_t$. So $Q \in \cap_{t' \sim \tau(Q)} S_{t'}$.

Let $\tau(Q) = t_0$. We would like to prove that S_{t_0} touches E at Q, that is, that $E \cap \mathfrak{M}(Q) = S_{t_0} \cap \mathfrak{M}(Q)$ (under a "dimension" assumption on T and the S_t, commented on below). We can in any case prove the inclusion $E \cap \mathfrak{M}(Q) \subseteq S_{t_0} \cap \mathfrak{M}(Q)$. For let $Q' \sim Q$ and $Q' \in E$. Then, $\tau(Q') \in T$ is defined, and since τ, as any function, preserves $\sim$, the assumption $Q \sim Q'$ implies $\tau(Q') \sim \tau(Q) = t_0$. Now $Q' \in \cap_{t \sim \tau(Q')} S_t$ by assumption, in particular, $Q' \in S_{t_0}$. Therefore

$$E \cap \mathfrak{M}(Q) \subseteq S_{t_0} \cap \mathfrak{M}(Q).$$

(To pass from the proved inclusion to the desired equality would involve a dimension argument like "the two sets have the same dimension, so the inclusion implies equality." We do not have such an argument available at this primitive stage; in Section 2.3, on wave fronts, this is part of the axiomatics.)

This leads to an alternative way to describe (but not construct) envelopes for such families S_t of subspaces of M. Namely, an envelope of such a family is *a* subspace $E \subseteq M$ such that each S_t touches E, and each point of E is touched by some S_t. This is also a classical definition, except that the word "touching" there is defined in terms of differential calculus, not available in the Stone Age. Note the indefinite article "*a* subspace." This "implicit" way of describing envelopes is the one we use in Section 2.3.

The primary notion in the explicit construction (2.2) of envelopes is that of *characteristic*; the envelope is derived from the characteristics. In the literature, based on analytic geometry, the characteristic C_{t_0} is sometimes, with some regret or reservation, defined as "the limit of the sets $S_{t_0} \cap S_t$ as t tends to t_0." In [7], Courant (talking about a one-parameter family of surfaces S_t in 3-space, where the intersection of any two of them therefore, in nondegenerate cases, is a curve) thus writes about a characteristic curve, say, C_{t_0}, for the family: "This curve is often referred to in a non-rigorous but intuitive way as the intersection of 'neighbouring' surfaces of the family" (p. 169) (offering instead "If we let h tend to zero, the curve of intersection will approach a definite limiting position" [p. 180]. What is the topology on the set of subsets which will justify the limit-position notion?).

We shall see (Section 4.2) that the axiomatics for SDG make the "limit" intersection curve rigorous by replacing the dubious *limit* with the simultaneous intersection of *all* neighbouring surfaces, now with "neighbouring S_t" in the strict sense of $t \sim t_0$. Thus, the "*nonrigorous but* intuitive" description in Courant's text now gets the status *rigorous and* intuitive.

2.3 Wave Fronts and Rays

Already with the neighbour relation as the only primitive concept, one can thus define the geometric notion (Huygens) of an *envelope* of families S_t of subspaces of a space M. Combined with a (weak) notion of *metric* (distance) on M, one can (cf. [23]), by less trivial synthetic reasoning (and under suitable axioms), recover some of Huygens's theory of *wave fronts* in geometrical optics: essentially, if B is a "hypersurface" (in a suitable sense) in M, one has an envelope $B \vdash s$ of the family of spheres of radius $s > 0$ and centre on B; the Huygens's principle states that (for s small enough), this is again a hypersurface, "the wave front which B becomes after time lapse s." (In particular, Huygens knew that if B is a sphere of radius r, then $B \vdash s$ is again a sphere, of radius $r + s$.)

To have a notion of metric, one needs a space of numbers to receive the values of the metric, that is, the distances. In the intended applications, this will be the strictly positive real numbers $R_{>0}$, but only its total strict order $>$ and the properties of the addition operation will be used in the following theory; so we are far from being in a situation where a coordinatization is used (still, we shall use $R_{>0}$ to denote the assumed object that receives the values of the metric). The fact that only strictly positive distances are considered means that we cannot talk about the distance from a point to itself; in fact, we cannot talk about the distance between a pair of neighbour points. (In the coordinatized model of our theory, this has to do with the fact that the square root function is not smooth at 0.) When we say that two points are *distinct*, we thus imply that their distance is defined (hence positive).

With a metric on M, we can define *spheres*: if $a \in M$ and $r \in R_{>0}$, the sphere $S(a, r)$ with centre a and radius $r > 0$ is the set $\{b \in M \mid ab = r\}$, where "ab" is short notation for the distance between a and b. So $ab = ba$. We assume that, as in Euclidean geometry, the centre a and the radius r can be reconstructed from the point set $S(a, r)$. No triangle inequality is used in the following.

Combining the two primitive notions, neighbours and metric, we can then define the notion of contact element P: a *contact element* at $b \in M$ is a subset of the form $P = \mathfrak{M}(b) \cap B$, where B is a sphere with $b \in B$. The same contact element may be presented in $\mathfrak{M}(b) \cap B'$ for many other spheres B', but all these spheres touch each other at b, since $\mathfrak{M}(b) \cap B = P = \mathfrak{M}(b) \cap B'$.

Since a contact element at b has $P \subseteq \mathfrak{M}(b)$, one has that b is a neighbour of all the points in P. We assume that b is the only point in P with this property. So b can be reconstructed from the point set P; we may call it the *focus* of P, to avoid saying "centre."

In the intended application, where M is a smooth manifold, the set of contact elements makes up the projectivized cotangent bundle of M.

Note that in the classical theory, any contact element P at b, say, $\mathfrak{M}(b) \cap B$ (where $b \in B$), is a one-point set, $P = \{b\}$, since $\mathfrak{M}(b)$ is so, whereas with a nontrivial $\sim$, there is much more information in P: it generates a nontrivial perpendicularity relation. Namely, for c distinct from b (and [hence]) from any $b' \sim b$), we say that c is *perpendicular to* P, or $c \perp P$, if for all $b' \in P$, we have $b'c = bc$ (where b is the focus of P). (For a trivial $\sim$, all points distinct from b are perpendicular to $P = \{b\}$.)

There are two basic structures in geometrical optics, (light-) *rays* and *wave fronts*. These can be described in the present framework. The *rays* in M are certain (open) half-lines, parameterized by $R_{>0}$, described more precisely below in terms of a collinearity condition. Wave fronts here occur in the present context as *(hyper-)surfaces*; the rude notion of *hypersurface* we are considering is the following: it is a subset of M "made up of contact elements," that is, it is a subset $B \subseteq M$ such that for each $b \in B$, the set $\mathfrak{M}(b) \cap B$ is a contact element (necessarily with focus b). So, in particular, a sphere is a hypersurface.

What makes synthetic reasoning about rays and wave fronts possible is an analysis about how spheres may touch. In the deductions in [23], this analysis takes the form of two axioms, one for "external" touching and one for "internal" touching. We state them below, noting that they are refinements of theorems of Euclidean geometry (in the sense that "*touch*" has a refined meaning). Thus, for external touching,

> Two spheres touch (externally) if the sum their radii is the distance between their centres.

Here is the picture for external touching; the spheres are $A = S(a,r)$, $C = S(c,s)$; they touch at b. The two other dots represent the centres a and c:

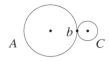

With the refined touching notion derived from $\sim$, here is how this basic fact gets formulated: given $A = S(a,r)$ and $C = S(c,s)$, if $r + s = ac$, then the spheres A and C touch at a unique point b, and this b is characterized by

$ab + bc = ac$; and for all $b' \sim b$, we have $(ab' = ab) \Leftrightarrow (b'c = bc)$. (2.3)

This is essentially the basic axiom (together with a similar axiom for internal touching), except that we weaken it by replacing the $\Leftrightarrow$ in (2.3) by $\Rightarrow$:

$ab + bc = ac$; and for all $b' \sim b$, we have $(ab' = ab) \Rightarrow (b'c = bc)$. (2.4)

This replacement is, in the intended models, justified by a dimension argument, as alluded to in Section 2.2. Note that (2.4) can also be expressed: c is a characteristic point (in the sense of Section 2.2), for parameter value b, of the family $S(b', s)$, as b' ranges over $P = \mathfrak{M}(b) \cap S(a, r)$ or as b' ranges over $S(a, r)$, for $b'c = s$ is equivalent to $c \in S(b', s)$.

We give an equivalent formulation of (2.3) and also of the corresponding way of writing the axiom for internal touching; a, b, and c denote points in M, and s denotes an element $\in R_{>0}$. Note that for internal touching, there is no restriction on $s \in R_{>0}$.

Given a, c, and s, with $s < ac$, $\exists! b$ such that $S(a, ac - s)$ and $S(c, s)$ touch at b. Given a, b, and s, $\exists! c$ such that $S(a, ab + s)$ and $S(b, s)$ touch at c.

Since touching of two spheres is either internal or external, it is straightforward that one can (transversally) *orient* a contact element P in two ways, and one then can divide the class of points perpendicular to P in two classes: those on the "outer" side and those on the "inner side." They are the two *rays* defined by P: given an orientation of P and an $s \in R_{>0}$, we let $P \vdash s$ denote the unique point on the outer side perpendicular to P and whose distance to b is s (where b denotes the focus of P). To jusitify the word "ray," note that the ray generated by P is a point set, bijectively parameterized by $R_{>0}$, and furthermore, any three distinct points (taken in suitable order) on this ray are *collinear*.

Collinearity is a notion which, when $\sim$ is trivial, may be formulated purely in terms of the metric, and it forms the basis of Busemann's theory of geodesics, cf. [5]. Three points a, b, c (say, distinct) are classically and in loc. cit. called collinear if $ab + bc = ac$. The stronger collinearity property that applies to the rays in the present theory is that $ab + bc = ac$ *and* that $S(a, ab)$ touches $S(c, bc)$ in b, equivalently, if (2.3) (or (2.4)) holds (with $r = ab$, $s = bc$).

To give a hypersurface an *orientation* is to give each of its contact elements an orientation. Let B be an oriented hypersurface, and let $s > 0$. Let us denote by $B \vdash s$ the set of points of the form $\mathfrak{M}(b) \vdash s$, that is, the envelope (in the explicit sense of (2.2)) of the spheres $S(b, s)$ as b ranges over B. Since the distance of b and $\mathfrak{M}(b) \vdash s$ is s, we would like to think of $B \vdash s$ as the parallel hypersurface to B at distance s; however, it may not be a hypersurface, as is well known in geometry, even when M is the Euclidean plane: there may be self-intersections, cusps, and so on if B is concave. But unless B is very crinkled, one will for sufficienty small s have that the map $s \mapsto \mathfrak{M}(b) \vdash s$ is a bijection $B \to B \vdash s$. The version of Huygens's principle we can prove synthetically (cf. [23]) is

Assume that B is an oriented hypersurface, and that $s > 0$ is so that the map $B \to B \vdash s$ described is a bijection. Then $B \vdash s$ is again a hypersurface.

If $B = S(a, r)$ is a sphere, $B \vdash s$ will be the sphere $B = S(a, r + s)$ for one orientation of B and will, for the other orientation, be the sphere $S(a, r - s)$ (provided $s < r$).

2.4 Geometric Distributions

A (geometric) *distribution* on M is a reflexive symmetric relation $\approx$ refining $\sim$ (i.e., $x \approx y$ implies $x \sim y$). It is called *involutive* if it satisfies, for all x, y, z in M,

$$(x \approx y) \wedge (x \approx z) \wedge (y \sim z) \quad \text{implies} \quad y \approx z.$$

A relevant picture is the following; single lines indicate the neighbour relation $\sim$, and double lines indicate the assumed "strong" neighbour relation $\approx$:

For instance, if $f : M \to N$ is any map between spaces, the relation on $\approx$ on M defined by $x \approx y$ iff $(x \sim y) \wedge (f(x) = f(y))$ is a distribution, in fact an involutive one.

An *integral subset* of a distribution $\approx$ is a subset $F \subseteq M$ such that on F, the relations $\sim$ and $\approx$ agree. An important integration theorem in differential geometry is Frobenius's theorem, whose conclusion is that for an *involutive* distribution, there exist maximal connected integral subsets (leaves).

Such integration results can usually not be *proved* in the context of SDG (even the very formulation may require some further primitive concepts), since they in a more serious way depend on limits and on completeness of the real number system. Sometimes, SDG can *reduce* one integration result to another; this is also an old endeavour in classical differential geometry, for example, Lie has many results about which differential equations can be solved *by quadrature*, that is, by reduction to existence of antiderivatives.

Example 2.1 The following is meant as a sketch of an (involutive) distribution in the plane. Consider

In this picture, the "line segments" are the $\approx$-monads $\mathfrak{M}_\approx(x) := \{y \mid y \approx x\}$ around (some of) the points x (drawn as dots) of M. But note that the notion of "line" has not yet entered in to our vocabulary, let alone coordinate systems like $R \times R$; when such things are present, an ordinary first-order differential equation

$$y' = F(x, y),$$

as in the Calculus Books, gives rise to such a picture, known as the "direction field" of the equation: through each point $(x, y) \in R \times R$, one draws a "little" line segment $S(x, y)$ with slope $F(x, y)$.

The "integral subsets" of a distribution of this kind are essentially (the graphs of) the solutions of the differential equation.

I cannot draw a good picture of a noninvolutive distribution: paper is two-dimensional. But in three dimensions, consider the scales of a ripe pinecone and extrapolate radially.

If M carries a metric (in the sense of Section 2.3), it makes sense to say that a distribution is "of codimension 1" if all the $\approx$-monads are contact elements. The two specific examples mentioned have this property.

2.5 Affine Connections

An *affine connection* on a space M is a law λ that completes any configuration (x, y, z) consisting of three points x, y, and z with $x \sim y$, $x \sim z$ by a fourth point $\lambda(x, y, z)$ with $y \sim \lambda(x, y, z)$ and $z \sim \lambda(x, y, z)$:

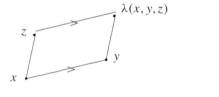

(2.5)

expressing "infinitesimal parallel transport of z along $\overline{xy}$" or "constructing an infinitesimal parallelogram." (We assume $\lambda(x,x,z) = z$ and $\lambda(x,y,x) = y$.) The connecting lines indicate the assumed neighbour relations. We use a different signature for the edges xy and xz, since we do not assume the symmetry condition $\lambda(x,y,z) = \lambda(x,z,y)$. If symmetry holds, λ is called a *symmetric* or *torsion-free* connection.

A *geodesic* for a given torsion-free affine connection on M is a subset $S \subseteq M$ that is stable under λ in the sense that if $x \sim y$ and $x \sim z$ with x, y and z in S, then $\lambda(x,y,z) \in S$.

The *curvature* of an affine connection may be described combinatorially by asking the question, what happens if we transport $z \sim x_0$ around a circuit from x_0 to x_1, then from x_1 to x_2, and finally from x_2 back to x_0? This makes sense whenever $x_0 \sim x_1 \sim x_2$ *and* $x_0 \sim x_2$ (the latter requirement is not automatic: the relation $\sim$ is not transitive). The result of such circuit transport gives a new point $z' \sim x_0$. Thus the "infinitesimal 2-simplex" (x_0,x_1,x_2) provides an automorphism $z \mapsto z'$, denoted $R(x_0,x_1,x_2)$, of the pointed set $\mathfrak{M}(x_0)$; this is the curvature of λ, or more precisely, the curvature of λ is the law which to an infinitesimal 2-simplex (x_0,x_1,x_2) associates the described automorphism of $\mathfrak{M}(x_0)$. If this automorphism is the identity map for all infinitesimal 2-simplices, the connection is called *flat*. (Any affine connection on a one-dimensional space M is flat. One may even experiment with this as a definition of "M is of dimension (at most) 1.")

2.6 Differential Forms

Differential forms are, in analytic differential geometry, certain functions taking values in a *ring* R of quantities (or in a *module* over R), but are in the present context (equivalent to) a special case of a more primitive, nonquantitative kind of thing. Thus, in SDG, one may, for any group G, define "(combinatorial) G-valued k-form on a space M" to mean a "function ω, that takes as input infinitesimal k-simplices ($k + 1$-tuples of mutual neighbour points in M) and returns as output elements in G." One imposes the normalization condition that $\omega(x_0, \ldots, x_k) = e$ whenever two of the x_is are equal (where e denotes the neutral element of G). A G-valued 0-form on M is then just a function $f : M \to G$; it has a "coboundary" df, which is a G-valued 1-form, defined by $df(x_0, x_1) := f(x_0)^{-1} \cdot f(x_1)$. A G-valued 1-form ω on M has a coboundary $d\omega$, which is a G-valued 2-form defined by

$$d\omega(x_0, x_1, x_2) := \omega(x_0, x_1) \cdot \omega(x_1, x_2) \cdot \omega(x_2, x_0).$$

The 1-form ω is *closed* if $d\omega$ is constant e. The 1-form df is always closed.

The group G carries a canonical closed G-valued 1-form, namely, df, where $f : G \to G$ is the identity function. This is the Maurer–Cartan form of G.

If there is given data identifying all the $\mathfrak{M}(x)$ of a given manifold M with each other, then the curvature of an affine connection λ on M may be seen as a 2-form with values in the automorphism group of the pointed set $\mathfrak{M}(x_0)$ (for some, hence any, $x_0 \in M$). (Alternatively, one gets a 2-form "with local coefficients"; then no identification data is needed.)

For most G, we have that G-valued differential forms are *alternating*: interchanging two of the input entries implies inversion of the value of the form. In particular,

$$\omega(x_0, x_1)^{-1} = \omega(x_1, x_0).$$

Such a 1-form ω on M then defines a geometric distribution on M by saying $x \approx y$ iff $\omega(x, y) = e$. If ω is closed, the $\approx$ that is defined by ω is involutive.

There is a relationship between combinatorial group valued 1-forms, on one hand, and the general notion of connection in a fibre bundle, or in a groupoid, on the other. This we expound in Section 6.1.

In case the value group G is commutative (additively written), there is for good M and G, a natural bijection between combinatorial G-valued forms, and the standard multilinear alternating forms on $T(M)$, the tangent bundle of M; see [15, I.18].

3 Neighbours in the Context of Euclidean Geometry

In this section, we move from the Stone Age into the era of civilization and assume that some classical Euclidean geometry (plane, say) is available in a space E (with a given neighbour relation $\sim$). In particular, there are given subsets called points and lines; they are *affine* subspaces of E (without yet assuming the existence of a "number" line $R \subseteq E$, i.e., a line equipped with a commutative ring structure).

Then we can be more explicit about our wishes for the compatibilities between the Euclidean notions and the combinatorics of the neighbour relation. We refrain from calling these wishes for "axioms," since they (for the coordinate spaces R^n built on R) lead to and are subsumed in a more complete comprehensive axiom scheme later on; so we call these wishes for "principles."

There are also some incompatibilities, essentially because in Euclid, the law of excluded middle is explicitly used. Thus, in Euclid, a curve, say, a circle, has *exactly one* point in common with any of its tangents, so that the picture (2.1) (with S as part of a circle) is an illusion for Euclid; already the contemporary Greek philosopher Protagoras is said to have ridiculed Euclidean geometry for insisting on the "only one point" idea, which seemed to him to go against experience. In Euclid's geometry, $\mathfrak{M}(x)$ is always just the one-point set $\{x\}$. Certainly, the following principle is incompatible with such a small $\mathfrak{M}(x)$; in the terminology of Section 2, this says that two lines that touch each other at some point are equal.

Principle Given two lines l_1 and l_2 in a plane E, let $x \in l_1 \cap l_2$. Then

$$\mathfrak{M}(x) \cap l_1 = \mathfrak{M}(x) \cap l_2 \text{ implies } l_1 = l_2.$$

A subspace $C \subset E$ is called a *curve* if, for each $x \in C$, there exists a line l such that $\mathfrak{M}(x) \cap l = \mathfrak{M}(x) \cap C$, that is, a line that touches C at x; such a line is unique, by the principle. This line then deserves the name the *tangent* of C in $x \in C$. In the picture (2.1), if T is a line (as the picture suggests), then the picture says that this line is the tangent to the curve S at x.

For any curve C, the family of its tangents T_x ($x \in C$) is a parameterized family, parameterized by the points of curve C.

Proposition 3.1 *Any curve C is contained in the envelope of its family of tangents.*

Proof: For $z \in C$, let T_z denote the tangent C at z. Let $x \sim y$ be points in C. So $y \in \mathfrak{M}(x) \cap C = \mathfrak{M}(x) \cap T_x$; so $y \in T_x$. Similarly, $x \in T_y$. So

$$x \in \bigcap_{y \in \mathfrak{M}(x) \cap C} T_y,$$

which is to say that x belongs to the characteristic set (for parameter value x) of the family of tangents. Hence it belongs to the envelope of the family. ∎

Let M and N be spaces (objects in $\mathcal{E}$). It will not in general be the case that $x \sim x'$ and $y \sim y'$ in N implies $(x, y) \sim (x', y')$ in $M \times N$ (although the converse implication will hold, since the projections, like any other map, preserve the assumed neighbour relations ∼). But if $f : M \to N$ is a map, then we also have that

$$x \sim x' \text{ in } M \text{ iff } (x, f(x)) \sim (x', f(x')) \text{ in } M \times N, \tag{3.1}$$

for the map $M \to M \times N$ given by $x \mapsto (x, f(x))$ preserves, like any map, the neighbour relation.

In the classical geometry of conics, consider a parabola. Then the tangent line at the apex is perpendicular to the axis of the parabola:

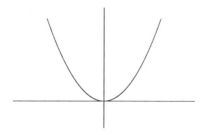

When coordinates are introduced in the plane, by making "the" geometric line R into a commutative ring, we may consider in particular the parabola P given as the graph of $y = x^2$. The axis of P is the y-axis, and the tangent line at the apex of P is the x-axis X. It follows from (3.1) that $x \sim 0$ implies $(x, x^2) \sim (0, 0)$. Since $\mathfrak{M}(0, 0) \cap P = \mathfrak{M}(0, 0) \cap X$, we conclude that $(x, x^2) \in X$, which implies that $x^2 = 0$. Thus

$$x \sim 0 \text{ implies } x^2 = 0, \tag{3.2}$$

or, writing $D \subseteq R$ for $\{x \in R \mid x^2 = 0\}$, this says that $\mathfrak{M}(0) \subseteq D$. (The other inclusion will be our definition of ∼ in this coordinate model.)

Next consider $(x, y) \sim (0, 0) \in R^2$. Since the projections $R^2 \to R$ preserve ∼, we conclude $x \sim 0$ and $y \sim 0$, so $(x, y) \in D \times D$, that is, $x^2 = y^2 = 0$; but we can say more, namely, that $x \cdot y = 0$, for the addition map $R \times R \to R$ preserves, like any map, the neighbour relation, so $(x, y) \sim (0, 0)$ implies $(x + y)^2 = 0$. But $(x + y)^2 = x^2 + y^2 + 2x \cdot y$. The first two terms we already know are 0, hence so is $2x \cdot y$, and since 2 is invertible, we conclude $x \cdot y = 0$; thus

$$(x, y) \sim (0, 0) \text{ implies } x^2 = y^2 = x \cdot y = 0. \tag{3.3}$$

We embark in the following section on a more serious investigation of how synthetic notions like ∼ can be conveniently coordinatized by suitable axiomatization of properties of the ring R.

4 Coordinate Geometry and the Axiomatics

It is not the intention of SDG to avoid using the wonderful tool of coordinates. So we now embark on the interplay between an assumed neighbour relation on the spaces and an assumed basic geometric line R with a commutative ring structure.

The reason we did not start there is to stress that the "arithmetization" in terms of R is a *tool*, not the *subject matter*, of geometry. This also applies in differential geometry, which has some important aspects without any R (as illustrated by the material in Section 2 and partly in Section 3); so in particular, it has a life without the ring $\mathbb{R}$ of real numbers, who sometimes thinks of himself as being the owner and boss of the company.

The scene of SDG in its present form is thus a category $\mathcal{E}$ (whose object we call *spaces* or *sets*), together with a commutative ring object R in it. But $\mathcal{E}$ is not the category of *discrete* sets, so some of the logical laws valid for the category of discrete sets, like the law of excluded middle, cannot be used. In differential geometry, whose maps are *smooth* maps, the law of excluded middle does anyway not apply; it would immediately lead out of the smooth world, like when one attempts to construct the absolute value function $x \mapsto |x|$ on the number line.

Nevertheless, we shall talk about the objects and maps of $\mathcal{E}$ as if they were sets; just recall that they are not *discrete* sets.[1] This is a basic technique in modern mathematics, more or less explicitly used in many other contexts. We shall not say more about it here. Basic concepts for making the technique explicit are cartesian closed categories, or, even better, locally cartesian closed categories, in particular, topoi (when talking about "families" of objects, as in the discussion above on envelopes). (There is some explicit description of the technique relevant for SDG in Part II of [15] and in Appendix A2 in [22].)

The axioms concern R; the category $\mathcal{E}$ should just have sufficiently good properties. The maximal thing wanted is that $\mathcal{E}$ is a topos, but less will often do. Thus, to get hold of an object like the unit circle $\{(x, y) \in R^2 \mid x^2 + y^2 = 1\}$, one needs only that $\mathcal{E}$ has finite limits; the circle then is a subobject of $R \times R$ given as the equalizer of two particular maps $R \times R \to R$. (In fact, the term *equalizer* came from such equational conditions as $x^2 + y^2 = 1$.)

For simplicity, we therefore in the following assume that $\mathcal{E}$ is a topos and that R is a commutative $\mathbb{Q}$-algebra in it. The intuition and terminology is as follows: R is the number line, and also, R is the ring of scalars.

---

[1] See the discussion in Section 8.

4.1 The Axiomatics

The axiom for such data, which is at the basis of the form of SDG considered here, is an axiom-*scheme*,[2] with one axiom for each Weil algebra; a Weil algebra is a finite-dimensional commutative algebra (over $\mathbb{Q}$, for the present purpose), where the nilpotent elements form an ideal of codimension 1. The name "Weil algebra" is used because they were introduced in the "Points proches" paper by A. Weil [42], whose aim was related to the one we present here. The simplest nontrivial Weil algebra is the "ring of dual numbers" $\mathbb{Q}[\epsilon] = \mathbb{Q}[X]/(X^2)$, which is two-dimensional over $\mathbb{Q}$.

Concerning R, we have already seen in (3.2) that $\mathfrak{M}(0) \subseteq \{x \in R \mid x^2 = 0\}$. The latter object we call D, as at the end of Section 3. To relate the combinatorics of $\sim$ with the algebra of R, we postulate the converse inclusion $D \subseteq \mathfrak{M}(0)$. It then follows that $x \sim y$ in R iff $(y - x)^2 = 0$.

The simplest instantiation of the axiom scheme concerns D. It can be seen as the instantiation of the axiom scheme for the two-dimensional Weil algebra

$$\mathbb{Q}[\epsilon] := \mathbb{Q}[X]/(x^2).$$

Axiom 1 *Every map $f : D \to R$ is of the form $d \mapsto a + d \cdot b$ for unique a and b in R.*

This has to be true *with parameters*; thus, if $f : I \times D \to R$ is an I-parameterized family of maps $D \to R$, then the a and b asserted by the axiom are likewise I-parameterized points of R, that is, they are maps $I \to R$. In a cartesian closed category $\mathcal{E}$, the "true with parameters" follows from a more succinct property, namely, the property that the map $R \times R \to R^D$, given by $(a,b) \mapsto [d \mapsto a + d \cdot b]$, is invertible. Thus, the axiomatics for SDG are simpler to state under the assumption that the category $\mathcal{E}$ is cartesian closed (although the idea and logic of parameterized families can also be made precise, even without cartesian closedness).

cartesian closedness of $\mathcal{E}$ is an aspect of talking about the objects of $\mathcal{E}$ as if they were sets.

We leave to the reader to prove (using D as a space of parameters) the following:

Corollary 4.1 *Every map $f : D \times D \to R$ is of the form $(d_1, d_2) \mapsto a + d_1 \cdot b_1 + d_2 \cdot b_2 + d_1 \cdot d_2 \cdot c$, for unique a, b_1, b_2, and c in R.*

In rough terms, since $R^D \cong R^2$, it follows that $(R^D)^D \cong (R^2)^D \cong (R^D)^2 \cong (R^2)^2 \cong R^4$. In itself, the corollary also appears as an instantiation

---

[2] Often referred to as the general KL axiom, for "Kock–Lawvere"; cf., e.g., [38] or [28].

of the axiom scheme, namely, for the four-dimensional Weil algebra

$$\mathbb{Q}[\epsilon_1, \epsilon_2] := \mathbb{Q}[X_1, X_2]/(X_1^2, X_2^2).$$

Remark 4.2 The relation $\sim$ defined in terms of D cannot be transitive, for transitivity is easily seen to be equivalent to D being stable under addition, and hence (using that 2 is invertible) that $d_1 \in D$ and $d_2 \in D$ imply $d_1 \cdot d_2 = 0$. But this contradicts the uniqueness of the coefficient c in the above corollary. So Axiom 1 implies that $\sim$ is not transitive.

We shall not be explicit how one goes from a (finite presentation of) a Weil algebra to the corresponding axiom (see [15, Section I.16]). The reader may guess the pattern from the examples given.

From the uniqueness assertion in Axiom 1, one derives the following:

Principle of cancelling universally quantified ds: let $r, s \in R$. Then *if $d \cdot r = d \cdot s$ for all $d \in D$, then $r = s$.*

In the classical treatment, any individual $x \neq 0$ in R is cancellable, that is, it has the property that it detects equality; $x \cdot r = x \cdot s$ implies $r = s$, for, in the classical treatment, R is a field, so $x \neq 0$ implies that x is invertible. On the other hand, in SDG, no individual $d \in D$ can be cancellable, for any such d is nilpotent. This, for some intuition, means that d is very small, "infinitesimal." So none of these small elements individually have the strength that they can detect equality, but when the small elements join hands, they can. Collective strength, of all the small together, replaces the strength of any individual.

Another consequence of Axiom 1 is that the beginnings of differential *calculus* become available: given $f : R \to R$, one applies, for each $x \in R$, the axiom to the function $d \mapsto f(x + d)$, so one gets for each x that there are unique a and b such that $f(x + d) = a + d \cdot b$ for all $d \in D$. The a and b depend on the x chosen, so write them $a(x)$ and $b(x)$, respectively. By setting $d = 0$, we conclude $a(x) = f(x)$. But $b(x)$ deserves a new name – we call it $f'(x)$, so for all $d \in D$, we have the exact "Taylor expansion"

$$f(x+d) = f(x) + d \cdot f'(x) \text{ for all } d \text{ with } d^2 = 0. \tag{4.1}$$

And this property characterizes $f'(x)$, by the principle of cancelling universally quantified ds.

Since such Taylor expansion holds also with parameters, one also gets partial derivatives for functions in several variables, by considering the variables, except one, as parameters. See (4.2) for an example.

Remark 4.3 For differential *calculus*, there are other synthetic/axiomatic theories available, for example, the "Fermat" axiom (suggested by Reyes) (see, e.g., [38, Section VII.2.3]); the axiomatics of "differential categories" (cf. [2, 6], and references therein); and the "topological differential calculus" (cf. [1], and references therein).

Corollary 4.1 could be seen as an instantiation of the general axiom scheme; a more interesting instantiation of the axiom scheme comes about by considering the three-dimensional Weil algebra

$$\mathbb{Q}[\epsilon_1, \epsilon_2]/(\epsilon_1 \cdot \epsilon_2) := \mathbb{Q}[X_1, X_2]/(X_1^2, X_2^2, X_1 \cdot X_2).$$

To state the axiom, let $D(2) \subseteq R^2$ be given as

$$\{(d_1, d_2) \in R^2 \mid d_1^2 = d_2^2 = d_1 \cdot d_2 = 0\}.$$

(Clearly, $D(2) \subseteq D \times D$. Note that $D(2)$ is defined by the equations occurring in (3.3).) Then,

Axiom 2 *Every map* $f: D(2) \to R$ *is of the form* $(d_1, d_2) \mapsto a + d_1 \cdot b_1 + d_2 \cdot b_2$ *for unique a, b_1, and b_2 in R.*

One may have deduced Axiom 2 from Corollary 4.1, *provided* one knew that any function $D(2) \to R$ may be extended to a function $D \times D \to R$. But this is not automatic – rather, this is guaranteed by Axiom 2.

Of course, there are similar axioms for $n = 3, 4, \ldots$, using

$$D(n) := \{(d_1, \ldots, d_n) \in R^n \mid d_i \cdot d_j = 0 \text{ for all } i, j = 1, \ldots n\}.$$

In short form, a general Axiom 2 says that *any map* $D(n) \to R$ *extends uniquely to an affine map* $R^n \to R$.

Another instantiation of the axiom scheme gives the following axiom (we shall not use it here): let $D_2 := \{x \in R \mid x^3 = 0\}$. Then *every function* $f: D_2 \to R$ *is uniquely of the form* $x \mapsto a_0 + a_1 \cdot x + a_2 \cdot x^2$, or every $f: D_2 \to R$ extends uniquely to a polynomial function $R \to R$ of degree ≤ 2. This axiom corresponds to the three-dimensional Weil algebra $\mathbb{Q}[X]/(X^3)$. More generally, let $D_k(n) := \{(x_1, \ldots, x_n) \in R^n \mid$ all products of $k + 1$ of the x_is is $0\}$. Then *every function* $D_k(n) \to R$ *extends uniquely to a polynomial function* $R^n \to R$ *of degree* $\leq k$. The polynomial functions occurring here are the Taylor polynomials at $0 \in R$) (resp. at $(0, \ldots, 0) \in R^n$) of f.

As a final example of an instantiation of the axiom scheme, let $D_L \subseteq R^2$ be given by

$$D_L := \{(x_1, x_2) \in R^2 \mid x_1^2 = x_2^2 \text{ and } x_1 \cdot x_2 = 0\}.$$

Then the following axiom is likewise an instantiation of the axiom scheme: *every function* $f\colon D_L \to R$ *is of the form* $f(x) = a + b_1 \cdot x_1 + b_2 \cdot x_2 + c \cdot (x_1^2 + x_2^2)$. The c occurring here can then be seen as (one-fourth of) the Laplacian $\Delta(f)$ of f at $(0,0)$. Note that $D(2) \subseteq D_L \subseteq D_2(2)$. The space D_L corresponds to a certain four-dimensional Weil algebra; see also [22, Section 8.3].

4.2 Envelopes Again

This section is to "justify" in classical terms the correctness of our description of envelopes in terms of characteristics, as in Section 2.2. For simplicity, we consider a one-parameter family of (unparameterized) curves S_t in R^2. We assume that there is some smooth function $F(x, y, t)$ such that the tth curve S_t is given as the zero set of $F(-, -, t)$. We then prove that the classical analytic "discriminant" description of the characteristics and the envelope agrees with the synthetic/geometric one that we have given; but note that our description is coordinate free, so in particular, it follows that the constructed envelope is independent of the analytic representation. To say that (x, y) belongs to the t_0 characteristic is by the synthetic definition to say that $F(x, y, t_0 + d) = 0$ for all $d \in D$ (the neighbours t of t_0 are of the form $t_0 + d$). Equivalently, by Taylor expansion,

$$F(x, y, t_0) + d \cdot \partial F/\partial t(x, y, t_0) = 0, \tag{4.2}$$

for all $d \in D$. By the principle of cancellation of universally quantified ds, this is equivalent to the conjunction of the two equations

$$F(x, y, t_0) = 0 \quad \text{and} \quad \partial F/\partial t(x, y, t_0) = 0, \tag{4.3}$$

which is how the t_0 characteristic, and hence the envelope, may be described by the discriminant method.

However, Courant gives an example ([7, Example 10 in III.3]) to show that "the envelope need not be the locus of the points of intersection of neighbouring[3] curves," in other words, the "*non-rigorous but intuitive*" description of characteristics suggested in loc. cit. is not only nonrigorous but is furthermore *wrong*. (So implicitly, do not believe in geometry!) The example is the following. Consider the family of curves in the plane given by $F(x, y, t) = y - (x - t)^3$. (This is the curve $y = x^3$, together with all its horizontal translates.) We leave to the reader to prove that the characteristic set at

---

[3] The word neighbouring here is not in the sense of the $\sim$ neighbour relation that we are using; in fact, it rather means "distinct."

parameter value t_0 (as calculated by (4.3)) is the subset $\{(t_0 + D, 0)\}$ of the x-axis; so the envelope is the x-axis, whereas the "limit intersection point" idea does not work here, since (to quote Courant) "no two of these curves intersect each other."

4.3 Defining $\sim$ in Terms of R?

We have already postulated that $x \sim y$ in R means $y - x \in D$, or $(y-x)^2 = 0$. A (first-order) neighbour relation $\sim$ on any object $M \in E$ can be defined by

$$x \sim y \text{ in } M \text{ iff } \alpha(x) \sim \alpha(y) \in D \text{ for all } \alpha : M \to R. \tag{4.4}$$

So $\sim$ is, for all objects M, defined in terms of the scalar valued functions on M. Trivially, any map $M' \to M$ preserves $\sim$. This is the "contravariant" or "weak" way of defining $\sim$. There is also a "covariant" or "strong" way of defining it (see [22, p. 31]). For good spaces, like R^n, they coincide. (The weak determination is not adequate in algebraic geometry, since projective space, and other important geometric objects, only admits constant scalar-valued functions. So one must here replace the consideration of scalar-valued functions by *locally defined* scalar-valued functions, and for this, one needs some notion of "local," as alluded to in Section 7.1.)

For the weak determination of $\sim$, we can identify the monad $\mathfrak{M}(\underline{0})$ around the origin in R^n:

Proposition 4.4 *We have* $\mathfrak{M}(\underline{0}) = D(n)$.

(For $n = 1$, this was postulated.) Let us prove it for $n = 2$. We have already seen in (3.3) that $\mathfrak{M}(\underline{0}) \subseteq D(2)$. For the converse, we have to consider an arbitrary map $\alpha : R^2 \to R$ and prove that $(d_1, d_2) \in D(2)$ implies $\alpha(d_1, d_2)^2 = 0$. By Axiom 2, $\alpha(d_1, d_2) = a + b_1 \cdot d_1 + b_2 \cdot d_2$, and so $\alpha(d_1, d_2) - \alpha(0, 0) = b_1 \cdot d_1 + b_2 \cdot d_2$, which has square 0 since $d_1^2 = d_2^2 = d_1 \cdot d_2 = 0$.

Remark 4.5 Since $D(2)$ is strictly smaller than $D \times D$, we therefore also have that $(d_1, d_2) \sim (0, 0)$ is stronger than the conjunction of $d_1 \sim 0$ and $d_2 \sim 0$.

The principle in the beginning of Section 3 may now be proved algebraically: we may assume that coordinates are chosen so that the considered common point $x \in l_1 \cap l_2$ is $(0, 0)$ and that l_1 and l_2 are graphs of the functions $x \mapsto b_1 \cdot x$ and $x \mapsto b_2 \cdot x$. We must prove that $b_1 = b_2$. For $d \in D$, we have $(d, b_i \cdot d) \in \mathfrak{M}(x) = D(2)$. So, by assumption, for all $d \in D$, we have $(d, b_1 \cdot d) \in \mathfrak{M}(x) \cap l_1 = \mathfrak{M}(x) \cap l_2 \subseteq l_2$, so for all $d \in D$, we have $b_1 \cdot d = b_2 \cdot d$; cancelling the universally quantified d then gives $b_1 = b_2$.

4.4 Contravariant and Covariant Hierarchy

The polynomial function $x: R \to R$ vanishes at 0; one also says that it vanishes to *first* order at 0 and that x^2 vanishes to *second* order at 0, and so on. More generally, $f: R \to R$ vanishes to second order at 0 if it may be written $f(x) = x^2 \cdot g(x)$ for some function $g: R \to R$, and similarly for kth-order vanishing. It generalizes to "order of vanishing" of $f: M \to R$ at a point $a \in M$. Note that kth-order vanishing is a *weaker* condition than $(k+1)$st-order vanishing. This (essentially classical) hierarchy of scalar-valued functions (quantities) is to be compared with the hierarchy of neighbours, applicable to points of spaces M, where kth-order neighbour is a *stronger* condition than $(k+1)$th-order neighbour. The neighbour relations are covariant notions, applicable to *points (elements)* of spaces (the assumed neighbour relations $\sim_1, \sim_2, \ldots$, are preserved by mappings and thus are *covariant*); the order of vanishing is a contravariant notion, applicable to *quantities* on M, that is, to R-valued functions $M \to R$.

The notions are related as follows, for a and b in M: $a \sim_k b$ iff, for any quantity, $f: M \to R$ vanishing to kth order at a, we have $f(b) = 0$; this is, for $k = 1$, just a reformulation of (4.4). Recall that $a \sim_k b$ on R is defined in terms of $(a-b)^{k+1} = 0$, that is, in terms of order of *nilpotency*.

A classical formulation, in certain contexts, is that we can "ignore" quantities of higher order in comparing a and b: "*Dabei sehen wir von unendlich kleinen Grössen höhere Ordnung ab.*" ("Here, we ignore infinitely small quantities of higher order"; [34, p. 523]). In rigorous mathematics, one cannot "ignore" anything, except 0. But one can certainly consider nilpotent elements in rings. Thus, an explicit theory of infinitesimals came in through the back door, namely, from algebraic geometry.

4.5 Wisdom from Algebraic Geometry

The development leading to the modern formulations of SDG began in French algebraic geometry in the mid-20th century by Grothendieck and his collaborators, with the notion (and category!) of *schemes*, as a generalization of the notion of algebraic varieties (over a field k, say).

In particular, the category $\mathcal{E}_k$ of affine schemes over k is by definition the dual of the category $\mathcal{A}_k$ of commutative k-algebras, suitably size restricted, say, of finite presentation. The algebras are allowed to have nilpotent elements. Such algebra A is seen as the ring of scalar-valued functions on the scheme M (geometric object, "space") that it defines. One writes $M = \mathrm{Spec}(A)$. (The "scalars" R is the scheme represented by $k[X]$.) Then the algebra $A \otimes A$ defines the space $M \times M$. If I is the kernel of the multiplication map $A \otimes A \to A$, then $(A \otimes A)/I \cong A$. Consider the ideal $I^2 \subseteq I$. The k-algebra $(A \otimes A)/I^2$ gives $M_{(1)}$, the first neighbourhood of the diagonal of M. So

$$M_{(1)} := \mathrm{Spec}((A \otimes A)/I^2).$$

The quotient map $(A \otimes A)/I^2 \to (A \otimes A)/I \cong A$ defines, in the category of schemes, the diagonal $M \to M_{(1)}$.

Note that $I/I^2 \subseteq (A \otimes A)/I^2$ consists of elements of square 0. It is in fact the module of *Kähler differentials of* A; $(A \otimes A)/I^2$ is the ring of scalar-valued functions on $M_{(1)}$, and the submodule I/I^2 consists of those functions that vanish on the diagonal $M \subseteq M_1$, that is, the combinatorial scalar-valued 1-forms, in the sense of Section 2.6. (Kähler introduced these differentials already in the 1930s.)

The simplest scheme that is not a variety is D, the affine scheme given by $k[\epsilon]$, the ring of dual numbers over k. The underlying variety of D has just one global point, since $k[\epsilon]$ has only one prime ideal, namely, (ϵ). Geometrically, D is a "thickened" version of its unique global point. Mumford [39, p. 338] describes D as "a sort of disembodied tangent vector," meaning that a map $D \to X$ may be identified with a tangent vector to X, for any scheme X.

The relationship between the infinitesimal objects like D and the neighbourhoods of diagonals may be exemplified by the isomorphism

$$R_{(1)} \cong R \times D,$$

given by $(x, y) \mapsto (x, y - x)$, for $x \sim y$ in R.

The crucial step in the formation of contemporary SDG was when Lawvere in 1967 combined this consideration of a "tangent vector representor" D with the idea of a cartesian-closed category $\mathcal{E}$. Thus, for any object X in $\mathcal{E}$, X^D is then the *object* (space) in $\mathcal{E}$ of all tangent vectors to X; in other words, it is the (total space of the) tangent bundle $T(X) \to X$.

To put this relationship into axiomatic form is most conveniently done by assuming a ring object R and describing D in terms of R (as is done in Section 4). There is a more radical approach, advocated by Lawvere in [31], where the ring R is to be constructed out of an infinitesimal object T ("an instant of time") (ultimately then proved to be isomorphic to D); see also [6, Section 5.3]

5 Models of the Axiomatics

For an axiomatic theory, models are useful but not crucial. Euclidean geometry has been useful for more than 2,000 years. When exactly was a model for it presented? Did it have to wait for the real numbers, or at least some subfields of it? Models are useful – they may guide the intuition and prevent inner contradictions. This also applies to SDG. The models for SDG come in two main groups: arising from algebraic geometry and from classical differential geometry over $\mathbb{R}$, respectively (and in fact, SDG serves to make explicit what the two groups have in common).

Models for the axiomatics of Section 2.3 may be built on the basis of some of the models of SDG mentioned above; see [23].

5.1 Algebraic Models

The category $\mathcal{E}_k$ of affine schemes over a commutative ring k (i.e., the dual of the category of [finitely presented, say] commutative k-algebras) is a model,[4] with $k[X]$ as R. $\mathcal{E}_k$ is not quite cartesian closed, but at least the scheme corresponding to $k[\epsilon]$ (or to any other Weil algebra) is exponentiable. The set-valued presheaves $\hat{\mathcal{E}}_k$ on $\mathcal{E}_k$ is a full-fledged topos model (with R represented by $k[X]$). The topos $\hat{\mathcal{E}}_k$ is of course the same as the category of covariant functors from the category of (finitely presentable) commutative k-algebras to sets, and R is in this setup just the forgetful functor, since $k[X]$ is the free k-algebra in one generator.

Many of the subtopoi of $\hat{\mathcal{E}}_k$ are likewise models; passing to suitable subtopoi, one may force R to have further properties. One may, for instance, force R to become a local ring; the subtopos forcing this is also known as the Zariski topos. These topoi are explicitly the main categories studied in [9].

5.2 Analytic Models Based on $\mathbb{R}$

There is, of course, a special interest in models $(\mathcal{E}, R)$ which contain the category Mf of smooth manifolds as a full subcategory, in a way that preserves known constructions and concepts from classical differential geometry. So one wants a full and faithful functor $i: Mf \to \mathcal{E}$, with $i(\mathbb{R}) = R$. Also *transversal* pullbacks should be preserved, and $i(T(M))$ should be $i(M)^D$. The properties of such a functor i have been axiomatized by Dubuc [10] under the name of "well-adapted models for SDG"; see also [14]. The book [38] is mainly devoted to the construction and study of such models.

The earliest well-adapted model (constructed by Dubuc [10]) is one now known as the "Cahiers topos." It can be proved to contain the category of convenient vector spaces (with smooth maps between them) as a full subcategory,

---

[4] If 2 is not invertible in k, there are things that work differently.

in a way that preserves the cartesian-closed structure (cf. [18, 25]). A more advanced topos $\mathcal{G}$, now called the "Dubuc topos" [11], even supports some "synthetic differential topology" (cf. [4]).

A main tool in the construction of analytic models is to take heed of the wisdom of algebraic geometry, but replacing the algebraic theory (in the sense of Lawvere) $\mathbb{T}$ of commutative rings with the richer algebraic theory $\mathbb{T}_\infty$, whose n-ary operations are not only the real polynomial functions but *all* the smooth maps $\mathbb{R}^n \to \mathbb{R}$. It contains the theory of commutative rings as a subtheory, since a polynomial in n variables defines a smooth function in n variables. The theory $\mathbb{T}_\infty$, and its importance for the project of categorical dynamics, was already in Lawvere's seminal 1967 lectures.

Note that any smooth manifold M gives rise to an algebra for this theory, namely, $C^\infty(M)$, the ring of smooth $\mathbb{R}$-valued functions on M. We may think of M as a "reduced" affine scheme corresponding to the ring $C^\infty(M)$, and then mimic the construction (described above) of set-valued presheaves on affine schemes, and subtopoi thereof. But note also that $\mathbb{R}[\epsilon]$ (and all other Weil algebras over $\mathbb{R}$) are algebras for $\mathbb{T}_\infty$ and define (nonreduced) affine schemes.

Modules of Kähler differentials for algebras for $\mathbb{T}_\infty$ were studied in [12].

(If one takes just the category of smooth manifolds [with open coverings] as the site of definition for a topos, one gets a topos already considered in SGA4, under the name of "the smooth topos"; it contains the category of diffeological spaces as a full subcategory, but lacks infinitesimal objects like D. These categories are models for the Fermat–Reyes axiomatics. See [15, Exercise III.8.1].)

6 New Spaces

Except for "infinitesimal" spaces like $D_k(n)$, the present account does not do justice to the *new* spaces that have emerged through the development of SDG. In particular, it has not capitalized on the unproblematic way in which function spaces exist in this context, by cartesian closedness of $\mathcal{E}$. These function spaces open the door to a synthetic treatment of calculus of variations, continuum mechanics, infinite-dimensional Lie groups, and so on. For such spaces, the neighbour relation (which has been my main focus here) is more problematic, however, and is not well exploited. Instead, one uses the (classical) method of encoding the infinitesimal information of a space X in terms of its tangent bundle $T(X) = X^D$, rather than in terms of $X_{(1)}$ (first neighbourhood of the diagonal). Notably, Nishimura has pushed the SDG-based theory far in this direction (cf., e.g., [41]).

Another type of new space comes from the observation that the functor $(-)^D$ in many of the models has a right adjoint, $(-)^{1/D}$ (Lawvere's notation, "fractional exponent"); the spaces $M^{1/D}$ are reminiscent of Eilenberg–Mac Lane spaces. There is some discussion of them in [15, Section I.20], in [26], and in [29].

Finally, the notion of *jets*, and jet bundles, as considered by Ehresmann in the 1950s forms, on one hand, one of the sources for SDG as presented here; on the other hand, the SDG method makes the consideration of jets and jet bundles simpler, since SDG makes the notion of a jet *representable*, in the sense that a k-jet at $x \in M$, with values in N, is here simply a map $\mathfrak{M}_k(x) \to N$, rather than an equivalence class of maps $U \to N$ (where $x \in U$). (In [4], the notion of the *germ* of a map is likewise representable.)

When jets are representable, Ehresmann's theory of differentiable groupoids, as a carrier of a general theory of connections, admits some simpler formulations.

6.1 Connections in Fibre Bundles and Groupoids

For the present purpose, a *fibre bundle* over a space M is just a map $\pi: E \to M$. (When it comes to proving things, one will need good exactness properties of π, such as being an effective descent map or being locally a projection $F \times M \to M$.) Then a combinatorial connection in the bundle $E \to M$ is an action ∇ of $M_{(1)}$ on E, in the sense that $(x, y) \in M_{(1)}$ and $e \in E_x$ define an element $\nabla(x, y)(e)$ in E_y. (Here, $E_x := \pi^{-1}(x)$, and similarly for E_y.) One requires the normalization condition $\nabla(x, x)(e) = e$. For good spaces, it then follows that $\nabla(y, x)\nabla(x, y)(e) = e$. The notion of affine connection λ considered above is a special case: the bundle $E \to M$ is in this case the first projection $M_{(1)} \to M$, and $\nabla(x, y)(z) = \lambda(x, y, z)$. If $E \to M$ is a vector bundle, say, a *linear* connection is a connection ∇ where the map $\nabla(x, y)(-): E_x \to E_y$ is linear for all $x \sim y$. For good spaces M, linear connections in the tangent bundle $T(M) \to M$ contain exactly the same information as affine connections λ on M.

There is also a notion of connection ∇ in a *groupoid* $\Phi \rightrightarrows M$. (This is closely related to the notion of *principal connection* in a principal fibre bundle $P \to M$; in fact, such P defines, according to C. Ehresmann, a groupoid $PP^{-1} \rightrightarrows M$, and a principal connection in $P \to M$ is then the same datum as a groupoid connection in $PP^{-1} \rightrightarrows M$.) Recall that a groupoid $\Phi \rightrightarrows M$ carries a reflexive symmetric structure: the reflexive structure picks out for every $x \in M$ the identity arrow at x, and the symmetric structure associates to an arrow $f: x \to y$ its inverse $f^{-1}: y \to x$. Then a connection in $\Phi \rightrightarrows M$

is simply a map $M_{(1)} \to \Phi$ preserving (the two projections to M and) the reflexive and symmetric structure,

$$\nabla(x,x) = \mathbf{1}_x \quad \text{and} \quad \nabla(y,x) = \nabla(x,y)^{-1},$$

for all $x \sim y$.

Given a bundle $E \to M$ in $\mathcal{E}$, if $\mathcal{E}$ is locally cartesian closed, one may form the groupoid $\Phi \rightrightarrows M$ where the arrows $x \to y$ are the invertible maps $f: E_x \to E_y$. Then a connection on $E \to M$, in the bundle sense, is equivalent to a connection, in the groupoid sense, of this groupoid $\Phi \rightrightarrows M$. If $E \to M$ is a vector bundle, there is a subgroupoid of $\Phi \rightrightarrows M$ consisting of the *linear* isomorphisms $E_x \to E_y$ (this groupoid deserves the name $GL(E)$), and similarly if $E \to M$ is a group bundle or has some other fibrewise structure.

The groupoid formulation of the notion of connection is well suited to formulating algebraic properties like curvature. We may observe that the curvature, as described in Section 2.5 for affine connections λ, is purely groupoid theoretical. Thus, if x, y, z form an infinitesimal 2-simplex in M, it makes sense to ask whether $\nabla(x, y)$ followed by $\nabla(y, z)$ equals $\nabla(x, z)$, or better, consider the arrow $R(x, y, z): x \to x$ given as the composite (composing from left to right)

$$R(x,y,z) := \nabla(x,y).\nabla(y,z).\nabla(z,x) \in \Phi(x,x).$$

This is the *curvature* of ∇; more precisely, the curvature R is a combinatorial 2-form with values in the group bundle $gauge(\Phi)$ of vertex groups $\Phi(x,x)$ of Φ. Now the connection ∇ in Φ gives rise to a connection $ad\nabla$ in the group bundle $gauge(\Phi)$: $ad\nabla(x, y)$ is the (group-) isomorphism $\Phi(x, x) \to \Phi(y, y)$ consisting in conjugation by $\nabla(x, y): x \to y$. This conjugation we write $(-)^{\nabla(x,y)}$. In terms of this, we have an identity, which deserves the name the *Bianchi* identity for (the curvature R of) the connection ∇; namely, for any infinitesimal 3-simplex (x, y, z, u), we have

$$id_x = R(yzu)^{\nabla(y,x)}.R(xyu).R(xuz).R(xzy), \tag{6.1}$$

verbally, *the covariant derivative of the $gauge(\Phi)$ valued 2-form R, with respect to the connection $ad\nabla$ in the group bundle, is "zero,"* that is, it takes only identity arrows as values.

The proof of (6.1) is trivial, in the sense that it is a case of Ph. Hall's 14-letter identity, which holds for any six elements in a group, or for the six arrows of a tetrahedron-shaped diagram in a groupoid; here, the six arrows are the $\nabla(x, y), \nabla(x, z), \ldots, \nabla(z, u)$ in Φ. See [22], and see [19] for how this implies the classical Bianchi identity for linear connections in vector bundles.

7 The Role of Analysis

7.1 Analysis in Geometry?

The phrase *analytic* geometry may be used in a wide sense: using coordinates and calculations. In this sense, SDG as presented here quickly becomes analytic (e.g., the basic axiomatics are formulated in such terms, as expounded in Chapter 4). But the more common use of the term *analytic* is that *limit* processes and topology are utilized.

Ultimately, topology and limits in real analysis have their origin in the strict order relation $<$ on $\mathbb{R}$. Then the partial order $\leq$ is defined by $x \leq y$ iff $\neg(y < x)$. The elements in $\mathbb{R}_{>0}$ are invertible. In SDG, it is also natural to have an order $<$ on R, given primitively, or in terms of the algebraic structure of R. (In well-adapted models $i : Mf \to \mathcal{E}$, the relation $<$ is definable in terms of the inclusion of the smooth manifold $\mathbb{R}_{>0}$ into $\mathbb{R}$, which by the embedding i defines a subobject $R_{>0} \subset R$, out of which a strict order $<$ can be defined.) Nilpotent elements d in a nontrivial ring cannot be invertible. It follows, for any nilpotent d, that $d \leq 0$, and hence also $-d \leq 0$ (since also $-d$ is nilpotent). So $0 \leq d \leq 0$. So if $\leq$ were a partial order (not just a preorder), this would imply that any nilpotent d is 0, which is incompatible with SDG. Thus, in SDG, $\leq$ is only a preorder, not a partial order. For preordered sets, a supremum is not uniquely defined; to have a *unique* number as supremum, one needs a partial order.

This is one reason why limit processes are not used in SDG at the present stage.

Topology comes in play, for example, when formulating statements about *local* existence of, say, solutions to particular differential equations. "Local" refers to some topology on a given object, and in SDG, there may be several natural choices (cf. in particular the recent [4]). The finest topology on an object (space) X is, in the context of SDG, the one where the open subsets are those $U \subseteq X$ which are closed under the neighbour relation $\sim$. For instance, a local solution f (for this fine topology) for a differential equation $f' = F(x, y)$ amounts to a formal power series solution and is therefore cheap. More serious existence statements are when stronger topologies, like the "intrinsic Zariski topology," are involved: a subset $U \subseteq X$ is open if it is of the form $f^{-1}(R^*)$, where $R^* \subseteq R$ consists of the invertible elements, or if it is of the form $f^{-1}(P)$, where $P \subseteq R$ consists in the strictly "positive" numbers – which then in turn have to be described or assumed (see [36] for some results in this direction).

SDG does not *prove* basic integration results, and even the formulation of such results does not come for free. Advances in this direction exist in what is now called synthetic differential topology. It builds on SDG, and its main model is the Dubuc topos $\mathcal{G}$; see [38, Chapter III] and, notably, [4], where also a synthetic theory of singuarity theory is considered.

The most basic integration result is the (essentially unique) existence of antiderivatives: for $f: R \to R$, there exists $F: R \to R$ with $F' = f$. In an axiomatic development, this has to be taken as an *axiom* – one that actually can be proved to hold in all the significant topos models $(\mathcal{E}, R)$ for SDG, and similarly, for many other basic results, such as a suitable version of the intermediate value theorem.[5] Thus, full-fledged analysis in axiomatic terms, incorporating SDG, quickly becomes overloaded with axioms and is better developed as a *descriptive* theory, describing what actually *holds* in *specific* models $(\mathcal{E}, R)$. This is the approach of [38], which, significantly, has the title "Models for Smooth Infinitesimal Analysis" (although also a full-fledged axiomatic theory is presented in loc. cit., Chapter 7). Note that the term *smooth*, in so far as SDG is concerned, is a void term, since unlimited differentiability is automatic in this context; and "smooth implies continuous" (equivalently, "all maps are continuous") is a *theorem* in the good, well-adapted models (see, e.g., Theorem III.3.5 in [38]).

I prefer not to think of SDG as a monolithic global theory but as a *method* to be used locally, in situations where it provides insight and simplification of a notion, of a construction, or of an argument. The assumptions, or axioms that are needed, may be taken from the valuable treasure chest of real anaysis.

Thus, the very construction of well-adapted models $Mf \to \mathcal{E}$ depends on the theory $\mathbb{T}_\infty$ whose n-ary operations are the smooth functions $\mathbb{R}^n \to \mathbb{R}$, so that, for example, the exponential function $\exp: \mathbb{R} \to \mathbb{R}$, or the trigonometric functions, are "imported" from the treasure chest (here, imported from Euler, say, much prior to the rigorous formulation of limit processes). In the context of SDG, it is possible to introduce existence of, say, these particular transcendental functions axiomatically, by functional equations, or by differential equations. This is what the Calculus Books in essence do.

7.2 Nonstandard Analysis?

Nonstandard analysis (NSA) is another theory where the notion of infinitesimals has an explicit and well-defined status. Therefore, one sometimes asks whether there is some relationship between SDG and NSA. There is very

---

[5] Significantly, the version valid in significant SDG models applies to functions f with a *transversality* condition, such as $f' > 0$ – like in constructive analysis.

little relationship; NSA is a descriptive, not an axiomatic, theory, dealing (at least insofar as differential geometry goes) with the real number field $\mathbb{R}$ and crucially capitalizing on its Cauchy completeness, since it is crucial that *every (bounded) nonstandard real number $\in \mathbb{R}^*$ have a unique standard part*. This is another expression of the completeness of the real number system. In this sense, NSA is a reformulation, with a richer vocabulary, of standard real analysis, and can, as such, cope with things defined in terms of limits, like definite integrals in terms of Riemann sums, say. SDG cannot do this; at best, it can introduce some integration by axioms (cf. the remark on the Frobenius integration theorem in Section 2.4).

In NSA, one has a neighbour notion for elements in $\mathbb{R}^*$; it is an *equivalence* relation, and the equivalence classes are called *monads* – a term that SDG has imported; but in SDG, it is crucial that the neighbour relations not be transitive and come in a hierarchy – first order, second order, ..., (hence first-order, second-order, ...kth-order monads $\mathfrak{M}_k(x)$, ...) – and this comes closer to important aspects of mathematical practice, where notably the first-order neighbour relation takes most of the work on its shoulders and has been the sole concern in this note. (The second-order monads in SDG play a role when discussing, e.g., dynamic or metric notions – thus a (pseudo-) Riemannian metric may be defined in terms of R-valued functions $f(x,y)$ defined for $x \sim_2 y$, and with $f(x,y) = 0$ if $x \sim_1 y$; see [22].)

NSA can also be axiomatized, but this amounts essentially to axiomatizing a further structure (an endofunctor) on the *category* of (discrete) sets [24] or a further primitive predicate in axiomatic (Zermelo Fraenkel) set theory [40].

8 The Continuum and the Discrete

A historically important problem in (the philosophy of) mathematics is the problem of understanding the nature of the continuum and its relationship to the discrete. Is the continuum just a discrete set of points (and motion therefore impossible, according to Parmenides)? In contrast, in Euclidean geometry, *line* (line segment) was a primitive notion and was not just the set of points in it. (And *time* was not a set of instants.) Even a contemporary geometer like Coxeter makes the distinction between a line and the "range of points" on it (cf. [8, p. 20]).

The principal side of the contradiction between continuum and discrete was, historically, the continuum. With the full arithmetization of the continuum, in the hands of, say, Dedekind, with the construction of the real number system $\mathbb{R}$, the continuum was reduced to a set of points, and the cohesion of the continuum was reduced to a topology on this point set. For mainstream differential geometry, synthetic axiomatic considerations became, in principle, redundant. Everything became reducible to real analysis.

Synthetic differential geometry refuses to take this one-sided reductionist view. (For one reason, $\mathbb{R}$, as a point set [set of global points], has no nontrivial nilpotent elements d.) Rather, SDG learns from (and possibly contributes to) *analysing* the relationship between the continuum and the discrete. Such analysis typically has the form of a functor $\gamma_* : \mathcal{E} \to \mathcal{S}$, with $\mathcal{S}$ some category of discrete sets, and with $\mathcal{E}$ some category of spaces with some kind of cohesion.[6] Preferably, both $\mathcal{E}$ and $\mathcal{S}$ are topoi and γ_* a geometric morphism, associating to a space $X \in \mathcal{E}$ its set of (global) points. The left adjoint γ^* of γ_* is a full embedding, so that discrete spaces form a full subcategory of $\mathcal{E}$. An example of such an $\mathcal{E}$-$\mathcal{S}$-pair is with $\mathcal{E}$ the topos of simplicial sets, with $\gamma_*(X)$ the set of 0-simplices (= global points) of X. This example is relevant to algebraic topology (cf., e.g., [35]), not to differential geometry, but it illustrates a phenomenon that is crucial also for SDG, namely, that there are nontrivial objects with only one global point (e.g., in the topos of simplicial sets: the simplicial n-sphere $\Delta(n)/\dot{\Delta}(n)$) – just like D in SDG has 0 as the only global point).

A well-adapted model $\mathcal{E}$ of SDG contains not only the category of discrete manifolds (sets) as a full subcategory but even the category of all smooth manifolds, in particular, $\mathbb{R}$. By the fullness, $\mathbb{R}$, when seen in $\mathcal{E}$ (and there denoted R), does not acquire any new global points (unlike the $\mathbb{R}^*$ of NSA). But it does acquire new subobjects, for example, $D \subseteq R$. When we talk about general elements $d \in D$, we are therefore not talking about *global* points $1 \to D \subseteq R$.

A space is an object in a category of spaces (Grothendieck, Lawvere). So what "is" the space $\mathbb{R}$? It depends on the category in which it is considered. In SDG, one considers $\mathbb{R}$ in certain ("well-adapted") topoi $\mathcal{E}$; $\mathbb{R} = R$ does not change; it is the ambient category that changes.

9 Looking Back

The discovery, by Huygens in the 17th century, of the notion of envelopes and their relatives (leading to a theory of waves, isochrones, etc.) was

---

[6] A situation axiomatized in Lawvere's [30] "Mengen" vs. "Kardinalen" and further elaborated in papers by Lawvere and by Menni (cf., e.g., [37] and [32]).

coined in geometric terms, without essential reference (so far as I know) to analytic considerations. When differential calculus, as we know it today, was developed, analytic methods became more dominant. A main treatise like Monge's in 1795 was entitled "L'application de l'analyse à la géométrie." But this treatise of Monge's goes also in the other direction: it forcefully uses geometric and synthetic reasoning for explaining the analytic theory of first-order PDEs of Lagrange – a thread taken up later by Sophus Lie; this comprises in particular the theory of *characteristics* of such PDEs, the curves, out of which the solutions of the PDE can be built. (They are built up from characteristics in the sense of Section 2.2, namely, intersections of families of surface elements.) Lie's 1896 book [34] on contact geometry has a chapter called "Die Theorie der partiellen Differentialgleichungen als Teil der Theorie der Flächenelemente" (Flächenelement = surface element = contact element $\mathfrak{M}(b) \cap B$, as in Section 2.3, or the sets $\mathfrak{M}_{\approx}(x)$ of suitable codimension 1 distribution, as in Section 2.4).

In one of Lie's early articles on the theory of differential equations, he wrote,

> The reason why I have postponed for so long these investigations, which are basic to my other work in this field, is essentially the following. I found these theories originally by synthetic considerations. But I soon realized that, as expedient [zweckmässig] the synthetic method is for discovery, as difficult it is to give a clear exposition on synthetic investigations, which deal with objects that till now have almost exclusively been considered analytically. After long vacillations, I have decided to use a half synthetic, half analytic form. I hope my work will serve to bring justification to the synthetic method besides the analytical one.
>
> (From Lie's "Allgemeine Theorie der partiellen Differentialgleichungen erster Ordnung," Math. Ann. 9 (1876); my translation)

Despite Lie's call for a synthetic language and logic, the differential geometry in the 20th century became more and more analytic and was removed from the geometric intuition – at the time of Einstein, the "débauche of indices," and rules for how the coordinates transform, later on more abstract and coordinate free, but still somewhat un-geometric – as it must be when explicit infinitesimals (neighbour points) have to be avoided.

The editors of the present volume asked me to address the question about the "advantages of SDG over other approaches." First of all, the neighbour notion, and synthetic reasoning and concept formation with it, is not an invention of present-day SDG; it has been, and is, used again and again by engineers and physicists, by Sophus Lie (cf. the above quotation), by David Hilbert [13], and (at least secretly) also by later mathematicians. However, explicit rules for such concept formation, construction, and reasoning have not been well formulated, and SDG is an attempt to provide such rules, so that the concepts, constructions, and reasoning can be clearly communicated and tested for rigour. What is the advantage of communication and rigour? It is a question not of "advantage" but of necessity.

References

[1] W. Bertram, *Calcul différentiel topologique élémantaire*, Calvage & Mounet (2010)

[2] R. Blute, J. R. B. Cockett, T. Porter, and R. A. G. Seely, *Kähler categories*, Cahiers de Topologie et Géométrie Différentielle 52 (2011) 253–68

[3] L. Breen and W. Messing, *Combinatorial differential forms*, Adv. in Math. 164 (2001) 203–82

[4] M. Bunge, F. Gago, and A. M. San Luis, *Synthetic Differential Topology*, London Mathematical Society Lecture Notes Series 448, Cambridge University Press (2018)

[5] H. Busemann, *On spaces in which two points determine a geodesic*, Trans. Amer. Math. Soc. 54 (1943) 171–84

[6] R. Cockett and G. Cruttwell, *Differential structure, tangent structure, and SDG*, Appl. Categor. Struct. 22 (2014) 331–417

[7] R. Courant, *Differential and Integral Calculus*, vol. II, Blackie & Son (1936)

[8] H. S. M. Coxeter, *The Real Projective Plane*, 2nd ed., Cambridge University Press (1955)

[9] M. Demazure and P. Gabriel, *Groupes Algébriques*, Tome I, Masson & Cie/North-Holland (1970)

[10] E. J. Dubuc, *Sur les modèles de la géométrie différentielle synthétique*, Cahiers de Top. et Géom. Diff. 20 (1979) 231–79

[11] E. J. Dubuc, C^∞-*schemes*, Amer. J. Math. 103–4 (1981) 683–90

[12] E. J. Dubuc and A. Kock, *On 1-form classifiers*, Commun. Algebra 12 (1984) 1471–531

[13] D. Hilbert and S. Cohn-Vossen, *Anschauliche Geometrie*, Grundlehren der Mathematischen Wissenschaften 37, Springer (1932)

[14] A. Kock, *Properties of well-adapted models for synthetic differential geometry*, J. Pure Appl. Alg. 20 (1981) 55–70

[15] A. Kock, *Synthetic Differential Geometry*, London Math. Soc. Lecture Notes Series 51, 2nd ed. (1981)

[16] A. Kock, *Differential forms with values in groups*, Bull. Austral. Math. Soc. 25 (1982) 357–86

[17] A. Kock, *A combinatorial theory of connections*, in J. Gray, ed., *Mathematical Applications of Category Theory*, Proceedings 1983, A.M.S. Contemporary Mathematics 30 (1984)

[18] A. Kock, *Convenient vector spaces embed into the Cahiers topos*, Cahiers de Top. et Géom. Diff. 27 (1986) 3–17. Corrections in [25]

[19] A. Kock, *Combinatorics of curvature and the Bianchi identity*, Theory Appl. Categories 2 (1996) 69–89
[20] A. Kock, *Principal bundles, groupoids, and connections*, in J. Kubarski, J. Pradines, T. Rybicki, and R. Wolak, eds., *Geometry and Topology of Manifolds*, Banach Center Publications 76 (2007) 185–200
[21] A. Kock, *Envelopes – notion and definiteness*, Beiträge Algebra Geometrie 48 (2007) 345–50
[22] A. Kock, *Synthetic Geometry of Manifolds*, Cambridge Tracts in Mathematics 180, Cambridge University Press (2010)
[23] A. Kock, *Metric spaces and SDG*, Theory Appl. Categories 32 (2017) 803–22
[24] A. Kock and C. J. Mikkelsen, *Topos theoretic factorization of non-standard extensions*, in A. Hurd and P. Loeb, eds., *Victoria Symposium on Nonstandard Analysis 1972*, Springer Lecture Notes in Mathematics 369, Springer (1974) 122–43
[25] A. Kock and G. E. Reyes, *Corrigendum and addenda to "Convenient vector spaces embed,"* Cahiers de Top. et Géom. Diff. 28 (1987) 69–89
[26] A. Kock and G. E. Reyes, *Aspects of fractional exponent functors*, Theory Appl. Categories 5 (1999) 251–65
[27] A. Kumpera and D. Spencer, *Lie equations*, Annals of Mathematics Studies 73, Princeton University Press (1972)
[28] R. Lavendhomme, *Basic Concepts of Synthetic Differential Geometry*, Kluwer Academic (1996)
[29] F. W. Lawvere, *Outline of synthetic differential geometry*, Notes Buffalo (1998), http://www.acsu.buffalo.edu/wlawvere/SDG_Outline.pdf
[30] F. W. Lawvere, *Axiomatic cohesion*, Theory Appl. Categories 19 (2007) 41–47
[31] F. W. Lawvere, *Euler's continuum functorially vindicated*, The Western Ontario Series in Philosophy of Science 75 (2011)
[32] F. W. Lawvere and M. Menni, *Internal choice holds in the discrete part of any cohesive topos satisfying stable connected codiscreteness*, Theory Appl. Categories 30 (2015) 909–32
[33] S. Lie, *Allgemeine Theorie der partiellen Differentialgleichungen erster Ordnung*, Math. Ann. 9 (1876) 245–96
[34] S. Lie, *Geometrie der Berührungstransformationen*, Leipzig 1896, reprint Chelsea Publ. Comp. (1977)
[35] P. May, *Simplicial Objects in Algebraic Topology*, van Nostrand Math. Studies 11, van Nostrand (1967)
[36] C. McLarty, *Local, and some global, results in synthetic differential geometry*, in A. Kock, ed., *Category Theoretic Methods in Geometry*, Aarhus Mat. Inst. Various Publ. Series 35 (1983) 226–56
[37] M. Menni, *Continuous cohesion over sets*, Theory Appl. Categories 29 (2014) 542–68
[38] I. Moerdijk and G. E. Reyes, *Models for Smooth Infinitesimal Analysis*, Springer (1991)
[39] D. Mumford, *The Red Book of Varieties and Schemes*, $\leq$ 1968, reprinted as Springer Lecture Notes in Mathematics 1358, Springer (1988)

[40] E. Nelson, *Internal set theory: A new approach to nonstandard analysis*, Bull. Amer. Math. Soc. 83 (1977) 1165–98
[41] H. Nishimura, *Higher-order preconnections in synthetic differential geometry of jet bundles*, Beiträge zur Algebra und Geometrie 45 (2004) 677–96
[42] A. Weil, *Théorie des points proches sur les variétés différentiables*, Colloq. Top. et Géom. Diff. (1953)

Anders Kock
Department of Mathematics, University of Aarhus, Denmark
kock@math.au.dk

3
Microlocal Analysis and Beyond

Pierre Schapira

Contents

Introduction		117
1	The Prehistory: Categories and Sheaves	118
2	Microlocal Analysis	127
3	Microlocal Sheaf Theory	134
4	Some Applications	143
References		150

Introduction

Mathematics often treats, in its own language, natural ideas and the concepts that one encounters in this discipline are frequently familiar. A good illustration of this fact is the dichotomy local/global. These notions appear almost everywhere in mathematics, and there is a tool to handle them: this is sheaf theory, a theory elaborated in the 1950s (see Section 1).

But another notion emerged in the 1970s, that of "microlocal," and being local on a manifold M becomes now a global property with respect to the fibers of the cotangent bundle $T^*M \to M$.

The microlocal point of view first appeared in analysis with Mikio Sato [34], soon followed by Lars Hörmander [17, 18], who both introduced, among others, the notion of the wave front set. The singularities of a hyperfunction or a distribution on a manifold M are viewed as the projection on M of singularities living in the cotangent bundle T^*M, more precisely, in $\sqrt{-1}T^*M$, and the geometry appearing in the cotangent bundle is in general much easier to analyze (see Section 2).

This microlocal point of view was then extended to sheaf theory by Masaki Kashiwara and the author in the 1980s (see [23–25]) who introduced the notion of microsupport of sheaves, giving rise to microlocal sheaf theory. To a sheaf F on a real manifold M, one associates its "microsupport" $\mu\mathrm{supp}(F)$[1] a closed conic subset of the cotangent bundle T^*M, which describes the codirections of nonpropagation of F (see Section 3).

Microlocal sheaf theory has many applications, and we will take a glance at some of them in Sections 4, first, in the study of linear partial differential equations (D-modules and their solutions), which was the original motivation of this theory, and next, in other branches of mathematics and in particular in symplectic topology (see in particular [31, 41]). The reason why microlocal sheaf theory is closely connected to symplectic topology is that the microsupport of a sheaf is a co-isotropic subset and the category of sheaves, localized on the cotangent bundle, is a homogeneous symplectic invariant playing a role similar to that of the Fukaya category, although the techniques in these two fields are extremely different.

Before entering the core of our subject, we shall briefly recall our basic language, categories, homological algebra, and sheaves. Then, following a historical point of view, we will recall the birth of algebraic analysis with Sato's hyperfunctions in 1959–60 and the birth of microlocal analysis with Sato's microfunctions around 1970. Then, we will describe some aspects of microlocal sheaf theory (1980–90) and some of its recent applications in symplectic topology.

1 The Prehistory: Categories and Sheaves

In everyday life, one often speaks of "local" or "global" phenomena (e.g., local wars, global warming), which are now common notions. These two concepts also exist in mathematics, in which they have a precise meaning. On a topological space X, a property is *locally satisfied* if there exists an open covering of X on which it is satisfied. But it can happen that a property is locally satisfied without being globally satisfied. For example, an equation may be locally solvable without being globally solvable. Or, more sophisticated, a manifold is always locally orientable but not always globally orientable, as shown by the example of the Möbius strip. And also, a manifold is locally isomorphic (as a topological space) to an open subset of the Euclidian space $\mathbb{R}^n$, but the two-dimensional sphere $\mathbb{S}^2$ is not globally isomorphic to any open

---

[1] $\mu\mathrm{supp}(F)$ was denoted $\mathrm{SS}(F)$ in loc. cit., a shortcut for "singular support."

subset of $\mathbb{R}^2$, and this is why, to recover the earth by planar maps, one needs at least two maps.

There is a wonderful tool that makes a precise link between local and global properties and that plays a prominent role in mathematics: sheaf theory. Sheaf theory was created by Jean Leray when he was a war prisoner in the 1940s. At the beginning, it was aimed at algebraic topology, but its scope goes far beyond, and this language is used almost everywhere – in algebraic geometry, representation theory, linear analysis, mathematical physics, and so on. It is a basic and universal language in mathematics. Leray's ideas were extremely difficult to follow but were clarified by Henri Cartan and Jean-Pierre Serre in the 1950s, who made sheaf theory an essential tool for analytic and algebraic geometry.[2]

But, as everyone knows, in mathematics and mathematical physics, the set theoretical point of view is often supplanted by the categorical perspective. Category theory was introduced by Eilenberg–Mac Lane [9], more or less at the same time as sheaf theory, and fantastically developed by Grothendieck, in particular in his famous Tohoku paper [12]. The underlying idea of category theory is that mathematical objects only take their full force in relation with other objects of the same type. This is part of a broader intellectual movement of which the structuralism of Claude Lévi-Strauss and the linguistics of Noam Chomsky are illustrations.

The link between categories and sheaves is due to Grothendieck. In his seminal work of 1957, he interprets a presheaf of sets F as a contravariant functor defined on the category of all open subsets of a topological space X with values in the category of sets, and a sheaf is a presheaf that satisfies natural gluing properties (see below). Later Grothendieck introduced what is now called "Grothendieck topologies" by remarking that there is no need of a topological space to develop sheaf theory. The objects of any category may perfectly play the role of the open sets, and it remains to define abstractly what the coverings are. But this is another story that we shall not develop here.

Note that instead of looking at a sheaf as a functor on the category of open sets, one can associate a functor $\Gamma(U; \bullet)$ to each open set U of X, from the category of sheaves to that of sets, namely, the functor that to a sheaf F associates $F(U)$, its value on U. When one considers sheaves with values in the category of modules over a given ring, the functor $\Gamma(U; \bullet)$ is left exact but in general not exact. This is precisely the translation of the fact that certain properties are satisfied locally and not globally. Then comes the dawn of modern homological algebra, with the introduction of derived functors, and

---

[2] For a short history of sheaf theory, see the historical notes by Christian Houzel in [25].

it appears that the cohomology of a sheaf F on U is recovered by the derived functor of $\Gamma(U;\bullet)$ applied to F. These cohomology objects are a kind of measure of the obstruction to pass from local to global.

Let us be a little more precise, referring to [27, 38] for an exhaustive treatment.

1.1 Categories

Let us briefly recall what a category is. A category $\mathscr{C}$ is the data of a set of objects, $\mathrm{Ob}(\mathscr{C})$, and given two objects $X, Y \in \mathrm{Ob}(\mathscr{C})$, a set $\mathrm{Hom}_{\mathscr{C}}(X, Y)$, called the set of morphisms from X to Y, and for any $X, Y, Z \in \mathrm{Ob}(\mathscr{C})$ a map $\circ: \mathrm{Hom}_{\mathscr{C}}(X, Y) \times \mathrm{Hom}_{\mathscr{C}}(Y, Z) \to \mathrm{Hom}_{\mathscr{C}}(X, Z)$, these data satisfying some axioms that become extremely natural as soon as one thinks of X, Y, Z as being, for example, sets, or topological spaces, or vector spaces, and $\mathrm{Hom}_{\mathscr{C}}(X, Y)$ as being the set of morphisms from X to Y, that is, maps in the case of sets, continuous maps in the case of topological spaces, and linear maps in the case of vector spaces. It is then natural to call the elements of $\mathrm{Hom}_{\mathscr{C}}(X, Y)$ the *morphisms* from X to Y and to use the notation $f: X \to Y$ for such a morphism. One shall be aware that, in general, the objects X, Y, Z, and so on are not sets and a fortiori the morphisms are not maps. One calls the map $\circ$ the *composition* of morphisms, and one simply asks two things: the composition is associative, $(f \circ g) \circ h = f \circ (g \circ h)$, and for each object X there exists a morphism $1_X: X \to X$ that plays the role of the identity morphism, that is, $f \circ 1_X = f$ and $1_X \circ g = g$ for any $f: X \to Y$ and $g: Z \to X$.

Category theory seems at first glance extremely simple and attractive, but there are traps. Indeed, the class of all sets is not a set, as noticed by Georg Cantor and later by Bertrand Russell, whose argument is a variant of the Greek paradox "all Cretans are liars." This leads to inextricable problems, unless one uses the concept of universe (or another equivalent notion, such as that of unaccessible cardinals) and adds an axiom to set theory that "any set belongs to a universe" – what Grothendieck did, but perhaps what scared Bourbaki, who never introduced category theory in his globalizing project.

Applying the philosophy of categories to themselves, we have to understand what a morphism $F: \mathscr{C} \to \mathscr{C}'$ from a category $\mathscr{C}$ to a category $\mathscr{C}'$ is. Such a morphism is called *a functor*. It sends $\mathrm{Ob}(\mathscr{C})$ to $\mathrm{Ob}(\mathscr{C}')$ and any morphism $f: X \to Y$ to a morphism $F(f): F(X) \to F(Y)$. Of course, one asks that F commutes with the composition of morphisms, $F(g \circ f) = F(g) \circ F(f)$, and sends identity morphisms in $\mathscr{C}$ to identity morphisms in $\mathscr{C}'$. In practice, one often encounters "contravariant functors," that is, functors that reverse the arrows, $F(g \circ f) = F(f) \circ F(g)$. It is better to consider them as usual functors from $\mathscr{C}^{op}$ to $\mathscr{C}'$, where $\mathscr{C}^{op}$, the opposite category of $\mathscr{C}$, is the category $\mathscr{C}$ in which the arrows are reversed: a morphism $f: X \to Y$ in $\mathscr{C}^{op}$ is a morphism $f: Y \to X$ in $\mathscr{C}$.

1.2 Homological Algebra

Homological algebra is essentially linear algebra, not over a field but over a ring, and, by extension, in any abelian categories, that is, categories that are modeled on the category of modules over a ring.

Consider first a finite system of linear equations over a (not necessarily commutative) ring **k**:

$$\sum_{i=1}^{N_0} a_{ji} u_i = v_j, \quad j = 1, \ldots, N_1. \tag{1.1}$$

Here u_i and v_j belong to some left **k**-module S and a_{ji} belongs to **k**. Denote by P_0 the matrix $(a_{ji})_{1 \leq i \leq N_0, 1 \leq j \leq N_1}$ and by $P_0 \cdot$ this matrix acting on the left on S^{N_0}:

$$S^{N_0} \xrightarrow{P_0 \cdot} S^{N_1}.$$

Now consider $\cdot P_0$, the matrix P_0 acting on the right on $\mathbf{k}^{N_0}$, and denote by M its cokernel, so that we have *an exact sequence*:

$$\mathbf{k}^{N_1} \xrightarrow{\cdot P_0} \mathbf{k}^{N_0} \to M \to 0. \tag{1.2}$$

Conversely, consider a **k**-module M, and assume that there exists an exact sequence (1.2). Then, one says that M admits a finite 1-presentation, but such a presentation is not unique, and different matrices with entries in **k** may give isomorphic modules. This is similar to the fact that a finite-dimensional vector space may have different systems of generators. As we shall see, when analyzing the system (1.1), the important (and "intrinsic") information is not the matrix P_0 but the module[3] M.

Applying the contravariant left exact functor $\mathrm{Hom}\,(\bullet, S)$ to (1.2), we find the exact sequence

$$0 \to \mathrm{Hom}\,(M, S) \to S^{N_0} \xrightarrow{P_0 \cdot} S^{N_1},$$

which shows that the kernel of $P_0 \cdot$ depends only on M (up to isomorphism) and not on the presentation (1.2).

---

[3] According to Sato (personal communication), at the origin of this idea is the mathematician and philosopher of the 17th century, E. W. von Tschirnhaus.

Assume now that **k** is right Noetherian. Then the kernel of $\cdot P_0$ in (1.2) is finitely generated, and one can extend the exact sequence (1.2) to an exact sequence

$$\mathbf{k}^{N_2} \xrightarrow{\cdot P_1} \mathbf{k}^{N_1} \xrightarrow{\cdot P_0} \mathbf{k}^{N_0} \to M \to 0. \tag{1.3}$$

By iterating this construction, one finds a long exact sequence

$$\cdots \to \mathbf{k}^{N_2} \xrightarrow{\cdot P_1} \mathbf{k}^{N_1} \xrightarrow{\cdot P_0} \mathbf{k}^{N_0} \to M \to 0. \tag{1.4}$$

Consider the complex $M^\bullet := \cdots \to \mathbf{k}^{N_2} \xrightarrow{\cdot P_1} \mathbf{k}^{N_1} \xrightarrow{\cdot P_0} \mathbf{k}^{N_0} \to 0$, and identify M with a complex *concentrated in degree* 0. We get a morphism of complexes $M^\bullet \to M$:

$$\begin{array}{ccccccccc} M^\bullet = & \cdots & \longrightarrow & \mathbf{k}^{N_2} & \xrightarrow{\cdot P_1} & \mathbf{k}^{N_1} & \xrightarrow{\cdot P_0} & \mathbf{k}^{N_0} & \longrightarrow 0 \\ & & & \downarrow & & \downarrow & & \downarrow & \\ M = & \cdots & \longrightarrow & 0 & \longrightarrow & 0 & \longrightarrow & M & \longrightarrow 0 \end{array}$$

and this morphism is a *qis*, a quasi-isomorphism, that is, it induces an isomorphism on the cohomology. Now one proves that, up to "canonical isomorphism," the complex $\operatorname{Hom}(M^\bullet, S)$ does not depend on the choice of the free resolution $M^\bullet$, and one sets

$$\operatorname{RHom}(M, S) = \operatorname{Hom}(M^\bullet, S). \tag{1.5}$$

This object is represented by the complex (which is no more an exact sequence, but the composition of two consecutive arrows is 0)

$$0 \to S^{N_0} \xrightarrow{P_0 \cdot} S^{N_1} \xrightarrow{P_1 \cdot} \cdots \tag{1.6}$$

One sets

$$\operatorname{Ext}^j(M, S) = H^j \operatorname{RHom}(M, S)$$
$$\simeq \operatorname{Ker}(P_j: S^{N_j} \to S^{N_{j+1}})/\operatorname{Im}(P_{j-1}: S^{N_{j-1}} \to S^{N_j}).$$

Hence, $\operatorname{Ext}^0(M, S) \simeq \operatorname{Hom}(M, S) \simeq \operatorname{Ker}(P_0)$, and for $j > 0$, $\operatorname{Ext}^j(M, S)$ is the obstruction for solving the equation $P_{j-1} u = v$, knowing that $P_j v = 0$.

One calls $\operatorname{RHom}(\bullet, S)$ the *right-derived functor* of the left-exact (contravariant) functor $\operatorname{Hom}(\bullet, S)$, and this construction is a toy model for the general construction of derived functors. Indeed, consider a left-exact functor of abelian categories $F: \mathscr{A} \to \mathscr{C}$. Its right-derived functor RF is defined as follows. Given X an object of $\mathscr{A}$, one first constructs (when it is possible) an exact sequence

$$0 \to X \to I^0 \to I^1 \to \cdots, \qquad (1.7)$$

where the I^js are *injective* objects of $\mathscr{A}$. Let us denote by $I^\bullet$ the complex $0 \to I^0 \to I^1 \to \cdots$, and set $RF(X) = F(I^\bullet)$. (The arrows in (1.3) go backward with respect to (1.7) because the functor Hom ($\bullet, S$) is contravariant.)

The construction of derived functors finds its natural place in the language of derived categories, again due to Grothendieck. The derived category $D(\mathscr{C})$ of an abelian category is obtained by considering complexes in $\mathscr{C}$ and identifying two complexes when they are quasi-isomorphic. When considering bounded complexes (those whose objects are all 0, except finitely many of them), one gets the bounded derived category $D^b(\mathscr{C})$.

Derived categories are particular cases of triangulated categories. In such categories, we have "distinguished triangles," which play the role of exact sequences in abelian categories. We shall not say more here.

1.3 Abelian Sheaves

Consider now a topological space X and the set Op_X of its open sets. This set may be considered a category by deciding that the morphisms are the inclusions (one morphism if $U \subset V$, and no morphisms otherwise). Denote by Mod(**k**) the abelian category of left **k**-modules. A presheaf of **k**-modules is nothing but a functor $F: \mathrm{Op}_X^{\mathrm{op}} \to \mathrm{Mod}(\mathbf{k})$. Hence, to any open set U, F associates a **k**-module $F(U)$, and for $U \subset V$, we get a **k**-linear map $F(V) \to F(U)$, called the *restriction morphism*. Of course, the composition of the restriction morphisms associated with inclusions $U \subset V \subset W$ is the restriction morphism associated with $U \subset W$, and the restriction associated with $U \subset U$ is the identity. If $s \in F(V)$ and $U \subset V$, one often simply denotes by $s|_U$ its image in $F(U)$ by the restriction morphism.

A sheaf F is a presheaf satisfying natural gluing conditions. Namely, for any open subset U of X, consider an open covering $U = \bigcup_{i \in I} U_i$. Then,
(i) if $s \in F(U)$ satisfies $s|_{U_i} = 0$ for all $i \in I$, then $s = 0$;
(ii) if one is given a family $\{s_i\}_{i \in I}$ with $s_i \in F(U_i)$ satisfying $s_i|_{U_i \cap U_j} = s_j|_{U_i \cap U_j}$ for all pairs (i, j), then there exists $s \in F(U)$ such that $s|_{U_i} = s_i$.

The presheaf C_X^0 of $\mathbb{R}$-valued continuous functions on X is a first example of a sheaf (here $\mathbf{k} = \mathbb{R}$), as well as the sheaf $\mathbf{k}_X$ of locally constant **k**-valued functions. More generally, if Z is a locally closed subset of X (the intersection of an open and a closed subset), there exists a unique sheaf $\mathbf{k}_{ZX}$ on X whose restriction to Z is the constant sheaf $\mathbf{k}_Z$ on Z and which is 0 on $X \setminus Z$. If there is no risk of confusion, we shall simply write $\mathbf{k}_Z$ instead of $\mathbf{k}_{ZX}$. On the other hand, the presheaf of *constant* functions on X is not a sheaf in general.

One easily proves that the category $\mathrm{Mod}(\mathbf{k}_X)$ of sheaves on X is an abelian category. One denotes by $\mathsf{D}^b(\mathbf{k}_X)$ its bounded derived category.

The open set U being given, consider the functor

$$\Gamma(U; \bullet): \mathrm{Mod}(\mathbf{k}_X) \to \mathrm{Mod}(\mathbf{k}), \quad F \mapsto F(U). \tag{1.8}$$

One easily shows the isomorphism of functors

$$\Gamma(U; \bullet) \simeq \mathrm{Hom}(\mathbf{k}_U, \bullet).$$

This functor is left exact but not right exact in general. For example, take for X the complex line, and consider the complex of sheaves

$$0 \to \mathbb{C}_X \to \mathscr{O}_X \xrightarrow{\partial_z} \mathscr{O}_X \to 0.$$

Here, $\mathscr{O}_X$ is the sheaf of holomorphic functions and ∂_z the holomorphic derivation. This sequence is exact because a holomorphic function is locally constant if and only if its derivative is 0 and, *locally* on X, one can solve the equation $\partial_z f = g$. For any nonempty connected open set U, the sequence

$$0 \to \mathbb{C} \to \mathscr{O}_X(U) \xrightarrow{\partial_z} \mathscr{O}_X(U) \tag{1.9}$$

remains exact, but one cannot solve the equation $\partial_z f = g$ globally on U when $U = \mathbb{C} \setminus \{0\}$ and $g(z) = 1/z$. Hence, the functor $\Gamma(U; \bullet)$ is left exact but is not right exact. Deriving it, one gets the functor $\mathrm{R}\Gamma(U; \bullet): \mathsf{D}^b(\mathbf{k}_X) \to \mathsf{D}^b(\mathbf{k})$ of derived categories.

We have chosen to describe the functor $\mathrm{R}\Gamma(U; \bullet)$, but it is a particular case of one of six natural functors, called the six Grothendieck operations. One has the internal hom functor $\mathrm{R}\mathscr{H}om$ and the tensor product functor $\overset{L}{\otimes}$, the functor of direct images $\mathrm{R}f_*$ and its left adjoint f^{-1}, and the functor of proper direct images $\mathrm{R}f_!$ and its right adjoint $f^!$. In this chapter, we shall make use of the duality functor $\mathrm{D}'_X := \mathrm{R}\mathscr{H}om(\bullet, \mathbf{k}_X)$.

Given an open set U and setting $S = X \setminus U$, we have an exact sequence of sheaves

$$0 \to \mathbf{k}_U \to \mathbf{k}_X \to \mathbf{k}_S \to 0.$$

For $F \in \mathsf{D}^b(\mathbf{k}_X)$, applying the functor $\mathrm{RHom}(\bullet, F)$, we get the distinguished triangle

$$\mathrm{R}\Gamma_S(X; F) \to \mathrm{R}\Gamma(X; F) \to \mathrm{R}\Gamma(U; F) \xrightarrow{+1} .$$

Hence, $\mathrm{R}\Gamma_S(X; F)$ is the obstruction to extend uniquely to X the cohomology classes of F defined on U. If F is a usual sheaf, then $H^0 \mathrm{R}\Gamma_S(X; F) = \Gamma_S(X; F)$ is the space of sections of F on X supported by S.

Now consider a sheaf of rings $\mathscr{R}$ on X. Replacing the constant ring **k** with $\mathscr{R}$, a system of linear equations on $\mathscr{R}$ becomes an $\mathscr{R}$-module that is *locally* of finite presentation. If $\mathscr{R}$ is *coherent*, whatever it means, such a module is called a coherent $\mathscr{R}$-module. Hence, the slogan is "a system of linear equations with coefficients in $\mathscr{R}$ is nothing but a coherent left $\mathscr{R}$-module." We denote by $\mathsf{D}^{\mathsf{b}}(\mathscr{R})$ the derived category of left $\mathscr{R}$-modules. We shall encounter such a situation in Section 4.1.

1.4 An Application: Generalized Functions

In the 1960s, people were used to working with various spaces of generalized functions constructed with the tools of functional analysis. Sato's construction of hyperfunctions in [33] is at the opposite of this practice: he uses purely algebraic tools and complex analysis. The importance of Sato's definition is twofold: first, it is purely algebraic (starting with the sheaf of holomorphic functions), and second, it highlights the link between real and complex geometry. Note that the sheaf $\mathscr{B}_M$ of hyperfunctions on a real analytic manifold M naturally contains the sheaf $\mathcal{D}b_M$ of distributions and has the nice property of being flabby (the restriction morphisms are surjective). We refer to [1, 36] for an exposition of Sato's work.

Consider first the case where M is an open interval of the real line $\mathbb{R}$, and let X be an open neighborhood of M in the complex line $\mathbb{C}$ satisfying $X \cap \mathbb{R} = M$. Denote by $\mathcal{O}(U)$ the space of holomorphic functions on an open set $U \subset X$. The space $\mathscr{B}(M)$ of hyperfunctions on M is given by

$$\mathscr{B}(M) = \mathcal{O}(X \setminus M)/\mathcal{O}(X). \tag{1.10}$$

In other words, a hyperfunction on M is a holomorphic function on $X \setminus M$, and such a function is considered as 0 if it extends to the whole of X.

It is easily proved, using the solution of the Cousin problem, that this space depends only on M, not on the choice of X, and that the correspondence $I \mapsto \mathscr{B}(I)$ (I open in M) defines a flabby sheaf $\mathscr{B}_M$ on M.

With Sato's definition, the boundary values always exist and are no more a limit in any classical sense.

Sato's definition is motivated by the well-known fact that the Dirac function at 0 is the "boundary value" of $\frac{1}{2i\pi} 1/z$. Indeed, if φ is a C^0-function on $\mathbb{R}$, one has

$$\varphi(0) = \lim_{\varepsilon \xrightarrow{>} 0} \frac{1}{2i\pi} \int_{\mathbb{R}} \left(\frac{\varphi(x)}{x - i\varepsilon} - \frac{\varphi(x)}{x + i\varepsilon} \right) dx,$$

and one can write formally

$$\delta(x) = \frac{1}{2i\pi}\left(\frac{1}{x-i0} - \frac{1}{x+i0}\right).$$

It follows that for any distribution u on $\mathbb{R}$ with compact support, say, $K \subset \mathbb{R}$, u is the boundary value of the function $u \star \frac{1}{2i\pi} 1/z$ holomorphic on $\mathbb{C} \setminus K$, and in fact any distribution on $\mathbb{R}$ is the boundary value of a holomorphic function on $\mathbb{C} \setminus \mathbb{R}$. However, there exist holomorphic functions on $\mathbb{C} \setminus \{0\}$ that have no boundary values as a distribution, such as the holomorphic function $\exp(1/z)$ defined on $\mathbb{C} \setminus \{0\}$.

To extend to the higher-dimensional case the definition of hyperfunctions, a natural idea would be as follows. Denote by M a real analytic manifold and by X a complexification of M. Then, it would be tempting to define $\mathscr{B}(M)$ by formula (1.10). Unfortunately, this does not work, since Hartog's theorem tells us that this space is 0 as soon as dim $M > 1$. (Here and in the sequel, dim denotes the dimension of a *real* manifold.)

Another way is to use local cohomology, that is, the derived functor of the functor $\Gamma_M(\bullet)$. Indeed, in dimension 1, one has

$$\mathscr{B}(M) = \mathscr{O}(X \setminus M)/\mathscr{O}(X) \simeq H^1_M(X; \mathscr{O}_X),$$

where $\mathscr{O}_X$ is the sheaf of holomorphic functions on X and the presheaf $H^1_M(\mathscr{O}_X)$ is a sheaf on M. This is the sheaf $\mathscr{B}_M$.

Note that Sato invented local cohomology independently from Grothendieck in the 1960s in order to define hyperfunctions. On a real analytic manifold M of dimension n, the sheaf $\mathscr{B}_M$ was originally defined as

$$\mathscr{B}_M = H^n_M(\mathscr{O}_X) \otimes \mathrm{or}_M,$$

after having proved that the groups $H^j_M(\mathscr{O}_X)$ are 0 for $j \neq n$. Here, or_M is the orientation sheaf on M. Since X is oriented, Poincaré's duality gives the isomorphism $\mathrm{D}'_X(\mathbb{C}_M) \simeq \mathrm{or}_M[-n]$, where D'_X is the duality functor for sheaves on X. An equivalent definition of the sheaf of hyperfunctions is thus given by

$$\mathscr{B}_M = \mathrm{R}\mathscr{H}om\,(\mathrm{D}'_X\mathbb{C}_M, \mathscr{O}_X). \tag{1.11}$$

The sheaf $\mathscr{A}_M$ of real analytic functions is given by

$$\mathscr{A}_M := \mathbb{C}_M \otimes \mathscr{O}_X.$$

Since $\mathbb{C}_M \simeq \mathrm{D}'_X \mathrm{D}'_X \mathbb{C}_M$, we get the natural morphism from real analytic functions to hyperfunctions:

$$\mathscr{A}_M \simeq \mathrm{R}\mathscr{H}om\,(\mathrm{D}'_X\mathbb{C}_M, \mathbb{C}_X) \otimes \mathscr{O}_X \to \mathrm{R}\mathscr{H}om\,(\mathrm{D}'_X\mathbb{C}_M, \mathscr{O}_X) \simeq \mathscr{B}_M.$$

Formula (1.11) opens the door to a vast generalization of distributions and hyperfunctions: one may consider the sheaves (in the derived sense) $\mathrm{R}\mathcal{H}om\,(F, \mathcal{O}_X)$ where now F is any sheaf on X. This is particularly interesting when F is $\mathbb{R}$-constructible (see Definition 3.6).

Similarly as in dimension 1, one can represent the sheaf $\mathcal{B}_M$ by using Čech cohomology of coverings of $X \setminus M$. For example, let X be a Stein open subset of $\mathbb{C}^n$ and set $M = \mathbb{R}^n \cap X$. Denote by x the coordinates on $\mathbb{R}^n$ and by $x + \sqrt{-1}y$ the coordinates on $\mathbb{C}^n$. One can recover $\mathbb{C}^n \setminus \mathbb{R}^n$ by $n+1$ open half-spaces

$$V_i = \{(x + \sqrt{-1}y; \langle y, \xi_i \rangle > 0\}, \quad \xi \in \mathbb{R}^n \setminus \{0\}, \quad i = 1, \ldots, n+1.$$

For $J \subset \{1, \ldots, n+1\}$, set $V_J = \bigcap_{j \in J} V_j$. Assuming $n > 1$, we have the isomorphism $H_M^n(X; \mathcal{O}_X) \simeq H^{n-1}(X \setminus M; \mathcal{O}_X)$. Therefore, setting $U_J = V_J \cap X$, we have

$$\mathcal{B}(M) \simeq \sum_{|J|=n} \mathcal{O}_X(U_J) / \sum_{|K|=n-1} \mathcal{O}_X(U_K).$$

Then comes naturally the following problem: how to recognize the directions associated with these U_Js? The answer is given by Sato's microlocalization functor that we shall describe in Section 2.

2 Microlocal Analysis

With any real manifold M (say, of dimension n) are naturally associated two important vector bundles, the tangent bundle $\tau : TM \to M$ and its dual, the cotangent bundle $\pi : T^*M \to M$. Classically, one interprets a vector $v \in T_{x_0}M$ as the speed at the point x_0 of something moving on M and passing at x_0. Up to the zero-section and to the action of $\mathbb{R}^+$ on these vector bundles, one may think of $T_{x_0}M$ as the space of all light rays issued from x_0 and of $T^*_{x_0}M$ as the space of all half-spaces, or walls, passing through x_0.

The tangent bundle is more intuitive, but it appears that the cotangent bundle is much more important. It is the *phase space* of the physicists and is endowed with a fundamental structure: it is a symplectic manifold. Symplectic geometry is a very classical subject whose origin perhaps goes back to Hamilton in the first half of the 19th century. Note that the duality tangent/cotangent reflects the duality observer/observed.

Analysts have known for a long time, after Petrowski, Hadamard, and Leray, that, given a linear differential operator P on a manifold X, its principal symbol $\sigma(P)$ is a well-defined function on the cotangent bundle and the geometry of its characteristic variety (the zeroes of $\sigma(P)$) plays a role in the behavior of its solutions. But the story of microlocal analysis really started in the 1970s with Sato.

As already mentioned, a hyperfunction may be represented (not uniquely) as a sum of boundary values of holomorphic functions defined on tuboids[4] in X, and an important problem is to understand from where these boundary values come. For that purpose, Sato defined the sheaf of microfunctions $\mathscr{C}_M$ on the conormal bundle T_M^*X to M in X (the dual of the normal bundle; see below) whose direct image on M is the sheaf $\mathscr{B}_M$. The reason for the success of this approach is that the study of partial differential equations is much easier in T_M^*X. Consider for example the wave operator $\square = \partial_t^2 - \sum_{i=1}^n \partial_{x_i}^2$ on $\mathbb{R}_t \times \mathbb{R}_x^n$. Its characteristic variety is a smooth manifold in $\dot{T}_M^*X$ (the bundle T_M^*X with the zero-section removed), and, locally on $\dot{T}_M^*X$, the equation $\square u = 0$ can be reduced after a "quantized contact transformation" to the equation $\partial_t u = 0$.

The breakthrough of microlocal analysis quickly spread from the analytic framework to the C^∞-framework under the impulse of Lars Hörmander, who replaced the use of holomorphic functions by that of the Fourier transform. Note that the $\sqrt{-1}$ that appears in the Fourier transform is related (in a precise sense, via the Laplace transform; see [26]) to the isomorphism of vector bundles $T_M^*X \simeq \sqrt{-1}T^*M$, where T^*M is the cotangent bundle to M.

Microfunctions are certainly an important tool of analysis, but in our opinion, their construction is still more important. Indeed, they are constructed by applying a new functor to the sheaf $\mathscr{O}_X$, the functor μ_M of microlocalization along M, and this functor is obtained as the "Fourier–Sato transform" of the specialization functor ν_M. We shall now describe these three functors, which are defined in a purely real setting. References are made to [25].

2.1 Specialization

Notation 2.1 Let $\mathbb{V} \to M$ be a real vector bundle. We identify M with the zero-section of $\mathbb{V}$. For $Z \subset \mathbb{V}$, we denote by Z^a its image by the antipodal map, $(x; v) \mapsto (x; -v)$. We say that a subset Z of $\mathbb{V}$ is $\mathbb{R}^+$-conic if it is invariant by the action of $\mathbb{R}^+$ on $\mathbb{V}$.

---

[4] A tuboid is an open subset of X which, in a local chart at $x_0 \in M$, contains $(\mathbb{R}^n \times \sqrt{-1}\Gamma) \cap W$ for an open nonempty convex cone Γ of $\mathbb{R}^n$ and an open neighborhood W of x_0 in X.

Let X be a *real* manifold and $\iota\colon M \hookrightarrow X$ the embedding of a closed submanifold M. Denote by $\tau_M\colon T_M X \to M$ the normal bundle to M in X defined by the exact sequence of vector bundles over M:

$$0 \to TM \to M \times_X TX \to T_M X \to 0.$$

If F is a sheaf on X, its restriction to M, denoted $F|_M$, may be viewed as a global object, namely, the direct image by τ_M of a conic sheaf $\nu_M F$ on $T_M X$, called the specialization of F along M. Here, *conic* means locally constant on the orbits of the action of $\mathbb{R}^+$. Intuitively, $T_M X/\mathbb{R}^+$ is the set of light rays issued from M in X, and the germ of $\nu_M F$ at a normal vector $(x;v) \in T_M X$ is the germ at x of the restriction of F along the light ray v.

In order to construct the specialization, one first constructs a commutative diagram of manifolds, called the normal deformation of X along M:

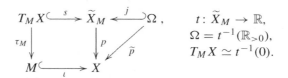

Locally, after choosing a local coordinate system (x',x'') on X such that $M = \{x' = 0\}$, we have $\widetilde{X}_M = X \times \mathbb{R}$, $t\colon \widetilde{X}_M \to \mathbb{R}$ is the projection, and $p(x',x'',t) = (tx',x'')$.

The specialization allows one to define intrinsically the notion of a Whitney normal cone. Let $S \subset X$ be a locally closed subset. The Whitney normal cone $C_M(S)$ is the closed conic subset of $T_M X$ given by

$$C_M(S) = \overline{\tilde{p}^{-1}(S)} \cap T_M X. \tag{2.1}$$

One also defines the normal cone for two subsets S_1 and S_2 by using the diagonal Δ of $X \times X$ and setting

$$C(S_1, S_2) = C_\Delta(S_1 \times S_2). \tag{2.2}$$

The Whitney normal cone $C_M(S)$ is given in a local coordinate system $(x) = (x',x'')$ on X with $M = \{x' = 0\}$ by

$$\begin{cases} (x_0''; v_0) \in C_M(S) \subset T_M X \text{ if and only if there exists a sequence} \\ \{(x_n, c_n)\}_n \subset S \times \mathbb{R}^+ \text{ with } x_n = (x_n', x_n'') \text{ such that } x_n' \xrightarrow{n} 0, x_n'' \xrightarrow{n} x_0'' \text{ and} \\ c_n(x_n') \xrightarrow{n} v_0. \end{cases}$$

Example 2.2 Assume that $M = \{x_0\}$. Then $C_{\{x_0\}}(S)$ is the tangent cone to S.

(i) If $X = \mathbb{C}^n$, $x_0 = 0$, and $S = \{x \in X; f(x) = 0\}$ for a holomorphic function f, then $C_{\{0\}}(S) = \{x \in X; g(x) = 0\}$, where g is the homogeneous polynomial of lowest degree in the Taylor expansion of f at 0.

(ii) Let $\mathbb{V}$ be a real finite-dimensional vector space, and let γ be a closed cone. Then $C_0(\gamma) = \gamma$ and $C_0(\gamma, \gamma)$ is the vector space generated by γ.

The specialization functor
$$\nu_M : \mathsf{D}^\mathsf{b}(\mathbf{k}_X) \to \mathsf{D}^\mathsf{b}(\mathbf{k}_{T_M X})$$
is then given by a formula mimicking (2.1):
$$\nu_M F := s^{-1} \mathrm{R} j_* \widetilde{p}^{-1} F.$$
Clearly, $\nu_M F \in \mathsf{D}^\mathsf{b}_{\mathbb{R}^+}(\mathbf{k}_{T_M X})$, that is, $\nu_M F$ is an $\mathbb{R}^+$-conic sheaf. Moreover,
$$\mathrm{R}\tau_{M*}\nu_M F \simeq \nu_M F|_M \simeq F|_M.$$
For an open cone $V \subset T_M X$, one finds that
$$H^j(V; \nu_M F) \simeq \varinjlim_U H^j(U; F), \tag{2.3}$$
where U ranges through the family of open subsets of X such that $C_M(X \setminus U) \cap V = \emptyset$.

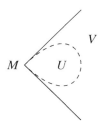

In other words, a section of $\nu_M F$ on a conic open set V of $T_M X$ is given by a section of F on a small open set U of X that is, in some sense, tangent to V near M.

2.2 Fourier–Sato Transform

The classical Fourier transform is an isomorphism between a space of (generalized) functions on a real vector space $\mathbb{V}$ and another space on the dual space $\mathbb{V}^*$. It is an integral transform associated with a kernel on $\mathbb{V} \times \mathbb{V}^*$. The Fourier–Sato transform is again an integral transform, but now in the language of the six

Grothendieck operations for sheaves. It induces an equivalence of categories between conic sheaves on a vector bundle and conic sheaves on the dual vector bundle. It seems to have been the first integral transform for sheaves, and, as its name may suggest, this construction is due to Sato.

Consider a diagram of real vector bundles over a real manifold M:

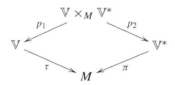

If γ is a cone in $\mathbb{V}$, its polar cone, or dual cone, is given by

$$\gamma^\circ = \{(x;\xi) \in \mathbb{V}^*; \langle \xi, v \rangle \geq 0 \text{ for all } v \in \gamma_x\}.$$

Then γ° is a closed convex cone of $\mathbb{V}^*$, and there is no hope to recover γ from γ° if γ is not convex. Things are different with sheaves since we can replace the usual functors on sheaves with their derived version. Define

$$P = \{(x, y) \in \mathbb{V} \times_M \mathbb{V}^*; \langle x, y \rangle \leq 0\}.$$

Denote by $\mathsf{D}^b_{\mathbb{R}_+}(\mathbf{k}_\mathbb{V})$ the subcategory of $\mathsf{D}^b(\mathbf{k}_\mathbb{V})$ consisting of conic sheaves. The Fourier–Sato transform of a conic sheaf F is a sheaf-theoretical version of the construction of the polar cone (see Example 2.4). It is given by the formula

$$F^\wedge = \mathrm{R}p_{2!}(p_1^{-1}F)_P.$$

Theorem 2.3 *The functor* $^\wedge$ *induces an equivalence of categories*

$$\wedge : \mathsf{D}^b_{\mathbb{R}_+}(\mathbf{k}_\mathbb{V}) \xrightarrow{\sim} \mathsf{D}^b_{\mathbb{R}_+}(\mathbf{k}_{\mathbb{V}^*}).$$

Example 2.4 Assume for short that $M = \mathrm{pt}$, and let $n = \dim \mathbb{V}$.

(i) Let γ be a closed proper[5] convex cone in $\mathbb{V}$. Then

$$(\mathbf{k}_\gamma)^\wedge \simeq \mathbf{k}_{\mathrm{Int}(\gamma^\circ)}.$$

Here $\mathrm{Int}\gamma^\circ$ denotes the interior of γ°.

(ii) Let γ be an open convex cone in $\mathbb{V}$. Then

$$(\mathbf{k}_\gamma)^\wedge \simeq \mathbf{k}_{\gamma^{\circ a}}[-n].$$

---

[5] A cone is proper if it contains no line.

(iii) Let $(x) = (x', x'')$ be coordinates on $\mathbb{R}^n$ with $(x') = (x_1, \ldots, x_p)$ and $(x'') = (x_{p+1}, \ldots, x_n)$, with $0 < p < n$. Denote by $(y) = (y', y'')$ the dual coordinates on $(\mathbb{R}^n)^*$. Set

$$\gamma = \{x; x'^2 - x''^2 \geq 0\}, \quad \lambda = \{y; y'^2 - y''^2 \leq 0\}.$$

Then $(\mathbf{k}_\gamma)^\wedge \simeq \mathbf{k}_\lambda[-p]$. (See [26].)

2.3 Microlocalization

Denote by $\pi_M : T_M^* X \to M$ the conormal bundle to M, the dual bundle to $T_M X$, given by the exact sequence of vector bundles over M:

$$0 \to T_M^* X \to M \times_X T^* X \to T^* M \to 0.$$

The microlocalization of F along M, denoted $\mu_M F$, is the Fourier–Sato transform of $\nu_M F$; hence it is an object of $D^b_{\mathbb{R}^+}(\mathbf{k}_{T_M^* X})$. It satisfies

$$R\pi_{M*}\mu_M F \simeq \mu_M F|_M \simeq R\Gamma_M F.$$

Roughly speaking, the sections of $\mu_M F$ on an open convex cone V of $T_M^* X$ are the sections of $\nu_M F$ supported by the polar cone to V in $T_M X$. More precisely, by using Theorem 2.3 and (2.3), we get

$$H^j(V; \mu_M F) \simeq \varinjlim_{U,Z} H^j_{Z \cap U}(U; F), \tag{2.4}$$

where U ranges through the family of open subsets of X such that $U \cap M = \pi(V)$ and Z ranges over the family of closed subsets of X such that $C_M(Z) \subset V^\circ$:

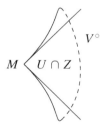

2.4 Application: Microfunctions and Wave Front Sets

Assume now that M is a real analytic manifold of dimension n and X is a complexification of M. First notice the isomorphisms

$$M \times_X T^* X \simeq \mathbb{C} \otimes_\mathbb{R} T^* M \simeq T^* M \oplus \sqrt{-1} T^* M.$$

One deduces the isomorphism

$$T_M^* X \simeq \sqrt{-1} T^* M. \qquad (2.5)$$

Sato introduced in [34] the sheaf $\mathscr{C}_M$ of microfunctions on $T_M^* X$ as

$$\mathscr{C}_M = H^n(\mu_M(\mathcal{O}_X)) \otimes \pi^{-1} \text{or}_M, \qquad (2.6)$$

after having proved that the other cohomology groups are 0. Thus $\mathscr{C}_M$ is a conic sheaf on $T_M^* X$ and one has a natural isomorphism

$$\mathscr{B}_M \xrightarrow{\sim} \pi_{M*} \mathscr{C}_M.$$

Denote by spec the natural map:

$$\text{spec}\colon \Gamma(M; \mathscr{B}_M) \xrightarrow{\sim} \Gamma(T_M^* X; \mathscr{C}_M).$$

Definition 2.5 The wave front set WF(u) of a hyperfunction $u \in \mathscr{B}(M)$ is the support of spec(u).

Example 2.6 Denote by (z_1, z_2) the coordinates on $X = \mathbb{C}^2$, with $z_j = x_j + \sqrt{-1} y_j$. Let $M = \mathbb{R}^2$, and let $(x_1, x_2; \sqrt{-1}\eta_1, \sqrt{-1}\eta_2)$ denote the coordinates on $T_M^* X$. The function $(z_1 + \sqrt{-1} z_2^2)^{-1}$ defines a holomorphic function f in the tuboid $\mathbb{R}^2 \times \sqrt{-1}\{y_1 > y_2^2\}$. The boundary value of f on $\mathbb{R}^2$ is a hyperfunction u, real analytic for $(x_1, x_2) \neq (0, 0)$, and whose wave front set above $(0, 0)$ is the half-line $\{\eta_1 \geq 0, \eta_2 = 0\}$.

Note that $(x_2 \partial_1 + \frac{\sqrt{-1}}{2} \partial_2) u = 0$, and this is in accordance with the result of Proposition 4.5. Also note that WF(u) is not co-isotropic (see Definition 3.3) after identifying $T_M^* X$ with $\sqrt{-1} T^* M$ and $\sqrt{-1} T^* M$ with the symplectic manifold $T^* M$.

Remark 2.7 Since the sheaf $\mathscr{B}_M$ contains the sheaf $\mathcal{D}b_M$ of distributions, one obtains what is called the *analytic wave front set* of distributions.

Consider a closed convex proper cone $Z \subset T_M^* X$ that contains the zero-section M. Then, WF(u) $\subset Z$ if and only if u is the boundary value b(f) of a holomorphic function f defined in a "tuboid" U with "profile" the interior of the polar tube to Z^a, that is, satisfying

$$C_M(X \setminus U) \cap \text{Int} Z^{\circ a} = \varnothing.$$

Moreover, the sheaf $\mathscr{C}_M$ is conically flabby. Therefore, any hyperfunction may be decomposed as a sum of boundary values of holomorphic functions f_is defined in suitable tuboids U_i, and if we have hyperfunctions u_i ($i = 1, \ldots N$) satisfying $\sum_j u_j = 0$, there exist hyperfunctions u_{ij} ($i, j = 1, \ldots N$) such that

$$u_{ij} = -u_{ji}, \quad u_i = \sum_{j=1}^{N} u_{ij} \text{ and } \mathrm{WF}(u_{ij}) \subset \mathrm{WF}(u_i) \cap \mathrm{WF}(u_j).$$

In other words, consider holomorphic functions g_i defined in tuboids U_i and assume that $\sum_i \mathrm{b}(g_i) = 0$. Then there exist holomorphic functions g_{ij} defined in tuboids U_{ij} whose profile is the convex hull of $U_i \cup U_j$ such that

$$g_{ij} = -g_{ji}, \quad g_i = \sum_{j=1}^{N} g_{ij}.$$

This is the so-called edge of the wedge theorem, which was intensively studied in the 1970s (see [29]).

Soon after Sato had defined the sheaf $\mathscr{C}_M$ and the analytic wave front set of hyperfunctions, Hörmander defined the C^∞-wave front set of distributions by using the classical Fourier transform. See [17, 18] and also [5, 40] for related constructions.

3 Microlocal Sheaf Theory

The idea of microlocal analysis was extended to sheaf theory by Masaki Kashiwara and the author (see [23–25]), giving rise to microlocal sheaf theory. With a sheaf F on a real manifold M, one associates its "microsupport" $\mu\mathrm{supp}(F)$,[6] a closed conic subset of the cotangent bundle T^*M. The microsupport describes the codirections of nonpropagation of F. Here we consider sheaves of **k**-modules for a commutative unital ring **k**. Roughly speaking, a codirection $(x_0; \xi_0) \in T^*M$ does not belong to $\mu\mathrm{supp}(F)$ if, for any smooth function $\varphi \colon M \to \mathbb{R}$ such that $\varphi(x_0) = 0$ and $d\varphi(x_0) = \xi_0$, any section of $H^j(\{\varphi < 0\}; F)$ extends uniquely in a neighborhood of x_0. In other words, $(R\Gamma_{\{\varphi \geq 0\}}(F))_{x_0} \simeq 0$. The microsupport of a sheaf describes the codirections in which it is not locally constant, and a sheaf whose microsupport is contained in the zero-section is nothing but a locally constant sheaf. Here again, a local notion (that of being locally constant) becomes a global notion with respect to the projection $T^*M \to M$.

One can give natural bounds to the microsupport of sheaves after the six operations, in particular after proper direct images and noncharacteristic inverse images. The formulas one obtains are formally similar to the classical

---

[6] Concerning the notation $\mu\mathrm{supp}$, see the footnote in the introduction to this chapter.

ones for D-modules. As an application, one gets a generalization of Morse theory. Indeed, in the classical setting, this theory asserts that, given a proper function $\varphi\colon M \to \mathbb{R}$, the topology of the set $M_t = \{x \in M; \varphi(x) < t\}$ does not change as far as t does not meet a critical value of φ; that is, $\mathrm{R}\Gamma(M_t; \mathbf{k}_M)$ is constant in t on an interval (t_0, t_1) in which there are no critical values. Now we can replace the constant sheaf $\mathbf{k}_M$ with any sheaf F on M, a critical value of φ becoming any $t = \varphi(x) \in \mathbb{R}$ such that $d\varphi(x) \in \mu\mathrm{supp}(F)$. Here, we shall interpret the new notion of "barcodes" (see [11]) in this setting.

The main property of the microsupport is that it is a co-isotropic (i.e., involutive) subset of the symplectic manifold T^*M. This is the reason why this theory has many applications in symplectic topology, as we shall see in Section 4.

3.1 Microsupport

Let M be a real manifold, say, of class C^∞.

Definition 3.1 ([25, Definition 5.1.2]) Let $F \in \mathsf{D}^\mathrm{b}(\mathbf{k}_M)$. One denotes by $\mu\mathrm{supp}(F)$ the closed subset of T^*M defined as follows. For an open subset $U \subset T^*M$, $U \cap \mu\mathrm{supp}(F) = \varnothing$ if and only if, for any $x_0 \in M$ and any real C^1-function φ on M defined in a neighborhood of x_0 satisfying $d\varphi(x_0) \in U$ and $\varphi(x_0) = 0$, one has $(\mathrm{R}\Gamma_{\{x; \varphi(x) \geq 0\}}(F))_{x_0} \simeq 0$. One calls $\mu\mathrm{supp}(F)$ the microsupport of F.

In other words, $U \cap \mu\mathrm{supp}(F) = \varnothing$ if the sheaf F has no cohomology supported by "half-spaces" whose conormals are contained in U.

In the sequel, we denote by $T^*_M M$ the zero-section of T^*M, identified to M.

- By its construction, the microsupport is closed and is conic, that is, invariant by the action of $\mathbb{R}^+$ on T^*M.
- $\mu\mathrm{supp}(F) \cap T^*_M M = \pi_M(\mu\mathrm{supp}(F)) = \mathrm{supp}(F)$.
- $\mu\mathrm{supp}(F) = \mu\mathrm{supp}(F[j])$ ($j \in \mathbb{Z}$).
- The microsupport satisfies the triangular inequality: if $F_1 \to F_2 \to F_3 \xrightarrow{+1}$ is a distinguished triangle in $\mathsf{D}^\mathrm{b}(\mathbf{k}_M)$, then $\mu\mathrm{supp}(F_i) \subset \mu\mathrm{supp}(F_j) \cup \mu\mathrm{supp}(F_k)$ for all $i, j, k \in \{1, 2, 3\}$ with $j \neq k$.

Example 3.2 (i) $\mu\mathrm{supp}(F) \subset T^*_M M$ if and only if $H^j(F)$ is locally constant on M for all $j \in \mathbb{Z}$.
(ii) If N is a smooth closed submanifold of M, then $\mu\mathrm{supp}(\mathbf{k}_N) = T^*_N M$, the conormal bundle to N in M.

(iii) The link between the microsupport of sheaves and the characteristic variety of D-modules will be given in Theorem 4.2.
(iv) Let φ be C^1-function with $d\varphi(x) \neq 0$ when $\varphi(x) = 0$. Let $U = \{x \in M; \varphi(x) > 0\}$, and let $Z = \{x \in M; \varphi(x) \geq 0\}$. Then,

$$\mu\text{supp}(\mathbf{k}_U) = U \times_M T_M^*M \cup \{(x; \lambda d\varphi(x)); \varphi(x) = 0, \lambda \leq 0\}$$
$$\mu\text{supp}(\mathbf{k}_Z) = Z \times_M T_M^*M \cup \{(x; \lambda d\varphi(x)); \varphi(x) = 0, \lambda \geq 0\}.$$

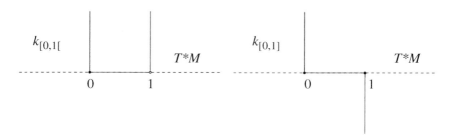

In these pictures, $M = \mathbb{R}$ and $T^*M = \mathbb{R}^2$.

Co-isotropic Subsets

The map $\pi_M : T^*M \to M$ induces the maps

$$T^*T^*M \xleftarrow{\pi_{Md}} T^*M \times_M T^*M \xrightarrow{\pi_{M\pi}} T^*M$$

(see Diagram 3.1). By composing the map π_{Md} with the diagonal map $T^*M \hookrightarrow T^*M \times_M T^*M$, we get a map $\alpha_M : T^*M \to T^*T^*M$, that is, a section of $T^*(T^*M)$. This is the Liouville 1-form, given in a local homogeneous symplectic coordinate system $(x; \xi)$ on T^*M, by

$$\alpha_M = \sum_{j=1}^n \xi_j \, dx_j.$$

The differential $d\alpha_M$ of the Liouville form is the symplectic form ω_M on T^*M given in a local symplectic coordinate system $(x; \xi)$ on T^*M by $\omega_M = \sum_{j=1}^n d\xi_j \wedge dx_j$. Hence T^*M is not only a symplectic manifold; it is a homogeneous (or exact) symplectic manifold.

The form ω_M induces an isomorphism $H : T^*(T^*M) \xrightarrow{\sim} T(T^*M)$ called the Hamiltonian isomorphism. In a local symplectic coordinate system $(x; \xi)$, this isomorphism is given by

$$H(\langle \lambda, dx \rangle + \langle \mu, d\xi \rangle) = -\langle \lambda, \partial_\xi \rangle + \langle \mu, \partial_x \rangle.$$

Definition 3.3 ([25, Definition 6.5.1]) A subset S of T^*M is co-isotropic (one also says involutive) at $p \in T^*M$ if $C_p(S, S)^\perp \subset C_p(S)$. Here we identify the orthogonal $C_p(S, S)^\perp$ to a subset of T_pT^*M via the Hamiltonian isomorphism.

When S is smooth, one recovers the usual notion.

Example 3.4 Let $M = \mathbb{R}$, and denote by $(t; \tau)$ the coordinates on T^*M. The set $\{(t; \tau); t \geq 0, \tau = 0\}$ is not co-isotropic, contrarily to the set $\{(t; \tau); t \geq 0, \tau = 0 \cup t = 0, \tau \geq 0\}$, which is co-isotropic.

An essential property of the microsupport is given by the next theorem.

Theorem 3.5 ([25, Theorem 6.5.4]) Let $F \in \mathsf{D}^b(\mathbf{k}_M)$. Then its microsupport $\mu\mathrm{supp}(F)$ is co-isotropic.

Constructible Sheaves

Assume that M is real analytic and that $\mathbf{k}$ is a field. We do not recall here the definition of a subanalytic subset and a subanalytic stratification, referring to [3].

Definition 3.6 A sheaf F is weakly $\mathbb{R}$-constructible if there exists a subanalytic stratification $M = \bigsqcup_\alpha M_\alpha$ such that, for each strata M_α, the restriction $F|_{M_\alpha}$ is locally constant. If, moreover, it is a local system (i.e., is locally constant of finite rank), then one says that F is $\mathbb{R}$-constructible.

One denotes by $\mathsf{D}^b_{\text{w-}\mathbb{R}\text{-c}}(\mathbf{k}_M)$ (resp. $\mathsf{D}^b_{\mathbb{R}\text{-c}}(\mathbf{k}_M)$) the full subcategory of $\mathsf{D}^b(\mathbf{k}_M)$ consisting of sheaves with weakly $\mathbb{R}$-constructible cohomology (resp. $\mathbb{R}$-constructible cohomology).

A subanalytic $\mathbb{R}^+$-conic subset Λ of T^*M is isotropic if the 1-form α_M vanishes on Λ. It is Lagrangian if it is both isotropic and co-isotropic.

Theorem 3.7 ([25, Theorem 8.4.2]) Assume that M is real analytic and $\mathbf{k}$ is a field. Let $F \in \mathsf{D}^b(\mathbf{k}_M)$. Then $F \in \mathsf{D}^b_{\text{w-}\mathbb{R}\text{-c}}(\mathbf{k}_M)$ if and only if $\mu\mathrm{supp}(F)$ is contained in a closed $\mathbb{R}^+$-conic subanalytic Lagrangian subset of T^*M, and this implies that $\mu\mathrm{supp}(F)$ is itself a closed $\mathbb{R}^+$-conic subanalytic Lagrangian subset of T^*M.

3.2 Microsupport and the Six Operations

Let $f: M \to N$ be a morphism of real manifolds. The tangent map $Tf: TM \to TN$ decomposes as $TM \to M \times_N TN \to TN$, and by duality, one gets the diagram

$$T^*M \xleftarrow{f_d} M \times_N T^*N \xrightarrow{f_\pi} T^*N \qquad (3.1)$$

with π_M, π, π_N projecting to $M \xrightarrow{f} N$.

One sets

$$T_M^*N := \operatorname{Ker} f_d = f_d^{-1}(T_M^*M).$$

Note that, denoting by Γ_f the graph of f in $M \times N$, the projection $T^*(M \times N) \to M \times T^*N$ identifies $T^*_{\Gamma_f}(M \times N)$ and $M \times_N T^*N$.

Definition 3.8 Let $\Lambda \subset T^*N$ be a closed $\mathbb{R}^+$-conic subset. One says that f is noncharacteristic for Λ if

$$f_\pi^{-1}\Lambda \cap T_M^*N \subset M \times_N T_N^*N.$$

This is equivalent to saying that f_d is proper on $f_\pi^{-1}\Lambda$.

Theorem 3.9 *Let $f : M \to N$ be a morphism of manifolds, let $F \in \mathsf{D}^b(\mathbf{k}_M)$, and assume that f is proper on* $\operatorname{supp}(F)$. *Then* $Rf_!F \xrightarrow{\sim} Rf_*F$ *and*

$$\mu\operatorname{supp}(Rf_*F) \subset f_\pi f_d^{-1} \mu\operatorname{supp}(F). \qquad (3.2)$$

Moreover, if f is a closed embedding, this inclusion is an equality.

This result may be interpreted as a "stationary phase lemma" for sheaves, or else as a generalization of Morse theory for sheaves (see below).

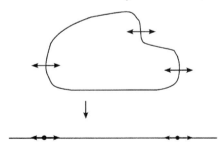

In this picture, one represents the direct image of the constant sheaf on the contour. The microsupport of the direct image is contained in the image of the "horizontal" conormal vectors. This shows that the inclusion in Theorem 3.9 may be strict.

Theorem 3.10 *Let $f : M \to N$ be a morphism of manifolds, let $G \in \mathsf{D}^b(\mathbf{k}_N)$, and assume that f is noncharacteristic with respect to $\mu\operatorname{supp}(G)$. Then the natural morphism $f^{-1}G \otimes \omega_{M/N} \to f^!G$ is an isomorphism and*

$$\mu\mathrm{supp}(f^{-1}G) \subset f_d f_\pi^{-1}(\mu\mathrm{supp}(G)). \tag{3.3}$$

Moreover, if f is submersive, this inclusion is an equality.

Note that the formulas for the microsupport of the direct or inverse images are analog to those for the direct or inverse images of D-modules.

Corollary 3.11 *Let $F_1, F_2 \in \mathsf{D}^\mathrm{b}(\mathbf{k}_M)$.*

(1) *Assume that $\mu\mathrm{supp}(F_1) \cap \mu\mathrm{supp}(F_2)^a \subset T_M^* M$. Then*

$$\mu\mathrm{supp}(F_1 \overset{L}{\otimes} F_2) \subset \mu\mathrm{supp}(F_1) + \mu\mathrm{supp}(F_2).$$

(2) *Assume that $\mu\mathrm{supp}(F_1) \cap \mu\mathrm{supp}(F_2) \subset T_M^* M$. Then*

$$\mu\mathrm{supp}(R\mathcal{H}om\,(F_2, F_1)) \subset \mu\mathrm{supp}(F_2)^a + \mu\mathrm{supp}(F_1).$$

Note that the formula for the microsupport of the tensor product is analog to that giving a bound to the wave front set of the product of two distributions u_1 and u_2 satisfying $\mathrm{WF}(u_1) \cap \mathrm{WF}(u_2)^a \subset \sqrt{-1}T_M^* M$.

Corollary 3.12 (A kind of Petrowsky theorem for sheaves) *Let $F_1, F_2 \in \mathsf{D}^\mathrm{b}(\mathbf{k}_M)$. Assume that F_2 is constructible and $\mu\mathrm{supp}(F_2) \cap \mu\mathrm{supp}(F_1) \subset T_M^* M$. Then the natural morphism*

$$\mathsf{D}'_M F_2 \otimes F_1 \to R\mathcal{H}om\,(F_2, F_1) \tag{3.4}$$

is an isomorphism.

The link with the classical Petrowsky theorem for elliptic operators will be given in Corollary 4.4.

Remark 3.13 (i) One can also give bounds to the microsupports of the sheaves obtained by the six operations without assuming any hypothesis of properness or transversality. See [25, Corollary 6.4.4, 6.4.5].

(ii) By applying the results on the behavior of the microsupport together with Theorem 3.7, one recovers easily the fact that the category of $\mathbb{R}$-constructible sheaves is stable with respect to the six operations (under suitable hypotheses of properness). See [25, Section 8.4].

Morse Theory

As an application of Theorem 3.9, one gets the following:

Theorem 3.14 *Let $F \in \mathsf{D}^\mathrm{b}(\mathbf{k}_M)$, let $\psi: M \to \mathbb{R}$ be a function of class C^1, and assume that ψ is proper on $\mathrm{supp}(F)$. Let $a, b \in \mathbb{R}$ with $a < b$, and assume that*

$d\psi(x) \notin \mu\mathrm{supp}(F)$ for $a \leq \psi(x) < b$. For $t \in \mathbb{R}$, set $M_t = \psi^{-1}(]-\infty, t[)$. Then the restriction morphism $R\Gamma(M_b; F) \to R\Gamma(M_a; F)$ is an isomorphism.

The classical Morse theorem corresponds to the constant sheaf $F = \mathbf{k}_M$. As an immediate corollary, one obtains the following:

Corollary 3.15 *Let $F \in \mathsf{D}^b(\mathbf{k}_M)$, and let $\psi : M \to \mathbb{R}$ be a function of class C^1. Set $\Lambda_\psi = \{(x; d\psi(x))\}$, a (nonconic in general) Lagrangian submanifold of T^*M. Assume that $\mathrm{supp}(F)$ is compact and $R\Gamma(M; F) \neq 0$. Then $\Lambda_\psi \cap \mu\mathrm{supp}(F) \neq \emptyset$.*

This corollary is an essential tool in a new proof of the Arnold nondisplaceability theorem (see Section 4.2).

Persistent Homology and Barcodes

Persistent homology and barcodes are recent concrete applications of algebraic topology. The link with sheaf theory has been noticed by several authors (see in particular [7, 11]), and the link with *microlocal* sheaf theory is developed in [28].

One has a finite subset S of $\mathbb{R}^N$, and one wants to understand its structure. For that purpose, one replaces each point $x \in S$ with a closed ball of center x and radius t and makes t go to infinity. The union of these balls gives a closed set $Z \subset X \times \mathbb{R}$, and one wants to understand how the homology of the union of the balls varies with t.

Denote by (x, t) the coordinates on $\mathbb{R}^{n+1}$, by $(x, t; \xi, \tau)$ the associated coordinates on $T^*\mathbb{R}^{n+1}$, and by $p : \mathbb{R}^{n+1} \to \mathbb{R}$ the projection $p(x, t) = t$.

Let $\mathbf{k}$ be a field, and consider the sheaf $\mathbf{k}_Z$. Clearly, this sheaf is $\mathbb{R}$-constructible. Since the map p is proper on Z, we get that $Rp_*\mathbf{k}_Z \in \mathsf{D}^b_{\mathbb{R}\text{-}c}(\mathbf{k}_\mathbb{R})$. Moreover, one has

$$\mu\mathrm{supp}(\mathbf{k}_Z) \subset \{(x, t; \xi, \tau); \tau \geq 0\}. \tag{3.5}$$

In fact, this result holds true in a more general situation, replacing X with a compact set and $\mathbb{R}^n$ with a Riemannian manifold. This follows from [25, Proposition 5.3.8].

Since p is proper on Z, we get by Theorem 3.9 that

$$\mu\mathrm{supp}(Rp_*\mathbf{k}_Z) \subset \{(t, \tau); \tau \geq 0\}. \tag{3.6}$$

A classical result of sheaf theory (see [6] or [15, Section 6]) asserts that any $G \in \mathsf{D}^b(\mathbf{k}_\mathbb{R})$ is a finite direct sum of constant sheaves on intervals. By (3.6), one gets that there are finitely many intervals $[a_j, b_j)$, $a_j \geq 0$, $a_j < b_j$, $0 < b_j \leq \infty$ and integers $d_j \in \mathbb{Z}$ ($j \in J$, J finite) such that

$$\mathrm{R}p_*\mathbf{k}_Z \simeq \bigoplus_{j \in J} \mathbf{k}_{[a_j, b_j)}[d_j]. \tag{3.7}$$

Here we may have $a_j = a_k, b_j = b_k, d_j = d_k$ for $j \neq k$. To the right-hand side of (3.7), we may associate a family of barcodes as follows. For each $d_j \in \mathbb{Z}$, replace $[a_j, b_j)[d_j]$ with the vertical interval $[0, b_j)$ centered at $a_j \in \mathbb{R}$.

Higher-dimensional persistent homology consists in replacing the poset $(\mathbb{R}, \leq)$ with a finite-dimensional vector space $\mathbb{V}$ endowed with a closed convex proper cone γ. Then barcodes are replaced with constructible sheaves on $\mathbb{V}$ whose microsupport is contained in $\mathbb{V} \times \gamma^{\circ a}$, and the study of such sheaves is rather delicate (see [28]).

3.3 The Functor μhom

In [25, Section IV.4], Sato's microlocalization functor is generalized as follows. The functor $\mu hom \colon \mathsf{D}^{\mathrm{b}}(\mathbf{k}_M)^{\mathrm{op}} \times \mathsf{D}^{\mathrm{b}}(\mathbf{k}_M) \to \mathsf{D}^{\mathrm{b}}(\mathbf{k}_{T^*M})$ is given by

$$\mu hom(F_2, F_1) = \tilde{\delta}^{-1} \mu_\Delta \mathrm{R}\mathcal{H}om(q_2^{-1} F_2, q_1^! F_1),$$

where q_i ($i = 1, 2$) denotes the ith projection on $M \times M$, Δ is the diagonal of $M \times M$, and $\tilde{\delta}$ is the isomorphism $\tilde{\delta} \colon T^*M \xrightarrow{\sim} T^*_M(M \times M)$, $(x; \xi) \mapsto (x, x; \xi, -\xi)$. One proves that

- $\mathrm{R}\pi_{M*}\mu hom(F_2, F_1) \simeq \mathrm{R}\mathcal{H}om(F_2, F_1)$;
- $\mu hom(\mathbf{k}_N, F) \simeq \mu_N(F)$ for N a closed submanifold of M;
- $\operatorname{supp} \mu hom(F_2, F_1) \subset \mu\operatorname{supp}(F_1) \cap \mu\operatorname{supp}(F_2)$.

The functor μhom is the functor of microlocal morphisms. Let us make this assertion more precise.

Let Z be a locally closed subset of T^*M. One denotes by $\mathsf{D}^{\mathrm{b}}(\mathbf{k}_M; Z)$ the localization of $\mathsf{D}^{\mathrm{b}}(\mathbf{k}_M)$ by its full triangulated subcategory consisting of objects F such that $\mu\operatorname{supp}(F) \cap Z = \emptyset$. The objects of $\mathsf{D}^{\mathrm{b}}(\mathbf{k}_M; Z)$ are those of $\mathsf{D}^{\mathrm{b}}(\mathbf{k}_M)$, but a morphism $u \colon F_1 \to F_2$ in $\mathsf{D}^{\mathrm{b}}(\mathbf{k}_M)$ becomes an isomorphism in $\mathsf{D}^{\mathrm{b}}(\mathbf{k}_M; Z)$ if, after embedding this morphism in a distinguished triangle $F_1 \to F_2 \to F_3 \xrightarrow{+1}$, one has $\mu\operatorname{supp}(F_3) \cap Z = \emptyset$.

One shall be aware that the prestack (a prestack is, roughly speaking, a presheaf of categories) $U \mapsto \mathsf{D}^{\mathrm{b}}(\mathbf{k}_M; U)$ (U open in T^*M) is not a stack – not even a separated prestack. The functor μhom induces a bifunctor (we keep the same notation),

$$\mu hom \colon \mathsf{D}^{\mathrm{b}}(\mathbf{k}_M; U)^{\mathrm{op}} \times \mathsf{D}^{\mathrm{b}}(\mathbf{k}_M; U) \to \mathsf{D}^{\mathrm{b}}(\mathbf{k}_U),$$

and for $F_1, F_2, F_3 \in \mathsf{D}^b(\mathbf{k}_M; U)$, there is a natural morphism

$$\mu hom(F_3, F_2) \overset{L}{\otimes} \mu hom(F_2, F_1) \to \mu hom(F_3, F_1). \tag{3.8}$$

Moreover, for $p \in T^*M$, one has an isomorphism

$$\mu hom(F_2, F_1)_p \simeq \mathrm{Hom}_{\mathsf{D}^b(\mathbf{k}_M; \{p\})}(F_2, F_1) \tag{3.9}$$

and (3.8) is compatible with the composition of morphisms in $\mathsf{D}^b(\mathbf{k}_M; \{p\})$. This shows that, in some sense, μhom is a kind of internal hom for $\mathsf{D}^b(\mathbf{k}_M; U)$.

Now let Λ be a smooth conic submanifold closed in U. Denote by $\mathsf{D}^b_\Lambda(\mathbf{k}_M; U)$ the full subcategory of $\mathsf{D}^b(\mathbf{k}_M; U)$ consisting of objects $F \in \mathsf{D}^b(\mathbf{k}_M)$ satisfying $\mu\mathrm{supp}(F) \cap U \subset \Lambda$.

Definition 3.16 Assume that $\mathbf{k}$ is a field. One says that $F \in \mathsf{D}^b_\Lambda(\mathbf{k}_M; U)$ is simple along Λ if the natural morphism $\mathbf{k}_\Lambda \to \mu hom(F, F)$ is an isomorphism. One denotes by $\mathrm{Simple}(\Lambda, \mathbf{k})$ the subcategory of $\mathsf{D}^b_\Lambda(\mathbf{k}_M; U)$ consisting of simple sheaves.

We shall see in Section 4.2 that simple sheaves play an important role in symplectic topology.

Microlocal Serre Functor

Recall first what a Serre functor is – a notion introduced in [4]. Consider linear triangulated category $\mathscr{T}$ over a field $\mathbf{k}$, and assume that the spaces $\bigoplus_{n \in \mathbb{Z}} \mathrm{Hom}_{\mathscr{T}}(A, B[n])$ are finite-dimensional for any $A, B \in \mathscr{T}$. A Serre functor S is an endofunctor S of $\mathscr{T}$ together with a functorial (in A and B) isomorphism:

$$\mathrm{Hom}_{\mathscr{T}}(A, B)^* \simeq \mathrm{Hom}_{\mathscr{T}}(B, S(A)).$$

Here $*$ denotes the duality functor for vector spaces.

This definition is motivated by the example of the category $\mathsf{D}^b_{\mathrm{coh}}(\mathscr{O}_X)$ of coherent $\mathscr{O}_X$-modules on a complex compact manifold X. In this case, a theorem of Serre asserts that the Serre functor is given by $\mathscr{F} \mapsto \mathscr{F} \otimes_{\mathscr{O}_X} \Omega_X[d_X]$, where Ω_X is the sheaf of holomorphic forms of maximal degree and d_X is the complex dimension of X.

There is an interesting phenomenon that holds with μhom and not with $R\mathscr{H}om$. Indeed, although the category $\mathsf{D}^b_{\mathbb{R}\text{-}c}(\mathbf{k}_M)$ does not admit a Serre functor, it admits a kind of microlocal Serre functor, as shown by the isomorphism, functorial in F_1 and F_2 (see [25, Proposition 8.4.14]):

$$\mathsf{D}_{T^*M} \mu hom(F_2, F_1) \simeq \mu hom(F_1, F_2) \otimes \pi_M^{-1} \omega_M.$$

Here, $\omega_M \simeq \mathrm{or}_M [\dim M]$ is the dualizing complex on M.

This confirms the fact that to fully understand $\mathbb{R}$-constructible sheaves, it is natural to look at them microlocally, that is, in T^*M. This is also in accordance with the "philosophy" of mirror symmetry, which interchanges the category of coherent $\mathcal{O}_X$-modules on a complex manifold X with the Fukaya category on a symplectic manifold Y (see Section 4.2).

4 Some Applications

4.1 Solutions of D-Modules

We shall first briefly present some applications of microlocal sheaf theory to systems of linear partial differential equations (LPDE), that is, (generalized) holomorphic solutions of D-modules. This was the original motivation of the theory. The main tool is a theorem that asserts that given a coherent D-module $\mathcal{M}$ on a complex manifold X, the microsupport of the complex of the holomorphic solutions of $\mathcal{M}$ is contained in the characteristic variety of the system. (In fact it is equal, but the other inclusion is not so useful.) This theorem is deduced from the Cauchy–Kowalevska theorem in its precise form given by Petrowsky and Leray and is the unique tool of analysis that is used thereafter. With this result, the study of generalized solutions of systems of LPDE reduces most of the time to a geometric study, the relations between the microsupport of the constructible sheaf associated with the space of generalized functions (e.g., the conormal bundle T_M^*X to a real manifold M) and the characteristic variety of the system. We shall only study here the particular case of elliptic systems.

D-Modules

We have seen at the end of Section 1.3 that a system of linear equations over a sheaf of rings $\mathcal{R}$ is nothing but an $\mathcal{R}$-module locally of finite presentation. We shall consider here the case where X is a complex manifold and $\mathcal{R}$ is the sheaf $\mathcal{D}_X$ of holomorphic differential operators. References for D-modules are made to [21, 22].

Let X be a complex manifold. One denotes by $\mathcal{D}_X$ the sheaf of rings of holomorphic (finite-order) differential operators. It is a right- and left-coherent ring. A system of linear partial differential equations on X is thus a left-coherent $\mathcal{D}_X$-module $\mathcal{M}$. Locally on X, $\mathcal{M}$ may be represented as the cokernel of a matrix $\cdot P_0$ of differential operators acting on the right:

$$\mathcal{M} \simeq \mathcal{D}_X^{N_0} / \mathcal{D}_X^{N_1} \cdot P_0.$$

By classical arguments of analytic geometry (Hilbert's syzygy theorem), one shows that $\mathcal{M}$ is locally isomorphic to the cohomology of a bounded complex

$$\mathcal{M}^\bullet := 0 \to \mathcal{D}_X^{N_r} \to \cdots \to \mathcal{D}_X^{N_1} \xrightarrow{\cdot P_0} \mathcal{D}_X^{N_0} \to 0. \tag{4.1}$$

Clearly, the sheaf $\mathcal{O}_X$ of holomorphic functions is a left $\mathcal{D}_X$-module. It is coherent since $\mathcal{O}_X \simeq \mathcal{D}_X / \mathcal{I}$, where $\mathcal{I}$ is the left ideal generated by the vector fields. For a coherent $\mathcal{D}_X$-module $\mathcal{M}$, one sets for short

$$Sol(\mathcal{M}) := R\mathcal{H}om_{\mathcal{D}_X}(\mathcal{M}, \mathcal{O}_X).$$

Representing (locally) $\mathcal{M}$ by a bounded complex $\mathcal{M}^\bullet$ as above, we get

$$Sol(\mathcal{M}) \simeq 0 \to \mathcal{O}_X^{N_0} \xrightarrow{P_0 \cdot} \mathcal{O}_X^{N_1} \to \cdots \mathcal{O}_X^{N_r} \to 0, \tag{4.2}$$

where now $P_0\cdot$ operates on the left.

Characteristic Variety

One defines naturally the characteristic variety of $\mathcal{M}$, denoted $\mathrm{char}(\mathcal{M})$, a closed complex analytic subset of T^*X, conic with respect to the action of $\mathbb{C}^\times$ on T^*X. For example, if $\mathcal{M}$ has a single generator u with relation $\mathcal{I}u = 0$, where $\mathcal{I}$ is a locally finitely generated left ideal of $\mathcal{D}_X$, then

$$\mathrm{char}(\mathcal{M}) = \{(z;\zeta) \in T^*X; \sigma(P)(z;\zeta) = 0 \text{ for all } P \in \mathcal{I}\},$$

where $\sigma(P)$ denotes the principal symbol of P.

The fundamental result below was first obtained in [35].

Theorem 4.1 *Let $\mathcal{M}$ be a coherent $\mathcal{D}_X$-module. Then $\mathrm{char}(\mathcal{M})$ is a closed conic complex analytic involutive (i.e., co-isotropic) subset of T^*X.*

The proof of the involutivity is really difficult: it uses microdifferential operators of infinite order and quantized contact transformations. Later, Gabber [10] gave a purely algebraic (and much simpler) proof of this result. Theorem 4.2 together with Theorem 3.5 gives another totally different proof of the involutivity.

Theorem 4.2 ([25, Theorem 11.3.3]) *Let $\mathcal{M}$ be a coherent $\mathcal{D}_X$-module. Then*

$$\mu\mathrm{supp}(Sol(\mathcal{M})) = \mathrm{char}(\mathcal{M}). \tag{4.3}$$

The only analytic tool in the proof of the inclusion $* \subset *$ in (4.3) is the classical Cauchy–Kowalevska theorem, in its precise form (see [18, Section 9.4]). To prove the reverse inclusion, one uses a theorem of [35] that asserts that the ring $\mathcal{D}_X^\infty$ of infinite-order differential operators is faithfully flat over $\mathcal{D}_X$.

Elliptic Pairs

Let us apply Corollary 3.12 when X is a complex manifold. For $G \in D^b_{\mathbb{R}\text{-}c}(\mathbb{C}_X)$, set

$$\mathscr{A}_G := \mathscr{O}_X \otimes G, \quad \mathscr{B}_G := R\mathscr{H}om(D'_X G, \mathscr{O}_X).$$

Note that if X is the complexification of a real analytic manifold M and we choose $G = \mathbb{C}_M$, we recover the sheaf of real analytic functions and the sheaf of hyperfunctions:

$$\mathscr{A}_{\mathbb{C}_M} = \mathscr{A}_M, \quad \mathscr{B}_{\mathbb{C}_M} = \mathscr{B}_M.$$

Now let $\mathscr{M} \in D^b_{\text{coh}}(\mathscr{D}_X)$. According to [37], one says that the pair $(G, \mathscr{M})$ is elliptic if $\text{char}(\mathscr{M}) \cap \mu\text{supp}(G) \subset T^*_X X$.

Theorem 4.3 ([37]) *Let $(\mathscr{M}, G)$ be an elliptic pair.*

(i) *We have the canonical isomorphism*

$$R\mathscr{H}om_{\mathscr{D}_X}(\mathscr{M}, \mathscr{A}_G) \xrightarrow{\sim} R\mathscr{H}om_{\mathscr{D}_X}(\mathscr{M}, \mathscr{B}_G). \quad (4.4)$$

(ii) *Assume moreover that* $\text{supp}(\mathscr{M}) \cap \text{supp}(G)$ *is compact and $\mathscr{M}$ admits a global presentation as in (4.1). Then the cohomology of the complex* $R\text{Hom}_{\mathscr{D}_X}(\mathscr{M}, \mathscr{A}_G)$ *is finite-dimensional.*

Proof: (i) This is a particular case of Corollary 3.12.

(ii) One first shows that one may represent G with a bounded complex whose components are a finite direct sum of sheaves of the type $\mathbf{k}_U$ for U open subanalytic relatively compact in X. Then one can represent the left-hand side of the global sections of (4.4) by a complex of topological vector spaces of type DFN and the right-hand side by a complex of topological vector spaces of type FN. The finiteness follows by classical results of functional analysis. ∎

Let us particularize Theorem 4.3 to the usual case of an elliptic system. Let M be a real anaytic manifold and X a complexification of M, and let us choose $G = D'_X \mathbb{C}_M$. Then $(G, \mathscr{M})$ is an elliptic pair if and only if

$$T^*_M X \cap \text{char}(\mathscr{M}) \subset T^*_X X. \quad (4.5)$$

In this case, one simply says that $\mathscr{M}$ is an elliptic system. Then one recovers a classical result:

Corollary 4.4 *Let $\mathscr{M}$ be an elliptic system.*

(i) *We have the canonical isomorphism*

$$R\mathscr{H}om_{\mathscr{D}_X}(\mathscr{M}, \mathscr{A}_M) \xrightarrow{\sim} R\mathscr{H}om_{\mathscr{D}_X}(\mathscr{M}, \mathscr{B}_M). \quad (4.6)$$

(ii) *Assume moreover that M is compact and $\mathcal{M}$ admits a global presentation as in (4.1). Then the cohomology of the complex* $\mathrm{RHom}_{\mathcal{D}_X}(\mathcal{M}, \mathcal{A}_M)$ *is finite-dimensional.*

There is a more precise result than Corollary 4.4 (a), due to Sato [34].

Proposition 4.5 *Let* $U \subset T_M^* X$ *be an open subset, let $\mathcal{M}$ be a coherent $\mathcal{D}_X$-module, let $j \in \mathbb{Z}$, and let $u \in \Gamma(U; \mathcal{E}xt^j_{\mathcal{D}_X}(\mathcal{M}, \mathcal{C}_M))$. Then* $\mathrm{supp}(u) \subset U \cap \mathrm{char}(\mathcal{M})$. *In particular, if $u \in \mathrm{Hom}_{\mathcal{D}_X}(\mathcal{M}, \mathcal{B}_M)$, then* $\mathrm{WF}(u) \subset T_M^* X \cap \mathrm{char}(\mathcal{M})$.

Proof: One has

$$\mathrm{R}\mathcal{H}om_{\pi_M^{-1}\mathcal{D}_X}(\pi_M^{-1}\mathcal{M}, \mathcal{C}_M) \simeq \mu hom(\mathrm{D}'\mathbb{C}_M, \mathrm{R}\mathcal{H}om(\mathcal{M}, \mathcal{O}_X)),$$

and the support of the right-hand side is contained in $\mu\mathrm{supp}(\mathrm{R}\mathcal{H}om(\mathcal{M}, \mathcal{O}_X)) \cap \mu\mathrm{supp}(\mathbb{C}_M)$, that is, in $T_M^* X \cap \mathrm{char}(\mathcal{M})$. ∎

In case $\mathcal{M}$ is elliptic, we get that $\mathrm{WF}(u)$ is contained in the zero-section, hence u is real analytic.

Remark 4.6 One can treat similarly hyperbolic systems, and in particular, one can solve globally the Cauchy problem on globally hyperbolic spacetimes, using only tools from sheaf theory (see [20]).

4.2 A Glance at Symplectic Topology

When a space is endowed with a certain structure, it is natural to associate to it a sheaf (or something similar, a stack, for example) that takes this structure into account. On a real manifold, one considers the sheaf of C^∞-functions, on a complex manifold, the sheaf of holomorphic functions, and on a complex symplectic manifold, the sheaves of holomorphic deformation-quantization algebras (which are sheaves only locally; globally, one has to replace the notion of a sheaf by that of an algebroid stack). Then, on a real symplectic manifold $\mathcal{X}$, what is, or what are, the candidate(s)? One answer is not given by a sheaf but by a category, the "Fukaya category." We shall not describe it here; let us only mention that its construction is not local (and cannot be so), its objects are smooth Lagrangian submanifolds (plus some data such as local systems), and the morphisms between two Lagrangians are only defined when the Lagrangians are transverse (which makes it difficult to define the identity morphisms!).

It is only quite recently that people realized that microlocal sheaf theory is an efficient tool to treat many questions of symplectic topology. In [41], Dmitri

Tamarkin gives a new proof of the Arnold nondisplaceability theorem, and in [30, 31], David Nadler and Eric Zaslow showed that the Fukaya category on T^*M is equivalent (in some sense) to the bounded derived category of $\mathbb{R}$-constructible sheaves on M. These works opened new perspectives in symplectic topology (see, in particular, [13–15]) and also in the study of Legendrian knots (see [39]).

Note that, for U open in T^*M, a substitute to the Fukaya category on U could be the triangulated category $\mathsf{D}^{\mathsf{b}}(\mathbf{k}_M; U)$ already encountered. However, if the objects of this category are associated with the microsupports of sheaves, which are co-isotropic, these microsupports are $\mathbb{R}^+$-conic. To treat nonconic Lagrangian submanifolds, Tamarkin develops a kind of no more conic microlocal sheaf theory by adding a variable t, with dual variable τ, and works in the category of sheaves on $M \times \mathbb{R}$ localized at $\tau > 0$.

Arnold Nondisplaceability Theorem

Let us explain the classical Arnold nondisplaceability conjecture, which has been a theorem for long.

A symplectic isotopy of a symplectic manifold $\mathscr{X}$ is a one-parameter family of isomorphisms of $\mathscr{X}$ that respect the symplectic structure. More precisely, I is an open interval containing 0, and $\Phi \colon \mathscr{X} \times I \to \mathscr{X}$ is a C^∞-map such that $\varphi_t := \Phi(\cdot, t) \colon \mathscr{X} \to \mathscr{X}$ is a symplectic isomorphism for each $t \in I$ and is the identity for $t = 0$. One says that the isotopy is Hamiltonian if the graph of Φ in $\mathscr{X} \times I \times \mathscr{X}$ is the projection of a Lagrangian submanifold Λ_Φ of $\mathscr{X} \times T^*I \times \mathscr{X}$. Then Arnold's conjecture says that for $\mathscr{X} = T^*N$ where N is a compact C^∞-manifold, $\varphi_t(T_N^*N) \cap T_N^*N \neq \varnothing$ for all $t \in I$.

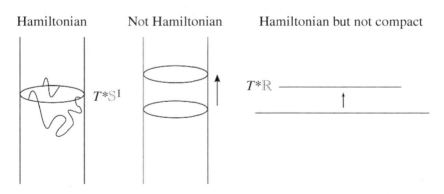

As already mentioned, Tamarkin [41] has given a totally new proof of this result by adapting microlocal sheaf theory to a nonconic setting. However, one can also deduce Arnold's conjecture from another conjecture that is itself "conic." This is the strategy of [16] that we shall expose.

Notice first that a homogeneous symplectic isotopy of an open conic subset $U \subset T^*M$ is a symplectic isotopy that commutes with the $\mathbb{R}^+$-action. In such a case, the isotopy is automatically Hamiltonian.

Recall that $\dot{T}^*M$ denotes the bundle T^*M with the zero-section removed. One denotes by $\mathsf{D}^{\mathrm{lb}}(\mathbf{k}_M)$ the full subcategory of $\mathsf{D}(\mathbf{k}_M)$ consisting of locally bounded objects. For $K \in \mathsf{D}^{\mathrm{lb}}(\mathbf{k}_{M \times M \times I})$ and $s \in I$, we shall denote by K_s the restriction of K to $M \times M \times \{s\}$.

Theorem 4.7 *(Quantization of Hamiltonian isotopies [16]). Let* $\Phi \colon \dot{T}^*M \times I \to \dot{T}^*M$ *be a homogeneous Hamiltonian isotopy. Then there exists* $K \in \mathsf{D}^{\mathrm{lb}}(\mathbf{k}_{M \times M \times I})$ *satisfying*

(a) $\mu\mathrm{supp}(K) \subset \Lambda_\Phi \cup T^*_{M \times M \times I}(M \times M \times I)$
(b) $K_0 \simeq \mathbf{k}_\Delta$.

Moreover,

(i) *both projections* $\mathrm{supp}(K) \rightrightarrows M \times I$ *are proper;*
(ii) *setting* $K_s^{-1} := v^{-1} R\mathcal{H}om(K_s, \omega_M \boxtimes \mathbf{k}_M)$ *with* $v \colon M \times M \to M \times M$, $v(x, y) = (y, x)$, *we have* $K_s \circ K_s^{-1} \simeq K_s^{-1} \circ K_s \simeq \mathbf{k}_\Delta$ *for all* $s \in I$;
(iii) *such a K satisfying the conditions (a) and (b) above is unique up to a unique isomorphism.*

Example 4.8 Let $M = \mathbb{R}^n$, and denote by $(x; \xi)$ the homogeneous symplectic coordinates on $T^*\mathbb{R}^n$. Consider the isotopy $\varphi_s(x; \xi) = (x - s\frac{\xi}{|\xi|}; \xi)$, $s \in I = \mathbb{R}$. One proves that there exists $K \in \mathsf{D}^{\mathrm{b}}(\mathbf{k}_{M \times M \times I})$ such that K_s denoting its restriction to $M \times M \times \{s\}$, one has $K_s \simeq \mathbf{k}_{\{|x-y| \leq s\}}$ for $s \geq 0$ and $K_s \simeq \mathbf{k}_{\{|x-y| < -s\}}[n]$ for $s < 0$.

Theorem 4.9 ([16]) *Consider a homogeneous Hamiltonian isotopy* $\Phi = \{\varphi_s\}_{s \in I} \colon \dot{T}^*M \times I \to \dot{T}^*M$ *and a C^1-map* $\psi \colon M \to \mathbb{R}$ *such that the differential $d\psi(x)$ never vanishes. Set*

$$\Lambda_\psi := \{(x; d\psi(x)); \ x \in M\} \subset \dot{T}^*M.$$

Let $F_0 \in \mathsf{D}^{\mathrm{b}}(\mathbf{k}_M)$ *with compact support and such that* $R\Gamma(M; F_0) \neq 0$. *Then for any* $s \in I$, $\varphi_s(\mu\mathrm{supp}(F) \cap \dot{T}^*M) \cap \Lambda_\psi \neq \varnothing$.

Sketch of proof: Set $F_s = K_s \circ F_0$. Then F_s has compact support and $\mu\mathrm{supp}(F_s) \cap \dot{T}^*M = \varphi_s(\mu\mathrm{supp}(F_0)) \cap \dot{T}^*M$. The direct image of $K \circ F_0$ on I is a constant sheaf, and this implies the isomorphism $R\Gamma(M; F_s) \simeq R\Gamma(M; F_0)$. Then the result follows from Corollary 3.15. ∎

It is possible to deduce (with some work; see loc. cit. Theorem 4.16) Arnold's conjecture from this result by choosing $M = N \times \mathbb{R}$ and $K = \mathbf{k}_{N \times \{0\}}$.

Legendrian Knots

We assume now that $\mathbf{k}$ is a field, and we consider a closed smooth conic Lagrangian submanifold Λ of $\dot{T}^*M$. Recall Definition 3.16.

It follows from Theorem 4.7 that the category Simple($\Lambda, \mathbf{k}$) is a Hamiltonian isotopy invariant. (Note that this category may be empty. Conditions that ensure that it is not empty are obtained in [13].)

Corollary 4.10 *Let Φ be a homogeneous Hamiltonian isotopy as in Theorem 4.7. Let Λ_0 be a smooth closed conic Lagrangian submanifold of $\dot{T}^*M$, and let $\Lambda_1 = \varphi_1(\Lambda_0)$. The categories* Simple($\Lambda_0; \mathbf{k}$) *and* Simple($\Lambda_1; \mathbf{k}$) *are equivalent.*

Example 4.11 In $\mathbb{R}^2$ with coordinates (x, y), we define the following locally closed subset with boundary the cusp

$$W = \{(x, y); x > 0, -x^{3/2} \le y < x^{3/2}\}. \tag{4.7}$$

Outside the zero section, $\mu\mathrm{supp}(\mathbf{k}_W)$ is the smooth Lagrangian submanifold

$$\Lambda_{\mathrm{cusp}} = \{(t^2, t^3; -3tu, 2u); t \in \mathbb{R}, u > 0\}. \tag{4.8}$$

Now assume that $M = N \times \mathbb{R}$ with $\dim N = 1$, and denote by $(t; \tau)$ the coordinates on $T^*\mathbb{R}$. Assume that Λ is a closed conic smooth Lagrangian submanifold contained in the open set $\{\tau > 0\}$ of T^*M and that its projection on M is compact. By considering the image of Λ in $T^*N \times \mathbb{R}$ by the map $(x, t; \xi, \tau) \mapsto (x, t; \xi/\tau)$, one gets a closed Legendrian smooth submanifold of the contact manifold $T^*N \times \mathbb{R}$. Its image by the projection π is called a Legendrian knot, or a link, in M. Generically, a link is a curve in M with ordinary double points and cusps as its only singularities. A natural and important problem is to find invariants that allow one to distinguish different links. The category Simple($\Lambda, \mathbf{k}$) is such an invariant, and in [39], the authors show that it allows us to distinguish the two so-called Chekanov knots, a result that was not obtained with the traditional methods.

4.3 Conclusion

We have discussed here a few applications of microlocal analysis or microlocal sheaf theory, first to systems of linear partial differential equations, next to symplectic topology.

However, there are other applications, in particular, in representation theory (see [19]), in singularity theory (see [8]), and, from quite recently, in algebraic geometry. Indeed, Alexander Beilinson has constructed the microsupport of constructible sheaves on schemes (see [2]), and this new theory is being developed by several authors (see, in particular, [32]).

References

[1] E. Andronikof, *Interview with Mikio Sato*, Notices AMS, 54 (2007) 208–22
[2] A. Beilinson, *Constructible sheaves are holonomic* Selecta Mathematica 22 (2016) 1797–1819
[3] E. Bierstone and P. D. Milman, *Semi-analytic and subanalytic sets*, Publ. Math. IHES 67 (1988) 5–42
[4] A. Bondal and M. Kapranov, *Representable functors, Serre functors, and mutations*, Izv. Akad. Nauk SSSR 53 (1989) 1183–1205
[5] J. Bros and D. Iagolnitzer, *Causality and local analyticity: Mathematical study*, Ann. Inst. Fourier 18 (1973) 147–84
[6] W. Crawley-Boevey, *Decomposition of pointwise finite-dimensional persistence modules*, J. Algebra Appl. 14 (2014) 1550066
[7] J. M. Curry, *Sheaves, cosheaves and applications* (2013), arXiv: 1303.3255v2
[8] A. Dimca, *Sheaves in topology*, Universitext, Springer (2004)
[9] S. Eilenberg and S. Mac Lane, *General theory of natural equivalences*, Trans. Amer. Math. Soc. 58 (1945) 231–94
[10] O. Gabber, *The integrability of the characteristic variety*, Amer. J. Math. 103 (1981) 445–68
[11] R. Ghrist, *Barcodes: The persistent topology of data*, Bull. Amer. Math. Soc. 45 (2008) 61–75
[12] A. Grothendieck, *Sur quelques points d'algèbre homologique*, Tohoku Math. J. (1957) 119–83
[13] S. Guillermou, *Quantization of conic Lagrangian submanifolds of cotangent bundles* (2012), arXiv:1212.5818
[14] ———, *The Gromov–Eliashberg theorem by microlocal sheaf theory* (2013), arXiv:1311.0187
[15] ———, *The three cusps conjecture* (2016), arXiv:1603.07876
[16] S. Guillermou, M. Kashiwara, and P. Schapira, *Sheaf quantization of Hamiltonian isotopies and applications to non-displaceability problems*, Duke Math. J. 161 (2012) 201245
[17] L. Hörmander, *Fourier integral operators. I.*, Acta Math. 127 (1971)
[18] ———, *The analysis of linear partial differential operators*, Grundlehren der Mathematischen Wissenschaften 256, 257, and 274, 275, Springer (1983–85)

[19] R. Hotta, K. Takeuchi, and T. Tanisaki, *D-modules, perverse sheaves, and representation theory*, Prog. Math 236, Birkhauser (2008)

[20] B. Jubin and P. Schapira, *Sheaves and D-modules on causal manifolds*, Lett. Math. Phys. 16 (2016) 607–48

[21] M. Kashiwara, *Algebraic study of systems of partial differential equations*, Mémoires SMF 63 (1995). Translated from author's thesis in Japanese, Tokyo 1970, by A. D'Agnolo and J.-P. Schneiders.

[22] _____, *D-modules and microlocal calculus*, Trans. of Math. Monogr. 217, AMS (2003)

[23] M. Kashiwara and P. Schapira, *Microsupport des faisceaux: applications aux modules différentiels*, C. R. Acad. Sci. Paris 295 (1982) 487–90

[24] _____, *Microlocal study of sheaves*, Astérisque 128, Mathematical Society of France (1985)

[25] _____, *Sheaves on manifolds*, Grundlehren der Mathematischen Wissenschaften 292, Springer (1990)

[26] _____, *Integral transforms with exponential kernels and Laplace transform*, J. AMS 10 (1997) 939–72

[27] _____, *Categories and sheaves*, Grundlehren der Mathematischen Wissenschaften, 332, Springer (2006)

[28] _____, *Persistent homology and microlocal sheaf theory*, Journal of Applied and Computational Topology, Vol. 2 (2018) 83–113, arXiv:1705.00955

[29] A. Martineau, *Théorèmes sur le prolongement analytique du type "Edge of the Wedge,"* Sem. Bourbaki 340 (1967–68) 340–417

[30] D. Nadler, *Microlocal branes are constructible sheaves*, Selecta Math. 15 (2009) 563–619

[31] D. Nadler and E. Zaslow, *Constructible sheaves and the Fukaya category*, J. Amer. Math. Soc. 22 (2009) 233–86

[32] T. Saito, *The characteristic cycle and the singular support of a constructible sheaf* (2015) arXiv:1510.03018

[33] M. Sato, *Theory of hyperfunctions, I & II*, J. Fac. Sci. Univ. Tokyo 8 (1959–60) 139–93, 387–436

[34] _____, *Regularity of hyperfunctions solutions of partial differential equations*, ICM 2 (1970) 785–94

[35] M. Sato, T. Kawai, and M. Kashiwara, *Microfunctions and pseudodifferential equations*. In Hyperfunctions and pseudo-differential equations (Proc. Conf., Katata, 1971; dedicated to the memory of André Martineau), Lecture Notes in Mathematics 287, Springer (1973) 265–529

[36] P. Schapira, *Mikio Sato, a visionary of mathematics*, Notices of the American Mathematical Society 54 (2007) 113

[37] P. Schapira and J.-P. Schneiders, *Index theorem for elliptic pairs*, Astérisque 224, Mathematical Society of France (1994)

[38] M. Artin, A. Grothendieck, and J.-L. Verdier, *Théorie des topos et cohomologie étale des schémas, Sém. Géom. Alg. (1963–64)*, Lecture Notes in Mathematics 269, 270, 305, Springer (1972–73)

[39] V. Shende, D. Treumann, and E. Zaslow, *Legendrian knots and constructible sheaves* (2014) arXiv:1402.0490
[40] J. Sjöstrand, *Singularités analytiques microlocales*, Astérisque 95, Mathematical Society of France (1982)
[41] D. Tamarkin, *Algebraic and analytic microlocal analysis, Microlocal conditions for non-displaceability*, Vol. 269, Springer, proceedings in Mathematics & Statistics (2012&2013) pp. 99–223, available at arXiv:0809.1584

Pierre Schapira
Sorbonne Universités, UPMC Univ. Paris 6,
Institut de Mathématiques de Jussieu, France
pierre.schapira@imj-prg.fr

PART II

Topology and Algebraic Topology

4
Topo-logie

Mathieu Anel and André Joyal*

En hommage aux auteurs de SGA 4

Contents

1	A Walk in the Garden of Topology	155
2	The Locale–Frame Duality	170
3	The Topos–Logos Duality	183
4	Higher Topos–Logos Duality	238
References		255

1 A Walk in the Garden of Topology

This text is an introduction to topos theory. Our purpose is to sketch some of the intuitive ideas underlying the theory, not to give a systematic exposition of it. It may serve as a complement to the formal expositions that can be found in the literature. We are using examples to illustrate many ideas.

We have tried to make the text both accessible to a reader unfamiliar with the theory and interesting for more familiar readers. Certain points of view presented here are nonstandard, even among experts, and we believe they should be more widely known.

The introduction explains how to compare topoi with more classical notions of spaces. It is aimed to be a summary of the rest of the text, where the same ideas will be detailed. In accordance with the theme of this book, we have limited this text to present topoi as a kind of spatial object. Unfortunately, the

---

* The first author gratefully acknowledges the support of the Air Force Office of Scientific Research through MURI grant FA9550-15-1-0053.

important relation of topoi with logic will not be dealt with as it should here. We have only made a few remarks here and there. Doing more would have required a much longer text.

1.1 Topoi as Spaces

1.1.1 From Sheaves to Topoi

The notion of topos was invented by Grothendieck's school of algebraic geometry in the 1960s. The motivation was Grothendieck's program for solving the Weil conjectures. An important step was the constructions of étale cohomology and l-adic cohomology for schemes. The methods to construct these relied heavily on sheaf theory as previously developed by Cartan and Serre after Leray's original work. A central notion was that of *étale sheaf*, a new notion of sheaf in two aspects:

1. an étale sheaf was defined as a contravariant functor on a *category* rather than on the partially ordered set of open subsets of a topological space;
2. the sheaf condition was formulated in term of *covering families* that could be chosen quite arbitrarily.

A *site* was defined to be a category equipped with a notion of covering families. Grothendieck and his collaborators eventually realized that the most important properties of a site depended only on the structure of the associated category of sheaves, for which sites were merely presentations by generators and relations [6, Section IV.0.1]. This structure was baptized *topos*, and an axiomatization was obtained by Jean Giraud. The name was chosen because a number of classical topological constructions (pasting, localizing, coverings, étale maps, bundles, fundamental groups, etc.) could be generalized from categories of sheaves on topological spaces to these abstract categories of sheaves. As a result, new objects, such as the category Set^G of actions of a group G or presheaves categories $\mathcal{P}r(C) = [C^{op}, \mathsf{Set}]$, could be thought as spatial objects. In the introduction of the chapter on topoi of [6], the authors wrote clearly their ambition for these new types of spaces:

> Exactly as the term topos itself suggests, it seems reasonable and legitimate to the authors of the current Seminar to consider that the object of Topology is the study of topoi (and not merely topological spaces).

It is the purpose of this text to explain how topoi can be thought as spaces. The following differences with topological spaces will be our starting point:

- The points of a topos are the objects of a *category* rather than the elements of a mere set. In particular, a central object of the theory is the topos $\mathbb{A}$ whose category of points $\mathcal{P}t\,(\mathbb{A})$ is the category Set of sets.
- A topos $\mathcal{X}$ is not defined by means of a "topology" structure on its category of points $\mathcal{P}t\,(\mathcal{X})$. Rather, it is defined by its *category of sheaves* $\mathcal{S}h\,(\mathcal{X})$, which are the continuous functions on $\mathcal{X}$ with values in $\mathbb{A}$.

1.1.2 A Category of Points

Recall that the set of points of a topological space can be enhanced into a preorder by the specialization relation.[1] The morphisms in the category of points of a topos must also be thought as specializations. Topological spaces with a nontrivial specialization order are nonseparated. Somehow, a topos with a nontrivial category of points corresponds to an even more extreme case of nonseparation since points can have several ways to be specializations of each other, or even be their own specializations!

We already mentioned that the theory contains a topos $\mathbb{A}$ whose category of points is the category Set of sets. Another example of a topos with a nontrivial category of points is given by the topos $\mathbb{B}G$ such that $\mathcal{S}h\,(\mathbb{B}G) = \text{Set}^G$ is the category of actions of a discrete group G. The category of points of this topos is the group G viewed as a groupoid with one object. A necessary condition for a topos to be a topological space is that its category of points be a poset. Both $\mathbb{A}$ and $\mathbb{B}G$ are then examples of topoi that are not topological spaces.

Having a category of points will allow the existence of topoi whose points can be the category of groups, or the category of rings, or of local rings, or many other algebraic structures. Topoi can be used to represent certain moduli spaces and this is an important source of topoi not corresponding to topological spaces. This relation to classifying spaces is also an important part of the relation with logic.

Let Topos be the category of topoi. Behind the fact of having a category of points is the more general fact that the collection of morphisms $\text{Hom}_{\text{Topos}}(\mathcal{Y}, \mathcal{X})$ between two topoi naturally forms a category. For example, when $\mathcal{Y} = \mathbb{1}$ is the terminal topos, we get back $\mathcal{P}t\,(\mathcal{X}) = \text{Hom}_{\text{Topos}}(\mathbb{1}, \mathcal{X})$, and when $\mathcal{X} = \mathbb{B}G$, the category $\text{Hom}_{\text{Topos}}(\mathcal{Y}, \mathbb{B}G)$ can be proven to be the groupoid of G-torsors over $\mathcal{Y}$. So categories of points go along with the fact that Topos is a 2-category.

---

[1] For two points x and y of a space X, x is a specialization of y if any open subset containing x contains y or, equivalently, if $\overline{x} \subset \overline{y}$, where $\overline{x}$ is the closure of $\{x\}$. This relation is a preorder $x \leq y$. A space X is called T_0 if this preorder is an order and T_1 if this preorder coincides with equality of points. Any Hausdorff space is T_1.

Table 4.1. *Types of spaces and their points*

Type of space	Top. space	Locale	Topos	∞-Topos
Points	a set	a preorder	a category	an (∞, 1)-category

The evolution of the collection of points from a set to a poset, and then to a category, and even to an ∞-category in the case of ∞-topoi, is part of a hierarchy of spatial notions (summarized in Table 4.1) that we are going to present.

1.1.3 Locales and Frames

In opposition to topological spaces, the points of topoi have in fact a secondary role. Topological spaces are defined by the structure of a *topology* on their set of points, but topoi are not defined in such a way.[2] In fact, we shall see that topos theory allows the existence of nonempty topoi with an empty category of points.

In order to understand the continuity of definition between topological spaces and topoi, we will require the slight change of perspective on what is a topological space given by the theory of locales. This theory is based on the fact that most features of topological spaces depend not so much on their set of points but only on their poset of open subsets (which we shall call *open domains* to remove the reference to the set of points). The open domains of a topological space X form a poset $\mathcal{O}(X)$ with arbitrary unions, finite intersections, and a distributivity relation between them. Such an algebraic structure is called a *frame*. A continuous map $f: X \to Y$ induces a morphism of frames $f^*: \mathcal{O}(Y) \to \mathcal{O}(X)$, that is, a map preserving order, unions, and finite intersections. The opposite of the category of frames is called the category of *locales*. The functor sending a topological space X to its frame $\mathcal{O}(X)$ produces a functor Top → Locale. We shall see in Section 2.2.13 how this functor corresponds in a precise way to forget the underlying set of points of the topological space. The theory of locales is sometimes called *point-free topology* for this reason.

The structure of a frame is akin to that of a commutative ring: the union plays the role of addition, the intersection that of multiplication, and there is a distributivity relation between the two. The definition of a locale as an object

---

[2] Topos theory has the notion of a Grothendieck topology on a category. It is unfortunate that the name suggests the notion of a topology on a set, but this is actually something of a completely different nature.

of the opposite category of frames is akin to the definition of an affine scheme as an object of the opposite category of commutative rings. The analogy goes even further since the frame $\mathcal{O}(X)$ can be realized as the set of continuous functions from X to the *Sierpiński space* $\mathbb{S}$.[3] This space plays a role analogous to that of the affine line $\mathbb{A}^1$ in algebraic geometry: $\mathbb{A}^1$ is dual to the free ring $\mathbb{Z}[x]$ on one generator and similarly $\mathbb{S}$ is dual to the free frame $\underline{2}[x]$ on one generator. This analogy shows that replacing topological spaces by locales is a way to define spaces as dual to some "algebras" of continuous functions.

1.1.4 Topoi and Logoi

Although this is not its classical presentation, we believe that topos theory is best understood similarly from a dual algebraic point of view. We shall use the term *logos* for the algebraic dual of a topos.[4] A logos is a category with (small) colimits and finite limits satisfying some compatibility relations akin to distributivity (see Section 3.3 for a detailed account on this idea). A morphism of logoi is a functor preserving colimits and finite limits. The category of topoi is defined to be the opposite of that of logoi (see Section 3.1 for a precise definition).

Table 4.2 presents the analogy of structure between the notions of logos, frame and commutative ring. The general idea of a duality between geometry and algebra goes back to Descartes in his *Geometry*, where geometric objects are constructed by algebraic operations. The locale–frame and topos–logos dualities are instances of many dualities of this kind, as shown in Table 4.3.[5]

1.1.5 Functions with Values in Sets

The analogue in the theory of topoi of the Sierpiński space $\mathbb{S}$, and of the affine line $\mathbb{A}^1$, is the *topos of sets* $\mathbb{A}$ (also known as the *object classifier*). The

---

[3] The Sierpiński space $\mathbb{S}$ is the topology on $\{0,1\}$, where $\{0\}$ is closed and $\{1\}$ is open. A continuous map $X \to \mathbb{S}$ is an open–closed partition of X. The correspondance $C^0(X,\mathbb{S}) = \mathcal{O}(X)$ associates to an open domain its characteristic function.

[4] The formal dual of a topos has been introduced by several authors. S. Vickers called the notion *geometric universes* in [34], and M. Bunge and J. Funk call them *topos frames* in [8]. Our choice of terminology is motivated by the play on the word *topo-logy*. It also resonates well with topos and with the idea that a logos is a kind of logical doctrine.
In practice, the manipulation of topoi forces one to jump between the categories Topos (where the morphisms are called *geometric morphisms*) and Topos^{op} (where the morphisms are called *inverse images* of geometric morphisms). It is a source of confusion that the same name of "topos" is used to refer to a spatial object and to the category of sheaves on this space. Rather than distinguishing the categories by different names for their morphisms, we have preferred to give different names for the objects.

[5] The structural analogy between topos/logos theory and affine schemes/commutative rings has been a folkloric knowledge among experts for a long time. However, this point of view is conspicuously absent from the main references of the theory. When it is mentioned in the literature, it is only as a small remark.

Table 4.2. *Ring-like structures*

Algebraic structure	Commutative ring	Frame	Logos
Addition	$(+, 0)$	$(\bigvee, 0)$	(colimits, initial object)
Product	$(\times, 1)$	$(\wedge, 1)$	(finite limits, terminal object)
Distrib.	$a(b+c) = ab + bc$	$a \wedge \bigvee b_i = \bigvee a \wedge b_i$	universality and effectivity of colimits
Initial algebra	$\mathbb{Z}$	$\underline{2}$	Set
Free algebra on one generator	$\mathbb{Z}[x] = \mathbb{Z}^{(\mathbb{N})}$	free frame $\underline{2}[x] = [\underline{2}, \underline{2}]$	free logos $\mathrm{Set}[X] = [\mathrm{Fin}, \mathrm{Set}]$
Corresponding geom. object	the affine line $\mathbb{A}^1$	the Sierpiński space $\mathbb{S}$	the topos classifying sets $\mathbb{A}$
General geom. objects	affine schemes	locales	topos

Table 4.3. *Some dualities*

Geometry	Algebra	Dualizing object (gauge **A**)
Stone spaces	boolean algebras	the boolean values $\underline{2} = \{0, 1\}$
compact spaces	commutative $\mathbb{C}^\star$-algebras	the complex numbers $\mathbb{C}$
affine schemes	commutative rings	the affine line $\mathbb{A}^1$
locales	frames	the Sierpiński space $\mathbb{S}$
topoi	logoi	the topos $\mathbb{A}$ of sets
∞-topoi	∞-logoi	the ∞-topos $\mathbb{A}_\infty$ of ∞-groupoids

corresponding logos is the functor category $\mathrm{Set}[X] := [\mathrm{Fin}, \mathrm{Set}]$, where Fin is the category of finite sets. We said that the category of points of $\mathbb{A}$ is the category of small sets. It is an object difficult to imagine geometrically, but, algebraically, it corresponds simply to the *free logos* on one generator, and we shall see in Table 4.10 that it has many similarities with the ring of polynomials in one variable $\mathbb{Z}[x]$.

The functions on a topos with values in $\mathbb{A}$ correspond to *sheaves of sets*. The notion of sheaf on a topological space depends only on the frame of open domains and can be generalized to any locale. The category of sheaves of sets $\mathrm{Sh}(X)$ on a locale X is a logos. This provides a functor Locale $\to$ Topos. This functor is fully faithful, and the topoi in its image are called *localic*. It can be proven that $\mathrm{Sh}(X)$ is equivalent to the category of morphisms of topoi $X \to \mathbb{A}$. Intuitively, the function corresponding to a sheaf F sends a point of X to the stalk of F at this point.[6] More generally, we shall see in (3.1) that the logos $\mathrm{Sh}(\mathcal{X})$ dual to a topos $\mathcal{X}$ can always be reconstructed as $\mathrm{Sh}(\mathcal{X}) = \mathrm{Hom}_{\mathrm{Topos}}(\mathcal{X}, \mathbb{A})$. The morphism $\chi_F \colon \mathcal{X} \to \mathbb{A}$ corresponding to a sheaf F in $\mathrm{Sh}(\mathcal{X})$ is called its *characteristic function*.

Finally, in the same way that locales are spatial objects defined by means of their frame of functions into the Sierpiński space, topoi can be described as those spatial objects that can be defined by means of their logos of functions into the topos of sets.

1.1.6 Etale Domains

Sheaves of sets have a nice geometric interpretation as *étale domains* (or local homeomorphisms). Given a topos $\mathcal{X}$ and an object F in the corresponding logos $\mathrm{Sh}(\mathcal{X})$, the slice category $\mathrm{Sh}(\mathcal{X})_{/F}$ is a logos and the pullback along $F \to 1$ defines a logos morphism $f^* \colon \mathrm{Sh}(\mathcal{X}) \to \mathrm{Sh}(\mathcal{X})_{/F}$. The corresponding morphism of topoi $\mathcal{X}_F \to \mathcal{X}$ is called *étale*. An étale domain of $\mathcal{X}$ is an étale morphism with codomain $\mathcal{X}$. We shall see in Section 3.2.6 that any morphism of topoi $F \colon \mathcal{X} \to \mathbb{A}$ corresponds uniquely to a morphism of topoi $\mathcal{X}_F \to \mathcal{X}$ (where $\mathrm{Sh}(\mathcal{X}_F) = \mathrm{Sh}(\mathcal{X})_{/F}$). This construction generalizes the construction of the *espace étalé* of a sheaf by Godement [13, Section II.1.2].

The Sierpiński space $\mathbb{S}$, when viewed as a topos, can be proven to be a subtopos of $\mathbb{A}$. At the level of points, the embedding $\mathbb{S} \hookrightarrow \mathbb{A}$ corresponds to the embedding of $\{\emptyset, \{\star\}\} \hookrightarrow$ Set. A particular kind of étale domain of a topos $\mathcal{X}$ are then the *open domains*: they are the one whose characteristic function takes values in $\mathbb{S}$. Intuitively, they are the sheaves whose stalks are either empty or a singleton. Table 4.4 summarizes the situation.

1.1.7 To Have or Have Not Enough Functions

Behind the idea to capture the structure of a space X by some algebra of functions into some fixed space $\mathbf{A}$, there is the idea that $\mathbf{A}$ is a kind of basic block from which X can be built. We shall say that a space X has *enough*

---

[6] This result a way to formalize the intuitive idea that a sheaf of sets on a space should be a continuous family of sets (the family of its stalks).

Table 4.4. *Sheaves on a topos*

Geometric interpretation	Algebraic interpretation
Etale domains $\mathcal{X}_F \to \mathcal{X}$	functions $\mathcal{X} \to \mathbb{A}$ to the topos of sets
Open domains $\mathcal{X}_U \to \mathcal{X}$	functions $\mathcal{X} \to \mathbb{S}$ to the Sierpiński subtopos $\mathbb{S} \subset \mathbb{A}$

functions into **A** if X can be written as a subspace $X \hookrightarrow \mathbf{A}^N$ of some power of **A**.[7]

This notion makes sense in a variety of contexts. For example, a locale X has always enough maps into the Sierpiński space $\mathbb{S}$: the canonical evaluation map $ev \colon X \times C^0(X, \mathbb{S}) \to \mathbb{S}$ defines a morphism of locales $X \to \mathbb{S}^{C^0(X,\mathbb{S})}$ that can be proven to be an embedding. Not every space (or locale) has enough maps into $\mathbb{R}$, but topological manifolds do and can be written as subspaces in some $\mathbb{R}^N$.[8] In the setting of algebraic geometry, affine schemes are precisely defined as the subobjects of affine spaces $\mathbb{A}^N$, that is, they are defined so that they have enough functions with values in $\mathbb{A}^1$. The fact that not every scheme is affine (like projective spaces) says that not all schemes have enough functions with values in $\mathbb{A}^1$. Finally, topoi can be proven to have enough maps in the topos $\mathbb{A}$.[9] However, not every topos has enough maps to the Sierpiński topos $\mathbb{S}$; only the localic topoi do.

This idea of having enough functions to some "gauge space" **A** is fundamental for all the dualities of Table 4.3. One of the main ideas behind the definition of topoi is that the Sierpiński gauge is not always enough: some spatial objects (such as the topoi $\mathbb{A}$ or $\mathbb{B}G$, or bad quotients such as $\mathbb{R}/\mathbb{Q}$) do not have enough open domains to be faithfully reconstructed from them. One needs to choose a larger gauge than $\mathbb{S}$ to capture those spaces. Topoi can – and must – be understood as those spatial objects that can be reconstructed from the gauge given by $\mathbb{A}$, that is, spaces with enough étale domains.[10]

Such a perspective on topoi raises the question of the existence of types of spaces beyond topoi, spaces that would not have enough étale domains. The answer is positive, and it is one of the motivations for the introduction of ∞-topoi and stacks (see Section 4 and Chapters 8 and 9). For now, let us only say that ∞-topoi and ∞-logoi are higher categorical analogues of topoi and

---

[7] The proper definition is that X can be written as the limit of some diagram of maps between copies of **A**, but the approximate definition will suffice for our purpose here.

[8] Since $\mathbb{R}$ is separated, nonseparated spaces (like the Sierpiński space) cannot embed faithfully in some $\mathbb{R}^N$. The locales with enough maps to $\mathbb{R}$ are the completely regular ones [17, Chapter IV].

[9] This is somehow the meaning of the statement that any topos is a subtopos of a presheaf topos. For a more precise statement, see the examples in Section 3.2.3.

[10] For more remarks on this line of ideas, see [1].

Table 4.5. *Types of spaces – 1*

Given a space X, maps $Y \to X$ which are	Are continuous functions on X with values in	They are also called	They form	Which is called a	A space with enough of them is called
Open immersions	the Sierpiński space $\mathbb{S}$	open domains	a poset	frame	a locale
Etale (local homeomorphisms)	the space $\mathbb{A}$ of sets	etale domains or sheaves	a 1-category	logos	a topos
∞-Etale	the space $\mathbb{A}_\infty$ of ∞-groupoids	∞-sheaves or stacks	an ∞-category	∞-logos	an ∞-topos

logoi where the role of the 1-category of sets is played by the ∞-category of ∞-groupoids. Table 4.5 summarizes different kinds of spaces.

1.1.8 To Have or Have Not Enough Points

The theory of locales is famous for providing nonempty locales that have an empty poset of points (we shall give examples in Section 2.6). A fortiori, there will exist nonempty topoi without any points.

The classical intuition of topological spaces, rooted in the ambient physical space, does not make it easy to imagine nonseparated spaces. But even more difficult is to imagine nonempty topoi or locales without any points. This seems to contradict all the common sense of topology. However, this phenomenon becomes understandable if we compare it with the more common fact of the existence of polynomials with no rational roots. We shall say more about this in Section 2.2.9.

A locale is said to have *enough points* if two open domains can be distinguished by the points they contain. A locale with enough points can be proven to be the same thing as a sober topological space. Similarly, a topos is said to have enough points if two sheaves can be distinguished by the family of their stalks (see Section 3.10). Intuitively, a topos $\mathcal{X}$ (or a locale) with enough points can be equipped with a surjection $\coprod_E 1 \twoheadrightarrow \mathcal{X}$ from a union of points.[11] In practice, most topoi have enough points. This is the case of $\mathbb{A}$, of $\mathbb{B}G$, of

---

[11] The two problems of having enough points $1 \to X$ or enough functions $X \to \mathbf{A}$ are somehow dual. In both cases, the question is how much of X can be "reconstructed" from some "gauge" given by mapping *from* a given objet (the point) or *to* a given object (the space of coordinates). An object has enough points if it admits a surjection from a union of points. An object has enough functions if it admits an embedding into a product of $\mathbf{A}$.

Table 4.6. *Types of spaces – 2*

Space with	Enough open domains	Enough etale domains	Enough higher etale domains	Maybe not enough higher etale domains
Enough points	topological space	topos with enough points	∞-topos with enough points	beyond ...
Maybe not enough points	locale	topos	∞-topos	

bad quotients such as $\mathbb{R}/\mathbb{Q}$, of presheaves topoi, of Zariski or étale spectra of rings, and of topoi classifying models of algebraic theories. Moreover, since any topos can always be embedded in a presheaf topos, any topos is always a subtopos of a topos with enough points.

1.1.9 Are Topoi Really Spaces?

Our excursion in the topological side of topoi has led us to distinguish different kinds of spatial objects summarized in Table 4.6. The discovery that topology is richer than the simple study of topological spaces is extraordinary. But after all these considerations, it is difficult not to question what a space is. Since we have removed points and open domains – the two fundamental features on which the notion of topological space is classically based – as defining characteristics of spaces, what is left of the intuition of what a space should be? And why should we agree to consider these news objects as spaces?

The best answer that we can propose – and that we will develop in the rest of this text – is that the intuition of space is in fact forged in a set of specific operations on spaces (e.g., covering, pasting, quotienting, localizing, intersecting, crossing, deforming, direct image, inverse image, homotopy, (co)homology), which leads to distinguishing some classes of spaces (compact, connected, contractible, etc.) and some classes of maps (open immersions, étale maps, submersions, proper maps, bundles, etc.). So far, all of these notions and the structural relations they have between them have been successfully generalized to topoi. Some of them, like quotienting or cohomology, have even gained more regular properties in the context of topoi. So, if all the tools, language, and structural relations of topology make sense for topoi, should not the question rather be, how can we afford not to think them as spaces?

1.2 Other Views

1.2.1 Topoi as Categories of Spaces

We have sketched how a logos can be thought dually as a single spatial object. But there exists also the point of view where a logos is thought as a *category* of spatial objects.[12] This point of view is justified by the following example. The category M of manifolds does not have certain quotients (e.g., leaf spaces of foliations are not manifolds in general). So it could be useful to embed M into a larger category where quotients could behave better. This is, for example, the idea behind the notion of *diffeology* (see Chapter 1). Another implementation is to consider the embedding $M \hookrightarrow \mathrm{Sh}(M)$ into sheaves of sets on M.[13] The embedding $M \hookrightarrow \mathrm{Sh}(M)$ suggests interpreting the objects of $\mathrm{Sh}(M)$ as some kind of generalized manifold. This is the so-called *functor of points* approach to geometry (Chapter 7). Within $\mathrm{Sh}(M)$, "bad quotients" such as the irrational torus $\mathbb{T}_\alpha^2 = \mathbb{T}^2/\mathbb{R}$ or even the more bizarre $\mathbb{R}/\mathbb{R}_{dis}$[14] do exist with nice properties. For example, it is possible to define a theory of fundamental groups for these objects and prove that $\pi_1(\mathbb{T}_\alpha^2) = \mathbb{Z}^2$ and $\pi_1(\mathbb{R}/\mathbb{R}_{dis}) = \mathbb{R}_{dis}$.

Other logoi exist in which to embed the category of manifolds M. Synthetic differential geometry uses sheaves on C^∞-rings (see Chapter 2 and [27]). Schreiber's approach to geometrization of gauge theories in physics relies on the same idea, but with sheaves of ∞-groupoids [5, Chapter 5]. The same idea has also been used in algebraic geometry (where it was actually invented), where the embedding {Affine schemes} $\hookrightarrow$ Sh({Affine schemes}, étale) provides a nice setting in which to define several kinds of gluing of affine schemes (general schemes, algebraic spaces). This setting has been useful for dealing with algebraic groups and constructing moduli spaces, such as Hilbert schemes. When sheaves of sets are replaced by sheaves of ∞-groupoids, the embedding {Affine schemes} $\hookrightarrow$ Sh_∞({Affine schemes}, étale) provides a nice setting in which to define Deligne–Mumford and Artin stacks. A variation on this setting involving ∞-logoi is also at the foundation of derived geometry (see Chapter 9).

---

[12] A logos $\mathrm{Sh}(X)$ can always be thought as a category of spaces étale over $\mathcal{X}$, but the interpretation we are talking about here is different.

[13] These two examples are actually related. The category Diff of diffeologies can be realized as a full subcategory of $\mathrm{Sh}(M)$, and the embedding $M \hookrightarrow \mathrm{Sh}(M)$ factors through Diff.

[14] The object $\mathbb{R}/\mathbb{R}_{dis}$ is the quotient of $\mathbb{R}$ by the discrete action of $\mathbb{R}$. Classically, it is a single point, but in $\mathrm{Sh}(M)$, a function from a manifold X to $\mathbb{R}/\mathbb{R}_{dis}$ is an equivalent class of families (U_i, f_i), where U_i is an open cover of X and $f_i : U_i \to \mathbb{R}$ are functions such that the differences $f_i - f_j$ are constant functions on U_{ij}. In more intrinsic terms, a morphism $X \to \mathbb{R}/\mathbb{R}_{dis}$ is the same thing as a closed differential 1-form on X, i.e., it represents the functor $X \mapsto Z^1_{dR}(X, \mathbb{R})$. In the embeddings $M \hookrightarrow \mathrm{Diff} \hookrightarrow \mathrm{Sh}(M)$, the object $\mathbb{R}/\mathbb{R}_{dis}$ is actually an example of a sheaf that is not a diffeology.

1.2.2 Topoi and Logic

The theory of topoi has a logical aspect, discovered by Lawvere and Tierney in the late 1960s, which has been developed into one of its most spectacular and fundamental features. A sheaf is intuitively a family of sets (the family of its stalks). Therefore, it should be clear enough that all the operations and language existing in the category of sets can be transported to sheaves with the idea that they are applied stalk-wise. This is the intuition behind the idea that a logos can be thought as a category of generalized sets.[15] From there, if **T** is a logical theory, the notion of model of **T** in sets can be extended into that of a model in the generalized sets/objects of a logos. This construction follows the spirit of the interpretation of propositional theories in frames of open domains of topological spaces (in fact, the latter can even be viewed as a particular case of the former). Logoi have provided a rich setting in which to interpret many features of logic; Table 4.7 gives a rough summary of some. The theory has notably led to independence proofs in set theory [25, Section VI.2].

If all the constructions of set theory make sense in a logos, the fact that a sheaf is a *continuous* family of sets leads to some differences of behavior. Such differences are already present in the frame semantics of propositional logic, where the logic ceases to be boolean and instead become intuitionist in the sense of Heyting. The logos semantics of logical theories is a fortiori intuitionistic, but there are new features. For example, the fact that not all covering maps have a section says that the axiom of choice can be false.

The logical use of logoi has also modified the notion a bit. The preference of logic for finite operations has led to replace SGA original definition by the so-called *elementary* definition of Lawvere and Tierney. The consideration of the internal homs and the subobject classifiers as being part of the structure of a logos has also led to considering notions of morphisms between logoi different than the original ones (morphisms of locally cartesian closed categories, logical morphisms). From this point of view, the logical notion of topos is not, strictly speaking, the same as the topological one.

Our priority in this chapter is to explain how topoi are spatial objects, and we will unfortunately not say much about the relationship with logic. We have only made a few remarks here and there about classifying topoi for some logical theories. We refer the reader to [19, 25] for a good treatment of classifying topoi and the intimate relationship between logoi and logic.

---

[15] The relation of this point of view with the previous one, where a logos is thought as a category of spatial objects, is the matter of Lawvere cohesion theory, central to Schreiber's geometrization of physics [5, Chapter 5].

Table 4.7. *Translation logic–logos*

Logic	Logos $\mathcal{E}$
Terms and types	*Objects and morphisms*
types/sorts $\mathbb{S}$	objects $[S]$
variable $s : S$	identity maps $[s] = [S] \xrightarrow{id} [S]$
context $s : S, t : T$	products $[S] \times [T]$
empty context	terminal object $[] = 1$
terms $f(s)$ of type T	maps $[f] : [S] \to [T]$
dependent types $T(s)$	object $[T] \to [S]$ in $\mathcal{E}_{/[S]}$
predicates (dependent booleans) $P(s)$	monomorphisms $[P] \rightarrowtail [S]$
propositions (booleans) p	subterminal object $[p] \rightarrowtail []$
Disjunctive operations	*Colimit constructions*
disjunction $P(s) \vee Q(s)$	union $[P] \cup [Q] \rightarrowtail [S]$
existential quantifier $\exists s\ f(s)$	image of a map $\mathrm{im}([f]) : \mathrm{Im}([f]) \to [S]$
dependent sums $\sum_{s:S} T(s)$	domain $[T]$ of the map $[T] \to [S]$ interpreting the dependent type $T(s)$
Conjunctive operations	*Limit constructions*
conjunction $P(s) \wedge Q(s)$	intersection $[P] \cap [Q] \rightarrowtail [S]$
implication $P(s) \Rightarrow Q(s)$	Heyting's right adjoint to $[P] \cap -$
universal quantifier $\forall s\ f(s)$	image by the right adjoint to base change of subobjects along $[S] \to []$
function type $S \to T$	internal hom $[T]^{[S]}$
dependent products $\prod_{x:S} T(x)$	image by the right adjoint to base change along $[S] \to []$
Specific types	*Specific objects*
the type of propositions	subobject classifier Ω
modalities on propositions	internal monads $j : \Omega \to \Omega$
the type of types	the object classifier/universe U (only in ∞-logoi)
modalities on types	internal monads $j : U \to U$ (only in ∞-logoi)

1.2.3 Higher Topoi

In the 1970s and 1980s, the construction of moduli spaces led geometers to enhance sheaves of sets into stacks, that is, sheaves valued in groupoids, which were objects of higher categories. Around the same time, it was gradually understood that the objects of algebraic topology (homotopy types, spectra, chain complexes, cobordisms, etc.) were also naturally objects of higher categories. Two types of higher categories have emerged from these considerations: ∞-*topoi* and *stable ∞-categories*. The first provides a setting for stacks, that is, sheaves in ∞-groupoids; the second provides a setting for stable homotopy theories, that is, sheaves of spectra.[16]

The theory of ∞-logoi is essentially similar to that of logoi, but with the replacement of the category Set of sets by the ∞-category $\mathcal{S}$ of ∞-groupoids, that is, homotopy types.[17] The category of points of an ∞-topos is an $(\infty, 1)$-category. This allows ∞-topoi to capture more spatial objects than topoi. For example, the analogue of the topos of sets $\mathbb{A}$ is the ∞-topos $\mathbb{A}_\infty$ whose points are ∞-groupoids. As for topoi, an ∞-topos $\mathcal{X}$ is defined dually by its ∞-*logos* $\mathrm{Sh}_\infty(\mathcal{X})$ of functions with values in $\mathbb{A}_\infty$ (see Section 4). Table 4.8 gives a few correspondences between notions of category and ∞-category theories.

Topos theory is actually having a tremendous renewal with the development of ∞-topos theory. In fact, we believe that, more than a simple higher categorical analogue, the notion of ∞-topos is actually an achievement of the notion of topos. Indeed, the theory of ∞-topoi/logoi turns out to be somehow simpler and more powerful than topos theory:

– it simplifies the descent properties of logoi (see Section 4.2.1);
– it simplifies the treatment of both homotopy theory and homology theory of logoi (see Sections 4.2.7 and 4.2.8);
– from a logical point of view, ∞-logoi provide a setting where quantification on arbitrary objects is allowed[18] (see Section 4.2.6).

But also, it contains a number of features absent from the classical theory. A central one is the notion of ∞-*connected objects* (see Section 4.2.4). To explain this, recall that, according to Whitehead's theorem, a homotopy type is contractible if and only if its homotopy groups are trivial. Roughly speaking, an object of an ∞-topos is ∞-connected if all its homotopy groups are trivial,

---

[16] A third kind of ∞-category has also emerged, ∞-categories with duals, which provides the proper setting for cobordism theories and extended field theories [7, 23]. We shall not talk about these categories.

[17] Some motivations for the enhancement Set $\hookrightarrow \mathcal{S}$ are explained in Chapter 9. See also Chapter 5 for some material on ∞-groupoids.

[18] Logoi only provide a setting in which to quantify on subobjects, a restriction that is arguably not natural.

Table 4.8. *Correspondence lower/higher category theories*

1-*Categories*	$(\infty, 1)$-*Categories*
Sets	∞-groupoids (homotopy types)
Property of equality $a = b$	*Structure* of the choice of an isomorphism (a homotopy) $\alpha : a \simeq b$
Presheaves of sets $\Pr(C) = [C^{op}, \text{Set}]$	Presheaves of ∞-groupoids $\Pr_\infty(C) = [C^{op}, \mathcal{S}]$
Logos = left exact localizations of $\Pr(C)$	∞-logos = left exact localizations of $\Pr_\infty(C)$
Topos of sets $\mathbb{A}$ dual to the free logos $\text{Set}[X] = [\text{Fin}, \text{Set}]$	∞-topos of ∞-groupoids $\mathbb{A}_\infty$ dual to the free ∞-logos $\mathcal{S}[X] = [\mathcal{S}_{\text{fin}}, \mathcal{S}]$
abelian groups	Spectra (reduced homology theories) or chain complexes
abelian categories	Stable ∞-categories

but such an object need not be a terminal object.[19] Their existence has several important consequences:

- they limit the power of Grothendieck topologies (not every ∞-logos can be defined from a site; see Section 4.2.5);
- they create unexpected links between unstable and stable homotopy theories (see Section 4.2.3);
- they give rise to a differential calculus for ∞-logoi related to Goodwillie theory.[20]

None of these properties have analogues, nor can any be seen in classical topos theory.

It is a good idea to compare the enhancement of Set into $\mathcal{S}$ to that of $\mathbb{R}$ into $\mathbb{C}$. This comparison illustrates both the simplification that is provided by ∞-groupoids (better regularity for some properties) and the new features that can appear (new objects, new methods, etc.), together with the price to pay to leave behind an ancient world of problems and points of view. As complex numbers, so do ∞-groupoids and ∞-logoi offer a new world, both in algebra and geometry. On the geometry side, the new features of ∞-topos theory push

---

[19] It is useful to compare them to nilpotent elements in a ring.
[20] This is an ongoing work of the authors and their collaborators [2, 3].

the notion of a spatial object further than anyone had anticipated (the situation compares to the enhancement of varieties into schemes with their singularities and nilpotent functions). On the algebra side, the interpretation of Goodwillie calculus in ∞-logoi provides a new "topological calculus" where spectra play the role of infinitesimal thickening of the point. These elements of the theory, which are ongoing work of the authors and others, are unfortunately too recent to be part of this report. We mention them only to give a glance at the future of the notion of space.

1.2.4 Further Reading

About locale theory, good books are [17, 28]. The article [21] contains nice elements of the theory not in the previous book. About topos theory, two very good books are [18, 25]. For the more experienced user, the two volumes of [19] are unavoidable. About ∞-topos theory, the note [29] contains essential ideas. The main references are [22] and also the appendix of [24]. For an approach closer to what we did here, some material is in [5]. About ∞-category theory, some ideas are explained in Chapters 5, 6, 8 and 9 of this volume; otherwise, we refer to the books [9, 20, 22].

2 The Locale–Frame Duality

The purpose of this section is to explain how topology, in parallel of being a theory of geometric objects, can also be understood as the study of some algebraic objects. To each space X is associated its frame $\mathcal{O}(X)$ of open domains, which is the same thing as $C^0(X, \mathbb{S})$, the set of continuous functions with values in the Sierpiński space. The frame $\mathcal{O}(X)$ is a ring-like object, and many of the geometric constructions about topological spaces can be formulated algebraically in terms of $\mathcal{O}(X)$. This easy model of an algebraic approach to geometry is a useful step in understanding the definition of a topos.

2.1 From Topological Spaces to Frames

The Sierpiński space $\mathbb{S}$ is defined as the topology on the set $\{0, 1\}$ such that 0 is a closed point and 1 an open point. The space $\mathbb{S}$ has an order on its points such that $0 < 1$. This makes it into a poset object in the category Top. If I is a set, then the map $\bigvee \colon \mathbb{S}^I \to \mathbb{S}$ sending a family to its supremum is continuous for the product topology. Moreover, when I is finite, the map $\wedge \colon \mathbb{S}^I \to \mathbb{S}$ sending a family to its infimum is also continuous. This presents $\mathbb{S}$ as a topological poset with all suprema and finite infima.

If X is a topological space, a continuous function $f: X \to \mathbb{S}$ is a partition of X into an open subset U (the inverse image of 1) and its closed complement (the inverse image of 0). We shall say that f is the *characteristic function* of U. The set $C^0(X, \mathbb{S})$ of characteristic functions inherits from $\mathbb{S}$ an order relation where $f \leq g$ if $f(x) \leq g(x)$ for all x in X. The resulting poset structure on $C^0(X, \mathbb{S})$ coincides with the poset $\mathcal{O}(X)$ of open subsets of X ordered by inclusion. Moreover, $C^0(X, \mathbb{S}) = \mathcal{O}(X)$ inherits also the algebraic operations of $\mathbb{S}$ where they coincide with the union and finite intersection in $\mathcal{O}(X)$: $(\bigvee f_i)(x) = \bigvee (f_i(x))$ and $(\bigwedge f_i)(x) = \bigwedge (f_i(x))$. This simple construction says an important thing: the algebra of open subsets of a space X can be thought as an algebra of continuous functions on X with values in the Sierpiński space.

The algebraic structure of $\mathcal{O}(X)$ is that of a *frame*: that is, a poset

- with arbitrary suprema $(\bigvee, 0)$,
- finite infima $(\wedge, 1)$,
- satisfying a distributivity condition $a \wedge \bigvee b_i = \bigvee (a \wedge b_i)$.

Given two frames F and F', a morphism of frames $u^*: F \to F'$ is a morphism of posets preserving all suprema and finite infima. The collection of frame morphisms $F \to F'$ is naturally a poset. This makes the category Frame of frames into a 2-category.

There exists a functor

$$\mathcal{O} = C^0(-, \mathbb{S}) : \text{Top}^{\text{op}} \longrightarrow \text{Frame},$$
$$X \longmapsto \mathcal{O}(X),$$
$$f: X \to Y \longmapsto f^*: \mathcal{O}(Y) \to \mathcal{O}(X).$$

The notion of *locale* is defined as an object of the category Locale = Frame^{op}.[21] This permits us to write the previous functor $\mathcal{O}$ as a covariant functor Top $\to$ Locale. If L is a locale, we denote by $\mathcal{O}(L)$ the corresponding frame. The objects of $\mathcal{O}(L)$ will be called the *open domains* of L. If $f: L \to L'$ is a morphism of locales, we shall denote by $f^*: \mathcal{O}(L') \to \mathcal{O}(L)$ the corresponding morphism of frames.

The functor Top $\to$ Locale is not faithful. If X is the indiscrete topology on a set E, then X and the one point space 1 have same image under $\mathcal{O}$. The spaces that can be faithfully represented in Frame are those spaces whose set of points can be reconstructed from the frame of open subsets. They are called *sober*

---

[21] When Frame is viewed as a 2-category, the 2-category Locale is defined by reversing the direction of 1-arrows only.

spaces.[22] This functor is not essentially surjective either. A frame F is the frame of open subset of a topological space if and only if there exists an injective frame morphism $F \hookrightarrow P(E)$ into the the power set of a certain set E. We shall see an example of a frame admitting no such embedding in Section 2.2.7 (7). We shall also see in Section 2.2.13 that the functor Top $\to$ Locale is in a very precise way the functor forgetting the underlying set of points.

2.2 Elements of Locale Geometry and Frame Algebra

The idea is that a locale is a formal geometric dual to the algebraic structure of frame. In other words, locales are spatial objects defined by an abstract algebra of open subsets, without reference to a set of points. The fact that Locale = Frame^{op} is indeed a category of geometric objects is justified by the fact that a number of topological notions and constructions can be transferred along Top $\to$ Locale. The mechanism is simple: take a topological notion, try to formulate it in terms of the frame of open subsets, then generalize it to any frame.

2.2.1 Punctual and Empty Locales

Let 1 be the one point space and $\emptyset$ the empty space. It is easy to prove that $\mathcal{O}(1) = \underline{2} := \{0 < 1\}$ is the initial object of the category Frame and that $\mathcal{O}(\emptyset) = \underline{1} := \{0\}$ is the terminal object. The corresponding objects in Locale are also denoted by 1 and $\emptyset$ and play the role of the point and the empty space. They are in the image of Top $\to$ Locale.

2.2.2 Free Frames and Affine Locales

The algebraic approach of topology that is locale theory distinguishes a class of topological objects corresponding to the freely generated algebraic objects. Given a poset P, there exists a notion of the free frame $\underline{2}[P]$ on P. The free frame on no generators ($P = \emptyset$) is $\underline{2} := \{0 < 1\}$. It is the initial object of the category Frame, the equivalent of $\mathbb{Z}$ in the category of commutative rings. The free frame on one generator x is $\underline{2}[x] := \{0 < x < 1\}$. It is the equivalent of $\mathbb{Z}[x]$ in the category of commutative rings.

More generally, the free frame on a poset P is constructed as follows: first, one constructs $P^\wedge$, the free completion of P for finite intersections, then one freely completes $P^\wedge$ for arbitrary unions into a poset $\underline{2}[P] := [(P^\wedge)^{op}, \underline{2}]$. This last construction is made by taking presheaves with values in $\underline{2}$. The construction of $\underline{2}[P]$ is analogous to that of the free ring on a set E by

---

[22] We shall not assume, as is sometimes the case when comparing topological spaces to locales, that our topological spaces are sober. We shall explain precisely in Section 2.2.13 how the two notions should be properly compared. We refer to the classical literature for more details on sober spaces [17, 28].

constructing first the free commutative monoid $M(E)$ on E and then the free abelian group $\mathbb{Z}.M(E)$ on $M(E)$ (see Section 3.4.1). A frame morphism $\underline{2}[P] \to F$ is then equivalent to a poset morphism $P \to F$.

We shall call $\mathbb{S}^P$ the locale dual to the free frame $\underline{2}[P]$. By analogy with algebraic geometry, the locales $\mathbb{S}^P$ can be called *affine spaces*. The algebraic result that any frame is a quotient of a free frame translates geometrically into the statement that any locale L has an embedding $L \hookrightarrow \mathbb{S}^P$ for some poset P.

Example 2.1 (Affine Locales)

1. The punctual locale is affine $1 = \mathbb{S}^0$. The free frame $\underline{2}$ on no generators is isomorphic to the frame $\mathcal{O}(1)$.
2. (The Sierpiński locale) The Sierpiński space is faithfully encoded by its corresponding locale. The frame $\mathcal{O}(\mathbb{S})$ has three elements $\{0 < \{1\} < \{0,1\}\}$. It is isomorphic to the free frame on one generator $\underline{2}[x] := \{0 < x < 1\}$.
3. If E is a set, then the frame $\underline{2}[E]$ is the poset of open subsets of the product $\mathbb{S}^E$ of E copies of the Sierpiński space $\mathbb{S}$.
4. If P is a poset, the locale dual to $\underline{2}[P]$ is $\mathbb{S}^P$, the "P-power" of $\mathbb{S}$. Recall that the category Locale is enriched over posets. It is in fact also cotensored over posets, and $\mathbb{S}^P$ is the cotensor of the Sierpiński space by P. It has the universal property that a morphism of locales $X \to \mathbb{S}^P$ is equivalent to a morphism of posets $P \to \mathrm{Hom}_{\mathsf{Locale}}(X, \mathbb{S}) = \mathcal{O}(X)$.

2.2.3 Alexandrov Locales

Let P be a poset. There exists a construction, due to Alexandrov, of a nonseparated topology on the set of elements of P such that the specialization order coincides with the order of P. The open subsets for this topology are the upward closed subsets of P, which can be also defined as order-preserving maps $P \to \underline{2}$. The Alexandrov locale of P is the locale $\mathrm{Alex}(P)$ defined by the frame $[P, \underline{2}]$ of poset morphisms from P to $\underline{2}$. There is a canonical map $P \to \mathcal{P}\mathrm{t}(\mathrm{Alex}(P))$ that is injective but not surjective in general.[23] This construction provides a functor $\mathrm{Alex}\colon \mathsf{Poset} \to \mathsf{Locale}$ that is left adjoint to the functor $\mathcal{P}\mathrm{t}\colon \mathsf{Locale} \to \mathsf{Poset}$. In other words, for a locale X, morphisms $\mathrm{Alex}(P) \to X$ are equivalent to morphisms of posets $P \to \mathcal{P}\mathrm{t}(X)$.

Example 2.2 (Alexandrov Locales)

1. Any discrete space defines an Alexandrov locale. The open subsets of the discrete topology on a set E do form the frame $P(E) = [E, \underline{2}]$.

---

[23] The poset $\mathcal{P}\mathrm{t}(A_P)$ is the completion of P for filtered unions, also called the poset of ideals of P (see [17]).

2. The Sierpiński space is the Alexandrov locale associated to $P = \underline{2} = \{0 < 1\}$, that is, $\mathcal{O}(\mathbb{S}) = \underline{2}[x] = [\underline{2}, \underline{2}]$.
3. Let $\underline{n}$ be the poset $\{0 < 1 < \cdots < n-1\}$. A morphism of locales $X \to \text{Alex}(\underline{n})$ is equivalent to a *stratification of depth n*, that is, a sequence $U_{n-1} \subset U_{n-2} \subset \ldots \subset U_0 = X$ of open domains of $\mathcal{X}$.
4. The poset $\left[\mathcal{O}(X)^{op}, \underline{2}\right]$ is an Alexandrov frame. The corresponding locale shall be denoted $\widehat{X}$. We shall see that there is an embedding $X \to \widehat{X}$ and that $\widehat{X}$ is a kind of compactification of X.

2.2.4 The Poset of Points

A point of a topological space X is the same thing as a continuous map $x: 1 \to X$. Such a map defines a morphism of frames $x^*: \mathcal{O}(X) \to \underline{2}$. Intuitively, this morphism sends an open subset to 1 if and only if it contains the point. Then, a *point* of a locale L is defined as a morphism $x: 1 \to L$, or equivalently, as a frame morphism $x^*: \mathcal{O}(L) \to \underline{2}$. Since the frame morphisms do form posets, the collection $\mathcal{P}\text{t}(L)$ of all the points is naturally a poset. For two points $x^*, y^*: \mathcal{O}(L) \to \underline{2}$, we shall say that x^* is a *specialization* of y^* when $x^* \leq y^*$. Intuitively, this says that any open domain containing x also contains y.

Example 2.3 (Points)

1. If X is a topological space and $\underline{X}$ the corresponding locale, there is a canonical map $\mathcal{P}\text{t}(X) \to \mathcal{P}\text{t}(\underline{X})$. This map is injective if and only if X is T_0-space and bijective if and only if X is a sober space.
2. For a locale L, let $|\mathcal{P}\text{t}(L)|$ be the underlying set of $\mathcal{P}\text{t}(L)$. There is a canonical morphism $\mathcal{O}(L) \to P(|\mathcal{P}\text{t}(L)|)$ that sends an open domain U to the set of points it contains. This defines a natural topology on the set $|\mathcal{P}\text{t}(L)|$. The corresponding functor Locale $\to$ Top is right adjoint to the functor Top $\to$ Locale. The image of this functor is the category of sober spaces. The map $\mathcal{O}(L) \to P(|\mathcal{P}\text{t}(L)|)$ is not injective in general, hence the functor Locale $\to$ Top is not fully faithful. When it is injective, the locale is said to have enough points; intuitively, this means that $\mathcal{O}(L)$ is the frame of open domains of a sober space.
3. The poset of points of $\widehat{X}$ is the poset of all filters in $\mathcal{O}(X)$. The embedding $X \to \widehat{X}$ sends a point of X to the filter of its neighborhoods.
4. We shall see in the examples of sublocales that there exist nonempty locales with an empty poset of points.

2.2.5 Open Domains

Let U be an open subset of a topological space X; then we have a canonical isomorphism of frames $\mathcal{O}(U) = \mathcal{O}(X)_{/U}$ (the slice of $\mathcal{O}(X)$

over U), and the inclusion $U \subset X$ corresponds to the frame morphism $U \cap - : \mathcal{O}(X) \to \mathcal{O}(X)_{/U}$. More generally, for any locale L and any U in $\mathcal{O}(L)$, the map $U \cap - : \mathcal{O}(L) \to \mathcal{O}(L)_{/U}$ is a frame morphism called an *open quotient* of frames. A map $U \to L$ of locales is called an *open embedding* if the corresponding map of frames is an open quotient. The class of open embeddings is compatible with the classical topological notion: if X is a topological space and $U \to X$ is an open embedding in Locale, then U can be proved to be an open topological subspace of X.

Example 2.4 (Open Domains)

1. The inclusion $\{1\} \hookrightarrow \mathbb{S}$ is an open embedding.
2. It is, in fact, the universal open embedding. Given an open embedding $U \hookrightarrow X$ of a locale X, there exists a unique morphism of locales $\chi_U : X \to \mathbb{S}$ inducing a cartesian square

$$\begin{array}{ccc} U & \longrightarrow & \{1\} \\ \downarrow & \ulcorner & \downarrow \\ X & \xrightarrow{\chi_U} & \mathbb{S} \end{array}$$

The morphism of frames $\underline{2}[x] \to \mathcal{O}(X)$ corresponding to the characteristic function $\chi_U : X \to \mathbb{S}$ is the unique frame morphism sending x to U.

2.2.6 Closed Embeddings

Let $U \subset X$ be an open subset of a topological space X and Z its closed complement. There is a canonical isomorphism of frames $\mathcal{O}(Z) = \mathcal{O}(X)_{U/}$ (the coslice of $\mathcal{O}(X)$ under U, that is, the poset of open domains containing U), and the inclusion $Z \subset X$ corresponds to the frame morphism $U \cup - : \mathcal{O}(X) \to \mathcal{O}(X)_{U/}$. In general, for any open domain U of a locale L, the map $U \cup - : \mathcal{O}(L) \to \mathcal{O}(L)_{U/}$ is a frame morphism called a *closed quotient* of frames. A map $U \to L$ of locales is called a *closed embedding* if the corresponding map of frames is a closed quotient.

Example 2.5 (Closed Embeddings)

1. The inclusion $\{0\} \hookrightarrow \mathbb{S}$ is an closed embedding.
2. It is, in fact, the universal closed embedding. Given a closed embedding $Z \to X$, there exists a unique morphism of locales $X \to \mathbb{S}$ inducing a cartesian square

$$\begin{array}{ccc} Z & \longrightarrow & \{0\} \\ \downarrow & \ulcorner & \downarrow \\ X & \xrightarrow{\chi_Z} & \mathbb{S} \end{array}$$

The morphism of frames $\underline{2}[x] \to \mathcal{O}(X)$ corresponding to the characteristic function $\chi_Z: X \to \mathbb{S}$ is the unique frame morphism sending x to the open complement U of Z.

2.2.7 Sublocales and Frame Quotients

Let $Y \subset X$ be an inclusion of topological spaces; then the corresponding frame morphism $\mathcal{O}(X) \to \mathcal{O}(Y)$ is surjective.[24] A morphism of frames is called a *quotient* if it is surjective. A morphism of locales $L' \to L$ is called an *embedding*, or a *sublocale*, if the corresponding map of frames is a quotient.

Quotients can be generated in several ways. For example, given any inequality $A \leq B$ in F, there exists a unique quotient $F \to F /\!/ (A = B)$ forcing the inclusion to become an identity. This is the analogue for a frame of the quotient of a commutative ring A by the relation $a = b$, for two elements a and b in A. Any quotient can be generated by forcing a set of inequalities to become equalities.[25]

For any frame quotient $q^*: F \to F'$, there exists a right adjoint $q_*: F' \to F$ that is injective (but this is only a poset morphism and not a frame morphism). Then the quotient is completely determined by the poset morphism $j: q_* q^* : F \to F$. Such morphisms are called *closure operators*, or *nuclei*, and they can be axiomatized by the properties $U \leq j(U)$, $j(j(U)) = j(U)$, and $j(U \wedge V) = j(U) \wedge j(V)$. A closure operator defines a unique quotient $q^*: F \to F /\!/ (1 = j)$ such that $j = q_* q^*$. The poset $F /\!/ (1 = j)$ is defined as the elements of F such that $U = j(U)$; in other terms, it is forcing all the canonical inequalities $U \leq j(U)$ to become identities. We refer to the literature for more details about those [17]. Table 4.9 compares the situation of quotients of frames and commutative rings.

If X is a topological space, not every sublocale is a topological subspace. This is one of the differences between topological spaces and the corresponding locale – the latter has more subobjects. We give an example below.

Example 2.6 (Sublocales)

1. Any open embedding of a locale X is an embedding. If U is the object of $\mathcal{O}(X)$ corresponding to the open embedding, the quotient $\mathcal{O}(X) \to \mathcal{O}(U) = \mathcal{O}(X)_{/U}$ is generated by forcing the inequality $U \leq 1$ to become an equality. The corresponding nucleus is $V \mapsto U \Rightarrow V$, where $U \Rightarrow V$ is a Heyting implication ($U \Rightarrow -$ is right adjoint to $U \cap -$).

---

[24] For topological spaces, the reciprocal is true only if X is T_0-separated.
[25] In terms of category theory, a frame quotient $F \to F'$ is a *left-exact localization* of F. The quotient $F \to F /\!/ (A = B)$ is then the left-exact localization generated by forcing $A \leq B$ to become an identity.

Table 4.9. *Quotients of frames and rings*

Comm. ring A	ideal $J \subseteq A$	generators a_i for J	projection $A \to A$ on a complement of J in A	quotient A/J
Frame F	the set J of inequalities $A \leq B$ that become equalities in the quotient	generating inequalities $A_i \leq B_i$	nucleus $j: F \to F$	quotient $F /\!/ (1 = j)$

2. Any closed embedding of a locale X is an embedding. Let U be the corresponding object of $\mathcal{O}(X)$; the quotient $\mathcal{O}(X) \to \mathcal{O}(Z) = \mathcal{O}(X)_{U/}$ is generated by forcing the inequality $0 \leq U$ to become an equality. The corresponding nucleus is $V \mapsto U \cup V$.
   The collection of all embeddings $L' \hookrightarrow L$ in a fixed locale L is a poset. It can be proven that the closed embedding $Z \hookrightarrow L$ is the maximal element of the poset of embeddings of L that is disjoint from $U \hookrightarrow L$. If X is a topological space, $Z \hookrightarrow X$ corresponds to the closed topological subspace that is the complement of U.

3. Recall the Alexandrov locale $\widehat{X}$ dual to the frame $[\mathcal{O}(X)^{op}, \underline{2}]$. There exists a unique frame morphism $[\mathcal{O}(X)^{op}, \underline{2}] \to \mathcal{O}(X)$ that is the identity on $\mathcal{O}(X) \hookrightarrow [\mathcal{O}(X)^{op}, \underline{2}]$. This frame morphism is surjective and defines the embedding $X \to \widehat{X}$ mentioned earlier.

4. The subposet $[\mathcal{O}(X)^{op}, \underline{2}]^{\text{lex}} \subset [\mathcal{O}(X)^{op}, \underline{2}]$ spanned by maps preserving finite infima is a frame, called the frame of *ideals* of the distributive lattice $\mathcal{O}(X)$. The dual locale shall be denoted X_{coh}. The previous frame quotient $[\mathcal{O}(X)^{op}, \underline{2}] \to \mathcal{O}(X)$ factors as $[\mathcal{O}(X)^{op}, \underline{2}] \to [\mathcal{O}(X)^{op}, \underline{2}]^{\text{lex}} \to \mathcal{O}(X)$. Dually, this defines embeddings $X \to X_{\text{coh}} \to \widehat{X}$. The locale X_{coh}, which is always spatial, is the so-called *coherent compactification* of X.

5. If E is a set viewed as a discrete locale, the *Stone–Čech compactification* βE of E can be defined as a sublocale of $\widehat{E}$. Let $[P(E)^{op}, \underline{2}]^{\text{ultra}} \subset [P(E)^{op}, \underline{2}]$ be the subposet spanned by maps $F: P(E)^{op} \to \underline{2}$ such that, for any subset $A \subset E$ and any partition $A = A_0 \coprod A_1$, we have $F(A) = F(A_0) \wedge F(A_1)$. Then $[P(E)^{op}, \underline{2}]^{\text{ultra}}$ is the frame of open domains of βE. Recall that the points of $\widehat{E}$ are the filters of $P(E)$. The points of βE are the ultrafilters.

6. Let x be a point of $\mathbb{R}$ and U_x be the complement of $\{x\}$. The open quotient $\mathcal{O}(X) \to \mathcal{O}(U_x)$ is generated by forcing the inclusion $]x - \epsilon, x[\cup]x, x + \epsilon[\subset]x - \epsilon, x + \epsilon[$ to become an equality.

The corresponding closure operator j_x is the following. For an open subset $V \subset \mathbb{R}$, we denote by V' its closed complement. If x is an isolated point of V', then $V \cup \{x\}$ is open and $j_x(V) = V \cup \{x\}$. If not, then $j_x(V) = V$. Hence, the image in the inclusion $\mathcal{O}(U_x) \to \mathcal{O}(X)$ is spanned by the open subsets V such that x is not an isolated point in V'.

7. Let x_i be an arbitrary family of points of $\mathbb{R}$ and U_i be the complement of $\{x_i\}$. The formalism of frames lets us describe in a simple way the frame corresponding to the intersection of all the U_i: it is the intersection of all the frames $\mathcal{O}(U_i)$ in $\mathcal{O}(X)$. By the previous example, this intersection is spanned by the open subsets V of X whose closed complement V' admits none of the x_i as isolated points.

This example becomes fun if we let x_i be the family of *all* points of $\mathbb{R}$. First, the intersection of all the U_x for all x identifies to the subframe of $\mathcal{O}(X)$ spanned by open subset V whose closed complement is *perfect*, that is, has no isolated points. Since nontrivial perfect subsets of $\mathbb{R}$ exist (e.g., closed intervals, Cantor sets), the resulting intersection is not trivial. Let $\mathbb{R}^\circ \subset \mathbb{R}$ be the corresponding sublocale of $\mathbb{R}$. Now the funny thing is that $\mathbb{R}^\circ$, even though it is not the empty locale, cannot have any points! Indeed, any such point would define a point of $\mathbb{R}$ through the inclusion $\mathbb{R}^\circ \subset \mathbb{R}$, but, by definition of $\mathbb{R}^\circ$, none of the points of $\mathbb{R}$ are in $\mathbb{R}^\circ$.

This is our first example of a locale without any points; we will see another one later. We shall call *thin* a subset of $\mathbb{R}$ with empty interior. Intuitively, a property is true on the locale $\mathbb{R}^\circ$ if it is true outside of a thin and perfect subset of numbers. The frame $\mathcal{O}(\mathbb{R}^\circ)$ is also an example of a frame without any injective frame morphism into a power set $P(E)$ (since any element of the set E would then be a point). This example can be generalized to any Hausdorff space.

2.2.8 Generators, Relations, and Classifying Locales

The algebraic notion of frame offers the means to define certain spaces by generators and relations for their frame. This fact is useful for constructing spaces classifying certain subsets of a given space. Let $\underline{2}[E]$ be the free frame on a set E. A point of $\underline{2}[E] \to \underline{2}$ is the same thing as a map $E \to \underline{2}$, which is a subset of E. From this point of view, the locale $\mathbb{S}^E$ is the classifying space of subsets of E.[26] If we impose relations on the free frame $\underline{2}[E]$, this corresponds to building a subspace of $\mathbb{S}^E$, which is to impose some constraints on the kind

---

[26] More precisely, if we define a family of subsets of E parameterized by a locale L as given by the data of a subobject of the trivial bundle $L \times E \to L$, then such data is equivalent to that of a morphism of locales $L \to \mathbb{S}^E$.

of subsets of E corresponding to the points. If $E = A \times B$, we can, for example, extract the subsets that are the graphs of functions $A \to B$. We shall denote by $[a \mapsto b]$ an element (a,b) in $A \times B$. The notation is chosen to suggest that this corresponds to the condition "a is sent to b." The relations to impose on $\underline{2}[A \times B]$ in order to classify graphs of functions are given by the following inequalities, which have to be forced to become equalities:

- (existence of image) for any a: $\bigvee_b [a \mapsto b] \leq 1$;
- (unicity of image) for any a and $b \neq b'$: $0 \leq [a \mapsto b] \wedge [a \mapsto b']$.

The frame classifying functions $A \to B$ is then the left-exact localization of $\underline{2}[A \times B]$ generated by those maps. To classify surjections or injections, we need to add the following further relations:

- (surjectivity) for any b: $\bigvee_a [a \mapsto b] \leq 1$;
- (injectivity) for any b and $a \neq a'$: $0 \leq [a \mapsto b] \wedge [a' \mapsto b]$.

One of the most intriguing facts about locales is that, when A is infinite and B is not empty, it can be proven that the sublocale of $\mathbb{S}^{A \times B}$ classifying surjections is never empty [21]. In particular, when $A = \mathbb{N}$ and $B = P(\mathbb{N}) \simeq \mathbb{R}$, there exists a nonempty locale of surjections $\mathbb{N} \to \mathbb{R}$. This produces another example of a locale without points since any point would construct an actual surjection $\mathbb{N} \to \mathbb{R}$ in set theory. There is also a nontrivial locale $\mathrm{Bij}(\mathbb{N}, \mathbb{R})$ classifying bijections between $\mathbb{N}$ and $\mathbb{R}$. From the point of view of this locale, the cardinals of $\mathbb{N}$ and $\mathbb{R}$ are then the same. More generally, any two infinite cardinals can be forced to be the same similarly. This kind of locale is useful in interpreting logical constructions, such as Cohen forcing [25].

2.2.9 Locales without Points

We mentioned a couple of examples of nonempty locales without any points. Another amusing example is given in [6, Section IV.7.4]. If $K = [0, 1]$ is the real interval equipped with a measure μ, the poset of measurable subsets of K is not a frame, but the poset of classes of measurable subsets of K up to null sets is. Since it is clearly nontrivial, it defines a nonempty locale K_μ. The points of this frame correspond to points of K with nonzero measure. If μ is the Lebesgue measure, no such points exist, and K_μ has no points.

These phenomena of locales without points can be nicely explained with the analogy of frame theory with commutative algebra. Let P be a polynomial in $\mathbb{Q}[x]$ and $A = \mathbb{Q}[x]/P$ the quotient ring. A root of P in $\mathbb{Q}$ is a ring morphism $A \to \mathbb{Q}$. Geometrically, such objects are called *rational points* of $Spec(A)$. Now if $P = x^2 + 1$, it does not have any root in $\mathbb{Q}$, and the corresponding scheme does not have enough rational points. To produce roots of P or points of $Spec(A)$, we need to take an extension of $\mathbb{Q}$.

The situation is similar for locales. The points of a locale X are defined as frame morphisms $\mathcal{O}(X) \to \underline{2}$. Given a presentation of $\mathcal{O}(X)$ by generators and relations, finding a point corresponds to interpreting the generators as 0 or 1 such that the relations are fulfilled. This might not be possible. However, this might become possible if the generators are interpreted as elements of larger frame than $\underline{2}$.

A locale is said to have *enough points* if two open domains can be distinguished by the points they contain. Recall that the set of points $|\mathcal{P}t(L)|$ of a locale L has a canonical topology. Then a locale has enough points precisely when the map $\mathcal{O}(L) \to P(E)$ is injective. A locale with enough points can be proved to be the same thing as a sober topological space. The affine locale $\mathbb{S}^P$ has enough points. Since any locale is a sublocale of some $\mathbb{S}^P$, any locale is a sublocale of a locale with enough points.

2.2.10 Product of Locales and Tensor Products of Frames

The product of two locales $X \times Y$ corresponds dually to a tensor product $\mathcal{O}(X) \otimes \mathcal{O}(Y)$ of their corresponding frames [21]. This tensor product is defined similarly to those of commutative rings and abelian groups.[27] Recall that a frame is in particular a sup-lattice, that is, a poset with arbitrary suprema. Sup-lattices play for frames the role played by abelian groups for commutative rings (see Table 4.18). A morphism of sup-lattices is defined to be a map preserving arbitrary suprema. For three sup-lattices A, B, C, a poset map $A \times B \to C$ is called *bilinear* if it preserves suprema in both variables. Then, it can be proven that such bilinear maps are equivalent to morphisms of sup-lattices $A \otimes B \to C$ for some sup-lattice $A \otimes B$ called the *tensor product* of A and B. There exists a canonical bilinear map $A \times B \to A \otimes B$.

Here are some properties of this tensor product. The unit is the poset $\underline{2}$. If P is a poset, the poset $[P^{op}, \underline{2}]$ is a sup-lattice,[28] and for two posets P and Q, we have $[P^{op}, \underline{2}] \otimes [Q^{op}, \underline{2}] = [(P \times Q)^{op}, \underline{2}]$. In other terms, the functor Alex : Poset $\to$ Locale preserves products.

In the same way that the tensor product $A \otimes B$ of two commutative rings is a commutative ring, the tensor product of two frames $F \otimes G$ is a frame.

---

[27] Recall that the coproduct of two commutative rings A and B is given by the tensor product $A \otimes B$ of the underlying abelian groups. This tensor product is defined by the universal property that maps of abelian groups $A \otimes B \to C$ are equivalent to bilinear maps $A \times B \to C$.

[28] We shall see in Section 3.4.1 that it is in fact the free sup-lattice generated by P.

Moreover, $A \otimes B$ is actually the sum of A and B in the category of commutative rings, and so is $F \otimes G$ the sum of F and G in the category of frames. Dually, the tensor operation corresponds to the cartesian product of locales. The canonical functor Top $\to$ Locale does not preserve cartesian products,[29] but products of locally compact spaces are preserved.

2.2.11 Surjections

If $X \to Y$ is a surjective map of topological spaces, the morphism of frames $\mathcal{O}(Y) \to \mathcal{O}(X)$ is injective. The reciprocal is not true, since surjective continuous maps need also to be surjective on the set of points. A morphism of locales $L' \to L$ is called a *surjective* if the corresponding morphism of frames is injective. If X is a topological space, then for any quotient $X \to L$ in Locale, there exists a surjective map $X \to Y$ in Top whose image under Top $\to$ Locale is $X \to L$.

Example 2.7 (Surjections)

1. Let X be a topological space and E its set of points. The canonical inclusion $\mathcal{O}(X) \subset P(E)$ is a frame morphism corresponding to a surjection $E \to X$, where E is viewed as a discrete locale. We shall see in Section 2.2.13 that the data of this surjection is precisely the difference between locales and topological spaces.
2. (Open covers) A collection $U_i \to L$ is an *open covering* if the resulting map $\coprod_i U_i \to L$ is surjective. This is equivalent to the condition that $\bigvee_i U_i = 1$ in $\mathcal{O}(L)$.
3. (Image factorization) Let $L' \to L$ be a map of locales; there exists a unique factorization $L' \to M \to L$ such that $L' \to M$ is a surjection and $M \to L$ is an embedding. This factorization is constructed dually by defining $\mathcal{O}(M)$ as the image of the frame map $\mathcal{O}(L) \to \mathcal{O}(L')$.

2.2.12 Compact Locales

A space X is compact if, for any directed union U_i of open subsets of X, the condition $X = \bigcup U_i$ implies that $X = U_i$ for some i. This property is a way to say that the maximal object 1 of the frame $\mathcal{O}(X)$ is *finitary*, or equivalently, that the poset morphism $\mathrm{Hom}_{\mathcal{O}(X)}(1, -) \colon \mathcal{O}(X) \to \underline{2}$ (the "global sections") preserves directed unions. Then, a locale L is called *compact* if the maximal object 1 of $\mathcal{O}(L)$ is finitary.

---

[29] $\mathbb{Q}^2$ is not the same computed in Top or in Locale (see [17, Section II.2.14]).

Example 2.8 (Compact Locales)

1. Any compact topological space is compact when viewed as a locale.
2. A frame $[P, \underline{2}]$ is dual to a compact locale if and only if the poset P is *filtering* (for any pair x, y of objects of P there exist $z \leq x$ and $z \leq y$). This is true in particular if P has a minimal element.
3. For X a locale or a topological space, the Alexandrov locale $\widehat{X}$ dual to the frame $[\mathcal{O}(X)^{op}, \underline{2}]$ is compact. This justifies the remark that it is a kind of compactification of X.
4. The coherent compactification X_{coh} of X, dual to the frame $[\mathcal{O}(X)^{op}, \underline{2}]^{\text{lex}}$, is also compact.

2.2.13 From Locales to Topological Spaces

We explained that the functor Top $\to$ Locale is not fully faithful, that is, that different spaces can have the same frame of open domains. Nonetheless, it is possible to reconstruct the category Top from Locale. For any set E, the power set $P(E)$ is a frame. A locale is called *discrete* if the corresponding frame is isomorphic to some $P(E)$. A locale L is said to have *enough points* if there exists a surjective morphism $E \to L$ from some discrete locale E. A choice of a set of points for a locale with enough points is a choice of such a surjection. Let X be a topological space and X_{dis} the discrete topology on the same set. The canonical embedding $\mathcal{O}(X) \subset P(X)$ is a frame morphism corresponding to a localic surjection $X_{dis} \to X$; that is, a topological space defines a locale together with a choice of a set of points.

Let Locale$^\to$ be the category whose objects are the morphisms of locales. The category of topological spaces is equivalent to the full subcategory of Locale$^\to$ spanned by maps $E \to L$, which are surjections with a discrete domain E. From this point of view, the functor Top $\to$ Locale is nothing but the functor sending a surjection $E \to L$ to L, that is, the functor forgetting the set of points. The image of this functor is the full subcategory of locales with enough points.

This simple result has two consequences. First, it should make clear the difference between the so-called *point-set topology* and *point-free topology*: topological spaces are locales with the extra structure of a fixed set of points. The second point is that the entire theory of topological spaces can be formulated in terms of the theory of locales, so the latter is in fact the most general one.

2.2.14 Concluding Remarks

Many other topological notions can be generalized to locales, such as connectedness, separation, pasting, or local homeomorphisms. Our purpose here was only to give a glance at the possibility of doing *point-free topology*, that is, topology without the prescription of a set of points. This step of forgetting the set of points is an essential one in the direction of the notion of topos. We refer to [17, 28] for a study of locales.

There are actually reasons to prefer the broader generality of locales to topological spaces. The most obvious reason is the nice duality Locale = Frame^{op}, that is, the fact that the spatial notion of locale can be equivalently manipulated in algebraic terms.[30] Another aspect is that the theory of locales is fundamentally constructive. For example, the proof that a product of a compact Hausdorff topological space is compact (Tychonov's theorem) depends on the axiom of choice, but not the proof that a product of compact Hausdorff locales is compact.

3 The Topos–Logos Duality

We have explained how the theory of topological spaces could be reformulated in terms of locale theory, a notion of spatial object dual to the algebraic structure of frame. The notion of topos can be similarly presented as dual to the algebraic notion of *logos*. We start in Section 3.1 by giving a first definition of logoi and topoi useful to give examples and play with them. Then, Section 3.2 defines a number of topological notions for topoi (and the corresponding algebraic notions for logoi) with the purpose of convincing the reader that topoi are indeed spatial objects. Finally, Section 3.3 has the purpose to explain Giraud and Lawvere definitions of logoi and topoi and their relation with a distributivity condition between limits and colimits in a logos. The explanation is given from the point of view of descent theory, aka the art of pasting. Section 3.3 is a more technical section that can be skipped at a first reading.

3.1 First Definition and Examples

Essentially, a logos is a category with colimits, finite limits, and a compatibility relation between them akin to distributivity. However, the precise formulation of this property demands the introduction of several concepts and will be postponed until Section 3.3. We shall start here with the simplest, albeit not the most intuitive, definition of a logos. Nonetheless, it is convenient to introduce

---

[30] The difference between topological spaces and locales is akin to that between algebraic varieties (over a nonalgebraically closed field) and schemes. The former have a prescription on the nature of their points that prevents them from being dual to some type of algebras, but the latter are designed to be perfectly dual to an algebraic structure; in particular, they can have no point in the sense of the former (rational points).

many examples to play with. The definitions by Giraud and Lawvere axioms will be given in Section 3.3.

We need a couple of preliminary notions. A *reflective localization* is a functor $L\colon \mathcal{E} \to \mathcal{F}$ admitting a fully faithful right adjoint. In particular, it is a cocontinuous functor. A *left-exact localization* is a reflective localization L that preserves finite limits.

A *logos* is a category $\mathcal{E}$ that can be presented as a left-exact localization of a presheaf category $\mathcal{Pr}(C) := [C^{op}, \mathsf{Set}]$ on a small category C. A *morphism of logoi* $f^*\colon \mathcal{E} \to \mathcal{F}$ is a functor preserving (small) colimits and finite limits. The category of logoi will be denoted Logos. It is a 2-category if we take into account the natural transformations $f^* \to g^*$ between the morphisms.[31] A *topos* is defined to be an object of the category Logos^{op}. The category of topoi is defined as

$$\mathsf{Topos} = \mathsf{Logos}^{op}.^{32}$$

We shall not use the classical terminology of *geometric morphisms* to refer to the morphisms in Topos, but simply talk about topos morphisms. If $\mathcal{X}$ is a topos, we shall denote by $\mathcal{Sh}(\mathcal{X})$ the corresponding logos. The objects of $\mathcal{Sh}(\mathcal{X})$ are called the *sheaves on* $\mathcal{X}$. For $u\colon \mathcal{Y} \to \mathcal{X}$, a topos morphism, we denote by $u^*\colon \mathcal{Sh}(\mathcal{X}) \to \mathcal{Sh}(\mathcal{Y})$ the corresponding logos morphism.

$$\mathsf{Logos}^{op} \underset{\mathcal{Sh}}{\overset{\text{dual}}{\rightleftarrows}} \mathsf{Topos}$$

Given F in $\mathcal{Sh}(\mathcal{X})$, the object u^*F in $\mathcal{Sh}(\mathcal{Y})$ is called the *pullback*, or *base change of F along u*. A logos $\mathcal{E}$ always has a terminal object 1; a map $1 \to F$ in $\mathcal{E}$ shall be called a *global section of F*. This geometric vocabulary will be justified in Section 3.2.6.

3.1.1 Sheaves on a Locale

The example motivating the definition of a logos is the category of sheaves of sets on a space. Let X be a topological space; the category $\mathcal{Sh}(X)$ of sheaves on X is a reflective subcategory of $\mathcal{Pr}(\mathcal{O}(X)) = [\mathcal{O}(X)^{op}, \mathsf{Set}]$. The localization $\mathcal{Pr}(\mathcal{O}(X)) \to \mathcal{Sh}(X)$ is the *sheafification functor* that happens to be left exact (we shall explain why below). Therefore, $\mathcal{Sh}(X)$ is a logos. The corresponding topos will be denoted simply by X. The construction of $\mathcal{Sh}(X)$ depends only

---

[31] Precisely, the category of morphisms of logoi is the full subcategory $[\mathcal{E}, \mathcal{F}]^{\text{lex}}_{\text{cc}} \subset [\mathcal{E}, \mathcal{F}]$ spanned by functors preserving small colimits and finite limits.

[32] When Logos is viewed as a 2-category, Topos is defined by reversing the direction of 1-arrows only. This definition of 2-cells in Topos is in accordance with most of the references but not with the original convention of [6].

on the frame $\mathcal{O}(X)$ and is therefore defined for any locale X. This produces a functor

$$\mathcal{S}\mathrm{h}: \mathsf{Locale}^{op} \longrightarrow \mathsf{Logos},$$
$$X \longmapsto \mathcal{S}\mathrm{h}(X),$$
$$f: X \to Y \longmapsto f^*: \mathcal{S}\mathrm{h}(Y) \to \mathcal{S}\mathrm{h}(X),$$

or equivalently a functor $\mathsf{Locale} \to \mathsf{Topos}$. This functor is faithful, and the topoi in the image of this functor are called *localic*. We shall see later the definition of the open domains of a topos and that the open domain of localic topos reconstructs the frame of open of the corresponding locale.

The fact that the sheafification functor $\mathcal{P}\mathrm{r}(\mathcal{O}(X)) \to \mathcal{S}\mathrm{h}(X)$ is left exact can be seen using the construction by Godement of this functor [13, Section II.1.2]. Let X be a topological space and $\mathcal{E}\mathrm{t}(X)$ be the full subcategory of $\mathsf{Top}_{/X}$ spanned by *local homeomorphisms*, or *étale maps*, $u: Y \to X$. Any such map $Y \to X$ defines a presheaf of local sections on X, which happens to be a sheaf. This produces a functor $\mathcal{E}\mathrm{t}(X) \to \mathcal{S}\mathrm{h}(X)$, which is an equivalence of categories. To prove this, Godement constructs a functor $\mathcal{P}\mathrm{r}(\mathcal{O}(X)) \to \mathcal{E}\mathrm{t}(X)$, which is the left adjoint to the functor $\mathcal{E}\mathrm{t}(X) \to \mathcal{P}\mathrm{r}(\mathcal{O}(X))$; hence it is the sheafification functor. The construction is based on the extraction of the stalks of a presheaf F. For any point x, let $U(x)$ be the filter of neighborhoods of x; the stalk of F at x is $F(x) = \mathrm{colim}_{V \in U(x)} F(V)$. The functor $F \mapsto F(x)$ is left exact because $U(x)$ is a filter and filtered colimits preserve finite limits. Let V be an open subset of X. Any point x in V defines a map $F(V) \to F(x)$, which sends a local section $\mathbb{S}$ of F to its *germ* $s(x)$ at x. Then, the underlying set of Y is $\coprod_{x \in X} F(x)$ and a basis for the topology is given by the sets $\{s(x) | x \in U\}$ for any s in $F(U)$. This geometric construction produces a functor $\mathcal{P}\mathrm{r}(\mathcal{O}(X)) \to \mathcal{E}\mathrm{t}(X)$, which is left exact because the construction of the stalks is

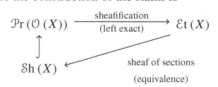

3.1.2 Presheaf Logoi and Alexandrov Topoi

The *Alexandrov logos* of a small category C is defined to be the category of set-valued C-diagrams $[C, \mathsf{Set}] = \mathcal{P}\mathrm{r}(C^{op})$. The *Alexandrov topos* of C is defined to be the dual topos, and we shall denote it by $\mathbb{B}C$. This defines a contravariant 2-functor $[-, \mathsf{Set}]: \mathsf{Cat}^{op} \to \mathsf{Logos}$ and a covariant 2-functor $\mathbb{B}: \mathsf{Cat} \to \mathsf{Topos}$, where Cat denotes the category of small categories. These 2-functors are not conservative since they take Morita-equivalent categories to equivalent Alexandrov logos/topos. Alexandrov topoi are analogues of Alexandrov locales (see Section 2.2.3). Many important examples of logoi/topoi are of this type.

Example 3.1 (Alexandrov Topoi)

1. When $C = \emptyset$, we get that the category 1 is a logos. It is the terminal object of Logos. Hence, the corresponding topos, denoted $\emptyset$, is the initial object of Topos and is called the *empty topos*. In the analogy logoi/commutative rings, this is the analogue of the zero ring.
2. When $C = 1$, the category Set is a logos. It is the initial object of Logos. In the analogy logoi/commutative rings, this is the analogue of the ring $\mathbb{Z}$. The corresponding topos, denoted $\mathbb{1}$, is the terminal object of Topos and will play the role of the *point*.
3. Let C be a small category; the presheaf category $\mathcal{P}r(C) := [C^{op}, \text{Set}]$ is a particular case of an Alexandrov logoi. The corresponding Alexandrov topos is $\mathbb{B}(C^{op})$. In particular, for a topological space X, the category $\mathcal{P}r(\mathcal{O}(X))$ is a logos and the sheafification $\mathcal{P}r(\mathcal{O}(X)) \to \mathcal{S}h(X)$ is a morphism of logoi. Recall the locale $\widehat{X}$ dual to the frame $[\mathcal{O}(X)^{op}, \underline{2}]$. Then we have in fact $\mathcal{P}r(\mathcal{O}(X)) = \mathcal{S}h\left(\widehat{X}\right)$. For this reason, we shall denote by $\widehat{X}$ the topos dual to $\mathcal{P}r(\mathcal{O}(X))$. We already saw the existence of an embedding $X \to \widehat{X}$, which is a kind of compactification of X. This will stay true in Topos.
4. The category of simplicial sets is a logos since it is defined as $\mathcal{P}r(\Delta)$, where Δ is the simplicial category, that is, the category of nonempty finite ordinals.
5. When C is a set E, that is, a discrete category, then $\mathcal{P}r(E) = \text{Set}^E$ is a logos. The corresponding Alexandrov topos $\mathbb{B}E$ is called *discrete*. In the analogy logoi/commutative rings, Set^E is analogue to $\prod_E \mathbb{Z}$.
6. Another example is the logos [Fin, Set], where Fin is the category of finite sets. This logos is arguably the central piece of the whole theory, and we are going to denote it by Set$[X]$. The notation is chosen to recall the free ring $\mathbb{Z}[x]$. The logos Set$[X]$ is in fact the free logos on one generator: for any logos $\mathcal{E}$, a logos morphism Set$[X] \to \mathcal{E}$ is the same thing as an object of $\mathcal{E}$. The "generic object" X in Set$[X]$ corresponds to the canonical inclusion Fin $\to$ Set. It is also the functor represented by the object 1 in Fin. The topos corresponding to Set$[X]$ will be denoted $\mathbb{A}$ and called the *topos of sets* or the *topos classifying objects*. It will play a role analogous to the affine line $\mathbb{A}^1$ in algebraic geometry. Table 4.10 details some aspects of the structural analogy between $\mathbb{Z}[x]$ and Set$[X]$.

Table 4.10. *Polynomial analogies*

	Commutative ring	*Logos*
Initial object	$\mathbb{Z}$	Set
Free on one generator	$\mathbb{Z}[x] = \mathbb{Z}^{(\mathbb{N})}$	$\text{Set}[X] = [\text{Fin}, \text{Set}]$
Monomials	x^n, for n in $\mathbb{N}$	X^N, for N in Fin (representable functors $X^N : \text{Fin} \to \text{Set}$ $E \mapsto E^N$)
Polynomial	$P(x) = \sum_n p_n x^n$	$F(X) = \int^N F(N) \times X^N$ (coend over Fin)
Polynomial function	for any ring A $P : A \to A$ $a \mapsto \sum_n p_n a^n$	for any logos $\mathcal{E}$ $F : \mathcal{E} \to \mathcal{E}$ $E \mapsto \int^N F(N) \times E^N$ (coend over Fin in $\mathcal{E}$)
Dual geometric object with an algebra structure	$\mathbb{A}^1$ is a ring object in Schemes	$\mathbb{A}$ is a logos object in Topos
Additive operation	$+ : \mathbb{A}^2 \to \mathbb{A}^1$ dual to $\mathbb{Z}[x] \to \mathbb{Z}[x, y]$ $x \mapsto x + y$	$\text{colim} : \mathbb{A}^C \to \mathbb{A}$ dual to $\text{Set}[X] \to \text{Set}[C]$ $X \mapsto \text{colim } c$
Multiplicative operation	$\times : \mathbb{A}^2 \to \mathbb{A}^1$ dual to $\mathbb{Z}[x] \to \mathbb{Z}[x, y]$ $x \mapsto xy$	$\lim : \mathbb{A}^C \to \mathbb{A}$ (C finite) dual to $\text{Set}[X] \to \text{Set}[C]$ $X \mapsto \lim c$

7. Let Fin$^\bullet$ be the category of pointed finite sets. The logos $\text{Set}[X^\bullet] := [\text{Fin}^\bullet, \text{Set}]$ is an important companion of $\text{Set}[X]$. A logos morphism $\text{Set}[X^\bullet] \to \mathcal{E}$ is the same thing as a *pointed object* in $\mathcal{E}$, that is, an object E with the choice of a global section $1 \to E$. The "generic pointed object" $X^\bullet$ in $\text{Set}[X^\bullet]$ corresponds to the functor $\text{Fin}^\bullet \to \text{Set}$, forgetting the base point. It is also the functor representable by the object $1 \to 1 \coprod 1$ in

Fin$^\bullet$. The topos corresponding to Set$[X^\bullet]$ will be denoted $\mathbb{A}^\bullet$ and called the *topos of pointed sets* or the *topos classifying pointed objects*. There is a distinguished topos morphism $\mathbb{A}^\bullet \to \mathbb{A}$ corresponding to the unique logos morphism Set$[X] \to$ Set$[X^\bullet]$ sending X to $X^\bullet$.

8. Let Fin$^\circ \subset$ Fin be the category of nonempty finite sets. The logos [Fin$^\circ$, Set] is denoted by Set$[X^\circ]$. The canonical object X° corresponds to the inclusion Fin$^\circ \subset$ Set. The corresponding logos is denoted $\mathbb{A}^\circ$. The inclusion Fin$^\circ \subset$ Fin produces a morphism of logoi Set$[X] \to$ Set$[X^\circ]$ sending X to X° and a morphism of topoi $\mathbb{A}^\circ \to \mathbb{A}$. The factorization Fin$^\bullet \to$ Fin$^\circ \subset$ Fin produces a factorization $\mathbb{A}^\bullet \to \mathbb{A}^\circ \to \mathbb{A}$. We shall see later that $\mathbb{A}^\circ$ classifies nonempty sets and that the factorization $\mathbb{A}^\bullet \to \mathbb{A}^\circ \to \mathbb{A}$ is the image factorization of $\mathbb{A}^\bullet \to \mathbb{A}$.

9. The logos of sheaves on the Sierpiński space is $\mathcal{S}h(\mathbb{S}) = [\underline{2}, \text{Set}] = \text{Set}^\to$, the arrow category of Set. The corresponding logos/topos is called the *Sierpiński logos/topos*. We shall see later that it plays the role of the Sierpiński space in classifying open domains of topoi, that is, that a morphism of topoi $\mathcal{X} \to \mathbb{S}$ is equivalent to an open subtopos of $\mathcal{X}$.

10. Let $[n]$ be the poset $\{0 < 1 < \cdots < n\}$. The category Set$^{[n]}$ is a logos. Morphisms of topoi $\mathcal{X} \to \mathbb{B}[n]$ can be proven to be equivalent to the data of a stratification of depth n on $\mathcal{X}$, that is, a sequence $U_n \subset U_{n-1} \subset \cdots \subset U_0 = \mathcal{X}$ of open subtopoi of $\mathcal{X}$. More generally, if P is a poset, morphisms $\mathcal{X} \to \mathbb{B}P$ can be proven to be stratifications on $\mathcal{X}$, whose strata are indexed by P.

11. Let G be a group; then the category Set^G of sets with a G-action is a logos since it can be described as the presheaf category $\mathcal{P}r(G)$, where G is viewed as a category with one object. The corresponding topos $\mathbb{B}G$ will play the role of a classifying space for G. A topos morphism $\mathcal{X} \to \mathbb{B}G$ can be proven to be the same thing as a G-torsor in the category $\mathcal{S}h(\mathcal{X})$ [25, Section VIII.2].

12. Let Ring$_{\text{fp}}$ be the category of commutative rings of finite presentations. The opposite category Ring$_{\text{fp}}^{\text{op}}$ is the category Aff$_{\text{fp}}$ of affine schemes of finite presentations. The Alexandrov logos [Ring$_{\text{fp}}$, Set] $= \mathcal{P}r\left(\text{Aff}_{\text{fp}}\right)$ and the dual topos $\mathbb{B}\left(\text{Ring}_{\text{fp}}\right)$ are *classifying rings*. A logos morphism $\mathcal{P}r\left(\text{Aff}_{\text{fp}}\right) \to \mathcal{E}$ is the same thing as a left-exact functor Aff$_{\text{fp}} \to \mathcal{E}$, which can be unraveled to be the same thing as a commutative ring object in $\mathcal{E}$, that is, a sheaf of rings.

13. Let **T** be a category with cartesian products, that is, a (multisorted) algebraic theory (aka a Lawvere theory). We denote by Mod(**T**) the category of models and by Mod(**T**)$_{\text{fp}}$ the subcategory of models of finite presentation. The Alexandrov logos Set$\langle$**T**$\rangle := \big[$Mod(**T**)$_{\text{fp}}$, Set$\big]$ has the

property that a logos morphism $\text{Set}\langle T\rangle \to \mathcal{E}$ is the same thing as a model of T in the logos $\mathcal{E}$. For this reason, the dual Alexandrov topos $\mathbb{B}\left(\text{Mod}(T)_{\text{fp}}\right)$ is called the *classifying topos of the algebraic theory* T and denoted $\mathbb{B}\langle T\rangle$

When T is the theory of rings, that is the full subcategory of Aff_{fp} spanned by affine spaces of finite dimension, $\text{Mod}(T)_{\text{fp}} = \text{Ring}_{\text{fp}}$, and we get back the previous example.

Let T be the theory of groups; then $\mathbb{B}\langle T\rangle$ is the topos classifying groups: one can prove that a topos morphism $\mathcal{X} \to \mathbb{B}\langle T\rangle$ is the same thing as a group object in $\mathcal{S}h(\mathcal{X})$, that is, a sheaf of groups on $\mathcal{X}$.

3.1.3 Other Examples

1. If $\mathcal{E}$ is a logos and E is an object of $\mathcal{E}$, then the category $\mathcal{E}_{/E}$ is again a logos. This is easy to see in the case $\mathcal{E} = \text{Set}$ since $\text{Set}_{/E} = \text{Set}^E = \mathcal{P}r(E)$. This is also easy to see in the case $\mathcal{E} = \mathcal{P}r(C)$ since $\mathcal{P}r(C)_{/E} = \mathcal{P}r(C_{/E})$, where $C_{/E}$ is the category of elements of the functor $E : C^{op} \to \text{Set}$. The base change along the map $e: E \to 1$ induces a functor $e^* : \mathcal{E} \to \mathcal{E}_{/E}$, which is a logos morphism. We shall see that such morphisms are étale maps.

2. Every logos $\mathcal{E}$ is a left-exact localization of a presheaf logos $\mathcal{P}r(C)$. The localization functor $\mathcal{P}r(C) \to \mathcal{E}$ is a surjective morphism of logoi. We shall see that the left-exact localizations of $\mathcal{P}r(C)$ are the "quotients" of $\mathcal{P}r(C)$ in the category of logoi.

3. Let G be a discrete group acting on a topological space X, and let $\mathcal{S}h(X, G)$ be the category of equivariant sheaves on X. Then $\mathcal{S}h(X, G)$ is a logos, and the corresponding topos $X/\!/G$ is the quotient of X by the action of G in the 2-category of topoi. The functor $q^* : \mathcal{S}h(X, G) \to \mathcal{S}h(X)$, forgetting the action, corresponds to the quotient map $q : X \to X/\!/G$.

4. Let G be a topological group, and let $\text{Set}^{(G)}$ be the category of sets equipped with a continuous action of G. Then $\text{Set}^{(G)}$ is a logos. If G is a connected group, then any continuous action of G on a set is trivial, and $\text{Set}^{(G)} = \text{Set}$. In fact, the logos $\text{Set}^{(G)}$ does depend only on the totally disconnected space of connected components of G, which is also a group. In particular, if G is locally connected, the connected components form a discrete group $\pi_0(G)$, and we have $\text{Set}^{(G)} = \text{Set}^{\pi_0(G)}$.

5. Let K be a profinite group (e.g., the Galois group of some field). Recall that K can be faithfully represented as a totally disconnected topological group. Then, by the previous example, the category $\text{Set}^{(K)}$ of continuous action of K is a logos.

3.2 Elements of Topos Geometry

As for locales, the fact that Topos = Logos^{op} is indeed a category of geometric objects is proved by the possibility to define there all the classical topological notions. The strategy to generalize topological notions to topoi is the same as before: first find a formulation in terms of sheaves, then generalize the notion to any logos.

3.2.1 Free Logoi and Affine Topoi

As with locales, the fact that topoi are defined as dual to some algebraic structure singularizes the class of topoi corresponding to the free algebras. Let C be a small category and C^{lex} the free completion of C for finite limits.[33] Then $\text{Set}[C] := \mathcal{P}r\left(C^{\text{lex}}\right) = \left[(C^{\text{lex}})^{op}, \text{Set}\right]$ is a logos called the *free logos* on C. The logos $\text{Set}[C]$ has the following fundamental property, which justifies its name: if $\mathcal{E}$ is a logos, then cocontinuous and left-exact functors $\text{Set}[C] \to \mathcal{E}$ are equivalent to functors $C \to \mathcal{E}$.[34] Inspired by algebraic geometry, the topos corresponding to $\text{Set}[C]$ will be denoted $\mathbb{A}^C$ and called an *affine topos*.

Example 3.2 (Free Logoi/Affine Topoi)

1. When $C = \emptyset$, we have $\emptyset^{\text{lex}} = 1$, and $\text{Set}[\emptyset] = \text{Set}$ is the initial logos, corresponding to the terminal topos $\mathbb{A}^0 = \mathbb{1}$.
2. When $C = 1$, we have $1^{\text{lex}} = \text{Fin}^{op}$, and $\text{Set}[1]$ is the logos $\text{Set}[X] = [\text{Fin}, \text{Set}]$ introduced before. The corresponding topos is $\mathbb{A}^1 = \mathbb{A}$. If $\mathcal{E}$ is a logos, a logos morphism $\text{Set}[X] \to \mathcal{E}$ is equivalent to choosing an object of $\mathcal{E}$. Geometrically, this gives the fundamental remark that the logos $\text{Sh}(\mathcal{X})$ of sheaves on a topos $\mathcal{X}$ can be described as topos morphisms into $\mathbb{A}$:
$$\text{Sh}(\mathcal{X}) = \text{Hom}_{\text{Topos}}(\mathcal{X}, \mathbb{A}). \tag{3.1}$$
This formula is analogous to $\mathcal{O}(X) = C^0(X, \mathbb{S})$ for locales. The morphism $\mathcal{X} \to \mathbb{A}$ corresponding to some F in $\text{Sh}(\mathcal{X})$ will be denoted χ_F and called the *classifying morphism* or *characteristic morphism* of F.
3. When $C = \{0 \to 1\}$, the category with one arrow, we have $C^{\text{lex}} = (\text{Fin}^{\to})^{op}$, where $\text{Fin}^{\to}$ is the arrow category of Fin, and $\text{Set}[\{0 \to 1\}] = [\text{Fin}^{\to}, \text{Set}]$. The corresponding topos is denoted $\mathbb{A}^{\to}$. A topos morphism $\mathcal{X} \to \mathbb{A}^{\to}$ is the same thing as a map $A \to B$ in $\text{Sh}(\mathcal{X})$. For this reason, $\mathbb{A}^{\to}$ is called the *topos classifying map*.

---

[33] This means that, if $\mathcal{E}$ is a category with finite limits, the data of a functor preserving finite limits $C^{\text{lex}} \to \mathcal{E}$ is equivalent to the data of a functor $C \to \mathcal{E}$.

[34] From a functor $C \to \mathcal{E}$, we get a functor $C^{\text{lex}} \to \mathcal{E}$ by right Kan extension and a function $\mathcal{P}r\left(C^{\text{lex}}\right) \to \mathcal{E}$ by left Kan extension. The fact that this last functor is cocontinuous and left exact is characteristic of logoi [12]. It would not be true if $\mathcal{E}$ were an arbitrary category with colimits and finite limits.

Table 4.11. *Classifying properties of affine and Alexandrov topoi*

	Topos morphism	Logos morphism	Interpretation
C small category	$\mathcal{X} \to \mathbb{A}^C$	$\mathsf{Set}[C] \to \mathsf{Sh}(\mathcal{X})$	diagram $C \to \mathsf{Sh}(\mathcal{X})$
E set	$\mathcal{X} \to \mathbb{A}^E$	$\mathsf{Set}[E] \to \mathsf{Sh}(\mathcal{X})$	family of sheaves $\mathcal{X}$ indexed by E
C small category	$\mathcal{X} \to \mathbb{B}C$	$\mathsf{Set}^C \to \mathsf{Sh}(\mathcal{X})$	flat C-diagram $C^{op} \to \mathsf{Sh}(\mathcal{X})$
D small category with finite colimits	$\mathcal{X} \to \mathbb{B}D$	$\mathsf{Set}^D \to \mathsf{Sh}(\mathcal{X})$	lex functor $D^{op} \to \mathsf{Sh}(\mathcal{X})$
E set	$\mathcal{X} \to \mathbb{B}E$	$\mathsf{Set}^E \to \mathsf{Sh}(\mathcal{X})$	partition of $\mathcal{X}$ indexed by E
P poset	$\mathcal{X} \to \mathbb{B}P$	$\mathsf{Set}^P \to \mathsf{Sh}(\mathcal{X})$	stratification of $\mathcal{X}$ indexed by P
G group	$\mathcal{X} \to \mathbb{B}G$	$\mathsf{Set}^G \to \mathsf{Sh}(\mathcal{X})$	G-torsor in $\mathcal{E}$

4. When $C = \{0 \simeq 1\}$, the category with one isomorphism, the affine topos $\mathbb{A}^{\{0 \simeq 1\}}$ is denoted $\mathbb{A}^\simeq$. A topos morphism $\mathcal{X} \to \mathbb{A}^\simeq$ is the same thing as an isomorphism $A \simeq B$ in $\mathsf{Sh}(\mathcal{X})$, and $\mathbb{A}^\simeq$ is called the *topos classifying isomorphisms*. The canonical functor $\{0 \to 1\} \to \{0 \simeq 1\}$ induces a map $\mathbb{A}^\simeq \to \mathbb{A}^\to$ of affine topoi. Intuitively, $\mathbb{A}^\simeq$ is the subtopos of $\mathbb{A}^\to$ classifying those maps that are isomorphisms.

Since $\{0 \simeq 1\}$ is equivalent to the punctual category 1, we have in fact $\mathbb{A}^\simeq = \mathbb{A}$. Intuitively, this says that the data of an isomorphism between two objects is equivalent to the data of a single object.

Table 4.11 summarizes some of the classifying properties of affine and Alexandrov topoi (some of these features will be explained later in the chapter).

3.2.2 The Category of Points

As mentioned in the introduction to this chapter, one of the differences between topological spaces and topoi is that the latter have a category of points instead of a mere set. The category of topoi has a terminal object $\mathbb{1}$ that corresponds to the logos Set. A *point* of a topos $\mathcal{X}$ is defined as a morphism of topoi $x: \mathbb{1} \to \mathcal{X}$. Equivalently, a point is a morphism of logoi $x^*: \mathsf{Sh}(\mathcal{X}) \to \mathsf{Set}$. The category of points of $\mathcal{X}$ is

$$\mathcal{P}t(\mathcal{X}) := \mathrm{Hom}_{\mathsf{Topos}}(\mathbb{1}, \mathcal{X}) = \mathrm{Hom}_{\mathsf{Logos}}(\mathcal{S}h(\mathcal{X}), \mathsf{Set}) = \left[\mathcal{S}h(\mathcal{X}), \mathsf{Set}\right]_{cc}^{lex},$$

which is the full subcategory of $\left[\mathcal{S}h(\mathcal{X}), \mathsf{Set}\right]$ spanned by functors preserving colimits and finite limits. Geometrically, a point x of $\mathcal{X}$ sends a sheaf F on $\mathcal{X}$ to its stalk $F(x) := x^*F$ at x.

Example 3.3 (Categories of Points)

1. When X is a locale, the category of points of $\mathcal{S}h(X)$ coincides with the poset $\mathcal{P}t(X)$ of points of X defined in Section 2.2.4.
2. By the universal property of free logoi, the category of points of $\mathbb{A}$ is the category Set. If E is a set, the logos morphism $\mathsf{Set}[X] \to \mathsf{Set}$ corresponding to E sends $X : \mathsf{Fin} \to \mathsf{Set}$ to E. More generally a functor $F : \mathsf{Fin} \to \mathsf{Set}$ is send to the coend $\int^{N \in \mathsf{Fin}} F(N) \times E^N$.
3. More generally, the category of points of $\mathbb{A}^C$ is the category $[C, \mathsf{Set}] = \mathcal{P}r(C^{op})$.
4. The classifying map $\chi_F : \mathcal{X} \to \mathbb{A}$ of some sheaf F on $\mathcal{X}$ induces a functor $\mathcal{P}t(\mathcal{X}) \to \mathcal{P}t(\mathbb{A}) = \mathsf{Set}$ that sends a point x to the stalk $F(x)$. In other words, the topos theory formalizes in a precise way the intuition that a sheaf is a continuous function with values in sets. In a sense, this fact is the whole point of topos theory.
5. The category of points of an Alexandrov topos $\mathbb{B}C$ is the category $\mathcal{I}nd(C)$, the free completion of C for filtered colimits.
6. In particular, for a topological space X, the points of the topos $\widehat{X}$, dual to the logos $\mathcal{P}r(\mathcal{O}(X))$, form the category $\mathcal{I}nd(\mathcal{O}(X))$. This category is equivalent to the poset of filters in $\mathcal{O}(X)$. We already mentioned that the inclusion $X \to \widehat{X}$ sends a point of X to the filter of its open neighborhoods.
7. When $C = \mathsf{fInj}$ the category of finite sets and injections, the category of points of $\mathbb{B}(\mathsf{fInj})$ is the category of all sets and injections.
8. Let $\mathbf{T}$ be an algebraic theory, that is, a category with cartesian products. The points of the topos $\mathbb{B}\langle \mathbf{T}\rangle$ do form the category $\mathcal{P}t(\mathbb{B}\langle \mathbf{T}\rangle) = [\mathbf{T}, \mathsf{Set}]^\times$ of functors preserving cartesian products. Such functors are also called the *models* of the theory $\mathbf{T}$. If $\mathbf{T}$ is the category opposite to the category of free groups on finite sets, then $\mathcal{P}t(\mathbb{B}\langle \mathbf{T}\rangle)$ is the category of all groups. If $\mathbf{T}$ is the category of affine spaces of finite dimension and algebraic maps, then $\mathcal{P}t(\mathbb{B}\langle \mathbf{T}\rangle)$ is the category of all commutative rings.
9. For a group G in Set, the category of points of $\mathbb{B}G$ is G itself viewed as a category with one object. This is a way to say that $\mathbb{B}G$ has essentially one point, but this point has G as its group of symmetries. The unique

point of $\mathbb{B}G$ is given by the functor $U: \text{Set}^G \to \text{Set}$ sending a G-set to its underlying set.
10. If G is a group acting on a space X, the category of points of the quotient topos $X/\!/G$ is the groupoid associated to the action of G on the points of X. In comparison, the points of the classical topological quotient X/G are only the isomorphism classes of objects of this groupoid. The difference is that the groupoid keeps the information about the stabilizers of each point. In the case of the quotient $\mathbb{R}/\!/\mathbb{Q}$, the category of points is the set of orbits of $\mathbb{Q}$ in $\mathbb{R}$. In the case of $\mathbb{R}/\!/\mathbb{R}_{dis}$ (where $\mathbb{R}_{dis}$ is $\mathbb{R}$ viewed as a discrete space), the category of points is a single point. Nonetheless, $\mathbb{R}/\!/\mathbb{R}_{dis}$ is not a point, and there exist many topos morphisms $\mathfrak{X} \to \mathbb{R}/\!/\mathbb{R}_{dis}$. For example, when X is a manifold, the set of closed differential forms embeds into the set of morphisms $X \to \mathbb{R}/\!/\mathbb{R}_{dis}$.
11. The category of points of $\mathbb{A}^\bullet$ is the category $\text{Set}^\bullet$ of pointed sets. The functor $\mathcal{P}t(\mathbb{A}^\bullet) \to \mathcal{P}t(\mathbb{A})$ induced by the topos morphism $\mathbb{A}^\bullet \to \mathbb{A}$ mentioned earlier is the forgetful functor $\text{Set}^\bullet \to \text{Set}$.
12. At the level of points, the embedding $\mathbb{A}^\circ \subset \mathbb{A}$ corresponds to the inclusion of nonempty sets into sets.
13. The category of points of $\mathbb{A}^\to$ is the arrow category $\text{Set}^\to$.
14. We define an *interval* to be a totally ordered set with a minimal and a maximal element that are distinct. For example, the real interval $[0,1]$ is an interval. A morphism of intervals is an increasing map preserving the minimal and maximal elements. It can be proven that the category of points of the topos $\mathcal{P}r(\Delta)$ of simplicial sets is the category of intervals.

    Recall that a simplicial set has a geometric realization that is a topological space. The functor $x^*: \mathcal{P}r(\Delta) \to \text{Set}$ corresponding to the interval $[0,1]$ sends a simplicial set to (the underlying set of) its geometric realization.

3.2.3 Quotient Logoi and Embeddings of Topoi

Let $u: Y \subset X$ be an embedding of topological spaces. We saw that $\mathcal{O}(X) \to \mathcal{O}(Y)$ was a surjective map of frames. The situation is the same for the corresponding map of logoi $u^*: \text{Sh}(\mathfrak{X}) \to \text{Sh}(\mathcal{Y})$, which is essentially surjective. In fact, more is true since u^* can be proven to have a fully faithful right adjoint u_*, that is, it is a left-exact localization. If Y is closed and F is a sheaf on Y, the sheaf u_*F is intuitively the extension of F to X obtained by declaring the fibers of u_*F outside of Y to be a single point.[35]

A morphism of logoi $\mathcal{E} \to \mathcal{F}$ shall be called a *quotient* if it is a left-exact localization. The corresponding morphism of topoi shall be called an

---
[35] When Y is not closed, the values of u_*F at the boundary of F are more involved.

embedding. If $\mathcal{Y} \hookrightarrow \mathcal{X}$ is an embedding, we shall also say that $\mathcal{Y}$ is a *subtopos* of $\mathcal{X}$. At the level of points, the functor $\mathcal{P}t(\mathcal{Y}) \to \mathcal{P}t(\mathcal{X})$ induced by an embedding is fully faithful. Classically, a quotient $\mathcal{E} \to \mathcal{F}$ is encoded by the data of a *Lawvere–Tierney topology* on $\mathcal{E}$. In the case where $\mathcal{E} = \mathcal{P}r(C)$ is a presheaf logos, this is also equivalent to the data of a *Grothendieck topology* on the category C. We shall come back to the notion of quotient of logoi in Section 3.4.3.

Example 3.4 (Embeddings)

1. From our definition of logoi, it is clear that every logos is a quotient of a presheaf logos, that is, that every topos $\mathcal{X}$ is a subtopos of an Alexandrov topos $\mathcal{X} \hookrightarrow \mathbb{B}C$. In fact, it can be proven that every logos is also a quotient of a free logos, that is, that every topos is a subtopos of an affine topos. This situation is similar to that of affine schemes.
2. If $Y \hookrightarrow X$ is an embedding of topological spaces or of locales, the corresponding map of topos is also an embedding. Moreover, any subtopos of a localic topos is localic.
3. For X a topological space or a locale, the logos morphism $\mathcal{P}r(\mathcal{O}(X)) \to \mathcal{S}h(X)$ is a quotient and the corresponding topos morphism $X \to \widehat{X}$ is an embedding of localic topoi. Recall that the points of $\widehat{X}$ are filters in $\mathcal{O}(X)$ and that the embedding $X \hookrightarrow \widehat{X}$ sends a point of X to the filters of its open neighborhoods.
4. Any fully faithful functor $C \hookrightarrow D$ between small categories induces a quotient $[D, \mathsf{Set}] \to [C, \mathsf{Set}]$ and an embedding $\mathbb{B}C \hookrightarrow \mathbb{B}D$. At the level of points, this embedding corresponds to the fully faithful functor $\mathfrak{Ind}(C) \hookrightarrow \mathfrak{Ind}(D)$.
5. In particular, the embedding $\underline{2} = \{\emptyset, \{\star\}\} \subset \mathsf{Fin}$ induces a quotient $\mathsf{Set}[X] = [\mathsf{Fin}, \mathsf{Set}] \to [\underline{2}, \mathsf{Set}] = \mathsf{Set}^\to$. Recall that $[\underline{2}, \mathsf{Set}] = \mathcal{S}h(\mathbb{S})$. We deduce that the Sierpiński space, when viewed as a topos, is a subtopos of the topos of sets: $\mathbb{S} \hookrightarrow \mathbb{A}$. At the level of points, this embedding corresponds to the inclusion $\{\emptyset, \{\star\}\} \subset \mathsf{Set}$. In other words, the Sierpiński topos can be said to classify sets with at most one element.
6. Another example is given by $\mathsf{Fin}^\circ \hookrightarrow \mathsf{Fin}$. This describes the topos $\mathbb{A}^\circ$ as a subtopos of $\mathbb{A}$. We already saw that at the level of points, this corresponds to the inclusion of nonempty sets in sets.
7. Yet another example is given by $C \hookrightarrow C_{\mathrm{rex}}$, where C_{rex} is the free completion of C for finite colimits. This builds a quotient of logoi $\mathsf{Set}\left[C^{op}\right] = [C_{\mathrm{rex}}, \mathsf{Set}] \to [C, \mathsf{Set}]$ and a dual embedding of topoi $\mathbb{B}C \to \mathbb{A}^{C^{op}}$. At the level of points, this embedding corresponds to the fully faithful functor

$\mathfrak{Ind}(C) \hookrightarrow \mathcal{Pr}(C)$. With the first example, this proves that any topos $\mathcal{X}$ can be embedded in some affine topos $\mathcal{X} \hookrightarrow \mathbb{B}C \hookrightarrow \mathbb{A}^{C^{op}}$.
8. The fully faithful inclusion $\text{Fin}^{\simeq} \hookrightarrow \text{Fin}^{\to}$ of isomorphisms into morphisms builds an embedding $\mathbb{A}^{\simeq} \hookrightarrow \mathbb{A}^{\to}$.

3.2.4 Products of Topoi

In analogy with locales/frames and commutative rings/schemes, the cartesian products of topoi correspond dually to a tensor product of logoi. If we forgot the existence of finite limits in a logos, the resulting category is a *presentable category*, that is, a reflective localization of a presheaf category. We shall say a few words about presentable categories in Section 3.3.3. The tensor product of logoi is defined at the level of their underlying presentable categories. A morphism of presentable categories is defined as a functor preserving all colimits. For three such categories $\mathcal{A}$, $\mathcal{B}$, and $\mathcal{C}$, a functor $\mathcal{A} \times \mathcal{B} \to \mathcal{C}$ is called *bilinear* if it preserves colimits in each variable. A bilinear functor $\mathcal{A} \times \mathcal{B} \to \mathcal{C}$ is determined by the same data as a morphism of presentable categories $\mathcal{A} \otimes \mathcal{B} \to \mathcal{C}$ for a certain presentable category $\mathcal{A} \otimes \mathcal{B}$. This category can be described as $\mathcal{A} \otimes \mathcal{B} = [\mathcal{A}^{op}, \mathcal{B}]^c$ (where $[\mathcal{A}^{op}, \mathcal{B}]^c$ is the category of functors preserving limits). This formula shows in particular that Set is the unit of this product. A comparison between this tensor product and that of abelian groups is sketched in Table 4.15.

Example 3.5 (Products)

1. The punctual topos $\mathbb{1}$ is the unit for the product. The equation $\mathbb{1} \times \mathcal{X} = \mathcal{X}$ for topoi is equivalent to $\text{Set} \otimes \mathcal{E} = \mathcal{E}$ for logoi.
2. The tensor product of presentable categories is such that $\mathcal{Pr}(C) \otimes \mathcal{Pr}(D) = \mathcal{Pr}(C \times D)$. We deduce that $\mathbb{B}C \times \mathbb{B}D = \mathbb{B}(C \times D)$.
3. The free nature of $\text{Set}[C]$ and the universal property of sums implies that $\text{Set}[C] \otimes \text{Set}[D] = \text{Set}[C \coprod D]$, that is, $\mathbb{A}^C \times \mathbb{A}^D = \mathbb{A}^{C \coprod D}$.
4. Given two topoi $\mathcal{X}$ and $\mathcal{Y}$, the logos corresponding to $\mathcal{X} \times \mathcal{Y}$ can be described as the category of sheaves on $\mathcal{X}$ with values in $\text{Sh}(Y)$ (or reciprocally):

$$\text{Sh}(\mathcal{X}) \otimes \text{Sh}(\mathcal{Y}) = [\text{Sh}(\mathcal{X})^{op}, \text{Sh}(\mathcal{Y})]^c = [\text{Sh}(\mathcal{Y})^{op}, \text{Sh}(\mathcal{X})]^c.$$

3.2.5 Fiber Products of Topoi

An important difference between topoi and topological space is the way fiber products are computed. The fact that topoi live in a 2-category requires the use of the so-called *pseudo fiber products*. We are only going to explain intuitively the situation. Let us consider a cartesian square

$$\begin{array}{ccc} \mathcal{X} \times_{\mathcal{Z}} \mathcal{Y} & \longrightarrow & \mathcal{X} \\ \downarrow {\scriptstyle \ulcorner} & & \downarrow {\scriptstyle f} \\ \mathcal{Y} & \xrightarrow{g} & \mathcal{Z} \end{array}$$

If $\mathcal{X}$, $\mathcal{Y}$, and $\mathcal{Z}$ were topological spaces or locales, $\mathcal{X} \times_{\mathcal{Z}} \mathcal{Y}$ would be the subspace of $\mathcal{X} \times \mathcal{Y}$ spanned by pairs (x, y) such that $f(x) = g(y)$ in $\mathcal{Z}$. The computation of fiber product of topoi is similar, but since the points of topoi leave in categories, the previous equality has to be replaced by an isomorphism. The choice of an isomorphism $f(x) \simeq g(y)$ being a structure and not a property, the map $\mathcal{X} \times_{\mathcal{Z}} \mathcal{Y} \to \mathcal{X} \times \mathcal{Y}$ will no longer be an embedding.[36] In the simplest case of the fiber product

$$\begin{array}{ccc} \mathbb{1} \times_{\mathbb{B}G} \mathbb{1} & \longrightarrow & \mathbb{1} \\ \downarrow {\scriptstyle \ulcorner} & & \downarrow {\scriptstyle b} \\ \mathbb{1} & \xrightarrow{b} & \mathbb{B}G \end{array}$$

we have $\mathbb{1} \times_{\mathbb{B}G} \mathbb{1} = G$, since the choice of an isomorphism $b \simeq b$ is the choice of an element of G.

More generally, let X be a space and G a discrete group acting on X. Recall from the examples that the quotient $X /\!/ G$ of X by G computed in the category of topoi is dual to the logos $\mathcal{S}h(X, G)$ of equivariant sheaves on X. It can be proved that the fibers of the quotient map $q: X \to X /\!/ G$ are isomorphic to G. Let x be a point of X and $\bar{x}$ be the corresponding point in $X /\!/ G$; then we have a cartesian square in the 2-category of topoi

$$\begin{array}{ccc} G = \mathbb{1} \times_{X /\!/ G} X & \xrightarrow{\mathrm{orbit}(x)} & X \\ \downarrow {\scriptstyle \ulcorner} & & \downarrow {\scriptstyle q} \\ \mathbb{1} & \xrightarrow{\bar{x}} & X /\!/ G \end{array}$$

where the top map sends G to the orbit of x. We mentioned that the category of points of $X /\!/ G$ is the groupoid associated to the action of G on the points of X. So an isomorphism $y \simeq x$ in this groupoid is equivalent to the choice of y in the orbit of x and of an element of g such that $g.x = y$. But this is just equivalent to just choosing an element $g \in G$. This is why the fiber is G. In fact, the morphism $X \to X /\!/ G$ can even be proven to be a principal G-cover. This is one of the nice features of quotients of discrete group actions in Topos – the quotient map is always a principal cover.

A variation on the same theme is the computation of fibers of the diagonal map $\mathcal{X} \to \mathcal{X} \times \mathcal{X}$ of a topos. Let $(x, y): \mathbb{1} \to \mathcal{X} \times \mathcal{X}$ be a pair of points

---

[36] The fiber of this map at a pair (x, y) being the choices of isomorphisms $f(x) \simeq g(y)$.

of $\mathcal{X}$. By a classical trick of category theory, the fiber product $\mathbb{1} \times_{\mathcal{X} \times \mathcal{X}} \mathcal{X}$ is equivalent to $\Omega_{x,y}\mathcal{X} := \mathbb{1} \times_{\mathcal{X}} \mathbb{1}$, that is, to "path space" between x and y in $\mathcal{X}$. If $\mathcal{X}$ is a topological space or even a locale, this intersection is empty if $x \neq y$ and a single point if $x = y$. But within a topos, points can have isomorphisms, and the topos $\mathbb{1} \times_{\mathcal{X}} \mathbb{1}$ is precisely the topos classifying the isomorphisms between x and y. It is empty if x and y are not isomorphic, but its category of points is the set $\mathrm{Iso}_{\mathcal{P}\mathrm{t}(\mathcal{X})}(x, y)$ if they are. It is possible to prove that $\Omega_{x,y}\mathcal{X}$ is always a localic topos. It follows from these observations that the diagonal map $\mathcal{X} \to \mathcal{X} \times \mathcal{X}$ of a topos is not necessarily an embedding!

Another important example of fiber products is the computation the fiber of the map $\mathbb{A}^\bullet \to \mathbb{A}$. Recall that this map sends a pointed space to its underlying set. Intuitively, the fiber over a set E should be the choice of a base point in E. One can prove that this is indeed the case: recall that $\mathbb{B}E$ is the discrete topos associated to a set E; then there exists a cartesian square

$$\begin{array}{ccc} \mathbb{B}E & \longrightarrow & \mathbb{A}^\bullet \\ \downarrow{\scriptstyle \Gamma} & & \downarrow \\ \mathbb{1} & \xrightarrow{\chi_E} & \mathbb{A} \end{array}$$

For this reason, $\mathbb{A}^\bullet \to \mathbb{A}$ is called the *universal family of sets*.

3.2.6 Etale Domains

We now turn to a central notion of topos theory. We explained in the introduction that, in the same way locales are based on the notion of open domain, the theory of topoi is based on the notion of étale morphism (see Table 4.4). Recall that an open embedding $U \to X$ was defined as an open quotient of frames $U \cap - : \mathcal{O}(X) \to \mathcal{O}(X)_{/U}$ for some U in $\mathcal{O}(X)$. The corresponding notion for logoi will correspond to étale maps. Let $\mathcal{E}$ be a logos and F an object of $\mathcal{E}$. The base change along the map $F \to 1$ provides a morphism of logoi $\epsilon_F^* : \mathcal{E} \to \mathcal{E}_{/F}$ called an *étale extension*. If $\mathcal{E} = \mathrm{Sh}(\mathcal{X})$, the corresponding morphisms of topoi will be denoted $\epsilon_F : \mathcal{X}_F \to \mathcal{X}$ and called an *étale morphism* or a *local homeomorphism*. Intuitively, an étale morphism is a morphism whose fibers are discrete. We are going to see that this is indeed the case. We are also going to explain the universal property of $\mathcal{E} \to \mathcal{E}_{/F}$.

Example 3.6 (Etale Morphisms)

1. The identity morphism of a topos $\mathcal{X}$ is étale.
2. The morphism $\emptyset \to \mathcal{X}$ from the empty topos is étale.

3. The morphism $\mathbb{A}^\bullet \to \mathbb{A}$ is étale. Recall that the object X in Set$[X] = $ [Fin, Set] is represented by the object 1 in Fin. Then the result is a consequence of the formula [Fin, Set]$_{/X} = $ [Fin$_{1/}$, Set] $= $ [Fin$^\bullet$, Set]. We shall see that it is the universal étale morphism.

4. The same proof shows that the morphism $\mathbb{A}^\bullet \to \mathbb{A}^\circ$ is étale. We shall see that it is also surjective.

5. The morphism $b\colon \mathbb{1} \to \mathbb{B}G$ is étale. Recall that it corresponds dually to the forgetful functor $U\colon \text{Set}^G \to \text{Set}$. Let G_λ be the action of G on itself by left translation. Then we have Set $= \left(\text{Set}^G\right)_{/G_\lambda}$.[37]
The morphism $b\colon \mathbb{1} \to \mathbb{B}G$ is moreover étale; it can be proved to be a principal covering with structure group G. It is in fact the universal cover of $\mathbb{B}G$.

The étale extension $\epsilon_F^*\colon \mathcal{E} \to \mathcal{E}_{/F}$ has an important universal property. The object $\epsilon_F^*(F)$ in $\mathcal{E}_{/F}$ corresponds to the map $p_1\colon F^2 \to F$, which admits a canonical section given by the diagonal $\Delta\colon F \to F^2$. Then pair (ϵ_F^*, Δ) is universal for creating a global section of F. More precisely, if $u^*\colon \mathcal{E} \to \mathcal{F}$ is a logos morphism and $\delta\colon 1 \to u^*F$ a global section of F in $\mathcal{F}$, there exists a unique factorization of u^* via $\mathcal{E}_{/F}$ such that $v^*(\Delta) = \delta$:

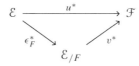

This property is to be compared with the splitting of a polynomial in commutative algebra, as shown in Table 4.12.

This property has also an important geometric interpretation. Suppose that $\mathcal{E} = \text{Sh}(\mathcal{X})$ and $\mathcal{F} = \text{Sh}(\mathcal{Y})$. Recall from the examples that a pointed object $\delta\colon 1 \to F$ in $\mathcal{F}$ determines a logos morphism Set$[X^\bullet] \to \mathcal{F}$. Then, the data of (u^*, δ) above is equivalent to a commutative square of logoi

$$\begin{array}{ccc} \text{Set}[X] & \xrightarrow{X \mapsto u^*F} & \mathcal{E} \\ \downarrow & & \downarrow u^* \\ \text{Set}[X^\bullet] & \xrightarrow{1 \to X^\bullet \mapsto 1 \xrightarrow{\delta} u^*F} & \mathcal{F} \end{array}$$

Geometrically, this corresponds to a square of topoi

---

[37] For a G-set F, an equivariant morphism $\varphi\colon F \to G_\lambda$ is equivalent to a trivialization of the action of G on F. Let $E \subset F$ be the elements of F sent to the unit of G by φ; then we have $G \times E \simeq F$ as G-sets.

Table 4.12. *Etale analogies*

Algebraic geometry	Topos theory
Ring A	logos $\mathcal{E}$
Separable polynomial $P(x)$ in $A[x]$	object F of $\mathcal{E}$
Separable (or étale) extension $A \to A[x]/P(x)$	étale extension $\mathcal{E} \to \mathcal{E}_{/F}$
Root of P in A = retraction of $A \to A[x]/P(x)$	global section $1 \to F$ = retraction of $\mathcal{E} \to \mathcal{E}_{/F}$

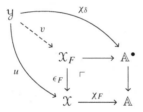

The universal property of $\mathcal{X}_F$ says exactly that it is the fiber product of $\mathcal{X} \to \mathbb{A} \leftarrow \mathbb{A}^\bullet$:

$$\begin{array}{ccc} \mathcal{Y} & \xrightarrow{\chi_\delta} & \\ & \searrow v & \\ u & \mathcal{X}_F & \longrightarrow \mathbb{A}^\bullet \\ & \epsilon_F \downarrow \ulcorner & \downarrow \\ & \mathcal{X} & \xrightarrow{\chi_F} \mathbb{A} \end{array}$$

The fact that any étale morphism is a pullback of the universal family of sets $\mathbb{A}^\bullet \to \mathbb{A}$ says that it is also the *universal étale morphism*. The previous computation of the fibers of $\mathbb{A}^\bullet \to \mathbb{A}$ gives a proof that the fiber of ϵ_F at a point x of $\mathcal{X}$ is the stalk $F(x)$ of F. If X is a topological space and F is a sheaf on X, one can prove that $X_F \to X$ is the *espace étalé* corresponding to the sheaf [13, Section II.1.2]. The construction $F \mapsto \mathcal{X}_F$ of the "topos étalé" of a sheaf builds a functor

$$\mathsf{Sh}(\mathcal{X}) \hookrightarrow \mathsf{Topos}_{/\mathcal{X}} \qquad (3.2)$$

whose image is spanned by étale morphisms over $\mathcal{X}$, or *étale domains of* $\mathcal{X}$. This functor is fully faithful and preserves colimits and finite limits. In other words, sheaves and their operations are faithfully represented as étale maps. Together with (3.1), this completes the algebraic/geometric interpretation of sheaves mentioned in Table 4.4.

3.2.7 Open Domains

In accordance with what is true for topological spaces, we define an *open embedding* of a topos $\mathcal{X}$ to be an étale morphism $\mathcal{Y} \to \mathcal{X}$ that is also an embedding. The corresponding morphisms of logoi will be called *open quotients*. For an object U in a logos $\mathrm{Sh}(\mathcal{X})$, the functor $\epsilon_U^* : \mathrm{Sh}(\mathcal{X}) \to \mathrm{Sh}(\mathcal{X})_{/U}$ is a quotient if and only if the canonical morphism $U \twoheadrightarrow 1$ is a monomorphism. This characterizes open domains as the étale domains $\mathcal{X}_U \to \mathcal{X}$ where U is a subterminal object. The étale domains of a topos $\mathcal{X}$ form a full subcategory $\mathcal{O}(\mathcal{X}) \subset \mathrm{Sh}(\mathcal{X})$ that coincides with the poset $Sub(1)$ of subobjects of 1 in $\mathrm{Sh}(\mathcal{X})$.

Intuitively, an étale morphism is an embedding if its fibers are either empty or a point. Recall the embedding $\mathbb{S} \subset \mathbb{A}$ of Sierpiński space into the topos of sets. It can be proven that an étale domain is open if and only if the classifying map $\mathcal{X} \to \mathbb{A}$ factors through $\mathbb{S} \subset \mathbb{A}$. This says that the Sierpiński space, when viewed as a topos, keeps the nice property of classifying open domains:

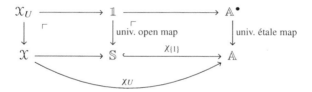

Example 3.7 (Open Embeddings)

1. The open embeddings of a localic topos coincide with the open domains of the corresponding locale.
2. Let $C \subset D$ be a full subcategory that is a *cosieve* (stable by postcomposition). Then the localization $[D, \mathrm{Set}] \to [C, \mathrm{Set}]$ is open and the embedding $\mathbb{B}C \to \mathbb{B}D$ is open. In fact, the poset of open quotients of $[D, \mathrm{Set}]$ can be proved to be exactly the poset of cosieves of D.
3. For any topos $\mathcal{X}$, the identity of $\mathcal{X}$ and the canonical morphism $\emptyset \to \mathcal{X}$ are always open embeddings.
4. The subtopos $\mathbb{A}^\circ \subset \mathbb{A}$ is open. This is the only nontrivial open subtopos of $\mathbb{A}$. The classifying morphism $\mathbb{A} \to \mathbb{S}$ of this open domain is a retraction of the embedding $\mathbb{S} \hookrightarrow \mathbb{A}$.

A topos $\mathcal{X}$ is said to have *enough open domains* if all sheaves on $\mathcal{X}$ can be written as pastings of open domains, that is, if the subcategory $\mathcal{O}(\mathcal{X}) \subset \mathrm{Sh}(\mathcal{X})$ generates by colimits. A topos has enough open domains if and only if it is localic, that is, in the image of the functor $\mathrm{Locale} \to \mathrm{Topos}$. Not every topos has enough open domains, and this is a very important fact of the theory. The topos $\mathbb{B}G$ does not have enough open domains. The computation shows that the only open domains of $\mathbb{B}G$ are the identity and $\emptyset \to \mathbb{B}G$, that is, $\mathbb{B}G$ has the same open domains as the point.

The intuitive explanation of what is going on is simple enough. Any morphism $\mathbb{B}G \to \mathbb{S}$ induces a functor $G = \mathcal{P}t(\mathbb{B}G) \to \mathcal{P}t(\mathbb{S}) = \{0 < 1\}$. Since the only isomorphisms in the poset $\{0 < 1\}$ are the identities, any functor from G has to be constant. This is why there are so few open domains. In other words, the Sierpiński space does not have "enough room" to reflect that some spaces have many morphisms between points. This is actually the source of the insufficiency of the notion topological space. In its essence, the theory of topoi proposes to enlarge the "gauge" poset $\{0 < 1\}$ by the "gauge" category Set. Doing so creates "enough room" to capture faithfully many spaces with a category of points.

3.2.8 Closed Embeddings

Let $\mathcal{X}_U \hookrightarrow \mathcal{X}$ be an open domain corresponding to an object U in $\mathrm{Sh}\,(\mathcal{X})$. It is possible to define a *closed complement* for $\mathcal{X}_U$, but we shall not detail this.

Example 3.8 (Closed Embeddings)

1. The closed embeddings of locales give closed embeddings of topoi.
2. We saw that cosieves $C \subset D$ correspond to open embeddings $\mathbb{B}C \to \mathbb{B}D$. Reciprocally, *sieves* (subcategories stable by precomposition) correspond to closed embeddings. If $C \subset D$ is a cosieve, the full subcategory C' of D spanned by the objects not in C is a sieve. Then $\mathbb{B}C \hookrightarrow \mathbb{B}D$ and $\mathbb{B}C' \hookrightarrow \mathbb{B}D$ are complementary open and closed embeddings.
3. The closed complement of the open embedding $\mathbb{A}^\circ \subset \mathbb{A}$ is the morphism $\chi_\emptyset : \mathbb{1} \hookrightarrow \mathbb{A}$ classifying the empty set.

3.2.9 Socle and Hyperconnected Topoi

For any topos $\mathcal{X}$, the poset $\mathcal{O}(\mathcal{X})$ of its open domains is a frame and defines a locale $\mathrm{Socle}(\mathcal{X})$. This provides a functor $\mathrm{Socle}: \mathsf{Topos} \to \mathsf{Locale}$, which is the left adjoint to the inclusion $\mathsf{Locale} \to \mathsf{Topos}$. The unit of this adjunction provides a canonical projection $\mathcal{X} \to \mathrm{Socle}(\mathcal{X})$. Intuitively, the socle of $\mathcal{X}$ is the best approximation of $\mathcal{X}$ that can be built out of open domains only.[38] A topos is called *hyperconnected* if its socle is a point. In other words, the hyperconnected topoi are exactly the kind of spatial object invisible from the usual point of view on topology (see [19] for more properties).

---

[38] The corresponding logos morphism $\mathrm{Sh}\,(\mathrm{Socle}(\mathcal{X})) \to \mathrm{Sh}\,(\mathcal{X})$ is full and faithful. Its image is the smallest full category containing $\mathcal{O}(\mathcal{X})$ and stable by colimits and finite limits. In other words, it is the subcategory of sheaves that can be generated by open domains.

Example 3.9 (Socles and Hyperconnected Topoi)

1. The inclusion of categories Poset → Cat has a left adjoint τ. The poset $\tau(C)$ has the same objects as C and $x \leq y$ if there exists an arrow $x \to y$ in C. The socle of $\mathbb{B}C$ is the Alexandrov locale associated to $\tau(C)$. Its frame of open domains is $[C, \underline{2}]$.
2. A category C is called hyperconnected if any two objects have arrows going both ways between them. This is equivalent to $\tau(C) = 1$. Then the corresponding Alexandrov topos $\mathbb{B}C$ is hyperconnected.
3. In particular, the topoi $\mathbb{A}^\bullet$, $\mathbb{A}^\circ$, $\mathbb{B}G$ are all hyperconnected, but not $\mathbb{A}$ (because of the strictness of $\emptyset$).
4. Examples of hyperconnected topoi are also given by the so-called bad quotients in topology. Let $\mathbb{Q}$, viewed as discrete group, act on $\mathbb{R}$ by translation. Every orbit is dense, and the topological quotient is an uncountable set with the discrete topology. The topos quotient $\mathbb{R}/\!/\mathbb{Q}$ is the topos corresponding to the logos of $\mathbb{Q}$-equivariant sheaves on $\mathbb{R}$. It stays true in the category of topoi that open domains of the quotient $\mathbb{R}/\!/\mathbb{Q}$ are equivalent to saturated open domains of $\mathbb{R}$, and this proves that $\mathbb{R}/\!/\mathbb{Q}$ is a hyperconnected topos. One can compute that its category of points is exactly the set of orbits of the action. So the topos $\mathbb{R}/\!/\mathbb{Q}$ has the same points and open domains as the topological quotient, but it has more sheaves! This topos enjoys many nice properties missing for the topological quotient. For example, it can be proved that its fundamental group is $\mathbb{Q}$. This is a good example of how defining a spatial object by its category of étale domains and not only its open domains leads to more regular objects.

3.2.10 Surjections

The notion of surjection of topoi is more subtle than the one of locales. The definition is based on the following property of surjection of spaces. Let $u\colon Y \to X$ be a continuous map and $f\colon F \to G$ a morphism of sheaves on X. Intuitively, f is an isomorphism if and only if all the maps $f(x)\colon F(x) \to G(x)$ between the stalks are bijections. If f is an isomorphism, then so is $u^*f\colon u^*F \to u^*G$ in $\mathcal{S}h(Y)$. If u is not surjective, the condition "u^*f is an isomorphism" is weaker than $F \simeq G$ because it does not say anything about the stalks that are not in the image of u. But if u is surjective, the condition "u^*f is an isomorphism" becomes equivalent to "f is an isomorphism."

A functor $f\colon C \to D$ is called *conservative* if it is true that "u is an isomorphism" ⇔ "$f(u)$ is an isomorphism." A morphism of topoi $f\colon \mathcal{Y} \to \mathcal{X}$ is called a *surjection* if the corresponding morphism of logoi $f^*\colon \mathcal{S}h(\mathcal{X}) \to \mathcal{S}h(\mathcal{Y})$ is conservative.

Example 3.10 (Surjections)

1. The morphism $\mathbb{1} \to \mathbb{B}G$ is a surjection. This is because the forgetful functor $\text{Set}^G \to \text{Set}$ is conservative.
2. The functor $[\text{Fin}^\circ, \text{Set}] \to [\text{Fin}^\bullet, \text{Set}]$ is conservative. Thus the morphism $\mathbb{A}^\bullet \to \mathbb{A}^\circ$ is surjective.
3. Let $\mathcal{X}$ be a topos and E be a set of points of $\mathcal{X}$. Then there exists a logos morphism $\text{Sh}(\mathcal{X}) \to [E, \text{Set}]$ sending a sheaf F to the family of its stalks corresponding to the points in E. Dually, this corresponds to a topos morphism $\mathbb{B}E \to \mathcal{X}$ where $\mathbb{B}E$ is the discrete topos associated to the set E. A topos is said to have *enough points* if there exists a set E such that the topos morphism $\mathbb{B}E \to \mathcal{X}$ is surjective. Intuitively, this means that a morphism $F \to G$ between sheaves on $\mathcal{X}$ is an isomorphism if and only if the morphism $F(x) \to G(x)$ is a bijection for all x in E.
Recall from Section 2.2.13 that topological spaces can be faithfully described as locales equipped with a surjective map from a discrete locale. The corresponding notion for topoi, which would be a categorification of topological spaces, is a topos equipped with a surjective morphism from a discrete topos. Such a notion has been studied in [11].

3.2.11 Image Factorization

With the notions of embedding and surjection, it is possible to define the image of a morphism of topoi $u: \mathcal{Y} \to \mathcal{X}$. From the corresponding morphism of logoi $f^*: \text{Sh}(\mathcal{X}) \to \text{Sh}(\mathcal{Y})$, we extract the class W of maps inverted by u^* and construct the left-exact localization of $\text{Sh}(\mathcal{X})/\!/W$ generated by W.[39] We deduce a factorization

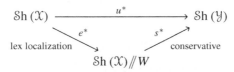

where e^* is a quotient and s^* is conservative by design. In the corresponding geometric factorization,

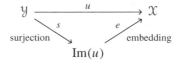

the subtopos $\text{Im}(u) \hookrightarrow \mathcal{X}$ is called the *image* of u.

---

[39] Technically, there is a size issue, and we need to prove that W can be generated by a single map $f: A \to B$ in $\text{Sh}(\mathcal{X})$. This is possible because f is an accessible functor between accessible categories.

Example 3.11 (Image Factorization)

1. Given a functor $C \to D$ between small categories, the image factorization of $\mathbb{B}C \to \mathbb{B}D$ is $\mathbb{B}C \to \mathbb{B}C' \to \mathbb{B}D$, where $C \to C' \to D$ is the essentially surjective/fully faithful factorization of $C \to D$.
2. In particular, the image of the morphism $\mathbb{A}^\bullet \to \mathbb{A}$ is the topos $\mathbb{A}^\circ$.
3. In the case of an object $x: 1 \to D$, the image $\mathbb{1} \to \mathbb{B}D$ is $\mathbb{B}(\text{End}(x))$ (dual to the logos of action of the monoid $\text{End}(x)$ on sets). The category of points of this topos consists in all the retracts of x in D.

3.2.12 Etale Covers

The image factorization in the category Topos echoes with another image factorization that exists *within* a given logos $\mathcal{E}$. Recall that for any map $f: A \to B$, the diagonal of f is the map $A \to A \times_B A$. The object $A \times_B A$ is a subobject of $A \times A$ that intuitively corresponds to the relation "having the same image by f." The coequalizer of $A \times_B A \rightrightarrows A$ is the quotient of A by this relation. The map f is called a *cover* if this coequalizer is B. This is a way to say that f is surjective. The map f is called a *monomorphism* if its diagonal $A \to A \times_B A$ is an isomorphism. This is a way to say that f is injective. We shall denote by $A \twoheadrightarrow B$ the covers and by $A \rightarrowtail B$ the monomorphisms. In the logos Set, the covers and monomorphisms are exactly the surjections and injections. In the logos $\mathcal{S}h(X)$ of sheaves on a topological space X, covers and monomorphisms are the maps that are surjective and injective stalk-wise.

Any map f in a logos can be factored uniquely in a cover followed by a monomorphism:

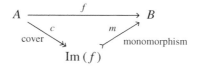

where the object $\text{Im}(f)$, called the *image* of f, is defined as the coequalizer of $A \times_B A \rightrightarrows A$.

If $\mathcal{E} = \mathcal{S}h(\mathcal{X})$, the correspondence (3.2) transforms the previous factorization into the surjection–embedding factorization:

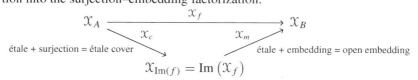

In other words, the correspondence (3.2) transforms covers into surjections and monomorphisms into embeddings. We saw that the class of monomorphisms produced this way, that is, monomorphisms that are étale, are the open embeddings. The class of surjections produced this way, that is, surjections that are étale, are called *étale covers*.

Example 3.12 (Etale Covers)

1. Any surjective local homeomorphism between topological spaces defines an étale cover between the associated topoi.
2. In particular, if U_i is an open covering of a space X, then $U = \coprod_i U_i \to X$ is an étale cover of the topos corresponding to X.
3. The étale covers of a topos $\mathcal{X}$ are equivalent to objects U in $\mathrm{Sh}(\mathcal{X})$ such that the map $U \to 1$ is a cover. Such objects are also called *inhabited* since they correspond intuitively to sheaves whose stalks are never empty. When viewed as a function, a sheaf $\mathcal{X} \to \mathbb{A}$ is inhabited if and only if it takes its values in the subtopos $\mathbb{A}^\circ \subset \mathbb{A}$. Finally, an étale cover of $\mathcal{X}$ is equivalent to a morphism $\mathcal{X} \to \mathbb{A}^\circ$.
4. The map $\mathbb{1} \to \mathbb{B}G$ is an étale cover since it is étale and surjective.
5. More generally, if a discrete group G acts on a space X, the quotient map $q: X \to X /\!/ G$ is also an étale cover. In particular, the map $\mathbb{R} \to \mathbb{R}/\!/\mathbb{Q}$ is étale.
6. The map $\mathbb{A}^\bullet \to \mathbb{A}^\circ$ is an étale cover since we saw that it was étale and surjective. Recall that it is given by $\mathsf{Set}[X^\circ] \to \mathsf{Set}[X^\circ]_{/X^\circ}$. The fact that X° is an inhabited object is the universal property of the logos $\mathsf{Set}[X^\circ]$. Any nonempty object E in a logos $\mathcal{E}$ defines a unique logos morphism $\mathsf{Set}[X^\circ] \to \mathcal{E}$ sending X° to E.
7. The factorization $\mathbb{A}^\bullet \to \mathbb{A}^\circ \to \mathbb{A}$ corresponds to the image factorization $X \to X^\circ \to 1$ of the map $X \to 1$ in $\mathsf{Set}[X]$. It is in fact the universal such factorization. Let F be a sheaf on $\mathcal{X}$ and let $F \to \mathrm{Im}(F) \to 1$ be the cover-monomorphism factorization of the canonical map $F \to 1$. Then the image factorization of $\mathcal{X}_F \to \mathcal{X}$ can be defined by the pullbacks

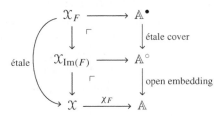

3.2.13 Constant Sheaves

Since Set is the initial logos, every logos $\mathcal{E}$ comes with a canonical morphism $e^*: \mathsf{Set} \to \mathcal{E}$. This functor is left adjoint to the *global section functor* $\Gamma = e_*: \mathcal{E} \to \mathsf{Set}$, which sends a sheaf F to $\Gamma(F) = \mathrm{Hom}_\mathcal{E}(1, F)$. The sheaves in the image of e^* are called *constant sheaves*. Geometrically, $e^*: \mathsf{Set} \to \mathrm{Sh}(\mathcal{X})$

corresponds to the unique morphism $X \to 1$. The interpretation of constant sheaves is that they are the pullback of sheaves on the point. In other words, they are the sheaves with a constant classifying morphism $X \to 1 \to A$.

3.2.14 Connected Topos

The previous functor $e^*\colon \text{Set} \to \mathcal{E}$ is not fully faithful in general. The only case where it is not faithful is when $\mathcal{E} = 1$ is the terminal logos (empty topos). But, when e^* is faithful, there might still be more morphisms between constant sheaves than between the corresponding sets. This is in fact characteristic of spaces with several connected components. For this reason, the logos $\mathcal{E}$ and the corresponding topos are called *connected* whenever e^* is fully faithful. More generally, a morphism of topos $u\colon \mathcal{Y} \to \mathcal{X}$ is called *connected* if the corresponding morphism of logoi $u^*\colon \text{Sh}(\mathcal{X}) \to \text{Sh}(\mathcal{Y})$ is fully faithful. The geometric intuition is that u has connected fibers. These definitions coincide with the existing notions for topological spaces.

Example 3.13 (Connected Topoi)

1. If X is a connected topological space or locale, then the corresponding topos is also.
2. An Alexandrov topos $\mathbb{B}C$ is connected if and only if the category C is connected (all objects can be linked by a zigzag of morphisms).
3. In particular, the topoi $\mathbb{1}, \mathbb{A}, \mathbb{A}^C, \mathbb{A}^\bullet, \mathbb{A}^\circ$, and $\mathbb{B}G$ are all connected.
4. Any hyperconnected topos is connected.

3.2.15 Connected–Disconnected Factorization

Given a morphism of topoi $u\colon \mathcal{Y} \to \mathcal{X}$, there exists a factorization related to connected morphisms. We define the *image* of $u^*\colon \text{Sh}(\mathcal{X}) \to \text{Sh}(\mathcal{Y})$ to be the smallest full subcategory $\mathcal{E}$ of $\text{Sh}(\mathcal{Y})$ containing the image of $\text{Sh}(\mathcal{X})$ and stable by colimits and finite limits.[40] It happens that $\mathcal{E}$ is a logos and that the functors $\mathcal{E} \to \text{Sh}(\mathcal{Y})$ and $\text{Sh}(\mathcal{X}) \to \mathcal{E}$ are logos morphisms. Let $\mathcal{Z}$ be the topos corresponding to $\mathcal{E}$. By design, the morphism $\text{Sh}(\mathcal{Z}) \to \text{Sh}(\mathcal{Y})$ is fully faithful, hence the corresponding topos morphism $\mathcal{Y} \to \mathcal{Z}$ has connected fibers. We shall call *dense* a morphism of logoi $\text{Sh}(\mathcal{Z}) \to \text{Sh}(\mathcal{Y})$ whose image is the whole of $\text{Sh}(\mathcal{Y})$ and *disconnected* the corresponding morphisms of topoi:

A topos $\mathcal{X}$ is called *disconnected* if the morphism $\mathcal{X} \to \mathbb{1}$ is. A disconnected topos $\mathcal{X}$ is such that the constant sheaves generate the whole of $\text{Sh}(\mathcal{X})$ by

---

[40] The construction is akin to that of the subring image of a ring morphism.

means of colimits and finite limits. Intuitively, it is easy to understand how this cannot be the case over a connected space like $\mathbb{R}$ of S^1: there is no way to build the open domains from constant sheaves since all morphisms between them are also constant. Therefore, the connected components of a disconnected topos must have "constant" trivial open domains and be points. In fact, it can be proved that disconnected topoi are totally disconnected spaces. Finally, the geometric intuition behind the connected–disconnected factorization $\mathcal{X} \to \mathcal{Z} \to \mathbb{1}$ is that $\mathcal{Z}$ is the disconnected space of connected components of the fiber. The intuition for the factorization of a morphism is the same fiber-wise.

Example 3.14 (Disconnected Morphisms)
1. Any discrete topos $\mathbb{B}E$ is disconnected over $\mathbb{1}$.
2. Any étale morphism, in particular, any open embedding, is disconnected. This is indeed the intuition of étale morphism, since we saw that the fibers are discrete topoi $\mathbb{B}E$.
3. Any limit of disconnected topoi is a disconnected topos. In fact, it can be proved that any disconnected morphism is, in a certain sense, a limit of étale maps.
4. Any embedding of topoi can be proved to be disconnected.
5. Let K be the Cantor set; then the topos morphism $K \to \mathbb{1}$ dual to the canonical functor Set $\to \mathcal{S}h(K)$ is disconnected. This is true essentially because K can be written as a limit of discrete spaces. Recall that the Cantor set is a profinite set. Let Pro(Fin) be the category of profinite sets. The functor Fin $\to$ Topos sending a finite set F to the discrete topos $\mathbb{B}F$ can be extended (by commutation to filtered limits) into a functor Pro(Fin) $\to$ Topos that is fully faithful. The image of this functor is inside disconnected topoi.
6. Let $\mathbb{Q}$ be the set of rational numbers with the topology induced by $\mathbb{R}$; then the logos morphism Set $\to \mathcal{S}h(\mathbb{Q})$ is dense. (It is sufficient to reconstruct from constant sheaves a basis of the topology of $\mathbb{Q}$. The open subsets (a, b) with a and b irrational numbers are a basis. Any such open subset can be written as the kernel of some maps $1 \rightrightarrows 2$ in $\mathcal{S}h(\mathbb{Q})$.)
7. The diagonal map $\mathcal{X} \to \mathcal{X} \times \mathcal{X}$ of a topos $\mathcal{X}$ can be proved to be a disconnected map. Recall that we saw that the fiber of this map at a pair of points (x, y) is a (localic) topos $\Omega_{x,y}\mathcal{X}$ whose points are the isomorphisms between x and y. The disconnection of the diagonal implies that $\Omega_{x,y}\mathcal{X}$ is a disconnected topos.[41]

---

[41] This result is actually a source of a limitation of the theory of topoi. Once the notion of a space with a category of points makes sense, it is reasonable to assume that the automorphisms of a given point do form a topological group. The answer is positive, but the disconnection of the diagonal of a topos says that the topology of these automorphism groups is at best disconnected.

8. Let G be a topological group and $\mathrm{Set}^{(G)}$ be the logos of continuous action of G on sets. Let $\mathcal{X}$ be the corresponding topos. Then $\mathcal{X}$ is a connected topos, and the fibers of its diagonal map are torsors over the totally disconnected space of connected components of G.

3.2.16 Locally Connected Maps and π_0 Theory

The simple definition of the connected–disconnected factorization in terms of sheaves shows that the theory of topoi is particularly suited to deal with connected components. This factorization can also be defined for topological spaces, but the definition of disconnected spaces and disconnected maps in terms of open domains only is more complex.

It is an important feature of topological spaces that not all spaces have a nice *set* of connected components (the easiest counterexamples being the Cantor set or $\mathbb{Q}$). This says that the functor $(-)_{dis}:$ Set $\to$ Top sending a set E to the corresponding discrete space E_{dis} does not have a globally defined left adjoint. The situation is a fortiori the same for topoi, and not every topos has a set of connected components. Somehow, the disconnected topoi enlarge the class of discrete topoi just by what is needed so that every space always has a disconnected topos of connected components.

Classically, the spaces whose connected components form a set are the locally connected spaces. Recall that a space X is locally connected if any open subset is a union of connected open subsets. In fact, more is true, and any étale domain $Y \to X$ is also a union of connected open domains. Let $\pi_0(Y)$ be the set of connected components of such a Y. This produces a functor $\pi_0:$ Sh $(X) \to$ Set that is left adjoint to the canonical logos morphism Set $\to$ Sh (X). The existence of this left adjoint is essentially the definition of a locally connected topos.[42] More generally, a morphism of topoi $u: \mathcal{Y} \to \mathcal{X}$ is *locally connected* if the functor $u^*:$ Sh $(\mathcal{X}) \to$ Sh $(\mathcal{Y})$ has a (local) left adjoint $u_!$.[43] Intuitively, this means that its fibers are locally connected topoi. When $u: \mathcal{Y} \to \mathcal{X}$ is locally connected, the disconnected part $\mathcal{Z} \to \mathcal{X}$ of its connected–disconnected factorization $u: \mathcal{Y} \to \mathcal{Z} \to \mathcal{X}$ is an étale morphism.[44]

In particular, it is impossible to obtain S^1 or other connected topological groups as such groups. Indeed, because S^1 is connected, any action on a set is constant, i.e., $\mathrm{Set}^{S^1} =$ Set. Hence, from the point of view of topoi and sheaves of sets, the classifying space of S^1 is indistinguishable from a point. This is an example of a space without enough étale domains, i.e., beyond the world of topoi. The theory of topological stacks is better suited to dealing with these objects.

[42] In fact, a stronger condition is required: the adjoint π_0 must be *local*, i.e., satisfy the technical assumption that, for any set E and any sheaf F, we have $\pi_0(E \times F) \simeq E \times \pi_0(F)$.

[43] Here again, $u_!$ must satisfy a locality condition: for any sheaf E in Sh $(\mathcal{X})$ and any sheaf F in Sh $(\mathcal{Y})$, we need to have $u_!(u^*E \times F) \simeq E \times u_!(F)$.

[44] In this case, we have Sh $(\mathcal{Z}) =$ Sh $(\mathcal{X})_{/u_!1}$.

Example 3.15 (Locally Connected Topoi)

1. Any locally connected space is a locally connected topos.
2. Any Alexandrov topos $\mathbb{B}C$ is a locally connected topos.
3. In particular, the topoi $\mathbb{1}$, $\mathbb{A}$, $\mathbb{A}^C$, $\mathbb{A}^\bullet$, $\mathbb{A}^\circ$, and $\mathbb{B}G$ are all locally connected.
4. The topoi corresponding to the Cantor set and $\mathbb{Q}$ are not locally connected.

3.2.17 Locally Constant Sheaves and π_1 Theory

Fundamental groupoids are related to locally constant sheaves, and the theory of topoi is also well suited to working with them. However, the resulting theory has a formulation that is more sophisticated than the π_0-theory [10]. The main difficulty is in fact the definition of locally constant sheaves and particularly of locally constant morphisms between them.[45] Another aspect is that the analogue of the connected–disconnected factorization system is difficult to define in terms of sheaves of sets only. If sheaves of sets are enhanced into sheaves of groupoids (i.e., 1-stacks), then the theory of fundamental groupoids can be nicely formulated in a way analogous to the theory of connected components. We shall see later how the notion of ∞-topos helps to have a nice theory for the whole homotopy type of topoi.

Example 3.16 (Fundamental Groupoids)

1. The fundamental groupoids of a locally simply connected space and of its corresponding topos are the same.
2. When $\mathbb{Q}$ is viewed as a discrete group, the quotient $\mathbb{R}/\!/\mathbb{Q}$ is a connected and locally simply connected topos, and its fundamental group is $\mathbb{Q}$. More amusing, if $\mathbb{R}_{dis}$ is $\mathbb{R}$ viewed as a discrete space, the quotient $\mathbb{R}/\!/\mathbb{R}_{dis}$ is connected and locally simply connected, with a single point but with $\mathbb{R}_{dis}$ as its fundamental group.
3. The fundamental groupoid of an Alexandrov topos $\mathbb{B}C$ is the groupoid G obtained from C by inverting all arrows.
4. In particular, the fundamental groupoid of $\mathbb{B}G$ is the group G viewed as a groupoid with one object. The map $\mathbb{1} \to \mathbb{B}G$ is an étale map from a connected space; it is then a universal cover of $\mathbb{B}G$. This is compatible with the earlier computation that the fibers of this map are copies of G.
5. We deduce also that $\mathbb{1}$, $\mathbb{A}$, $\mathbb{A}^\circ$, and $\mathbb{A}^\bullet$ have trivial fundamental groupoids. (They are in fact examples of topoi with trivial homotopy type.)

---

[45] When a space X (or a topos) is not locally 1-connected, the category of locally constant sheaves is not a full subcategory of $\mathrm{Sh}(X)$.

3.2.18 Compact Topoi

We mention briefly how to define a condition of compactness on topoi. Recall that a locale X is called *compact* if the functor $\mathrm{Hom}_{\mathcal{O}(X)}(1, -) \colon \mathcal{O}(X) \to \underline{2}$ preserves directed unions. The corresponding property for a topos is to ask that the global section functor $\Gamma \colon \mathrm{Sh}(\mathfrak{X}) \to \mathrm{Set}$ preserve filtered colimits. A topos is called *tidy* if this is the case. As it happens, the condition to be tidy on a topological space or on a locale is a bit stronger than the compactness condition. More details can be found in [19].

Example 3.17 (Tidy Topoi)

1. Any compact Hausdorff space is tidy as a topos.
2. When G is a group of finite generation, the topos $\mathbb{B}G$ is tidy.
3. All $\mathbb{A}^C$ are tidy. The global section $\Gamma \colon [C^{\mathrm{lex}}, \mathrm{Set}] \to \mathrm{Set}$ is simply the evaluation at the terminal object 1 in C^{lex}. In particular, this is a cocontinuous functor.
4. An Alexandrov topos $\mathbb{B}C$ is tidy if C is a cofiltered category. This is true as soon as C has a terminal object.
5. In particular, $\mathbb{A}^\circ$ and $\mathbb{A}^\bullet$ are tidy.
6. For any locale, we saw that the topos $\widehat{X}$, dual to the presheaf logos $\mathcal{P}r(\mathcal{O}(X))$, is a localic and compact as a locale. It is in fact tidy as a topos. The coherent envelope $X_{\mathrm{coh}} \hookrightarrow \widehat{X}$ is also tidy as a topos.

3.2.19 Cohomology

It should not be a surprise that the setting of topoi is convenient for sheaf cohomology. This includes cohomology with constant coefficients or locally constant coefficients. This has actually been a motivation for the theory. We shall not develop this and refer to the literature for details [6]. However, as for the theory of fundamental groupoids and higher homotopy invariants, the notion of topos turns out to be less suited than that of ∞-topos for the purposes of cohomology theory (see Section 4.2.8).

3.2.20 Topoi as Groupoids

Topoi turned out to have a close relationship with stacks on the category of locales. A *localic groupoid* G is a groupoid $G_1 \rightrightarrows G_0$ where G_0 and G_1 are locales. The category of such groupoids is denoted GpdLocale. To any such groupoid, we can associate a logos $\mathrm{Sh}(G)$ of equivariant sheaves on G_0. This produces a functor GpdLocale $\to$ Topos between 2-categories. The main theorem of [21] proves that this functor is essentially surjective. However, this functor is not fully faithful (see [26] for a study of its fully faithfulness).

3.3 Descent and Other Definitions of Logoi/Topoi

The previous section explained how a number of topological features could be extended to topoi. We focus now more on the algebraic side of topos theory, that is, logos theory. The basic idea we have laid out is that a logos is a category $\mathcal{E}$ with finite limits, (small) colimits, and a compatibility relation between them akin to distributivity. There exist several ways to formulate this relation, and this is essentially the difference between the several definitions of topoi. We are going to present a unified view on the structure of logoi based in the geometric theory of descent, that is, the art of gluing. Such a path will also make it clear what is gained with the notion of ∞-logos/topos.

We start by some recollections on descent. Then, we formulate descent in a way that makes it closer to a distributivity condition. This will help us to explain Giraud and Lawvere axioms. Finally, we will sketch the deep analogy of structure between logoi, frames, and commutative rings.

3.3.1 Descent for Sheaves

We first recall some facts about the gluing of sheaves. Let $U_i \to X$ be an open covering of a space X, and let $U_{ij} = U_i \cap U_j$ and $U_{ijk} = U_i \cap U_j \cap U_k$. Let F be a sheaf on X. We define F_i, F_{ij}, and F_{ijk} to be the pullbacks of F along $U_i \to X$, $U_{ij} \to X$, and $U_{ijk} \to X$. All these data organize into a diagram[46]

$$\begin{array}{ccccccc}
\coprod_{ijk} F_{ijk} & \rightrightarrows & \coprod_{ij} F_{ij} & \rightrightarrows & \coprod_i F_i & \longrightarrow & F \\
\downarrow & & \downarrow & & \downarrow & & \downarrow \\
\coprod_{ijk} U_{ijk} & \rightrightarrows & \coprod_{ij} U_{ij} & \rightrightarrows & \coprod_i U_i & \longrightarrow & X
\end{array}$$

where the vertical maps are the étale maps corresponding to the sheaves. By construction of this diagram by pullback, all the squares of the diagram are cartesian. The cartesian property of this diagram is a clever way to encode the data of the *cocycle* gluing the F_i together to get back F. The cartesianness of the middle square says that the two pullbacks of F_i and F_j along $U_{ij} \to U_i$ and $U_{ij} \to U_j$ are isomorphic and gives $\phi_{ij} : F_{i|ij} \simeq F_{j|ij}$. The cartesianness of the left square says that these isomorphisms satisfy a coherence condition on U_{ijk}: $\phi_{ki}\phi_{jk}\phi_{ij} = id$.[47]

---

[46] This diagram is technically a truncated simplicial diagram. We have not drawn the degeneracy arrows to facilitate reading, but they are part of the diagram.

[47] The degeneracy maps not drawn in the diagram also give conditions on the ϕ_{ij}. In the middle square, we get the condition $\phi_{ii} = id$. In the left square, we get the conditions $\phi_{ij}\phi_{ji} = id = \phi_{ji}\phi_{ij}$.

We define a *descent data* relative to the covering $\{U_i\}$ as a cartesian diagram of sheaves

$$\begin{array}{ccccc} \coprod_{ijk} F_{ijk} & \rightrightarrows & \coprod_{ij} F_{ij} & \rightrightarrows & \coprod_i F_i \\ \downarrow & \ulcorner & \downarrow & \ulcorner & \downarrow \\ \coprod_{ijk} U_{ijk} & \rightrightarrows & \coprod_{ij} U_{ij} & \rightrightarrows & \coprod_i U_i \end{array} \qquad (3.3)$$

Morphisms of descent data are defined as morphisms of diagrams. The category of descent data is denoted $\mathrm{Desc}(\{U_i\})$ and called the *descent category* of the covering U_i.

This category has a conceptual definition. The vertical maps of (3.3) define objects in the categories $\prod_i \mathrm{Sh}(U_i)$, $\prod_{ij} \mathrm{Sh}(U_{ij})$, and $\prod_{ijk} \mathrm{Sh}(U_{ijk})$. These categories are related by pullback functors:

$$\prod_i \mathrm{Sh}(U_i) \rightrightarrows \prod_{ij} \mathrm{Sh}(U_{ij}) \rightrightarrows \prod_{ijk} \mathrm{Sh}(U_{ijk}).$$

Then, a descent data is the same thing as an object in the limit of this diagram of categories.[48] In other terms, we can define the descent category as

$$\mathrm{Desc}(\{U_i\}) = \lim \left(\prod_i \mathrm{Sh}(U_i) \rightrightarrows \prod_{ij} \mathrm{Sh}(U_{ij}) \rightrightarrows \prod_{ijk} \mathrm{Sh}(U_{ijk}) \right). \qquad (3.4)$$

The construction of the beginning builds a *restriction functor*:

$$\mathrm{rest}_{\{U_i\}} : \mathrm{Sh}(X) \longrightarrow \mathrm{Desc}(\{U_i\})$$
$$F \longmapsto (F_i, F_{ij}, F_{ijk}).$$

It is a classical result about sheaves that, reciprocally, it is possible to define a sheaf F on X by gluing a descent data (F_i, F_{ij}, F_{ijk}) relative to a covering U_i. In terms of category theory, this gluing is nothing but the colimit of the diagram

$$\coprod_{ijk} F_{ijk} \rightrightarrows \coprod_{ij} F_{ij} \rightrightarrows \coprod_i F_i.$$

This constructs a functor

$$\mathrm{glue}_{\{U_i\}} : \mathrm{Desc}(\{U_i\}) \longrightarrow \mathrm{Sh}(X)$$

that is left adjoint to the restriction functor.

---

[48] More precisely, it is a *pseudo-limit* computed in the 2-category of categories.

We shall say that a descent data along the covering $\{U_i\}$ is *faithful* if the functor $\mathrm{rest}_{\{U_i\}}$ is fully faithful and *effective* if the functor $\mathrm{colim}_{\{U_i\}}$ is fully faithful. Intuitively, the faithfulness of a descent data means that, given a sheaf F, its decomposition into (F_i, F_{ij}, F_{ijk}) followed by the gluing of the (F_i, F_{ij}, F_{ijk}) reconstructs F. The effectivity of a descent data says that the gluing of (F_i, F_{ij}, F_{ijk}) into some F followed by the decomposition of F reconstructs the diagram (F_i, F_{ij}, F_{ijk}). We shall say that the *descent property holds* along the covering $\{U_i\}$ if the descent data is effective and faithful, that is, if the adjunction $\mathrm{colim}_{\{U_i\}} \dashv \mathrm{rest}_{\{U_i\}}$ is an equivalence of categories,[49]

$$\mathrm{Sh}(X) \simeq \mathrm{Desc}(\{U_i\}).$$

These considerations can be extended to a topos $\mathfrak{X}$ in a straightforward way. The only difference is that the open embeddings $U_i \to X$ can be enhanced into étale maps $\mathcal{U}_i \to \mathfrak{X}$. Then, the $\mathcal{U}_{ij}$ are defined by the fiber products $\mathcal{U}_i \times_{\mathfrak{X}} \mathcal{U}_j$, and so on. Let U_i be the object of $\mathrm{Sh}(\mathfrak{X})$ corresponding to the étale morphisms $\mathcal{U}_i \to \mathfrak{X}$ by the correspondence (3.2). Recall that this correspondence preserves finite limits. This says that the fiber products $\mathcal{U}_i \times_{\mathfrak{X}} \mathcal{U}_j$ can be dealt with by means of the corresponding object $U_{ij} = U_i \times U_j$ in $\mathrm{Sh}(\mathfrak{X})$. The category $\mathrm{Desc}(\{\mathcal{U}_i\})$ is defined by the same diagrams as in (3.4), the restriction and gluing functors $\mathrm{rest}_{\{\mathcal{U}_i\}}$ and $\mathrm{colim}_{\{\mathcal{U}_i\}}$ are defined similarly, and the same vocabulary makes sense.

Example 3.18 (Descent Data)

1. Recall the étale cover $\mathbb{1} \to \mathbb{B}G$. Using the computation of $G = \mathbb{1} \times_{\mathbb{B}G} \mathbb{1}$ made earlier, a descent data with respect to this map is an object in the limit of the diagram

$$\mathrm{Sh}(\mathbb{1}) \rightrightarrows \mathrm{Sh}(\mathbb{1} \times_{\mathbb{B}G} \mathbb{1}) \Rrightarrow \mathrm{Sh}(\mathbb{1} \times_{\mathbb{B}G} \mathbb{1} \times_{\mathbb{B}G} \mathbb{1})$$

$$= \mathrm{Set} \rightrightarrows \mathrm{Set}_{/G} \Rrightarrow \mathrm{Set}_{/G \times G},$$

---

[49] Given an adjoint pair of functors $L \dashv R$, recall that L is fully faithful if and only if the unit $1 \to RL$ is an isomorphism, and R is fully faithful if and only if the counit $LR \to 1$ is an isomorphism. Then, L and R are inverse equivalences of categories if and only if they are both fully faithful.

that is, a diagram of sets of the type

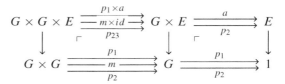

Such a data is the same thing as an action of the group G over a set E. The action is given by the map $a: G \times E \to E$, and the diagram relations ensure that it is unital and associative.

2. More generally, if a discrete group G acts on a space X, the quotient map $q: X \to X /\!/ G$ is also an étale cover of topoi. A descent data with respect to this cover is the same thing as a sheaf on X with an equivariant action of G.

3.3.2 Descent and Distributivity

We abstract from the previous section the structure of descent. This will lead us to conditions with a flavor of distributivity, summarized in Table 4.14.

The distributivity relation $c(a + b) = ca + cb$ has an obvious analogue in terms of colimits and limits, which is the property of *universality of colimits*. Let A_i be a diagram $I \to \mathcal{E}$, $u: C \to B$ be a map in $\mathcal{E}$, and $\mathrm{colim}_i A_i \to B$ be another map. Then, the universality of colimits is the condition that the base change along u preserves the colimit of A_i:

$$C \times_B (\underset{i}{\mathrm{colim}}\, A_i) = \underset{i}{\mathrm{colim}}(C \times_B A_i).$$

The analogy with the distribution of products over sums should be clear.

There exist a number of equivalent formulations for this condition. For example, this is equivalent to saying that the pullback, or base change, functor

$$u^*: \mathcal{E}_{/B} \to \mathcal{E}_{/C}$$

preserves colimits. Geometrically, this says that the pullback of sheaves along étale maps preserves the colimits. By symmetry of the fiber product, this says also that, for any B in $\mathcal{E}$, the fiber product $- \times_B -$ preserves colimits in both variables. This is somehow analogue to the bilinearity of the product $m: R^2 \to R$ of a commutative ring R.

The universality of colimits will be one of the conditions to hold in a logos, but to formulate the other conditions, we need to reformulate it. Let us assume that $B = A$ is the colimit of the A_i and let $C_i = A_i \times_A C$; then we have two cocones $A_i \to A$ and $C_i \to C$ and a morphism between them (represented vertically):

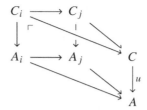

By construction, all the square faces of this diagram are cartesian. Then, the universality of colimits is the condition for C to be the colimit of the diagram C_i.

The other condition we are looking for is a kind of reciprocal statement. We are going to need a few prior steps to be able to formulate it properly. Let us assume that we have a natural transformation of diagrams $C_i \to A_i$ such that, for any map $u: i \to j$ in the indexing category I, the corresponding square is cartesian:

$$\begin{array}{ccc} C_i & \longrightarrow & C_j \\ \downarrow \ulcorner & & \downarrow \\ A_i & \longrightarrow & A_j \end{array} \qquad (3.5)$$

An example of such a cartesian natural transformation is given by a descent data along a covering (see (3.3) and the following examples). In this case, the role of the diagram A_i is played by the so-called *nerve* of the covering family $U_i \to X$, which is the truncated simplicial diagram[50]

$$\coprod_{ijk} U_i \times_X U_j \times_X U_k \rightrightarrows \coprod_{ij} U_i \times_X U_j \rightrightarrows \coprod_i U_i. \qquad (3.6)$$

Intuitively, the cartesian transformations between diagrams correspond also to a descent data, but relative to an arbitrary diagram A_i instead of the nerve of a covering family.

From there, the situation is very similar to what we did with descent. For a diagram $A_\bullet : I \to \mathcal{E}$, let $\mathrm{Desc}(A_\bullet)$ be the category of cartesian natural transformations $C_i \to A_i$, as above. For each map $i \to j$ in I, we have a map $A_i \to A_j$ and a base change functor $\mathcal{E}_{/A_j} \to \mathcal{E}_{/A_i}$. Then, the category $\mathrm{Desc}(A_\bullet)$ can be described as the limit diagram of $\mathcal{E}_{/A_i}$:[51]

---

[50] Precisely, the indexing category is $(\Delta_{\leq 2})^{op}$, where $\Delta_{\leq 2}$ is the full subcategory of the simplex category Δ spanned by simplices of dimensions 0, 1, and 2 only.

[51] This limit is a pseudo-limit in the 2-category of categories. It can be computed as the category of cartesian sections of a certain fibered category over the indexing category I.

$$\mathrm{Desc}(A_\bullet) = \lim_i \mathcal{E}_{/A_i}. \tag{3.7}$$

Let A be the colimit of A_i; then we have a natural "restriction" functor (pullback along the maps $A_i \to A$) and a "gluing" functor (colimit of the diagram)

$$\mathcal{E}_{/A} \underset{\mathrm{rest}_{A_\bullet}}{\overset{\mathrm{glue}_{A_\bullet}}{\longleftrightarrow}} \mathrm{Desc}(A_\bullet) = \lim_i \mathcal{E}_{/A_i}. \tag{3.8}$$

We shall say that the colimits of A_i are *faithful* if the functor $\mathrm{rest}_{A_\bullet}$ is fully faithful and that they are *effective*[52] if the functor $\mathrm{glue}_{A_\bullet}$ is fully faithful. The faithfulness condition says that, given $C \to A$, C can be decomposed into the pieces $C_i = A_i \times_A C$ and recomposed as the colimit of this diagram. The effectivity condition says that, given a cartesian morphism $C_i \to A_i$, we can compose the diagram C_i into its colimit $C = \mathrm{colim}\, C_i$ and then decompose the resulting object C into its original pieces by $C_i = A_i \times_A C$. In other words, the effectivity of the colimit of A_i is equivalent to the following squares being cartesian for all i:

$$\begin{array}{ccc} C_i & \longrightarrow & C = \mathrm{colim}\, C_i \\ \downarrow & & \downarrow \\ A_i & \longrightarrow & A = \mathrm{colim}\, A_i \end{array}$$

The descent property along the diagram A_i is then formulated by the equivalence of categories

$$\mathcal{E}_{/\mathrm{colim}\, A_i} \simeq \lim_i \mathcal{E}_{/A_i}. \tag{3.9}$$

We have finally arrived at the end of the formulation of the descent property. The slice categories $\mathcal{E}_{/A}$ and the base change functors define a functor, called the *universe*, with values in the 2-category of categories:

$$\begin{array}{rcl} \mathbb{U} : \mathcal{E}^{op} & \longrightarrow & \mathrm{Cat}, \\ A & \longmapsto & \mathcal{E}_{/A}, \\ f : A \to B & \longmapsto & f^* : \mathcal{E}_{/B} \to \mathcal{E}_{/A}. \end{array} \tag{3.10}$$

By the formula (3.9), the diagrams for which the descent property holds are precisely those for which their colimit is sent to a limit by the functor $\mathbb{U}$.

For example, let G be a sheaf of groups acting on a sheaf F over some space X. The group action defined a simplicial diagram in $\mathrm{Sh}(\mathcal{X})$

$$\ldots G \times G \times F \underset{p_{23}}{\overset{p_1 \times a}{\underset{m \times \mathrm{id}}{\rightrightarrows}}} G \times F \underset{p_2}{\overset{a}{\rightrightarrows}} F.$$

---

[52] This condition on a colimit is also called to be Van Kampen.

Table 4.13. *Descent conditions for a diagram $A_\bullet : I \to \mathcal{E}$*

Descent category	
$\mathrm{Desc}(A_\bullet) = \lim_i \mathcal{E}_{/A_i}$	
Descent property	
$\mathcal{E}_{/\operatorname{colim}_i A_i} \simeq \lim_i \mathcal{E}_{/A_i}$	
Faithfulness	*Effectivity*
$\mathrm{rest}_{A_\bullet} : \mathcal{E}_{/\operatorname{colim}_i A_i} \to \lim_i \mathcal{E}_{/A_i}$ is fully faithful	$\mathrm{glue}_{A_\bullet} : \lim_i \mathcal{E}_{/A_i} \to \mathcal{E}_{/\operatorname{colim}_i A_i}$ is fully faithful
$C = \operatorname{colim}_i (C \times_{\operatorname{colim}_i A_i} A_i)$ decomposition-then-composition identity	$C_i = (\operatorname{colim}_i C_i) \times_{\operatorname{colim}_i A_i} A_i$ composition-then-decomposition identity
Case of a group action $F /\!/ G$	
Faithfulness	*Effectivity*
a sheaf on $F /\!/ G$ can be described faithfully by an equivariant sheaf on F	any equivariant sheaf of F describes faithfully a sheaf on $F /\!/ G$

The quotient of the action $F /\!/ G$ is the colimit of this diagram in $\mathrm{Sh}(\mathcal{X})$. The descent data associated with this diagram is equivalent to a sheaf E with an action of G and an equivariant map of sheaves $E \to F$. Then, the descent property says that a sheaf over the quotient $F /\!/ G$ is equivalent to an equivariant sheaf over F. This equivalence does not hold for a general group action, but it holds when the action is free. The general descent condition can be understood intuitively in the same way: a diagram has the descent property if working over its colimit is equivalent to working "equivariantly" over the diagram.[53]

Table 4.13 summarizes all the descent conditions, and Table 4.14 sets up the comparison with the distributivity relation in a commutative ring.[54] The descent conditions make sense in any category $\mathcal{E}$ with colimits and finite limits, but they do not hold in general. Whether they hold or not is going to define logoi. As it happens, every diagram in a logos is going to be of faithful descent, but not every diagram is going to be of effective descent.[55] There are two

---

[53] We shall see that in sheaves of ∞-groupoids, within an ∞-logos, all diagrams have the descent property. In particular, any group action will be qualified for working equivariantly. This property is one of the motivations to define ∞-logoi/topoi.

[54] The conditions of Table 4.14 do have a flavor of distributivity, but a better formulation would be a general relation of commutation of finite limits and colimits, like $\lim_i \operatorname{colim}_j X_{ij} = \operatorname{colim}_k \lim_i X_{i,k(i)}$. However, we do not know any such formulation.

[55] For a counterexample, see [29]. The condition for every diagram to be of effective descent is going to be the definition of an ∞-logos.

Table 4.14. *Descent and distributivity*

	Logos	Commutative ring
Faithfulness (decomposition-then-composition condition)	$C = \underset{i}{\mathrm{colim}} \left(C \times_{\mathrm{colim}_i A_i} A_i \right)$	distributivity relation $c \sum_i a_i = \sum_i c a_i$
Effectivity (composition-then-decomposition condition)	given $\begin{array}{c} C_i \to C_j \\ \downarrow \ulcorner \quad \downarrow \\ A_i \to A_j \end{array}$ $C_i = (\underset{j}{\mathrm{colim}}\, C_j) \times_{\mathrm{colim}_j A_j} A_i$ (not a consequence of faithfulness)	given elements a_i and c_i such that $c_i a_j = a_i c_j$ $c_i \sum_j a_j = a_i \sum_j c_j$ (consequence of distributivity)

natural ways to restrict the effectivity condition: either we ask that a specific class of diagrams be of effective descent, or we can ask that all diagrams be of effective descent, but for a restricted class of descent data. The first condition will lead us to Giraud axioms, the second to Lawvere–Tierney axioms.

3.3.3 Presentable Categories

The last ingredient before we are able to state the definitions of a logos is the notion of presentable category, which, in the analogy between logoi and commutative rings, plays the role of abelian groups. The structural analogy between presentable categories and abelian groups is presented in Table 4.15.

The notion of presentable category is one of the most crucial notions of category theory. They are a particularly nice class of categories with all colimits (or cocomplete categories) for which a technical problem of size is tamed. Let $\mathcal{C}$ be a cocomplete category and R be a class of arrows in $\mathcal{C}$. We denote by $\mathcal{C}/\!/R$ the localization of $\mathcal{C}$ forcing all the arrows in R to become isomorphisms.[56] We call it the *quotient* of $\mathcal{C}$ by R.[57] A category $\mathcal{C}$ is called

---

[56] This localization is taken in the category of cocomplete categories and functors preserving colimits. This forces $\mathcal{C}/\!/R$ to have all colimits and the canonical functor $\mathcal{C} \to \mathcal{C}/\!/R$ to preserve them.

[57] The vocabulary is a bit awkward here – the classical name of the operation $\mathcal{C} \to \mathcal{C}/\!/R$ is *localization* because the operation is thought from the point of view of the arrows of $\mathcal{C}$, but, from the point of view of the objects of $\mathcal{C}$, this operation is in fact a *quotient* of $\mathcal{C}$ identifying

Table 4.15. *Presentable categories versus abelian groups*

	Presentable categories	abelian groups
Operations	colimits $\mathcal{A}^I \to \mathcal{A}$	sums $A^n \to A$
Morphisms	functors $\mathcal{A} \to \mathcal{B}$ preserving colimits (cc functors)	linear maps $A \to B$
Initial object	Set	$\mathbb{Z}$
Free objects	$\mathcal{P}\mathrm{r}\,(C)$	$\mathbb{Z}.E := \oplus_E \mathbb{Z}$
Quotients	$\mathcal{P}\mathrm{r}\,(C)/\!/(A_i \to B_i \text{ iso})$	$\mathbb{Z}.E/(a_i - b_i = 0)$
Additivity	$\mathcal{A} \oplus \mathcal{B} = \mathcal{A} \times \mathcal{B}$	$A \oplus B = A \times B$
Self-enrichment	the category of cc functors $[\mathcal{A},\mathcal{B}]_{cc}$ is presentable	the set of group maps $\mathrm{Hom}(A,B)$ is an abelian group
Tensor product	functor preserving colimits in each variable $\mathcal{A} \times \mathcal{B} \to \mathcal{C}$ = functor preserving colimits $\mathcal{A} \otimes \mathcal{B} \to \mathcal{C}$ $\mathcal{A} \otimes \mathcal{B} = [\mathcal{A}^{op}, \mathcal{B}]^c$ $\mathcal{P}\mathrm{r}\,(C) \otimes \mathcal{P}\mathrm{r}\,(D) = \mathcal{P}\mathrm{r}\,(C \times D)$	bilinear map $A \times B \to C$ = linear map $A \otimes B \to C$ $\mathbb{Z}.E \otimes \mathbb{Z}.F = \mathbb{Z}.(E \times F)$
Closure of the tensor product	$[\mathcal{A} \otimes \mathcal{B}, \mathcal{C}]_{cc} = [\mathcal{A}, [\mathcal{B}, \mathcal{C}]_{cc}]_{cc}$	$\mathrm{Hom}(A \otimes B, C) = \mathrm{Hom}(A, \mathrm{Hom}(B,C))$
Dual objects	$\mathcal{A}^\star = [\mathcal{A}, \mathrm{Set}]_{cc}$ $\mathcal{P}\mathrm{r}\,(C)^\star = \mathcal{P}\mathrm{r}\,(C^{op})$	$A^\star = \mathrm{Hom}(A, \mathbb{Z})$ $(\mathbb{Z}.E)^\star = \mathbb{Z}.E$
Dualizable objects	retracts of $\mathcal{P}\mathrm{r}\,(C)$	retracts of $\mathbb{Z}.E$

presentable if it is equivalent to some quotient $\mathcal{P}\mathrm{r}\,(C)/\!/R$, where C is a small category and R a *set* (rather than a class). The intuitive idea is that, even though presentable categories are not small, they are still controlled by the small data (C, R).

Here follows a list of some properties for which presentable categories are so nice. Let $\mathcal{C}$ be a presentable category; then

the domain and codomain of the maps $A \to B$ in R. This second point of view is better for our purposes. The notation $\mathcal{C}/\!/R$ is intended to be more evocative of this fact than the classical notation $\mathcal{C}[R^{-1}]$.

1. $\mathcal{C}$ has (small) limits in addition to (small) colimits;
2. (special adjoint functor theorem) if $\mathcal{D}$ is a cocomplete category, a functor $\mathcal{C} \to \mathcal{D}$ has a right adjoint if and only if it preserves (small) colimits;
3. (representability theorem) in particular, a functor $\mathcal{C}^{op} \to$ Set is representable by an object X in $\mathcal{C}$ if and only if it sends colimits to limits;
4. (quotients as full subcategories) if R is a set of maps in $\mathcal{C}$, the quotient $\mathcal{C}/\!/R$ is again presentable, and the right adjoint to the quotient functor $\mathcal{C} \to \mathcal{C}/\!/R$ is fully faithful.

The last property is the one we need now. The existence of a fully faithful right adjoint $q_*: \mathcal{C}/\!/R \to \mathcal{C}$ to the quotient functor $q^*: \mathcal{C} \to \mathcal{C}/\!/R$ means that any quotient of $\mathcal{C}$ can be identified canonically to a full subcategory of $\mathcal{C}$ (however, this embedding does not preserves colimits). An object X of $\mathcal{C}$ is called *orthogonal* to R if, for any $f: A \to B$ in R, the map $\mathrm{Hom}(B,X) \to \mathrm{Hom}(A,X)$ is a bijection. This relation is denoted $R \perp X$. Intuitively, this says that, from the point of view of X, the maps in R are isomorphisms. Then, the image of $q_*: \mathcal{C}/\!/R \to \mathcal{C}$ is the full subcategory $R^\perp$ spanned by the objects orthogonal to all maps in R.[58]

Example 3.19 (Presentable Categories)

1. The categories Set, $\mathcal{P}r(C)$, Set$[C]$ are presentable. Set^{op} is not a presentable category.
2. An important example of a quotient is the construction of categories of sheaves. Let C be a small category with finite limits, and for each object X in C, let $J(X)$ be a set of covering families $U_i \to X$. A presheaf F in $\mathcal{P}r(C)$ is a *sheaf* if and only if, for each covering family, we have

$$F(X) = \lim \left(\prod_i F(U_i) \rightrightarrows \prod_{ij} F(U_i \times_X U_j) \right).$$

Let $U = \mathrm{colim} \left(\coprod_{ij} U_i \times_X U_j \rightrightarrows \coprod U_i \right)$ computed in $\mathcal{P}r(C)$. The canonical map $U \to X$ is a monomorphism in $\mathcal{P}r(C)$, called the *covering sieve* associated to the covering family $U_i \to X$. Let J be the set of all the covering sieves. Then, the previous condition can be reformulated as follows: F is a sheaf if and only if $J \perp F$. In other words, $\mathcal{S}h(C, J) = J^\perp \subset \mathcal{P}r(C)$. The property that $J^\perp = \mathcal{P}r(C)/\!/J$ says that the category of sheaves can be thought as the quotient of $\mathcal{P}r(C)$ by the relations given by the topology J. This is actually the proper way to think about it.

---

[58] This is how quotients are dealt with in practice: they are defined as categories $R^\perp$ (see the example of sheaves below). The quotient functor $\mathcal{C} \to R^\perp$ is then constructed by a small object argument from the set R.

3.3.4 Definitions of a Logos/Topos

We are now ready to present several definitions of logoi. We are going to explain in detail the ones of Giraud and Lawvere. The comparison between these definitions is summarized in Table 4.17.

A presentable category $\mathcal{E}$ is a *logos* if

1. (Our first definition) it is a left-exact localization of some presheaf category $\mathcal{P}r(C)$;
2. (Original definition in [6, Chapter IV]) it is a category of sheaves on a site;
3. (Giraud) it has universal colimits, disjoint sums, and effective equivalence relations;
4. (Lawvere) it is locally cartesian closed and has a subobject classifier Ω.[59]

3.3.5 Universality of Colimits and Local Cartesian Closeness

We defined the universality of colimits as the condition that, for any map $u : C \to B$ in $\mathcal{E}$, the base change functor

$$u^* : \mathcal{E}_{/B} \longrightarrow \mathcal{E}_{/C}$$

preserves colimits. When the category $\mathcal{E}$ is assumed presentable, this condition is also equivalent to the existence of a right adjoint for this functor,

$$u_* = \prod_u : \mathcal{E}_{/C} \longrightarrow \mathcal{E}_{/B}.$$

This functor is called the *relative limit*, the *multiplicative direct image*, or the *depend product*, along u. A category $\mathcal{E}$ such that, for every map u in $\mathcal{E}$, the adjoint pair $u^* \dashv u_*$ exists is called *locally cartesian closed*. These conditions are also equivalent to the condition that every diagram be of faithful descent. Hence, although they are stated differently, Giraud's and Lawvere's definitions both assume this half of the descent property.

3.3.6 Giraud Definition

The first condition of Giraud axioms is that all diagrams be of faithful descent. The idea behind the other axioms is to ask for the effectivity of descent for some diagrams only. Intuitively, these diagrams are going to be the nerves of covering families (3.6). But such a characterization of these diagrams will be

---

[59] Lawvere's original definition does not in fact require the category $\mathcal{E}$ to be presentable. Without this hypothesis, we get the notion of an *elementary topos* (but we shall say *elementary logos*). This notion is not equivalent to the other definitions. By comparison, the other notion is called a *Grothendieck topos* (but we shall say *Grothendieck logos*). To view topoi as spatial objects, as is the purpose of this chapter, we need to use Grothendieck's definition, not Lawvere's. This is why we have chosen not to present Lawvere's definition in full generality but to restrict it to the case of a presentable category only.

true only if the Giraud axiom holds. So we need to define them without the fact that they correspond to nerves of covering families. There are going to be two cases. The first case is that of unions. The second case is that of the quotient of an object by an equivalence relation.

Let A_i be a set of objects; the descent property for the sum of the A_i is the condition

$$\mathcal{E}_{/\coprod_i A_i} \simeq \prod_i \mathcal{E}_{/A_i}.$$

This is sometimes called *extensivity of sums*. As it happens, this whole condition boils down to a single simpler condition called the *disjointness of sums*. Sums are said to be disjoint if, for any $i \neq j$, the following square is cartesian:

$$\begin{array}{ccc} \emptyset & \longrightarrow & A_i \\ \downarrow\scriptstyle{\ulcorner} & & \downarrow \\ A_j & \longrightarrow & \coprod_i A_i \end{array}$$

The second condition concerns equivalence relations within the category $\mathcal{E}$ that we now define. Let A_0 be an object in $\mathcal{E}$. An *equivalence relation* on A_0 is a relation $A_1 \rightarrowtail A_0 \times A_0$ (a monomorphism) satisfying the following:

1. (reflexivity) the diagonal of $A_0 \rightarrowtail A_0 \times A_0$ factors through A_1 ($A_0 \subset A_1$ as subobjects of $A_0 \times A_0$);
2. (transitivity) for $A_2 = A_1 \times_{p_2, A_0, p_1} A_1$, we have $A_2 \subset A_1$ as subobjects of $A_0 \times A_0$;
3. (symmetry) $A_1 \rightarrowtail A_0 \times A_0 \stackrel{\sigma}{\simeq} A_0 \times A_0$ is A_1.

Such data provides a truncated simplicial diagram[60]

$$A_2 \Rrightarrow A_1 \rightrightarrows A_0.$$

The equivalence relation $A_1 \rightrightarrows A_0$ is said to be of effective descent if the previous diagram is. As with sums, this condition boils down to a single simpler condition, called the *effectivity of equivalence relations*. The quotient A of the equivalence relation is defined to be the colimit of the previous diagram.[61] Then, the equivalence relation is of effective descent if and only if the following square is cartesian:

$$\begin{array}{ccc} A_1 & \stackrel{p_1}{\longrightarrow} & A_0 \\ \scriptstyle{p_2}\downarrow\scriptstyle{\ulcorner} & & \downarrow \\ A_0 & \longrightarrow & A \end{array}$$

---

[60] The indexing category is $(\Delta_{\leq 2})^{op}$. Again we are drawing only the face maps.
[61] Or equivalently, the coequalizer of $A_1 \rightrightarrows A_0$.

Table 4.16. *Giraud axioms*

Under assumption of universality of colimits	
descent for sums $$\mathcal{E}_{/\coprod_i A_i} \simeq \prod_i \mathcal{E}_{/A_i}$$ $\Longleftrightarrow$	disjointness of sums $$\begin{array}{ccc} \emptyset & \longrightarrow & A_1 \\ \downarrow {\scriptstyle \ulcorner} & & \downarrow \\ A_j & \longrightarrow & \coprod_i A_i \end{array}$$
descent for equivalence relations $$\mathcal{E}_{/A} = \lim \left(\mathcal{E}_{/A_0} \rightrightarrows \mathcal{E}_{/A_1} \Rrightarrow \mathcal{E}_{/A_2} \right) \Longleftrightarrow$$	effectivity of equivalence relations $$\begin{array}{ccc} A_1 & \longrightarrow & A_0 \\ \downarrow {\scriptstyle \ulcorner} & & \downarrow \\ A_0 & \longrightarrow & \mathrm{colim}\,(A_1 \rightrightarrows A_0) \end{array}$$

Table 4.16 summarizes the Giraud axioms and the descent conditions they correspond to.

We have already said that the descent condition is not true for all diagrams within a logos. This raises the question to characterize the diagrams for which it holds. Giraud axioms give a family of diagrams (sums and equivalence relations) sufficient to define the structure of logoi, but more diagrams have the descent property. They are the π_1-*acyclic* diagrams, that is, the diagrams A_i for which the ∞-colimit, computed in sheaves of ∞-groupoids, have trivial fundamental group.

3.3.7 Lawvere Definition

We already explain the local cartesian closure property of the Lawvere definition. The definition of Lawvere of a logos emphasizes the so-called subobject classifier Ω. For an object A in $\mathcal{E}$, a subobject of A is a monomorphism $B \rightarrowtail A$.[62] The subobjects of A span a full subcategory $\mathrm{Sub}(A) \subset \mathcal{E}_{/A}$, which is equivalent to a poset. We denote by $\mathrm{sub}(A)$ the set of objects of this poset. Since monomorphisms are preserved by base change, the family of all $\mathrm{sub}(A)$ defines a functor[63]

$$\begin{array}{rcl} \mathrm{sub}: \mathcal{E}^{op} & \longrightarrow & \mathrm{Set} \\ A & \longmapsto & \mathrm{sub}(A). \end{array}$$

---

[62] Recall that a monomorphism is a morphism $f: B \rightarrowtail A$ such that the diagonal $\Delta f: B \rightarrowtail B \times_A B$ is an isomorphism.

[63] The family of all $\mathrm{Sub}(A)$ defines also a functor Sub with values in Poset, which is a subfunctor of the universe $\mathbb{U}$, but we shall not need this functor.

Table 4.17. *Definitions of logoi/topoi*

	Giraud	Lawvere–Tierney
Decomposition-then-composition condition	universality of colimits ($\Leftrightarrow$ all diagrams are of faithful descent)	
Composition-then-decomposition condition	only π_1-acyclic diagrams are of effective descent	all diagrams are of effective descent, but for subobjects only
	sums are disjoint $\emptyset = X_i \times_{\coprod_k X_k} X_j$	equivalence relations are effective $X_1 \simeq X_0 \times_{X_{-1}} X_0$	the functor Sub: $\mathcal{E}^{op} \to$ Set of subobjects is representable by an object Ω

Since we have assumed $\mathcal{E}$ to be a presentable category, the property (3) of such categories says that this functor is representable by an object Ω, that is, $\operatorname{sub}(A) = \operatorname{Hom}(A, \Omega)$, if and only if it sends colimits in $\mathcal{E}$ to limits in Set. This condition is exactly a descent condition,[64] but for the class of diagrams (3.5) where the vertical maps are monomorphisms only:

$$\begin{array}{ccc} C_i & \longrightarrow & C_j \\ \downarrow & \ulcorner & \downarrow \\ A_i & \longrightarrow & A_j \end{array}$$

In other words, Lawvere's axiom of existence of Ω is a way to impose a general descent property but for a restricted class of descent data.

3.4 Elements of Logos Algebra

3.4.1 Structural Analogies

In this section, we sketch the structural analogy between the theories of logoi, frames, and commutative rings. We already saw the analogy between presentable categories and abelian groups in Table 4.15. We are going to continue along the same spirit.

---

[64] Strictly speaking, the descent condition would be for the functor Sub defined in the previous footnote. We are smoothing things out a bit here.

The theory of commutative rings is related in a fundamental way to that of abelian groups and that of commutative monoids. Between these structures, there exist forgetful functors and their left adjoints, or free constructions.

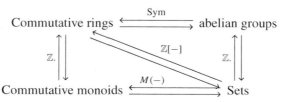

The functor $\mathbb{Z}.$ constructs the free abelian group on a set. The functor M constructs the free commutative monoid. The functor Sym constructs the symmetric tensor algebra. The functor $\mathbb{Z}[-]$ constructs the free commutative ring on a set. The commutativity of the square says that this last construction can be obtained either by taking first the free abelian group and then the symmetric algebra, or first the free monoid and then the free abelian group on the resulting set.

The analogues of these structures for locales and topoi are summarized in Table 4.18 (we have included also ∞-topoi for future reference). The notion of sup-lattice is a poset with arbitrary suprema. The notion of meet-lattice is a poset with finite infima. The notion of lex category is a category with finite limits. And we already saw the notion of presentable category. These structures are also related by a number of forgetful and free functors, presented in Figure 4.1.[65] In the diagram for frames, the functor $\underline{2}[-]$ is the free frame functor, mentioned earlier. The functor $\bigvee$ is the free sup-lattice functor; if P is a small poset, $\bigvee P = [P^{op}, \underline{2}]$. The functor $(-)^\wedge$ is the free meet-lattice functor. For a poset P $(P^\wedge)^{op}$ is the sub-poset of $[P, \underline{2}]$ generated by finite unions of elements of P. The functor Sym is an analogue of the symmetric algebra functor. In the diagram for logoi, the functor $\mathcal{P}$ is the free cocompletion functor. It is defined only for small categories C, where it is given by the presheaves $\mathcal{P}r(C) = [C^{op}, \mathsf{Set}]$. The functor $(-)^{\mathrm{lex}}$ is the free finite limit completion functor. The functor Sym is an analogue of the symmetric algebra functor; we refer to [8] for details. The functor $\mathsf{Set}[-]$ is the free logos functor. It is defined only for small categories C by the formula that we have seen already:

$$\mathsf{Set}[C] = \mathcal{P}r\left(C^{\mathrm{lex}}\right) = \left[(C^{\mathrm{lex}})^{op}, \mathsf{Set}\right].$$

---

[65] In the right diagram of Figure 4.1, the left-adjoint functors going up do not, strictly speaking, exist for problems of size. This is why we put them in dashed arrows. They are only defined for small categories and small lex categories.

Table 4.18. *Analogies of structure*

Algebraic geometry	Locale theory	Topos theory	∞-Topos theory
Set	Poset	Category	∞-Category
abelian group	Sup-lattice	Presentable category	Presentable ∞-category
addition $(+,0)$	suprema $(\bigvee, \bot)$	colimits, initial object	colimits, initial object
$\mathbb{Z}$	$\underline{2} = \{0 < 1\}$	Set	$\mathcal{S}$
Commutative monoid	Meet lattice	Lex category	Lex ∞-category
multiplication $(\times, 1)$	finite infima $(\wedge, \top)$	finite limits, terminal object	finite limits, terminal object
$x^{\mathbb{N}}$	$\underline{2}^{op}$	Fin^{op}	$\mathcal{S}_{\text{fin}}^{op}$
Commutative ring	Frame	Logos	∞-Logos
$\mathbb{Z}[x] = \mathbb{Z}.x^{\mathbb{N}}$	$\underline{2}[x] = [\underline{2},\underline{2}]$	$\text{Set}[X] = [\text{Fin}, \text{Set}]$	$\mathcal{S}[X] = [\mathcal{S}_{\text{fin}}, \mathcal{S}]$
Distributivity relation $c\sum a_i = \sum ca_i$	Distributivity relation $c \wedge \bigvee_i a_i = \bigvee_i c \wedge a_i$	Distributivity relations (see Tables 4.14, 4.17)	Distributivity relation (all colimits have the descent property)
Affine scheme	Locale	Topos	∞-Topos
affine line $\mathbb{A}^1$	Sierpiński space $\mathbb{S}$	topos of sets $\mathbb{A}$	∞-topos $\mathbb{A}_\infty$ of ∞-groupoids

3.4.2 Presentation of Logoi by Generators and Relations

The previous paragraph essentially detailed the construction of the free logos. As is true for any kind of algebraic structure, any logos is a quotient of a free logos. This leads to the possibility to define logoi by generators and relations. This is a key feature in the connection of logoi with classifying problems and logic.

3.4.3 Relations and Quotients of Logoi

The computation of quotients of logoi is one of the most fundamental pieces of technology of the theory. The collection of quotients of a given logos $\mathcal{E}$ is a poset. Given any family R of maps in a logos $\mathcal{E}$, the class of all quotients of $\mathcal{E}$ where all maps in R become an invertible map has a minimal

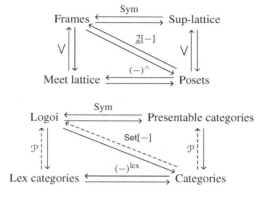

Figure 4.1 Free constructions

element $\mathcal{E} \to \mathcal{E}/\!/R$ called the *quotient generated by* R.[66] Any quotient can be generated this way. Geometrically, the situation is clear: in the case of a single map, if $f \colon A \to B$ is a map of sheaves on a topos $\mathcal{X}$, the subtopos $\mathcal{X}^f$ corresponding to $\mathrm{Sh}(\mathcal{X})/\!/f$ is intuitively the subspace of points x where the map $f(x) \colon A(x) \to B(x)$ between the stalks of A and B is a bijection.[67]

The construction $\mathcal{E}/\!/R$ has the following universal property: given a logos morphism $u^* \colon \mathcal{E} \to \mathcal{F}$ such that, for any f in R, $u^*(f)$ is an isomorphism in $\mathcal{F}$, there exists a unique logos morphism $\mathcal{E}/\!/R \to \mathcal{F}$ and a factorization $u^* \colon \mathcal{E} \to \mathcal{E}/\!/R \to \mathcal{F}$. Geometrically, this factorization says that if $u \colon \mathcal{Y} \to \mathcal{X}$ is such that the pullback of the maps $f \colon A \to B$ of R on $\mathcal{Y}$ are isomorphisms, then the image of u is within the subtopos of $\mathcal{X}$ where all maps in R are isomorphisms.

Recall from Section 2.2.7 that the quotients of a frame F were encoded by nuclei $j \colon F \to F$. There exists an analogue notion for quotient of logoi called a *left-exact idempotent monad* (we shall say *lex reflector* for short). Such an object is an (accessible) endofunctor $j \colon \mathcal{E} \to \mathcal{E}$ with a natural transformation $1 \to j$ such that the induced transformation $j \to j \circ j$ is an isomorphism and j is a left-exact functor. Recall that quotients of logoi $q^* \colon \mathcal{E} \to \mathcal{F}$ are reflective, that is, they have a fully faithful right adjoint $q_* \colon \mathcal{F} \to \mathcal{E}$. In this situation, the

---

[66] Technically, $\mathcal{E} \to \mathcal{E}/\!/R$ is the left-exact localization generated by the family of maps R. The detailed construction is given in the examples. There exists the same problem of vocabulary (localization or quotient) as with presentable categories (see Footnote 57). Again, thinking of a logos in terms of its objects and not its arrows, the term *quotient* is more appropriate.

[67] This construction is what becomes the construction of a subspace $Y \subset X$ as equalizer of two maps $a, b \colon X \rightrightarrows A$ ($Y = \{x | a(x) = b(x)\}$). When sets of points are replaced by categories of points, the equality of two objects has to be replaced by an isomorphism.

Table 4.19. *Quotients of logoi and commutative rings*

Comm. ring A	ideal $J \subseteq A$	generators a_i for J	projection $\pi : A \to A$ on a complement of J in A	quotient A/J in bijection with the set of fixed points $a = \pi(a)$
Logos $\mathcal{E}$	the class W of all maps $A \to B$ inverted by the quotient	a generating set R of maps $A_i \to B_i$	left-exact idempotent monad $j : \mathcal{E} \to \mathcal{E}$	quotient $\mathcal{E}/\!\!/ W$ equivalent to the category of fixed points $F \simeq j(F)$

endofunctor j is q_*q^* and projects $\mathcal{E}$ to the full subcategory equivalent to $\mathcal{F}$. Reciprocally, any lex reflector j determines a quotient $\mathcal{E} \to \mathcal{F}$ where $\mathcal{F}$ is the full subcategory of *fixed points* of j (objects F such that the map $F \to j(F)$ is an isomorphism). Table 4.19 presents a comparison of the theory of quotients of logoi and commutative rings.

Example 3.20 (Quotients and Reflectors)

1. For X a topological space or a locale, the lex reflector associated to the quotient $\mathcal{P}r(X) \to \mathcal{S}h(X)$ is the sheafification endofunctor.
2. (Open reflector) Let $\mathcal{Y} \to \mathcal{X}$ be the open embedding associated to the subterminal object U in $\mathcal{S}h(\mathcal{X})$. The associated lex reflector is the functor $\mathcal{S}h(\mathcal{X}) \to \mathcal{S}h(\mathcal{X})$ sending F to $U \times F$. Intuitively, this functor replaces the stalks of F outside U by a point, leaving the others unchanged.
3. (Closed reflector) Let $\mathcal{Y} \to \mathcal{X}$ be the closed embedding associated to the subterminal object U in $\mathcal{S}h(\mathcal{X})$. For F in $\mathcal{S}h(\mathcal{X})$, we define $U \star F$ as the pushout of the diagram $U \leftarrow U \times F \to F$. The associated lex reflector is the functor sending F to $U \star F$. Intuitively, this functor replaces the stalks of F in U by a point, leaving the others unchanged.
4. We detail the general construction of the $\mathcal{E} \to \mathcal{E}/\!\!/ R$. Thanks to the reflectivity of localizations, $\mathcal{E}/\!\!/ R$ can be described as the full subcategory $\mathcal{E}^R$ of $\mathcal{E}$ of objects X satisfying the following condition. Let G be a small category of generators for $\mathcal{E}$. We define R' to be the smallest class of maps in $\mathcal{E}$ containing R that is (1) stable by diagonals (if $f : A \to B$ is in R', then $\Delta f : A \to A \times_B A$ is in R') and (2) stable by all base change along maps in G (if $f : A \to B$ is in R', then for any $g : C \to B$ in G, the map $f' : C \times_B A \to C$ is in R'). Then, X is in $\mathcal{E}^R$ if, for any map $u : C \to D$ in R', the canonical map of sets $\text{Hom}(D, X) \to \text{Hom}(C, X)$ is a bijection. With the notation introduced for quotients of presentable categories, we

have $\mathcal{E}^R = (R')^\perp$. The corresponding reflector and the localization functor $\mathcal{E} \to \mathcal{E}/\!/R$ are then constructed with a small object argument.
5. If R is made of monomorphisms only, the previous description simplifies. It is enough to define the class R' to satisfy condition (2) only, that is, that R' be stable by base change (along generators). Then, an object X is in $\mathcal{E}^R$ if $\mathrm{Hom}(D, X) \to \mathrm{Hom}(C, X)$ is a bijection for any map $u\colon C \to D$ that is a base change of some map in R. The reflector is again constructed with a small object argument.

3.4.4 Presentations

We define a *logos presentation* as a pair (G, R), where G is a small category and R a set of maps in $\mathsf{Set}\,[G]$. The objects of G are called the *generators* and the maps in R the *relations*. A *presentation of a logos* $\mathcal{E}$ is a triple (G, R, p), where (G, R) is a presentation and p is a functor $p\colon G \to \mathcal{E}$ inducing an equivalence $\mathsf{Set}\,[G]/\!/R \simeq \mathcal{E}$. Every logos admits a presentation.

Recall that a logos morphism $\mathsf{Set}\,[G] \to \mathcal{E}$ is equivalent to a diagram $G \to \mathcal{E}$. Then, a morphism $\mathsf{Set}\,[G]/\!/R \to \mathcal{E}$ corresponds to a diagram $G \to \mathcal{E}$ satisfying extra conditions. It is useful to introduce the vocabulary that $\mathsf{Set}\,[G]$ is the *logos classifying G-diagrams* and that $\mathsf{Set}\,[G]/\!/R$ is the logos classifying G-diagrams that are R-exact.[68] Any structure that can be described diagrammatically (such as groups; rings, as we saw; and also local rings, as we will see) can be classified in this way by a topos. And since every logos admits a presentation, every logos can be thought as classifying some kind of exact diagram. This fact is important in the relationship of logoi with logical theories (see Section 3.4.6).

Recall from the examples of affine topoi the topos $\mathbb{A}^\to$ classifying maps and its subtopos $\mathbb{A} \simeq \mathbb{A}^{\simeq} \subset \mathbb{A}^\to$ classifying isomorphisms. Geometrically, the data of a map f in $\mathbb{A}^G$ corresponds to a topos morphism $\mathbb{A}^G \to \mathbb{A}^\to$. For R a family of maps in $\mathcal{E}$, the topos $\mathcal{X}$ corresponding to $\mathsf{Set}\,[G]/\!/R$ is defined by the fiber product in Topos (or the corresponding pushout in Logos)[69]

$$\left(\begin{array}{ccc} \mathcal{X} & \longrightarrow & (\mathbb{A}^{\simeq})^R \\ \downarrow{}^{\ulcorner} & & \downarrow \\ \mathbb{A}^G & \xrightarrow{R} & (\mathbb{A}^\to)^R \end{array}\right. \qquad \left.\begin{array}{ccc} \mathsf{Set}\,[G]/\!/R & \longleftarrow & \mathsf{Set}\,[R] \\ \uparrow^{\ulcorner} & & \uparrow \\ \mathsf{Set}\,[G] & \longleftarrow & \mathsf{Set}\,[\underline{2} \times R] \end{array}\right).$$

---

[68] Recall that any ring can be presented as classifying the solutions to some polynomial equations. Classifying R-exact diagrams is the analogue for logoi.

[69] Notice the analogy with the definition of affine schemes as zeros of a set of polynomials.

Example 3.21 (Presentations)

1. (Flat diagrams) Let C be a small category with finite limits. We already mentioned that the logos $\Pr(C)$ classifies diagrams $C \to \mathcal{E}$ that preserve finite limits. Let us compute a presentation of this topos. For a finite diagram c_i in C, let $\lim_i^{(C)} c_i$ be the limit of the diagram in C, and let $\lim_i^{\text{free}} c_i$ be the limit of the same diagram in $\text{Set}[C]$. There is a canonical map $f_c : \lim_i^{(C)} c_i \to \lim_i^{(\text{free})} c_i$ in $\text{Set}[C]$. Let Λ be the collection of all these maps. Then, the logos quotient $\text{Set}[C] /\!/ \Lambda$ is the logos $\Pr(C)$.
   A logos morphism $\text{Set}[C] \to \mathcal{E}$ is the same thing as a diagram $C \to \mathcal{E}$. The logos morphisms $\Pr(C) \to \mathcal{E}$ correspond to those diagrams $C \to \mathcal{E}$ that are *flat*, or *filtering* in the sense of [25, Section VII.8]. In the case where C has finite limits, a diagram $C \to \mathcal{E}$ is flat if and only if it is a left-exact functor.

2. (Torsors) In the case where $C = G$ is a group viewed as a category with one object, a diagram $G \to \mathcal{E}$ corresponds to a sheaf with an action of G. Such a diagram is flat if and only if the action is free and transitive, that is, if and only if it is a G-torsor [25, Section VIII]. Moreover, natural transformations between logos morphisms $\text{Set}^G \to \mathcal{E}$ correspond to morphisms of G-torsors. This says that $\mathbb{B}G$ is the topos classifying G-torsors.

3. Let C be a small category with finite sums; then there exists a topos classifying diagrams $C \to \mathcal{E}$ that preserve sums. For a finite family (c_i) of objects in C, let $\coprod_i^{(C)} c_i$ be the sum of the family in C, and let $\coprod_i^{\text{free}} c_i$ be the sum of the family in $\text{Set}[C]$. There is a canonical map $\coprod_i^{\text{free}} c_i \to \coprod_i^{(C)} c_i$ in $\text{Set}[C]$. Let Σ be the collection of all these maps. Then, the logos $\text{Set}[C] /\!/ \Sigma$ is the logos classifying diagrams $C \to \mathcal{E}$ preserving sums.
   More generally, the same construction works for any class of colimits existing on C and leads to a topos classifying diagrams $C \to \mathcal{E}$ preserving any set of colimits.

4. (Inhabited sets revisited) The left-exact localizations of the logos $\text{Set}[X]$ classify objects satisfying some conditions. For example, one can ask that the canonical map $X \to 1$ be a cover (see Section 3.2.12). This condition is equivalent to the exactness of the diagram $X \times X \rightrightarrows X \to 1$. One can prove that $\text{Set}[X]/\!/(\text{colim}\,(X \times X \rightrightarrows X) \to 1) = [\text{Fin}^\circ, \text{Set}] = \text{Sh}(\mathbb{A}^\circ)$. That is, the topos classifying inhabited objects is the topos classifying nonempty sets.

5. (Sierpiński revisited) Another example is to ask that the canonical map $X \to 1$ be a monomorphism, that is, X is subterminal. This condition is equivalent to the diagonal $X \to X \times X$ being an isomorphism. One can prove that $\text{Set}[X]/\!/(X \simeq X \times X) = \text{Sh}(\mathbb{S})$, that is, that subterminal

objects are classified by the Sierpiński topos. We already saw this, since subterminal objects are equivalent to open domains.

6. (Arrow classifier) Let $C = \{Y \to X\} \simeq \underline{2}$ be the category with one arrow. Then Set $[Y \to X]$ is the logos classifying arrows. It can be proven to be $[\text{Fin}^\to, \text{Set}]$. We can impose the condition that $Y = 1$; this is equivalent to inverting the canonical map $Y \to 1$. The resulting logos is Set $[Y \to X]/\!/(Y \simeq 1) = \text{Set}[X^\bullet]$.

7. (Mono classifier) A monomorphism in a logos is defined as a map $A \to B$ such that the diagonal $A \to A \times_B A$ is an isomorphism. Intuitively, a monomorphism of sheaves on a space X is a map $f: A \to B$ that is injective stalk-wise. Let $\text{Fin}^{\rightarrowtail}$ be the full subcategory of $\text{Fin}^\to$ whose objects are monomorphisms between finite sets. It can be proven that the $[\text{Fin}^{\rightarrowtail}, \text{Set}]$ is the logos classifying monomorphisms Set $[Y \rightarrowtail X]$. The corresponding subtopos of $\mathbb{A}^\to$ will be denoted $\mathbb{A}^{\rightarrowtail}$.

If we further force the map $X \to 1$ to be an isomorphism, we get back the Sierpiński logos.

8. (Cover classifier) Let $f: A \to B$ be a map in a logos $\mathcal{E}$. Recall from Section 3.2.12 that the image factorization of f is $A \to \text{im}(f) \to B$, where $\text{Im}(f) = \text{colim}(A \times_B A \rightrightarrows A)$. The map f is a cover if and only if the monomorphim im $(f): \text{Im}(f) \to B$ is an isomorphism. Let $\text{Fin}^\twoheadrightarrow$ be the full subcategory of $\text{Fin}^\to$ whose objects are surjections between finite sets. It can be proven that the $[\text{Fin}^\twoheadrightarrow, \text{Set}]$ is the logos classifying surjections Set $[Y \twoheadrightarrow X]$. The corresponding subtopos of $\mathbb{A}^\to$ will be denoted $\mathbb{A}^\twoheadrightarrow$.

The image factorization of maps gives a topos morphism $\mathbb{A}^\to \to \mathbb{A}^{\rightarrowtail}$ and a cartesian square

$$\begin{array}{ccc} \mathbb{A}^\twoheadrightarrow & \longrightarrow & \mathbb{A}^\simeq \\ \downarrow & \ulcorner & \downarrow \\ \mathbb{A}^\to & \xrightarrow{\text{image}} & \mathbb{A}^{\rightarrowtail} \end{array}$$

The fact that a map is an isomorphism if and only if it is a cover and a monomorphism gives a cartesian square

$$\begin{array}{ccc} \mathbb{A}^\simeq & \hookrightarrow & \mathbb{A}^{\rightarrowtail} \\ \downarrow & \ulcorner & \downarrow \\ \mathbb{A}^\twoheadrightarrow & \hookrightarrow & \mathbb{A}^\to \end{array}$$

3.4.5 Topologies and Sites

Although presentations may be the most natural way to define logoi by generators and relation, history and practice have imposed another way to do it: the notion of site. In a presentation by means of a site, the free logoi $\mathrm{Set}\,[G]$ are replaced by presheaf logoi $\mathcal{P}r\,(C)$, and the relations are replaced by the data of a *topology*. Recall that the quotient of a logos $\mathcal{E}$ generated by a map $f: A \to B$ forces f to become an isomorphism. A variation on this is to force f to become a cover instead. This is the main idea behind the notion of a topology. The comparison between sites and presentations is summarized in Table 4.21.

Let $A \to \mathrm{Im}(f) \to B$ be the image factorization of f. The image factorizations are built using colimits and finite limits, so they are preserved by any morphism of logoi $\mathcal{E} \to \mathcal{F}$. The map f becomes a cover in $\mathcal{F}$ if and only if the monomorphism $\mathrm{im}(f): \mathrm{Im}(f) \to B$ becomes an isomorphism in $\mathcal{F}$. Thus, forcing a map to become a cover is equivalent to forcing some monomorphism to become an isomorphism, which is a particular case of a quotient. Topological relations on a logos $\mathcal{E}$ are defined to be a family J of maps that are meant to become covers. Equivalently, the data of topological relations can be given as a family J of monomorphisms to be inverted.

Let us see how this is related to the so-called sheaf condition. Recall from the examples of quotients the construction of the quotient $\mathcal{E}/\!/(\mathrm{im}(f)) \simeq \mathcal{E}^{\mathrm{im}(f)} \hookrightarrow \mathcal{E}$ as a full subcategory of $\mathcal{E}$. A necessary condition for an object F of $\mathcal{E}$ to be in $\mathcal{E}^{\mathrm{im}(f)}$ is that $\mathrm{Hom}(B, F) \simeq \mathrm{Hom}(\mathrm{Im}(f), F)$. Using the fact that $\mathrm{Im}(f) = \mathrm{colim}\,(A \times_B A \rightrightarrows A)$, this condition becomes the *sheaf condition*:

$$\mathrm{Hom}(B, F) = \lim \Big(\mathrm{Hom}(A, F) \rightrightarrows \mathrm{Hom}(A \times_B A, F) \Big).$$

Then, one can prove that F is in $\mathcal{E}^{\mathrm{im}(f)}$ if and only if it satisfies the same condition not only for f but for all base changes of f.

A *site* is the data of a small category C and a set J of topological relations on $\mathcal{P}r\,(C)$ satisfying some extra conditions (stability by base change, composition, etc.). We shall not detail them since most of them are superfluous in order to characterize the corresponding reflective subcategory. Only the stability by base change is crucial.[70]

As for presentations, the notion of a site can be interpreted geometrically in Topos. Recall the subtopos $\mathbb{A}^{\twoheadrightarrow} \hookrightarrow \mathbb{A}^{\to}$ classifying arrows that are covers. Let $\mathbb{B}(C^{op})$ be the Alexandrov topos dual to $\mathcal{P}r\,(C)$ and J a topological relation in $\mathcal{P}r\,(C)$. The subtopos $\mathcal{X}$ of $\mathbb{B}(C^{op})$ defined by J can be defined as the following pullback in Topos:

---

[70] The situation compares to a more classical one. Recall that any relation R on a set E generates an equivalence relation. But, to compute the quotient E/R, it is not necessary for R to be an actual equivalence relation. Similarly, any set of monomorphisms in a logos $\mathcal{E}$ can be completed into a topology, but the characterization of the quotient reflective subcategory can be done directly from the generators.

Table 4.20. *Quotient and topologies*

Forcing condition	*Formulation in terms of monomorphisms*
Inverting a map $f: A \to B$	inverting the two monomorphisms $\text{im}(f): \text{Im}(f) \rightarrowtail B$ and $\Delta f: A \to A \times_B A$.
Forcing a map $c: U \to X$ to become a cover	inverting the monomorphism $\text{im}(c): \text{im}(c) \rightarrowtail X$
Forcing a family $c_i: U_i \to X$ to become covering	inverting the monomorphism $\left(\bigcup_i \text{im}(c_i)\right) \rightarrowtail X$

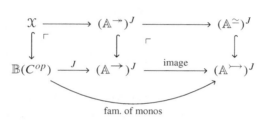

fam. of monos

It is a very important feature of logoi that the two conditions of forcing some maps to become isomorphisms and forcing some maps to become surjective are in fact equivalent, that is, every quotient can be described in terms of topological relations.[71] Recall that the diagonal of $f: A \to B$ is the map $\Delta f: A \to A \times_B A$, which is always a monomorphism. The map $f: A \to B$ is a monomorphism if and only if Δf is an isomorphism. Then a map f is an isomorphism if and only if it is a cover and a monomorphism if and only if both monomorphisms $\text{im}(f)$ and Δf are isomorphisms. As a consequence, any logos can be presented by means of topological relations. Table 4.20 recalls how to translate some conditions in terms of topologies, that is, of monomorphisms, and Table 4.21 summarizes the comparison between sites and presentations.

Example 3.22 (Topological Relations and Sites)

1. (Canonical and coherent topologies) Let X be a space. Let J_{can} be the collection of all open covers $U_i \to X$. Then the logos $\text{Sh}(X)$ is the quotient of the logos $\text{Pr}(\mathcal{O}(X))$ forcing the families in J_{can} to be covering families. If we consider instead the class J_{fin} of all finite open covers $U_i \to X$, then the quotient is the logos $\text{Sh}(X_{\text{coh}})$.

---

[71] We shall see that this property fails for ∞-logoi.

Table 4.21. *Comparison of sites and presentations*

	Site	Presentation
Generators	a category C of *representables*	a category G of *generators*
"Free" objet	$\mathcal{P}r(C)$ (presheaf logos/ Alexandrov topos)	$\mathrm{Set}[G] = \mathcal{P}r\left(G^{\mathrm{lex}}\right)$ (free logos/affine topos)
Relations	a topology J on C (forcing some maps to become covers)	a set R of maps in $\mathrm{Set}[G]$ (forcing some maps to become isomorphisms)
convenient for conditions of the type	colim of representables = representable	colim of lim of generators = lim of colim of generators
Quotient	$\mathcal{P}r(C)/\!/J = \mathrm{Sh}(C, J)$	$\mathrm{Set}[G]/\!/R$

2. (Stone–Čech) Let E be a set. Recall that the Stone–Čech compactification βE of E is a subtopos of $\widehat{E}$. Let J be the collection of all partitions $E_1 \coprod E_2 \to E$ of E. Then the logos $\mathrm{Sh}(\beta E)$ is the quotient of the logos $\mathrm{Sh}\left(\widehat{E}\right) = \mathcal{P}r(P(E))$ forcing the families in J to be covering families.

3. (Zariski spectrum) Let fLoc_A be the poset of finitely generated localizations of a ring A. Every finitely generated localization of A is of the form $A_f = A[f^{-1}]$ for some element f in A. If f and g are in A, let us write $f \leq g$ to mean that g is invertible in A_f. The relation $f \leq g$ is a preorder (it is transitive and reflexive). Let P_A be the associated poset, and let us write $D(f)$ for the image of $f \in A$ in P_A. The poset P_A is an inf-semi-lattice with $D(f) \wedge D(g) = D(fg)$ and $D(1) = 1$. The points of the Alexandrov logos $[P_A, \mathrm{Set}]$ form the poset $\mathrm{Loc}_A = \mathrm{Ind}(\mathrm{fLoc}_A)$ of all localizations $A \to B$. If $D(f_i) \leq D(f)$ $(1 \leq i \leq n)$ and $f_1 + \cdots + f_n = f$, let us declare that the family $D(f_i)$ $(1 \leq i \leq n)$ covers $D(f)$. For example, the pair $(D(f), D(1-f))$ covers $D(1) = 1$ for every $f \in A$. Also, $D(0)$ is covered by the empty family. This defines a topology on the presheaf logos $[P_A, \mathrm{Set}]$. The corresponding topos is the *Zariski spectrum* $Spec_{Zar}(A)$ of A. The topos $Spec_{Zar}(A)$ is localic, and its poset of points is the subposet of Loc_A spanned by localizations $A \to B$, where B is a local ring. This poset is the opposite of the poset of prime ideals of A.

4. (Actions of a Galois group) Let fSep_k be the category of finite separable field extensions of a field k. We consider the Alexandrov logos $[\mathrm{fSep}_k, \mathrm{Set}]$. A point of the corresponding topos is a separable field extension of k. Then,

we can construct the localization forcing all maps in (fSep$_k$)^{op} to become covers in $[\text{fSep}_k, \text{Set}]$. The resulting quotient is the logos $\mathcal{S}$h (fEt$_k$, étale) of sheaves for the étale topology on fEt$_k$. The corresponding topos is the so-called *étale spectrum* of k.

Recall that the Galois group Gal(k) of k is defined as a pro-finite group. We mentioned that the category Set$^{(\text{Gal}(k))}$ of sets equipped with a continuous action of Gal(k) is a logos. The logos $\mathcal{S}$h (Et$_k$, étale) can be proven to be equivalent to Set$^{(\text{Gal}(k))}$.

5. (Schanuel logos) Let fInj be the category of finite sets and injective maps. The category of points of the logos $\mathbb{B}$fInj is the category of all sets and injective maps. Then, we can construct the localization forcing every map in fInj^{op} to become cover in [fInj, Set]. The resulting category of sheaves $\mathcal{S}$h (fInj^{op}) is called the *Schanuel logos*. Its category of points is the category of infinite sets and injective maps. Let $G := \text{Aut}(\mathbb{N})$ be the group of automorphisms of $\mathbb{N}$ with the topology induced from the infinite product $\mathbb{N}^{\mathbb{N}}$, and let Set$^{(G)}$ be the category of continuous G-sets. It can be proven that the logos $\mathcal{E}$ is equivalent to the category Set$^{(G)}$.

6. (Etale spectrum of a commutative ring) Let fSep$_A$ be the category of finite separable extensions of a ring A. The opposite category is the category fEt$_A$ of finite étale extensions of the scheme dual to A. We consider the Alexandrov logos $[\text{fSep}_A, \text{Set}] = \mathcal{P}\text{r}(\text{fEt}_A)$. Its category of points is the category Sep$_A = \mathcal{I}\text{nd}(\text{fSep}_A)$ of all separable extensions $A \to B$.

The Yoneda embedding fEt$_A \hookrightarrow \mathcal{P}\text{r}(\text{fEt}_A)$ does not send étale coverings in fEt$_A$ to covering families $\mathcal{P}\text{r}(\text{fEt}_A)$. Forcing this, define the *étale spectrum* $Spec_{Et}(A)$ of A. The category of points of $Spec_{Et}(A)$ is the subcategory of Sep$_A$ spanned by separable extensions $A \to B$ such that B is a strictly Henselian local ring. The isomorphism classes of $\mathcal{P}\text{t}(Spec_{Et}(A))$ are in bijection with prime ideals of A. For an ideal p, the symmetries of the corresponding strict Henselianization $A \to A_p^h$ are given by the Galois group of the residue field of p. This category is not a poset, and this proves that the topos $Spec_{Et}(A)$ is not localic. However, its localic reflection, that is, the socle of $Spec_{Et}(A)$, is $Spec_{Zar}(A)$. Intuitively, $Spec_{Et}(A)$ is the space $Spec_{Zar}(A)$, but with the extra information of Galois groups at each point. The construction of étale spectra was the original motivation to develop topos theory. Its most important property is that the functor $Spec_{Et}$: Ring$^{op} \to$ Topos sends étale maps of schemes to étale maps of topoi. This is what allows us to interpret the algebraic Galoisian or étale descent as an actual topological descent and permits the construction of ℓ-adic cohomology theories.

7. (Nisnevich spectrum) In the previous example, if we force only the Nisnevich covering families to become covering families $\mathcal{P}r\,(\mathsf{fEt}_A)$, this defines the subtopos *Nisnevich spectrum* $Spec_{Nis}(A)$ of A.

Geometrically, the Nisnevich spectrum is further from the classical intuition of the Zariski spectrum of A than the étale spectrum is. The category of points of $Spec_{Nis}(A)$ is the subcategory of Sep_A spanned by separable extensions $A \to B$ such that B is a Henselian ring. There exists an inclusion $Spec_{Et}(A) \hookrightarrow Spec_{Nis}(A)$, which at the level of points corresponds to that of strict Henselian rings. Since not every Henselian ring is strict, the set of isomorphism classes of $\mathcal{P}t\,(Spec_{Nis}(A))$ is strictly larger than the set of prime ideals of A. For example, in the case of field k, the Nisnevich topology is trivial, and $Spec_{Nis}(k) = \mathcal{P}r\,(\mathsf{fEt}_A)$, whose points are all separable extensions of fields $k \to k'$. The poset reflection of this category is the poset of conjugacy classes of intermediate fields between k and some separable closure $\bar{k}$. This proves that the socle of $Spec_{Nis}(A)$ is not $Spec_{Zar}(A)$. There exist two morphisms of topoi

$$Spec_{Et}(A) \hookrightarrow Spec_{Nis}(A) \twoheadrightarrow Spec_{Zar}(A)$$

where the first one is an embedding and the second a surjection, and the composite is the socle projection of $Spec_{Et}(A)$. Intuitively, the Nisnevich spectrum is a sort of "mapping cone" (in the sense of homotopy theory) interpolating between the étale and Zariski spectra.

8. (Zariski sheaves) Let $\mathsf{Ring}_{\mathsf{fp}}$ be the category of commutative rings of finite presentation and $\mathsf{Aff}_{\mathsf{fp}} = \mathsf{Ring}_{\mathsf{fp}}^{\mathsf{op}}$ be the category of affine schemes of finite presentation. We consider the Alexandrov logos $\mathcal{P}r\left(\mathsf{Aff}_{\mathsf{fp}}\right) = \left[\mathsf{Ring}_{\mathsf{fp}}, \mathsf{Set}\right]$. The Yoneda embedding $\mathsf{Aff}_{\mathsf{fp}} \hookrightarrow \mathcal{P}r\left(\mathsf{Aff}_{\mathsf{fp}}\right)$ sends $\mathbb{A}^1$ to the forgetful functor

$$\mathbb{A}^1 : \mathrm{Hom}_{\mathsf{Ring}_{\mathsf{fp}}}(\mathbb{Z}[x], -) : \mathsf{Ring}_{\mathsf{fp}} \to \mathsf{Set}.$$

Recall that $\mathbb{A}^1$ is a ring object in the category of affine schemes with addition and multiplication given by maps $+, \times \colon \mathbb{A}^2 \to \mathbb{A}^1$. The Yoneda embedding preserves products, and $\mathbb{A}^1$ is also a ring object in $\left[\mathsf{Ring}_{\mathsf{fp}}, \mathsf{Set}\right]$. If $f^* \colon \left[\mathsf{Ring}_{\mathsf{fp}}, \mathsf{Set}\right] \to \mathcal{E}$ is a morphism of logoi, then $f^*(\mathbb{A}^1)$ is a ring object in $\mathcal{E}$. This defines an equivalence between the category of logos morphisms $\left[\mathsf{Ring}_{\mathsf{fp}}, \mathsf{Set}\right] \to \mathcal{E}$ and the category of ring objects in $\mathcal{E}$. Thus, the logos $\left[\mathsf{Ring}_{\mathsf{fp}}, \mathsf{Set}\right]$ classifies commutative rings.

Recall that a ring A is *nonzero* if $0 \neq 1$ in A. Let $\mathsf{Ring}_{\mathsf{fp}}^{\circ} \subset \mathsf{Ring}_{\mathsf{fp}}$ be the full category of nonzero rings. The forgetful functor $\mathbb{A}^1 \colon \mathsf{Ring}_{\mathsf{fp}}^{\circ} \to \mathsf{Set}$ is a nonzero ring object in the logos $\left[\mathsf{Ring}_{\mathsf{fp}}^{\circ}, \mathsf{Set}\right]$. The fully faithful inclusion $\mathsf{Ring}_{\mathsf{fp}}^{\circ} \hookrightarrow \mathsf{Ring}_{\mathsf{fp}}$ induces a left-exact localization

$[\mathrm{Ring}_{\mathrm{fp}}, \mathrm{Set}] \to [\mathrm{Ring}_{\mathrm{fp}}^\circ, \mathrm{Set}]$ that presents $[\mathrm{Ring}_{\mathrm{fp}}^\circ, \mathrm{Set}]$ as the logos classifying nonzero rings.

Recall that a commutative ring A is a *local ring* if $0 \neq 1$, and for every element a in A, either a or $1 - a$ is invertible. An element a in A is the same thing as a map $\mathbb{Z}[x] \to A$. This element is invertible if and only if the classifying map can be factored as $\mathbb{Z}[x] \to \mathbb{Z}[x, x^{-1}] \to A$. The definition of a nonzero local ring can be encoded by saying that, in the following diagram, one of the two dashed arrows has to exist:

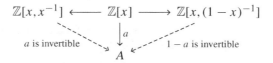

Let $A^\times = \mathrm{Hom}(\mathbb{Z}[x, x^{-1}], A)$ be the subset of invertible elements in A. The two horizontal maps define two maps $A^\times \to A \leftarrow A^\times$, and a nonzero ring A is local if they are jointly surjective. The two horizontal maps of the diagram above correspond to two maps in the opposite category $\mathrm{Aff}_{\mathrm{fp}}^\circ$:

$$\mathbb{G}_m \xrightarrow{\iota} \mathbb{A} \xleftarrow{1-\iota} \mathbb{G}_m$$

The two maps define a single map $\mathbb{G}_m \coprod \mathbb{G}_m \to \mathbb{A}$ in $\mathcal{P}r\left(\mathrm{Aff}_{\mathrm{fp}}^\circ\right)$. This map is not a cover, but it can be forced to be. And this is exactly the condition that defines local rings. The quotient of $\mathcal{P}r\left(\mathrm{Aff}_{\mathrm{fp}}^\circ\right)$ generated by the condition "$\mathbb{G}_m \coprod \mathbb{G}_m \to \mathbb{A}$ is a cover" is the logos $\mathcal{S}h\left(\mathrm{Aff}_{\mathrm{fp}}^\circ\right)$ that classifies local rings. The image of $\mathbb{A}$ in $\mathcal{S}h\left(\mathrm{Aff}_{\mathrm{fp}}^\circ\right)$ is the generic local ring, and it is often denoted $\mathbb{A}^1$. The category $\mathcal{S}h\left(\mathrm{Aff}_{\mathrm{fp}}^\circ\right)$ can be proven to be the category $\mathcal{S}h\left(\mathrm{Aff}_{\mathrm{fp}}, \mathrm{Zar}\right)$ of sheaves on $\mathrm{Aff}_{\mathrm{fp}}$ for the Zariski topology.

Similar considerations apply to defining the topoi classifying Henselian rings (with the Nisnevich topology) and strict Henselian rings (with the étale topology). However, these topologies are not nicely generated by a single map, as is the Zariski topology.

3.4.6 Presentations from Logical Theories

We mentioned in the introduction that logoi could be thought as categories of generalized sets and were suited to producing semantics for all sorts of logical theories. A particular aspect of this relationship with logic is that logical theories can be used as generating data for logoi. Roughly presented, a logical theory has sorts (or types), formulas, and axioms. Intuitively, the sorts and formulas generate the objects and morphisms of a category G, and the axioms distinguish a set of maps R in $\mathrm{Set}\,[G]$ (using the dictionary sketched in

Table 4.7). A model of the logical theory in a logos $\mathcal{E}$ is an interpretation of sorts and formulas such that the axioms are validated. In terms of category, this is a functor $G \to \mathcal{E}$ such that the canonical extension $\mathrm{Set}[G] \to \mathcal{E}$ sends the maps of R to isomorphisms. In other terms, a model in $\mathcal{E}$ is a logos morphism $\mathrm{Set}[G]/\!/R \to \mathcal{E}$. For this reason, the logos $\mathrm{Set}[G]/\!/R$ is called the *classifying logos* of the theory. Details about this construction can be found in [25, Sections VI, VIII, and X]. The previous construction of the logos of Zariski sheaves is an example of this construction. The quotient forcing the map $\mathbb{G}_m \coprod \mathbb{G}_m \to \mathbb{A}^1$ to become a cover corresponds to the axiom that the ring must be local.

However, such a construction is not pertinent for all logical theories. It relies implicitly on the fact that morphisms of logoi preserve the logical constructions, but this is mostly false. Logoi morphisms preserve all colimits, but only finite limits. This means that, in the dictionary of Table 4.7, they will only be compatible with logical theories involving *finite* conjunctive conditions, that is, only finite conjunctions of propositions and no function type, no universal quantification, no implication, and no subobject classifier. Logical theories compatible with logoi morphisms are called *geometric* (see [19, 25]).

A particular instance of the dictionary of Table 4.7 is that an existential statement translates into the image of a morphism. This gives an elegant logical interpretation to the presentation of logoi by sites: topological relations correspond to forcing some statements of existence. Again, the previous construction of the logos of Zariski is an example: the axiom forcing a ring to be local is existential.[72]

4 Higher Topos–Logos Duality

4.1 Definitions and Examples

4.1.1 Enhancing Set into $\mathcal{S}$

Our presentation should have made it clear that the theory of topoi is essentially what becomes locale theory when the "basic coefficients" are enhanced from the poset $\{0 < 1\}$ to the category Set. Similarly, the theory of ∞-topoi is what become topos theory when the category Set is enhanced into the ∞-category $\mathcal{S}$ of ∞-groupoids (e.g., homotopy types of spaces). Intuitively, an ∞-logos is an ∞-category of sheaves with values in ∞-groupoids.[73]

---

[72] $\vdash_a \exists b, (ab = 1) \vee ((1-a)b = 1)$.

[73] Sheaves of ∞-groupoids are also called *stacks* in ∞-groupoids. However, the usage in ∞-topos theory has simplified the vocabulary and kept only the name of sheaves.

The replacements of $\{0 < 1\}$ by Set and then by $\mathcal{S}$ follow a precise logic. In posets, $\underline{2}$ is the free sup-lattice on one generator. In categories, Set $= \mathcal{P}r(1)$ is the free cocomplete category on one generator. And in ∞-categories, $\mathcal{S} = \mathcal{P}r_\infty(1)$ is the free cocomplete ∞-category on one generator. These universal properties are the reason why $\underline{2}$, Set, and $\mathcal{S}$ are so important. This may explain also why, in the setting of ∞-categories, $\mathcal{S}$ is a more fundamental object than Set: the category Set is still cocomplete as an ∞-category, but it is no longer freely generated.[74]

The manipulation of ∞-groupoids is, in practice, remarkably similar to that of sets. The main operations of manipulation of ∞-groupoids are still limits and colimits, but their behavior in the ∞-categorical setting is different. For example, the diagonal $\Delta f : A \to A \times_B A$ of a map $f : A \to B$ need not be a monomorphism any longer. Also, using the embedding Set $\hookrightarrow \mathcal{S}$ whose image is discrete ∞-groupoids, the colimit of a diagram of sets computed in $\mathcal{S}$ need not be discrete.[75] Otherwise, the theory of ∞-logoi is very similar in its structure to that of logoi (see Table 4.18). Essentially, it suffices to replace Set by $\mathcal{S}$ everywhere and to interpret all constructions (limits, colimits, adjunctions, commutativity of diagrams) in the ∞-categorical sense. For example, the free cocompletion of an ∞-category C is now given by the ∞-category of presheaves of ∞-groupoids $\mathcal{P}r_\infty(C) = [C^{op}, \mathcal{S}]$ rather than presheaves with values in Set. An ∞-logos can then be defined as an (accessible) left-exact localization of some $\mathcal{P}r_\infty(C)$. Morphisms of ∞-logoi are defined as functors preserving colimits and finite limits in the ∞-categorical sense. This defines an ∞-category Logos$_\infty$, and the category Topos$_\infty$ is then defined to be (Logos$_\infty$)^{op}.[76] We shall denote by $\mathcal{S}h_\infty(\mathcal{X})$ the ∞-logos dual of an ∞-topos $\mathcal{X}$.

Affine topoi, Alexandrov topoi, points, subtopoi, étale morphisms, and so on, are all defined the same way as in topos theory. For this reason, we shall not present the theory of ∞-topoi systematically, as in the case of topoi (see [5, 22]). We will just underline the important new features of the theory. Before we do this, we are going to introduce some examples to play with.

4.1.2 First Examples
1. (Point) The ∞-category $\mathcal{S}$ is the initial ∞-logos. Any ∞-logos $\mathcal{E}$ has a canonical logos morphism $\mathcal{S} \to \mathcal{E}$. The ∞-topos $\mathbb{1}$ dual to $\mathcal{S}$ is terminal.

---

[74] Other motivations to enhance sets into ∞-groupoids are given in Chapter 9.
[75] This new colimit is the so-called homotopy colimit. For a description of the notion of homotopy colimit, see Chapter 9.
[76] When Logos$_\infty$ is viewed as an $(\infty, 2)$-category, we defined the $(\infty, 2)$-category of ∞-topoi as Topos$_\infty$ = (Logos$_\infty$)^{1op}, i.e., by reversing the direction of 1-arrows only.

A *point* of an ∞-topos $\mathcal{X}$ is a morphism $\mathbb{1} \to \mathcal{X}$, that is, a logos morphism $\mathrm{Sh}_\infty(\mathcal{X}) \to \mathcal{S}$. The ∞-category of points of a topos $\mathcal{X}$ is $\mathcal{P}\mathrm{t}(\mathcal{X}) := \mathrm{Hom}_{\mathsf{Topos}_\infty}(\mathbb{1}, \mathcal{X}) = \mathrm{Hom}_{\mathsf{Logos}_\infty}(\mathrm{Sh}_\infty(\mathcal{X}), \mathcal{S})$.

2. (The ∞-topos of a topos) In the same way that any frame $\mathcal{O}(X)$ defines a logos $\mathrm{Sh}(X)$ of sheaves of sets, any logos $\mathrm{Sh}(\mathcal{X})$ defines an ∞-logos $\mathrm{Sh}_\infty(\mathcal{X})$ of sheaves of ∞-groupoids. The ∞-category $\mathrm{Sh}_\infty(\mathcal{X})$ is defined at the full sub-∞-category of $[\mathrm{Sh}(\mathcal{X})^{op}, \mathcal{S}]$ spanned by functors F satisfying the *higher sheaf condition*: for any covering family $U_i \to U$ in $\mathrm{Sh}(\mathcal{X})$, we must have

$$F(U) \simeq \lim \left(\prod_i F(U_i) \rightrightarrows \prod_{ij} F(U_{ij}) \underset{\longrightarrow}{\overset{\longrightarrow}{\longrightarrow}} \prod_{ijk} F(U_{ijk}) \cdots \right),$$

where the diagram is now a *full* cosimplicial diagram. This defines a functor $\mathrm{Sh}_\infty : \mathsf{Logos} \to \mathsf{Logos}_\infty$, and dually a functor $\mathsf{Topos} \to \mathsf{Topos}_\infty$, which are both fully faithful. In particular, the ∞-category of points of a topos $\mathcal{X}$ does not change when it is viewed as an ∞-topoi and stays a 1-category.

3. (Quasi-discrete ∞-topos) For K an ∞-groupoid, the ∞-category $\mathcal{S}_{/K}$ is an ∞-logos. The dual ∞-topos is denoted $\mathbb{B}_\infty K$ and called *quasi-discrete*. An ∞-topos is called *discrete* if it is of the type $\mathbb{B}_\infty E$ for E a set. This construction defines a fully faithful functor $\mathbb{B}_\infty : \mathcal{S} \to \mathsf{Topos}_\infty$, which is analogue to the "discrete topos" functor $\mathsf{Set} \to \mathsf{Topos}$. The ∞-category of points of $\mathbb{B}_\infty K$ is K. In particular, when K is not a 1-groupoid (e.g., the homotopy type $K(\mathbb{Z}, 2)$ of $\mathbb{CP}^\infty$, which is a nontrivial 2-groupoid), the quasi-discrete topos $\mathbb{B}_\infty K$ is not in the image of $\mathsf{Topos} \hookrightarrow \mathsf{Topos}_\infty$. This proves that there are more ∞-topoi than topoi.

4. (Alexandrov ∞-topos) For C a small ∞-category, the diagram ∞-category $[C, \mathcal{S}] = \mathcal{P}\mathrm{r}_\infty(C^{op})$ is an ∞-logoi. The dual *Alexandrov ∞-topos* is denoted $\mathbb{B}_\infty C$. This construction defines a functor $\mathbb{B}_\infty : \mathsf{Cat}_\infty \to \mathsf{Topos}_\infty$ that is not fully faithful.[77] The restriction of this functor to ∞-groupoids via $\mathcal{S} \hookrightarrow \mathsf{Cat}_\infty$ gives back the previous example. The ∞-category of points of $\mathbb{B}_\infty C$ is $\mathcal{P}\mathrm{t}(\mathbb{B}_\infty C) = [C^{op}, \mathcal{S}]^{\mathrm{lex}} = \mathrm{Ind}(C)$. Quasi-discrete ∞-topoi are examples of Alexandrov ∞-topoi. This is a consequence of the *Galoisian interpretation of homotopy theory* [30, 32] that provides the important equivalence of ∞-categories $\mathcal{S}^K \simeq \mathcal{S}_{/K}$. In the case where $K = BG$ is the classifying space of some group G, this equivalence encodes the statement that a homotopy type with an action of

---

[77] Two Morita-equivalent ∞-categories define the same Alexandrov ∞-topos.

G is the same thing as a homotopy type over BG. In this case, we shall denote simply by $\mathbb{B}_\infty G$ the quasi-discrete ∞-topos $\mathbb{B}_\infty(BG)$.

5. (Affine ∞-topos) For C a small ∞-category, the *free ∞-logoi* on C is $\mathcal{S}[C] := \mathcal{P}r_\infty(C^{\text{lex}}) = [(C^{\text{lex}})^{op}, \mathcal{S}]$, where the lex completion is taken in the ∞-categorical sense. It satisfies the expected property that an ∞-logos morphism $\mathcal{S}[C] \to \mathcal{E}$ is equivalent to a diagram $C \to \mathcal{E}$. The dual *affine ∞-topos* is denoted $\mathbb{A}_\infty^C$.

6. (The ∞-topos of ∞-groupoids) In particular, the free ∞-logos on one generator is $\mathcal{S}[X] = [\mathcal{S}_{\text{fin}}, \mathcal{S}]$, where $\mathcal{S}_{\text{fin}}$ is the ∞-category of finite ∞-groupoids (homotopy types of finite cell complexes). The object X corresponds to the canonical inclusion $\mathcal{S}_{\text{fin}} \to \mathcal{S}$. The corresponding ∞-topos shall be denoted simply by $\mathbb{A}_\infty$. Its ∞-category of points is $\mathcal{P}t(\mathbb{A}_\infty) = \mathcal{S}$. The universal property of $\mathcal{S}[X]$ translates geometrically into the result that

$$\mathcal{S}h_\infty(\mathcal{X}) = \text{Hom}_{\text{Topos}_\infty}(\mathcal{X}, \mathbb{A}_\infty).$$

7. (∞-Etale morphisms) If $\mathcal{E}$ is an ∞-logos, then so is the slice $\mathcal{E}_{/E}$ for any object E of $\mathcal{E}$. Moreover, the base change along $E \to 1$ in $\mathcal{E}$ provides an ∞-logos morphism $\epsilon_E^* : \mathcal{E} \to \mathcal{E}_{/E}$ called an *∞-étale extension*. Let $\mathcal{X}$ and $\mathcal{X}_E$ be the ∞-topoi dual to $\mathcal{E}$ and $\mathcal{E}_{/E}$. Observe that the diagonal map $\delta_E : E \to E \times E$ is defining a global section of the object $\epsilon_E^*(E) := (E \times E, p_1)$. The pair $(\epsilon_E^*(E), \delta_E)$ is universal in the sense that for any morphism of ∞-logoi $u^* : \mathcal{E} \to \mathcal{F}$ and any global section $s : 1 \to u^*E$, there exists a morphism of ∞-logoi $v^* : \mathcal{E}_{/E} \to \mathcal{F}$ such that $v^* \circ \epsilon_E^* = u^*$ and $u^*(\delta_E) = s$; moreover, the morphism u^* is essentially unique:

In other words, the ∞-logos $\mathcal{E}_{/E}$ is obtained from $\mathcal{E}$ by adding freely a global section δ_E to the object E.

The corresponding morphism $\mathcal{X}_E \to \mathcal{X}$ is called *∞-étale*, and $\mathcal{X}_E$ is called an *∞-étale domain* of $\mathcal{X}$. Intuitively, in the same way that an étale morphism of topoi has discrete fibers, an ∞-étale morphism of ∞-topoi has quasi-discrete fibers.

8. (Pointed objects) Let $\mathcal{S}_{\text{fin}}^\bullet$ be the ∞-category of pointed finite ∞-groupoids (pointed finite cell complexes). The Alexandrov ∞-topos $\mathbb{A}_\infty^\bullet$ is defined to be the dual of $\mathcal{S}[X^\bullet] := [\mathcal{S}_{\text{fin}}^\bullet, \mathcal{S}]$. It has the classifying property that

an ∞-logos morphism $S[X^\bullet] \to \mathcal{E}$ is equivalent to a pointed object of $\mathcal{E}$, that is, an object E together with a global section $1 \to E$. There exists an equivalence $S[X^\bullet] = S[X]_{/X}$ that gives an étale morphism $\mathbb{A}_\infty^\bullet \to \mathbb{A}_\infty$. This map is the universal ∞-étale morphism: for any ∞-topos $\mathcal{X}$ and any object E in $Sh_\infty(\mathcal{X})$, there exists a unique cartesian square

$$\begin{array}{ccc} \mathcal{X}_E & \longrightarrow & \mathbb{A}_\infty^\bullet \\ \epsilon_E \downarrow & \ulcorner & \downarrow \\ \mathcal{X} & \xrightarrow{\chi_E} & \mathbb{A}_\infty \end{array}$$

The argument is the same as in Section 3.2.6.

9. (Quotient) Let R be a set of maps in an ∞-logos $\mathcal{E}$. The quotient $\mathcal{E}/\!/R$ is defined to be the left-exact localization of $\mathcal{E}$ generated by R. It is equivalent to the sub-∞-category $\mathcal{E}^R$ of $\mathcal{E}$ spanned by objects E satisfying the following condition. Recall that for a map $f \colon A \to B$, the iterated diagonals of f are defined by $\Delta^0 f := f$ and $\Delta^n f := \Delta(\Delta^{n-1} f)$. Let $C \to D$ be a base change of some $\Delta^n f$ for f in R; then E must satisfy that $\mathrm{Hom}(D, E) \to \mathrm{Hom}(C, E)$ is an invertible map in S.

10. (Truncated objects) For $-2 \leq n \leq \infty$, a morphism $f \colon A \to B$ of $\mathcal{E}$ is said to be n-*truncated* if $\Delta^{n+2} f$ is invertible. A (-1)-truncated morphism is the same thing as a monomorphism. An object E is called n-truncated if the map $E \to 1$ is. In this case, we simply put $\Delta^n E = \Delta^n(E \to 1)$. In the ∞-logos S, the n-truncated objects are the n-groupoids. Intuitively, the n-truncated objects in $\mathcal{E}$ are sheaves with values in n-groupoids. In particular, 0-truncated objects are sheaves with discrete fibers, and (-1)-truncated objects are sheaves with fibers an empty set or a singleton. Given an ∞-logos $\mathcal{E}$, we denote by $\mathcal{E}^{\leq n}$ the full sub-∞-category spanned by n-truncated objects. A morphism of ∞-logoi $\mathcal{E} \to \mathcal{F}$ induces a functor $\mathcal{E}^{\leq n} \to \mathcal{F}^{\leq n}$.

The ∞-logos $S[X^{\leq n}] := S[X]/\!/(\Delta^{n+2} X)$ is the classifier for n-truncated objects. This means that $\mathrm{Hom}_{\mathrm{Logos}_\infty}(S[X^{\leq n}], \mathcal{E}) = \mathcal{E}^{\leq n}$. In particular, the ∞-category of points of $S[X^{\leq n}]$ is the ∞-category $S^{\leq n}$ of n-groupoids. Since any n-truncated object is also $(n+1)$-truncated, we have a tower of quotients of ∞-logoi:

$$S\left[X^{\leq -1}\right] \longleftarrow S\left[X^{\leq 0}\right] \longleftarrow S\left[X^{\leq 1}\right] \longleftarrow \ldots \longleftarrow S[X].$$

We denote by $\mathbb{A}_\infty^{\leq n}$ the ∞-topos dual to $S\left[X^{\leq n}\right]$. It is a sub-∞-topos of $\mathbb{A}_\infty$. We have $S\left[X^{\leq 0}\right] = Sh_\infty(\mathbb{A})$ and $S\left[X^{\leq -1}\right] = Sh_\infty(\mathbb{S})$, hence $\mathbb{A}_\infty^{\leq 0}$ and $\mathbb{A}_\infty^{\leq -1}$ are respectively the ∞-topos corresponding to the topos of sets and the Sierpiński space though the embeddings Locale ↪ Topos ↪ Topos$_\infty$. Altogether, we have an increasing sequence of sub-∞-topoi:

$$\mathbb{S} = \mathbb{A}_\infty^{\leq -1} \longhookrightarrow \mathbb{A} = \mathbb{A}_\infty^{\leq 0} \longhookrightarrow \mathbb{A}_\infty^{\leq 1} \longhookrightarrow \ldots \longhookrightarrow \mathbb{A}_\infty.$$

4.1.3 Extension and Restriction of Scalars

For $\mathfrak{X}$ an ∞-topos, the ∞-category $\mathcal{O}(\mathfrak{X}) := \mathrm{Sh}_\infty(\mathfrak{X})^{\leq -1}$ of (-1)-truncated objects is a frame, called the frame of *open domains* of $\mathfrak{X}$. The corresponding locale is denoted $\tau_{-1}(\mathfrak{X})$ and called the *socle* of $\mathfrak{X}$. The ∞-category $\mathrm{Sh}_\infty(\mathfrak{X})^{\leq 0}$ of 0-truncated objects is a logos called the *discrete truncation* of $\mathrm{Sh}_\infty(\mathfrak{X})$. The corresponding topos is denoted $\tau_0(\mathfrak{X})$. The socle of $\tau_0 \mathfrak{X}$ in the sense of ordinary topoi is the socle of $\mathfrak{X}$ in the sense of ∞-topoi.[78] These constructions build left adjoints to the inclusion functors:

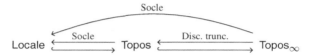

At this point, it is perhaps useful to make an analogy with commutative algebra. The embedding $\underline{2} \simeq \{\emptyset, \{\star\}\} \hookrightarrow$ Set compares somehow with the inclusion $\{0,1\} \subset \mathbb{Z}$. Schemes over $\mathbb{Z}$ are defined as zeros of polynomial with coefficients in $\mathbb{Z}$. Among them are those that can be defined as zeros of polynomials with coefficients in $\{0,1\}$ (e.g., toric varieties). There are more of the former than the latter. The relation between locales and topoi can be thought the same way: there are more topoi than locales because the latter are allowed to be defined only by equations involving a restricted class of functions. And there are more ∞-topoi than topoi for the same reason. Table 4.22 details a bit this analogy.

Moreover, the above truncation functors Topos$_\infty$ → Topos → Locale can be formalized as actual base change along the coefficient morphisms $\mathcal{S} \xrightarrow{\pi_0}$ Set $\xrightarrow{\pi_{-1}} \underline{2}$. Presentable ∞-categories have a tensor product, denoted $\otimes_\mathcal{S}$, defined similarly to the one of presentable categories (which we rename $\otimes_{\mathsf{Set}}$ here). We shall not expand on it here. We shall only give the computation formula $\mathcal{A} \otimes_\mathcal{S} \mathcal{B} = [\mathcal{A}^{op}, \mathcal{B}]^c$, where $[-,-]^c$ refer to the ∞-category of functors preserving limits. All structural relations of Table 4.15 make sense also for presentable ∞-categories, provided Set is replaced by $\mathcal{S}$. Using this tensor product, the truncation functor can be written as base change formula

---

[78] There exists a notion of n-logos corresponding to the categories $\mathrm{Sh}(\mathfrak{X})^{\leq n}$, but, once in the paradigm of ∞-categories, the notion of ∞-logos/topos encompasses all the others, and it is also the one with the most regular features. For these reasons, we shall not say much about n-logoi/topoi (see [22]).

Table 4.22. *Coefficient analogies*

Degree	Commutative algebra		Logos theory	
	Coefficient k	k-Algebra	Coefficient $\mathcal{K}$	$\mathcal{K}$-Logos
-1	$\mathbb{Z}/2\mathbb{Z}$	$\mathbb{Z}/2\mathbb{Z}$-algebra	$\{0 \to 1\} =$ $\mathcal{S}^{\leq -1}$ (-1-groupoids)	frame = 0-logos
0	$\mathbb{Z}$	$\mathbb{Z}$-algebra	Set $= \mathcal{S}^{\leq 0}$ (0-groupoids)	logos = 1-logos
1	$\mathbb{Z}[\epsilon] = \mathbb{Z}[x]/(x^2)$	$\mathbb{Z}[\epsilon]$-algebra	$\mathcal{S}^{\leq 1}$ (1-groupoids)	2-logos
n	$\mathbb{Z}[x]/(x^{n+1})$	$\mathbb{Z}[x]/(x^{n+1})$-algebra	$\mathcal{S}^{\leq n}$ $(n\text{-groupoids})$	$(n+1)$-logos
∞	$\mathbb{Z}[\![s]\!]$	$\mathbb{Z}[\![s]\!]$-algebra	$\mathcal{S}$ $(\infty\text{-groupoids})$	∞-logos

and

$$\mathcal{S}h_\infty(\mathfrak{X})^{\leq 0} = \mathcal{S}h_\infty(\mathfrak{X}) \otimes_\mathcal{S} \text{Set}$$

$$\mathcal{S}h_\infty(\mathfrak{X})^{\leq -1} = \mathcal{S}h_\infty(\mathfrak{X})^{\leq 0} \otimes_{\text{Set}} \underline{2} = \mathcal{S}h_\infty(\mathfrak{X}) \otimes_\mathcal{S} \underline{2}.$$

4.2 New Features

4.2.1 Simplification of Descent Properties

Although the use of ∞-groupoids instead of sets might look like a sophistication, it happens that the characterization of ∞-logoi by their descent properties is actually simpler than the one of logoi. Recall from Section 3.3.4 and Table 4.17 that not every colimit had the descent property in a logos and that we had to restrict this condition in order to characterize logoi. It is a remarkable fact that *all* colimits have the descent property in an ∞-logos. This leads to a very compact characterization first proposed by Rezk [29]: a presentable ∞-category $\mathcal{E}$ is an ∞-logos if and only if, for any diagram $X \colon I \to \mathcal{E}$, we have

$$\mathcal{E}_{/\operatorname{colim}_i X_i} \simeq \lim_i \mathcal{E}_{/X_i}. \tag{4.1}$$

In the case of $\mathcal{E} = \mathcal{S}$, this property is equivalent to the Galoisian interpretation of homotopy theory, $\mathcal{S}^K = \mathcal{S}_{/K}$ (see Example 4.1.2(4) and [30, 32]).[79]

---

[79] Essentially, if $\mathcal{S}^K = \mathcal{S}_{/K}$, we deduce $\mathcal{E}_{/\operatorname{colim}_i X_i} = \mathcal{E}^{\operatorname{colim}_i X_i} = \lim_i \mathcal{E}^{X_i} = \lim_i \mathcal{E}_{/X_i}$. Reciprocally, we use $K = \operatorname{colim}_K 1$ to get $\mathcal{E}_{/K} = \mathcal{E}_{/\operatorname{colim}_K 1} = \lim_K \mathcal{E} = \mathcal{E}^K$.

Definitions à la Giraud or Lawvere can also be given, but we shall not detail them here (see [22, 31, 33]).

This property is equivalent to another one that we will need below. Let $\mathcal{E}_{/E}^{(\text{core})}$ be the *core* of $\mathcal{E}_{/E}$, that is, the sub-∞-groupoid containing all objects and only invertible maps. The core functor $(-)^{(\text{core})}$: $\mathsf{Cat}_\infty \to \mathcal{S}$ is right adjoint to the inclusion $\mathcal{S} \to \mathsf{Cat}_\infty$. In particular, it preserves limits, and we get from the descent property of the ∞-logos $\mathcal{E}$ that

$$\mathcal{E}_{/\operatorname{colim}_i X_i} \simeq \lim_i \mathcal{E}_{/X_i}. \tag{4.2}$$

Under the assumption that $\mathcal{E}$ has universal colimits, the core descent property, written in terms of ∞-groupoids, turns out to be equivalent to the previous one in terms of ∞-categories.

4.2.2 The Universe

One of the reasons to deal with ∞-groupoids instead of sets is the failure of sets to classify themselves. Letting aside size issues for now, the problem is that sets do not so much form a set as a category, or a groupoid if we are only interested in classifying them up to isomorphism. Only ∞-groupoids have a self-classification property: there exists naturally an ∞-groupoid of ∞-groupoids.[80]

The only sets that are classified by an actual set are those without symmetries, that is, the empty set and singletons. This singles out the set $\{\emptyset, 1\}$ as a classifier for these "rigid" sets. In a logos $\mathcal{E}$, thought as a category of generalized sets, the role of $\{\emptyset, 1\}$ is played by the subobject classifier Ω. A map $E \to \Omega$ is intuitively the same thing as a family of empty or singleton sets parameterized by E, that is, a subobject $F \hookrightarrow E$.

To classify more general families, that is, general maps $f : F \to E$, by some characteristic map $\chi_f : E \to U$, the codomain U needs to be able to classify sets of all sizes, that is, sets with symmetries. The symmetries are a well-known obstruction to construct any kind of classifying (or moduli) space with the property that χ_f is *uniquely* determined by f. The solution was found with the idea that the classifying object U need not only classify sets up to symmetries but sets *and* their symmetries. That is, U needs to have a groupoid of points and not only a set. This is the beginning of stack theory (see Chapters 8 and 9).

The formalism of presheaves is actually of great help to formalize classification problems. Let a family of objects of a logos $\mathcal{E}$ parameterized by an object

---

[80] Notice that, because n-groupoids form an $(n + 1)$-groupoid, we need to go to infinity in order to have this property.

E be a map $F \to E$ in $\mathcal{E}$, that is, an object of $\mathcal{E}_{/E}$ (intuitively, the family is that of the fibers of this map). A morphism of families is a morphism $F \to F'$ compatible with the projections to E, that is, a morphism in $\mathcal{E}_{/E}$. Since we are only interested in classifying objects of $\mathcal{E}$ up to isomorphisms, we are going to consider only the subgroupoid $\mathcal{E}^{(\text{core})}_{/E} \hookrightarrow \mathcal{E}_{/E}$ containing all objects, but only isomorphisms. If $E' \to E$ is a map, any family on E can be pulled back on E'. This builds the *functor of families*, called also the *universe* of the logos $\mathcal{E}$:

$$\begin{aligned}
\mathbb{U} : \mathcal{E}^{op} &\longrightarrow \mathsf{Gpd}, \\
E &\longmapsto \mathbb{U}(E) := \mathcal{E}^{(\text{core})}_{/E}, \\
f : E' \to E &\longmapsto f^* : \mathcal{E}^{(\text{core})}_{/E} \longrightarrow \mathcal{E}^{(\text{core})}_{/E'}.
\end{aligned} \quad (4.3)$$

There exists a Yoneda embedding $\mathcal{E} \hookrightarrow [\mathcal{E}^{op}, \mathsf{Gpd}]$ sending an object E to the functor $\widehat{E} := \text{Hom}(-, E)$ with values in $\mathsf{Set} \hookrightarrow \mathsf{Gpd}$, in particular, the groupoid of natural transformations $\text{Hom}(\widehat{E}, \mathbb{U})$ is $\mathbb{U}(E) = \mathcal{E}^{(\text{core})}_{/E}$. This equivalence implies that, in the category of presheaves of groupoids, the object $\mathbb{U}$ has the property that a map $F \to E$ in $\mathcal{E}$ corresponds uniquely to a map $\widehat{E} \to \mathbb{U}$. In other words, the presheaf $\mathbb{U}$ is the formal solution to the classification of families of objects of $\mathcal{E}$.

Now, the classification problem can be formulated properly as the problem of finding an object U in $\mathcal{E}$ such that $\widehat{U} \simeq \mathbb{U}$. There are two obstructions to this:

1. $\text{Hom}(-, U)$ takes values in sets and not groupoids.
2. (size issue) the values of $\text{Hom}(-, U)$ are small, but those of $\mathbb{U}$ are large.

In logos theory, the first obstruction is handled by restricting the functor $\mathbb{U}$. If we limit ourselves to families $F \to E$ that are monomorphisms, then the groupoid of such $F \rightarrowtail E$ is actually a set. This defines a subfunctor $\mathbb{U}^{\leq -1} \hookrightarrow \mathbb{U}$ with values in sets and can be represented by an object of $\mathcal{E}$. This is actually the universal property of subobject classifier $\mathbb{U}^{\leq -1} = \text{Hom}(-, \Omega)$. But the first obstruction is better dealt with by enhancing sets into ∞-groupoids and logoi into ∞-logoi. When $\mathcal{E}$ is an ∞-logos, both the functor of points $\text{Hom}(-, U)$ of an object U and the core universe $\mathbb{U}$ take values in the ∞-category $\mathcal{S}$ of (large) ∞-groupoids. Moreover, since $\mathcal{E}$ is assumed a presentable ∞-category, a functor $\mathcal{E}^{op} \to \mathcal{S}$ is representable if and only if it sends colimits in $\mathcal{E}$ to limits in $\mathcal{S}$. But this is exactly the descent property of (4.2) characterizing ∞-logoi. So the object U would exist if it were not for the second obstruction.

This second obstruction is dealt with by considering only partial universes, that is, universe, that classified uniquely *some* families. We shall say that an object U of an ∞-logos $\mathcal{E}$ is a *partial universe* if it is equipped with a monomorphism $\widehat{U} \rightarrowtail \mathbb{U}$.[81] This means that, for an object E in $\mathcal{E}$, the ∞-groupoid $\text{Hom}(E, U)$ is a full sub-∞-groupoid of $\mathcal{E}_{/E}^{(\text{core})}$. For example, the subobject classifier Ω classifies only families $F \to E$ that are monomorphisms. Now, a fundamental property of ∞-logoi is that, even though the universe is too big to be an actual object of $\mathcal{E}$, there exist always partial universes. In other words, given any map $f: F \to E$, there exists always a partial universe U such that f is classified by a unique map $\chi_f: E \to U$. Moreover, there are always enough partial universes in the sense that $\mathbb{U}$ is the union of all the partial universes of $\mathcal{E}$. This last property has the practical effect that, for the most part, one can manipulate the universe as if it were an actual object of the ∞-logos.

4.2.3 ∞-Topoi from Homology Theories

Eilenberg–Steenrod axioms for homology theories have a modern formulation in terms of ∞-category theory. Let $\mathcal{S}_{\text{fin}}^{\bullet}$ be the category of pointed finite ∞-groupoids. A functor $H: \mathcal{S}_{\text{fin}}^{\bullet} \to \mathcal{S}$ is a *homology theory* if it satisfies the *excision property*, that is, if it sends pushout squares to pullback squares:

$$\begin{array}{ccc} A & \longrightarrow & B \\ \downarrow & \ulcorner & \downarrow \\ C & \longrightarrow & D \end{array} \quad \longmapsto \quad \begin{array}{ccc} H(A) & \longrightarrow & H(B) \\ \downarrow & \urcorner & \downarrow \\ H(C) & \longrightarrow & H(D) \end{array} \qquad (4.4)$$

A homology theory H is called *reduced* if, moreover, $H(1) = 1$.[82]

Homology theories define a full sub-∞-category $\left[\mathcal{S}_{\text{fin}}^{\bullet}, \mathcal{S}\right]^{(1)}$ of $\mathcal{S}[X^{\bullet}] = \left[\mathcal{S}_{\text{fin}}^{\bullet}, \mathcal{S}\right]$. The sub-$\infty$-category of reduced homology theories can be proven to be equivalent to the ∞-category Sp of *spectra* (in the sense of algebraic topology) and the ∞-category $\left[\mathcal{S}_{\text{fin}}^{\bullet}, \mathcal{S}\right]^{(1)} = \mathcal{P}\text{Sp}$ the ∞-category $\mathcal{P}\text{Sp}$ of *parameterized spectra*.[83] Moreover, Goodwillie's calculus of functors proves

---

[81] Partial universes are equivalent to codomains of Voevodsky's notion of *univalent maps*.
[82] For H a reduced homology theory and $B = C = 1$, the excision condition says $H(\Sigma A) = \Omega H(A)$. Passing to the homotopy groups $H_i(A) := \pi_i(H(A))$, we get the more classical form of the excision $H_i(\Sigma A) = H_{i+1}(A)$.
[83] A *spectrum* is a collection of pointed spaces $(E_n)_{n \in \mathbb{N}}$ and of homotopy equivalences $E_n = \Omega E_{n+1}$. Let S^n be the sphere of dimension n viewed as an object of $\mathcal{S}$. A reduced homology theory defines such a sequence by $E_n = H(S^n)$.
  A *parameterized spectrum* is the data of an object B of $\mathcal{S}$ (the space of parameters), of a collection of pointed objects $(E_n)_{n \in \mathbb{N}}$ in $\mathcal{S}_{/B}$, and of homotopy equivalences $E_n = \Omega_B E_{n+1}$. Equivalently, spectra parameterized by B can be defined as diagrams $B \to \text{Sp}$. Intuitively, they

that $[\mathcal{S}_{\text{fin}}^\bullet, \mathcal{S}]^{(1)}$ is in fact a left-exact localization of $\mathcal{S}[X^\bullet]$ (see [2]). Let $\mathbb{A}_\infty^{(1)}$ be the dual ∞-topos; we have an embedding $\mathbb{A}_\infty^{(1)} \hookrightarrow \mathbb{A}_\infty^\bullet$.

It is possible to give a presentation of the ∞-logos $[\mathcal{S}_{\text{fin}}^\bullet, \mathcal{S}]^{(1)} = \mathcal{PSp}$. Let us say that a pointed object $1 \to E$ in a logos is *additive* if sums and products of this object coincide, that is, if the canonical map $E \vee E \to E \times E$ is invertible. An additive pointed object is called *stably additive* if the additivity property extends to all its loop objects, that is, if, for all m, n, $\Omega^m X^\bullet \vee \Omega^n X^\bullet \simeq \Omega^m X^\bullet \times \Omega^n X^\bullet$. The logos classifying stably additive objects is $\mathcal{S}\left[X^{(1)}\right] := \mathcal{S}[X^\bullet]/\!/(\Omega^m X^\bullet \vee \Omega^n X^\bullet \to \Omega^m X^\bullet \times \Omega^n X^\bullet, m, n \in \mathbb{N})$. In [3], we prove that $[\mathcal{S}_{\text{fin}}^\bullet, \mathcal{S}]^{(1)} = \mathcal{S}\left[X^{(1)}\right]$. Under the equivalence $[\mathcal{S}_{\text{fin}}^\bullet, \mathcal{S}]^{(1)} = \mathcal{PSp}$, the universal stably additive object $X^{(1)}$ corresponds to the sphere spectrum $\mathbb{S}$ in $\mathcal{PSp}$.

The fact that $\mathcal{PSp}$ is an ∞-logos has been a surprise for everybody in the higher category community. The category $\mathcal{Sp}$ is an example of a *stable* ∞-category.[84] Another example is the ∞-category $C(k)$ of chain complexes over a ring k. It is a result of Hoyois that the parameterized version of $C(k)$ (or of any stable ∞-category $\mathcal{C}$) is an ∞-logos [15].[85] Intuitively, if ∞-topoi are ∞-categories of generalized homotopy types, stable ∞-categories are ∞-categories of generalized homology theories (aka generalized stable homotopy types). The two worlds used to be thought as quite different (stable homotopy types behave very differently than their unstable counterpart), but the result of Hoyois shows that they are closer than expected.

4.2.4 ∞-Connected Objects

The ∞-connected objects are arguably the most important new feature of ∞-topoi. They provide an unexpected bridge between stable and unstable homotopy theories. They are also responsible for the failure of the notion of site to present ∞-logoi by generators and relations.

In the same way that sheaves are continuous families of sets, sheaves of ∞-groupoids are continuous families of ∞-groupoids (their stalks). Therefore,

---

can be thought as locally constant families of spectra parameterized by B (local systems of spectra). A homology theory defines such data by putting $B = H(1)$ and $E_n = H(S^n)$.

[84] A presentable ∞-category is called *stable* if its colimits commute with finite limits. In particular, it is an additive category: initial and terminal objects coincide, and so do finite sums and products. Stable categories are the proper higher notion replacing abelian categories. Another example is the ∞-category $C(k)$ obtained by localizing the 1-category of chain complexes over a ring k by quasi-isomorphism.

[85] Parameterized chain complexes are the same thing as local systems of chain complexes. If $\mathcal{C}$ is an ∞-category, the ∞-category $\mathcal{PC}$ of parameterized objects of $\mathcal{C}$ is defined in the following way. Its objects are diagrams $x: K \to \mathcal{C}$, where K is an ∞-groupoid. The 1-morphisms $x' \to x$ are pairs (u, υ) where $u: K' \to K$ is a map of ∞-groupoids and $\upsilon: x' \to x \circ u$ is a natural transformation of diagrams $K' \to \mathcal{C}$. Higher morphisms are defined in the obvious way. There is a canonical embedding $\mathcal{C} \to \mathcal{PC}$ induced by the choice $K = 1$.

∞-logoi can be understood as generalized categories of ∞-groupoids, that is, generalized homotopy theories. The operations of manipulation of these generalized homotopy types are the same as for homotopy types, but their behavior is different. The most important difference is arguably the failure of the Whitehead theorem to ensure that a homotopy type with trivial homotopy groups is the point. To explain this, we need some definitions.

Given a map $f: A \to B$, the *nerve of f* is the simplicial diagram

$$\ldots \rightrightarrows A \times_B A \times_B A \rightrightarrows A \times_B A \rightrightarrows A. \qquad (4.5)$$

The image of f, denoted $\text{Im}(f)$, is the colimit of this diagram.[86] The map f is called a *cover*, or a (-1)-*connected* maps if its image is B. Intuitively, a map is a cover if its fibers are not empty. Recall that a map $f: A \to B$ is a monomorphism if $\Delta f: A \to A \times_B A$ is an invertible map (in $\mathcal{S}$, this corresponds to a fully faithful functor between ∞-groupoids). The construction of the image produces a factorization of any map $f: A \to B$ into a cover followed by a monomorphism $A \to \text{Im}(f) \to C$.

More generally, f is called n-*connected* if all its iterated diagonals $\Delta^k f$ are covers for $0 \leq k \leq n+1$. An object E is called n-*connected* if the map $E \to 1$ is. An object E of $\mathcal{S}$ is n-connected if and only if $\pi_k(E) = 0$ for all $k \leq n$. Intuitively, an object in an ∞-logos is n-connected if it is a sheaf with n-connected stalks, and a map between sheaves is n-connected if its fibers are. The definition makes sense for $n = \infty$. In $\mathcal{S}$, an ∞-connected object corresponds to an ∞-groupoid with trivial homotopy groups. By the Whitehead theorem, only the point satisfies this. However, there exist ∞-logoi with nontrivial ∞-connected objects.

Example 4.1 (∞-Connected Objects)

1. Recall the ∞-logos $\mathcal{PSp}$ of parameterized spectra and the canonical inclusion $\mathcal{Sp} \hookrightarrow \mathcal{PSp}$ of reduced homology theories into homology theories. There exists a canonical functor red: $\mathcal{PSp} \to \mathcal{S}$, called the *reduction*, sending a parameterized spectrum $B \to \mathcal{Sp}$ to its indexing ∞-groupoid B. This functor is a logos morphism that happens to be the only point of the topos $\mathbb{A}_\infty^{(1)}$. It is possible to prove that an object of $\mathcal{PSp}$ is ∞-connected if and only if it is in the image of $\mathcal{Sp} \to \mathcal{PSp}$, that is, a reduced homology

---

[86] In a 1-category, the beginning of this diagram $A \times_B A \rightrightarrows A$ is sufficient to define covers. It is the graph of the equivalence relation on A "having the same image by f." But in higher categories, in $\mathcal{S}$, for example, "having the same image by f" is no longer a relation but a *structure* on the pairs (a, a') in A: that of the choice of a homotopy $\alpha: f(a) \simeq f(a')$ in B. This is why the higher part of the simplicial diagram is needed. The nerve of f defines a groupoidal relation in $\mathcal{S}$ that encodes the coherent compositions of the homotopies α.

theory. More generally, a morphism in $\mathcal{P}\mathcal{S}\mathrm{p}$ is ∞-connected if and only if its image under the reduction red: $\mathcal{P}\mathcal{S}\mathrm{p} \to \mathcal{S}$ is an invertible map in $\mathcal{S}$. This proves that there are plenty of ∞-connected morphisms in $\mathcal{P}\mathcal{S}\mathrm{p}$.

It is possible to think the situation intuitively in the following way. The objects of $\mathcal{P}\mathcal{S}\mathrm{p}$ are sorts of infinitesimal thickenings of the objects of $\mathcal{S}$. In particular, spectra are infinitesimal thickenings of the point. From this point of view, the morphism red: $\mathcal{P}\mathcal{S}\mathrm{p} \to \mathcal{S}$ is indeed a reduction, forgetting the infinitesimal thickening.[87]

2. Another source of examples of ∞-connected objects is the *hypercovers* in the ∞-logos $\mathrm{Sh}_\infty(X)$ associated to a space X, but we shall not detail this here (see [22, Section 6.5.3]). Because of this example, an ∞-logos such that the only ∞-connected maps are the invertible maps is called *hypercomplete*. This is the case of $\mathcal{S}$ and any diagram category $[C, \mathcal{S}]$. In particular, free ∞-logoi are hypercomplete. The ∞-logos $\mathrm{Sh}_\infty(X)$ of a space of "finite dimension" (like a manifold) is hypercomplete (see [22, Section 6.5.4]).

An ∞-topos $\mathcal{X}$ is said to have enough points if a map $A \to B$ in $\mathrm{Sh}_\infty(\mathcal{X})$ is invertible if and only if, for any point x of $\mathcal{X}$, the map $A(x) \to B(x)$ between the stalks is invertible in $\mathcal{S}$. Intuitively, this means that a sheaf is faithfully represented by the diagram of its stalks. If $\mathrm{Sh}_\infty(\mathcal{X})$ has some hyperconnected maps, then it cannot have enough points. This creates the bizarre situation that a topological space X such that $\mathrm{Sh}_\infty(X)$ is nonhypercomplete does not have enough points![88]

3. In homotopy theory, the construction of the free group on a pointed homotopy type X is given by $\Omega\Sigma X$, where Σ is the suspension functor. There exists a canonical map $X \to \Omega\Sigma X$ (the inclusion of generators). In $\mathcal{S}$, this map is invertible if and only if X is the point. But there exist examples of topoi where $X = \Omega\Sigma X$ for some nontrivial object. This is the case in $\mathcal{P}\mathcal{S}\mathrm{p}$. The embedding $\mathcal{S}\mathrm{p} \hookrightarrow \mathcal{P}\mathcal{S}\mathrm{p}$ preserves pushout and fiber products, hence if E is a spectrum, the object $\Omega\Sigma E$ is the same computed in $\mathcal{S}\mathrm{p}$ or in $\mathcal{P}\mathcal{S}\mathrm{p}$. But in the first case, we have trivially $E = \Omega\Sigma E$. In other terms, any spectrum viewed in $\mathcal{P}\mathcal{S}\mathrm{p}$ provides a pointed object that is its own free group.

---

[87] There again, the situation compares to algebraic geometry. Recall that in algebraic geometry, the connected components of a scheme depend only on its reduction. In particular, the spectrum of a local artinian ring is connected. Similarly, the homotopy invariants of an object E of $\mathcal{P}\mathcal{S}\mathrm{p}$ are those if its reduction $B = \mathrm{red}(E)$ in $\mathcal{S}$.

[88] The situation is comparable with a well-known fact in algebraic geometry. Let a be an element of a ring A viewed as a function $Spec(A) \to \mathbb{A}$. The value of this function at a point p is the residue of a in the field $\kappa(p)$. Then, because of nilpotent elements, an element a of a ring A is not completely determined by its set of values. In fact, it seems a good idea to compare the subcategory spanned by ∞-connected objects of an ∞-logos to the radical of a ring.

The logos classifying these *self-free-groups* is $\mathsf{Set}\,[X^\bullet]/\!\!/(X^\bullet \simeq \Omega\Sigma X^\bullet)$. Any self-free-group is ∞-connected. This explains why there are not more of them in $\mathcal{S}$.

4.2.5 Insufficiency of Topologies

We saw that any quotient of a logos $\mathcal{E}$ could be generated by a set of monomorphisms. This property fails drastically for ∞-logoi since there exist quotients of logoi inverting no monomorphisms at all. An example is given by the reduction morphism red: $\mathcal{PS}\mathrm{p} \to \mathcal{S}$. It is a localization because its right adjoint is the canonical ∞-logos morphism $\mathcal{S} \to \mathcal{PS}\mathrm{p}$, which is fully faithful.[89] We saw that a map is inverted by red if and only if it is ∞-connected. So we need to prove that no proper monomorphism can be ∞-connected. This is because an ∞-connected map is in particular a cover, and a map that is both a cover and a monomorphism is always invertible.

We now analyze why the trick that worked with logoi no longer works for ∞-logoi. Recall that a map is a monomorphism if and only if its diagonal is invertible. Let $f: A \to B$ be a map in an ∞-logos. We have that "$f: A \to B$ is invertible" if and only if "f is both a cover and a monomorphism" if and only if "f is a cover and Δf is invertible." In the context of logoi, the map Δf is a monomorphism, and the reformulation stops there. But in the context of ∞-logoi, Δf is no longer a monomorphism, so the equivalence of conditions continues into "f is invertible" if and only if "f and Δf are covers and $\Delta^2 f$ is invertible" if and only if "f, Δf, $\Delta^2 f$ are covers and $\Delta^3 f$ is invertible," and so on. At the limit of this process, we get the condition "$\forall n$, $\Delta^n f$ is a cover." But this condition is not equivalent to "f is invertible"; it is equivalent to "f is ∞-connected." This explains the failure of being able to write the invertibility of a map f by means of a topological relation. The best one can do with topological relations for an arbitrary map is to force it to become ∞-connected. This is in fact the new meaning of topological relations in the setting of ∞-logoi. The following conditions of generation are equivalent for a quotient of ∞-logoi:

- inverting some monomorphisms;
- forcing some maps to become covers;
- forcing some maps to become ∞-connected.

We shall say that a quotient is *topological* if it satisfies the above conditions and that a quotient is *cotopological* if it can be presented by inverting a set R of

---

[89] This functor sends an object B in $\mathcal{S}$ to the constant diagram $B \to \mathcal{S}\mathrm{p}$ with value the null spectrum.

∞-connected maps. An example of a cotopological relation is red: $\mathcal{PSp} \to \mathcal{S}$, where all ∞-connected maps are inverted. Any quotient $\mathcal{E} \to \mathcal{E}/\!\!/ R$ of ∞-logoi can be factored into a topological quotient followed by a cotopological one: the topological quotient forces the relations to become ∞-connected maps, then the cotopological quotient finishes the job by inverting these ∞-connected maps [22, Section 6.5.2]. Finally, we see that even though the notion of site, that is, topological quotients, is insufficient to present all ∞-logoi, it is nonetheless a meaningful notion of the theory.

Example 4.2 (Topological Relations and Factorizations)

1. The ∞-logos classifying n-connected objects is defined by

$$\mathcal{S}[X_{>n}] := \mathcal{S}[X]/\!\!/ \left(\forall k \leq n+1, \Delta^n X \text{ is a cover}\right).$$

   In particular, the ∞-logos classifying ∞-connected objects is

$$\mathcal{S}[X_{>\infty}] := \mathcal{S}[X]/\!\!/ \left(\forall n, \Delta^n X \text{ is a cover}\right).$$

   A variation is the ∞-logos classifying *pointed* ∞-connected objects defined by

$$\mathcal{S}\left[X^\bullet_{>\infty}\right] := \mathcal{S}\left[X^\bullet\right]/\!\!/ \left(\forall n, \Delta^n X^\bullet \text{ is a cover}\right).$$

   All of these are examples of topological quotients of $\mathcal{S}[X]$ or $\mathcal{S}[X^\bullet]$.

2. Recall the quotient

$$\mathcal{S}\left[X^\bullet\right] \to \mathcal{S}[X^{(1)}] := \mathsf{Set}\left[X^\bullet\right]/\!\!/ \left(\forall m, n, \Omega^m X^\bullet \vee \Omega^n X^\bullet \simeq \Omega^m X^\bullet \times \Omega^n X^\bullet\right)$$

   classifying stably additive objects. Any stably additive object can be proved to be ∞-connected. This gives a factorization $\mathcal{S}[X^\bullet] \to \mathcal{S}\left[X^\bullet_{>\infty}\right] \to \mathcal{S}\left[X^{(1)}\right]$ that is the topological/cotopological factorization.

3. Recall the logos classifying *self-free groups* is $\mathsf{Set}[X^\bullet]/\!\!/(X^\bullet \simeq \Omega\Sigma X^\bullet)$. Any self-free group is ∞-connected, and the factorization $\mathcal{S}[X^\bullet] \to \mathcal{S}\left[X^\bullet_{>\infty}\right] \to \mathcal{S}[X^\bullet]/\!\!/(X^\bullet \simeq \Omega\Sigma X^\bullet)$ is the topological/cotopological factorization.

4. Recall that $\mathsf{Set}[X^\bullet] = \left[\mathcal{S}^\bullet_{\mathrm{fin}}, \mathcal{S}\right]$. In particular, $\mathcal{S}\left[X^{(1)}\right]$ and $\mathsf{Set}[X^\bullet]/\!\!/(X^\bullet \simeq \Omega\Sigma X^\bullet)$ are examples of ∞-logoi that cannot be presented by a topology on $\mathcal{S}^\bullet_{\mathrm{fin}}{}^{op}$.

4.2.6 New Relations with Logic

In the line of what we said in Section 3.4.6, ∞-logoi provide several important new elements. The almost representability of the universe $\mathbb{U}$ and the existence of enough partial universes authorize semantics for logical theories having a type of types, quantification on objects, or modalities on types. This feature is somehow behind the whole homotopical semantics of Martin–Löf type theory with identity types [14].

The existence of ∞-connected objects also has consequences from the logical point of view. Recall from Section 3.4.6 that topological relations correspond logically to forcing some existential statements. Then logical meaning of the impossibility to present all quotients of ∞-logoi by topological relations is the surprising fact that it is impossible to describe the invertibility of a map by means of geometric formulas. Related to this, the ∞-connected objects are also responsible for the failure of the Deligne completion theorem for coherent topoi [24, Appendix A].

The notion of ∞-logoi also leads to the construction of classifying objects for some nontrivial theories with only the point as a model in S, namely, theories where the underlying objects are ∞-connected. We saw examples with stably additive objects and self-free group objects. These theories are somehow akin to theories without any models in Set or S.

4.2.7 Homotopy Theory of ∞-Logoi

We have explained in Section 3.2.15 how topos theory provides a nice theory of connectedness with the connected–disconnected factorization. The same definitions make sense in the setting of ∞-topoi, but changing the coefficients from Set to S has the effect of enhancing the theory of connectedness into a theory of contractibility. A morphism of ∞-topoi $\mathcal{Y} \to \mathcal{X}$ is called *contractible* if the corresponding morphism of ∞-logoi $\mathrm{Sh}_\infty(\mathcal{X}) \to \mathrm{Sh}_\infty(\mathcal{Y})$ is fully faithful. An ∞-topos $\mathcal{X}$ is contractible if the morphism $\mathcal{X} \to \mathbb{1}$ is. The *image* of a morphism of ∞-logoi $u^* \colon \mathcal{E} \to \mathcal{F}$ is defined as the smallest full sub-∞-category of $\mathrm{Sh}(\mathcal{Y})$ containing the image of $\mathcal{F}$ and stable under finite limits and colimits. The morphism u^* is said to be *dense* if its image is the whole of $\mathcal{F}$. A morphism of ∞-topoi $\mathcal{Y} \to \mathcal{X}$ is *uncontractible* if the corresponding morphism of ∞-logoi $\mathrm{Sh}_\infty(\mathcal{X}) \to \mathrm{Sh}_\infty(\mathcal{Y})$ is dense.[90] Any morphism of ∞-topoi $u \colon \mathcal{Y} \to \mathcal{X}$ factors as a contractible morphism followed by an uncontractible morphism:

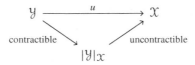

---

[90] These morphisms are called *algebraic* in [22, Section 6.3.6].

Table 4.23. *Degrees of homotopy theory*

	Locale (0-topos)	Topos	∞-Topos
Coefficients	$\{0 \leq 1\} = \mathcal{S}^{\leq -1}$	Set $= \mathcal{S}^{\leq 0}$	$\mathcal{S}$
Algebraic morphism	$\mathcal{O}(X) \xrightarrow{u^*} \mathcal{O}(Y)$	$\mathrm{Sh}(X) \xrightarrow{u^*} \mathrm{Sh}(Y)$	$\mathrm{Sh}_\infty(X) \xrightarrow{u^*} \mathrm{Sh}_\infty(Y)$
u^* fully faithful	surjective morphisms	connected morphisms	contractible morphisms
u^* dense	embeddings	disconnected morphisms	uncontractible morphisms
u^* has a local left adjoint	open morphisms	locally connected morphisms	locally contractible morphisms
Convenient for	image theory (π_{-1})	connected components theory (π_0)	full homotopy type

We call the morphism $|\mathcal{Y}|_X \to X$ the *residue of the contraction* of $\mathcal{Y} \to X$. This construction is an analogue for the whole homotopy type of the π_1 construction of Dubuc for topoi [10].

A morphism $u: \mathcal{Y} \to X$ is *locally contractible* when u^* has a local left adjoint. In this case, the residue $|\mathcal{Y}|_X \to X$ is ∞-étale and associated to an object of $\mathrm{Sh}_\infty(X)$. When $X = \mathbb{1}$, this object is called the *homotopy type* of the topos $\mathcal{Y}$. This generalizes to the whole homotopy type the situation of connected components of topoi. The set of connected components of a topos does not always exist as a set but always exists as totally disconnected space. Similarly, the whole homotopy type of an ∞-topos does not always exist as an ∞-groupoid but always exists as an uncontractible ∞-topos.[91]

From locales to topoi to ∞-topoi, there is a progression in the kind of homotopy features for which the theory is convenient. Table 4.23 summarizes the situation.

4.2.8 Cohomology Theory of ∞-Topoi

The theory of ∞-topoi is also well suited for cohomology theory with coefficient in sheaves. The modern formulation of derived functors as functors between ∞-categories has reformulated sheaf cohomology as the computation of the global sections of sheaves of spectra. The cohomology of an ∞-topos

---

[91] This point of view goes around the theory of shape of [16, 22].

$\mathfrak{X}$ is then dependent on the ∞-category of sheaves of spectra $\mathcal{S}h_\infty(\mathfrak{X}, \mathcal{S}p)$. The nice descent properties of ∞-logoi provide a simple description of this category as a tensor product of presentable ∞-categories:[92]

$$\mathcal{S}h_\infty(\mathfrak{X}, \mathcal{S}p) := \mathcal{S}h_\infty(\mathfrak{X}) \otimes_\mathcal{S} \mathcal{S}p = [\mathcal{S}h_\infty(\mathfrak{X}), \mathcal{S}p]^c.$$

The cohomology spectrum of $\mathfrak{X}$ with values in a sheaf of spectra E is given simply by the global sections

$$\Gamma : \mathcal{S}h(\mathfrak{X}, \mathcal{S}p) \longrightarrow \mathcal{S}p$$
$$E \longmapsto \Gamma(\mathfrak{X}, E).$$

Then, the cohomology groups of $\mathfrak{X}$ with coefficients in E are defined as the stable homotopy groups of the spectra $H^i(\mathfrak{X}, A) := \pi_{-i}(\Gamma(\mathfrak{X}, H(A)))$.

In terms of the analogy of logos theory with commutative algebra, the formula $\mathcal{S}h(\mathfrak{X}, \mathcal{S}p) = \mathcal{S}h(\mathfrak{X}) \otimes \mathcal{S}p$ says that the stabilization operation is a change of scalar from $\mathcal{S}$ to $\mathcal{S}p$ along the canonical stabilization map $\Sigma_+^\infty : \mathcal{S} \to \mathcal{S}p$. The resulting ∞-category is not a logos, though, but a stable ∞-category.

References

[1] M. Anel, *What is a space?* Slides from a talk at the 6th Workshop on Formal Topology, Birmingham (2019), http://mathieu.anel.free.fr/mat/doc/Anel-2019-Birmingham.pdf

[2] M. Anel, G. Biedermann, E. Finster, and A. Joyal, *Goodwillie's calculus of functors and higher topos theory*, J. Topol. 11 (2018) 1100–1132

[3] M. Anel, G. Biedermann, E. Finster, and A. Joyal, *Left-exact modalities and localizations of higher topoi*, Ongoing work

[4] M. Anel and G. Catren, eds., *New Spaces in Physics: Formal and Conceptual Reflections*, Cambridge University Press (2021)

[5] M. Anel and D. Lejay, *Exponentiable ∞-topoi*, Preprint Version 2, arXiv:1802.10425

[6] M. Artin, A. Grothendieck, and J.-L. Verdier, *Théorie des topos et cohomologie étale des schémas*, Lecture Notes in Mathematics 269 (Tome 1) 270 (Tome 2) 305 (Tome 3), Springer (1972–73)

[7] J. Baez and J. Dolan, *Higher-dimensional algebra and topological quantum field theory*. J. Math. Phys. 36 (1995) 6073–6105

[8] M. Bunge and J. Funk, *Singular Coverings of Toposes*, Lecture Notes in Mathematics 1890, Springer (2006)

---

[92] Such a presentation does not work if ∞-logoi are replaced by logoi. It relies on the fact that a sheaf on ∞-topos with values in a category $\mathcal{C}$ is a functor $\mathcal{S}h_\infty(\mathfrak{X})^{op} \to \mathcal{C}$ sending colimits to limits. For logoi $\mathcal{S}h(\mathfrak{X})$, the exactness condition involves, rather, covering families.

[9] D.-C. Cisinksi, *Higher Categories and Homotopical Algebra*, Cambridge Studies in Advanced Mathematics 180, Cambridge University Press (2019)
[10] E. Dubuc, *The fundamental progroupoid of a general topos*, J. Pure Appl. Algebra 212 (2008) 2479–92
[11] R. Garner, *Ionads*, J. Pure Appl. Algebra 216 (2012) 1734–47
[12] R. Garner and S. Lack, *Lex colimits*, J. Pure Appl. Algebra 216 (2012) 1372–96
[13] M. Godement, *Théorie des faisceaux*, Actualités Scientifiques et Industrielles 1252, Publications de l'Institut de Mathematique de l'Universite de Strasbourg, Hermann & Cie (1958)
[14] The Univalent Foundations Program, *Homotopy Type Theory: Univalent Foundations of Mathematics*, Institute for Advanced Study (2013)
[15] M. Hoyois, *Topoi of parametrized objects*, Theory Appl. Categ. 34, no. 9 (2019)
[16] M. Hoyois, *Higher Galois theory*, J. Pure Appl. Algebra 222 (2018)
[17] P. T. Johnstone, *Stone Spaces*, Cambridge Studies in Advanced Mathematics 3, Cambridge University Press (1982)
[18] P. T. Johnstone, *Topos Theory*, Academic Press (1977)
[19] P. T. Johnstone, *Sketches of an Elephant*, 2 vols., Oxford Logic Guide, 43 and 44 (2002)
[20] A. Joyal, *Notes on quasi-categories*, http://www.math.uchicago.edu/~may/IMA/Joyal.pdf
[21] A. Joyal and M. Tierney, *An extension of the Galois theory of Grothendieck*, Mem. AMS 51 (1984)
[22] J. Lurie, *Higher topos theory*, Ann. Math. Stud. 170 (2009)
[23] J. Lurie, *On the classification of topological field theories*, Preprint, http://people.math.harvard.edu/~lurie/papers/cobordism.pdf
[24] J. Lurie, *Spectral Algebraic Geometry*, Book in preparation, http://people.math.harvard.edu/~lurie/papers/SAG-rootfile.pdf
[25] S. Mac Lane and I. Moerdijk, *Sheaves in Geometry and Logic: A First Introduction to Topos Theory*, Springer (1992)
[26] I. Moerdijk, *The Classifying Topos of a Continuous Groupoid. I*, Transactions of the American Mathematical Society, Vol. 310, No. 2 (Dec., 1988) 629–668
[27] I. Moerdijk and G. Reyes, *Models for Smooth Infinitesimal Analysis*, Springer (1991)
[28] J. Picado and A. Pultr, *Frames and Locales: Topology without Points*, Birkhäuser (2012)
[29] C. Rezk, *Toposes and homotopy toposes*, https://faculty.math.illinois.edu/~rezk/homotopy-topos-sketch.pdf
[30] M. Shulman, *Parametrized spaces as model for locally constant homotopy sheaves*, Topol. Appl. 155 (2008) 412–32
[31] C. Simpson, *A Giraud-type characterization of the simplicial categories associated to closed model categories as 1-pretopoi*, Preprint, arXiv:math.AT/9903167
[32] B. Toën, *Vers une interprétation galoisienne de la théorie de l'homotopie*, Cahiers de topologie et géométrie différentielle catégoriques 43 (2002) 257–312

[33] B. Toën and G. Vezzosi, *Homotopical algebraic geometry, I. Topos theory*, Adv. Math. 193 (2005) 257–372

[34] S. Vickers, *Topical categories of domains*, Math. Struct. Comp. Sci. 9 (1999) 569–616

Mathieu Anel
Department of Philosophy, Carnegie Mellon University
mathieu.anel@gmail.com

André Joyal
Département de mathématiques, Université du Québec
joyal.andre@uqam.ca

5
Spaces as Infinity-Groupoids

Timothy Porter

Contents

1	Introduction	258
2	The Beginnings: Recollections of Poincaré's Fundamental Groupoid	260
3	Whitehead's Algebraic and Combinatorial Homotopy	273
4	Higher Homotopy Groups, Weak Homotopy Types, Truncation, and Connectedness	274
5	Simplicial Sets, Higher Combinatorics, and ∞-Groupoids	297
6	Higher Galois Theory and Locally Constant Stacks	312
7	Concluding Discussion	317
	References	318

1 Introduction

As the title of this chapter suggests, there are aspects of "spaces" mirrored by things called "∞-groupoids." There are also such ∞-groupoids that arise in "nature" from other contexts. This raises a whole lot of questions. Some of these are obvious, for example, What "aspects"? What on earth are ∞-groupoids? What job do they do? What sorts of "spaces" are we considering? These questions are, formally, easy to answer, but they leave deeper, harder questions still to be considered. Informally, the initial vague idea is that an infinity-groupoid model of a space should generalise the classical idea of a fundamental group or groupoid of a space that you may have met from algebraic topology textbooks, working not just with paths but with higher-dimensional analogues, "paths between paths," "paths between those things," and so on, all being considered up to some appropriate idea of homotopy or deformation. What it should do is, thus, to generalise the sorts of things that the fundamental groupoid is good at doing, such as classifying covering spaces and other types of bundle-like objects.[1] That being said, how to generalise those properties is not always obvious, and their formalisation can be tricky.

To continue with our questions, What is the conceptual advantage of working with things such as ∞-groupoids – whatever they might be (and there is more than one answer to that question)? Very importantly, what are the intuitions underpinning this formalisation? What are the limitations of the formalisation? How do these objects fit into the general scheme of things, say, from the perspective of algebraic topology? How "practical" is it to try to "calculate" with such models of spaces? If we model "spaces" by ∞-groupoids, do we gain some new insights on the spaces? Turning that around, the opposite question is: How good are "spaces" themselves as models for the perhaps naive notions of ∞-groupoid that arise in other areas of mathematics? Is the spatial intuition, thus being invoked, a good one to use, or is it too constraining or, alternatively, too wide?

We will attempt to answer some of these questions. To answer them all fully would take a lot more space. We will attempt to do this by looking back at the developments that led to the perception that there was a link between spatial phenomena of a more or less geometric nature and something that would be eventually called an "∞-groupoid" and which can be thought of as being more algebraic or "categorical" in its inspiration rather than inherently "geometric." We will go right back to the beginnings of the use of algebraic tools to study topology so as to look, very briefly, at some of the ways that Poincaré, and some of those who came after him, looked at the fundamental groupoid of a space. (Here, thankfully, we can keep things quite brief, as the detailed historical analysis of how the algebraic structure of paths in a space was first encoded has been initiated by Krömer [41]. That article also contains some very relevant passages quoted from Poincaré and others.)

Right from the origins of the fundamental group(oid), the link with covering spaces was recognised as one of the important aspects, and that will be seen also in the motivation for the idea behind an ∞-groupoid approach to analogous higher-order structures. We will briefly mention this with respect to Grothendieck's approach to the fundamental group in SGA1, [33]. That will lead us, naturally, to the letters from Alexander Grothendieck to Larry Breen,

---

[1] We will be looking at precisely that very shortly, as there are several types of property that we will be seeking to generalise. This will also serve as an *aide mémoire* on the fundamental group(oid) construction.

[27–29], of which, for us, the first is probably the most central for our purposes here, and also to the subsequent "letter to Quillen" [30], in which a strong relationship between spaces and ∞-groupoids is more explicitly mentioned.

Before that, we will need to give some indication of what ∞-groupoids are, as there are several possible manifestations of the notion. (We will look at only two of them in any detail.)

In another thread of the chapter, we will try to show how J. H. C. Whitehead's "combinatorial homotopy" or "algebraic homotopy" fits into this theme. One of the test problems he considered was to model polyhedra by algebraic data. Here dimension is crucial, yet within the corresponding area of ∞-groupoid theory, this geometric aspect is less evident.[2] In such applications, it would seem that the combinatorial approach, via simplicial complexes and their combinatorial (Whitehead) homotopy theory, may be more directly useful than the fully ∞-groupoid one.

As the area is a huge one, we will tend to give brief descriptions rather than detailed definitions, directing the reader to the original literature where needs be.

2 The Beginnings: Recollections of Poincaré's Fundamental Groupoid

The fundamental group of a space was introduced by Poincaré in 1895 [56]. At this point in time, the "spaces" concerned arose as Riemann surfaces and thus naturally came with the insights and problems of analytic continuation of functions, integration along paths, and integration over regions bounded by paths or collections of paths. As noted by Sarkaria [62] (in [37, Chapter 6]), Poincaré gave four approaches to the fundamental group(oid) of such a space, M; see also Krömer [41]. These were, in modern terminology,

1. as the group of deck transformations of covering spaces over M, thus implicitly involving a form of "multiple valued function";
2. as the holonomy group of what would now be called an integrable connection on a vector bundle;[3]
3. as the set of homotopy classes of loops at a base point (or, more generally, of paths) in M; and finally,
4. on any such space, M, obtained from a polytope or simplicial complex, as a group given by explicit generators and relations.

Here we will initially be looking at the third of these but will also need to consider the covering space approach and that using combinatorial group

---

[2] But see Ara and Maltsiniotis [2].
[3] We will not be following up on this approach here.

theoretic methods, corresponding, usually, to some simplicial or CW-complex structure and thus, more or less, to triangulations.[4]

2.1 Homotopy Classes of Paths

Although this subsection will consist of well-known standard material, in order to increase the accessibility of the account, it will be useful to recall and comment on some of the basic definitions and terminology relating to homotopy and the fundamental group(oid). This will also draw attention to certain aspects of this basic theory that we will need but that are perhaps understressed in the standard accounts:

- Given a space, X, a *path* in X is a continuous map, $\alpha\colon I \to X$, where $I = [0, 1]$. The path has *source*, $\alpha(0)$, and *target*, $\alpha(1)$.
- Given two continuous maps, $f_0, f_1 \colon X \to Y$, a *homotopy* between them is a continuous map, $h\colon X \times I \to Y$, such that, for all $x \in X$, $h(x, 0) = f_0(x)$ and $h(x, 1) = f_1(x)$. The two maps are said to be *homotopic* if there is a homotopy between them and then are said "*to be in the same homotopy class.*" We write $h\colon f_0 \simeq f_1$ in this case.

Intuitions. With two homotopic maps, the interpretation is that each can be deformed continuously to the other. For instance, in the case of X being just a single point, each function from X to Y gives a point in Y, and vice versa. A homotopy between two such maps gives a path "deforming" one point in Y to the other. If X is a unit interval, then the two maps will just be paths and the homotopy deforms one into the other. An important case in this situation is when the paths share the same source and also the same target, $x_0 = f_0(0) = f_1(0)$ and $x_1 = f_0(1) = f_1(1)$, then the homotopy may *fix endpoints* (see Figure 5.1)

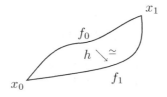

Figure 5.1

---

[4] This latter situation relates strongly to certain constructions within theoretical physics as well as raising the question as to whether, in our discussion, "spaces" should be just "topological spaces" or should come with additional structure, such as that of a simplicial or CW-complex or that of a manifold. We will meet this several times later on and will give a more detailed account then.

so that the map $h(0, -)$ is constant at $f_0(0)$ and $h(1, -)$ is constant at $f_0(1)$. By a *path class*, we will mean a fixed endpoint homotopy class of paths:

- Two spaces, X and Y, have *the same homotopy type* if there are continuous maps, $f : X \to Y$ and $g : Y \to X$, such that there are homotopies, $gf \simeq id_X$ and $fg \simeq id_Y$. We also say X and Y are *homotopy equivalent* and that f is a *homotopy equivalence* between them. When we refer to a *homotopy type*, we thus mean a maximal family of "spaces" all of which are homotopically equivalent to each other.

Intuitions. The idea is that if two spaces are homeomorphic (i.e., are essentially "the same"), then they will be homotopically equivalent. Often, however, the aim is to decide if two spaces are essentially different and so should have different properties, behavior, and so on, and for that, one tries to find "quantities" that are invariant under homeomorphism to test if the two spaces are, or are not, "the same." It is much easier to find invariants of homotopy type, however, and if X and Y can be shown to differ on some homotopy invariant quantity, then, as they can then not be of the same homotopy type, they must also not be homeomorphic. Equally importantly, that same basic methodology often can be adapted to show whether some mapping has, or has not, some particular property. To do this, one searches for "algebraic" invariants of homotopy types – but we are getting ahead of ourselves here!

To return to describing Poincaré's constructions:

- The paths in X can be composed (concatenated) in more or less the same way as when integrating along paths in $\mathbb{R}^n$. A fairly obvious formula for this corresponds to concatenation (which defines the composite on an interval of length 2) followed by "rescaling." This gives the following:

    If $\alpha, \beta : I \to X$ are two paths such that $\alpha(1) = \beta(0)$, then $\alpha \cdot \beta : I \to X$ is defined by

$$\alpha \cdot \beta(t) = \begin{cases} \alpha(2t) & 0 \leq t \leq \tfrac{1}{2} \\ \beta(2t - 1) & \tfrac{1}{2} \leq t \leq 1. \end{cases}$$

Although this is the "obvious" composition, corresponding to the subdivision, $\{[0, \tfrac{1}{2}], [\tfrac{1}{2}, 1]\}$ of $[0, 1]$, it is not "God given." There are a whole lot of others that could have been used. For any subdivision $\{[0, r], [r, 1]\}$ of $[0, 1]$, we could have scaled α to fit on the first subinterval and scaled and shifted β to fit on the second one, before concatenating. All these composites would give homotopic paths, however. (Even that does not exhaust the possible compositions, as we could have scaled nonlinearly.)

- The composition we have chosen to give is not associative. If $\gamma: I \to X$ is such that $\gamma(0) = \beta(1)$, we can form both $(\alpha \cdot \beta) \cdot \gamma$ and $\alpha \cdot (\beta \cdot \gamma)$, but they are clearly not equal. They are, however, homotopic.[5] In fact, the usual homotopy given in texts just slides the "middle" subinterval, $[\frac{1}{4}, \frac{1}{2}]$, on which β is used, along to the corresponding position, $[\frac{1}{2}, \frac{3}{4}]$, rescaling the other two subintervals accordingly. This means that it takes place *within* the "track" of the composite, that is, the image of the composite function within X. It is, thus, very "thin" in the sense that while most homotopies can be thought of as "sweeping out an area" within the space, here what is happening is more like a continuous reparameterisation of the function from one using the subdivision of $[0, 1]$ given by $\{[0, \frac{1}{4}], [\frac{1}{4}, \frac{1}{2}], [\frac{1}{2}, 1]\}$ to one using $\{[0, \frac{1}{2}], [\frac{1}{2}, \frac{3}{4}], [\frac{3}{4}, 1]\}$. We will return to ideas of thinness somewhat later.
- If we use the notation $[\alpha]$ for the class of a path, α, in X under *fixed endpoint* homotopies, then the set of such classes has an algebraic structure given by a partially defined composition, $[\alpha] \cdot [\beta] := [\alpha \cdot \beta]$, defined just on those pairs $([\alpha], [\beta])$ such that $\alpha(1) = \beta(0)$ – which is, of course, why we used fixed endpoint homotopies rather than "free" homotopies in defining the path classes, $[\alpha]$. The algebraic structure we have here is a groupoid, that is, a (small) category in which every morphism is invertible. It is the *fundamental groupoid*[6] $\Pi_1 X$ of X. That this all works is standard, but any reader who has not seen it spelled out in some detail should consult standard texts or look at Krömer's article that was mentioned earlier for a historical viewpoint. (It is of interest that the groupoid version is explicitly given by Schreier in 1927 and then later by Reidemeister, but then was not used for some time; see Krömer [41] again. Their work is an early example of a structure that could be considered algebraically, being thought of as a space.)

That completes a description of Poincaré's path-based definition, except to note that he actually defines the fundamental *group* and not the more general, and more natural,[7] groupoid version. For this "fundamental group," one needs

---

[5] And this is where things start being interesting, as it is a natural occurrence of "weak" structure rather than "strict."

[6] The objects of $\Pi_1 X$ are the points of X; the morphisms are the path classes, so that, between two points, x_0 and x_1, in X, the set $\Pi_1 X(x_0, x_1) = \{[\alpha] \mid \alpha(1) = x_0, \alpha(1) = x_1\}$. Composition is given as above and is associative; identities are given by the classes of constant paths at the points of X, and the inverse of $[\alpha]$ is $[\alpha^{(r)}]$, where $\alpha^{(r)}(t) = \alpha(1 - t)$, the path, α traversed in the opposite direction.

[7] Recall that a monoid can be thought of as a category with a single object, and in the same way, a group can be thought of as a groupoid again having just a single object. If G is a group, the corresponding groupoid, $\mathbb{G}$, has one object, denoted $*$, for the moment, and $\mathbb{G}(*, *)$ is the group G, with composition in $\mathbb{G}$ being just the multiplication in G. Note that we will not, in fact, use a different notation for a group and the corresponding one-object-groupoid in the main text.

to choose a "*base-point*," $x_0 \in X$, and then the fundamental group, $\pi_1(X, x_0)$, of the *pointed space*, (X, x_0), is the vertex group of $\Pi_1 X$ at x_0, that is, $\Pi_1 X(x_0, x_0)$, so is the group of path classes of loops in X, based at x_0.

2.2 Covering Spaces

If we now assume that X is "sufficiently locally nice,"[8] we can pass to another of Poincaré's definitions (again we will only sketch the theory and more briefly than above, leaving more "for the reader to check"). We will assume the space, X, is connected and will choose a base point, x_0, in X. Furthermore, let $p: \tilde{X} \to X$ be a universal covering space for X. In other words, p is a local homeomorphism, so given any $y \in \tilde{X}$, if we look near enough to y, that is, in a small enough neighbourhood of it, p behaves as a homeomorphism, mapping that neighbourhood to a neighbourhood of $p(y)$, and, moreover, a universality condition holds (which we will skate over; see standard algebraic topology texts for this, also for conditions on X for such a universal cover to exist, and, once again, Krömer [41] for a historical perspective). This p will have a unique path lifting property: if we have a path, α, in X and pick a point, $x \in \tilde{X}$, such that $p(x) = \alpha(0)$, then α lifts uniquely to a path, $\tilde{\alpha}$, in $\tilde{X}$, so that $p\tilde{\alpha} = \alpha$ and $\tilde{\alpha}(0) = x$.

Using this, one shows that the category of covering spaces of X is equivalent to the category of $\pi_1(X, x_0)$-sets, that is, sets with an action of $\pi_1(X, x_0)$ on them:

$$Fibre: Cov/X \xrightarrow{\simeq} \pi_1(X, x_0) - Sets,$$

where $Fibre(q: Y \to X)$ will be the set, $q^{-1}(x_0)$, with the action of $\pi_1(X, x_0)$ given by lifting paths.[9] This is almost Poincaré's deck transformation "definition" of $\pi_1(X, x_0)$. (A deck transformation is simply an automorphism of a covering space, hence is compatible with the covering map.) Deck transformations of the universal cover of X give a group that is isomorphic to $\pi_1(X, x_0)$. To see why, we note that $Fibre$ sends the universal cover to the *set* of elements of $\pi_1(X, x_0)$ with the action of that same *group* given by multiplication. Any automorphism of the universal cover goes via the equivalence, $Fibre$, to an automorphism of that $\pi_1(X, x_0)$-set, and that gives an element of $\pi_1(X, x_0)$.

---

[8] Meaning that small enough neighbourhoods of each point are "homotopically trivial," so, intuitively, nothing "interesting" is happening at the very small scale!

[9] This easily generalises to nonconnected spaces. Replace $\pi_1(X, x_0)$ by $\Pi_1(X)$ and $\pi_1(X, x_0) - Sets$ by the category, $Sets^{\Pi_1(X)}$, of functors from $\Pi_1(X)$ to $Sets$. This also frees up the construction from needing to choose a base point, x_0.

Comment. The above correspondence works for any "nice topological space," but its importance for us is, also, that it acted as a key starting point for the definition by Grothendieck of the fundamental group of a scheme, the main algebraic geometric version of "space." This, in turn, used the exciting insight that this is a version of the fundamental theorem of Galois theory relating extension fields with actions of a Galois group; (see SGA1 [33] for the basic source and Douady and Douady [24] for a neat treatment), but then there is an enormous literature on this theory, as you would expect. Because of that link, it is useful to think of the above correspondence as being part of some more encompassing Galois–Poincaré theory. The use of a *topos*, $\pi_1(X) - Sets$ or $Sets^{\Pi_1(X)}$, to model aspects of the *topological properties* of X is, perhaps, to be noted for use elsewhere.[10]

2.3 Complexes as "Presentations" of Spaces and of Groupoids

For the above theory, all that was needed was that X was a "sufficiently nice" topological space. For the final approach to the fundamental group that we will look at, Poincaré assumed, in addition, that the space was specified as a "complex" of some sort. When using "path classes," one has to face the initially very large number of paths that there are in the "usual" spaces that are considered. Although the ideas that Poincaré used were later extended and made much more exact, the intuitive ideas remain clear in what he introduced. These ideas were somewhat later applied by Schreier, Neilsen, and others to problems in group theory. This illustrates well the somewhat symbiotic relationship between spaces and algebra, which relates to the main theme of this chapter. The initial development was topological and allowed one to encode spatial information in an algebraic form. The work in (combinatorial) group theory first encoded algebraic structure in combinatorial, and then in spatial, form, where methods derived from algebraic topology could be applied to give new insights and results.

To start with, the "complexes" that we will need are *simplicial complexes*, a notion that we will recall below. Later we will need simplicial sets, CW-complexes, and various variants of such. We will not give fully formal definitions of these, as they are readily available elsewhere, but the intuition for the simplicial setting is that of a triangulation of the space being studied.[11]

---

[10] An action of a group, G, on a set, S, is often called a *representation* of the group, as it *represents* the elements of the group as permutations of S. This also leads on to the subject of representation theory and the various categorical approaches to that subject area.

[11] Think "wire frame" image!

Note, however, that each of these is an instance of a (topological) space *plus instructions on how it was built*, so is *not just a space*.

Simplicial complexes come in two flavors, one "abstract" or combinatorial, the other "geometric," or, perhaps in better terminology, "spatial." Abstractly, a simplicial complex, K, is specified by a set, $V(K)$, of "vertices" and a set, $S(K)$, of "simplices," which are nonempty finite subsets of vertices. Not every finite subset need be in $S(K)$, but all singletons, $\{v\}$, $v \in V(K)$, are considered to be simplices of K, and, very importantly, if a set of vertices gives a simplex, σ, of K, then *all* its nonempty subsets are also to be in $S(K)$. These subsets give the *faces* of σ.

As a simple example, suppose $V(K) = \{0, 1, 2, 3, 4\}$, while $S(K)$ consists of $\{0, 1, 2\}$, $\{2, 3\}$, $\{3, 4\}$, and all the nonempty subsets of these. We draw this schematically as shown in Figure 5.2:

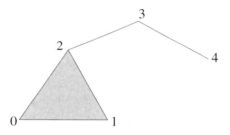

Figure 5.2

The simplex, $\{0, 1, 2\}$, is a "2-simplex" and is pictured as being two-dimensional. In general, an n-simplex of K is a subset in $S(K)$ with $n + 1$ elements. We will write K_n for the set of n-simplices. Note that with the above example, K_n is empty for $n \geq 3$, and that any graph/network corresponds to a simplicial complex with no simplices in dimensions 2 or larger.

That gives a brief outline of the abstract form of the notion of simplicial complex. We can use such a specification as a plan for building a space, roughly as follows.[12] For each $n \geq 0$, we have a space, Δ^n, which is an n-simplex.[13] The idea is now to take, for each n, and each n-simplex, σ, in K, a copy, $K(\sigma)$, of Δ^n; then if τ is an $(n - 1)$-simplex in K, which is a face of σ, we identify $K(\tau)$ with the corresponding face of $K(\sigma)$. Doing this

---

[12] This can be done in various equivalent ways, but we will just look at one that generalises easily to simplicial sets.

[13] The space, Δ^n, is given, for example, by

$$\{\underline{x} = (x_0, ..., x_n) \in \mathbb{R}^{n+1} \mid \sum_0^n x_i = 1; \text{ all } x_i \geq 0\}.$$

for all the simplices gives a space, the geometric realisation of K, but this description suffers from being too informal.[14] Of course, for our example, we need five Δ^0s and also five Δ^1s, that is, geometric segments, and one Δ^2, and the resulting space, clearly, looks like we have pictured it! We think of the combinatorial gadget, K, as "presenting" the space given by its "geometric realisation." Not all spaces can be represented in this way. Those that can are sometimes called *polyhedra*.

We will return to this below but now need to give how to go from such a "presentation" of a space to a presentation of its fundamental groupoid.

Let K be a simplicial complex (and our notation will not distinguish between the combinatorial object and the corresponding space). If we restrict attention to the vertices and the 1-simplices of K, we obtain a graph, K^1, that forms the *1-skeleton*[15] of K. For our simple example, this has the same set of vertices, but $S(K^1)$ does not contain $\{0, 1, 2\}$, so the picture/space is as shown in Figure 5.3, in which there is now a hole where, in Figure 5.2, there was the 2-simplex corresponding to $\{0, 1, 2\}$.

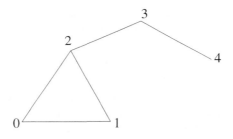

Figure 5.3

Going back to the general abstract case, we can form the free groupoid, $F(K^1)$, on this graph. This has the vertices of K as its objects and between two such vertices, morphisms are reduced edge paths between them. (An edge path is just a list of edges or their inverses in the graph, which are "composable" so the target of each is the source of the next.) We omit the detailed construction of this free groupoid as it is relatively well known and can easily be found in the literature. We note that it is important to choose a direction on each edge in K^1 and we will need a notation for such a directed edge. We will write $\langle v, v' \rangle$ for the edge with source, v, and target, v'. We will usually do more than just ordering the edges, rather we will pick a total order on $V(K)$ and then write $\langle v_0, \ldots, v_n \rangle$ for an n-simplex, $\{v_0, \ldots, v_n\} \in S(K)$, in which

---

[14] Discussion of formal definitions of the geometric realisation of a simplicial complex can be found in many books on algebraic topology.
[15] This is not just the set of edges but also involves the information on two ends of each edge.

$v_0 < v_1 < \cdots < v_n$.[16] Returning to the groupoid, we now form the quotient of $F(K^1)$ by relations that come from the 2-simplices of K:
For each 2-simplex, $\langle v_0, v_1, v_2 \rangle$, in K, which we picture as

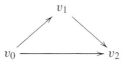

we form the relation

$$\langle v_0, v_1 \rangle \cdot \langle v_1, v_2 \rangle \cdot \langle v_0, v_2 \rangle^{-1} = id_{v_0}.$$

If we write $R(K)$ for the set of such relations, then the groupoid, $\Pi_1^{comb}(K)$, with presentation $\langle K^1 : R(K) \rangle$, is the combinatorial version of the fundamental groupoid of K. It will be isomorphic to the full subgroupoid of $\Pi_1(K)$ formed by the objects corresponding to vertices of the complex, K, and therefore is equivalent, as a groupoid, to $\Pi_1(K)$, but with far fewer objects. (To see, geometrically, why it is equivalent to $\Pi_1(K)$, first note that in $\Pi_1(K)$, any object is a point of K, so is in some simplex of K, and hence can be joined to a vertex of K by a path. We thus have that every such object is isomorphic to one in $\Pi_1^{comb}(K)$. Next given any path in K between two vertices, it will be homotopic[17] to one whose image is within the 1-skeleton of K, and which can be represented by an edge path, thus by a morphism in $\Pi_1^{comb}(K)$. Finally any homotopy between paths can be replaced,[18] up to a second level homotopy, that is, a homotopy between homotopies, by one within the 2-skeleton of K, that is, the subcomplex given just by the 0-, 1-, and 2-simplices of K. What this implies is that homotopy can be mirrored, algebraically, by "moves across 2-simplices" and thus by the *rewriting process* associated to the presentation that we gave.) Restricting attention to a single vertex, v_0, the vertex group of $\Pi_1^{comb}(K)$ at v_0 is isomorphic to $\pi_1(K, v_0)$ and gives us Poincaré's combinatorial form of his fundamental group.

---

[16] This gives a unique "ordered simplex" representing each element of $S(K)$. It has the additional benefit of allowing us to talk of the kth face, $d_k(\sigma)$, of a simplex, σ, just by deleting v_k from the simplex; thus, in our example, if we take the obvious order on the vertices, $d_1 \langle 0, 1, 2 \rangle = \langle 0, 2 \rangle$, and so on, but note that if we change the total order, this will change the way the faces turn out. If we had ordered the vertices, $\{3 < 2 < 0 < 4 < 1\}$, then although $\{0, 1, 2\}$ still would be a simplex, now written $\langle 2, 0, 1 \rangle$, we would have $d_1 \langle 2, 0, 1 \rangle = \langle 2, 1 \rangle$.
[17] To visualise this, think of a path in our example, going from 1 to 2, perhaps wandering around within the 2-simplex given by 0, 1, and 2. It could be pushed out (and thus "homotoped") to the 1-skeleton, and this could be done in several different ways.
[18] The key to all this is a *simplicial approximation theorem*, which can be found in most books on basic algebraic topology.

This process shows some inadequacies in the simple combinatorial language given to us by simplicial complexes, at least when we want to build an algebraic object from it. For example, in the above description, we wrote id_{v_0} for the identity element at the vertex v_0, that is, the empty edge path starting at v_0. In the simplicial complex, K, we do *not* have an edge $\langle v_0, v_0 \rangle$. One way to get around this is to relax the condition $v_0 < v_1 < \cdots < v_n$ for this ordered set of vertices to be a simplex, replacing $<$ by $\leq$, thus allowing a vertex label to be repeated. For instance, if we take our example and order the vertices in the obvious way, then, as well as the 1-simplices, $\langle 0, 1 \rangle$, and so on, we would have "degenerate" 1-simplices, such as $\langle 2, 2 \rangle$, and degenerate 2-simplices, such as $\langle 2, 2, 3 \rangle$ and $\langle 2, 3, 3 \rangle$. We would have simplices in all (positive) dimensions, as a string with n copies of 2, followed by m copies of 3, would give us a degenerate $(n + m - 1)$-simplex. The rule for defining the faces of a simplex would still apply, so $d_0 \langle 2, 2, 3 \rangle = \langle 2, 3 \rangle = d_1 \langle 2, 2, 3 \rangle$, whilst $d_2 \langle 2, 2, 3 \rangle = \langle 2, 2 \rangle$, a degenerate 1-simplex, so an "identity edge" at 2.

Let us write K_n for the set of *all* n-simplices, now including the degenerate ones as well.[19] We now not only have the face operators, $d_k \colon K_n \to K_{n-1}$, $k = 0, \ldots, n$, but also some degeneracy operators, which, in the usual notation, are denoted $s_i \colon K_n \to K_{n+1}$, where, for instance,

$$s_1 \langle v_0, v_1, \ldots, v_n \rangle = \langle v_0, v_1, v_1, \ldots, v_n \rangle,$$

so repeats the vertex label in position 1, while

$$s_0 \langle v_0, v_1, \ldots, v_n \rangle = \langle v_0, v_0, v_1, \ldots, v_n \rangle,$$

and so on. In our calculation of the faces of $\langle 2, 2, 3 \rangle$, which is, of course, $s_0 \langle 2, 3 \rangle$, we verified that, at least in this case, $d_0 s_0 = d_1 s_0$, but, of course, that is true in general as deleting either copy of a repeated label gets you the same result. Similar reasoning gives other such "simplicial identities,"[20] such as if $i < j$, then $d_i d_j = d_{j-1} d_i$.

We need to abstract from this example. We here have a structure consisting of a family, $\{K_n\}_{n \geq 0}$, of sets, plus face and degeneracy operations which satisfy the simplicial identities. Such a structure, in general, is called a *simplicial set*[21] and is the second of our ways of "presenting a space." Simplicial complexes with a total order on their set of vertices gives just one example of such

---

[19] So, slightly more formally, the basic setup of $V(K)$ and $S(K)$ is still the same, but now $\sigma = \langle v_0, \cdots, v_n \rangle$ stands for a set, after deleting any repetitions, $\{v_0, \ldots v_n\}$ in $S(K)$, with $v_0 \leq v_1 \leq \cdots \leq v_n$, and K_n is the set of all such σ.

[20] We will not give the usual complete list of these *simplicial identities* here but refer the reader to the standard texts on simplicial sets and related homotopy theory. A useful, if slightly old, brief introduction to this theory is to be found in Curtis's survey [23].

[21] The category of simplicial sets will be denoted $\mathcal{S}$.

things, but there are other important examples that do not come from simplicial complexes. We will consider two such.

Example 2.1 (Nerves of Small Categories) In the above, we could have used any partial order on $V(K)$ for which each simplex of each dimension was ordered. We did not really need a total order for the construction to work, although that is the simplest type to work with. More generally, if we have a partially ordered set, $\mathcal{X} = (X, \leq)$, then we can form a simplicial set, $Ner(\mathcal{X})$, by taking its set of n-simplices to consist of all sequences, $x_0 \leq x_1 \leq \cdots \leq x_n$, and with face and degeneracy operators much as in the simplicial complex case we looked at before. A particular case that is very useful is $[n] = \{0, 1, \ldots, n\}$ with the usual order. The simplicial set, $Ner[n]$, is the standard model for the n-simplex, as is evident if you look at low values of n. This is usually written $\Delta[n]$.

The definition of $Ner(\mathcal{X})$ is a special case of the nerve of a (small) category.[22] If $\mathcal{C}$ is an arbitrary small category, we can define a simplicial set, $Ner(\mathcal{C})$, by taking its set of n-simplices to consist of all *composable* sequences of n-arrows in $\mathcal{C}$, that is, of form

$$\sigma = (x_0 \xrightarrow{c_1} x_1 \xrightarrow{c_2} \cdots \xrightarrow{c_{n-1}} x_{n-1} \xrightarrow{c_n} x_n).$$

The set of 0-simplices is simply the set of objects of $\mathcal{C}$. For the face and degeneracy operators, we will leave the details for the reader to search for in the literature, but will rather look at the faces of a typical 2-simplex:

$$\sigma := (x_0 \xrightarrow{c_1} x_1 \xrightarrow{c_2} x_2),$$

or, sometimes, more conveniently, in the opposite order and for a general n, $(c_n, \ldots, c_2, c_1)$, recording just the morphisms[23] and which we draw as a triangle:

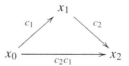

(and the reversal of order, above, allows this to avoid a reversal here to get to $c_2 c_1$). From this perspective, there is a clear idea of what the faces should be: $d_0(\sigma) = (c_2)$, the face opposite the vertex, 0; $d_2(\sigma) = (c_1)$, the face opposite

---

[22] Remember that any partially ordered set can be considered as a category with X being the set of objects and there being a single morphism from x to y if and only if $x \leq y$.
[23] The reason for the change in order in the symbols will be clear in a short while.

vertex 2; and $d_1(\sigma) = (c_2c_1)$, so given by the composition[24] of the two arrows, and giving the face opposite vertex 1. The degeneracies insert identity maps in a fairly obvious way.

We can give another brief equivalent description of $Ner(\mathcal{C})$. We let $Ner(\mathcal{C})_n = Cat([n],\mathcal{C})$, which is easily seen to be the same as before, but in different language. The face and degeneracy maps are derived from functors/order-preserving maps between the various $[n]$.

Example 2.2 (The Singular Complex Functor) We can use the same idea as above to obtain a simplicial set associated to a topological space, X. This is the classical *singular complex*, $Sing(X)$, of X. We use the topological simplices, Δ^n, that we have met earlier. There are face inclusions, $\delta_k \colon \Delta^{n-1} \to \Delta^n$, for $0 \leq k \leq n$, and some squashing maps, $\sigma_i \colon \Delta^{n+1} \to \Delta^n$, here given in a footnote.[25] These induce face maps,

$$d_i \colon Sing(X)_n \to Sing(X)_{n-1}, \qquad 0 \leq i \leq n,$$

and degeneracy maps,

$$s_i \colon Sing(X)_n \to Sing(X)_{n+1}, \qquad 0 \leq i \leq n.$$

Remarks (i) The singular complex construction is one of the key examples for our "narrative" about ∞-groupoids and spaces. It is one of the main candidates for something worth calling an ∞-groupoid, and, most importantly, it is easy to construct from a space. We still have a way to go before giving a better idea of what an ∞-groupoid is, but we will be revisiting $Sing(X)$ several times later on.

This singular complex construction is one of several used to encode the results of "probing" a space by nice "test objects." These test objects, in this case the topological simplices, Δ^n, are "spaces" that are well understood, both in themselves individually and also in their relationship between each other.[26]

(ii) We note that the notion of geometric realisation that we sketched for simplicial complexes extends to one for simplicial sets, for which see any of the standard texts on simplicial homotopy theory. The geometric

---

[24] A very important observation here is that the algebraically defined composition in $\mathcal{C}$ corresponds to the "geometric" process of filling the (up-side down) V-shaped diagram consisting of the given two arrows. This V-shaped diagram is usually called a $(2,1)$-horn. It looks like the one-dimensional skeleton of a 2-simplex with the d_1-face omitted.

[25] The face inclusion δ_k, sends $(x_0, \ldots, x_{n-1})$ to $(x_0, \ldots, 0, \ldots, x_{n-1})$, putting a 0 in the kth position. The squashing map, σ_i, adds x_i and x_{i+1} together before placing the result in the ith position then shifting each of the subsequent entries one place to the left.

[26] We could have based the discussion on other categories of test objects, for instance, n-cubes or n-globes, of more generally multiprisms, that is, products of topological simplices, Δ^n. Each has nice properties, but we will more or less restrict attention to the simplices, partially for historical and expositional reasons but also because that theory is the most developed one.

realisation of a simplicial set is an example of a type of space called a *CW-complex*.[27]

(iii) We can also extend the idea of an *n*-skeleton from simplicial complexes to simplicial sets, but the construction is a little bit more subtle. Given an n and a simplicial set, K, we can form its *n*-skeleton, $sk_n(K)$, by throwing away the nondegenerate simplices in dimensions greater than n. There will still be simplices in those higher dimensions, but, if $\sigma \in sk_n(K)_m \subseteq K_m$ for $m > n$, then there will be an *n*-simplex, $\tau \in K$, and a sequence of degeneracy operators whose composite sends τ to σ.

(iv) We met in the footnote to the previous page the idea of a (2, 1)-horn. This generalises to an (n, k)-horn in a simplicial set, K, for $0 \leq k \leq n$. The (2, 1)-horn that we considered consisted of two 1-simplices that fitted together as if they formed all but one of the faces of a 2-simplex. (In the case $K = Ner(\mathcal{C})$ that we looked at, there was a 2-simplex there "filling the horn," but it is easy to see if we just had a simplicial set coming from a simplicial complex, for instance, the "horn" might not have such a "filler.") A (n, k)-horn in K consists of a collection, $\underline{x} = (x_0, \ldots, x_{k-1}, -, x_{k+1}, \ldots, x_n)$, of $(n-1)$-simplices of K that fit together like all but the *k*th face of a *n*-simplex. An *n*-simplex $x \in K_n$ "fills" the horn if, for $j \neq k$, $d_j x = x_j$. The *k*-horn of a topological *n*-simplex, Δ^n, is defined in an analogous way to the *k*-horns in $\Delta[n]$. In each case one takes the $(n-1)$-skeleton and then removes the *k*th face. We will write $\Lambda[k, n]$ for the corresponding simplicial subset of $\Delta[n]$ and note that $\underline{x}$ can be thought of as a simplicial morphism from $\Lambda[k, n]$ to K.

In the singular complex, $Sing(X)$, of a space, X, all horns have fillers since a topological *n*-simplex can easily be shown to retract onto any of its horns; see any introduction to simplicial sets for more discussion. (This means not only that there is a filler but that such fillers are "thin" in the same intuitive sense as we mentioned earlier.) Those simplicial sets that satisfy the property of having fillers for all horns are called *Kan complexes* and will be very important later in our discussion. The nerve of a category is a Kan complex if, and only if, the category is a groupoid.[28]

---

[27] A CW-complex is built up from a discrete set of "vertices" by progressively attaching *n*-dimensional cells to lower-dimensional parts of the space, so we get a filtered space $X_0 \subseteq X_1 \subseteq \cdots \subseteq X_n \subseteq \cdots \subseteq X$ where X_0 is a discrete space, and, for each n, X_n is obtained from X_{n-1} by "gluing" in some *n*-disks.

[28] In the nerve of an arbitrary category, all (n, k)-horns for $0 < k < n$ (the so-called inner horns) have fillers given by the composition in the category, but the $(n, 0)$- and (n, n)-horns, the "outer horns," may not have fillers in general. Simplicial sets satisfying the weaker condition that all "inner horns" have fillers were originally called "weak Kan complexes," but now the term *quasi-category* is perhaps more often used. They are also one of the models for a class of ∞-category – but that is getting ahead of ourselves for the moment.

(v) An important related idea is that of a simplicial object in a category, $\mathcal{C}$. For the case of $\mathcal{C}$ being the category of sets, the simplicial objects are just the simplicial sets, of course, but taking $\mathcal{C}$ to be the category of groups, or abelian groups, will give simplicial groups[29] and simplicial abelian groups.[30]

3 Whitehead's Algebraic and Combinatorial Homotopy

On a seemingly tangential note, we will now consider the more general question of modelling homotopy types with algebraic data. We will see that this general idea lies in the background of our central theme, the point being that the idea of ∞-groupoids is "algebraic" in some sense, at least in some of the interpretations of the term. This also asks when a term such as "algebraically given data" can reasonably be thought of in spatial terms, as, for instance, group presentations can lead to spaces. Here we will very briefly set the scene for a fuller presentation of some of the ideas, but before that more detailed treatment, we will have to do some more groundwork introducing notions in the next section that illustrate the ideas here more fully and giving sketches of definitions.

In his 1950 ICM address, J. H. C. Whitehead summarised his vision of what he called *algebraic homotopy*:

> *The ultimate aim of algebraic homotopy is to construct a purely algebraic theory, which is equivalent to homotopy theory in the same sort of way that "analytic" is equivalent to "pure" projective geometry.*
>
> *([70], quoted in [4])*

A statement of the aims of algebraic homotopy might thus include the following homotopy classification problems [70]:

> *Classify the homotopy types of polyhedra, $X, Y, \ldots$, by algebraic data.*
> *Compute the set of homotopy classes of maps, $[X, Y]$, in terms of the classifying data for X, Y.*

This nicely sets up the idea of modelling (nice) spaces by "algebra," ..., but does not make precise what "algebraic data" is to mean. In fact, when modelling spaces by algebraic data, there is nearly always a balance to be

---

[29] In which each K_n is a group and each face and degeneracy map is a group homomorphism.
[30] These latter objects form a category equivalent to that of chain complexes of abelian groups, by the Dold–Kan theorem. This uses the Moore complex, which is an intersection of all the kernels of the face maps, d_k, $k > 0$, giving a chain complex from a simplicial abelian group. This is closely related to the way we get from 2-groupoids to crossed modules (see Footnote 67) and will be briefly examined in Section 5.3.

struck. More finely structured models are better for classifying the spaces and morphisms, but often the more structure there is, the harder it is to handle it all.[31] One question is, thus, what an analysis of this set of problems looks like in the context of the putative comparison

$$spaces \leftrightarrow \infty\text{-}groupoids,$$

and another query, again inspired by Whitehead's own list of problems, would be, if we have a finite-dimensional space, say, a k-dimensional CW, or simplicial, complex, how might that finite dimensionality be reflected in any associated ∞-groupoid?

As we said, detailed algebraic invariants become harder to calculate the more detailed they are! The exact sense of *calculate* here is quite hard to pin down! Some of the meanings of *detailed* are easier to explore, and we will endeavor to do so. The overall aim is to find algebraic models for homotopy types in the above sense and in particular, here, to come up with a notion of ∞-groupoid that will fit into this Whitehead program as a suitable form of algebraic data. As a step in that direction, we will look at various types of algebraic data and explore their connection with this setting.

Clearly, from today's perspective, the assignment of algebraic data to "spaces" that Whitehead was proposing has to be functorial.[32] From that viewpoint, the overall aim of Whitehead's algebraic homotopy program is to find natural algebraic models for homotopy types in such a way that the resulting "functor" is as close to an equivalence of categories as possible. Any functorial homotopy invariant, $F: Top \to AlgebraicData$, will, however, determine a class of morphisms between spaces that become isomorphisms on application of F. By assumption, this class will contain homotopy equivalences but may be much bigger. Controlling such a class for a given modelling functor F is where the more structured notions of homotopy theory come in.

4 Higher Homotopy Groups, Weak Homotopy Types, Truncation, and Connectedness

As an example of a type of algebraic data that are typical for Whitehead's setting, but that are very "minimal," in some sense, we could take a set to represent the set of arcwise connected components of the space, plus, for each element in that set, an $\mathbb{N}$-indexed family of groups, and loosely take

---

[31] This is very well discussed in the early sections of Baues [5].
[32] And he was one of the first to adopt an overtly categorical view of such situations.

$F(X) = \{\pi_n(X, x_0) \mid x_0 \in X, n \in \mathbb{N}, n \geq 0\}$. The corresponding class of morphisms will be that of *weak homotopy equivalences* (often just called *weak equivalences*).

4.1 Higher Homotopy Groups and Weak Homotopy Types

To make sense of this, and it *has* several bits of terminology and notation that we have not yet formally met, we first need to set out the main ideas on the higher homotopy groups, $\pi_n(X, x_0)$, of a pointed space, (X, x_0). We first recall that one of the definitions of the fundamental group of a pointed topological space was as homotopy classes of loops at the base point. We can think of a loop as being a map from the circle, S^1, to X, and, as we want the loop "at the base point," x_0, of X, we make S^1 into a pointed space by realising it as $\{\underline{x} = (x, y) \in \mathbb{R}^2 \mid ||\underline{x}|| = 1\}$ and then choosing $\underline{1} = (1, 0)$ as its base point. A loop at the base point of X is then a continuous function, $\gamma: S^1 \to X$, satisfying $\gamma(\underline{1}) = x_0$. The fundamental group, $\pi_1(X, x_0)$, is thus $[(S^1, \underline{1}), (X, x_0)]$, that is, the set of pointed homotopy classes of pointed maps from the pointed circle to the pointed space, (X, x_0), and as, intuitively, this corresponds to studying one-dimensional holes in X, it is natural to consider the n-sphere, $S^n = \{\underline{x} \in \mathbb{R}^{n+1} \mid ||\underline{x}|| = 1|\}$, based at $\underline{1} = (1, 0, \ldots, 0)$ and to look at $\pi_n(X, x_0) := [(S^n, \underline{1}), (X, x_0)]$, in an attempt to capture something of the behavior of the n-dimensional holes in X. This set has a natural group structure if $n > 0$, and that structure is abelian if $n > 1$. The resulting group is called the nth *homotopy group* of the pointed space, (X, x_0). Of course, if $f: X \to Y$ is a continuous map, then there is an induced group homomorphism,

$$\pi_n(f): \pi_n(X, x_0) \to \pi_n(Y, f(x_0)).$$

If X is connected, then $\pi_n(X, x_0)$ does not really depend on x_0. If we choose a path from x_0 to some other point x_1, then there is an induced isomorphism from $\pi_n(X, x_0)$ to $\pi_n(X, x_1)$, which depends only on the homotopy class of the path. In fact, in this way, we get a functor, $\pi_n(X)$, from the fundamental groupoid, $\Pi_1(X)$, to the category of groups (if $n > 0$) sending the point x, thought of as an object of $\Pi_1(X)$, to $\pi_n(X, x)$. Note that, as well as the case $n = 1$ corresponding to the fundamental group of (X, x_0), the case $n = 0$ gives the pointed set of connected components of (X, x_0), since S^0 is the two-point discrete space, $\{-1, 1\}$.

Returning, now, to *weak homotopy equivalences*, these are central to a lot of what follows, so we will give a slightly more formal definition.

A continuous map, $f: X \to Y$, between topological spaces is said to be a *weak equivalence* if f induces a bijection, $\pi_0(f): \pi_0(X) \to \pi_0(Y)$, between the sets of arcwise connected components of the two spaces, and also, for each $x_0 \in X$ and each $n \geq 1$, the induced homomorphism, $\pi_n(f)\colon \pi_n(X, x_0) \to \pi_n(Y, f(x_0))$, is an isomorphism of groups.

Two spaces are said to have the same *weak homotopy type* if there is a zigzag of maps between them, all of which maps being weak equivalences.[33]

The loose interpretation is that, if there is a weak equivalence between X and Y, then the set of invariants, π_n, cannot tell the two spaces, X and Y, apart. We note that any homotopy equivalence is a weak homotopy equivalence. In fact, if we restrict to CW-complexes, then weak equivalences *are* exactly homotopy equivalences, by a famous result of J. H. C. Whitehead, but, in general, there are pairs of topological spaces that are weakly equivalent without being homotopy equivalent.[34]

Given any space, X, we can form $Sing(X)$ and then take its geometric realisation. The two constructions are adjoint functors, and there is a natural map, $|Sing(X)| \to X$, that is a weak homotopy equivalence.[35] As $|Sing(X)|$ is a CW-space, we have any space is weakly equivalent to a CW-space; see below.

Note that spaces with the extra "combinatorial" information of being a CW-complex[36] make for a fairly well-behaved setting; however, even with such CW-complexes, and weak equivalences, this does not completely give the answer to Whitehead's idea of algebraic homotopy, since we do not know that there might not be two spaces, X and Y, with an isomorphism, $\theta\colon F(X) \to F(Y)$, yet there would be no continuous map, $f\colon X \to Y$, that satisfies $F(f) = \theta$. This is *Whitehead's realisation problem*, and we will discuss it in slightly more detail later on. One has to realise both spaces *and morphisms*.[37] This is important for our main theme, as it asks whether the algebra *accurately models* homotopy aspects of the spatial structure!

---

[33] We can form a new category from our category of spaces by "formally inverting" the weak equivalences to form the corresponding *homotopy category*. Two spaces have the same weak homotopy type if they are isomorphic in that homotopy category.

[34] The space known as the Warsaw circle has the same weak homotopy type as the discrete space with two elements, but not the same homotopy type. The study of the algebraic topology of such more general spaces is known as *shape theory*, and that term is also applied, by generalisation, to handle topoi; see the various relevant entries in the *n*Lab [54].

[35] Strong homotopy equivalence, although useful, does not so readily lead to simply defined, good algebraic invariants. General spaces, i.e., ones that are not "locally nice" need additional machinery for any effective study. This does relate to methods of both topos theory and of noncommutative geometry but will not concern us in this chapter, where we will almost always be restricting to CW-spaces.

[36] Or rather a *CW-space*, so the existence of a CW-complex structure on the space is required, but no choice of that structure is specified.

[37] And perhaps homotopies between morphisms, and higher homotopies between them etc.

Given any specific topological situation, the amount of information encoded in the weak homotopy type may be "unnecessary" or "unnatural" for the application that is in mind, or it may be impractical to calculate it,[38] so it is quite natural to work with a subset of the possible dimensions, thus looking at the homotopy groups, say, from 1 up to some given n, or, alternatively, from some integer n onward to infinity, or over some other suitable range of values, say, a segment, $[n, n+k]$, from n to $n+k$ for some integers n and k.[39] With regard to the original query of modelling "spaces" by "∞-groupoids," these classes of weak homotopy types "should" correspond to restricted classes of ∞-groupoids – ones, it is hoped, that help one understand the general picture better as well as shedding light on the specific situation.

We will look at several such situations in a bit more detail. In each case, the corresponding F will be different, and hence the notion of "equivalence" being used will change. First, in Section 4.2, we concentrate on the homotopy groups, π_k, for $k \in [n, \infty)$ and, in fact, will look at spaces that only have nontrivial homotopy groups in that range. The following section to that will look at the complementary setting, that is, where the only nontrivial homotopy groups are concentrated in the range $[1, n]$, and then we will look at one or two classical situations in which there is some information giving the homotopy groups in all dimensions but not enough to determine the homotopy type. In each case, there is a suitable choice of functor F, leading to a corresponding class of spaces.

4.2 n-Connectedness

As was said above, when studying "spaces" as (weak) homotopy types,[40] one of the evident simplifications to make about the spaces is that some of that structure is trivial. This assumption can then be the starting point for attempts to decompose a given (general) homotopy type somehow into a part for which the assumption holds plus "the rest." To a minor extent, we can already see this idea in our restriction to considering arcwise connected spaces. For any space, X, we can form $\pi_0(X)$ as a quotient of X, so getting $q: X \to \pi_0(X)$, and, trivially, each fibre of q is arcwise connected as it is a component of X.

There are higher forms of connectedness that can be used to get more useful and interesting instances of this sort of idea:

a space, X, is *n-connected* if $\pi_k(X)$ is trivial for $k = 0, 1, \ldots, n$.

---

[38] For example, even for spheres, it is not known how to work out a general formula for the homotopy groups or to what extent there is one.
[39] We will not follow up on these "segments" here, but the interested reader can find some useful results in [38], which also puts this into the context of this chapter.
[40] Via the information encoded in their homotopy groups.

What this means is that in an n-connected space, if we have any continuous map, $f: S^k \to X$, for $0 \leq k \leq n$, then as f must be homotopic to a constant map (at the unmentioned base point), we can extend f over the $(k + 1)$-disc, D^{k+1}, in the sense that the $(k + 1)$-disc, $D^{k+1} = \{\underline{x} \in \mathbb{R}^{k+1} \mid ||\underline{x}|| \leq 1\}$, has the k-sphere as its boundary, and there will be a map, $g: D^{k+1} \to X$, that restricts to f on the subspace S^k.

For $n = 0$, we retrieve the original idea, as this is just saying that X is arcwise connected, since if we have any two points $x_{-1}, x_1 \in X$, then we define a continuous map, $x: S^0 \to X$, by setting $x(-1) = x_{-1}$ and $x(1) = x_1$. If X is 0-connected, then this x extends to $y: D^1 \to X$, that is, as $D^1 \cong [0,1]$, to an arc joining x_{-1}, and $x_1 \in X$, and conversely, thus "0-connected" = "arcwise connected".[41]

For $n = 1$, 1-connectedness is the same as what is often called *simple connectedness*. It interprets as saying that any loop in X extends to a map of the disc, D^2. This notion is very important since, as we noted earlier, $\pi_1(X)$ acts on all the $\pi_n(X)$, $n \geq 2$, in fact making them into $\pi_1(X)$-modules. If X is simply connected, $\pi_1(X)$ is trivial, so the $\pi_n(X)$ are "merely" abelian groups and so are much easier to classify and use. Simple connectedness connects up with universal covers as, if X has a universal cover, $\tilde{X}$, then the space, $\tilde{X}$, is simply connected.[42] More than this is true, in fact. If $p: Y \to X$ is any connected covering space of X, and Y is simply connected, then Y is a universal covering space for X.

This type of example is the $n = 1$ case of a much more general phenomenon that is quite central to our overall story but, once again, needs expanding a little first. Recall that a covering space, $Y \to X$, has nice lifting properties. Generalising these is the notion of a fibration; see Footnote[43] for the definition. In a fibration, if $b \in B$, the *fibre over b* is the subspace, $F_b := p^{-1}(b)$.[44]

---

[41] To make the exposition slightly easier, we will usually assume in what follows that the spaces considered are arcwise connected, so we will not need to mention the information on π_0.

[42] The fundamental group of X is "still around" in this covering space as it acts on $\tilde{X}$ with X being the quotient space of "orbits." This is, of course, another of the classical Poincaré viewpoints.

[43] There are several versions, but roughly, a map $p: E \to B$ is a fibration if, for any CW-complex, X, and subcomplex, $A \subset X$, and for any commutative diagram

there is a map, f, that "lifts" v to E (so $pf = v$) and extends u from A to X (so $fi = u$).

[44] It is important to remember that changing b along a path makes the fibre change, so that, omitting the details, there is a "homotopy" action of the fundamental groupoid of the base, B, on the set of fibres. In particular, $\pi_1(B,b)$ acts on $\pi_k(F_b)$ for all $k > 0$. The homotopy groups

Example 4.1 (a) A covering space is a fibration and its fibres will be discrete spaces.

(b) Given any pointed space, $\underline{X} = (X, x_0)$, the set, $P(X)$, of all paths, $\alpha: [0, 1] \to X$, which start at x_0, so $\alpha(0) = x_0$, can be given a natural topology so that the map, $p: P(X) \to X$, given by $p(\alpha) = \alpha(1)$, is a fibration, called the *path fibration of* $\underline{X}$, and the fibre at x_0 is the space, $\Omega(\underline{X})$, of loops at the base point of X. (It is usual to write $\Omega(X)$, for simplicity.)

(c) The third example is rather a way of making more examples than one itself. If $p: E \to B$ is a fibration and $f: X \to B$ a continuous map, then in the pullback square

$$\begin{array}{ccc} f^*(E) & \longrightarrow & E \\ f^*(p) \downarrow & & \downarrow p \\ X & \underset{f}{\longrightarrow} & B \end{array}$$

$f^*(p): f^*(E) \to X$ is a fibration.

Using these, we can shed more light on one version of higher-dimensional analogs of covering spaces. Given any CW-space, X, and any $n > 0$, we can construct a space, $X(n)$, containing X as a (closed) subcomplex, and such that (i) for all k, $0 \leq k \leq n$, the induced homomorphism from $\pi_k(X)$ to $\pi_k(X(n))$ is an isomorphism[45] and (ii) $\pi_k(X(n))$ is trivial for $k > n$.[46] Now take the path fibration, $P(X(n)) \to X(n)$, and restrict it to X by pulling back along $X \to X(n)$. The resulting fibration will be written $X\langle n\rangle \to X$, and it is easy to show that $X\langle n\rangle$ is n-connected. Its other higher homotopy groups are the same as those of X.

If we go back to the case $n = 1$ and assume that X is connected, then $X\langle 1\rangle$ will have just one nontrivial homotopy group, $\pi_1(X\langle 1\rangle) \cong \pi_1(X)$, and $\pi_k(X\langle 1\rangle) = 1$ if $k = 1$ and is isomorphic to $\pi_k(X)$ for $k > 1$. In other words, $X\langle 1\rangle \to X$ is like the universal cover, and, in fact, its fibre is homotopically equivalent to the underlying set of $\pi_1(X)$.

Often $X\langle n\rangle \to X$ is called the *n-connected cover* of X, but it is not a covering space, although it has many properties that are analogous to those

---

of a fibre, the total space, and the base of a fibration are linked by a long exact sequence; see, for instance, Hatcher [35, p. 376].

[45] We will meet such maps shortly and in more detail.
[46] Such spaces are said to be *n-coconnected* in some of the classical literature but are also called *n*-truncated.

of classical covering spaces, so perhaps a little care has to be taken with the terminology.[47]

4.3 n-Truncation

We now turn to the type of morphism exemplified by $X \to X(n)$ above. This mapping kills off any information encoded in the homotopy groups of X above level n. It *truncates* the homotopy type at dimension n.

Adapting the definition and terminology of weak equivalences, we say a continuous map, $f: X \to Y$, is a *homotopy n-equivalence* (or simply an *n-equivalence*) if it induces an isomorphism, $\pi_k(f): \pi_k(X) \to \pi_k(Y)$ for $k = 1, \ldots, n$. Two spaces, X and Y, are said to *have the same n-type* if there is a zigzag of n-equivalences joining them. By this we mean that there is a diagram of form

$$X = X^{(0)} \to X^{(1)} \leftarrow \ldots \leftarrow X^{(2k)} = Y$$

with all the maps n-equivalences.[48]

As we mentioned in the previous section, for any space X, we can build another space, which we denoted $X(n)$, together with an n-equivalence, $X \to X(n)$, such that $\pi_k(X(n)) = 0$ for $k > n$.[49] For this sort of space, it can be very much easier to understand what the "spaces as ∞-groupoids" paradigm looks like, while our previous discussion indicates that there are ways of "decomposing" a general CW-space, X, into an n-connected piece and an n-type, via a fibration. These ideas, in the main, were already present in Whitehead's conception of algebraic homotopy and do not, initially, seem that connected to the ideas of higher category theory. Turning to the models for homotopy n-types, however, we will start to see the beginnings of a link with ∞-groupoids via various types of higher-dimensional groupoids encoding more information on a homotopy type, and, to aid this, we will explore n-types and their algebraic models in a bit more detail.

---

[47] The above construction is quite tricky to make into a functorial one, as it involves constructing $X(n)$, which usually involves choices, but the corresponding problem for simplicial sets has a neat functorial solution using the idea of a coskeleton that can be found in standard texts on simplicial homotopy.

[48] A weak equivalence is an n-equivalence for all n, so we can think of any n-type as being made up of lots of weak homotopy types. Each of these will consist of spaces with the same homotopy groups up to the nth one, but after that, they will, in general, be different.

[49] Such a space with vanishing homotopy groups above the nth one is itself often called a homotopy n-type, although strictly speaking from the definition that we have given, that is an abuse of terminology.

4.3.1 1-Types and Groupoids

For $n = 1$, the only nontrivial homotopy group would be $\pi_1(X)$. A 1-type thus corresponds to an isomorphism class of groups. This is not quite all, however. The spaces we are looking at are connected, so if we choose any base point, we will get the fundamental group of that pointed space, and if we change base point, we will get an isomorphic fundamental group. The isomorphism between them will be given by a path from one base point to the other, and different paths will usually give different isomorphisms. Because of this, it is better to use the fundamental *groupoid* of the space, $\Pi_1(X)$, even if the space is connected.

For any group (or groupoid), G, we can find a space, a *classifying space*, BG, with that group as its fundamental group and no other nontrivial homotopy groups, so a group(oid) yields a 1-type, modelling algebra by topology;[50] see, for instance, the treatment by Baues [6, p. 18]. This illustrates an important aspect/theme of this general program: *to extract algebra from a space and to build a space from algebraic data*.

Looking back, we can see a vague idea emerging when going from the case $n = 0$ to $n = 1$. In forming $\pi_0(X)$, we put an equivalence relation on X, relating two points if there is a path joining them. It is thus the *existence* of a path that counts at this stage, not the actual paths. In forming $\Pi_1(X)$, the paths joining the points are promoted to being of value *in themselves*. They become more centre stage. We do not just consider two points of X equivalent because a path *exists*; rather, we seriously look at all the paths between them[51] and ask when those paths are, themselves, to be thought of as being "equivalent" by some "higher path of paths," that is, by the existence of a homotopy.

In general, an equivalence relation *encodes* a (special type of) groupoid, but general groupoids encode more information "stored" in their vertex groups, telling us about the "automorphisms" of the object at which one is looking.[52] Of course, in an equivalence relation thought of as a groupoid, those vertex groups are trivial.

We can probe that "vague idea" a bit more. What would be the result if we were to take the equivalences between paths seriously as well, more or less considering them as "reasons" that two paths are equivalent and thus would

---

[50] The classifying space of a groupoid is most usually taken to the geometric realisation of the nerve of that groupoid but can also be constructed starting from a *presentation* of the group(oid).
[51] Perhaps thinking of the paths as the different "reasons" that the points are to be "equivalent."
[52] This idea of a groupoid is thus very useful in classifying situations in which objects have important local symmetries. Often it is useful to replace a quotienting operation by the formation of a groupoid for this reason.

be "identified" in the fundamental groupoid? We can find this sort of idea in algebra to some extent. In the theory of groups, it is not equivalence relations on a group, G, that are useful but rather congruences. These have compatibility with the multiplication built in. They are "internal" equivalences within the category of groups and, of course, can be usefully handled by looking at the subgroup of G consisting of those elements that are "congruent" to the identity. In that way, the congruence is encoded by a *normal* subgroup. Our "vague idea" about "reasons between reasons" and also of replacing "equivalence relations" by "groupoids," so as to encode nontrivial automorphisms of objects suggests that it is natural to extend "internal equivalence relations" to "internal groupoids" – but note that this is *quite natural* when coming from the situation with paths and is not just some abstract generalisation for the sake of it.

That "natural" progression was not quite the way that the theory developed in the 1940s and 1950s. The reason would seem to be that the notion of groupoid was not that obviously useful for researchers having a good working knowledge of group theory. Groupoids had been introduced by Brandt in 1926, but his use was in other areas of algebra than those adjacent to topology. Schreier did make explicit use of them in topology in 1927, and Reidemeister included the construction of the fundamental groupoid in his book of 1932, but they were still not considered that useful by other topologists. The interaction of the fundamental group(oid) concept and spatial aspects of combinatorial *group theory*, as developed by Reidemeister for applications in knot theory,[53] seems to have been central to that work. (This is explored by Krömer [41].) This work in combinatorial group theory mirrored, in an algebraic context, ideas that were emerging[54] in homotopy theory that directly related to the idea of calculating homotopy groups from combinatorial models of a space. These advances, however, used the "normal subgroup" side of the picture rather than the "congruence" one.

4.3.2 2-Types, Crossed Modules, and 2-Group(oid)s

The developments that led to an understanding of models for 2-types came from three closely related areas: combinatorial group theory, homotopy theory itself, and group cohomology. (We will not go into the third here, other than to say it relates to the homotopy theory of the classifying space of a group.) The first two of these correspond to ideas that were already there in Poincaré's approaches to the fundamental group that we

---

[53] Many early invariants of knots were derived from homotopical invariants of the complement of the knot, via group presentations of its fundamental group, such as those developed by Dehn and Wirtinger.

[54] E.g., Whitehead [67].

mentioned earlier. We will explore a little the path that led from the case of "1-types/groups" to "2-types/crossed modules and 2-group(oid)s," as this shows the start of a shift of focus toward groupoid methods and then to "higher-dimensional groupoids." The historical development of the "new" concepts of crossed modules and 2-groupoids starting from more "classical" notions shows clearly the beginnings of the progression from a low-dimensional notion of "space" encoding simple relationships between "points" to one encompassing many dimensional relationships, that is, higher categories and groupoids.

The nearest of these ideas to the further development of the "vague idea" comes from Reidemeister, and later Peiffer,[55] working on "identities among relations" of presentations of groups. Given a group, G, it is often usual and useful to label the elements by some alphabet of generators and then to say which "words" in the symbols of the alphabet correspond to the same element of the group. In more usual terms, one gives a presentation, $\mathcal{P} = \langle X : R \rangle$, of G, where X is a set of generators, often thought of as a subset[56] of G, and R, a subset[57] of the free group, $F(X)$, on X. The elements of R are often called "relators" or "relations."[58] Reidemeister started looking at "identities among relations" in the following sense. There is a morphism, $\varphi \colon F(X) \to G$, given by evaluating each generator as an element of the group and the kernel of φ is the normal closure of R, so as was said above, the elements of $Ker\varphi$ are "consequences" of R, and thus are words made up of conjugates of relators and their inverses. To study these, we could form a free group on symbols for these conjugates and then see if there were relations between them, that is, "relations between the relations." That sounds like our "vague idea" coming near the surface again, and it is. It also has a topological aspect, to which we turn next.

We will first need another classical definition, namely, that of the relative homotopy groups of a (pointed) pair of spaces. We postpone a more detailed description of these to an appendix to this section, but for the moment it suffices to say they give homotopy groups for *pairs* of base pointed spaces, (X, A), together with natural homomorphisms linking them with the homotopy groups

---

[55] This research would seem to have been done early in the 1940s, but publication was delayed until 1949; see [55, 61].
[56] But that can sometimes be inconvenient,
[57] The congruence on $F(X)$ that identifies words, i.e., elements of $F(X)$, if they represent the same element of G, corresponds to the normal subgroup of $F(X)$ generated by the elements of R and all their conjugates. The words in the elements of R and their conjugates are called *consequences* of R. The "relators" in R thus help us to understand the congruence.
[58] Although that latter term is a bit confusing in our context.

of X and A; for a full treatment, see Hatcher [35, p. 343] or many other texts on homotopy theory.

In 1941, Whitehead examined a problem that is clearly related to the idea of specifying a space, or building it, by iteratively attaching cells to "lower-dimensional" parts as in a simplicial or CW-complex. He asked what would happen to the invariants, and in particular the homotopy groups, if extra cells were added to such a complex. The key setup is thus a space, A, to which one attaches ("glues on") some cells to get a new space, X. Knowing information on the homotopy groups of A, and how the new cells were attached, what can be said about the homotopy type of X and the corresponding invariants? Before discussing the structure that he revealed, let us see why this is related to the question of identities among relations.

Suppose $\mathcal{P} = \langle Y : R \rangle$ is a presentation of a group, G; then we can form a CW-complex, $K(\mathcal{P})$, having a single vertex, with 1-skeleton consisting of a collection of pointed loops or circles, one for each generator (and indexed by the set, Y), all attached at their base points to that single vertex. (This will make up our space A, in this case.) The fundamental group of A is a free group[59] on the set Y. Each relation, $r \in R$, corresponds to (the homotopy class of) some map $f_r : S^1 \to A$, and we use f_r to attach a 2-cell to A. Doing this for all r gives us a two-dimensional complex, $K(\mathcal{P})$, which is the X in our statement of the general problem that Whitehead considered. (This is the start of a process that can lead to a *small* complex having the homotopy type of the classifying space,[60] BG, of G.) The relative homotopy group, $\pi_2(X, A)$, and the boundary map, $\partial : \pi_2(X, A) \to \pi_1(A)$, encode interesting and useful information about the presentation *and* the group, G. The kernel of ∂ is isomorphic to $\pi_2(X)$ and can be interpreted in the case of $K(\mathcal{P})$ as the G-module of identities among the relations of the presentation.

Whitehead's 1941 paper examined the general case, and, in the situation, as here, of attaching *two-dimensional* discs to a complex A, he identified the structure of the algebraic object, $(\pi_2(X, A), \pi_1(A), \partial)$, encoding the way the cells were attached.[61] This algebraic structure was what is called a *crossed module*.[62] Just as replacing an equivalence relation by a groupoid encoded more of the spatial structure of paths in a space, so replacing a congruence

---

[59] The elements of $\pi_1(A)$ are homotopy classes of paths that go around the loops, and a word in Y can be used to encode the order in which this happens.

[60] The data needed to build such a small complex and thus to encode the homotopy type of BG may be finite, even when the group G is infinite. This is important for understanding properties of G.

[61] In [67], Whitehead considers and solves the problem for the case of general n-dimensional discs, but, for the immediate story here, the $n = 2$ case is the most important.

[62] The term seems first to have been used by Whitehead about that time.

in the form of a normal subgroup by a crossed module continues that "vague idea" corresponding to replacing the "congruence," thought of as an internal "equivalence relation" by an internal "groupoid," and yes, crossed modules *do* correspond to a form of "2-groupoid," as we will see.

In 1949–50, Mac Lane and Whitehead [49] combined these ideas with some from group cohomology to show that 2-types corresponded to these *crossed modules*.[63] We will give the definition and some basic, but important, examples but will not develop the theory here.

Definition 4.2 A *crossed module*, (C, G, δ), consists of groups, C and G, with a (left) action of G on C, written $(g, c) \to {}^g c$ for $g \in G$, $c \in C$, and a group homomorphism, $\delta \colon C \to G$, satisfying the following conditions:

(1) for all $c \in C$ and $g \in G$, $\delta({}^g c) = g\delta(c)g^{-1}$,
(2) for all $c_1, c_2 \in C$, ${}^{\delta(c_2)} c_1 = c_2 c_1 c_2^{-1}$.[64]

Example 4.3[65]

(1) Suppose that δ is just the inclusion of a subgroup of G and that the action of G on that subgroup is by conjugation; then, to say (C, G, ∂) is a crossed module just says that C is a *normal* subgroup of G, thus a normal subgroup yields a crossed module. (As we suggested above, a crossed module is, from this viewpoint, a generalisation of a normal subgroup in which we ditch the requirement of "being a subgroup"!)
(2) At the other extreme, if δ is the trivial homomorphism, then C will be abelian and is just a G-module.

4.3.3 2-Groupoids, Crossed Modules, and the Mac Lane–Whitehead Theorem

Much later than the Mac Lane–Whitehead result on 2-types, some time about 1965, Verdier noticed that crossed modules corresponded to what would now be called 2-*groupoids*, that is, 2-categories[66] in which every 1-arrow and every 2-arrow is invertible. Crossed modules of groups, as we have defined them above, correspond to 2-*groups*, that is, 2-groupoids with exactly one object.

---

[63] Warning: the definition of n-types has changed since that date, so their "3-types" are now usually called 2-types.
[64] There is a fairly obvious many object/nonconnected version of this.
[65] We direct the reader to the literature for more examples and the development of the elementary theory of crossed modules. These can be found in [11] and in numerous other sources, both surveys and original articles.
[66] We will look at 2-categories and 2-groupoids in a bit more detail very shortly in Section 4.5. Here we only need the idea that 2-categories are like categories, but with objects, morphisms (= 1-arrows) between them, and, in addition, 2-arrows between (parallel) 1-arrows.

This relationship was rediscovered by Brown and Spencer in 1972 (see the introduction to [12]), and a closely related result appeared, at about the same time, in the thesis [63] of Grothendieck's student Hoàng Xuân Sính. Her result and Verdier's original one relate to non-abelian cohomology and the representation of cohomology classes, and this theme comes in later in some of Grothendieck's letters to Breen.

The method of going from crossed modules to 2-groups is quite simple. It is a simple extension of the way one replaces a congruence on a group by a normal subgroup, without loosing information. Here it will be relegated to the footnotes[67] so as not to interrupt the main flow of ideas. It does depend on having some idea of what a 2-category is but is otherwise quite simple. We will discuss various simple ideas about 2-categories and related structures slightly later (see Section 4.5).

The result of Mac Lane and Whitehead thus says that 2-types correspond to 2-groupoids. There is, however, a "but." The classification of crossed modules/2-groupoids that it needs is not that of "up to isomorphism," rather, it uses an algebraic form of homotopy equivalence, adapted for crossed modules. In fact, an even better way to think of it is that the category of 2-types should have a 2-category structure, as should that of crossed modules, in which the 2-cells encode the homotopies.

We can use the relative homotopy groups that we met earlier to give the explicit functor from simplicial (or CW-) complexes to crossed modules, and which is the basis for the Mac Lane–Whitehead result. We assume that K is a simplicial or CW-complex, and in our relative crossed module of a pair (X, A),

---

[67] Given a crossed module, $\mathsf{C} = (C, G, \delta)$, (of groups), the corresponding 2-category, denoted $\mathcal{X}(\mathsf{C})$ here, has a single object, which we will denote $*$, the set of 1-arrows from $*$ to itself is the set of elements of G, with composition being its multiplication, the set of 2-arrows is the set of pairs (c, g), but with horizontal composition given by the multiplication of the semi-direct product group, $C \rtimes G$. The 1-source of a 2-arrow (c, g) is g, while its 1-target will be $\delta c.g$. This makes it look a bit like the following:

(That this picture looks like two paths/loops and a homotopy between them is, of course, more than coincidental!)

Coming back from 2-group(oid)s to crossed modules is now easy. You look at the 1-arrows as the bottom group of the crossed module, and then the top group will be the group of 2-arrows with source the identity element of the group of 1-arrows. (That looks hopeful for a generalisation to higher dimensions!)

we take A to be $K^{(1)}$, the 1-skeleton of K, and so get a crossed module, $(\pi_2(K, K^{(1)}), \pi_1(K^{(1)}), \partial)$, from the complex. This looks very good as an invariant of the CW-space, K, until one realises that it depends on the specified combinatorial structure of the complex and thus on the "triangulation" or "cellular decomposition" of the space, rather than just on the *topological* structure of the "space," K. It is an analogue of Poincaré's combinatorial definition of the fundamental group but does depend more on the combinatorial structure. A subdivision of the (simplicial) complex structure will give another *nonisomorphic* crossed module. What Mac Lane and Whitehead do in [49] is to analyse how such a combinatorial change is reflected by a "combinatorial homotopy" of crossed modules.[68]

Crossed modules are, thus, the "slimmed-down" encoding of a 2-group(oid), and we have associated a 2-groupoid to a (CW-) space, albeit by choosing a CW-structure on it.

4.3.4 n-Types for $n \geq 3$

The story of "modelling" n-types was really only continued in the 1980s by Loday [46] (see also [13, 58]). Loday introduced generalisations of crossed modules and 2-groupoids valid "for all n." His models were *strict* n-fold groupoids, a slightly different form of "multiple groupoid" than we will be considering, but still an indication of that somewhat elusive link between spaces and ∞-groupoids. Loday was able to use "strict" objects because his models are very "spread out."[69] We have not the space here to describe Loday's construction in more detail, although it leads to some interesting points relevant to our themes.[70]

---

[68] Note the date and title of Whitehead's two papers [68, 69]. He envisaged, as part of algebraic homotopy, a "combinatorial homotopy" that would extend ideas and methods of *combinatorial group theory* to higher dimensions. The introductions to these papers contain important reflections on homotopy types and algebraic models for them.

[69] For instance, for 3-types, the structure corresponding to the extra data required for encoding "weak" 3-groupoids is given by his h-maps, which are derived from commutators in the group structures.

Two "criticisms" of Loday's models are that (1) from the point of view of "spaces," their interpretation is, perhaps, less intuitive than one initially might hope for, and (2) it is not clear if there is a way of adapting the theory to handle the case of $n = \infty$. (This may be just a question of looking at the structures in the "right way," but that "right way" is not yet obvious.)

[70] There *is* a topological interpretation of the construction that Loday uses and that provides interesting insight into the whole question of what "spaces" are. Loday does not work with a pair, (X, A), of spaces as such; rather, he converts the inclusion $A \hookrightarrow X$ into a fibration, $\overline{A} \to X$, with $\overline{A}$ being the space of paths in X that start in A.

Given any (pointed) fibration $p \colon E \to B$, with fibre $F = p^{-1}(b_0)$, it is not hard to see that there is an action of $\pi_1(E)$ on $\pi_1(F)$ and that the inclusion $inc \colon F \hookrightarrow E$ induces a morphism, $\pi_1(inc) \colon \pi_1(F) \to \pi_1(E)$, that satisfies the crossed module axioms. Better than that, on converting this to the corresponding 2-groupoid, you get that there is a structure of

We should also mention Conduché's notion of 2-crossed module – see Conduché [17] and the closely related notion of quadratic module, due to Baues [5] – but we will not give details, as that would require some background that we have not assumed. Both provide models of connected 3-types.

4.4 Beyond 2-Types towards Infinity Groupoids, from a Classical/Strict Viewpoint

From our point of view, the Mac Lane–Whitehead result shows that 2-types and 2-group(oid)s are closely linked. Whitehead's combinatorial homotopy papers [68, 69] also talked of algebraic models that provide (usually incomplete) information in all dimensions. Surprisingly enough, these also have an interpretation in terms of ∞-groupoids, but we will take a slightly leisurely approach using Whitehead's classical homotopy theoretic machinery rather than going directly to the ∞-groupoid model.[71]

The use of the relative homotopy groups by Mac Lane and Whitehead, and the resulting crossed module used to model a 2-type, fits into another sequence of models that link into a classical construction due to Blakers (1948) and which was developed further by Whitehead (1949) and then by Brown and Higgins, Baues, and others[72] from the 1970s onward.

As a first step toward them, we look at the chains on the universal cover of a CW-complex. This was one of the classical tools used by Whitehead in his key papers. This allows us to state a result that illustrates some of Whitehead's ideas simply and clearly.

First a little background. We mentioned the universal covering space, $\tilde{X}$, of a (nice) space, X. If X has a CW-complex structure, then the local homeomorphism property of covering maps allows one to obtain a CW-complex structure on $\tilde{X}$ for which the covering map, $p\colon \tilde{X} \to X$, is a cellular map.

Any CW-complex, X, gives rise to a complex, $C(X)$, of "cellular chains," (see [35]), and for the universal cover, $\tilde{X}$, the action of $\pi_1(X)$ on that space transfers to the chain complex, $C(\tilde{X})$, of chains on the universal cover, giving it the structure of a chain complex of modules over $\pi_1(X)$, which we will again use shortly. (We will denote the homology of this complex by $H_*(\tilde{X})$.)

---

a weak 2-groupoid on the pullback of E with itself (over B). (This seems to have been first noticed by Deligne; see Friedlander's paper [26].)

Loday's general construction takes a (fibrant) $(n+1)$-cube of fibrations and constructs an n-fold groupoid from it.

[71] Note that this is algebraic topology from the end of the 1940s but is not that well known.

[72] For more detail on the history of this, see [11, p.255, note 96]. These models, which are called *crossed complexes*, are equivalent to a special class of ∞-groupoids.

To show the relevance of this for our "quest" for algebra mirroring homotopy, we note the following theorem of Whitehead:

Theorem 4.4 *A map, $f: X \to Y$, of (pointed connected) CW-complexes is a homotopy equivalence if, and only if, the induced homomorphisms, $\pi_1(f): \pi_1(X) \to \pi_1(Y)$ and $H_n(\tilde{f}): H_n(\tilde{X}) \to H_n(\tilde{Y})$, for all $n \geq 2$, are isomorphisms.*

We thus have that not only do the homotopy groups constitute a system of algebraic invariants sufficiently powerful to characterise the homotopy type of a CW-space but so does the combination of π_1 and the homology of the universal covering space. Importantly, however, this does *not* mean that it solves all the basic problems of algebraic homotopy. We could have isomorphisms, $\phi_n: \pi_n(X) \to \pi_n(Y)$, of all the homotopy groups of two spaces, or of the homology of their universal covers, but would not know if there was an $f: X \to Y$ realising these ϕ_n.

On the other hand, the question about how dimension of a complex might be reflected in the models is here very easy to resolve. If X is a CW-complex of dimension k, then so is $\tilde{X}$, and as the generators of $C(\tilde{X})$ in dimension n are the n-cells, we immediately have that $C(\tilde{X})$ is trivial in dimensions greater than k. Whitehead used this in his two papers on combinatorial homotopy [68, 69] to show that the cellular chain complex of the universal covering does act as an "algebraic equivalent" of a three-dimensional polyhedron, so here the algebra does reflect a lot of the geometry of the space.[73]

The compatibility conditions that would be needed between the input data[74] must thus be part of the key to understanding the structure of homotopy types. Baues [6] calls the problem of finding necessary and sufficient conditions for this the *realisation problem of Whitehead*. For any system of algebraic invariants, there will be a similar realisation problem.[75] In an attempt to study these, Whitehead introduced another algebraic model, which is nowadays called a *crossed complex*. These crossed complexes are equivalent to a special class of ∞-groupoids in the same way that crossed modules are an equivalent algebraic model to 2-groupoids. We will merely give the idea of the definition, referring to [11] for much fuller information.

A *crossed complex* is a chain complex of groups (or groupoids), (C_n, ∂), where C_n is defined for $n > 0$,

---

[73] It is an interesting question to see the explicit relationship between these results and models of such spaces by ∞-groupoids. (Perhaps this is known, but I cannot recall seeing such a study.)
[74] In the above case, the ϕ_n and the actions.
[75] The problem is discussed in detail in that source, so we will not repeat here what is said there.

$$\ldots C_n \xrightarrow{\partial} C_{n-1} \xrightarrow{\partial} \ldots \xrightarrow{\partial} C_3 \xrightarrow{\partial} C_2 \xrightarrow{\partial} C_1,$$

in which there is an action of C_1 on all the terms, with ∂ respecting that action, (C_2, C_1, ∂) is a crossed module, and, for $n \geq 3$, C_n is abelian and, in fact, is a $C_1/\partial C_2$-module.

This "model" will therefore have something of a crossed module/2-type in it, and as that crossed module has "fundamental group" $C_1/\partial C_2$, it has something of the "chains on the universal cover" model that we saw just now. It is a natural abstraction from the following motivating example in which the two parts fit exactly as required.

Example 4.5 Let K be a CW-complex, and, for each n, let $K^{(n)}$ denote its n-skeleton.[76] We set, for $n \geq 2$, $C_n = \pi_n(K^{(n)}, K^{(n-1)}, x_0)$, and when $n = 1$, we just take C_1 to be the fundamental group of $K^{(1)}$. The boundary ∂ will be the composite

$$\pi_n(K^{(n)}, K^{(n-1)}, x_0) \to \pi_{n-1}(K^{(n-1)}, x_0) \to \pi_{n-1}(K^{(n-1)}, K^{(n-2)}, x_0).$$

As we have indicated earlier, $\pi_1(K^{(1)}, x_0)$ will act on all the higher relative homotopy groups[77] in the complex, and this gives a crossed complex.

This still seems very far from ∞-groupoids. (Historically, we are still in the late 1940s or early 1950s!) The category of crossed complexes (over groupoids) is, however, equivalent to a category of *strict*[78] ∞-groupoids, [10].[79] We can think of these "∞-groupoids" as having not only objects and arrows, which will have "inverses," as with an ordinary groupoid, but also invertible 2-arrows/2-cells joining the (1-)arrows, invertible 3-arrows joining certain pairs of 2-arrows, and so on. Once again, our "vague idea" is coming into evidence.

We cannot give the detailed description of the equivalence between these objects and crossed complexes here, but note that if we have a strict ∞-groupoid, H, then the module of n-cells in the associated crossed complex is the group of those n-cells of H, whose $(n-1)$-source is an $(n-1)$-fold identity. This thus generalises the way we mentioned of getting from a 2-group to a crossed module.[80] Note that these strict ∞-groupoids do *not* answer the

---

[76] For simplicity, we will assume that K is "reduced" and so is connected and, in fact, $K^{(0)}$ consists of just one point, x_0. This means we can limit ourselves to groups rather than groupoids, which, although more natural, do require a bit more "setting up" if they are to be handled well.
[77] Recall these are looked at in more detail in the appendix to this section.
[78] The meaning of *strict* should become apparent shortly.
[79] This was discovered by Brown and Higgins in their study of higher-dimensional analogs of van Kampen's theorem; see [11] for a full treatment and discussion of that theorem.
[80] Or from a congruence to the corresponding normal subgroup.

general Whitehead's "algebraic homotopy" problem, however. The point is well made in the book by Brown, Higgins, and Sivera [11, p. xxvii]. Quoting that source:

> Crossed complexes give a kind of 'linear model' of homotopy types which includes all 2-types. Thus although they are not the most general model by any means (they do not contain quadratic information such as Whitehead products), this simplicity makes them easier to handle and to relate to classical tools. The new methods and results obtained for crossed complexes can be used as a model for more complicated situations.

It is this very "linearity" which means crossed complexes do not trap the whole of the information on the homotopy type. It also means that they correspond to a strict ∞-groupoid and not the fully general type of ∞-groupoids. For the relevance of this to Whitehead's realisation problem, this means that just as the chains on the universal cover do not capture the whole of the 2-type structure, although the fundamental crossed complex captures the 2-type, it does not capture the 3-type. Going further, Baues [5] defines a "quadratic" model (quadratic complexes), but while it captures the 3-type, and so on.[81]

4.5 From "2-" Heading to "Infinity," from Grothendieck's Viewpoint

To go from the fundamental group and covering spaces, or from crossed complexes, all the way to

$$spaces \leftrightarrow \infty\text{-groupoids}$$

seems a very high step to jump. The next stage in understanding that link, both historically and conceptually, is via various intermediate stages that indicate that *there is something to jump to*! As a first small step, we need to gain some idea about infinity categories and groupoids in the sense needed later. In so doing, we will also encounter some other useful ideas.

The ∞-groupoids that we will be meeting later will not usually be "strict," that is, they will be infinite-dimensional analogues of *bicategories*, with all arrows in all dimensions being "weakly invertible," so we should first glance at bicategories as being the simplest such "weak" context. (We will only give a sketch as usual, leaving the reader to follow up links to the literature. Introductions can be found in Leinster [44] and Lack [42], for instance.)

---

[81] What is not always that clear is how the extra structure at each stage corresponds to adding "weakness" into a notion of ∞-groupoid, bit by bit.

A (strict) 2-category, $\mathbb{A}$, is a category *enriched over* the category of small categories, so each "hom" $\mathbb{A}(x, y)$ is a small category, and the composition

$$\mathbb{A}(x, y) \times \mathbb{A}(y, z) \to \mathbb{A}(x, z)$$

is a functor. Composition is associative, so given objects, w, x, y, z, the square of functors,

$$\begin{array}{ccc} \mathbb{A}(w, x) \times \mathbb{A}(x, y) \times \mathbb{A}(y, z) & \longrightarrow & \mathbb{A}(w, x) \times \mathbb{A}(x, z) \\ \downarrow & & \downarrow \\ \mathbb{A}(w, y) \times \mathbb{A}(y, z) & \longrightarrow & \mathbb{A}(w, z), \end{array}$$

is commutative; similarly for the identities. We have already seen some *2-groupoids* above, and they are, of course, the corresponding structures in which the $\mathbb{A}(x, y)$ are groupoids, and also there is an inversion operation from each $\mathbb{A}(x, y)$ to the corresponding $\mathbb{A}(y, x)$ satisfying some, it is hoped, obvious axioms.

In a *bicategory*,[82] $\mathbb{B}$, although the basic structure looks the same, the corresponding diagrams are only required to commute up to specified natural isomorphisms. An important special case is that in which $\mathbb{B}$ has just one object. Such bicategories correspond to the monoidal categories that are widespread in algebraic contexts.[83]

It is fairly easy to define the notion of a strict ∞-category, generalising that of strict 2-categories, and thus to define strict ∞-groupoids as being a special class of such. Such strict ∞-groupoids correspond to crossed complexes.[84] We will need the weak form of ∞-category, and that, in some sense, is much harder to "get right." There are, perhaps, two main problems here: (1) How "weak" should this be? Do we just weaken things like associativity, and the need for composition to be a functor, or do we also weaken composition to

---

[82] Note that the terminology is perhaps slowly changing from that used initially in this area, and as these bicategories are seen as being more natural and significant than the strict form, the term *bicategory* is now often replaced by *2-category*, with the older meaning of that latter term corresponding to "strict 2-category."

[83] As was mentioned earlier, a category with one object is essentially just a monoid. A bicategory, $\mathbb{B}$, with just one object, $*$, will have a category, $\mathbb{B}(*, *)$, forming the endomorphisms of $*$ and then a functor

$$\mathbb{B}(*, *) \times \mathbb{B}(*, *) \to \mathbb{B}(*, *),$$

which is then the tensor product, $\otimes$, of the corresponding monoidal category. This argument can be reversed, so any monoidal category can be thought of as a single object bicategory. The monoidal category is *strict* exactly when the bicategory, $\mathbb{B}$, is a 2-category.

[84] The process of passing from a strict ∞-groupoid to a crossed complex is a natural generalisation of the one we sketched earlier for 2-groupoids. One takes kernels of the source maps.

being "existence of a composite" up to higher homotopies? and (2) Do we base things on globes, simplices, cubes, or what?

We must now look to the 1970s, when Grothendieck exchanged a series of letters [27–29] with Larry Breen. In these he sketched out a theory of objects that he called *n-stacks* and of the possible analogue of the Galois–Poincaré theory (see [33]) in higher dimensions. We will return to that later, but for the moment, we will just note that this series of letters resurfaced in February 1983, being mentioned in the "letter to Quillen" that formed the first six pages of Grothendieck's epic manuscript *Pursuing Stacks* [32].[85]

To set the scene for that, Ronnie Brown had just sent Grothendieck some of the preprints produced in Bangor, in which ideas on (strict) infinity groupoid models (thus crossed complexes) for some aspects of homotopy types were discussed. Grothendieck wrote to Quillen:

At first sight, it seemed to me that the Bangor group had indeed come to work out (quite independently) one basic intuition of the program I had envisaged in those letters to Larry Breen – namely the study of n-truncated homotopy types (of semi-simplicial sets, or of topological spaces) was essentially equivalent to the study of so-called n-groupoids (where n is a natural integer). This is expected to be achieved by associating to any space (say) X its "fundamental n-groupoid" $\Pi_n(X)$, generalizing the familiar Poincaré fundamental groupoid for $n = 1$. The obvious idea is that 0-objects of $\Pi_n(X)$ should be points of X, 1-objects should be "homotopies" or paths between points, 2-objects should be homotopies between 1-objects, etc. This $\Pi_n(X)$ should embody the n-truncated homotopy type of X in much the same way as for $n = 1$ the usual fundamental groupoid embodies the 1-truncated homotopy type. For two spaces X, Y, the set of homotopy classes of maps $X \to Y$ (more correctly, for general X, Y, the maps of X into Y, in the homotopy category) should correspond to n-equivalence classes of n-functors from $\Pi_n(X)$ to $\Pi_n(Y)$ –, etc. There are very strong suggestions for a nice formalism including a notion of geometric realization of an n-groupoid, which should imply that any n-groupoid is n-equivalent to a $\Pi_n(X)$. Moreover when the notion of n-groupoid (or more generally of an n-category) is relativised over an arbitrary topos to the notion of an n-gerbe (or more generally, an n-stack) these become the natural "coefficients" for a formalism of non-commutative cohomological algebra, in the spirit of Giraud's thesis.

Later in the letter, he noted that these *n*-categories would have nonassociative compositions, including "whiskerings" (as they are now called) of all types, but that the nonassociativity would be up to a cell in the next higher dimension. This is one of the points of the idea of "weak" as against "strict"

---

[85] It is worth remarking that, both here and later, Grothendieck refers to some ideas coming from algebraic geometry and non-abelian cohomology, but our use of these texts will not require any real knowledge of that area. It just feels strange to cut up the quotation in an attempt to make a text without such mentions.

in these n-categories and groupoids. The composition would be associative "up to higher cells," just as path composition is associative up to specified homotopies. Likewise, he points out that n-objects would be invertible up to an $(n + 1)$-object.

4.6 Appendix: More Technical Comments

We will pick up the main "narrative" in the next full section but, before that, will collect up here a few longer comments that may provide more detail or insight into things that we have been looking at but that can safely be skipped or skimmed on a first reading.

1. *Relative homotopy groups.* We will give this in general as it can be useful later on, even though, for our main use, it was only the case $n = 2$ that is required.[86]

   The nth homotopy group of a pointed space, (X, x_0), can be defined in another way from which we used above. In this alternative definition, it consists of homotopy classes of maps of pairs, $f: (I^n, \partial I^n) \to (X, \{x_0\})$, the homotopies being through maps of the same form. In other words, the maps, f, send the n-cube, I^n, to X in such a way as to send its boundary, ∂I^n, to the single point x_0, and the homotopies used must deform the way the interior is mapped while not changing the behavior on the boundary.

   For the relative homotopy groups of a pair, (X, A), with $A \subseteq X$, and a base point $x_0 \in A$, we again use the n-cube, I^n, and think of it as $I \times I^{n-1}$. Let $J^{n-1} = \{1\} \times I^{n-1} \cup I \times \partial I^{n-1}$, so consisting of all the faces of I^n except (the interior of) $\{0\} \times I^{n-1}$; then $\pi_n(X, A, x_0)$ is defined to be the set of homotopy classes of maps, $f: (I^n, \partial I^n, J^{n-1}) \to (X, A, \{x_0\})$, with homotopies being through maps of the same form.[87]

   Restricting such a map to $\{0\} \times I^{n-1}$, we get an element of $\pi_{n-1}(A, x_0)$, and this assignment gives a homomorphism

   $$\partial: \pi_n(X, A, x_0) \to \pi_{n-1}(A, x_0).$$

   The above groups "depend on the choice of base point" in the same way as we saw earlier. This means that, for a given n and a choice of base point, x_0, the groups, $\pi_n(X, A, x_0)$, do not depend on x_0 if A is pathwise connected,

---

[86] Again, this will be a sketch, and for fuller details, the reader should "consult the literature." Fuller expositions can be found, for instance, in Hatcher's book [35, p. 343], and, especially relevantly for us, in the book by Brown, Higgins, and Sivera [11, p. 35].

[87] We thus have that all but one of the faces of the n-cube are sent to the base point, while the last face is sent into A. We will see a similar situation later when looking at simplicial groups.

and there is an action of the fundamental groupoid of A on the family of groups, $(\pi_n(X, A, x_0) \mid x_0 \in A)$, that can contain valuable information.

2. *From chain complexes to crossed complexes.* We have met, above, the fundamental crossed complex of a CW-complex. Another type of crossed complex comes from chain complexes of modules over a group G. Suppose G is a group and (M, δ) is a (positively graded) chain complex of left G-modules. We can form a crossed complex by taking $C_1 = M_1 \rtimes G$ and $C_n = M_n$ for $n \geq 2$. We then take all the "boundary" maps, ∂ to be δ, except for $\partial \colon M_2 \to M_1 \rtimes G$, which sends m to $(\partial m, 1_G)$. Finally, we make $(m, g) \in C_1$ act on higher dimensions using just the action of g.

This construction, in fact, gives a functor from the category consisting of such pairs, (G, M), to the category, $CrsComp$, of crossed complexes. This functor has a left adjoint.

An especially important instance of this is when we apply that adjoint to the fundamental crossed complex of a CW-complex, X. This gives a chain complex of $\pi_1(X)$-modules, which is isomorphic to the complex of chains on the universal cover of X. There is, thus, a direct functorial construction going from the fundamental crossed complex, $\pi(X)$, of a CW-complex, X, to this crossed complex.[88] Applying this, *the crossed complex of a CW-complex must be at least as good at distinguishing spaces and homotopy equivalences as is the cellular homology of the universal covering, because we can functorially derive the latter from the former.*

3. *What kind of homotopy types are completely captured by crossed complexes?* This is a very natural question to ask. Each time we find a "model" such as these, we have two related questions: How good a model is it? What kind of homotopy types are captured by the models, that is, are completely modelled by a model of that type? That the two questions are related is clear if we look at 1-types. The model of a 1-type is given by a group. Two spaces cannot be distinguished by their "models" if they have the same fundamental group, so one might say the model is not that good (but in fact it is still very useful!). A space is completely modelled by its fundamental group if, and only if, all its other homotopy groups are trivial, but any space has the same 1-type as a space whose higher homotopy groups, π_k, for $k > 1$, are trivial,[89] which shows how the two answers are connected.

---

[88] This construction relates to certain well-known notions (Fox derivative, Reidemeister–Fox Jacobian of a presentation) from combinatorial group theory.
[89] X has the same 1-type as the classifying space of $\pi_1(X)$.

That is quite easy and can, with a bit of work, be extended without much change to 2-types and even to general n-types. A bit more work is needed to describe the homotopy types that correspond to crossed complexes. It is relatively easy to show that there is a classifying space construction for crossed complexes and that there is a continuous map, $K \to B\pi(K)$, from a complex, K, to the classifying space of its fundamental crossed complex. The question is thus of determining when this map is a weak equivalence. We know that a homotopy type leads to a crossed complex. What we hope for is to be able to say what properties a homotopy type has if this encoding in terms of crossed complexes is enough to give all the possible information, that is, that the crossed complex "characterises" the homotopy type.

There are higher-order "pairings" or "actions" within the homotopy groups of a space given by the Whitehead products, for which see Hatcher [35], for example. It is not too difficult to see that the classifying space of a crossed complex must have vanishing Whitehead products, but there are spaces that do have vanishing Whitehead products yet require more structure than that encoded in a crossed complex to completely model their weak homotopy type. The key to answering the problem lies elsewhere, within Whitehead's theory.[90]

In [68], Whitehead introduces the notion of a J_m-complex as follows:

Let K be a (connected) CW-complex (with a 0-cell chosen as base point for all the homotopy groups concerned). Let $\rho_n := \pi_n(K^{(n)}, K^{(n-1)})$ be the nth relative homotopy group of the n-skeleton of K, relative to the $(n-1)$-skeleton.[91] There is a natural homomorphism

$$j_n : \pi_n(K^{(n)}) \to \rho_n.$$

Definition 4.6 The complex, K, is said to be a J_m-*complex* if j_n is a monomorphism[92] for each $n = 2, \ldots, m$. The complex is said to be a J-*complex* if it is a J_m-complex for all $m > 2$.

He introduced the notation $\Gamma_n(K)$ for the kernel of j_n, so K is a J_m-complex if $\Gamma_n(K)$ is trivial for all $n = 2, \ldots, m$ and is a J-complex is $\Gamma_n(K)$ is trivial for all $n \geq 2$.

---

[90] But he did not, in fact, prove the characterisation needed for our question.
[91] N.B. This is the nth-group, $\pi(K)_n$, of the fundamental (reduced) crossed complex of the space, K, filtered by skeleta that was mentioned above.
[92] Whitehead actually uses the term *isomorphism into* rather than *monomorphism*.

Theorem 4.7 *Let K be a connected CW-complex. The natural morphism*

$$K \to B\pi(K)$$

is a weak equivalence if, and only if, K is a J-complex.

This result explicitly gives a characterisation of those homotopy types representable by crossed complexes. It has appeared in several places in the literature, for example, in [15, Corollary 2.2.7], or, more recently, in [11], but these have usually been slightly submerged in a mass of other results and so are not that well known. Carrasco and Cegarra's proof uses simplicial groups and is thus quite algebraic.[93]

5 Simplicial Sets, Higher Combinatorics, and ∞-Groupoids

We have given quite a lot of historical background in the earlier sections on the lowest-level "classical" viewpoint. For the most part, that left us in the first half of the 20th century. We have mentioned quite a few ideas that arose from work by Whitehead in the 1940s and 1950s.[94] Later on, these gave various algebraic structures equivalent to some forms of (strict) ∞-groupoid.

5.1 Kan Complexes and ∞-Groupoids

Grothendieck's sketch, which we saw earlier, is reminiscent of the construction of the singular complex of a space, the difference being that it was based on a globular intuition while the latter was simplicial, corresponding to the two sides of Grothendieck's "yoga" from letters [27–29] to Larry Breen, which will be revisited shortly. I mentioned this idea in a letter [57] to Grothendieck in June 1983. Explicitly, I said,

> *I believe, in fact, the ultimate in non-strict or lax ∞-groupoid structures is already essentially well known (even well loved) although not by that name. The objects to*

---

[93] There is an important slightly technical point of interest here, relating to the fact that, from a crossed complex, one can build a *strict* ∞-groupoid by a construction generalising the way one constructs a 2-groupoid from a crossed module. We thus do get some infinity groupoids using the relative homotopy groups and classical constructions from Whitehead's 1950s papers, but they are *strict* and correspond to J-complexes. A question is, how does one relate the J-complex condition *explicitly* to that categorical "strictness"? In low dimensions, weak versions of ∞-groupoids are well known and relatively well understood, and some of them model, for instance, 3-types: see Joyal's letter to Grothendieck [38]. How does the "weakness" of these models, expressed in purely categorical terms, explicitly correspond to the nonvanishing of Whitehead's Γ functors?

[94] These ideas have been developed a lot further by Brown and Higgins, and also by Baues in the work cited in the bibliography.

which I am referring are Kan complexes, [...]. Here composition is not even strictly defined – given $\alpha, \beta \in X_1$, X a Kan complex, one forms a composite by filling the horn,

in any way whatsoever. Two such fillers are homotopic, associativity is only defined up to homotopy and so on.

Grothendieck's reply, a couple of weeks later [31], raised several objections. He felt, among other things, that composition should be defined as a function and that there should be just two boundary maps from each dimension giving the source and target of each n-object. (This then led on to a discussion of "test categories" that will not be explored here but does relate to the problem of the choice of "test objects" that we looked at earlier.)

These objections are at the same time serious and not that difficult to counter, at least in part.

(i) First, the fact that "filling" is not defined as a function could be met by making it one! We could then make composition "algebraic," adding enough formal composites into the Kan complex, then 2-simplices between the formal composite and the corresponding horn, and so on. One way of making this idea explicit has been done by Thomas Nikolaus [53]. He has introduced a form of Kan complex with given fillers, which he calls *algebraic Kan complexes*. There is also another way of doing this that links in with more classical constructions of simplicial homotopy, namely, by generating composites freely in a certain sense, using a construction of Dwyer and Kan, in such a way as to be able to compare "formal composites" with "geometric composites." We will look at this in the next section.

(ii) The second objection is partially handled by exploring Grothendieck's sketched construction in detail, as is done by Georges Maltsiniotis in [50, 51] and by Dimitri Ara [1].[95]. This approach uses n-globes as the test objects.

This raises the interesting question of which test objects one should use. Grothendieck preferred n-globes in some form. The difficulty is that although n-globes correspond to the intuition most neatly and to the occurrences of low-dimensional versions in homological algebra and algebraic geometry, the idea of (weak) composition is much more difficult to handle for them. The use

---

[95] More generally, useful references for higher category theory include [16, 45].

of n-cubes has a lot to recommend it (see the discussion in [11]) especially for composition, but the resulting theory of cubical sets is less well known and less well developed and so needs some development work to get to where we are going. In both cases, one would also like nicely behaved globular or cubical nerves of categories, while in that context, the simplicial case seems to be very neat, as we will see in the coming pages.

Picking up the point about composition, another reason for claiming a strong link between Kan complexes and ∞-groupoids is the neat observation, usually attributed to Grothendieck himself and that we made earlier, that the nerve of a small category is a Kan complex if, and only if, the category is a groupoid. The simplicial filling properties of the nerve do correspond *precisely* to the algebraic structure of the composition in the category. The clearest and simplest example of this is that, given a $(2,1)$-horn, the filler of that horn *is* the sequence formed by that pair of arrows, viewed as a 2-simplex of $Ner(\mathcal{C})$, and the missing one-dimensional face *is* the composite of that pair. There is a *unique* filler for the horn, and this uniqueness is unusual (and significant). The related property, often known as the Segal condition, is also important but will not be discussed here.[96]

All this suggests that we *could* decide to take "Kan complex" to be the idea of "∞-groupoid" that we would work with. In that case, the process of the passage from spaces to ∞-groupoids would just be the classical singular complex functor, while the functor going in the other direction would be the geometric realisation. Certainly this gives an equivalence of homotopy categories and, even better, that equivalence is given by an equivalence of the homotopy theoretic structure, interpreting that in the sense of Quillen's theory of model categories, either in its original form (from [60]) or in almost any of the refinements and variants made since his initial idea. This however is not the whole picture. It would not advance the study very much if that was all. The missing perspectives include the objections that Grothendieck raised and the result of the interaction of Grothendieck's ideas and those of Whitehead's algebraic/combinatorial homotopy. Examining this brings up the "homotopy hypothesis" (which we will often call the HH), that, in the case of Kan complexes, is a classical theorem of Milnor. This "hypothesis" is really not one, but rather is a test for any putative definition of ∞-groupoid together with assorted "baggage" of homotopies, and of higher homotopy structure, including fibrations, cofibrations, and other similar concepts. The test is that *there should be, at very least, an equivalence of homotopy categories between*

---

[96] Segal "spaces," which involve an abstraction of this condition, are another important model for ∞-categories and groupoids.

that of some category of "spaces" and that of that notion of ∞-groupoid being "tested" and this takes us right back to the questions with which we started.

If a candidate notion of ∞-groupoid is to be tested against the HH, then it must be able to model the various (relevant) structures of the category of "spaces." A key example of such structure would be the idea of fibrations, and if that works, then to translate some of the ideas of covering space, n-connected covering, and so on that we met earlier to that ∞-groupoidal setting. The simplest context in which to try out this extra structural test is, of course, that of Kan complexes, and there everything works well.[97] There is a simple notion of (Kan) fibration, $p \colon E \to B$; see the footnote below.[98] For the case $B = \Delta[0]$, p is the unique map to that object[99] and is a Kan fibration exactly if E is a Kan complex.[100] If B is connected, and the fibre of p over any vertex of B is discrete,[101] then p is the simplicial set analogue of a covering space. In general, if p is a Kan fibration, its fibres are Kan complexes (and so a Kan fibration is "fibred in ∞-groupoids" if you are using Kan complexes as models for ∞-groupoids).[102]

5.2 Simplicially Enriched Categories and Groupoids

(We will tacitly assume a bit more basic knowledge of simplicial objects in this section.) We next need to go toward understanding the Dwyer–Kan construction that we mentioned above. This takes a simplicial set and gives a

---

[97] This is a good point to remember that the best-known meaning of *works well* in this context involves the idea of a Quillen model category structure in one of its forms. As we said earlier, for this we merely direct the reader to the extensive literature on that theory.

[98] Similarly defined to the topological notion, a morphism, $p \colon E \to B$, of simplicial sets is a *Kan fibration* if, in any commutative square of the form

$$\begin{array}{ccc} \Lambda[k,n] & \xrightarrow{u} & E \\ {\scriptstyle i}\downarrow & {\scriptstyle f}\nearrow & \downarrow{\scriptstyle p} \\ \Delta[n] & \xrightarrow{v} & B \end{array}$$

there exists a (dotted) arrow, as shown, so for any n-simplex, v, in B and a (n,k)-horn in E, which maps down to the correspond (n,k)-horn of v, the horn in E can be filled to an n-simplex mapping down to v.

[99] As $\Delta[0]$ is the terminal object of $\mathcal{S}$.

[100] In the language of model category theory, this says that the Kan complexes are the fibrant objects in the category, $\mathcal{S}$, of simplicial sets. In general, with a good notion of fibration, a *fibrant object* is one for which the unique morphism from the object to the terminal object is a fibration.

[101] That is, is a constant simplicial set.

[102] Given a lot more space, we could have described the fibrations in both cubical and globular ∞-groupoid models. In each case, the models correspond to the fibrant objects.

simplicially enriched groupoid. To understand those objects, it will help not only to meet *simplicially enriched categories* but to see how these ideas fill in some of the blanks in a simplicially based version of Grothendieck's idea.

In the picture that we sketched above, there is a lot more structure around, namely, that which we have loosely termed the *assorted baggage* of homotopies and of higher homotopy structure. For instance, the category of spaces can be "enriched" over the category of simplicial sets. It has a structure that merits being called an ∞-category, as we will see shortly. Grothendieck's suggestion, implied by his comments in his letters to Larry Breen, was that any sensible category of ∞-groupoids should have an ∞-category structure and the hypothetical correspondence

$$spaces \leftrightarrow \infty\text{-}groupoids$$

should be an equivalence of ∞-categories, in an appropriate sense. His sketch left a lot of ideas at the "intuitive" level, that is, the theory seems to need certain ideas so as to work well, and moreover, coming from the examples, the intuitions behind those ideas, about homotopies and higher homotopies, would seem to fit, but that still left a lot of detailed exploration, checking, and so on to be done.

Let us give a few more details. If our category of spaces has function space objects,[103] then we can take their singular complexes to get simplicial sets of morphisms from X to Y. In fact, even if the particular category of spaces being considered does not have nice function spaces, we can still form up this simplicial set of maps by defining Spaces(X, Y) to be the simplicial set having $Spaces(X \times \Delta^n, Y)$ as its set of n-simplices.[104] There are "composition maps,"

$$\text{Spaces}(X, Y) \times \text{Spaces}(Y, Z) \to \text{Spaces}(X, Z),$$

and a little routine checking shows that this gives *Spaces* the structure of a *simplicially enriched category*, a term often abbreviated to $\mathcal{S}$-category.[105]

Remark 5.1 (i) There are lots of other examples of $\mathcal{S}$-categories in the areas we have been discussing. The category, $\mathcal{S}$, itself is an $\mathcal{S}$-category with

---

[103] So if we have X and Y, we also have a space, Y^X, of continuous functions from X to Y and there are nice isomorphisms, such as $Spaces(X \times Y, Z) \cong Spaces(X, Z^Y)$.

[104] N.B. This is isomorphic to the singular complex of Y^X if that latter space exists, but this definition does not depend on the existence of that function space. There is a very slight condition on the category of spaces that is needed. The category of spaces being used must contain $X \times \Delta^n$ for all X and all $n \geq 0$.

[105] An $\mathcal{S}$-category, A, is a category enriched over the category of simplicial sets, so each "hom" $A(x, y)$ is a simplicial set, and the composition,

$$A(x, y) \times A(y, z) \to A(x, z)$$

"function spaces," S(K, L), given by S(K, L)$_n$ = $\mathcal{S}(K \times \Delta[n], L)$. Other examples include the category of simplicial abelian groups, that of chain complexes of modules over a commutative ring, that of simplicial groups, and so on. All these are "large" $\mathcal{S}$-categories. There are small ones as well. Any simplicial monoid is essentially a small $\mathcal{S}$-category with a single object, and conversely, for any $\mathcal{S}$-category, A, and any object, x, the simplicial set, A(x, x), of endomorphisms of x is a simplicial monoid. Another class of examples come from 2-categories. If $\mathbb{A}$ is a (strict) 2-category, then we can define a corresponding $\mathcal{S}$-category, A, by taking A(x, y) := $Ner(\mathbb{A}(x, y))$ with the induced face and degeneracy maps. As any crossed module, C, gives a 2-group(oid), $\mathcal{X}$(C), it also gives a small $\mathcal{S}$-category, $Ner\mathcal{X}$(C), which is, in fact, an $\mathcal{S}$-groupoid. For that $\mathcal{S}$-groupoid, as its set of objects consists just of the single object, $*$, of C, we have that $Ner\mathcal{X}$(C) is a simplicial group.[106] Any simplicial group likewise gives a crossed module.[107]

(ii) In the $\mathcal{S}$-category structure on the category of spaces, each simplicial function space, Spaces(X, Y), is a Kan complex.[108] Such "Kan-enriched" categories include the category of Kan complexes itself. If Kan complexes are taken to be a possible model for ∞-groupoids, then these Kan-enriched categories are also enriched over ∞-groupoids and would form a particular class of the structures that would warrant being called ∞-categories. In such an ∞-category, any k-arrow for $k > 1$ will be invertible. Of course, Grothendieck's criticisms would still apply to these, and perhaps one needs something a bit more "algebraic."

is a simplicial map. Composition is associative, so given w, x, y, z, the square of simplicial maps,

$$\begin{array}{ccc} A(w,x) \times A(x,y) \times A(y,z) & \longrightarrow & A(w,x) \times A(x,z) \\ \downarrow & & \downarrow \\ A(w,y) \times A(y,z) & \longrightarrow & A(w,z) \end{array}$$

is commutative, and similarly for the identities, which are zero simplices in the various A(x, x).

[106] There are several simple ways to write down this simplicial group in terms of the data of C, which can be found in the literature but will not be explored here. The way back from simplicial groups to crossed modules using the Moore complex construction is also very neat.

[107] Simplicial groups represent all connected homotopy types, and the purely algebraic way from a simplicial group to the corresponding crossed module is an algebraic form of the Mac Lane–Whitehead model for the 2-type of a general (CW-)space.

[108] The proof of this follows the same route as that of proving that each $Sing(X)$ is a Kan complex. One uses that the horns of topological n-simplices are retracts of the simplex.

The criticisms by Grothendieck of the idea of using Kan complexes as a model for ∞-groupoids included that, for that model, composition is not determined uniquely. It used a filler of a horn, and that filler need not be unique; hence, for instance, a composable pair of 1-arrows might have a whole lot of "composites." Of course, that does mirror the topological context, but, in that setting, there is a reasonably natural choice of a filler. We have already mentioned one way to get around this, namely, by choosing one "formal" filler for each horn in a Kan complex, K. Another way would be to freely combine simplices in a horn in some way to give a formal composite, which, then, could be linked to the "composites defined by the Kan condition" via some homotopies. This vague idea is somewhat of an analogue, in all dimensions, of the way in which Poincaré formed the edge-path groupoid of a simplicial complex. There one formed a free groupoid on the 1-skeleton and then used the 2-simplices to link the formal composite of two faces of the 2-simplex to the generator in the third face. Technically, the "all dimensions" version of this is quite difficult to do, and the formulas that result[109] are not "self-evident." The result, however, is an S-groupoid, $\mathcal{G}(K)$, that does work well. This construction is due to Dwyer and Kan in [25] (but beware of some silly typographic errors in that source, and look for the details elsewhere). Writing S-$Grpd$ for the category of S-groupoids, the Dwyer–Kan functor, $\mathcal{G}: S \to S$-$Grpd$, has a right adjoint, $\overline{W}$, and, for any S-groupoid, G, its *classifying space*, $\overline{W}G$, is a Kan complex.[110] If we think of a groupoid as an S-groupoid that is simplicially trivial in a fairly obvious sense, then its classifying space, as a simplicially enriched groupoid, is the same as the classifying space given by its nerve as a groupoid. As the adjoint pair, $\mathcal{G} \dashv \overline{W}$, induces an equivalence of homotopy categories between that of S and that of S-$Grpd$, we could have taken S-groupoids as our choice for ∞-groupoids.

These simplicial models are much more "algebraic" than the simple Kan complex ones. Restricting to connected homotopy types, and thus to simplicial groups, we can gain some additional insight into the models of the special kinds of homotopy types that we mentioned earlier, for example, those corresponding to strict infinity groupoids. They are related to

---

[109] One takes K_0 as the set of objects and then considers each $n + 1$-simplex, σ, as being an n-dimensional edge going from the zeroth to the first vertex of σ. This gives a graph, and one then takes the free groupoid on that graph to be $\mathcal{G}(K)_n$. There is then a way to define induced face and degeneracy maps.

[110] If we think of this Kan complex as an ∞-groupoid, we can analyse the composition *algebraically*. To see why, at least in the case of a simplicial group, G, note that the underlying simplicial set of a simplicial group is a Kan complex and that there are algorithms giving fillers for horns. These, in turn, give algorithmic fillers for the "classifying space," $\overline{W}G$, with a complete analysis of possible choices being feasible!

special algebraic and combinatorial properties of the corresponding simplicial groups.[111]

Here is a good point at which to mention the role simplicial groups play in the theory of simplicial fibrations, fibre bundles, and related ideas in the theory of simplicial homotopy as it connects up not only with the adjunction, $\mathcal{G} \dashv \overline{W}$, but with another of Poincaré's ways of considering the fundamental group. We saw how he considered it as a group of deck transformation of the universal covering of the space. In other words, he looked at the group of automorphisms of the object $\tilde{X} \to X$ in the "slice category" of spaces over X. When we are working in an $\mathcal{S}$-enriched category, $\mathbb{A}$, the automorphisms of an object, y, naturally form a simplicial group, $\mathrm{aut}(y)$, and if we have a specified map, $p \colon y \to x$ in $\mathbb{A}$, then the automorphisms of y *over* x form a simplicial subgroup of $\mathrm{aut}(y)$, generalising the group of deck transformations of a covering. This suggests that some, at least, of the rôle of "$\pi_1(X)$ as automorphisms of $\tilde{X} \to X$" might be transferred to the automorphisms of the fibrations that generalise the universal cover, and this *is* the case.[112]

Another related link between simplicial groups and fibred things comes about in the adjunction, $\mathcal{G} \dashv \overline{W}$. This gives, for any (reduced) simplicial set, K, a unit morphism, $\eta_K \colon K \to \overline{W}\mathcal{G}K$. The next ingredient that we need is that for a simplicial group,[113] G, its classifying space, $\overline{W}\mathrm{G}$, comes with a natural Kan fibration, $p_\mathrm{G} \colon W\mathrm{G} \to \overline{W}\mathrm{G}$, whose fibre is the underlying simplicial set, $U(\mathrm{G})$, of G, and whose simplicial group of automorphisms will be isomorphic to G. Now any morphism $f \colon K \to \overline{W}\mathrm{G}$ will induce a fibration over K by pullback of p_G along it. This induced fibration will have fibre $U(\mathrm{G})$.[114] We direct the reader to the literature for more on this. (A fairly brief treatment can be found in Curtis [23].)

Some examples of this are of note for our earlier theme of factorising a homotopy type into simpler bits. If G is a group, then we can construct a constant simplicial group, $K(G,0)$, with $K(G,0)_n = G$ for all n and all face and degeneracy maps the identity isomorphism on G. Taking $G = \pi_1(K)$,

---

[111] A word of caution here is needed. The transition from simplicial sets to $\mathcal{S}$-groupoids involves a shift in the usual dimension of simplices, so a 1-simplex in K becomes a zero simplex in one of the "hom-sets" of $\mathcal{G}(K)$. Because of this, for a connected K, $\pi_n(K) \cong \pi_{n-1}(\mathcal{G}(K))$, which initially can lead to some confusion. The same goes for "n-type."

[112] Of particular interest would be the n-connected cover of a homotopy type considered as a fibration $X\langle n \rangle \to X$ in the simplicial category, Spaces, or in the analogous simplicial setting.

[113] We asked for K to be reduced to ensure that the $\mathcal{G}K$ was a simplicial group and so could apply this fact to it. This is mostly for the sake of the exposition.

[114] And is technically a *principal* G*-fibration* or G*-torsor*, and any such fibration will correspond to some f.

there is a natural map, $K \to K(G,0)$, related to the unit we mentioned above.[115] If we pull p_G along this, we get the universal (simplicial) cover of K. More generally, we could also use the unit composed with other quotients of $\mathcal{G}(K)$ and pullback $p_{\mathcal{G}(K)}$ along that. In each case, one is approximating to the information on the homotopy type of K by means of a truncation or similar. Note that if we start with K being a Kan complex, then this can be interpreted as being a series of operations on ∞-groupoids.

This rich structure suggests the extent to which the classical Kan complex/Kan fibration setup gives a useful first example of a situation satisfying the expectations of the HH. It also sets up a family of results and subtheories that operate as a model for how any good general abstract theory will look.

5.3 Chain Complexes, Globular and Simplicial Models

Here we will look at globular models and how they relate via chain complexes to simplicial ones.

Chain complexes (of abelian groups) form very simple models of a class of homotopy types.[116] As mentioned earlier, they correspond to simplicial *abelian* groups. We also mentioned that strict ∞-groupoids corresponded to crossed complexes and that such objects are also chain complexes, but with not all the groups involved being abelian, and there being some extra structure in the shape of some well-behaved actions. The functor from strict ∞-groupoids to crossed complexes extends that from 2-group(oid)s to crossed modules, and the idea there is very simple. Given a 2-group(oid), $\mathbb{G}$, you take those 2-arrows, x, which have their 1-source at an identity, so they look like

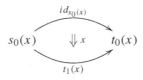

where $s_0(x)$ is the object that is the source of x, and similarly, $t_0(x)$ is its target.[117]

---

[115] As $\pi_1(K) = \pi_0(\mathcal{G}(K))$, there is a quotient map $\mathcal{G} \to K(\pi_1(K), 0)$; now apply $\overline{W}$ to that and compose with the unit η_K.
[116] And also correspond to a fairly simple class of (strict) ∞-groupoids.
[117] So $x \in \mathbb{G}(s_0(x), t_0(x))$, while $s_1(x)$, (resp $t_1(x)$) denotes the 1-arrow that is the source (resp. target) of x *within* the groupoid $\mathbb{G}(s_0(x), t_0(x))$. In our particular situation, $s_1(x) = id_{s_0(x)}$, so $s_0(x) = t_0(x)$, and x actually is a vertex 2-loop of the 2-groupoid.

In a strict ∞-groupoid, $\mathbb{G}$, based on a globular model, each n-arrow, x, would have sources, $s_k(x)$, and targets, $t_k(x)$, in all lower dimensions.[118] The group in the nth-dimension of the corresponding crossed complex consists of those n-arrows, x, such that $s_{n-1}(x) = id_{s_{n-2}(x)}$, that is, the source trivial n-arrows.

If one goes to the more general form of ∞-groupoid, again in a globular form, then one can again consider the source trivial n-arrows in each dimension (but they may not form a group). The result will look somewhat like a chain complex, as the target map will provide a boundary operator.[119]

If one chooses Kan complexes as the model for ∞-groupoids, then handling the analogous construction seems hard, as there is little or no algebra to help, but if we work with a S-groupoid, not only does it work much better, after a bit of "simplicial adjustment," but it relates to known structures that we have seen already.

For ease of exposition, we will assume that the S-groupoid is "reduced," that is, has a single object, which makes it essentially just a simplicial group. Suppose therefore that $\mathbb{G}$ is a simplicial group. We want, in some sense, to look at the group of "source-trivial" n-simplices, but there is no clear sense of a single source for an n-simplex, at least for $n \geq 2$. Of course, if $g \in \mathbb{G}_1$, it is thought of as an arrow, and the source will be $d_1 g$, so the group of source trivial 1-simplices is simply $Ker\, d_1$ – but what about higher dimensions? One way to see a solution to this is to take an idea from the construction of the relative homotopy groups of a CW-complex, K, as used when constructing its fundamental crossed complex, $\pi(K)$. That construction used $\pi_n(K^{(n)}, K^{(n-1)}, x_0)$, so the elements can be represented by cellular singular n-cubes in K, all but one of whose faces is at the base point. We adapt that to a simplicial group context and look at, for each n, the subgroup, $N\mathbb{G}_n$, of $\mathbb{G}_n$ consisting of those $g \in \mathbb{G}_n$ having all but their d_0-face at the identity. In other words, $N\mathbb{G}_n = \bigcap_{k=1}^{n} Ker\, d_k$. This subgroup forms the n-dimensional part of a chain complex as the remaining face, d_0, restricts to give a "boundary" morphism, $\partial_n : N\mathbb{G}_n \to N\mathbb{G}_{n-1}$. It is easily checked that $\partial_{n-1} \partial_n$ is the trivial morphism, as $d_0 d_0 = d_0 d_1$ is a consequence of the simplicial identities that encode how face maps and degeneracy maps interact.

---

[118] Think of x as being an n-globe, with a boundary consisting of two $(n - 1)$-dimensional hemispheres, $s_{n-1}(x)$ and $t_{n-1}(x)$, which are thus in $\mathbb{G}_{n-1}$, meeting "at the equator" in two $(n - 2)$-dimensional hemispheres, $s_{n-2}(x)$ and $t_{n-2}(x)$, and so on.

[119] Chain complex models for these ∞-groupoidal structures have the advantage of many years of experience in handling chain complexes in algebraic topology and homological algebra, although those latter chain complexes are usually of abelian groups, which is not the case here. Some of the algebraic information encoded in them is thus easily accessible, although other parts of the structure may be less so.

Those simplicial identities also easily show that the image of ∂_n is a normal subgroup of $N\mathbb{G}_{n-1}$, so $N\mathbb{G}$ is not just a chain complex of possibly non-abelian groups, it has the extra property that boundaries form normal subgroups, and that means that we can form the homology, $H_*(N\mathbb{G})$, of such a complex, even though we are in a non-abelian setting.[120] The complex, $(N\mathbb{G}, \partial)$, is well known from simplicial homotopy theory. It is often called the *Moore complex*, as many of its properties were developed by John Moore in his seminar [52] in the late 1950s. This complex relates to very many of the invariants of homotopy types represented by the simplicial group, $\mathbb{G}$. As the form of $\mathbb{G}$ is sufficiently general to represent *any* connected homotopy type, and it *is* evidently a form of ∞-groupoid, the way in which the more classical invariants of a homotopy type depend *algebraically* on the Moore complex of $\mathbb{G}$ provides some more intuitive interpretation of the model and of the way well-known homotopy invariants of a homotopy type correspond to properties of an ∞-groupoid.

5.4 Appendix: Some Moore Complex Properties

To help in this process, we will look at how properties of $(N\mathbb{G}, \partial)$ relate to structures that we have already met here.

1. If $\mathbb{G}$ is a simplicial group, the homotopy groups of the underlying simplicial set of $\mathbb{G}$ are isomorphic to the homology groups of $(N\mathbb{G}, \partial)$. If $\mathbb{G} = \mathcal{G}(K)$ for a (reduced) simplicial set K, then $\pi_k(K) = \pi_{k-1}(\mathbb{G}) = H_{k-1}(N\mathbb{G}, \partial)$. (In particular, we note the shift in dimension so that, for instance, $\pi_0 \mathbb{G}$ is a group.) We thus have that the homotopy groups of the connected homotopy type represented by K are very simply related to the Moore complex of $\mathcal{G}(K)$ and hence to the "source-trivial" part of an ∞-groupoid model for the homotopy type.[121]

2. If $\mathbb{G}$ is a simplicial abelian group, then $N\mathbb{G}$ is just a "standard" chain complex, and it is relatively easy, given an arbitrary chain complex, C, to build a simplicial abelian group whose Moore complex is isomorphic to C. This gives the classical Dold–Kan theorem.[122]

---

[120] Although we have seen this just in the simplicial group setting, this fact encodes facts about "whiskering," which is the general form of conjugation in ∞-groupoid models and so is an expected feature.

[121] Of course, we should also look at the structure of homotopies through the eyes of this process. That is more tricky – and more revealing – but cannot be handled here.

[122] For an arbitrary simplicial group, $N\mathbb{G}$ has considerably more structure than being merely a chain complex of groups. (That structure is trivial in the abelian case.) That extra structure corresponds, in part, to the extra structure that an arbitrary homotopy type may have (Whitehead products, actions, etc.) and so corresponds to the *weak* ∞-groupoid structure,

3. If $\mathbb{G}$ is such that $N\mathbb{G}_n = 1$ for $n \geq 2$, then, of course, all the higher homotopy groups, $\pi_k(\mathbb{G})$, for $k \geq 1$, will be trivial.[123] In this case, the only nontrivial part of the Moore complex, $\partial_1 : N\mathbb{G}_1 \to N\mathbb{G}_0$, will be a crossed module. For a general $\mathbb{G}$, the 2-type model that it represents is given by the crossed module

$$\partial : \frac{N\mathbb{G}_1}{\partial N\mathbb{G}_2} \to N\mathbb{G}_0.$$

4. If $\mathbb{G}$ is such that $N\mathbb{G}_n = 1$ for $n \geq 3$, then the three remaining terms of $N\mathbb{G}$ form what is called a *2-crossed module*.[124] We mention this more technical model because the corresponding (globular) ∞-groupoid is a weak 3-groupoid in which, in general, the interchange law fails to hold in the underlying 2-groupoid. This can be seen in the 2-crossed module, as the Peiffer identity does not, in general, hold in the structure encoded by the bottom two terms.[125]

5. As a final example, we will look at what stops a general Moore complex, $N\mathbb{G}$, from being a crossed complex, as this is the simplicial version of what stops a general ∞-groupoid from being a strict ∞-groupoid.[126]

Let D_n be the subgroup of $\mathbb{G}_n$ generated by the degenerate elements.[127]

A Moore complex is a crossed complex if, and only if, for each $n \geq 2$, $N\mathbb{G}_n \cap D_n = 1$.

Turning this around, with a bit more work, it also provides a functor from simplicial groups to crossed complexes, but we will not explore that here. It also allows one to give a description of the property of "being a crossed complex" in terms of "having a unique thin filler for each horn"; see Ashley,

---

which is "weak" as against "strict." The analogue of the Dold–Kan theorem with regard to this extra structure was given by Carrasco and Cegarra [15].

[123] If $\mathbb{G} = \mathcal{G}(K)$, we have to remember that $\pi_0(\mathbb{G}) \cong \pi_1(K)$, and $\pi_1(\mathbb{G}) \cong \pi_2(K)$, so there is a shift in dimension, and so, generally, if $N\mathbb{G}_n = 1$ for $n \geq 2$, then $\mathbb{G}$ is a simplicial groupoid model of a 2-type *and not of a 1-type as one might think*.

[124] Cf. Conduché's paper [17] that was mentioned earlier.

[125] There is a pairing $N\mathbb{G}_1 \times N\mathbb{G}_1 \to N\mathbb{G}_2$ that "lifts" the difference of the two sides of the Peiffer rule, making it a boundary of a higher element. The weakening thus replaces an equality by a "reason" why the two sides of the equation are to be thought equivalent. The corresponding structure for higher values of n is related to the notion of hypercrossed complex, found in Carrasco and Cegarra [15].

[126] This is easily related to the simplicial group theoretic analogue of a J-complex, as mentioned earlier.

[127] These elements are sometimes called "thin" elements, in a sense that intuitively extends our earlier use of "thin" to describe certain homotopies.

[3].[128] Note that, for comparison, in the nerve of a groupoid, there is a unique filler for each horn.

5.5 Higher-Dimensional Combinatorial Homotopy

Poincaré's combinatorial approach to the fundamental group of a polytope deserves to be explored up to higher dimensions. Let us put this as follows.

Pick a notion of ∞-groupoid, denoted $\Pi_\infty(X)$ for a "space" X. We will work on the assumption here that $\Pi_\infty(X)$ is the singular complex of X, as that is a Kan complex, and we have been exploring the use of Kan complexes as models of ∞-groupoids.

Suppose now that X is a polytope/simplicial complex (or more generally a CW-complex, that is, with explicit *CW-structure). Can we use the extra structure of a complex to produce an ∞-groupoid, $P_\infty(X)$, presented in some sense by the combinatorial structure, smaller, we hope, than $\Pi_\infty(X)$, and an ∞-equivalence*

$$P_\infty(X) \leftrightarrow \Pi_\infty(X)?$$

As usual, we could rewrite this question with ∞ replaced by n.

This is almost Whitehead's algebraic homotopy problem in disguise, and a start on it is made in his papers [68, 69]. There has been a lot of progress on it in the work that we have already mentioned by Brown and Higgins [11] and of Baues [5], but *this work is using models that are not explicitly linked to ∞-groupoids*. There is a complication that the necessary weakness of the ∞-groupoids needs encoding in an *economical* way. It is not clear how to do this with some of the models[129] of ∞-groupoids. It would need a combinatorial ∞-groupoid theory to mirror combinatorial homotopy theory and to extend combinatorial group theory.

Other aspects of that overall combinatorial approach are worth mentioning. We have already pointed out the original link between combinatorial group theory and Whitehead's combinatorial homotopy. To some extent, working with simplicial sets or S-groupoids can be thought of as an extension of that link, but, although it is relatively easy to define step-by-step constructions of simplicial sets (or simplicial objects in algebraic categories) having desired properties, this is not the usual method. Likewise, if one has a naturally

---

[128] In some way, the "weakness" of a simplicial group modelling a given connected homotopy type is related to the nonuniqueness of such thin fillers and the size of the various subgroups, $N\mathbb{G}_n \cap D_n$, of the Moore complex terms.

[129] Which we have not discussed here.

occurring ∞-groupoid, the idea of working with a small "presentation" of that object, perhaps reflecting some geometry of how it occurs "in nature," is not one that has yet been explored to any great extent. Of course, 2-groups of *symmetries* have occurred in work on various properties in non-abelian cohomology, but there are relatively few studies of explicit presentations of such as yet, although the cohomology of 2-groups/crossed modules has begun to be applied to problems in algebraic and differential geometry and related areas of theoretical physics. We thus have less evidence of "2-groups as spaces," at least for that interpretation of our initial query. Crossed modules *do* yield interesting "classifying spaces," but the models used in their construction usually give "big" presentations, having a lot of cells, and are often obtained from a nerve construction followed by geometric realisation.

One block to constructing the "combinatorial homotopy" of an algebraically defined ∞-groupoid is the second objection of Grothendieck that we mentioned. Algebraically occurring n- or ∞-categories are often "globular" rather than simplicial. Likewise, group presentations are more globular in their "feeling," so what is needed is more a globular, abstract combinatorial homotopy rather than a simplicial one. In part, such a theory is given by the theory of computads/polygraphs introduced by Street [64] for $n = 2$, and later by Burroni [14]. These have wider application than merely presenting certain strict ∞-groupoids, as they seem very useful for studying rewriting systems and for operads rather than merely spaces and then questions of finite presentation.[130] They usually are used to present strict ∞-categories, but skilled use allows some use of explicit "weakness." Their usefulness and power come, in part, from being able to be *analogues* of "spaces." They lead to a "folk" homotopy structure (see [43]). The "cofibrant" objects are the free strict ∞-categories on polygraphs and so are in close analogy to CW-spaces, which are the cofibrant objects in one of the usual homotopy structures on the category of topological spaces. We are not here able to explore that line of development as deeply as it deserves.

5.6 ∞-Categories?

We will briefly need to mention ∞-categories in the next section. Grothendieck, in his letters to Breen and to Quilllen, to which we will be returning shortly, seems to make the assumption that not only would certain features of spaces be modelled by n-groupoids for any n including ∞ but that the right context would be to have some sort of n-equivalence of some

---

[130] Some idea of this application can be found in the work of Guiraud and Malbos; cf. [34].

Spaces as Infinity-Groupoids   311

corresponding n-categories. These n-categories, moreover, would be "weak" rather than "strict."

There are many different models for what are now usually called $(\infty, 1)$-categories.[131] These are the ∞-categories in which all 2-arrows are invertible "up to 3-arrows," and similarly, 3-arrows are invertible "up to 4-arrows," and so on. They are, at present, the best understood class of ∞-categories. The different models correspond in part to the models for ∞-groupoids that we have met.[132]

From the point of view that we have been exploring, the easiest potential approach to explaining what ∞-categories are is probably via S-categories, since we know that the categories both of spaces and of simplicial sets have S-category structures. Of course, Grothendieck's objection regarding simplicial as against globular approaches still applies here, but if S-groupoids are acceptable as ∞-groupoids, then probably S-categories should be considered as, at least, a working alternative to a globular version – and the preexisting extensive theory of homotopy in some of the key examples gives them some advantages. One caveat is that the really nice S-categories seem to have $\mathbb{A}(x, y)$ being a Kan complex.[133]

(A bit of terminology may help here in bridging between natural, intuitive ∞-category terms and their analogues for S-categories. If $\mathbb{A}$ is an S-category that we are thinking of as an ∞-category, it is often useful to think of, and to speak of, the 0-simplices in $\mathbb{A}(x, y)$ as being the $(1-)$morphisms/1-arrows of the infinity category from x to y. Then this continues with the elements of $\mathbb{A}(x, y)_1$ being the 2-arrows or even "homotopies," the 2-simplices as 3-arrows or "homotopies between homotopies," and, in general, of "higher homotopies." This terminology is in some ways not strictly accurate but can be useful in helping intuition.)

If the $\mathbb{A}(x, y)$ are Kan complexes, then the homotopies in $\mathbb{A}$ will be "reversible up to higher homotopies."[134]

---

[131] A useful survey of $(\infty, 1)$-categories is [8].
[132] More generally, an (∞, r)-category is one in which the n-cells are invertible "up to $(n + 1)$-cells" for $n > r$, so in this terminology, ∞-groupoids are $(\infty, 0)$-categories.
[133] There is a Quillen model category structure on the catgory $S-Cat$ of (small) S-categories in which these "locally Kan" S-categories are exactly the fibrant objects; see Bergner [7].
[134] To see this, let $\alpha: f_0 \to f_1$ be in $\mathbb{A}(x, y)_1$; then you can use it to build horns,

$$\begin{array}{ccc} & f_1 & \\ \alpha \nearrow & \searrow \beta & \\ f_0 \xrightarrow{s_0(f_0)} & & f_0 \end{array} \qquad \begin{array}{ccc} & f_0 & \\ \gamma \nearrow & \searrow \alpha & \\ f_1 \xrightarrow{s_0(f_1)} & & f_1 \end{array}$$

in $\mathbb{A}(x, y)$, shown with "full" arrows, which can be filled since $\mathbb{A}(x, y)$ is a Kan complex. This gives left and right inverses "up to homotopy" for α, namely, β and γ. You then show

There are directly simplicial models of $(\infty, 1)$-categories, namely, the quasi-categories, that were mentioned in Footnote 28. These are very like Kan complexes but have a slightly, but significantly, weaker filling requirement on horns. As the nerve of a category is a quasi-category, there is a fairly clear intuition as to how to develop quasi-categorical analogues of many categorical properties. This gives one the most developed versions of ∞-category theory.[135]

Globular approaches are also known – see Maltsiniotis [51] and Ara's thesis [1] – but are to some extent less developed than the simplicial ones as that can draw on the classical theory of simplicial homotopy, which can be an advantage. There are also treatments that are "model independent," that is, they do not choose between simplicial, globular, operadic, etc. models but look at the structures from the point of view of Quillen model category theory.

6 Higher Galois Theory and Locally Constant Stacks

When we started our discussion of spaces and groupoids, we mentioned three of the ways that Poincaré had of thinking of the fundamental group(oid). The first was as the algebraic structure of path classes in the space. The second was related to deck transformations and, via SGA1, to Galois theory. The last was a combinatorial group-theoretic approach given a simplicial or CW-complex structure on the space, which we briefly revisited in Section 5.5. We saw how Grothendieck's "letter to Quillen" in *Pursuing Stacks* sketched a higher-dimensional version of the path class idea, so what about deck transformations and Poincaré–Galois theory? How does this interpret for a Kan complex/simplicial model?

6.1 "Stacks"

We first need to explore briefly some notion of "stack" and its relationship with covering spaces. (Other related notions of stack as occur in geometric contexts are discussed in other chapters in this volume. A set of lectures giving a perspective linked to that taken here were given by Bertrand Toën [65, 66],

---

using higher horns that these are themselves homotopic, in a homotopy coherent variant of the classic argument for inverses in a group.

[135] The initial idea is due to Boardman and Vogt [9]; the simplicial version was developed by Cordier [18] and then with the author [19–22], and this basic theory was then pushed much further by Joyal [40] and Lurie [47, 48].

while there are many other treatments available perhaps more optimised for applications of stacks in other areas of mathematics.)

A covering space, $q: Y \to X$, is equivalently a locally constant sheaf.[136] A reader who is "new" to sheaves might initially think of them both as a continuously varying family of "spaces," namely, the fibres of q, indexed by the points of X, and also as a presheaf with nice gluing properties. A presheaf on X is just a functor from the opposite of the partially ordered set of open sets of the space, X, with inclusions as morphisms to, in our case here, the category of sets. Given q, as above, we get for each open set, U, in X, the set of maps, $s: U \to Y$, such that $qs(x) = x$ for all $x \in U$. We will denote this $F(U)$. For any open $V \subset U$, such "local sections," s, of course, restrict to local sections on V, so we get a presheaf, $F: Open(X)^{op} \to Sets$. That would work for any "space over X," $f: X' \to X$, as local sections of f restrict to subsets giving a presheaf on X. If one has an arbitrary presheaf, $F: Open(X)^{op} \to Sets$, on X, it need not come from a space over X. Presheaves of local sections have a special "gluing" property generalising that met in elementary calculus.

If we have a presheaf, F, of local sections of some map, $f: X' \to X$, then, for open sets, U_0, and U_1, in X and $U = U_0 \cup U_1$, and local sections s_i, $i = 0, 1$, each over the corresponding U_i, if $s_0(x) = s_1(x)$ for all $x \in U_0 \cap U_1$, then, clearly, we can define a function $s: U \to X'$ by $s(x) = s_0(x)$ if $x \in U_0$ and $s(x) = s_1(x)$ if $x \in U_1$, and this is a continuous section of f over U, because the two sets, U_0 and U_1, are open. To state the obvious, it is the unique local section over U that restricts to the given ones over the given two open sets.

This is the condition that the presheaf is a sheaf. It is also called the *descent condition*.

The local homeomorphism aspect of covering spaces means that, for small enough open sets, the restriction morphisms are, in fact, bijections, so the sets, $F(U)$, are "really all the same" for small enough U, and F is, as we said, "locally constant." (Notice that $F(X)$ can be empty, yet $F(U)$ may be nonempty for many open sets, U. "Local sections" may not be restrictions of "global" ones, but if a family of local sections is compatible over intersections of their domain, then it can be built up into local sections on the union of their domains.) We thus have that covering spaces "are" locally constant families of sets, and as sets "are" homotopy 0-types, they are "locally constant families of homotopy 0-types."

The idea of a sheaf as a special form of presheaf on a space generalises in a useful way to general functors, $F: \mathcal{C}^{op} \to Sets$, and corresponds to giving an

---

[136] We will not give a detailed introduction to sheaf theory here, but we direct the reader to Michel Vaquié's chapter "Sheaves and Functors of Points" in this volume.

abstract analogue of "open covering," or, more exactly, of a "covering family of maps." This leads on to the idea of a Grothendieck topology explored here in the chapter "Sheaves and Functors of Points" by Michel Vaquié. A category of sheaves on such a "site"[137] is called a *(Grothendieck) topos*.

It is easy to see how one can extend the idea of presheaves of sets to that of presheaves of other objects – you just replace *Sets* by the category of "other objects." To define stacks, of n-stacks, ∞-stacks, and so on, one approach is to start with presheaves of, perhaps, categories, groupoids, n-groupoids, or simplicial sets, depending on what "flavor" of stacky objects you need. We will not give a detailed treatment as it would take too long and would be quite technical, and it is not really necessary for the limited use we have for it. We will thus be a bit "vague"! Perhaps the best way here is to take "prestacks" to be presheaves of $(\infty, 1)$-categories on a site (which may be a topological space) or alternatively on $(\infty, 1)$-category objects internal to the corresponding topos.[138] We could also replace the basic category by an S-category, or an ∞-category, if the potential application merited that extra structure.[139] The second stage is then to find a suitable $(\infty, 1)$-categorical analogue of the gluing condition. Not surprisingly, that is often seen in terms of a notion of fibration with the ∞-stacks being the fibrant objects. If a (basic) simplicial model is taken for $(\infty, 1)$-categories, then the resulting fibrant objects, when considered as presheaves, take values in the subcategory consisting of Kan complexes.[140] The gluing condition then takes the form of local elements combining up to a form of coherent homotopy with higher homotopies linking the various levels. This leads to a simplicial set of "descent data" (relative to a covering of an object) which are to be compared with the given value of the presheaf on that object. As this is quite a bit more technically challenging than earlier sections, we will leave this deliberately vague.[141] (For more precision, the reader may want to look at the nLab pages on stacks descent and presheaves of simplicial sets, with subsequent following up of the references given there.)

6.2 And Their Pursuit

This allows for a very useful generalisation and leads us to a short quote from the letter of Grothendieck to Larry Breen, dated February 17, 1975 (in a

---

[137] "Site = category together with a Grothendieck topology."
[138] In a sense that we leave you to work out or look up.
[139] We here are deviating from having simply a classical space on which things are happening.
[140] A detailed treatment of this would require the introduction of model category theory and the discussion of the model category structures on categories of simplicial presheaves.
[141] But hope that some of the intuition gets through the vagueness!

translation [27], annotated for use here) on the "yoga of homotopy," which led on to the manuscript "Pursuing Stacks," [32]:

> In other terms, the constructions on a topos[142] X which one can make in terms of $(n-1)$-stacks which are locally constant, depend only of its "n-truncated prohomotopy[143] type" and define it. In the case where X is locally homotopically trivial in dim $\leq n$ and so defines an n-truncated ordinary homotopy type, one can interpret these last as an n-groupoid C_n, defined up to n-equivalence. In terms of these.[144]
>
> The $(n-1)$-stacks on X should be able to be identified with the n-functors from the category C_n to the category $((n-1)-Cat)$ of all $(n-1)$-categories.
>
> In the case $n = 1$, this is nothing other than the Poincaré theory of the classification of coverings of X in terms of the "fundamental groupoid" C of X. By extension, C_n merits the name of fundamental n-groupoid of X, which I propose to write $\Pi_n(X)$. Knowledge of this includes knowledge of the $\pi_i(X)$, $(0 \leq i \leq n$ and the Postnikoff invariants of all orders up to $H^{n+1}(\Pi_{n-1}(X), \pi_n)$

We will stop the quotation there as it does contain the point that we will be needing, but would also suggest that that letter and the following one [28], of July 17, 1975, which continues some of the same themes, are well worth looking at. The main point for us is that Grothendieck's conception of an ∞-groupoidal model for "spaces" not only includes the direct "equivalence" between "spaces" and some kind of weak ∞-groupoid but also continues and extends the covering space formulation started by Poincaré and continued by Grothendieck himself in SGA1. The idea is that if we have a locally constant $(n-1)$-stack, then the fibre over any point should be an $(n-1)$-groupoid, and the automorphisms of that fibre should give an n-groupoid. The stack would then correspond to a morphism of n-groupoids from $\Pi_n(X)$ to that automorphism object, that is, a representation of $\Pi_n(X)$. Doing a small reality check on this possibility for low values of n, for the case $n = 1$, a 0-groupoid is a set, and the automorphisms of a set form a group, thus a 1-groupoid (with one object); for $n = 2$, a 1-groupoid has an automorphism gadget that is a 2-groupoid, so that fits, and for $n = \infty$, the automorphism gadget of a Kan complex (considered as model for an ∞-groupoid) is a simplicial group, and that can also be considered to be an ∞-groupoid.[145]

---

[142] For more on the idea of a topos, see other chapters in this collection. We think of this just as being the category of sheaves on a (topological) space or on a site. Remember that Grothendieck was mainly interested in applications to problems in algebraic geometry.

[143] See later in the text.

[144] Two earlier forms of the idea were explored in the letter but are more technically stated and so will not be included here.

[145] In fact, the classical theory of twisted cartesian products in simplicial homotopy theory can be thought of as being one simple form of the correspondence that Grothendieck is talking about.

We need, however, to give some extra "notes" on various points in the quotation. These by necessity are slightly more technical but are only intended to make the quotation slightly easier to approach for the more general reader.

1. The idea of a pro-homotopy type is that of a set of interrelated approximations to a general homotopy type. Such things are needed, for instance, for handling the more general objects found in algebraic geometry. Here Grothendieck is slightly simplifying things, but when he talks of "locally homotopically trivial," *that* is the analogue of "locally contractible," so then X looks a bit like a CW-space and the "pro-homotopy type" simplifies to being a "homotopy type." This is relevant for the question: *What sort of spaces are we considering?* For spaces and concepts of spaces (general topological spaces, schemes, topoi, etc.) that are more general than CW-spaces, probing the "space" even via "points" (i.e., singular 0-simplices) is very problematic. There may be very few points and very few singular simplices, hence the approach via paths and their generalisations will not be adequate. The pro-homotopy type can lead to a pro-∞-groupoid, in some sense, regardless of whether there are enough points.

2. An important point to note once again is that this sketch by Grothendieck also needs a working theory of ∞-categories, or rather of n-categories, n-functors, "categorical" n-equivalences, and so on, for all n, including $n = \infty$. Such a theory existed for small values of n at the time the letter was written but has involved much effort to formulate it precisely[146] for $n = \infty$.

Given that, we can restate Grothendieck's idea as a form of higher Poincaré–Galois theory, namely, there should be an "∞-equivalence":

locally constant stacks of ∞-groupoids on X $\leftrightarrow$ representations of $\Pi_\infty(X)$, the fundamental ∞-groupoid of X.

Such a "Galois theorem" is, then, the ∞-version of a hierarchy of results for n-groupoids, which looks as follows:

- representations of the fundamental 1-groupoid classify coverings;
- representations of the fundamental 2-groupoid classify locally constant 1-stacks (in particular, connected 1-stacks, usually called *gerbes*);
- (for subsequent ns);

---

[146] References for this include Maltsiniotis [51], Ara's thesis [1] and, for a simplicially based theory using *quasi-categories*, Joyal [39, 40], Lurie [47, 48], and below, here. Hoyois [36] has now completed a detailed quasi-categorical attack on this aspect of Grothendieck's ideas.

and then representations of the fundamental ∞-groupoid classify locally constant ∞-stacks (in particular, the higher gerbes, as mentioned elsewhere in this collection).

7 Concluding Discussion

It seems a good idea to try to bring together some of the themes we have been following in an attempt to answer some of the questions with which we started this chapter. The title of the chapter suggested that some aspects of "spaces" could be encoded by ∞-groupoids. We saw that probing a space first by paths and then by higher-dimensional analogues allowed an algebraic/combinatorial model of the space to be built. For both historical reasons and expositional purposes, we looked in detail only at the structures encoded this way using the simplices, Δ^n, but we could have used n-globes or n-cubes.[147] So "what aspects?" initially has to be limited to those features that are available via probing with some test models. This works best with spaces that are sufficiently "locally homotopically trivial," and thus *spaces* ends up being limited in meaning to being "topological spaces" and, more often than not, "CW-spaces," as those are constructible from the usual test objects. Luckily, many spaces[148] occurring in geometric and theoretical physical contexts are CW-spaces, so this gives a large degree of applicability to the resulting theory.[149] It also provides the simple answer to "what spaces?."

The relationship between (topological) spaces and ∞-groupoids is symbiotic. The aspects of a space that are coded up in its singular complex are "∞-groupoidal" and as such provide a (classical) somewhat combinatorial model for the space. Applying algebraic constructions allows one to mix the combinatorial structure with suitable algebra and extract useful and usable information. If no CW-structure is available, then things are more tricky, but if there is one, then useful information of an ∞-categorical nature can be obtained with much smaller simpler models, but often at the expense of a loss of detailed structure. Mixing Whitehead's and Grothendieck's viewpoints,

---

[147] These two come with some advantages from the intuitive viewpoint of having a simpler compositional structure. Simplices, however, lead to simplicial sets, and their very well-understood homotopy theory makes them a natural first choice, at least at this point in time.

[148] But not all.

[149] "Spaces" by themselves are unable to handle some important situations. Sometimes objects that one expects to be modelled by points of a space have naturally symmetries, which does not accord well with just a directly spatial model. Such things as these "orbifolds," although not "directly spatial," do correspond to certain forms of stack and so fit into the overall ∞-groupoid model quite easily.

∞-groupoids are a good first step to analysing more of the structure of a spatial homotopy type.[150]

In many parts of mathematics, there are naturally occurring objects that look like ∞-groupoids or one of their avatars, such as crossed complexes, crossed n-cubes, or chain complexes, descending from the lofty heights of "∞-groupoidheim" to live among us by strictifying or nullifying some of the structure from the general case. Typically, such structures arise because they enhance the notion of "identity" (of sets) into that of "identification."[151] By this we mean that when there is some reason to "identify" two situations in mathematics, it is often, nay, *nearly always*, beneficial to record the collection of "reasons" for so doing (cf. page 282 for some brief development of this idea[152]). This leads to a quite constructive form[153] of mathematics. As this often gives a type of ∞-groupoid, the "spaces as ∞-groupoids" analogy can be turned around to provide a collection of spatial insights and tools in many other settings in mathematics.

Is the spatial intuition, thus being invoked, a good one to use, or is it too constraining or, alternatively, too wide?

My own feeling on this is that only time will tell. Some of the ways of thought involved seem quite difficult to handle in some instances. Even in the theory of group presentations, it is fair to ask what the module of identities of a presentation tells one about the group itself, yet that is, perhaps, still not clear. Working out higher invariants of situations like that does tell one something, but it is not always clear how to use that.

References

[1] D. Ara, *Sur les ∞-groupoïdes de Grothendieck et une variante ∞-catégorique*, PhD thesis, Université Paris 7 (2010)

[2] D. Ara and G. Maltsiniotis, *The homotopy type of the ∞-category associated to a simplicial complex* (2015), arXiv:1503.02720

[3] N. Ashley, *Simplicial T-complexes: A non-Abelian version of a theorem of Dold-Kan*, Diss. Math. 165 (1989) 11–58

---

[150] Using n-stacks of such things allows interpretation of non-abelian forms of cohomology and a transfer of the ∞-groupoid technology to other types of space.

[151] This situation is even more explicit in homotopy type theory, since that theory is built around the idea of multiple identification.

[152] And see the short article [59] for a fairly informal discussion of some of its consequences.

[153] In fact, something along these lines was pointed out by Mike Shulman in a discussion in the *n-cat café* about slogans in category theory. He mentioned that Errett Bishop *defined* a set as follows: *a set is defined by describing exactly what must be done in order to construct an element of the set and what must be done in order to show that two elements are equal*. In other words, "equality" as a structure needs to be taken seriously!

[4] H. J. Baues, *Algebraic Homotopy*, Cambridge Studies in Advanced Mathematics 15, Cambridge University Press (1989)
[5] H. J. Baues, *Combinatorial Homotopy and 4-Dimensional Complexes*, de Gruyter Expositions in Mathematics 2, Walter de Gruyter (1991)
[6] H. J. Baues, *Homotopy types*, in I. M. James, ed., *Handbook of Algebraic Topology*, Elsevier (1995) 1–72
[7] J. E. Bergner, *A model category structure on the category of simplicial categories*, Trans. Amer. Math. Soc. 359 (2007) 2043–58
[8] J. E. Bergner, *A survey of $(\infty, 1)$-categories*, in *Towards Higher Categories*, The IMA Volumes in Mathematics and Its Applications 152, Springer (2010) 69–83
[9] J. M. Boardman and R. M. Vogt, *Homotopy Invariant Algebraic Structures on Topological Spaces*, Lecture Notes in Mathematics 347, Springer (1973)
[10] R. Brown and P. J. Higgins, *The equivalence of ∞-groupoids and crossed complexes*, Cahiers Top. Géom. Diff. 22 (1981) 371–86
[11] R. Brown, P. J. Higgins, and R. Sivera, *Nonabelian Algebraic Topology: Filtered Spaces, Crossed Complexes, Cubical Homotopy Groupoids*, EMS Tracts in Mathematics 15, European Mathematical Society (2010)
[12] R. Brown and C. Spencer, *G-groupoids, crossed modules and the fundamental groupoid of a topological group*, Proc. Kon. Ned. Akad. v. Wet 79 (1976) 296–302
[13] M. Bullejos, A. M. Cegarra, and J. Duskin, *On cat^n-groups and homotopy types*, J. Pure Appl. Alg. 86 (1993) 135–54
[14] A. Burroni, *Higher-dimensional word problems with applications to equational logic*, Theor. Comput. Sci. 115 (1993) 43–62
[15] P. Carrasco and A. M. Cegarra, *Group-theoretic algebraic models for homotopy types*, J. Pure Appl. Alg. 75 (1991) 195–235
[16] E. Cheng and A. Lauda, *Higher-dimensional categories: An illustrated guide book*, Preprint (2004)
[17] D. Conduché, *Modules croisés généralisés de longueur 2*, J. Pure Appl. Alg. 34 (1984) 155–78
[18] J.-M. Cordier, *Sur la notion de diagramme homotopiquement cohérent*, Cahiers de Top. Géom. Diff. 23 (1982) 93–112
[19] J.-M. Cordier and T. Porter, *Vogt's theorem on categories of homotopy coherent diagrams*, Math. Proc. Cambridge Philos. Soc. 100 (1986) 65–90
[20] J.-M. Cordier and T. Porter, *Maps between homotopy coherent diagrams*, Top. Appl. 28 (1988) 255–75
[21] J.-M. Cordier and T. Porter, *Fibrant diagrams, rectifications and a construction of Loday*, J. Pure. Appl. Alg. 67 (1990) 111–24
[22] J.-M. Cordier and T. Porter, *Homotopy coherent category theory*, Trans. Amer. Math. Soc. 349 (1997) 1–54
[23] E. Curtis, *Simplicial homotopy theory*, Adv. Math. (1971) 107–209
[24] R. Douady and A. Douady, *Algèbre et Théories Galoisiennes*, 2 vols., CEDIC (1979)
[25] W. G. Dwyer and D. M. Kan, *Homotopy theory and simplicial groupoids*, Nederl. Akad. Wetensch. Indag. Math. 46 (1984) 379–85
[26] E. M. Friedlander, *The etale homotopy theory of a geometric fibration*, Manuscripta Math. 10 (1973) 209–44

[27] A. Grothendieck, *Letter 1 from Grothendieck to Larry Breen, (17/2/1975)*, Grothendieck circle (1975), http://webusers.imj-prg.fr/~leila.schneps/grothendieckcircle/Letters/breen1.html
[28] A. Grothendieck, *Letter 2 from Grothendieck to Larry Breen, (17/7/1975)*, Grothendieck circle (1975), http://webusers.imj-prg.fr/~leila.schneps/grothendieckcircle/Letters/breen2.html
[29] A. Grothendieck, *Letter 3 from Grothendieck to Larry Breen, (1975)*, Grothendieck circle (1975), http://webusers.imj-prg.fr/~leila.schneps/grothendieckcircle/Letters/breen3.html
[30] A. Grothendieck, *Letter from Grothendieck to Dan Quillen, (19/2/1983)*, in *Pursuing Stacks*, Manuscript (1983) 12–17 (1983)
[31] A. Grothendieck, *Letter to Tim Porter (28/06/83)*, Doc. Math. (in press)
[32] A. Grothendieck, *Pursuing Stacks*, Manuscript (1983)
[33] A. Grothendieck, *Revêtements étales et groupe fondamental (SGA1)*, Doc. Math. 3 (2003). Directed by A. Grothendieck, with two papers by M. Raynaud. Updated and annotated reprint of the 1971 original from Lecture Notes in Mathematics 224, Springer
[34] Y. Guiraud and P. Malbos, *Higher-dimensional categories with finite derivation type*, Theory Appl. Cat. 22 (2009) 420–78
[35] A. Hatcher, *Algebraic Topology*, Cambridge University Press (2002)
[36] M. Hoyois, *Higher Galois theory*, Preprint, arXiv:1506.07155v3.pdf
[37] I. M. James, *History of Topology*, Elsevier (1999)
[38] A. Joyal, *Lettre d'Andé Joyal à Alexandre Grothendieck*, G. Maltsiniotis (ed.) (1984), https://webusers.imj-prg.fr/~georges.maltsiniotis/ps/lettreJoyal.pdf
[39] A. Joyal, *Quasi-categories and Kan complexes*, J. Pure Appl. Alg. 175 (2002) 207–22
[40] A. Joyal, *The theory of quasi-categories and its applications*, in C. Casacuberta and J. Kock, eds., *Advanced Course on Simplicial Methods in Higher Categories*, *i*-Math 2, Centre de Recerca Matemàtica (2008), www.crm.cat/HigherCategories/hc2.pdf
[41] R. Krömer, *The set of paths in a space and its algebraic structure. A historical account*, Ann. Math. Fac. Sci. Toulouse 22 (2013) 915–68
[42] S. Lack, *A 2-categories companion*, in J. Baez and P. May, eds., *Towards Higher Categories*, Springer (2010) 105–91
[43] Y. Lafont, F. Métayer, and K. Worytkiewicz, *A folk model structure on omega-cat*, Adv. Math. 224 (2010) 1183–1231
[44] T. Leinster, *Basic bicategories*, Preprint, arXiv:9810017
[45] T. Leinster, *Higher Operads, Higher Categories*, London Mathematical Society Lecture Note Series 298, Cambridge University Press (2004)
[46] J.-L. Loday, *Spaces with finitely many homotopy groups*, J. Pure Appl. Alg. 24 (1982) 179–202
[47] J. Lurie, *Higher Topos Theory*, Annals of Mathematics Studies 170, Princeton University Press (2009), arXiv:0608040
[48] J. Lurie, *Higher algebra*, Unpublished manuscript (2011), http://www.math.harvard.edu/~lurie/papers/higheralgebra.pdf
[49] S. Mac Lane and J. H. C. Whitehead, *On the 3-type of a complex*, Proc. Nat. Acad. Sci. U.S.A. 36 (1950) 41–48

[50] G. Maltsiniotis, *Infini groupoïdes non stricts, d'après Grothendieck*, Preprint (2007)
[51] G. Maltsiniotis, *Grothendieck ∞-groupoids, and still another definition of ∞-categories*, Preprint, arXiv:1009.2331
[52] J. C. Moore, *Seminar in Algebraic Homotopy*, Princeton University (1956)
[53] T. Nikolaus, *Algebraic models for higher categories*, Indag. Math. (N.S.) 21 (2011) 52–75
[54] nLab team, *nLab*, http://ncatlab.org/nlab/show/HomePage
[55] R. Peiffer, *Über Identitäten zwischen Relationen*, Math. Ann. 121 (1949) 67–99
[56] H. Poincaré, *Analysis situs*, J. l'École Polytechnique (1895) 1–121
[57] T. Porter, *Letter to Grothendieck (16/06/1983)*, in press
[58] T. Porter, *n-types of simplicial groups and crossed n-cubes*, Topology 32 (1993) 5–24
[59] T. Porter, *Variations on a theme of homotopy*, in Annales de la faculté des sciences de Toulouse Mathématiques 22, Université Paul Sabatier (2013) 1045–89
[60] D. G. Quillen, *Homotopical Algebra*, in Lecture Notes in Mathematics 53, Springer (1967)
[61] K. Reidemeister, *Über Identitäten von Relationen*, Abh. Math. Sem. Hamburg 16 (1949) 114–18
[62] K. S. Sarkaria, *The topological work of Henri Poincaré*, in I. M. James, ed., History of Topology [37], Elsevier (1999) 123–39
[63] H. X. Sinh, *Gr-catégories*, Thèse d'État, Paris VII (1975)
[64] R. Street, *Limits indexed by category-valued 2-functors*, J. Pure Appl. Alg. 8 (1976) 149–81
[65] B. Toën, *Notes on non-Abelian cohomology I: Stacks and non-Abelian cohomology*, Lectures in MSRI (January 2002), https://perso.math.univ-toulouse.fr/btoen/files/2015/02/msri2002.pdf
[66] B. Toën, *Notes on non-Abelian cohomology II: Affine stacks, schematic homotopy types and Hodge theory*, Lectures in MSRI (January 2002), https://perso.math.univ-toulouse.fr/btoen/files/2015/02/msri2002-2.pdf
[67] J. H. C. Whitehead, *On adding relations to homotopy groups*, Ann. Math. 409–28
[68] J. H. C. Whitehead, *Combinatorial homotopy I*, Bull. Amer. Math. Soc. 55 (1949) 213–45
[69] J. H. C. Whitehead, *Combinatorial homotopy II*, Bull. Amer. Math. Soc. 55 (1949) 453–96
[70] J. H. C. Whitehead, *Algebraic homotopy theory*, Proc. Int. Cong. Math. Harvard 2 (1950) 354–57

Timothy Porter
Emeritus Professor, University of Wales, Bangor
t.porter.maths@gmail.com

6
Homotopy Type Theory: The Logic of Space

Michael Shulman*

Contents

1	Introduction to Synthetic Spaces	322
2	Type Theory	325
3	Toward Synthetic Topology	354
4	Homotopy Type Theory	367
5	Cohesive Homotopy Type Theory	388
6	Conclusion	393
References		398

1 Introduction to Synthetic Spaces

There are so many different notions of "space" (topological spaces, manifolds, schemes, stacks, and so on, as discussed in various other chapters of this book and its companion volume [2]) that one might despair of finding any common thread tying them together. However, one property shared by many notions of space is that they can be "background" structure. For instance, many kinds of algebraic objects, such as groups, rings, lattices, and boolean algebras, often

---

* This material is based on research sponsored by the US Air Force Research Laboratory under agreement FA9550-15-1-0053. The US government is authorized to reproduce and distribute reprints for governmental purposes notwithstanding any copyright notation thereon. The views and conclusions contained herein are those of the author and should not be interpreted as necessarily representing the official policies or endorsements, either expressed or implied, of the US Air Force Research Laboratory, the US government, or Carnegie Mellon University. I would like to thank Peter LeFanu Lumsdaine, Steve Awodey, Urs Schreiber, and Daniel Grayson as well as the editors of the volume for careful reading and helpful comments.

come with "extra space structure" that is respected by all their operations. In the case of groups, we have topological groups, Lie groups, sheaves of groups, ∞-groups, and so on.

For each kind of "spatial group," much of the theory of ordinary groups generalizes directly, with the "extra space structure" simply "coming along for the ride." Additionally, many naturally arising groups, such as the real and complex numbers, matrix groups, the p-adic numbers, profinite groups, and loop spaces, come "naturally" with spatial structure, and usually it would be ridiculous to study them without taking that spatial structure into account. On the other hand, "ordinary" groups are the special case of "spatial groups" whose spatial structure is trivial (e.g., discrete); but certain natural constructions on groups, such as the Pontryagin dual, profinite completion, or delooping, take us out of the discrete world. Thus, the theory of "groups with spatial structure" subsumes, and in a sense "completes," the study of ordinary groups. Similar statements can be made about many other kinds of algebraic structure.

With this in mind, the idea of *synthetic spaces* can be summarized as follows: if all objects in mathematics come naturally with spatial structure, then it is perverse to insist on defining them first in terms of bare sets, as is the official foundational position of most mathematicians, and only *later* equipping them with spatial structure. Instead, we can replace set theory with a different formal system whose basic objects are *spaces*. Since spaces admit most of the same constructions that sets do (products, disjoint unions, exponential objects, and so on), we can develop mathematics in such a system with very few changes to its outward appearance, but all the desired spatial structure will automatically be present and preserved. (In fact, as we will see in Section 2.8, this can even be regarded as an explanation of *why* many objects in mathematics come naturally with spatial structure.) Moreover, if our formal system is sufficiently general, then its objects will be interpretable as many *different* kinds of space; thus the same theorems about "groups" will apply to topological groups, Lie groups, sheaves of groups, ∞-groups, and so on.

A formal system with these properties is *Martin-Löf dependent type theory* [84, 85]. Originally conceived as a constructive foundation for mathematics where everything has "computational" content, it turns out also to admit "spatial" interpretations. This connection between constructivity/computability and topology/continuity goes back at least to Brouwer and was gradually developed by many people.[1] It was originally restricted to

---

[1] Escardó [40] cites "Kleene, Kreisel, Myhill/Shepherdson, Rice/Shapiro, Nerode, Scott, Ershov, Plotkin, Smyth, Abramsky, Vickers, Weihrauch and no doubt many others."

particular topologies on computational data types but eventually broadened to the realization that types could be interpreted as almost any kind of space. Categorically speaking, each "kind of space" forms a *topos* or something like it (a category that shares many properties of the category of sets), and the interpretation proceeds by way of constructing a "free topos" from the syntax of type theory (see Section 2).

There are other formal systems that can be interpreted in toposes, such as intuitionistic higher-order logic. Dependent type theory has some minor advantages of convenience, but more importantly, it has recently been recognized [11, 64] also to admit interpretations in *higher* toposes. More concretely, this means we can also interpret its basic objects as *homotopy spaces*, aka ∞-groupoids. The resulting collection of new axioms and techniques is known as *homotopy type theory* [112] or *univalent foundations* [117]. It includes *synthetic homotopy theory*, which studies homotopical objects "directly" without the need for topological spaces, simplicial sets, or any other combinatorial gadget. Like any new perspective on a subject, synthetic homotopy theory suggests new ways to attack problems; it has already led to new proofs of known theorems.

Classically, ∞-groupoids arose to prominence gradually, as repositories for the homotopy-theoretic information contained in a topological space; see Chapter 5 for an extensive survey. As we will see, however, the synthetic viewpoint emphasizes that this structure of a "homotopy space" is essentially *orthogonal* to other kinds of space structure, so that an object can be both "homotopical" and (for example) "topological" or "smooth" in unrelated ways. This sort of mixed structure is visible in many other chapters of the present volume, such as those about toposes (Chapter 4) and stacks (Chapter 8). It is also central to many applications, such as differential cohomology and gauge field theory (e.g., Schreiber's chapter in the companion volume [2]). Finally, it describes cleanly how topological and smooth spaces give rise to homotopy ones (see Section 5).

This chapter is intended as a brief introduction to the above ideas: type theory, synthetic spaces, and homotopy type theory. Of course, many details will be left out, but I hope to convey a flavor of the subject and leave the reader with some idea of what it means to talk about the *logic of space*.

It should be emphasized that homotopy type theory, in particular, is a very new subject. Many of its basic definitions are still in flux, and some of its expected fundamental theorems have not yet been completely proven. In general I will focus on what is *expected* to be true, in order to emphasize the possibilities opened up by these ideas; but I will endeavor not to lie and to include some remarks on the current state of the art as well.

I will begin in Section 2 with an introduction to type theory. Then in Sections 3 and 4, I will discuss its spatial and homotopical aspects, respectively, and some of their applications. Finally, in Section 5, I will briefly mention how these aspects are combined. For further reading, I recommend [7, 10, 90, 106] and [112].

2 Type Theory

2.1 On Syntax

Mathematicians often have a lot of difficulty understanding type theory (and the author was no exception). One reason is that the usual presentation of type theory is heavy on *syntax*, which most mathematicians are not used to thinking about. Thus, it is appropriate to begin with a few general remarks about syntax, what it is, and its role in type theory and in mathematics more generally.[2]

In general, *syntax* refers to a system of formal symbols of some sort, whereas *semantics* means the interpretation of those symbols as "things." In the language of category theory, we can generally think of syntax as describing a *free* (or *presented*) object and semantics as the morphisms out of that object determined by its universal property.

For instance, in group theory, we may write a sequence of equations such as

$$(gh)h^{-1} = g(hh^{-1}) = ge = g. \tag{2.1}$$

Where does this computation take place? One obvious answer is "in an arbitrary group." But another is "in the free group $F\langle g,h\rangle$ generated by two symbols g and h." Since the elements of $F\langle g,h\rangle$ are literally strings of symbols ("words") produced by multiplication and inversion from g and h, strings such as "$(gh)h^{-1}$" *are themselves* elements of $F\langle g,h\rangle$, and (2.1) holds as an equality between these elements, that is, a statement *in syntax*. Now if we have any other group G and two elements of it, there is a unique group homomorphism from $F\langle g,h\rangle$ to G sending the letters g and h to the chosen elements of G. This is the *semantics* of our syntax, and it carries the equation (2.1) in $F\langle g,h\rangle$ to the analogous equation in G. Such reasoning can be applied to arguments involving hypotheses, such as "if $g^2 = e$, then $g^4 = (g^2)^2 = e^2 = e$," by considering (in this case) the group $F\langle g \mid g^2 = e\rangle$ *presented* by one generator g and one equation $g^2 = e$. (A free group, of course, has a presentation with no equations.)

---

[2] The "algebraic" perspective I will present is only one of many valid ways to look at type theory. It has been developed by [39, 68, 108, 119], among others.

In other words, we can regard an argument such as (2.1) either as a "semantic" statement about "all groups" or as a "syntactic" statement about a particular free or presented group. The former is a consequence of the latter, by the universal property of free groups and presentations.

This may seem like mere playing with words,[3] and the reader may wonder how such a viewpoint could ever gain us anything. The reason is that often, we can say more about a free object than is expressed tautologically by its universal property.[4] Usually, this takes the form of an explicit and tractable construction of an object that is then *proven* to be free. Of course, *any* construction of a free object must be proven correct, but such a proof can range from tautological to highly nontrivial. The less trivial it is, the more potential benefit there is from working syntactically with the free object.

For instance, a "tautological" way to define $F\langle g,h \rangle$ is by "throwing in freely" the group operations of multiplication and inversion, obtaining formal "words" such as $(gg^{-1})(h^{-1}(hg))$, and then quotienting by an equivalence relation generated by the axioms of a group. The universal property of a free group is then essentially immediate. But a more interesting and useful construction of $F\langle g,h \rangle$ consists of "reduced words" in g, h, and their formal inverses (finite sequences in which no cancellation is possible), such as $ghg^{-1}g^{-1}hhgh^{-1}$, with multiplication by concatenation and cancellation. The proof that this yields a free group is not entirely trivial (indeed, even the definition of the group multiplication is not completely trivial); but once we know it, it can simplify our lives.

As a fairly banal example of such a simplification, recall that the *conjugation* of h by g is defined by $h^g = ghg^{-1}$. Here is a proof that conjugation by g is a group homomorphism:

$$h^g k^g = (ghg^{-1})(gkg^{-1}) = ghkg^{-1} = (hk)^g \qquad (2.2)$$

As straightforward as it is, this is not, technically, a complete proof from the usual axioms of a group. For that, we would have to choose parenthesizations and use the associativity and unit axioms explicitly:

$$h^g k^g = ((gh)g^{-1})((gk)g^{-1}) = (g(hg^{-1}))((gk)g^{-1}) = ((g(hg^{-1}))(gk))g^{-1}$$
$$= (g((hg^{-1})(gk)))g^{-1} = (g(h(g^{-1}(gk))))g^{-1} = (g(h((g^{-1}g)k)))g^{-1}$$
$$= (g(h(ek)))g^{-1} = (g(hk))g^{-1} = (hk)^g \qquad (2.3)$$

---

[3] No pun intended.
[4] This is dual to the familiar fact that studying a "classifying space" can yield insights about the objects it classifies – a classifying space being a representing object for a contravariant functor, while a free or presented object represents a covariant one.

Of course, this would be horrific, so no one ever does it. If mathematicians think about this sort of question at all, they usually call (2.2) an "acceptable abuse of notation." But with the above explicit description of free groups, we can make formal sense of (2.2) as a calculation in $F\langle g,h,k\rangle$, wherein ghg^{-1} and gkg^{-1} are specific elements whose product is $ghkg^{-1}$. Then we can extend this conclusion to every other group by freeness. Note that if we tried to do the same thing with the "tautological" presentation of a free group, we would be forced to write down (2.3) instead, so no simplification would result.

In general, there are several ways that a presentation of a free object might make our lives easier. One is if its elements are "canonical forms," as for free groups (e.g., $ghkg^{-1}$ is the canonical form of $((gh)g^{-1})((gk)g^{-1})$). This eliminates (or simplifies) the quotient by an equivalence relation required for "tautological" constructions. Often there is a "reduction" algorithm to compute canonical forms, making equality in the free object computationally decidable.

Another potential advantage is if we obtain a "version" of a free object that is *actually* simpler. For instance, it might be stricter than the one given by a tautological construction. This is particularly common in category theory and higher category theory, where it can be called a *coherence theorem*.

Finally, a particular construction of a free object might also be *psychologically* easier to work with, or at least suggest a different viewpoint that may lead to new insights. The best example of this is type theory itself: though it also offers the advantages of canonical forms and strictness (see Sections 3.3 and 4.6), arguably its most important benefit is a way of thinking.

2.2 Universes of Mathematics

What, then, *is* type theory?[5] Roughly speaking, it is a particularly convenient construction of free objects for the theory of *all of mathematics*. Just as a group presented by g, h and $gh = hg$ admits a unique homomorphism to any other group equipped with two commuting elements, type theory with certain structures presents "a universe of mathematics" with a unique "mathematics-homomorphism" to any other such universe of mathematics.

---

[5] Unfortunately, the phrase "type theory" has many different meanings. On one hand, type theory is a *discipline* lying at the boundary of mathematics and computer science. This discipline studies deductive systems that are themselves also known as type theories. But in the context of mathematical foundations, such as here, "type theory" generally refers to a particular subclass of these deductive systems, which are more precisely called *dependent type theories* (because they admit "dependent types"; see below). The type theory we are interested in here is also sometimes called "formal type theory" to distinguish it from "computational type theory," which is about assigning types to untyped computations (see, e.g., [3, 4]).

This discussion of "universes of mathematics" may sound odd; surely there is only one universe of mathematics? Well, yes, mathematics is a whole; but it has been known since the early 20th century that some formal systems, such as Zermelo–Fraenkel set theory, can encode almost all of mathematics. To first approximation, by a "universe of mathematics" I mean a model of a formal system in which mathematics can be encoded. Note that Gödel's incompleteness theorem ensures that any such system has many different models.[6] Thus, there are many "universes of mathematics" in this sense.

Often the incompleteness theorem is seen as a bug, but from our point of view it is actually a feature! We can make positive use of it by recognizing that certain mathematical structures, like notions of space, happen to form new universes of mathematics by themselves. In other words, starting from one universe of mathematics,[7] we can construct another universe whose objects are, from the point of view of our original universe, "spaces" of some sort. Thus, when a mathematician living in this new world constructs a bare function $A \to B$ between sets, the mathematician in the old world sees that it is in fact a *continuous* function between *spaces*.

This is admittedly a bit vague, so let me pass to a second approximation of what I mean by a "universe of mathematics": a *category*, or $(\infty, 1)$-category, with certain structure. Our starting universe is then the category of sets (or perhaps the $(\infty, 1)$-category of ∞-groupoids). Thus, type theory gives a way to construct free or presented objects in some *category of structured categories*. Such a free object is sometimes called the "syntactic category" or "classifying category" of the type theory. In the words of Scott [100]:

> [...] a category represents the "algebra of types," just as abstract rings give us the algebra of polynomials, originally understood to concern only integers or rationals.

Now the usual way of working "inside" a particular category is to write all arguments in diagrammatic language. For instance, if G is a group object in a category (such as a topological group in the category of topological spaces, or a Lie group in the category of smooth manifolds), then the argument analogous to (2.1) would be the commutativity of the following diagram:

---

[6] Specifically, it shows that any sufficiently powerful formal system contains statements that are neither provable nor disprovable. The *completeness* theorem then implies that there must be some models in which these statements are true and some in which they are false.

[7] We may regard the starting universe as the "true" one, but there is no formal justification for this. We will come back to this in Section 6.

$$G \times G \xrightarrow{1 \times \Delta} G \times G \times G \xrightarrow{1 \times 1 \times \text{inv}} G \times G \times G \xrightarrow{\text{mult} \times 1} G \times G$$

(with proj_1 down from $G \times G$ to G; $1 \times \text{id}$ from G to $G \times G$; $1 \times \text{mult}$ down to $G \times G$; mult to G; mult down from $G \times G$ to G; and 1 from G to G)

(2.4)

Categorically trained mathematicians become quite adept at translating calculations like (2.1) into diagrams like (2.4). However, objectively I think it is hard to deny the relative simplicity of (2.1) compared to (2.4). The benefits are magnified further when we include additional simplifications like those in (2.2).

Type theory allows us to use equations like (2.1) and (2.2) to prove things about all group objects in *all* categories. Its syntax involves elements with operations and equations, so we can speak and think as if we were talking about ordinary sets.[8] But it is nevertheless a description of a free category of a certain sort,[9] so that its theorems can be uniquely mapped into any other similar category. Thus, type theory supplies a different perspective on categories that is often more familiar and easier to work with.

To be a little more precise, the benefit here comes from the interplay between two modes of interacting with type theory. On one hand, we can define and study the formal system of type theory *inside* mathematics. This enables us to talk about its having multiple models, and hence functioning as a syntax for categories, as described above. But on the other hand, because type theory is sufficiently powerful to encode all of mathematics, we are also free to regard it as the "ambient foundation" for any mathematical theory. Most modern mathematicians implicitly assume set theory as a foundation, but for the most part type theory is just as good (and, as we will see in Section 4, it makes "new kinds of mathematics" possible as well). Of course, real-world mathematics is rarely "fully encoded" into *any* foundational system, but experience shows that

---

[8] Another approach to this problem, which enables us to *literally* talk about ordinary sets, is to speak about "generalized elements" of an object G (meaning arbitrary morphisms with codomain G) or, equivalently, to apply the Yoneda embedding. This works for structures defined using only limits, such as group objects, but it breaks down when colimits, images, exponentials, and so on come into play. It can be enhanced to deal with some such cases using "Kripke–Joyal semantics," but this is essentially equivalent to type theory.

[9] See Sections 2.5 and 4.6 for some caveats to this statement.

it is always possible in principle, and nowadays with computer proof assistants it is becoming more common and feasible to do explicitly.

The point, then, is that any theorem in "ordinary" mathematics can be encoded using the second "foundational" point of view, obtaining a derivation in the formal system of type theory; but then we can switch to the first "semantic" point of view and conclude that *that theorem* is actually true (suitably interpreted) in all categories with appropriate structure. In this way, *any* mathematical theorem is actually much more general than it appears.[10]

2.3 Types versus Sets

With those lengthy preliminaries out of the way, let us move on to what type theory actually looks like. If it is to describe the free "universe of mathematics," type theory should be a formal system into which mathematics can be encoded. The currently accepted formal system for encoding mathematics is Zermelo–Fraenkel set theory (ZFC), and mathematicians have a great deal of practice representing structures as sets. Thus it makes sense that the basic objects of type theory, called *types*, are very set-like – with one important difference.

In ZFC, an assertion of membership like "$x \in A$" is a statement about two previously given objects x and A, which might be true or false, and can be hypothesized, proven, or disproven. In other words, the universe of ZFC is a vast undifferentiated collection of things called "sets," with a relation called "membership" that can be applied to any two of them. By contrast, in type theory, the type to which an element belongs is "part of its nature," rather than something we can ask about and prove or disprove; two distinct types can never[11] share any elements in common. To emphasize this difference, we write $x : A$, rather than $x \in A$, to mean that x is an element of the type A.

This perspective on sets is like that of categorical or "structural" set theory, such as Lawvere's ETCS [72, 75], which axiomatizes the *category* of sets and functions. It contrasts with membership-based or "material" set theory such as ZFC, which axiomatizes the *class* of sets and its membership relation. The structural approach generalizes better when thinking of the basic objects as spaces rather than bare sets, since the spatial relationships between points are specified by an ambient space: it does not make sense to ask whether

---

[10] However, as we will see in Section 3.2, it requires some care on the side of ordinary mathematics – specifically, avoiding certain restrictive logical axioms – to maximize this resulting generality.

[11] As with almost any general statement about type theory, there are exceptions to this, but for the most part, it is true.

two points are "nearby" unless we have fixed some space in which they both reside.

In principle it may be possible to use a more ZFC-like formal system for at least some of the same purposes as type theory (see, e.g., [9]), but the connection to spaces would become rather more tenuous. Moreover, the structural perspective matches the usage of "sets" in most of mathematics. Outside the formal theory of ZFC, the primary place where one element can belong to more than one set, or where elements of distinct sets are compared, is when the given sets are *subsets* of some ambient set. This situation is encoded in type theory by a notion of "subset of A" that, like "element of A," is a basic notion not reducible to something like "set that happens to be a subset of A"; see Section 3.1.

While we are talking about ZFC and set theory, it is worth mentioning another reason type theory is often difficult for mathematicians. Any formal system for encoding mathematics, be it ZFC, ETCS, or type theory, must by its nature be careful about many things that mathematicians usually gloss over. Ordinary mathematical notation and writing is, technically speaking, ambiguous and full of gaps, trusting the human reader to draw "obvious" conclusions. But to give a mathematical *theory* of mathematics (and in particular, to prove things like "type theory presents a free structured category"), we have to remove all such ambiguity and fill in all the gaps. This causes the syntactic formulas of the formal system to appear quite verbose, and often barely comprehensible to a mathematician accustomed to informal mathematical language.

The important points are that this is true for *all* formal systems and that it should not bother us when doing ordinary mathematics. The process of "encoding" mathematics into a formal system such as ZFC, ETCS, or type theory looks somewhat different depending on which formal system is chosen, but it is generally well understood. In particular, no matter what formal system we choose, there is no need for its verbosity to infect ordinary mathematics; we remain free to "abuse notation" in the usual way.

I stress this point because one sometimes encounters a false impression that type theory requires "heavier syntax" than set-based mathematics or that it forbids "abuse of notation." This is probably partly because type theory is often presented in a very formal and syntactic way – perhaps because many type theorists are logicians or computer scientists – whereas most mathematicians' exposure to set theory has been fairly informal and intuitive. Moreover, the particular notations used in type theory are somewhat unfamiliar to mathematicians, and take some practice to learn to read correctly. But the syntax of type theory is *intrinsically* no heavier or unabusable than that of set theory. (Promoting a style of informal mathematics that matches the formal system of type theory was one of the explicit goals of [112].)

2.4 Judgments and the Classifying Category

Finally, we are ready to describe the syntax of type theory and how it generates a category (which we will call the *classifying category*; it is also called the *syntactic category* and the *category of contexts*). In Sections 4 and 5 we will be concerned with "homotopy" type theory, whose classifying category is an $(\infty, 1)$-category; but for simplicity, in Sections 2 and 3 we begin with so-called extensional type theory, whose classifying category is an ordinary 1-category. Most of what we say here will remain true in the homotopy case with only minor modifications.

Like the elements of a free group, the syntactic objects of type theory are "words" built out of operations. In a free group, there is only one sort of word, since a group involves only one collection of "things" (its elements). But since type theory presents a category with both objects and morphisms, it has at least two sorts of "words." Type theorists call a "sort of word" a *judgment form* and a particular word a *judgment*.

The first judgment form is a *type judgment*; it is written "B type" and pronounced "B is a type." Here B is a syntactic expression like $\mathbb{N} \times (\mathbb{R} + \mathbb{Q})$, in which $\times$ and $+$ are operations on types, formally analogous to the multiplication of elements represented by concatenation of words in a free group; we will come back to them in Section 2.5. The objects of the "classifying category" generated by a type theory are[12] the syntactic expressions B for which the judgment B type can be produced by the rules (i.e., operations) to be described in Section 2.5. For clarity, we will write $[\![B]\!]$ when B is regarded as an object of this category, and say that B *presents* the object $[\![B]\!]$.

The second judgment form is a *term judgment*, written "$b \colon B$." Here B is a syntactic expression for a type (i.e., we must also have "B type"). For instance, we might have $(3 \cdot 2 + 1, \mathrm{inr}(\frac{3}{4} - 17)) \colon \mathbb{N} \times (\mathbb{R} + \mathbb{Q})$. Here again, $\cdot, +, -, \mathrm{inr}$ and so on denote operations that will be described in Section 2.5. We pronounce $b \colon B$ as "b is an element of B" or "b is a point of B" or "b is a term of type B," emphasizing respectively the set-like, space-like, or syntactic character of B.

More generally, a term judgment can include a *context*, consisting of a list of variables, each with a specified type, that may occur in the term b. For instance, we might also write $(3x + 1, \mathrm{inr}(\frac{3}{4} - y)) \colon \mathbb{N} \times (\mathbb{R} + \mathbb{Q})$, which only makes sense

---

[12] Well, not exactly; see below.

in the context of $x\colon \mathbb{N}$ and $y\colon \mathbb{Q}$. The traditional notation in type theory is to write the context as a list of variables with their types, joined by commas, and separate it from the judgment with the symbol $\vdash$ (called a turnstile). Thus, the above judgment would be written

$$x\colon \mathbb{N}, y\colon \mathbb{Q} \vdash (3x + 1, \mathsf{inr}(\tfrac{3}{4} - y))\colon \mathbb{N} \times (\mathbb{R} + \mathbb{Q}).$$

Here the $\vdash$ is the "outer relation" that binds most loosely; then the commas on the left-hand side bind next most loosely, separating the (variable: type) pairs. Thus, for emphasis it could be bracketed as

$$((x\colon \mathbb{N}), (y\colon \mathbb{Q})) \vdash ((3x + 1, \mathsf{inr}(\tfrac{3}{4} - y))\colon (\mathbb{N} \times (\mathbb{R} + \mathbb{Q}))).$$

Often the Greek letter Γ denotes an arbitrary context, so that $\Gamma, x\colon A \vdash b\colon B$ (to be parsed as $(\Gamma, (x\colon A)) \vdash (b\colon B)$) means that in some arbitrary context together with a variable x of type A, we have a term b of type B.[13]

Term judgments $\Gamma \vdash a\colon A$ present *morphisms* in the classifying category. In the simplest case, Γ contains only one variable, such as $x\colon A \vdash b\colon B$, and this morphism $[\![b]\!]$ is from $[\![A]\!]$ to $[\![B]\!]$. For the general case $\Gamma \vdash a\colon A$, we have to modify our definition of the classifying category by taking its objects to be *contexts* rather than types, with our previous $[\![A]\!]$ corresponding to $[\![x\colon A]\!]$; then $\Gamma \vdash a\colon A$ presents a morphism from $[\![\Gamma]\!]$ to $[\![A]\!]$. For this reason, the classifying category is also known as the *category of contexts*; we denote it by **Ctx**.

We stipulate that $[\![\Gamma]\!]$ is a product of the types in Γ, so $[\![x\colon A, y\colon B]\!] \cong [\![A]\!] \times [\![B]\!]$ and so on. (In particular, the empty context yields a terminal object $[\![\ ]\!]$.) Thus, for instance, $x\colon \mathbb{R}, y\colon \mathbb{R} \vdash xy\colon \mathbb{R}$ yields the multiplication map $\mathbb{R} \times \mathbb{R} \to \mathbb{R}$. The universal property of products implies that for contexts Γ and Δ, a morphism in **Ctx** from $[\![\Gamma]\!]$ to a general context $[\![\Delta]\!]$ must consist of a tuple of term judgments $\Gamma \vdash b_i\colon B_i$ for all variables $y_i\colon B_i$ occurring in Δ. If we also have $\Delta \vdash c\colon C$, we get another term judgment denoted

$$\Gamma \vdash c[b_1/y_1, \ldots, b_m/y_m]\colon C$$

by *substituting* each b_i for y_i in c; this presents the composite $[\![\Gamma]\!] \to [\![\Delta]\!] \to [\![C]\!]$. For instance, we have a morphism from $[\![x\colon \mathbb{R}]\!]$ to $[\![z\colon \mathbb{R}, w\colon \mathbb{R}]\!]$ defined by the terms $x\colon \mathbb{R} \vdash (x-1)\colon \mathbb{R}$ and $x\colon \mathbb{R} \vdash (x+1)\colon \mathbb{R}$; substituting it into $z\colon \mathbb{R}, w\colon \mathbb{R} \vdash zw\colon \mathbb{R}$ gives $x\colon \mathbb{R} \vdash (x-1)(x+1)\colon \mathbb{R}$. That is,

$$(zw)\big[(x-1)/z, (x+1)/w\big] = (x-1)(x+1).$$

---

[13] Technically, Γ, b, A, and B here are "metavariables," not to be confused with "variables," such as x in a context. We will come back to this in Section 2.5.

So far, we have described *simple* type theory. Next we allow type judgments "B type" to have a context as well, making B into a *dependent type* or *type family*. Intuitively, a dependent type "$\Gamma \vdash B$ type" presents an object of the *slice category* $\mathbf{Ctx}_{/[\![\Gamma]\!]}$, that is, an object with a morphism to $[\![\Gamma]\!]$. We think of the "fiber" over a point of $[\![\Gamma]\!]$ as the instance of B corresponding to that point.

For instance, in informal mathematics, we might speak of "an arbitrary finite cyclic group C_n," for $n\colon \mathbb{N}_+$. In type theory, this becomes $n\colon \mathbb{N}_+ \vdash C_n$ type, corresponding categorically to $\coprod_{n\in\mathbb{N}_+} C_n$ with its projection to $\mathbb{N}_+$. Topologically, this is a *bundle* over $\mathbb{N}_+$, with the C_n as its fibers. Working in a category in the style of (2.4) requires manually translating from "arbitrary objects" to bundles; this is one of the least convenient aspects of categorical set theories and of the traditional way of doing mathematics in a topos. The ability to talk directly about families of types and have them interpreted automatically as bundles is one of the most significant advantages of type theory.

A crucial fact is that *substitution into a dependent type* presents the *pullback* functor between slice categories. For instance, we have a judgment $\vdash 3\colon \mathbb{N}_+$ with no variables, yielding a context morphism from the terminal object $[\![\,]\!]$ to $[\![\mathbb{N}_+]\!]$. Substitution into $n\colon \mathbb{N}_+ \vdash C_n$ type yields the nondependent type $\vdash C_3$ type, which is the pullback of $\coprod_{n\in\mathbb{N}_+} C_n$ along the inclusion $3\colon [\![\,]\!] \to [\![\mathbb{N}_+]\!]$:

$$\begin{array}{ccc} C_3 & \longrightarrow & \coprod_{n\in\mathbb{N}_+} C_n \\ \downarrow & \lrcorner & \downarrow \\ 1 & \longrightarrow & \mathbb{N}_+ \end{array}$$

As an even simpler example, if $\vdash B$ type is a nondependent type, we can substitute it along the unique context morphism from any Γ to the empty context, yielding a "trivially dependent type" $\Gamma \vdash B$ type. This presents the pullback of $[\![B]\!]$ to the slice over $[\![\Gamma]\!]$, that is, the projection $[\![\Gamma]\!] \times [\![B]\!] \to [\![\Gamma]\!]$ (a "trivial bundle"):

$$\begin{array}{ccc} [\![\Gamma]\!] \times [\![B]\!] & \longrightarrow & [\![B]\!] \\ \downarrow & \lrcorner & \downarrow \\ [\![\Gamma]\!] & \longrightarrow & [\![\,]\!] = 1 \end{array}$$

With dependent types, we can allow the type B in a term judgment $\Gamma \vdash b\colon B$ to also depend on Γ. For instance, the generators of the cyclic groups form a term judgment $n\colon \mathbb{N}_+ \vdash g_n\colon C_n$. Such a judgment $\Gamma \vdash b\colon B$ represents a *section* of the projection represented by the dependent type $\Gamma \vdash B$ type: we "select one point in each fiber." This includes the nondependent case because morphisms $[\![\Gamma]\!] \to [\![B]\!]$ are equivalent to sections of the projection $[\![\Gamma]\!] \times [\![B]\!] \to [\![\Gamma]\!]$.

An example that will be central to the story of this entire chapter is the *diagonal* map $\Delta_{[\![A]\!]} : [\![A]\!] \to [\![A]\!] \times [\![A]\!]$. We can regard this as an object of the slice category $\mathbf{Ctx}_{/([\![A]\!] \times [\![A]\!])}$, or equivalently $\mathbf{Ctx}_{/[\![x:A,\,y:A]\!]}$; it is then presented by a dependent type called the *equality type* or *identity type*, written

$$x : A, y : A \vdash (x = y) \text{ type} \quad \text{or} \quad x : A, y : A \vdash \mathsf{Id}(x, y) \text{ type}.$$

We will explain this type in more detail in Section 4.1. For the moment, we observe that it reduces equalities of terms to existence of terms. For instance, given $\Gamma \vdash a : A$ and $\Gamma \vdash b : A$ representing morphisms $[\![a]\!], [\![b]\!] : [\![\Gamma]\!] \to [\![A]\!]$, substituting them into the equality type we get a dependent type $\Gamma \vdash (a = b)$ type that presents the pullback of $\Delta_{[\![A]\!]}$ along $([\![a]\!], [\![b]\!]) : [\![\Gamma]\!] \to [\![A]\!] \times [\![A]\!]$, or equivalently the *equalizer* of $[\![a]\!]$ and $[\![b]\!]$. Thus, a judgment $\Gamma \vdash e : a = b$ says that this equalizer has a section, or equivalently that $[\![a]\!] = [\![b]\!]$. So our type and term judgments also suffice to present *equality* of morphisms.

To be precise, in the presence of dependent types we extend our previous definition of the classifying category $\mathbf{Ctx}$ as follows. First, we also allow the types in a context to depend on the variables occurring earlier in the same context. For instance, we can form the context $(n : \mathbb{N}_+, x : C_n)$, and then in this context write $n : \mathbb{N}_+, x : C_n \vdash x^2 : C_n$ for the operation that squares an arbitrary element of an arbitrary cyclic group. Categorically, if $\Gamma \vdash B$ type presents an object of the slice over $[\![\Gamma]\!]$, that is, a morphism with codomain $[\![\Gamma]\!]$, then the extended context $[\![\Gamma, x : B]\!]$ is the *domain* of this morphism. This reduces to our previous $[\![\Gamma, x : B]\!] = [\![\Gamma]\!] \times [\![B]\!]$ if B is nondependent.

Second, we take the objects of $\mathbf{Ctx}$ to be contexts in this generalized sense, and a morphism from $[\![\Gamma]\!]$ to $[\![\Delta]\!]$ to consist of term judgments for all $1 \leq i \leq m$:

$$\Gamma \vdash b_i : B_i[b_1/y_1, \ldots, b_{i-1}/y_{i-1}],$$

where $\Delta = (y_1 : B_1, y_2 : B_2, \ldots, y_m : B_m)$, with y_j potentially occurring in B_i for $j < i$. That is, we first give $\Gamma \vdash b_1 : B_1$, presenting a morphism

$$[\![b_1]\!] : [\![\Gamma]\!] \to [\![B_1]\!].$$

Then we substitute b_1 for y_1 in B_2, obtaining a type $\Gamma \vdash B_2[b_1/y_1]$ and a corresponding extended context that presents the pullback

$$\begin{array}{ccc} [\![\Gamma, y_2 : B_2[b_1/y_1]]\!] & \longrightarrow & [\![y_1 : B_1, y_2 : B_2]\!] \\ \downarrow & & \downarrow \\ [\![\Gamma]\!] & \xrightarrow{[\![b_1]\!]} & [\![B_1]\!]. \end{array}$$

Next we give $\Gamma \vdash b_2 : B_2[b_1/y_1]$, which presents a section of this pullback, or equivalently a morphism $\Gamma \to [\![y_1 : B_1, y_2 : B_2]\!]$ making this triangle commute:

$$\begin{array}{ccc} & & [\![y_0 : B_0, y_1 : B_1]\!] \\ & \overset{[\![b_1]\!]}{\nearrow} & \downarrow \\ [\![\Gamma]\!] & \underset{[\![b_0]\!]}{\longrightarrow} & [\![y_0 : B_0]\!]. \end{array}$$

Continuing in this way, the sequence of terms $(b_0, b_1, \ldots, b_m)$ that represent a morphism $[\![\Gamma]\!] \to [\![\Delta]\!]$ individually present a tower of sections

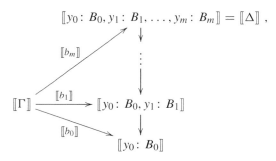

with $[\![b_m]\!]$ being the overall morphism $[\![\Gamma]\!] \to [\![\Delta]\!]$. For instance, the "squaring" injections $i_n : C_n \hookrightarrow C_{2n}$, represented by term judgments

$$n : \mathbb{N}_+ \vdash 2n : \mathbb{N}_+ \quad \text{and} \quad n : \mathbb{N}_+, x : C_n \vdash i_n(x) : C_{2n},$$

assemble into a morphism $[\![n : \mathbb{N}_+, x : C_n]\!] \to [\![m : \mathbb{N}_+, y : C_m]\!]$. Categorically, this is a morphism $\coprod_n C_n \to \coprod_m C_m$ that sends the nth summand to the $2n$th summand.

Finally, we quotient these morphisms by an equivalence relation arising from the identity type. In the simplest case where each context has only one type, we identify the morphisms presented by $x : A \vdash b_1 : B$ and $x : A \vdash b_2 : B$ if there is a term $x : A \vdash p : b_1 = b_2$. The case of morphisms between arbitary contexts is a generalization of this. (We will reconsider this last step in Sections 4.6 and 4.7.)

This completes our definition of the classifying category of a type theory. We can now *define* the projection morphism $[\![\Gamma, z : C]\!] \to [\![\Gamma]\!]$ associated to a dependent type $\Gamma \vdash C$ type, exhibiting $[\![\Gamma, z : C]\!]$ as an object of the slice category over $[\![\Gamma]\!]$, as we intended. According to the above description of morphisms, this projection morphism should consist of a term in context $\Gamma, z : C$ for each type in Γ; we take these to be just the variables in Γ, ignoring z.

For instance, the projection map $[\![x\colon A, y\colon B, z\colon C]\!] \to [\![x\colon A, y\colon B]\!]$ is determined by the terms $x\colon A, y\colon B, z\colon C \vdash x\colon A$ and $x\colon A, y\colon B, z\colon C \vdash y\colon B$. Similarly, a section of this projection consists of terms

$$x\colon A, y\colon B \vdash a\colon A,$$
$$x\colon A, y\colon B \vdash b\colon B[a/x],$$
$$x\colon A, y\colon B \vdash c\colon C[a/x, b/y],$$

such that the composite $[\![x\colon A, y\colon B]\!] \to [\![x\colon A, y\colon B, z\colon C]\!] \to [\![x\colon A, y\colon B]\!]$ is the identity, that is, that a and b are the same as x and y. Thus, such a section is simply determined by a term $x\colon A, y\colon B \vdash c\colon C$, as we intended.

Of course, not *every* object of the slice category $\mathbf{Ctx}_{/[\![\Gamma]\!]}$ is of this form, but every object of $\mathbf{Ctx}_{/[\![\Gamma]\!]}$ is *isomorphic* to one of this form. Consider the simplest case when Γ is a single type B, and we have an object of $\mathbf{Ctx}_{/[\![B]\!]}$ whose domain is also a single type A, equipped with a term $x\colon A \vdash f(x)\colon B$. Let Ψ denote the context $(y\colon B, x\colon A, p\colon f(x) = y)$; then $[\![\Psi]\!]$ is the pullback

$$\begin{array}{ccc} [\![\Psi]\!] & \longrightarrow & [\![B]\!] \\ \downarrow & & \downarrow {\scriptstyle \Delta_{[\![B]\!]}} \\ [\![B]\!] \times [\![A]\!] & \xrightarrow{1 \times [\![f]\!]} & [\![B]\!] \times [\![B]\!] \end{array}$$

using the identity type $y_1\colon B, y_2\colon B \vdash (y_1 = y_2)$ type mentioned above to present the diagonal $\Delta_{[\![B]\!]}$. It is easy to see categorically that such a pullback is isomorphic to $[\![A]\!]$. Thus, every object of $\mathbf{Ctx}_{/[\![B]\!]}$ is at least isomorphic to a composite of *two* projections from dependent types

$$[\![y\colon B, x\colon A, p\colon f(x) = y]\!] \to [\![y\colon B, x\colon A]\!] \to [\![y\colon B]\!].$$

Using the Σ-type to be defined in Section 2.5, we can reduce this to one such projection:

$$[\![y\colon B, z\colon \textstyle\sum_{x:A}(f(x) = y)]\!] \to [\![y\colon B]\!].$$

A similar argument works with B replaced by any context Γ. Thus we can assume that any object of a slice category is determined by a dependent type.

2.5 Rules and Universal Properties

In the previous section we described the judgment forms of type theory and how they present the classifying category, claiming that each judgment is analogous to a word in a free group. In this section we will describe *what* the judgments are for each judgment form, or more precisely how we can *generate* them.

The words in a free group are generated by successive application of "operations." For the tautological description of a free group, these operations are just the operations of a group: multiplication, inversion, and the identity (a nullary operation). When describing an arbitrary group, we think of these operations as defined on a fixed underlying set; but when generating a free group, we instead think of each of them as a "way to produce new elements," usually represented as syntactic strings of symbols. That is, the elements of the free group are a quotient of the set of all the syntactic strings obtainable by successive application of the following rules:

1. Given elements X and Y, we have an element (XY).
2. Given an element X, we have an element (X^{-1}).
3. We have an element e.

Formally, this is an *inductive definition*: we consider the smallest set of syntactic strings closed under the rules. Usually we think of applying these operations starting with a set of *generators*, but an equivalent description that generalizes better is to include each generator as another nullary operation:

4. For any generator g, we have an element g.

That is, generators are a special case of operations. This enables us to similarly define the "reduced words" version of a free group using inductive operations:

1. We have an element e,
2. for any generator g and any element X not ending with g^{-1}, we have an element Xg, and
3. for any generator g and any element X not ending with g, we have an element Xg^{-1}.

Defining sets inductively in this way makes it easy to define operations on them by recursion. For instance, the group multiplication in the reduced-words case is defined by recursion on the second word as follows:

1. The product of Y and e is Y.
2. To multiply Y and Xg, first multiply Y by X. If the result ends with a g^{-1}, remove it to get the answer; otherwise, concatenate a g at the end.
3. Similarly, to multiply Y and Xg^{-1}, first multiply Y by X. If the result ends with a g, remove it; otherwise, concatenate a g^{-1} at the end.

Note that in the tautological version, the group multiplication is one of the inductive clauses *defining* the elements of the free group, and hence is automatically present, whereas in the reduced-words version, the inductive clauses tell us how to multiply *by generators only*, and then we have to define multiplication of arbitrary elements afterward. As mentioned in Section 2.1, this extra work pays dividends. For instance, in this case, there is no need to quotient by any equivalence relation; we can *prove* (by induction) that the above-defined multiplication is already associative, is unital, and has inverses.

Now, the judgments of type theory, like the words in a free group, are generated inductively by operations, which in this case are usually called *rules*. Categorically, these rules build new objects and morphisms from old ones, generally according to them some universal property. For example, we might have a rule saying that any two types have a coproduct (disjoint union). This rule applies in any context (i.e., we have coproducts in every slice category); type theorists write it as

$$\frac{\Gamma \vdash A \text{ type} \qquad \Gamma \vdash B \text{ type}}{\Gamma \vdash (A+B) \text{ type}}. \tag{2.5}$$

As with judgments, this notation takes practice to read. The horizontal bar separates the "inputs" (called *premises*), on top, from the "output" (or *conclusion*), on the bottom. Each input or output is a judgment-in-context, and the inputs are separated by wide spaces or linebreaks. If the operations for the tautological description of a free group were written analogously, they would be

$$\frac{X \text{ elt} \qquad Y \text{ elt}}{(XY) \text{ elt}} \qquad \frac{X \text{ elt}}{(X^{-1}) \text{ elt}} \qquad \frac{}{e \text{ elt}} \qquad \frac{g \text{ is a generator}}{g \text{ elt}}.$$

Here "X elt" is the judgment that X is an element of the free group, analogous to the judgments "A type" and "$x: A \vdash b: B$" that A is an object and b a morphism in a free category. Note that the identity (a nullary operation) has *no* premises. Similarly, the operations for the reduced-words description are

$$\frac{}{e \text{ elt}} \qquad \frac{X \text{ elt} \qquad g \text{ is a generator} \qquad X \text{ does not end with } g^{-1}}{(Xg) \text{ elt}}$$

$$\frac{X \text{ elt} \qquad g \text{ is a generator} \qquad X \text{ does not end with } g}{(Xg^{-1}) \text{ elt}}.$$

The variables X and Y are analogous to Γ, A, B, and b in type theory. We call the latter *metavariables* to distinguish them from the variables $x: A$ occurring *in* a context Γ, which have no analog in group theory.

Returning to the coproduct type $A+B$, for it to be worthy of the name "coproduct," it needs to have certain structure. For simplicity, let us consider first the case when A and B are not dependent types; that is, the context Γ

in (2.5) is empty, so that $[\![A]\!]$ and $[\![B]\!]$ are objects of **Ctx** rather than one of its slices. Now to start with, we need injections from $[\![A]\!]$ and $[\![B]\!]$ into $[\![A+B]\!]$. It may seem natural to write these as

$$\frac{}{x\colon A \vdash \mathsf{inl}(x)\colon (A+B)} \quad \text{and} \quad \frac{}{y\colon B \vdash \mathsf{inr}(y)\colon (A+B)}. \tag{2.6}$$

(We omit the judgments $\vdash A$ type and $\vdash B$ type from the premises, since these are implied[14] by mention of $A+B$.) However, instead, one usually uses the following rules, with an arbitrary context Δ:

$$\frac{\Delta \vdash a\colon A}{\Delta \vdash \mathsf{inl}(a)\colon (A+B)} \quad \text{and} \quad \frac{\Delta \vdash b\colon B}{\Delta \vdash \mathsf{inr}(b)\colon (A+B)}. \tag{2.7}$$

The naïve rules (2.6) say that "there is a morphism $[\![\mathsf{inl}]\!]\colon [\![A]\!] \to [\![A+B]\!]$" (and similarly for $[\![\mathsf{inr}]\!]$), whereas the rules (2.7) say that "for any morphism $[\![a]\!]\colon [\![\Delta]\!] \to [\![A]\!]$, there is an induced morphism $[\![\mathsf{inl}(a)]\!]\colon [\![\Delta]\!] \to [\![A+B]\!]$." Intuitively, the latter should be thought of as describing $[\![\mathsf{inl}]\!]$ indirectly in terms of its image under the Yoneda embedding, that is, as a natural family of operations $\mathbf{Ctx}([\![\Delta]\!], [\![A]\!]) \to \mathbf{Ctx}([\![\Delta]\!], [\![A+B]\!])$.

The reason type theorists choose (2.7) over (2.6) is somewhat esoteric and is not necessary to understand in order to use type theory successfully. However, I will spend a little time explaining it, because it has to do with the specifics of how type theory presents a free category and hence with the question (already mentioned in Section 2.1) of *why* one might use the complicated syntax of type theory rather than arguing directly in the language of category theory.

The advantage of (2.7) over (2.6) is closely analogous to that of the reduced-words description of a free group over the tautological one. More precisely, the rules (2.6) are analogous to the generator rule

$$\frac{g \text{ is a generator}}{g \text{ elt}},$$

whereas the rules (2.7) are analogous instead to

$$\frac{X \text{ elt} \quad g \text{ is a generator} \quad X \text{ does not end with } g^{-1}}{(Xg) \text{ elt}}.$$

That is, we regard the morphism $[\![\mathsf{inl}]\!]$ as a "generator," and composition in a category as analogous to group multiplication. Just as the tautological description of a free group requires an explicit multiplication operation

---

[14] Depending on technical details far beyond our present scope, this implication might be a theorem about type theory or it might be just an unproblematic abuse of notation.

$$\frac{X \text{ elt} \qquad Y \text{ elt}}{(XY) \text{ elt}},$$

if we used (2.6), then our type theory would need an explicit rule allowing us to compose $[\![a]\!] \colon [\![\Delta]\!] \to [\![A]\!]$ and $[\![b]\!] \colon [\![A]\!] \to [\![B]\!]$. As we saw in Section 2.4, composition is represented type-theoretically by *substitution*,[15]

$$\frac{\Delta \vdash a \colon A \qquad x \colon A \vdash b \colon B}{\Delta \vdash b[a/x] \colon B}, \tag{2.8}$$

and we would have to quotient by an equivalence relation forcing this composition to be associative and unital. But if we instead use (2.7), which incorporates "postcomposition with generators" into the inductive definition of the set of judgments, then just as we defined multiplication as an operation on reduced words, with one recursive clause for each "postmultiply with a generator" rule, we can define composition as an operation on judgments, with one recursive clause for each "postcompose with a generator" rule such as (2.7). We can then prove that this composition is associative and so on, rather than having to quotient by an equivalence relation to enforce associativity. (This is called the *admissibility of substitution*; it is closely related to *cut-elimination*.) Unlike in the simple case of free groups, we cannot generally entirely eliminate the quotient, but we can significantly reduce the complexity of the necessary equivalence relation; see Sections 3.3, 4.6, and 4.7. This is one of the advantages of presenting free categories using type theory, rather than a more "tautological" category-theoretic syntax.

When A and B depend on a nonempty context Γ, the analogue of (2.6) is

$$\frac{}{\Gamma, x \colon A \vdash \mathsf{inl}(x) \colon (A+B)} \quad \text{and} \quad \frac{}{\Gamma, y \colon B \vdash \mathsf{inr}(y) \colon (A+B)}. \tag{2.9}$$

The former literally means a section of the pullback of $[\![\Gamma, A+B]\!]$ to $[\![\Gamma, A]\!]$:

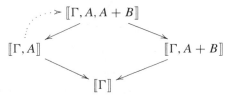

although by the universal property of pullback, this is equivalent to a morphism $[\![\Gamma, A]\!] \to [\![\Gamma, A+B]\!]$ over $[\![\Gamma]\!]$.

---

[15] The actual substitution rule is more general, allowing A, B, and b to depend on Δ as well. But this simple version more obviously represents categorical composition.

The analogue of (2.7) for dependent types actually looks no different from (2.7) itself, although I will write it with Γ instead of Δ for the context, to emphasize that this is the same Γ on which A and B depend:

$$\frac{\Gamma \vdash a : A}{\Gamma \vdash \mathsf{inl}(a) : (A + B)} \quad \text{and} \quad \frac{\Gamma \vdash b : B}{\Gamma \vdash \mathsf{inr}(b) : (A + B)}. \quad (2.10)$$

This seems only to yield an operation taking every *section* of the projection $[\![\Gamma, A]\!] \to [\![\Gamma]\!]$ to a section of $[\![\Gamma, A + B]\!] \to [\![\Gamma]\!]$, but there is a trick that gives more. For any morphism $[\![\theta]\!] : [\![\Delta]\!] \to [\![\Gamma]\!]$ in **Ctx**, we can substitute θ into A and B, obtaining types $A[\theta]$ and $B[\theta]$ representing the pullbacks of $[\![A]\!]$ and $[\![B]\!]$ along $[\![\theta]\!]$. Moreover, the definition of substitution ensures that $(A + B)[\theta]$ is $A[\theta] + B[\theta]$, so we can also apply (2.10) in context Δ. This gives an operation taking any section of any *pullback* of the projection $[\![\Gamma, A]\!] \to [\![\Gamma]\!]$ along some morphism $[\![\theta]\!]$ to a section of the corresponding pullback of $[\![\Gamma, A + B]\!] \to [\![\Gamma]\!]$. But by the universal property of pullback, such a section corresponds to a map $[\![\Delta]\!] \to [\![\Gamma, A]\!]$ over $[\![\theta]\!]$, and likewise for $[\![\Gamma, A + B]\!]$. Since $[\![\Delta]\!]$ is an arbitrary object of **Ctx**, we have a natural transformation

$$\mathbf{Ctx}_{/[\![\Gamma]\!]}([\![\Delta]\!], [\![\Gamma, A]\!]) \cong \mathbf{Ctx}_{/[\![\Gamma]\!]}([\![\Delta]\!], [\![\Gamma, A + B]\!])$$

and hence a morphism $[\![A]\!] \to [\![A + B]\!]$ over $[\![\Gamma]\!]$. Thus, (2.10) and (2.9) are also different ways of describing the same categorical structure, and there are technical but real advantages to (2.10). However, (2.9) (plus substitution rules like (2.8)) still yields a respectable type theory (called "explicit substitution calculus") that has its uses.

Whichever of (2.9) or (2.10) we use, they are called the *introduction* rules for the coproduct (they "introduce" elements of $A + B$), whereas (2.5) is called the *formation* rule. The coproduct also has an *elimination* rule, which expresses the "existence" part of its universal property:

$$\frac{\Gamma, x : A \vdash c_A : C \quad \Gamma, y : B \vdash c_B : C \quad \Gamma \vdash s : A + B}{\Gamma \vdash \mathsf{case}(C, c_A, c_B, s) : C} \quad (2.11)$$

That is, given morphisms $[\![A]\!] \to [\![C]\!]$ and $[\![B]\!] \to [\![C]\!]$, we have a morphism $[\![A + B]\!] \to [\![C]\!]$. The notation $\mathsf{case}(C, c_A, c_B, s)$ suggests that it is defined by inspecting the element s of $A + B$ and dividing into cases: if it is of the form $\mathsf{inl}(x)$, then we use c_A, whereas if it is of the form $\mathsf{inr}(y)$, then we use c_B.[16]

---

[16] Technically, we should really write something like $\mathsf{case}(C, x.c_A, y.c_B, s)$, to indicate which variables x and y are being used in the terms c_A and c_B.

More generally, we allow C to be a dependent type:

$$\frac{\Gamma, z \colon A + B \vdash C \text{ type} \qquad \Gamma \vdash s \colon A + B}{\Gamma \vdash \mathsf{case}(C, c_A, c_B, s) \colon C[s/z]} \qquad (2.12)$$

Categorically, this says that given a map $[\![C]\!] \to [\![A + B]\!]$ and sections of its pullbacks to $[\![A]\!]$ and $[\![B]\!]$, we can define a section over $[\![A + B]\!]$ by the universal property of $[\![A + B]\!]$. This generalization of the existence part of the universal property is actually an equivalent way to include the *uniqueness* part of it. On one hand, categorically, uniqueness is what tells us that the induced map $[\![A + B]\!] \to [\![C]\!]$ is in fact a section. On the other hand, assuming (2.12), if $z \colon A + B \vdash c \colon C$ and $z \colon A + B \vdash d \colon C$ have equal composites with inl and inr, then we can express this using the "equality type" from Section 2.4,

$$x \colon A \vdash e_A \colon c[\mathsf{inl}(x)/z] = d[\mathsf{inl}(x)/z]$$

$$y \colon B \vdash e_B \colon c[\mathsf{inr}(y)/z] = d[\mathsf{inr}(y)/z],$$

and then use (2.12) to construct $z \colon A + B \vdash e \colon (c = d)$.

Finally, the universal property also requires that $[\![\mathsf{case}(C, c_A, c_B)]\!] \circ [\![\mathsf{inl}]\!]$ equals $[\![c_A]\!]$, and similarly for inr. These are called *computation rules*. Here "equals" is usually taken to mean something stronger than the equality type, analogous to the equivalence relation imposed on words in a free group, and written with $\equiv$; we will return to this in Sections 3.3 and 4.7:

$$\frac{\Gamma, z \colon A + B \vdash C \text{ type}}{\Gamma, x \colon A \vdash c_A \colon C[\mathsf{inl}(x)/z] \qquad \Gamma, y \colon B \vdash c_B \colon C[\mathsf{inr}(y)/z]}{\Gamma \vdash \mathsf{case}(C, c_A, c_B, \mathsf{inl}(a)) \equiv c_A[a/x]}$$

In conclusion, we have four groups of rules relating to coproducts: formation (how to build types), introduction (how to build elements of those types), elimination (how to use elements of those types to build elements of other types), and computation (how to combine introduction and elimination). Most rules of type theory come in packages like this, associated to one "type constructor" (here the coproduct) and expressing some universal property. Given any class of *structured categories* determined by universal properties, we can obtain a corresponding type theory by choosing all the corresponding packages of rules. By and large, the rules for each type constructor are self-contained, allowing them to be "mixed and matched"; thus unlike ZFC, type theory is not a fixed system of axioms or rules, but a "modular" framework for such systems.

Table 6.1. *Type constructors and their semantics*

Type constructor	Universal property	See
Coproduct types $A + B$	binary coproducts	Section 2.5
Empty type $\emptyset$	initial object	
Product types $A \times B$	binary products	
Unit type $\mathbf{1}$	terminal object	
Natural numbers $\mathbb{N}$	natural numbers object	Section 4.4
Identity type $(x = y)$	diagonal/equalizer	Section 4.1
Function type $A \to B$	exponential object (cartesian closure)	Section 2.5
Dependent sum $\sum_{x:A} B$	left adjoint to pullback	
Dependent product $\prod_{x:A} B$	right adjoint to pullback (lcc)	
Proposition type Ω	subobject classifier (elementary topos)	Section 2.6
Universe type $\mathcal{U}$	object classifier (∞-topos)	Section 2.6
Coequalizer type $\mathsf{coeq}(f, g)$	coequalizer	Section 4.4

$$\frac{\Gamma \vdash A \text{ type} \quad \Gamma \vdash B \text{ type}}{\Gamma \vdash (A \to B) \text{ type}} \qquad \frac{\Gamma, x : A \vdash b : B}{\Gamma \vdash \lambda x.M : A \to B}$$

$$\frac{\Gamma \vdash f : A \to B \quad \Gamma \vdash a : A}{\Gamma \vdash f(a) : B} \qquad \frac{\Gamma, x : A \vdash b : B \quad \Gamma \vdash a : A}{\Gamma \vdash (\lambda x.b)(a) \equiv b[a/x]}$$

$$\frac{\Gamma \vdash f : A \to B}{\Gamma \vdash f \equiv (\lambda x.f(x))}$$

Figure 6.1 The rules for function types.

The most common type constructors and their corresponding universal properties are shown in Table 6.1. We will discuss some of these further in later sections, as indicated. Here we give the rules explicitly only for *function types*, which correspond to categorical exponentials; see Figure 6.1. The exponential object from A to B is often denoted by B^A or $^A B$, but in type theory, we denote it by $A \to B$; this way, the notation for its elements is $f : A \to B$, matching the usual notation for "f is a function from A to B." Again we have a formation rule saying when $A \to B$ is a type, an introduction rule saying how to produce terms in $A \to B$, an elimination rule saying how to use such terms (by applying them to an argument), and two computation rules. Categorically, the elimination rule yields an "evaluation" morphism $[\![A \to B]\!] \times [\![A]\!] \to [\![B]\!]$, while the introduction rule says that any map $[\![\Gamma]\!] \times [\![A]\!] \to [\![B]\!]$ has a "transpose" $[\![\Gamma]\!] \to [\![A \to B]\!]$. The computation rules say that these compose correctly.[17]

An important generalization of this is the *dependent function type*, where the codomain B is allowed to depend on the domain A. For instance, the family of generators of cyclic groups $n\colon \mathbb{N}_+ \vdash g_n\colon C_n$ yields a dependent function, assigning to each n the generator of C_n:

$$\lambda n.g_n\colon \ \prod_{n:\mathbb{N}_+} C_n.$$

Categorically, given $\Gamma \vdash A$ type and $\Gamma, x\colon A \vdash B$ type, the type $\Gamma \vdash \prod_{n:A} B$ type is obtained from $[\![\Gamma, x\colon A, y\colon B]\!] \to [\![\Gamma, x\colon A]\!]$ by applying the right adjoint of pullback along $[\![\Gamma, x\colon A]\!] \to [\![\Gamma]\!]$. Such a right adjoint exists exactly when the category is locally cartesian closed (LCC).

Of the type constructors in Table 6.1, LCC categories also have products $A \times B$, a terminal object **1**, diagonals represented by the identity type ($x = y$),[18] and left adjoints to pullback represented by dependent sums $\sum_{x:A} B$. The latter generalizes $A \times B$; its elements are pairs (x, y) where the type of y can depend on x. This collection of type constructors, corresponding to locally cartesian closed categories, is one of the most "standard" type theories, sometimes called *extensional Martin-Löf type theory* (EMLTT) without universes (though that phrase also sometimes includes coproducts and the empty type, and sometimes the "reflection rule" to be discussed in Section 4.7).

In general, choosing a particular collection of type constructors specifies the rules for a particular type theory $\mathfrak{T}$, and thereby the collection of derivable judgments. From this we construct a classifying category **Ctx**($\mathfrak{T}$) as in Section 2.4, which one can prove to be *initial* among categories with the corresponding structure. For instance, the classifying category of extensional Martin-Löf type theory, as above, is the initial locally cartesian closed category. It follows that the types and terms we construct in type theory have unique interpretations as objects and morphisms in *any* category with appropriate structure, by applying the unique structure-preserving functor out of the initial object.

As in the case of groups, we often want to generalize this by including "generators" in addition to operations, allowing us to reason about *arbitrary* objects and morphisms that may not necessarily be constructible "from nothing" using the categorical structure present. And as we did for groups, we can do this by adding stand-alone rules to our type theory. For instance, if we add the rules

---

[17] We need two computation rules because we cannot prove the uniqueness part of the universal property for functions the way we did for coproducts. See also Section 4.3.
[18] When we discuss the identity type in Section 4.1, we will see that there are multiple choices that can be made for its rules. For present purposes, the rules include "UIP" (see Section 4.1).

$$\frac{}{\Gamma \vdash \mathsf{X}\ \text{type}} \qquad \frac{}{\Gamma \vdash \mathsf{Y}\ \text{type}} \qquad \frac{}{\Gamma \vdash \mathsf{f}\colon \mathsf{X} \to \mathsf{Y}} \qquad (2.13)$$

to EMLTT, we obtain a type theory whose classifying category is the free locally cartesian closed category generated by two objects X and Y and a morphism $\mathsf{f}\colon \mathsf{X} \to \mathsf{Y}$. (Here X, Y, and f are "constants," distinct from both variables and metavariables.) Thus, given any other locally cartesian closed category $\mathscr{C}$ in which we have chosen two objects A, B and a morphism $g\colon A \to B$, there is a unique map from this classifying category sending X, Y, and f to A, B, and g respectively. Thus, anything constructable in type theory with the additional rules (2.13) can be interpreted in $\mathscr{C}$ *and yield a result relative to A, B, and g.*

Rules like (2.13) appear to correspond only to *free* groups, whereas we generally also consider *presented* groups, with relations in addition to generators. However, as we saw in Section 2.4, equalities can be represented by elements of identity types; thus here the "presented" case is subsumed by the "free" case.

Constant rules like (2.13) allow us to reason about small collections of data in arbitrary structured categories. In addition, given a category $\mathscr{C}$, there is a way to reason about "all of $\mathscr{C}$ and nothing else," by adding a constant for *every* object, morphism, and equality in $\mathscr{C}$. This yields a type theory $\mathfrak{Lang}(\mathscr{C})$ called the *internal language* of $\mathscr{C}$. It is closely related to the "Mitchell–Benabou language" and "Kripke–Joyal semantics" of a topos [18, 47, 59, 81], which are type theories of a sort but, unlike ours, do not include general dependent types.

If we fix some collection of type constructors, corresponding to a notion of structured category, then we can define a "category of type theories" based on these type constructors (with varying choices of constants). Then **Ctx** becomes a functor from this category to an appropriate category of structured categories and $\mathfrak{Lang}$ into a right adjoint (or perhaps, depending on how we define the category of type theories, an inverse equivalence) to **Ctx**; see Figure 6.2. The counit of this adjunction is a functor $\mathbf{Ctx}(\mathfrak{Lang}(\mathscr{C})) \to \mathscr{C}$ that interprets the internal language of $\mathscr{C}$ in $\mathscr{C}$ itself; this gives a "complete" syntax for constructions in $\mathscr{C}$, analogous to the canonical presentation of a group G involving one generator for each element and one relation for each equality.

I have just sketched an appealing general picture of the correspondence between type theories and categories. However, proving the correctness of this picture can be exceedingly technical. Seely's original proposal [101] contained a subtle technical flaw, later fixed by Hofmann and others [31, 36, 50, 51]. But

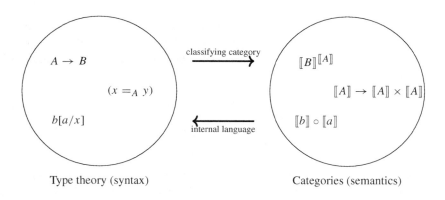

Figure 6.2 Syntax and semantics.

even now, complete proofs of the freeness of **Ctx**($\mathfrak{T}$) are quite involved and hard to find: they exist for some collections of type constructors [31, 56, 59, 69, 108], and everyone expects all other cases to be analogous, but at present there is no general theorem. Indeed, even a precise definition of "any type constructor" is still lacking in the literature. This is a current research problem, but I expect it to be solved one day; so I will say no more about this issue, except for a brief discussion of the (even more difficult) higher-categorical situation in Section 4.6.

2.6 Subobject Classifiers and Universes

There is one class of type constructors, called *universes*, that merits some individual discussion. The simplest of these is a *subobject classifier*, which categorically is a monomorphism T: $1 \to \Omega$ of which every other monomorphism is uniquely a pullback. In the category of sets, $\Omega = \{T, F\}$ is the set of truth values, and for a subset $A \subseteq B$ we have $A = \chi_A^{-1}(T)$, where $\chi_A \colon B \to \Omega$ is the characteristic function of A.

If we identify T and F with a 1-element set and a 0-element set, respectively, then up to isomorphism, the characteristic function of $A \subseteq B$ sends each $b \in B$ to its preimage under the inclusion $A \hookrightarrow B$. This leads us to represent Ω by a type whose elements *are themselves types*, with a rule like

$$\frac{\Gamma \vdash P \colon \Omega}{\Gamma \vdash P \ \text{type}} \qquad (2.14)$$

In particular, we have $x \colon \Omega \vdash x$ type, and any other instance of (2.14) can be obtained from this "universal case" by substitution. Semantically, the interpretation of $\Gamma \vdash P : \Omega$ is a morphism $[\![P]\!] : [\![\Gamma]\!] \to [\![\Omega]\!]$, while the interpretation of $\Gamma \vdash P$ type is an object of the slice over $[\![\Gamma]\!]$, that is, a projection morphism $[\![\Gamma, x : P]\!] \to [\![\Gamma]\!]$. These two morphisms fit into a pullback square:

$$\begin{array}{ccc} [\![\Gamma, x : P]\!] & \longrightarrow & [\![x : \Omega, y : x]\!] \\ \downarrow & \lrcorner & \downarrow \\ [\![\Gamma]\!] & \xrightarrow{[\![P]\!]} & [\![\Omega]\!] \end{array}$$

Thus, the morphism on the right (the interpretation of $x : \Omega \vdash x$ type) is the universal monomorphism $T : 1 \to [\![\Omega]\!]$ for a subobject classifier (and in particular, $[\![x : \Omega, y : x]\!]$ is a terminal object).

The fact that any $\Gamma \vdash P : \Omega$ classifies a monomorphism means equivalently that the types in Ω (the fibers of the corresponding objects of $\mathbf{Ctx}_{/[\![\Gamma]\!]}$) should "have at most one element." We thus express it by the following rule:

$$\frac{\Gamma \vdash P : \Omega \quad \Gamma \vdash a : P \quad \Gamma \vdash b : P}{\Gamma \vdash \mathsf{tr}_P(a,b) : a = b}. \tag{2.15}$$

This says that the diagonal $[\![\Gamma, x : P]\!] \to [\![\Gamma, x : P, y : P]\!]$ has a section; hence it is an isomorphism and so $[\![\Gamma, x : P]\!] \to [\![\Gamma]\!]$ is mono. (The notation tr_P stands for "truncation"; see Section 3.1.) The universality of $T : 1 \to [\![\Omega]\!]$ means that any type with "at most one element" is equivalent to one in Ω:[19]

$$\frac{\Gamma \vdash P \text{ type} \quad \Gamma, x : P, y : P \vdash p : x = y}{\Gamma \vdash \mathsf{Rsz}(P, xy.p) : \Omega} \tag{2.16}$$

$$\frac{\Gamma \vdash P \text{ type} \quad \Gamma, x : P, y : P \vdash p : x = y}{\Gamma \vdash \mathsf{rsz}_{P,xy.p} : (\mathsf{Rsz}(P, xy.p) \to P) \times (P \to \mathsf{Rsz}(P, xy.p))}. \tag{2.17}$$

The notations Rsz and rsz stand for "resize," indicating that P may be "too big to fit inside" Ω, but there is an equivalent type that does.

More generally, we can consider a *universe type* $\mathscr{U}$, whose elements are types without any monomorphy restriction.[20] That is, we have the analogue of (2.14),

$$\frac{\Gamma \vdash P : \mathscr{U}}{\Gamma \vdash P \text{ type}},$$

---

[19] The astute reader may notice that something is missing; we will return to this in Section 4.
[20] It is common in type theory to denote $\mathscr{U}$ by "Type," and similarly to denote Ω by "Prop." The latter will make more sense in Section 3.1.

but no analogue of (2.15). The direct analogue of (2.16) would yield in particular $\mathsf{Rsz}(\mathcal{U})\colon \mathcal{U}$, making the theory inconsistent due to Cantorian-type paradoxes. Instead we assert that $\mathcal{U}$ is closed under the *other* type constructors, for example,

$$\frac{\Gamma \vdash P\colon \mathcal{U} \quad \Gamma, x\colon P \vdash Q\colon \mathcal{U}}{\Gamma \vdash \prod_{x:P} Q\colon \mathcal{U}}, \qquad \frac{\Gamma \vdash P\colon \mathcal{U} \quad \Gamma, x\colon P \vdash Q\colon \mathcal{U}}{\Gamma \vdash \sum_{x:P} Q\colon \mathcal{U}},$$

$$\frac{\Gamma \vdash P\colon \mathcal{U} \quad \Gamma \vdash a\colon P \quad \Gamma \vdash b\colon P}{\Gamma \vdash (a = b)\colon \mathcal{U}}.$$

Thus $\mathcal{U}$ is similar to a set-theoretic "Grothendieck universe" or inaccessible cardinal. Of course, we can also have many universes $\mathcal{U}_i$ of different sizes.

Categorically, subobject classifiers are characteristic of elementary toposes, while universe objects arise in algebraic set theory [6, 61]. But we will see in Section 4 that universes really come into their own when we pass to $(\infty, 1)$-categories and incorporate Voevodsky's *univalence axiom*.

2.7 Toposes of Spaces

The preceding general theory tells us that given any category of spaces, if we choose a collection of type constructor "packages" corresponding to universal properties that exist in that category, the resulting type theory can be used to reason "internally" about that category. Turning this around, for each "package" we want to include in our type theory, there is a corresponding restriction on the categories of spaces in which we can model it.

Starting with the least restrictive case, *simple* type theory (i.e., no dependent types) requires only a category with finite products, which includes practically any category of spaces. Nothing further is required to interpret binary product types $A \times B$ and the unit type $\mathbf{1}$. To interpret the coproduct type $A + B$ and the empty type $\emptyset$, we need a category with finite products and finite coproducts, with the former distributing over the latter (this is because the elimination rule for coproducts can be applied anywhere in a context, so that for instance $[\![x\colon A+B, z\colon C]\!]$ also has the universal property of $[\![x\colon A, z\colon C]\!] + [\![y\colon B, z\colon C]\!]$).

To interpret dependent type theory, we require at least finite limits (since substitution into dependent types is interpreted by pullback). This rules out a few examples, such as smooth manifolds, but these can generally be embedded into larger categories having limits, such as "generalized smooth spaces" (see, e.g., Chapter 1). Nothing further is required to interpret the dependent sum type $\sum_{x:A} B$ and the identity type $(x = y)$.

The function type $A \to B$ in simple type theory can be interpreted in any cartesian closed category, but in *dependent* type theory it requires a *locally* cartesian closed category, since type constructors can be applied in any context (i.e., in any slice category). Local cartesian closure also allows us to interpret the dependent function type $\prod_{x:A} B$.

Ordinary topological spaces are not cartesian closed, but various slight modifications of them are. The best-known of these, such as compactly generated spaces, are not *locally* cartesian closed, but there are others that are, such as subsequential spaces [58] (sets with a convergence relation between sequences and points) or pseudotopological spaces [28, 122] (similar, but using filters instead of sequences). In such categories, a sequence of functions $f_n \colon X \to Y$ converges to $f_\infty \colon X \to Y$ in Y^X if for any convergent sequence $x_n \rightsquigarrow x_\infty$ in X, the sequence $f_n(x_n)$ converges to $f_\infty(x_\infty)$ in Y; this is sometimes called *continuous convergence*. In fact, subsequential and pseudotopological spaces both form *quasi-toposes* [123], as do various kinds of generalized smooth spaces (see [14] and also Iglesias-Zemmour's chapter on diffeologies, Chapter 1).

It is much less clear what sort of "space" could function as a subobject classifier or a universe. One guess for a subobject classifier that does not work is the Sierpinski space Σ (the set $\{T, F\}$ where $\{T\}$ is open but $\{F\}$ is not). Continuous maps $X \to \Sigma$ classify *open* subspaces of X, but not every mono is open. If instead we give the set $\{T, F\}$ the indiscrete topology, then it classifies arbitrary subspaces, but the monos of topological spaces (and their relatives) include all *injective continuous functions*, which need not be subspace inclusions.

Thus, a subobject classifier Ω has to have sufficient structure that maps into it can encode chosen topologies on subsets. For instance, if $Y \to X$ is mono and a sequence (x_n) lying in Y converges (in X) to a point x_∞ also lying in Y, then it might or might not also converge to x in the topology of Y. Thus, in defining a map $\chi_Y \colon X \to \Omega$ classifying Y, even after we know that x_n and x_∞ are sent to T (hence lie in Y), we need an additional degree of freedom in defining χ_Y to specify whether or not the convergence $x_n \to x_\infty$ is still "present" in Y.

Obviously this is impossible with classical topological spaces, but there are categories of spaces in which such an object exists: the trick is to make the "spatial structure" into *data* rather than a *property*. For instance, the quasi-topos of subsequential spaces sits inside the topos of *consequential spaces*. A consequential space is a set equipped with, for every sequence (x_n) and point x_∞, a *set* of "reasons why" or "ways in which" (x_n) converges to x_∞. (Of course, this set might be empty, i.e., (x_n) might not converge to x_∞ at all.) The axioms of a subsequential space are then promoted to operations on these "witnesses of convergence," which then have to satisfy their own axioms. See [58] for details.[21]

The category of consequential spaces is a topos, so it is locally cartesian closed and has a subobject classifier. The latter has two points {T,F}, but many different "witnesses" that the constant sequence at T converges to T, allowing the characteristic function of a mono $Y \to X$ to retain information about the topology of Y. One might think we only need *two* such witnesses, to record whether a convergent sequence $x_n \rightsquigarrow x_\infty$ in X also converges in Y; but in fact we need to record which *subsequences* of $\{x_n\}$ also converge to x_∞ in Y. We can exactly determine the witnesses of convergence in Ω by its universal property: they must be the sub–consequential spaces of the "universal convergent sequence" $\mathbb{N}_\infty$ (the one-point compactification of $\mathbb{N}$). See [58, Corollary 4.2].

The category of consequential spaces also has universe objects $\mathcal{U}$. Roughly speaking, this means that the collection of all consequential spaces (bounded in size by some cardinality) can be made into a consequential space. By the desired universal property of $\mathcal{U}$, a witness that a sequence of spaces (X_n) converges to X_∞ consists of a consistent way to make $\left(\coprod_n X_n\right) \sqcup X_\infty$ into a consequential space over $\mathbb{N}_\infty$, which roughly means giving a consistent collection of witnesses for convergence of sequences $x_n \in X_n$ to points $x_\infty \in X_\infty$. In [43] it is shown that for any (X_n) and X_∞ there is at least one such witness, so the topology of $\mathcal{U}$ is "indiscrete" in some sense (though unlike for a classical indiscrete space, interesting information is still carried by *how many* such witnesses there are).

Generalizing from consequential spaces, for any small collection of spaces $\mathcal{T}$, we can build a topos whose objects are "spaces" whose "topology" is determined by "probing" them with maps out of $\mathcal{T}$. More precisely, we take the category of sheaves for some Grothendieck topology on $\mathcal{T}$. Consequential spaces are the case when $\mathcal{T} = \{\mathbb{N}_\infty\}$, so that a space is determined by its "convergent sequences."

Another reasonable choice is $\mathcal{T} = \{\mathbb{R}^n\}_{n\in\mathbb{N}}$, yielding spaces determined by a notion of "continuous paths, homotopies, and higher homotopies."[22] Urs Schreiber has suggested to call these *continuous sets*. Just as a sequence in a

---

[21] The term "consequential space" should not be blamed on [58], who considered it but then discarded it. I have chosen to use it anyway, since I know no other term for such spaces.

[22] To be precise, we take the full subcategory of spaces $\mathcal{T} = \{\mathbb{R}^n\}_{n\in\mathbb{N}}$ with the Grothendieck topology of open covers, whereas consequential spaces are obtained from the one-object full

consequential space can "converge to a point" in more than one way, a path in a continuous set can "be continuous" in more than one way. By the adjunction, continuous paths $\mathbb{R} \to Y^X$ are equivalently homotopies $\mathbb{R} \times X \to Y$. Similarly, in $\mathcal{U}$, the witnesses to continuity of a "path of spaces" $\{X_t\}_{t \in \mathbb{R}}$ are the ways to make $\coprod_{t \in \mathbb{R}} X_t$ into a space over $\mathbb{R}$, which roughly means a consistent collection of witnesses for the continuity of "paths" consisting of points $x_t \in X_t$ for all $t \in \mathbb{R}$. Consequential spaces and continuous sets are similar in many ways, but different in others, and each has advantages and disadvantages.

If we restrict the morphisms of $\mathcal{T}$ to preserve some additional structure on its objects, we obtain a topos whose objects have a version of that structure. For instance, if we choose $\mathcal{T} = \{\mathbb{R}^n\}_{n \in \mathbb{N}}$ but with only the *smooth* functions between them, then we obtain a topos whose objects are a kind of generalized smooth space, in which any given path or homotopy has a set of "witnesses to smoothness" rather than witnesses to continuity; we call these *smooth sets*. Just as the quasi-topos of subsequential spaces sits inside the topos of consequential spaces, the quasi-topos of diffeological spaces (and hence also the usual category of smooth manifolds) sits inside the topos of smooth sets.

There are likewise toposes of "algebraic" spaces, as well as toposes containing "infinitesimal" or "super" spaces; see [14, 89] and [2, Chapter 5].

2.8 Why Spaces?

We can now address the question mentioned in Section 1 of *why* so many objects in mathematics come naturally with spatial structure. The short answer is that since type theory can be a language for all of mathematics, any construction of a mathematical object can be phrased in type theory and then interpreted into any of these toposes of spaces, thereby yielding not just a set but a space.

However, this answer is missing something important: to say that a set X comes naturally with spatial structure is to say that we have a canonically defined space *whose underlying set is* X, and simply being able to interpret the construction of X in a topos of spaces does not ensure this. If the "underlying set" functor from our topos of spaces to the category of sets preserved all the structure used to interpret type theory (limits, colimits, dependent exponentials, universes, and so on – this is called a *logical functor*), then

subcategory $\{\mathbb{N}_\infty\}$ with the "canonical" Grothendieck topology. This difference in topology is the actual source of a significant amount of the differences between the two toposes.

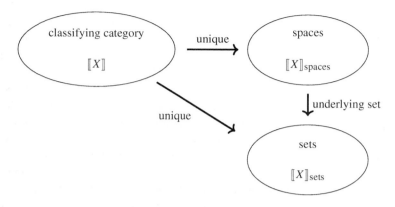

Figure 6.3 A hypothetical logical underlying-set functor.

this would follow from the initiality of the classifying category. Namely, the unique functor from the classifying category to the category of sets would factor uniquely through our topos of spaces, implying that the image of X in sets would be the underlying set of the image of X in spaces, as shown in Figure 6.3.

In general, the underlying-set functor does not preserve *all* the structure. But it does preserve quite a lot of it. In fact, for all the categories of spaces considered above, the underlying-set functor fits into a string of adjunctions:

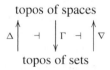

where Γ is the underlying-set functor, Δ equips a set with a "discrete topology," and ∇ equips it with an "indiscrete topology." (In consequential spaces, the discrete topology says that only eventually constant sequences converge, while the indiscrete one says that every sequence converges [uniquely] to every point.) Moreover, Δ preserves finite limits, and Δ and ∇ are fully faithful; this makes the category of spaces into a *local topos* over sets (see [57] or [59, Section C3.6]).

In particular, this string of adjunctions implies that Γ preserves all limits and colimits. In the above examples, Γ also preserves the subobject classifier and the universes – this is clear from our explicit descriptions of these above, and categorically it implies that the spaces are also a *hyperconnected topos*.

The main thing that Γ does not preserve is function-spaces. However, it does preserve function-spaces whose domain is discrete: for any set X and space Y we have $\Gamma(Y^{\Delta X}) \cong (\Gamma Y)^X$, and likewise for dependent exponentials. (This follows formally from the fact that Δ preserves finite limits.)

The upshot is that any mathematical construction can be interpreted as a space, and as long as the only functions it uses have a domain that (when interpreted as a space) is discrete, the resulting space will be a topology on the set that we originally thought we were defining. This restriction makes sense: once we start using functions with nondiscrete domain, in the world of spaces we have to consider only *continuous* functions, causing a divergence from the world of sets. Moreover, in this case often the world of spaces is the "correct" one: we *should* restrict to continuous maps between profinite groups, and use only continuous homomorphisms in the Pontryagin dual.

The next natural question to ask is, when *does* a given mathematical construction inherit a nondiscrete topology? Categorically, this means asking which constructions are (not) preserved by Δ. Since Δ is a left adjoint, it preserves colimits, and as remarked it preserves finite limits; but in general it does not preserve much more than this. Thus, nontrivial topologies can arise from (1) infinite limits, (2) function-spaces (which, in the category of sets, are just particular infinite products), or (3) the subobject classifier or a universe.

Many constructions that automatically inherit spatial structure are obtained by infinite limits. For instance, the profinite completion is a limit of finite quotients. In the topos of consequential spaces, this infinite limit generally has its expected topology, since "take the convergent sequences" is a limit-preserving functor from classical topological spaces to consequential spaces.

By contrast, for the topos of continuous sets (and similarly smooth sets), Δ *does* preserve infinite limits, and indeed it has a further left adjoint (see Section 5). Thus, the profinite completion in these toposes gets the discrete topology (intuitively, profinite topologies are incompatible with the "manifold-like" structure of a continuous set). More generally, nontrivial continuous-set structures can arise *only* in constructions that include the subobject classifier or a universe. Perhaps the most important example of such a space is the real numbers $\mathbb{R}$, but before discussing them, we need to talk about logic in type theory.

3 Toward Synthetic Topology

In Section 2.1 I claimed that type theory presents a free "universe of mathematics." So far, we have seen that type theory contains type constructors such as products, coproducts, and exponentials that look like the standard operations on sets. Moreover, when taken literally (rather than as "code" for objects and morphisms in a category), the syntax of type theory talks about *elements* of types, and the rules stipulate that the elements of product types, exponential types, and so on are exactly what we would expect. Thus, any mathematical construction that is classically performed with sets, such as building the rational and real numbers out of the integers, can be performed with types instead.

However, there is more to mathematics than constructing things: we also like to *prove* things about them. The formal system of ZFC set theory is formulated inside of first-order logic, so that proving is the "basic act of mathematics": what are called "constructions" are actually just existence (or existence-and-uniqueness) proofs. But in type theory, *constructions* are the "basic act of mathematics" – so what has happened to proofs?

3.1 Propositions as Types

In set theory, a property P of elements of a set A can be equivalently expressed as a subset of A, namely $\{x \in A \mid P(x)\}$. Categorically, this is a monomorphism into A, or equivalently its characteristic function $A \to \Omega$. This provides us with the means to *define logic inside of type theory*: we declare that by a **property** of elements of a type A we *mean* a judgment $x : A \vdash P : \Omega$. Similarly, by a *proposition*[23] we mean simply an element of Ω: that is, a type having *at most one element*. We regard such a proposition as "true" when it *does* have an element. That is, the proposition corresponding to $P : \Omega$ is "P has an element." Similarly, by a *proof* of $P : \Omega$ we mean a construction of an element of P.

We have already seen an example of this way of representing properties: the equality type $x : A, y : A \vdash (x = y)$ type, which in Section 2.4 we said represents the diagonal $\triangle_A : [\![A]\!] \to [\![A]\!] \times [\![A]\!]$ as an object of $\mathbf{Ctx}_{/([\![A]\!] \times [\![A]\!])}$. Since $\triangle_A$ is a monomorphism, it is classified by a map $[\![A]\!] \times [\![A]\!] \to \Omega$, which is the binary relation of equality. In the category of sets, this map sends (x, x) to T, since the fiber of $\triangle_A$ over (x, x) has one element, and sends (x, y) to F if $x \neq y$, since then the fiber of $\triangle_A$ over (x, y) is empty. (But see Section 4.)

This identification of propositions with "types having at most one element" is close to, but not quite, the usual meaning of the phrase *propositions as*

---

[23] We use proposition in the logician's sense of something that one might *try* to prove, rather than the other common meaning of something that *has been* proven. Note that in [112], types with at most one element are called *mere propositions* rather than just "propositions."

$$\frac{\Gamma \vdash A \text{ type}}{\Gamma \vdash \|A\| : \Omega} \qquad \frac{\Gamma \vdash A \text{ type} \qquad \Gamma \vdash a : A}{\Gamma \vdash |a| : \|A\|}$$

$$\frac{\Gamma \vdash A \text{ type} \qquad \Gamma \vdash B : \Omega \qquad \Gamma, x : A \vdash b : B \qquad \Gamma \vdash a : \|A\|}{\Gamma \vdash \text{ptr}(x.b, a) : B}$$

$$\frac{\Gamma \vdash A \text{ type} \qquad \Gamma \vdash B : \Omega \qquad \Gamma, x : A \vdash b : B \qquad \Gamma \vdash a : A}{\Gamma \vdash \text{ptr}(x.b, |a|) \equiv b[a/x]}$$

Figure 6.4 Propositional truncation.

types. The latter refers to allowing *any* type to be called a proposition, rather than only those with at most one element. However, our choice (which is increasingly common) relates better to the standard practice in mathematics whereby once a proposition has been proven, the particular proof given has no further mathematical (as opposed to aesthetic or conceptual) importance.

One reason the identification of propositions with (certain) types is so convenient is an observation called the *Curry–Howard correspondence* [37, 53, 84, 120]: the operations of logic are *already present* in type theory as constructions on types. For instance, if P and Q are propositions, then so is $P \times Q$; and since it has an element just when P and Q do, it is natural to call it "P and Q." Similarly, $P \to Q$ is "if P then Q," since a function $f : P \to Q$ transforms the truth of P into the truth of Q, while $\prod_{x:A} P(x)$ is "for all $x : A$, $P(x)$," since a dependent function $f : \prod_{x:A} P(x)$ assigns to any $x : A$ a proof of $P(x)$. It is also reasonable to regard $P \to \emptyset$ as "not P," since a function $f : P \to \emptyset$ can only exist if P is empty (i.e., false).

We might expect $P + Q$ to be "P or Q," but $P + Q$ may not be a proposition even if P and Q are. For this purpose we introduce a new type constructor called the *propositional truncation*,[24] whose rules are shown in Figure 6.4. (The notation ptr just stands for "propositional truncation.") Intuitively, $\|A\|$ is the proposition "A has at least one element"; while categorically, $\Gamma \vdash \|A\| : \Omega$ presents the *image* of the projection $[\![\Gamma, x : A]\!] \to [\![\Gamma]\!]$. The introduction rule says that if we have an element of A, then A has at least one element. The elimination rule says that if we know that A has at least one element, then when proving a proposition we may assume *given* an element of A. (Removing the hypothesis that B is a proposition would imply a choice principle that is too strong even for classical mathematics; see Footnote 36.)

---

[24] Propositional truncation has a long history and many variations, with names such as *squash type*, *mono-type*, and *bracket type* (see, e.g., [12, 33, 82, 87, 112]).

Now we define "P or Q" to be $\|P+Q\|$, and similarly "there exists an $x: A$ such that $P(x)$" to be $\|\sum_{x:A} P(x)\|$. As observed by Lawvere [73], this definition of the existential quantifier can be described categorically as the left adjoint to pullback between posets of subobjects $\mathrm{Sub}(\llbracket \Gamma \rrbracket) \to \mathrm{Sub}(\llbracket \Gamma, x: A \rrbracket)$. The untruncated $\sum_{x:A}$ gives the left adjoint to the pullback between slice categories $\mathbf{Ctx}_{/\llbracket \Gamma \rrbracket} \to \mathbf{Ctx}_{/\llbracket \Gamma, x:A \rrbracket}$, and the truncation reflects it back into monomorphisms. Similarly, the universal quantifier "for all $x: A$, $P(x)$" is the right adjoint of the same functor: since the right adjoint $\prod_{x:A}$ between slice categories already preserves monomorphisms, no truncation is necessary.

Rather than being the existential quantifier, the untruncated $\sum_{x:A} P(x)$ plays the role of the subset $\{x \in A \mid P(x)\}$. Its elements are pairs of an element $x: A$ and a proof that $P(x)$ holds, and since $P(x)$ is a proposition, to give an element of $P(x)$ contains no more information than that $P(x)$ "is true." Thus we may consider the elements of $\sum_{x:A} P(x)$ to be "the elements of A such that $P(x)$ is true." The type of *all* subsets of A, mentioned in Section 2.4, is just "$A \to \Omega$."

In conclusion, instead of describing mathematics inside of logic, as is done by ZFC set theory, in type theory, we "define logic inside of mathematics." One advantage of this is that, as we have seen, type theory can be interpreted in suitably structured categories, yielding an intrinsic "logic" internal to any such category. Another is that it allows us to draw finer distinctions, thereby actually representing informal mathematics *more* faithfully, in the following way.

We mentioned above that there is a tradition in type theory that interprets "there exists an $x: A$ such that $P(x)$" as $\sum_{x:A} P(x)$ rather than $\|\sum_{x:A} P(x)\|$. In addition to its mismatch with the standard practice of mathematics, when done indiscriminately, this can actually lead to inconsistencies; see [112, Section 3.2] and [42]. However, there *are* places in ordinary mathematics where a "there exists" statement is more naturally interpreted as $\sum_{x:A} P(x)$. For instance, the Yoneda lemma $\mathrm{Nat}(\mathscr{C}(-,a),F) \cong F(a)$ does not mean that there merely *exists* such an isomorphism but that we have specified a *particular* one. ZFC set theory, being formulated inside first-order logic, forces every "theorem" to be a "mere existence" statement; type theory frees us from this straitjacket, allowing us to directly express "constructions" in addition to "proofs."

3.2 Constructive Logic

When the logical connectives are defined according to the Curry–Howard correspondence, most of the basic laws of logic can be derived from the rules for the basic type constructors. For instance, one of de Morgan's laws,

$$(\neg P \wedge \neg Q) \to \neg(P \vee Q)$$

(where as usual $\wedge$, $\vee$, $\neg$ mean "and," "or," and "not"), has the following proof:

$$\lambda x.\lambda y.\text{ptr}(\text{case}(\emptyset, \text{pr}_1(x)(y), \text{pr}_2(x)(y), y), y) : (\neg P \wedge \neg Q) \to \neg(P \vee Q).$$

(Of course, it would be a heavy burden to carry around such long terms whenever we want to use de Morgan's law. But writing out a fully formalized proof in ZFC is no easier – in fact, often it is much harder! In both cases the formalism simply justifies our ordinary mode of mathematical writing.)

However, we cannot derive from the rules of type theory any proofs of the following classical tautologies of logic:

$$\neg(P \wedge Q) \to (\neg P \vee \neg Q), \qquad \neg\neg P \to P, \qquad P \vee \neg P,$$

$$\neg(\forall x, P(x)) \to (\exists x, \neg P(x)).$$

In other words, the logic we obtain from propositions-as-types is *constructive* or *intuitionistic logic*. Constructive logic acquired a bad name due to some fundamentalists in the early 20th century, but in fact it is natural and unavoidable once we recognize that type theory is a syntax for categories, and that sets are not the only category in the world (see [16]). The above tautologies are simply *not true* in most categories of spaces.

For example, consider the proposition $\prod_{x:A} P(x) \vee \neg P(x)$, where A is a type and $P: A \to \Omega$ a property, and let us interpret this in a "topological" topos such as those discussed in Section 2.7. Now P classifies a monomorphism $B \rightarrowtail A$, which as mentioned previously need not be a subspace embedding. Similarly, $\neg P$ classifies a different monomorphism $\neg B \rightarrowtail A$, which turns out to be the "maximal mono disjoint from B." In other words, $\neg B$ contains all the points of A that are absent from B, and also "all the topology" on those points that is absent from B (e.g., all the convergent sequences, or all the continuous paths). However, $(\lambda x.P(x) \vee \neg P(x)): A \to \Omega$ classifies their union $B \cup \neg B$ as monos into A, which is not generally a subspace even if B is: it contains all the points of A, but its topology is that of the disjoint union $B \sqcup \neg B$. Thus the mono $B \sqcup \neg B \to A$ has no *continuous* section, and so we cannot assert $\prod_{x:A} P(x) \vee \neg P(x)$.

In other words, constructive logic is simply more general than classical logic. As always, using fewer assumptions – here, assumptions about *logic* – leads to a more general conclusion – here, one that applies to more categories.

There is also another sense in which constructive logic is more general: classical logic can be *embedded* in constructive logic. Specifically, the subset $\Omega_{\neg\neg}$ of Ω consisting of those P such that $\neg\neg P \to P$ (formally, the Σ-type

$\sum_{P:\Omega}(\neg\neg P \to P))$ admits logical operations satisfying the laws of classical logic. In fact, $\Omega_{\neg\neg}$ is closed under all the ordinary logical operations except for "or" and "there exists," and we can define $P \vee' Q$ to be $\neg\neg(P \vee Q)$ and $\exists' x : A$ to be $\neg\neg\exists x : A$. (Note the similarity to how in Section 3.1 we applied $\|-\|$ to $+$ and $\sum$ to get $\vee$ and $\exists$.) In categories of spaces, the subtypes whose classifying map factors through $\Omega_{\neg\neg}$ generally coincide with the subspace embeddings.

Using $\Omega_{\neg\neg}$ gives us a different "logic" that is always classical but is not as well behaved in other ways. For instance, it fails to satisfy *function comprehension* (aka the *principle of unique choice*): "if for all $x : A$ there is a unique $y : B$ such that $P(x, y)$, then there is $f : A \to B$ such that $P(x, f(x))$ for all $x : A$." However, there is a subuniverse of types where $\Omega_{\neg\neg}$-logic does behave well. Define a type A to be a $\neg\neg$-*sheaf* if the "constant functions" map $(\lambda x.\lambda p.x): A \to (P \to A)$ is an isomorphism for any $P : \Omega$ such that $\neg\neg P$. The world of $\neg\neg$-sheaves in constructive mathematics behaves just like the world of classical mathematics, with both classical logic and function comprehension. In categories of spaces, the $\neg\neg$-sheaves are generally the *indiscrete* spaces. (If you were expecting to hear "discrete" instead of "indiscrete," wait for Section 5.3.)

On the other hand, it is always possible to add classical logic *globally* to type theory as a rule:

$$\frac{\Gamma \vdash P : \Omega}{\Gamma \vdash \mathsf{lem}(P) : P \vee \neg P}.$$

(This rule, called the *law of excluded middle* (LEM), suffices to prove all the other classical tautologies.) This would mean restricting our syntax to apply only to "boolean" categories, such as sets, and excluding most topological ones, just as adding the relations $gh = hg$ to a free group turns it into a free *abelian* group, with a more restricted universal property. Similarly, we can add a type-theoretic version of the *axiom of choice* (AC), which is not provable[25] even after adding LEM, and whose precise formulation we leave to the reader.[26]

---

[25] The reader may have heard rumors that the axiom of choice is actually provable in type theory *without* any added axioms. It is true that one can prove a statement that looks like the axiom of choice *if* arbitrary types are allowed to play the role of "propositions," i.e., if all propositional truncations are removed from the definitions of the logical operations in Section 3.1. But from our perspective, this provable statement is not at all the axiom of choice, since its hypothesis already essentially carries along the data of a choice function.

[26] There is also a subuniverse like the $\neg\neg$-sheaves that satisfies AC as well as LEM: one can build Gödel's "constructible universe" L (no relation to "constructive logic") inside the $\neg\neg$-sheaves. However, this relies on first-order logic and ZFC-style membership-based set theory; no category-theoretic or type-theoretic construction of such a subuniverse is known.

If we wanted only to use type theory merely as a foundation for classical mathematics, there would be no problem with this.[27] But in this chapter our focus is on type theory as a syntax for categories of *spaces*, which frequently means that we must learn to live with constructive logic (though we will see in Section 4 that "homotopy spaces" can be compatible with classical logic). Fortunately, this is usually not very difficult; often it suffices to rephrase things carefully, avoiding unnecessary negations. For instance, constructively, it is not very useful to say that a type is nonempty ($\neg(A \cong \emptyset)$ or equivalently $\neg\neg A$); instead, we use the positive statement that it is "inhabited" ($\exists x : A$, i.e., $\|A\|$). It also often happens that a group of classically equivalent definitions are no longer the same constructively, and we have to judiciously choose the "correct" one. For instance, classically a set is finite just when it is not bijective to any proper subset of itself, but constructively this is a weaker and less useful condition; the correct definition of "finite" is "bijective to $\{k : \mathbb{N} \mid k < n\}$ for some $n : \mathbb{N}$."

Once we get over this minor hurdle, we can develop mathematics on top of type theory in basically the same way as usual. Formally, type-theoretic proofs and constructions involve heavy manipulation of syntax (just as in any other formal foundational system like ZFC); but as mentioned above, when actually *doing* mathematics there is usually no reason to bother about this.

As a simple example, once we have the natural numbers type $\mathbb{N}$, we can define the integers $\mathbb{Z}$ as $\mathbb{N}+\mathbb{N}$ (with appropriate structure), the rational numbers $\mathbb{Q}$ as a subtype of $\mathbb{Z} \times \mathbb{N}$ (the "fractions a/b in lowest terms"), and the real numbers $\mathbb{R}$ as a subtype of $(\mathbb{Q} \to \Omega) \times (\mathbb{Q} \to \Omega)$ (the two-sided Dedekind cuts). Recall that by a "subtype of A" we mean a type of the form $\sum_{x:A} P(x)$ where $P : A \to \Omega$ is a property; for instance, more formally we have

$$\mathbb{Q} = \sum_{r:\mathbb{Z}\times\mathbb{N}} \left(\mathsf{pr}_2(r) > 0 \wedge \prod_{n:\mathbb{N}} \left((n \mid \mathsf{pr}_1(r)) \wedge (n \mid \mathsf{pr}_2(r)) \to n = 1 \right) \right).$$

Here pr_1 and pr_2 are the projections out of a cartesian product, and $>$ and $\mid$ are relations that we have to define previously, for example,

$$(n \mid m) = \left(\exists p : \mathbb{N}, (p \cdot n = m) \right)$$
$$= \left\| \sum_{p:\mathbb{N}} (p \cdot n = m) \right\|,$$

where multiplication has been previously defined, and so on. We can then proceed to define all the usual functions and properties of numbers of all sorts and build the rest of mathematics on top of them.

---

[27] There *are* many toposes, other than the category of sets that satisfy both LEM and AC. They are roughly the same as forcing models of ZFC. However, none of them are "spatial" in the sense we care about here.

When this syntax is interpreted into the category of sets, it of course yields the usual sets of numbers. The point, however, is that if we instead interpret it into a category of spaces, the types of numbers automatically inherit a spatial structure, and usually that spatial structure is the *intended* one! For instance, in the toposes of consequential spaces or continuous sets from Section 2.7, the real numbers type $\mathbb{R}$ is interpreted by the real numbers *with their usual Euclidean topology* (see [58, Proposition 4.4] and [81, Theorem VI.9.2]).

Note that the definition of $\mathbb{R}$ using Dedekind cuts (which we may denote $\mathbb{R}_d$ for emphasis) fulfills the requirement for a nontrivial continuous-set structure, since it uses Ω. However, not all the classically equivalent definitions of $\mathbb{R}$ remain equivalent constructively. For instance, the *Cauchy reals* $\mathbb{R}_c$, defined by taking equivalence classes of Cauchy sequences, come with an inclusion $\mathbb{R}_c \to \mathbb{R}_d$ that is not generally surjective. In consequential spaces, we have $\mathbb{R}_c \cong \mathbb{R}_d$; but in continuous sets, $\mathbb{R}_c$ gets the discrete topology. Thus, it is usually better to regard the Dedekind real numbers as "the" real numbers.

Since the Dedekind reals $\mathbb{R}$ have their usual topology in our toposes, other types built from them, such as the circle $S^1 = \{\, (x, y) \colon \mathbb{R} \times \mathbb{R} \mid x^2 + y^2 = 1 \,\}$, the complex numbers $\mathbb{C} \cong \mathbb{R} \times \mathbb{R}$, or matrix groups $\mathrm{GL}_n(\mathbb{R}) \subseteq \mathbb{R}^{n^2}$, also have their usual topologies. Furthermore, all functions definable in type theory are interpreted by continuous maps; so the fact that we can define addition of real numbers in type theory tells us automatically that $\mathbb{R}$ is a topological group, and so on. Analogous facts are true for the other constructions leading to nontrivial topologies mentioned in Section 2.8, such as profinite completion. Thus, we have finally made good on our promise from Section 1 to provide a formal system for describing "groups with background spatial structure" that is sufficiently flexible to include all different kinds of spatial structure at once, with the added benefit of a uniform way of constructing the standard examples.

The fact that "the real numbers" defined in type theory are interpreted in some categories by the usual *space* of real numbers has an interesting consequence: *constructively, we cannot define*[28] *any discontinuous function* $\mathbb{R} \to \mathbb{R}$. In particular, the usual examples of discontinuous "piecewise" functions $\mathbb{R} \to \mathbb{R}$, such as the Heaviside step function

$$f(x) = \begin{cases} 0 & \text{if } x < 0 \\ 1 & \text{if } x \geq 0 \end{cases}$$

cannot be defined constructively – or, more precisely, their domain cannot be shown constructively to be all of $\mathbb{R}$ (that being tantamount to the assertion

---

[28] To be precise, we cannot define such a function "in the empty context," i.e., without any ambient assumptions.

that every real number is either < 0 or ≥ 0, which is essentially an instance of LEM). That is, restricting ourselves to constructive logic automatically "notices," and forces us to respect, a canonical and implicit topological structure on types such as $\mathbb{R}$. In the next section we briefly discuss another such implicit structure.

3.3 A Digression on Computation

While our primary concern here is with the suitability of type theory as a "logic of space," historically, it developed rather differently. The first type theory, which bore little resemblance to its modern descendants, was introduced by Russell to avoid his eponymous paradox. After other logicians, such as Gödel, refined Russell's type theory in various ways, Church [29, 30] combined it with his "λ-calculus" to obtain what today we can see as a typed functional programming language. The dependent type theory we are using here is mainly due to Martin-Löf [84, 85], whose intent was to give "a full scale system for formalizing intuitionistic mathematics" in the sense of Bishop [20, 21]. Bishop, in turn, wanted to develop a form of mathematics in which all statements would have computational meaning, so that for instance whenever we assert something to *exist* we must have a method for finding it. This led him, following the earlier pioneering work of Brouwer, to reject the law of excluded middle, since in general there can be no *method* for deciding which of P or $\neg P$ holds.

Thus, type theory was originally conceived as a formal basis for a mathematics that would be "constructive" in this computational sense. It is remarkable that it turned out to also be a flexible system for reasoning in arbitrary categories! The existence of internal languages of categories was apparently first recognized for elementary toposes [22, 68, 88]; the "Mitchell–Benabou language" of a topos is a sort of type theory in which the only dependent types are those in Ω. The generalization to Martin-Löf's type theory was first written down by Seely [101], and corrected and refined by others as discussed in Section 2.5.

From our category-theoretic point of view, the computational aspect of type theory is partly explained by a different class of models. In addition to categories of "spaces" where every map is continuous, there are categories of "computable objects" [54, 115] in which every map is computable. Thus, everything in type theory must be "potentially computable" in addition to "potentially continuous." In particular, just as any constructively definable function $\mathbb{R} \to \mathbb{R}$ must be continuous, any constructively definable function $\mathbb{N} \to \mathbb{N}$ must be computable.

However, there is more to the computational side of type theory than this: its syntax *is* actually a programming language that can be executed. That is, not only does every term $f : \mathbb{N} \to \mathbb{N}$ represent a computable function, but its definition is an algorithm for computing that function. The "execution" of such a program is essentially the type-theoretic version of the "reduction" algorithm for free groups that simplifies $(g((hg)^{-1}((hh)g^{-1}))) h^{-1}$ to $hg^{-1}h^{-1}$.

Specifically, these reductions implement the "computation rules" mentioned in Section 2.5 (hence the name). For instance, the first computation rule in Figure 6.1 can be interpreted as a "reduction" or "normalization" step allowing us to "simplify" $(\lambda x.M)(N)$ to $M[N/x]$. (This partially explains why we used a different equality symbol $\equiv$; see also Section 4.7.) In good cases, these reduction steps are guaranteed to terminate at a unique "value" or "normal form," analogous to the reduced words for elements of a free group. Reduction in a free group is a fairly simple process, but since type theory is complicated enough to encode all of mathematics, its notion of "reduction" can serve as a general-purpose programming language.[29]

This makes type theory a convenient language for reasoning about computer programs, and as such it has many adherents in computer science. Moreover, we can implement a "compiler" for type theory on a physical computer, which then also serves as a *proof checker* for mathematical arguments. Thus, mathematics done in type theory not only can be interpreted in many categories but can have its correctness formally verified in this way. Such computerized "proof assistants" built on type theory play an increasingly important role in computer science, and are slowly growing in importance in mathematics.

One thing to note is that the computational interpretation of type theory is rather "fragile" and depends on the details and interactions of all the rules. Adding new rules can cause the reduction algorithm to loop, diverge or give up; while adding axioms (such as LEM and AC) can cause it to stop before reaching a "value." The latter is not surprising, since LEM and AC assert that certain things exist or are true without giving any way to construct them, thereby allowing us to define terms of type $\mathbb{N}$ like "1 if the Riemann hypothesis is true and 0 if it is false" that no terminating algorithm can possibly simplify to an "answer." (This does not prevent us from using computer proof assistants to check proofs in such type theories, however.)

---

[29] This programming language is not technically "Turing-complete," since all its "programs" must terminate; otherwise, we could prove a contradiction with a divergent computation. But it can still encode all computable functions, e.g., with a partiality monad [25, 38].

Since this book is about space rather than computation, I will not say much more about the computational side of type theory. However, it is worth pointing out that computation and topology are actually closely related. As first recognized by Scott [98–100], computational objects often come naturally with, or are represented by, topologies representing the fact that a finite computation can only consume a finite amount of data. In fact, Brouwer's original intuitionism was also arguably more "topological" than computational. For further discussion of these ideas, see for instance [40, 116].

3.4 Synthetic Topology

So far we have seen that the types in type theory admit interpretations as various different kinds of space. Thus, one might say that they have *latent* or *potential* spatial structure: they might be spaces, but they also might not have any nontrivial spatiality, depending on where we interpret them. Moreover, they also have other latent structures, such as computability.

Until now we have considered mainly the aspects of types that are independent of their potential spatial structure, where the topology simply comes along for the ride. However, some spatial aspects of types are visible inside of type theory, without needing to interpret them first in some category. This leads to subjects called *synthetic topology* and *synthetic differential geometry*.

One important observation is that in many cases we can detect topology using structures that already exist in type theory. For instance, following [43], define $\mathbb{N}_\infty$ to be the type of nonincreasing binary sequences:

$$\mathbb{N}_\infty = \sum\nolimits_{a:\mathbb{N}\to 2} \prod\nolimits_{n:\mathbb{N}} (a_{n+1} \leq a_n)$$

where **2** is the "boolean" type with two elements 0 and 1. Then we have an injection $i: \mathbb{N} \to \mathbb{N}_\infty$ where $i(m)_n = 1$ if $m < n$ and 0 otherwise, and we also have an element "$\infty: \mathbb{N}_\infty$" defined by $\infty_n = 1$ for all n. In the topos of consequential spaces, $\mathbb{N}_\infty$ is interpreted by the "actual" one-point compactification of $\mathbb{N}$; thus it is sensible to *define* a *convergent sequence* in a type A to be a map $\mathbb{N}_\infty \to A$. In this way, without assuming any axioms, we see that every type automatically has a structure like a consequential space, and every function is automatically "continuous" in the sense of preserving convergent sequences.

Of course, the actual interpretation of $\mathbb{N}_\infty$ depends on the category: in the topos of continuous sets, it yields $\mathbb{N} \sqcup \{\infty\}$ with the discrete topology, so that every sequence "converges" uniquely to every point. However, the structure of continuous sets is detectable internally in a different way: define a *continuous path* in a type A to be a map $\mathbb{R} \to A$ out of the (Dedekind) real numbers. This gives the expected answer for both consequential spaces and continuous sets, since in both cases $\mathbb{R}$ has its usual topology. We will come back to this in Section 5.

These internally defined "topologies" are only "potentially nontrivial": for example, if we assume LEM, then every sequence converges uniquely to every point and every path is uniquely continuous. If we want to ensure that they definitely *are* nontrivial, we can assert "nonclassical axioms" that contradict LEM, excluding the category of sets but retaining topological models. For instance, we could assert that the only "convergent sequences" in $\mathbb{R}$ are those that converge in the ϵ-N sense, or that every "continuous path" in $\mathbb{R}$ is continuous in the ϵ-δ sense.[30]

Convergent sequences and continuous paths are "covariant" notions of topology, that is, they are defined using maps *into* a type. We can also describe "contravariant" notions of topology synthetically, involving maps *out* of a type. For instance, with continuous sets in mind, we can define *open subsets* as the preimages of open intervals under functions $A \to \mathbb{R}$. Alternatively, we can construct or postulate a subtype $\Sigma \subseteq \Omega$ behaving like the Sierpinski space (usually called a *dominance*), and define an *open subset* to be one whose classifying map $A \to \Omega$ factors through Σ. In a topos of sheaves on a category $\mathcal{T}$ of spaces, there is an obvious choice of such a Σ, namely, the sheaf represented by the actual Sierpinski space (whether or not it is in $\mathcal{T}$). On the other hand, if we want to construct a particular Σ *inside* type theory, one possibility is the *Rosolini dominance* [95]:

$$\Sigma_{\text{Ros}} = \left(\sum_{P:\Omega} \exists f : \mathbb{N} \to \mathbf{2}, \left(P \cong \exists n : \mathbb{N}, (f(n) = 1) \right) \right).$$

That is, Σ_{Ros} is the type of propositions of the form $\exists n : \mathbb{N}, (f(n) = 1)$ for some $f : \mathbb{N} \to \mathbf{2}$. In consequential spaces, Σ_{Ros} is the Sierpinski space, so the resulting "open subsets" are as we would expect. (But in continuous sets, $\Sigma_{\text{Ros}} \cong \mathbf{2}$, so the only "open subsets" in this sense are unions of connected components.) However we choose Σ, once chosen we can develop much of classical point-set topology with these "synthetic open sets," including compactness, Hausdorffness, and so on. See [17, 40, 41, 111] for more examples of this sort of "synthetic topology." There are also other ways to define open sets synthetically; for instance, [91] defines $U \subseteq A$ to be open if $\forall x : A, \forall y : A, (x \in U \to (y \in U \vee x \neq y))$.

---

[30] The latter statement is sometimes known as *Brouwer's theorem*, since Brouwer proved it in his "intuitionistic" mathematics using principles derived from "choice sequences." It is *not* a theorem of pure constructive mathematics, which, unlike Brouwer's "intuitionism," is fully compatible with classical principles like excluded middle, though it does not assume them.

It is more subtle to obtain a synthetic theory of *smoothness*, since smoothness does not arise automatically from constructive logic the way that continuity and computation do. Indeed, it is easy to define nondifferentiable functions $\mathbb{R} \to \mathbb{R}$ in constructive mathematics, such as the absolute value. Semantically, the type $\mathbb{R}_d$ of Dedekind reals interpreted in the topos of *smooth* sets actually yields the sheaf of *continuous* (not necessarily smooth) real-valued functions on the domain spaces.[31]

Clearly a more interesting smooth set than this is the sheaf of *smooth* real-valued functions, which is equivalently the usual smooth manifold $\mathbb{R}$ regarded as a diffeological space. In the internal language of smooth sets, this appears as a type $\mathbb{R}_s$ of "smooth reals" living strictly in between the "discrete (Cauchy) reals" $\mathbb{R}_c$ and the "continuous (Dedekind) reals" $\mathbb{R}_d$. It seems unlikely that there is any type definable in type theory whose interpretation in smooth sets is $\mathbb{R}_s$, but we can at least write down some axioms that $\mathbb{R}_s$ satisfies, such as being a subring of $\mathbb{R}_d$, or more generally closed under the action of all "standard smooth functions" (see [45] for one way to make this precise).

A more transformative approach is to make the notion of "smoothness" synthetic as well, rather than relying on the classical limit definition of derivative. Following Grothendieck's insight into the importance of nilpotent elements in algebraic geometry, we can enhance the category $\mathcal{T} = \{\mathbb{R}^n\}_{n \in \mathbb{N}}$ by replacing each $\mathbb{R}^n$ by its algebra of smooth functions $C^\infty(\mathbb{R}^n)$ and turning the arrows around to obtain a category of $\mathbb{R}$-algebras, then adding new algebras that are "deformations" of some $C^\infty(\mathbb{R}^n)$ containing nilpotents. Whatever the details, the resulting topos will contain an internal ring $\mathbf{R}$ that enhances $\mathbb{R}_s$ to include nilpotent "infinitesimals," with $\mathbb{R}_s$ the quotient by these:

$$\begin{array}{c} \mathbf{R} \\ \downarrow \\ \mathbb{R}_c \to \mathbb{R}_s \to \mathbb{R}_d \end{array}$$

Nilpotents allow a synthetic definition of differentiation: if $D \subseteq \mathbf{R}$ is defined by $D = \{d : \mathbf{R} \mid d^2 = 0\}$, then for any $f : D \to \mathbf{R}$ there is a unique $f'(0) : \mathbf{R}$ such that $f(d) = f(0) + f'(0) \cdot d$ for all $d : D$. (This is sometimes called the *Kock–Lawvere axiom*.) In particular, all functions $f : \mathbf{R} \to \mathbf{R}$ are differentiable in this synthetic sense. The theory resulting from this and similar axioms that hold in the above sheaf toposes is called *synthetic differential geometry*; for further reading, see [19, 66, 70, 89] and Chapter 2.

---

[31] In particular, while it is true in a sense that "everything is smooth" in this topos, what this actually means is that each object comes with a "notion of smoothness," that could in some cases happen to coincide with mere continuity.

This section has been a very brief sketch of some ways to access the spatial structure of types internally. Rather than pursue any of these avenues in detail, I want to describe (in Section 5) a newer approach to synthetic topology that leverages more of the categorical and type-theoretic structure, involving *higher modal operators*. However, first we must move sideways to consider another very different latent structure in type theory: that of *homotopy spaces* or ∞-*groupoids*.

4 Homotopy Type Theory

4.1 The Mystery of Identity Types

For many years, the most mysterious part of Martin-Löf's type theory was the identity types "$x = y$." As mentioned in Sections 2.4 and 3.1, the semantic idea is that the dependent type $x: A, y: A \vdash (x = y)$ type represents the diagonal $\triangle_A : [\![A]\!] \to [\![A]\!] \times [\![A]\!]$, regarded as an object of $\mathbf{Ctx}_{/([\![A]\!] \times [\![A]\!])}$. Of course, $\triangle_A$ is automatically present in the classifying category, as defined in Section 2.4; but without the identity type it is not represented by any dependent type.

The rules of the identity type are difficult to understand at first, but essentially they use a universal property to express the fact that it is the diagonal. In fact, any object $f : B \to A$ of a slice category $\mathscr{C}/A$ has a universal property: it is the image of the terminal object $1_B : B \to B$ of $\mathscr{C}/B$ under the left adjoint $f_! : \mathscr{C}/B \to \mathscr{C}/A$ to pullback along f. In other words, for any object $g : C \to A$ of $\mathscr{C}/A$, morphisms $f \to g$ in $\mathscr{C}/A$ are in natural bijection with sections of the pullback $f^*(g) : f^*C \to B$:

$$\begin{array}{ccc} f^*C & \longrightarrow & C \\ \downarrow & \lrcorner & \downarrow g \\ B & \xrightarrow{f} & A \end{array} \quad \Longleftrightarrow \quad \begin{array}{c} B \cdots\cdots\cdots\cdots\!\!\!\!\!\!\!\!\to C \\ {}_f\searrow \;\; \swarrow_g \\ A \end{array}$$

Thus, $\triangle_A : A \to A \times A$ is characterized in $\mathscr{C}/(A \times A)$ by saying that for $g : C \to A \times A$, morphisms $\triangle_A \to g$ are naturally bijective to sections of $\triangle_A^*(g)$:

$$\begin{array}{ccc} \triangle_A^* C & \longrightarrow & C \\ \downarrow & \lrcorner & \downarrow g \\ A & \xrightarrow{\triangle_A} & A \times A \end{array} \quad \Longleftrightarrow \quad \begin{array}{c} A \cdots\cdots\cdots\cdots\!\!\!\!\!\!\!\!\to C \\ {}_{\triangle_A}\searrow \;\; \swarrow_g \\ A \times A \end{array}$$

If we represent g by a dependent type $x: A, y: A \vdash C$ type, then the pullback on the left corresponds to substitution of the same variable for both x and y, for example, $C[w/x, w/y]$ in context $w: A$. The section on the left is then a term $w: A \vdash c: C[w/x, w/y]$, whereas the induced map on the right corresponds to a term $x: A, y: A, p: x = y \vdash d: C$. If we represent the latter in Yoneda form, as we did for (2.11), we obtain the following rule:

$$\frac{\Gamma, x: A, y: A \vdash C \text{ type} \qquad \Gamma, w: A \vdash c: C[w/x, w/y]}{\Gamma \vdash \mathsf{J}(C, c, p): C[a/x, b/y]} \quad \Gamma \vdash a: A \quad \Gamma \vdash b: A \quad \Gamma \vdash p: a = b \tag{4.1}$$

The corresponding introduction rule is just the unit of this adjunction, saying that the diagonal of A has a specified section when pulled back along itself. Logically, it expresses the reflexivity of equality:

$$\frac{\Gamma \vdash A \text{ type} \qquad \Gamma \vdash a: A}{\Gamma \vdash \mathsf{refl}_a: a = a}$$

In the classifying category, this rule is a section as on the left below:

where Id denotes $[\![x: A, y: A, p: x = y]\!]$. Thus, refl gives a morphism as on the right above; the universal property should make this an isomorphism $[\![A]\!] \cong \mathsf{Id}_A$.

This adjoint characterization of equality is due to Lawvere [71], but it is closely related to Leibniz's "indiscernibility of identicals." Specifically, given a property $x: A \vdash P(x): \Omega$, we can form $x: A, y: A \vdash (P(x) \to P(y)): \Omega$. Taking this as C, we have the second hypothesis of (4.1) given by $w: A \vdash (\lambda p.p): P(w) \to P(w)$. Thus, from (4.1) we get

$$x: A, y: A, p: x = y \vdash \mathsf{J}(\cdots): P(x) \to P(y).$$

This says that if x and y are equal ("identical") then any property that holds of x also holds of y ("indiscernible"). The function $P(x) \to P(y)$ induced by p is often denoted p_* and called *substitution* or *transport*.

In dependent type theory, we need to enhance (4.1) to allow C to depend on a "witness of equality" $p: x = y$ as well, and also add a computation rule relating it to the introduction rule refl_a. This yields the rules shown in Figure 6.5, which are due to Martin-Löf [84, 85]. Just as for coproducts, we expect this

$$\frac{\Gamma \vdash A \text{ type} \quad \Gamma \vdash a\colon A \quad \Gamma \vdash b\colon A}{\Gamma \vdash (a = b) \text{ type}} \qquad \frac{\Gamma \vdash A \text{ type} \quad \Gamma \vdash a\colon A}{\Gamma \vdash \mathsf{refl}_a \colon a = a}$$

$$\frac{\Gamma, x\colon A, y\colon A, e\colon x = y \vdash C \text{ type}}{\Gamma, w\colon A \vdash c\colon C[w/x, w/y, \mathsf{refl}_w/e] \quad \Gamma \vdash a\colon A \quad \Gamma \vdash b\colon A \quad \Gamma \vdash p\colon a = b}{\Gamma \vdash \mathsf{J}(C, c, p)\colon C[a/x, b/y, p/e]}$$

$$\frac{\Gamma, x\colon A, y\colon A, e\colon x = y \vdash C \text{ type}}{\Gamma, x\colon A \vdash c\colon C[x/y, \mathsf{refl}_x/e] \quad \Gamma \vdash a\colon A}{\Gamma \vdash \mathsf{J}(C, c, \mathsf{refl}_a) \equiv c[a/x]}$$

Figure 6.5 The rules for identity types.

enhancement to ensure the full universal property of the desired adjunction. When phrased informally in terms of points, the stronger elimination rule says that if we want to perform a construction or proof involving a general element $p\colon x = y$ (for general x and y), it suffices to consider the case when y is x and p is refl_x. This is formally analogous to the elimination rule for (say) coproducts, which says that to perform a construction or proof involving a general element $z\colon A + B$, it suffices to consider the two cases when z is $\mathsf{inl}(x)$ and $\mathsf{inl}(y)$.

It may seem odd to introduce a new notation and system of rules for an object (the identity type) that turns out to be isomorphic to something we already had (the diagonal map). The point is that dependent type theory has two different ways of representing a morphism $A \to B$, depending on whether we view it as simply a morphism between two objects or as an object of the slice category over its codomain B. The same is true in set theory: we can have a function $f\colon A \to B$ between sets, or we can have a B-indexed family of sets $\{A_b\}_{b \in B}$, and up to isomorphism the two are equivalent by $A = \coprod_b A_b$ and $A_b = \{a \in A \mid f(a) = b\}$. As mentioned in Section 2.4, enabling us to think in terms of "indexed families" in an arbitrary category is actually one of the *strengths* of dependent type theory, and the identity type and Σ-type are exactly what supply the isomorphism between functions and families in type theory.

However, although the rules in Figure 6.5 are well motivated categorically from a universal property, they appear to have a serious problem. Specifically, if the identity type satisfies these rules only, and we define the classifying category as in Section 2.4, then it does not have pullbacks!

Consider, for instance, how we might try to pull back a dependent type $[\![y\colon B, z\colon C(y)]\!] \to [\![B]\!]$ along a morphism $[\![f]\!]\colon [\![A]\!] \to [\![B]\!]$. In Section 2.4

we claimed that such a pullback should be obtained by substituting $f(x)$ for y in $C(y)$. To check the universal property, we would consider a diagram as below:

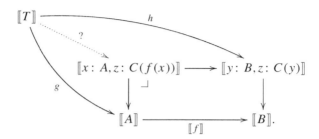

Then g is a term $t: T \vdash g(t): A$, while h is determined by two terms $t: T \vdash h_1(x): B$ and $t: T \vdash h_2(t): C(h_1(x))$. Now it seems as though the commutativity of the square would force $h_1(t)$ to be $f(g(t))$, so that h_2 would be a term $t: T \vdash h_2(t): C[f(g(t))/y]$, or equivalently $t: T \vdash h_2(t): C[f(x)/y][g(t)/x]$, inducing the dotted morphism.

This appealing argument stumbles on the fact that we quotiented the morphisms in **Ctx** by an equivalence relation induced by the identity type. Thus, to say that the above square commutes does not mean that $h_1(t)$ is literally $f(g(t))$, only that we have a term $t: T \vdash p: h_1(t) = f(g(t))$. This is not by itself the end of the world, because p induces a transport function $p_*: C(h_1(t)) \to C(f(g(t)))$, so we can define the dotted morphism as $t: T \vdash p_*(h_2(t)): C(f(g(t)))$. The real problem is that this morphism depends on the choice of the term p, but the term p is not *specified* by the mere *fact* that the outer square commutes; thus the dotted factorization is not unique.

I have stated this problem in a form that may make its solution seem obvious to a modern reader with a certain background. However, for many years after Martin-Löf, no one with this background looked at the problem; nor was it stated in this way. Instead, the question was whether we can prove *inside type theory* that the identity type $x: B, y: B \vdash (x = y)$ type is a "proposition" in the sense of Sections 2.6 and 3.1, that is, any two of its elements are equal. Categorically, this would mean proving that the projection $[\![x: B, y: B, p: x = y]\!] \to [\![x: B, y: B]\!]$ is a monomorphism (as we expect, if it is to be the diagonal); while syntactically it would mean constructing, given $p: x = y$ and $q: x = y$, a term of $p = q$. If this were the case, different choices of p would result in terms $p_*(h_2(t))$ that are equal, so that uniqueness for the pullback would be restored.

It turns out, however, that we *cannot* prove that the identity type is always a proposition. Thus, people (starting with Martin-Löf) considered adding this statement as an extra stand-alone axiom,

$$\frac{\Gamma \vdash p: a = b \qquad \Gamma \vdash q: a = b}{\Gamma \vdash e: p = q}, \qquad (4.2)$$

called *uniqueness of identity proofs* (UIP). Type theory with UIP is sometimes called *extensional*,[32] while type theory without UIP is called *intensional*. Thus, as mentioned in Section 2.5, it is *extensional* Martin-Löf type theory for which our previous construction of **Ctx**($\mathfrak{T}$) presents free locally cartesian closed categories (or other kinds of structured categories). However, baldly assuming UIP is unsatisfying, since it does not fit into the system of rule packages motivated by universal properties, as described in Section 2.5. Moreover, this approach provides no insight into why UIP might be true, or why it is not provable.

But to a modern reader with a background in homotopy theory, the above problem looks familiar: it is the same reason why the homotopy category of spaces does not have pullbacks. In that case, we consider instead *homotopy* pullbacks, where the factorization morphism is *required* to depend on a choice of homotopy filling the square. This suggests we should regard **Ctx** as a "homotopy theory" or $(\infty, 1)$-category, thereby explaining why we cannot prove that $x = y$ is a proposition: diagonals in an $(\infty, 1)$-category are *not* in general monic. For instance, in the 2-category of groupoids, the monomorphisms are the fully faithful functors, but a diagonal $G \to G \times G$ is not generally full: its functorial action on hom-sets

$$\hom_G(x, y) \longrightarrow \hom_{G \times G}((x, x), (y, y)) = \hom_G(x, y) \times \hom_G(x, y)$$

is not an isomorphism if $\hom_G(x, y)$ has more than one element. The first model of type theory using this idea was constructed by [52] using groupoids; later authors [11, 64] generalized it using homotopy theory.

This situation should be compared with the remarks about constructive logic in Section 3.2. In both cases we have a rule (LEM or UIP) that seems reasonable given one model or class of models (the category of sets, or all 1-categories). But this rule turns out not to be provable, because type theory admits more general models, in some of which the rule is false. This provides us with the proper attitude towards the rule: assuming it simply means restricting the class of categories we are interested in, whereas declining to assume it allows us to use type theory as a syntax for a wider class of models.

---

[32] More precisely, "propositionally extensional"; see Section 4.7.

Unfortunately, there is much unresolved subtlety in an $(\infty, 1)$-categorical interpretation of type theory. One hopes for an analogue of the 1-categorical situation, with a "classifying $(\infty, 1)$-category" that is free in some $(\infty, 1)$-category of structured $(\infty, 1)$-categories, setting up an $(\infty, 1)$-adjunction, but it is quite difficult to make this precise. In Section 4.6 I will sketch the current state of the art; until then I will just assume that the problem will be solved somehow, as I believe it will be.

4.2 Types as ∞-Groupoids

In Section 3.4 we saw that if we decline to assume LEM, we can detect the potential spatial structure of types internally. Similarly, if we decline to assume UIP, then types have potential "homotopy space" or "∞-groupoid" structure, and the natural way to try to detect this is by using the identity types. But what does $x = y$ mean when it is not just a proposition?

In higher category theory, we have a notion of *n-groupoid*, which is an ∞-groupoid containing no interesting information above dimension n. This can be defined inductively: a 0-groupoid is an ∞-groupoid that is equivalent to a discrete set, while an $(n + 1)$-groupoid is one all of whose hom-∞-groupoids $\hom_A(u, v)$ are n-groupoids. Moreover, we can extend the induction downwards two more steps: an ∞-groupoid is a 0-groupoid just when each $\hom_A(u, v)$ is empty or contractible, so it makes sense to define a (-1)-*groupoid* to be an ∞-groupoid that is either empty or contractible. Similarly, an ∞-groupoid is a (-1)-groupoid just when its homs are *all* contractible, so we can define a (-2)-*groupoid* to be a contractible one. (See, for instance, [15, Section 2].)

In particular, when we regard a set as an ∞-groupoid, the proposition that two elements u, v are equal turns into the (-1)-groupoid $\hom_A(u, v)$. Thus, the homs of an ∞-groupoid generalize the notion of equality for elements of a set, so it is natural to expect the *type* $u = v$ to behave like $\hom_A(u, v)$. This is correct: we can derive all the composition structure on these hom-objects that should be present in an ∞-groupoid from the rules in Figure 6.5 [78, 113]. For instance, we can construct the composition law

$$(x = y) \times (y = z) \to (x = z)$$

by applying the eliminator to $p \colon x = y$ to assume that y is x and p is refl_x, in which case the other given $q \colon y = z$ has the same type as the goal $x = z$. (This is the same as the proof of transitivity of equality in extensional type theory.)

Other aspects of homotopy theory can also be defined using the identity type. For instance, the *loop space* $\Omega(A, a)$ of a type A at a point a is just the identity type $(a = a)$. Voevodsky also showed that we can mimic the above inductive definition of n-groupoids, also called *homotopy n-types*:[33] a type A is an $(n + 1)$-type if for all $x \colon A$ and $y \colon A$ the type $(x = y)$ is a n-type. We can start at $n = -1$ with the propositions as defined in Section 3.1, that is, types A such that for all $x \colon A$ and $y \colon A$ we have $x = y$. We can also start at $n = -2$ with the *contractible types*, which are just the propositions that have an element. Note that the homotopy 0-types, also called *sets*, are those that satisfy UIP; so UIP could equivalently be phrased as "all types are sets."

Two types are *homotopy equivalent* if we have $f \colon A \to B$ and $g \colon B \to A$ such that $g \circ f = 1_A$ and $f \circ g = 1_B$. However, the type of such data

$$\sum_{f:A \to B} \sum_{g:B \to A} (g \circ f = 1_A) \times (f \circ g = 1_B) \qquad (4.3)$$

is not a correct definition of *the type of homotopy equivalences*. (It is correct if A and B are sets, in which case we generally say *isomorphism* or *bijection* rather than *equivalence*.) The problem is that, given $f \colon A \to B$, the rest of (4.3),

$$\sum_{g:B \to A} (g \circ f = 1_A) \times (f \circ g = 1_B), \qquad (4.4)$$

may not be a proposition, whereas we want "being an equivalence" to be a mere *property* of a morphism. For instance, if f is the identity map of the homotopical circle S^1 (see Section 4.5), then (4.4) is equivalent to $\mathbb{Z}$. Thus if we took (4.3) as our definition of equivalence, there would be "infinitely many self-equivalences of S^1," which is not correct: up to homotopy, there should be only two.

Many equivalent ways to correct (4.3) are now known; here are a few:

$$\sum_{f:A \to B} \sum_{g:B \to A} \sum_{h:B \to A} (g \circ f = 1_A) \times (f \circ h = 1_B), \qquad (4.5)$$

$$\sum_{f:A \to B} \sum_{g:B \to A} \sum_{\eta: g \circ f = 1_A} \sum_{\epsilon: f \circ g = 1_B} f(\eta) = \epsilon_f, \qquad (4.6)$$

$$\sum_{f:A \to B} \left\| \sum_{g:B \to A} (g \circ f = 1_A) \times (f \circ g = 1_B) \right\|, \qquad (4.7)$$

$$\sum_{f:A \to B} \prod_{y:B} \left(\left(\sum_{x:A} f(x) = y \right) \text{ is contractible} \right). \qquad (4.8)$$

Each of these admits maps back and forth from (4.3), while the data after the $\sum_{f:A \to B}$ form a proposition. We can think of these as building contractible cell complexes. For instance, in (4.5), we glue on two 1-cells g, h each with a contracting 2-cell, giving a contractible space, whereas in (4.3), we glue on

---

[33] Voevodsky's terminology [117] is "type of h-level $n + 2$."

one 1-cell with *two* contracting 2-cells, giving a noncontractible 2-sphere. And in (4.6), we add to (4.3) a 3-cell filler, getting a contractible 3-ball.

When making definitions of this sort, we generally think of types as ∞-groupoids (or homotopy spaces), just as in Section 3.2 we thought of types as sets. However, since type theory presents an initial structured $(\infty, 1)$-category, these definitions can also be interpreted in any structured $(\infty, 1)$-category, yielding "classifying spaces" for n-types and equivalences. For example, given $\Gamma \vdash A$ type and $\Gamma \vdash B$ type, if $\Gamma \vdash \mathsf{Equiv}(A, B)$ type denotes the type of equivalences (with any of the corrected definitions above), then the object $[\![\mathsf{Equiv}(A, B)]\!] \to [\![\Gamma]\!]$ of the slice category has the universal property that, for any map $f \colon X \to [\![\Gamma]\!]$, lifts of f to $[\![\mathsf{Equiv}(A, B)]\!]$ are equivalent to homotopy equivalences $f^*[\![A]\!] \simeq f^*[\![B]\!]$ over X. (See [64, Section 3] or [103, Section 4].) In other words, $[\![\mathsf{Equiv}(A, B)]\!]$ is a "classifying space for equivalences between $[\![A]\!]$ and $[\![B]\!]$."

4.3 Extensionality and Univalence

In Section 3.4 we "actualized" the potential spatial structure of types by adding axioms such as a dominance or a type of smooth reals. Similarly, we can add axioms ensuring that there really are types with higher groupoid structure – that is, that not all types are sets, or that UIP fails. The only serious contender for such an axiom at present is Voevodsky's *univalence axiom*.

To explain univalence, let us return to the type constructors discussed in Sections 2.5 and 2.6. We claimed that the rules for coproduct types, function types, subobject classifiers, and so on express their desired categorical universal properties. For most types, this is literally true, but there are a couple of cases[34] in which something is missing from our discussion so far.

First, the universal property of an exponential object requires that any map $X \times [\![A]\!] \to [\![B]\!]$ factor through a *unique* map $X \to [\![B]\!]^{[\![A]\!]}$. For $[\![A \to B]\!]$ to be $[\![B]\!]^{[\![A]\!]}$, therefore, requires that if $\Gamma \vdash f \colon A \to B$ and $\Gamma \vdash g \colon A \to B$ and $\Gamma, x \colon A \vdash h \colon f(x) = g(x)$, then also $\Gamma \vdash e \colon f = g$ (because elements of the identity type induce equalities of morphisms in **Ctx**, or homotopies in $\mathscr{C}\!tx$). This is not derivable from the rules in Figure 6.1; it is an extra

---

[34] What these cases have in common is that they are "mapping in" universal properties. "Mapping out" universal properties, like for coproducts, can be expressed more powerfully in type theory using a dependent output such as in (2.12), enabling us to derive their full universal property from the basic rules. (Of course, the cartesian product also has a "mapping in" universal property, but it does not have this problem; formally, this is because the classifying category is better described as a sort of "cartesian multicategory" in which the cartesian product also has a "mapping out" universal property.)

axiom called *function extensionality*. Informally, it says that two functions are equal if they take equal values. (Dependent function types require a similar axiom.)

Second, the universal property of a subobject classifier requires that a mono $M \to [\![\Gamma]\!]$ is classified by a *unique* map $[\![\Gamma]\!] \to \Omega$, or equivalently, two maps $[\![\Gamma]\!] \to \Omega$ classifying the same subobject of $[\![\Gamma]\!]$ are equal. Here "the same" means *isomorphism* in $\mathbf{Ctx}_{/[\![\Gamma]\!]}$; a classifying map only determines a mono up to isomorphism anyway. Type-theoretically, this means that if $\Gamma \vdash P \colon \Omega$ and $\Gamma \vdash Q \colon \Omega$ and $\Gamma \vdash h \colon \mathsf{Equiv}(P, Q)$, then $\Gamma \vdash e \colon P = Q$. This is not derivable from the rules in Section 2.6; it is an extra axiom called *propositional extensionality*. (Note that two propositions are equivalent as soon as each implies the other.)

When we try to generalize propositional extensionality for Ω to a statement about type universes $\mathcal{U}$, things become more subtle. We still expect classifying maps to classify only up to isomorphism – or better, up to homotopy equivalence, which leads us towards homotopical classifying spaces. In traditional homotopy theory (e.g., [86]), *homotopy classes* of maps into a classifying space correspond to *homotopy equivalence classes* of fibrations over it. But in an $(\infty, 1)$-category, it is more natural to ask directly that the ∞-groupoid $\hom(X, \mathcal{U})$ is equivalent to a full sub-∞-groupoid of the slice category over X; this gives the notion of an *object classifier* [80, Section 6.1.6]. Type-theoretically, the corresponding condition is that for $\Gamma \vdash A \colon \mathcal{U}$ and $\Gamma \vdash B \colon \mathcal{U}$, the type $\Gamma \vdash (A = B)$ type (i.e., the ∞-groupoid of homotopies between classifying maps) is *equivalent* to the type $\mathsf{Equiv}(A, B)$ of homotopy equivalences as in Section 4.2. More precisely, identity-elimination yields a function $\mathsf{idtoeqv}_{A, B} \colon (A = B) \to \mathsf{Equiv}(A, B)$, and we should require that this map is itself an equivalence.

This axiom is due to Voevodsky, who dubbed it *univalence*.[35] Univalence clearly implies propositional extensionality, while Voevodsky showed [117] that it also implies function extensionality (see, e.g., [112, Section 4.9]). Univalence also does indeed ensure that not all types are sets (i.e., homotopy 0-types). For instance, if $B \colon \mathcal{U}$ has a nontrivial automorphism, such as $B = \mathbf{1} + \mathbf{1}$, then $\mathsf{Equiv}(B, B)$ is not a proposition. Hence, neither is the equality

---

[35] By analogy with function extensionality and propositional extensionality, univalence could be called *typal extensionality*. In particular, like function extensionality and propositional extensionality, univalence is an "extensionality" property, meaning that "types are determined by their behavior." For this reason, it is unfortunate that the phrase "extensional type theory" has come to refer to type theory with UIP, which is incompatible with univalence. Historically, the special case of univalence when A and B are sets, in which case one can use (4.3) without "correction," was proposed by Hofmann and Streicher [52] under the name "universe extensionality," but it did not attract much attention.

type $B = B$ in $\mathcal{U}$, so $\mathcal{U}$ is not a set. More generally, with a hierarchy of universes $\mathcal{U}_n$ with $\mathcal{U}_n : \mathcal{U}_{n+1}$, each $\mathcal{U}_n$ is not an n-type [67].

In particular, for models in the category of sets, or more generally in any 1-category, univalence must be *false*. For instance, any "Grothendieck universe" in ZFC set theory can be used as a type-theoretic universe $\mathcal{U}$ in **Set**; but it is not univalent, since it would be a set (a 0-type), whereas the above argument shows no univalent universe containing a 2-element set can be a set.

Formally, univalence is an axiom like UIP and LEM that cuts down our collection of models, only now in a way that *excludes* all 1-categories. Just as the topological axioms from Section 3.4 are incompatible with LEM, univalence is incompatible with UIP. These two oppositions are essentially independent: the topos of sets satisfies both LEM and UIP, the toposes of consequential spaces and continuous sets satisfy UIP together with topological axioms instead of LEM, and the $(\infty, 1)$-category of ∞-groupoids satisfies univalence instead of UIP but still satisfies LEM. (In particular, LEM does not rule out *all* "spatial" interpretations of type theory, at least if we regard ∞-groupoids as a kind of "space.") Finally, in Section 5, we will mention some $(\infty, 1)$-categories that combine topological axioms with univalence, thus satisfying neither LEM nor UIP.[36]

To a homotopy theorist or higher category theorist, assuming univalence instead of UIP is obviously the right move; but it can be a difficult step for those used to thinking of types as sets. However, univalence can also be motivated from purely type-theoretic considerations, as giving a "correct" answer to the question "what are the identity types of a universe?" just as function extensionality answers "what are the identity types of a function type?" And from a philosophical point of view, univalence says that all properties of types are invariant under equivalence, since we can make any equivalence into an equality and apply transport; thus it expresses a strong "structural" nature of type theory [8, 35], in contrast to ZFC-style set theory.

For the homotopy theorist, univalence is one place where we start to see the advantage of the type-theoretic syntax. Inside type theory, the "elements" of a universe type $\mathcal{U}$ *are themselves types*, in contrast to the classical construction of classifying spaces whose "points" often lack a meaning directly connected to the things being classified. This enables us to define other classifying spaces and operations between them in a very intuitive way. For instance, if G is

---

[36] Most "classicality" properties like the axiom of choice behave similarly to LEM in this way, but some very strong choice principles do conflict with univalence, such as the existence of a global "Hilbert choice" operator (see [112, Section 3.2 and Exercise 3.11]).

a group (meaning a *set*, a 0-type, with a group structure), we can define its classifying space to be "the type of free transitive G-sets":

$$\mathbf{B}G = \sum_{A:\mathcal{U}} \sum_{a:G \times A \to A} (A \text{ is a set and } a \text{ is a free transitive action}).$$

That is, an element of $\mathbf{B}G$ is a tuple $(A, a, \ldots)$ consisting of a type, an action of G on that type, and witnesses of the truth of the necessary axioms. It turns out that $\mathbf{B}G$ is a connected 1-type with $\Omega(\mathbf{B}G) = G$. If G is abelian, we can define an operation $\mathbf{B}G \times \mathbf{B}G \to \mathbf{B}G$ by taking the "tensor product" of G-sets, and so on.

This definition of $\mathbf{B}G$ also immediately defines the objects it classifies: a "torsor" over a type X is just a function $X \to \mathbf{B}G$. The first component of such a function is a map $X \to \mathcal{U}$, corresponding to a dependent type $x: X \vdash A$ type, and hence a map $[\![A]\!] \to [\![X]\!]$. The rest of the classifying map equips this with the usual structure of a torsor over X.

In fact, *any* definition of a structure in type theory automatically defines the classifying space for such structures and therefore also automatically the corresponding notion of "bundle of structures." For instance, a group can be considered a tuple $(G, e, m, \ldots)$ of a set, an identity, a multiplication, and proofs of the axioms, giving a definition of "the type of groups":

$$\mathsf{Group} = \sum_{G:\mathcal{U}} \sum_{e:G} \sum_{m:G \times G \to G} (G \text{ is a set and } (m, e) \text{ is a group structure}).$$

We then automatically obtain a notion of a "family of groups," namely, a function $X \to \mathsf{Group}$. This turns out to correspond precisely to a *local system* of groups in the sense of classical homotopy theory. Similarly, we can define a *spectrum* to be a sequence of pointed types (X_n, x_n) each of which is the loop space of the next; thus the "type of spectra" is

$$\mathsf{Spectrum} = \sum_{X:\mathbb{N} \to \mathcal{U}} \sum_{x:\prod_{n:\mathbb{N}} X_n} \prod_{n:\mathbb{N}} \big((X_n, x_n) = \Omega(X_{n+1}, x_{n+1})\big).$$

This yields automatically a notion of "parameterized spectrum," namely, a function $X \to \mathsf{Spectra}$. The homotopy groups of a spectrum are functions $\pi_n: \mathsf{Spectrum} \to \mathsf{AbGroup}$, while the Eilenberg–Mac Lane construction is a function $H: \mathsf{AbGroup} \to \mathsf{Spectrum}$; these then act by simple composition to relate parameterized spectra and local systems. Thus, type theory automatically handles generalizations to "parameterized spaces," which in classical homotopy theory and category theory have to be done by hand.

4.4 Higher Inductive Types

In Section 2.5 we mentioned type constructors corresponding to coproducts, products, exponentials, initial and terminal objects, diagonals, and natural

$$\dfrac{\Gamma \vdash f : A \to B \qquad \Gamma \vdash g : A \to B}{\Gamma \vdash \mathsf{coeq}(f,g)\ \text{type}} \qquad \dfrac{\Gamma \vdash N : B}{\Gamma \vdash \langle N \rangle : \mathsf{coeq}(f,g)}$$

$$\dfrac{\Gamma \vdash M : A}{\Gamma \vdash \mathsf{ceq}(M) : \langle f(M) \rangle = \langle g(M) \rangle}$$

$$\dfrac{\Gamma, z : \mathsf{coeq}(f,g) \vdash C\ \text{type} \qquad \Gamma, y : B \vdash c_B : C[\langle y \rangle / z]}{\Gamma, x : A \vdash c_A : \mathsf{ceq}(x)_*(c_B[f(x)/y]) = c_B[g(x)/y] \qquad \Gamma \vdash P : \mathsf{coeq}(f,g)}{\Gamma \vdash \mathsf{cind}(C, c_B, c_A, P) : C[P/z]}$$

$$\vdots$$

$$\Gamma \vdash \mathsf{cind}(C, c_B, c_A, \langle N \rangle) = c_B[N/y]$$

$$\vdots$$

$$\Gamma \vdash \mathsf{ap}_{\mathsf{cind}(C, c_B, c_A)}(\mathsf{ceq}(M)) = c_A[M/x]$$

Figure 6.6 The rules for coequalizer types.

numbers objects. Combining dependent sum types with the identity type yields all finite limits; for instance, the pullback of $f : A \to C$ and $g : B \to C$ is

$$\textstyle\sum_{x:A} \sum_{y:B} (f(x) = g(y)).$$

Moreover, with the natural numbers type we can express certain infinite limits, for example, the limit of a sequence $\cdots \xrightarrow{p_2} A_2 \xrightarrow{p_1} A_1 \xrightarrow{p_0} A_0$ is

$$\textstyle\sum_{f : \prod_{n:\mathbb{N}} A_n} \prod_{n:\mathbb{N}} p_n(f(n+1)) = f(n).$$

However, to represent *colimits* other than coproducts we need new type constructors. For instance, the rules for the coequalizer type $\mathsf{coeq}(f,g)$ are shown in Figure 6.6. The first is formation: any $f, g : A \to B$ have a coequalizer. The next two are introduction: there is a map $B \to \mathsf{coeq}(f,g)$, and the two composites $A \rightrightarrows B \to \mathsf{coeq}(f,g)$ are equal. The third is the elimination rule, which is analogous to the case analysis rule (2.12). To understand this, consider first the simpler version analogous to (2.11), where C does not depend on $\mathsf{coeq}(f,g)$:

$$\dfrac{\Gamma \vdash C\ \text{type} \qquad \Gamma, y : B \vdash c_B : C}{\Gamma, x : A \vdash c_A : c_B[f(x)/y] = c_B[g(x)/y] \qquad \Gamma \vdash P : \mathsf{coeq}(f,g)}{\Gamma \vdash \mathsf{cind}(C, c_B, c_A, P) : C}.$$

This expresses the existence part of the universal property of a coequalizer: given a map $B \to C$ such that the composites $A \rightrightarrows B \to C$ are equal, there is an induced map $\operatorname{coeq}(f, g) \to C$.

As with coproducts, the more general version in Figure 6.6 also implies the uniqueness part of the universal property. It contains one new aspect: if C depends on $\operatorname{coeq}(f, g)$, then $c_B[f(x)/y]$ and $c_B[g(x)/y]$ have different types $C[f(x)/y]$ and $C[g(x)/y]$, so we cannot write "$c_B[f(x)/y] = c_B[g(x)/y]$." But we have $\operatorname{ceq}(x): f(x) = g(x)$, so the types $C[f(x)/y]$ and $C[g(x)/y]$ ought to be "the same"; but formally we need to "transport" $c_B[f(x)/y]$ along $\operatorname{ceq}(x)$ (using identity-elimination) to get an element of $C[g(x)/y]$ that we can compare to $c_B[g(x)/y]$. This is what the notation $\operatorname{ceq}(x)_*(c_B[f(x)/y])$ means.

Finally, the last two rules (in which I have omitted the premises for brevity) are the computation rules. The first says that when a map $\operatorname{coeq}(f, g) \to C$ is induced by the universal property, the composite $B \to \operatorname{coeq}(f, g) \to C$ is indeed the original map $B \to C$. The second says similarly that the "induced equality" between the composites $A \rightrightarrows B \to \operatorname{coeq}(f, g) \to C$ is the originally given one. (Do not worry about the notation; it is not important for us.)

If this "equality of equalities" sounds weird, recall that in *homotopy* type theory, the type $x = y$ represents the hom-∞-groupoid and hence can have many different elements. Thus, it makes sense to ask whether two such "equalities" are equal. In fact, when we regard type theory as presenting an $(\infty, 1)$-category rather than a 1-category, the type $\operatorname{coeq}(f, g)$ represents an ∞-categorical coequalizer, aka, a homotopy coequalizer. From this we can build all finite (homotopy) colimits, as in [80, Corollary 4.4.2.4]. We also obtain certain infinite colimits: for example, the coproduct of a countably infinite family $A : \mathbb{N} \to \mathcal{U}$ is just $\sum_{n:\mathbb{N}} A_n$, and the colimit of a sequence $A_0 \xrightarrow{f_0} A_1 \xrightarrow{f_1} A_2 \xrightarrow{f_2} \cdots$ is the coequalizer of two maps $(\sum_{n:\mathbb{N}} A_n) \rightrightarrows (\sum_{n:\mathbb{N}} A_n)$.

The rules for $\operatorname{coeq}(f, g)$ do not *require* any ∞-categorical behavior, and are perfectly consistent with UIP. In particular, adding them to *extensional* MLTT yields a type theory for locally cartesian closed categories with finite colimits. Nevertheless, types such as $\operatorname{coeq}(f, g)$ were not widely studied prior to the advent of homotopy type theory; they are known as *higher inductive types*.

In general, an *inductive type* W is specified by a list of *constructors*, which are (possibly dependent) functions into W. For instance, the coproduct $A + B$ is the inductive type specified by two constructors $\operatorname{inl}: A \to A + B$ and $\operatorname{inr}: B \to A + B$. (The empty type $\emptyset$ is inductively specified by *no* constructors.) The constructors are the introduction rules of the resulting type, while the elimination rule says that to define a map out of the inductive type it is sufficient to specify its behavior on the constructors.

A *higher inductive type* (HIT) is similar, but the constructors can also be functions into equality types of the HIT. For instance, $\text{coeq}(f, g)$ is specified by two constructors $\langle - \rangle : B \to \text{coeq}(f, g)$ and $\text{ceq} : \prod_{x:A}(\langle f(x) \rangle = \langle g(x) \rangle)$.

The word "inductive" comes from the fact that in general, the type being defined is allowed to appear in the *domains* of its constructors in certain limited ways. For instance, the natural numbers are the inductive type specified by two constructors $0 : \mathbb{N}$ (a 0-ary function) and $\text{succ} : \mathbb{N} \to \mathbb{N}$. Informally, this means that the elements of $\mathbb{N}$ are generated by applying the constructors successively any number of times; thus we have 0, $\text{succ}(0)$, $\text{succ}(\text{succ}(0))$, and so on.

When combined with higher constructors, this additional feature is quite powerful; for instance, the propositional truncation $\|A\|$ from Figure 6.4 is the HIT specified by two constructors $|-| : A \to \|A\|$ and $\text{tprp} : \prod_{x,y:\|A\|}(x = y)$. We can similarly construct an *n-truncation* that is the universal map $A \to \|A\|_n$ into a homotopy n-type (i.e., its nth Postnikov section). In particular, the 0-truncation $\|A\|_0$ is the "set of connected components."

"Recursive" HITs of this sort can also be used to construct more exotic objects, such as homotopical localizations. Given a map $f : S \to T$, we say that a type A is f-*local* if the map $(- \circ f) : (T \to A) \to (S \to A)$ is an equivalence. The f-*localization* is the universal map from a type X into an f-local type $L_f X$. In classical homotopy theory, constructing localizations in general requires a fairly elaborate transfinite composition. But in homotopy type theory, we can simply define $L_f X$ to be the HIT generated by the following constructors:

- A map $\eta : X \to L_f X$.
- For each $g : S \to L_f X$ and $t : T$, an element $\text{ext}(g, t) : L_f X$.
- For each $g : S \to L_f X$ and $s : S$, an equality $\text{ext}(g, f(s)) = g(s)$.
- For each $g : S \to L_f X$ and $t : T$, an element $\text{ext}'(g, t) : L_f X$.
- For each $h : T \to L_f X$ and $t : T$, an equality $\text{ext}'(h \circ f, t) = h(t)$.

The last four constructors combine to lift $(- \circ f)$ to an element of (4.5). (This is why we have both ext and ext'; if we collapsed them into one we would only get (4.3).) This is one example of how homotopy type theory gives a "direct" way of working with objects and constructions that in classical homotopy theory must be laboriously built up out of sets. For more examples and theory of higher inductive types and their applications, see [112, Chapter 6].

4.5 Synthetic Homotopy Theory

With HITs we can define many familiar spaces from classical homotopy theory. For instance, in the ∞-category of ∞-groupoids, the circle S^1 is the homotopy coequalizer of $\mathbf{1} \rightrightarrows \mathbf{1}$; thus we expect the corresponding coequalizer type to behave like an "internal S^1" in homotopy type theory. It is equivalently the HIT generated by two constructors base: S^1 and loop: base = base; its elimination rule (universal property) says roughly that to give a map $S^1 \to C$ is equivalent to giving a point $c\colon C$ and a loop $l\colon c = c$.

Since HITs are consistent with UIP, this "circle" may not behave as expected: in a 1-category, the coequalizer of $\mathbf{1} \rightrightarrows \mathbf{1}$ is just $\mathbf{1}$. But if we also assume univalence, type theory becomes a powerful tool for working directly with ∞-groupoids such as S^1. By the universal property of S^1, to give a dependent type $C\colon S^1 \to \mathcal{U}$, we must give a type $B\colon \mathcal{U}$ and an equality $B = B$; but by univalence, the latter is the same as an *autoequivalence* of B. For instance, if B is $\mathbb{Z}$, we can use the autoequivalence "+1"; the resulting dependent type is then a version of the *universal cover* of S^1. With a little extra work [77], we can adapt the classical calculation of $\pi_1(S^1)$ to show, in type theory, that $\Omega S^1 \simeq \mathbb{Z}$.

This is the first theorem of a growing field known as *synthetic homotopy theory*, more of which can be found in [112, Chapter 8] and recent work such as [24, 44, 76]. Just as in the *synthetic topology* of Section 3.4, the types come automatically with topological structure, which we can then study "synthetically" rather than breaking it down into a set equipped with a topology, in synthetic homotopy theory the types come automatically with homotopical or ∞-groupoid structure, which we can then study synthetically rather than breaking it down into any explicit definition of an ∞-groupoid. Thus it is a "model-independent" language for homotopy theory, avoiding the need to choose (say) topological spaces or simplicial sets as a definition of "∞-groupoid."

It is too early to say how useful this will be to classical homotopy theory. In its very short existence so far, synthetic homotopy theory has not led to proofs of any new theorems, but it has shown an impressive ability to produce new proofs of old theorems: as of this writing, synthetic homotopy theorists have calculated $\pi_n(S^n) = \mathbb{Z}$, $\pi_k(S^n) = 0$ for $k < n$, $\pi_3(S^2) = \mathbb{Z}$, and $\pi_4(S^3) = \mathbb{Z}/2\mathbb{Z}$, and proven numerous foundational results, such as the Freudenthal suspension theorem, the Blakers–Massey connectivity theorem, and the Serre spectral sequence.

More importantly, the theorems of synthetic homotopy theory are more general than those of classical homotopy theory, because (modulo subtleties to be mentioned in Section 4.6) they apply in any well-behaved

(∞, 1)-category, including any ∞-*topos* [80, 92]. (Some ∞-toposes of interest to classical homotopy theorists include equivariant and parameterized homotopy theory.) A particularly interesting example is the Blakers–Massey theorem, for which no purely homotopical proof applicable to ∞-toposes was known prior to the synthetic one [44]; the latter has now been translated back into categorical language [94].

Finally, synthetic homotopy theory gives a new way to think about the "homotopy hypothesis" of Grothendieck [13] that ∞-groupoids describe the homotopy theory of spaces. Rather than looking for an equivalence between some notions of ∞-groupoid and space, we have a *synthetic* theory of ∞-groupoids that is *modeled by* classical homotopy spaces – but also other things. (In fact, Brunerie has observed that the ∞-groupoid structure of types in homotopy type theory looks almost exactly as it was envisioned by Grothendieck [83], rather than like any of the definitions of ∞-groupoid used more commonly today.) In Section 5 I will sketch a particular context in which this extra generality is useful.

4.6 The Classifying (∞, 1)-Category

In this appendix to Section 4 I will describe the "classifying (∞, 1)-category" informally, then give a precise definition of it, and end with some remarks about the current state of knowledge as regards its freeness. This appendix and Section 4.7 are provided to satisfy the curious reader, but can be skipped without consequence.

Let $\mathfrak{T}$ be an intensional type theory. We define the objects and morphisms of its classifying (∞, 1)-category $\mathscr{C}tx(\mathfrak{T})$ just as we did for the classifying 1-category **Ctx**($\mathfrak{T}$) in Section 2.4. However, we do *not* quotient the morphisms by terms in the equality type. Instead we will use those to define the 2-morphisms, as well as 3-morphisms, 4-morphisms, and so on.

The idea is to generalize the representation of equalities using diagonals to a characterization of 2-morphisms. Given morphisms $f, g: A \to B$ in an (∞, 1)-category, their "equalizer" is a morphism $e: E \to A$ equipped with a 2-morphism $fe \cong ge$ that is universal among such 2-morphisms. In particular, to give a 2-morphism $f \cong g$ is equivalent to giving a section of e (that is, a morphism $s: A \to E$ and a 2-morphism $es \cong 1_A$). As in the 1-categorical case, this equalizer can be constructed, up to equivalence, as the pullback of the diagonal $B \to B \times B$ along $(f, g): A \to B \times B$. Thus, assuming that the identity type of B still presents the diagonal (up to the appropriate sort of (∞, 1)-categorical equivalence), and substitution still presents pullback, 2-morphisms $f \cong g$ should be equivalent to terms of the form

Homotopy Type Theory

$$x: A \vdash p(x): f(x) =_B g(x). \tag{4.9}$$

Hence we simply *define* a 2-morphism in $\mathscr{C}tx(\mathfrak{T})$ to be a term of this sort. Similarly, we define a 3-morphism $p \cong q$ to be a term in an iterated identity type $x: A \vdash h(x): p(x) =_{(f(x)=_B g(x))} q(x)$, and so on.

To make this precise, we need to choose a method of presenting $(\infty, 1)$-categories. In principle, there are many options; but at present, the method of choice for defining $\mathscr{C}tx(\mathfrak{T})$ is to use *fibration categories* [23]. A fibration category (or "category of fibrant objects") is a 1-category with two classes of morphisms called *weak equivalences* and *fibrations*, satisfying certain axioms, for example, pullbacks of fibrations exist and preserve weak equivalences. The most important axiom is that every diagonal $A \to A \times A$ factors as a weak equivalence followed by a fibration, with the intermediate object called a *path object PA* for A.

Of course, this is an abstraction of a common situation from homotopy theory: fibrations of topological spaces, Kan simplicial sets, or chain complexes (and more generally the fibrations between fibrant objects in any Quillen model category) all have these properties. Generalizing these examples, in any fibration category, we define a *homotopy* between $f, g \colon A \rightrightarrows B$ to be a lift of $(f, g) \colon A \to B \times B$ to a map $A \to PB$. We can similarly define higher homotopies and thereby construct a more explicit notion of $(\infty, 1)$-category (such as a quasi-category), although the combinatorics are somewhat involved; see [109].

Now, if in the definition of $\mathbf{Ctx}(\mathfrak{T})$ from Section 2.4 we omit the quotient of morphisms,[37] we obtain a fibration category $\underline{\mathsf{Ctx}}(\mathfrak{T})$. Its fibrations are the composites of projections $[\![\Gamma, x\colon A]\!] \to [\![\Gamma]\!]$, its weak equivalences are the homotopy equivalences defined in Section 4.2, and its path objects are the identity types $P[\![A]\!] = [\![x\colon A, y\colon A, p\colon x = y]\!]$. (The fact that identity types satisfy the axioms of path objects was one of the central insights of Awodey and Warren [11, 121].) With this definition, homotopies in the fibration-category sense correspond bijectively to terms of the form (4.9) – the former are lifts as on the left below, whereas the latter are sections as on the right:

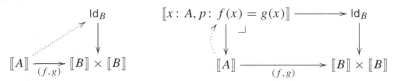

---

[37] Technically, we replace it with a different quotient; see Section 4.7.

Thus, we may define $\mathscr{C}tx(\mathfrak{T})$ to be the $(\infty, 1)$-category presented by $\mathsf{Ctx}(\mathfrak{T})$. Here we see the second advantage of syntax mentioned in Section 2.1: giving a presentation of a free object (here, an $(\infty, 1)$-category) that is actually stricter (here, a fibration category) than one would expect from only its universal property.

However, although this $\mathscr{C}tx(\mathfrak{T})$ has some of the expected structure [63, 65], no one has yet proven its $(\infty, 1)$-categorical freeness. Instead, to interpret type theory in $(\infty, 1)$-categories, we use the fact that $\mathsf{Ctx}(\mathfrak{T})$ is free in a category of structured fibration categories. The latter have various names like "contextual categories" [27], "comprehension categories" [55], "categories with families" [39], "categories with attributes" [27], "display map categories" [110, Section 8.3], "type-theoretic fibration categories" [104], "tribes" [62], "C-systems" [118], and so on. Although this approach has proven more tractable, it is still quite difficult, for two reasons. One is that, as mentioned for the 1-categorical case in Section 2.5, complete proofs of the freeness of $\mathsf{Ctx}(\mathfrak{T})$ have been given only for a few particular type theories [108]. Everyone expects these proofs to generalize to all other type theories, but actually writing down such a generalization, and in a useful amount of generality, is a current research problem.

Another difficulty is that this approach incurs a new proof obligation. In principle, a type theory $\mathfrak{T}$ should be interpreted in an $(\infty, 1)$-category $\mathscr{C}$ by means of the unique functor $\mathscr{C}tx(\mathfrak{T}) \to \mathscr{C}$ determined by the universal property of $\mathscr{C}tx(\mathfrak{T})$. If we stick with the 1-categorical universal property of $\mathsf{Ctx}(\mathfrak{T})$, then to interpret $\mathfrak{T}$ in $\mathscr{C}$ we need to also present $\mathscr{C}$ by a fibration category of the appropriate sort. This is a sort of "coherence theorem" for structured $(\infty, 1)$-categories – which, again, is known in some particular cases, but a fully general version of which is a current research problem; the state of the art includes [5, 46, 63, 64, 79, 102–104, 107, 114]. (Part of this coherence theorem is showing that pullbacks of fibrations can be made strictly functorial and preserve all the type operations strictly, which is nontrivial even for 1-categories [31, 36, 50, 51].)

I have chosen not to dwell on these issues because I have faith that they will eventually be resolved. Instead, I want to focus on the picture that such a resolution will make possible (and which is *substantially* achievable even with current technology). Thus one might call this chapter a "program" for homotopy type theory and its higher-categorical semantics. In Section 4.7 I will briefly discuss another technical detail; in Section 5 we will return to the program.

4.7 Judgmental Equality

In Section 2.1 we described both the "tautological" and the "reduced-words" presentation of a free group using "rules" in the style of type theory. For the reduced-words description, this is the end of the definition; but for the tautological description, we need to describe the equivalence relation to quotient by. This can also be defined inductively by the rules shown in Figure 6.7, which essentially say that it is the smallest equivalence relation imposing the group axioms and compatible with the operations. We also remarked that there is an algorithm for "reducing" any word from the tautological presentation, so that two terms are related by $\equiv$ precisely when they reduce to the same result. Finally, in Section 3.3, we mentioned that type theory includes an analogous "reduction algorithm" making it into a general-purpose programming language.

Taken together, these remarks suggest that there should be two forms of type theory, one involving an equivalence relation $\equiv$ and one not, with a "reduction algorithm" mapping the first to the second. This is more or less correct, but it turns out to be quite fiddly to describe the second type theory without reference to the first. It is sometimes possible [49], but more common is to describe only a type theory involving $\equiv$, with the reduction algorithm an endofunction of its terms, and then *define* the "canonical forms" to be those that are "fully reduced." This is also more flexible, since we can add new $\equiv$ axioms without knowing whether there is a corresponding reduction algorithm that terminates at a canonical form (or even knowing that there is not!). The relation $\equiv$ is known as *judgmental equality* or *definitional equality* or *substitutional equality*.[38]

$$\frac{X \text{ elt}}{X \equiv X} \qquad \frac{X \equiv Y}{Y \equiv X} \qquad \frac{X \equiv Y \quad Y \equiv Z}{X \equiv Z} \qquad \frac{X \equiv X' \quad Y \equiv Y'}{(XY) \equiv (X'Y')}$$

$$\frac{X \equiv Y}{X^{-1} \equiv Y^{-1}} \qquad \frac{X \text{ elt} \quad Y \text{ elt} \quad Z \text{ elt}}{(X(YZ)) \equiv ((XY)Z)} \qquad \frac{X \text{ elt}}{(Xe) \equiv X} \qquad \frac{X \text{ elt}}{(eX) \equiv X}$$

$$\frac{X \text{ elt}}{(XX^{-1}) \equiv e} \qquad \frac{X \text{ elt}}{(X^{-1}X) \equiv e}$$

Figure 6.7 Equality rules for free groups.

---

[38] Technically, these three terms have slightly different meanings, but in the most common type theories, they all turn out to refer to the same thing.

Just as for free groups, when defining the corresponding free object, we have to quotient by the relation ≡. For the classifying 1-category **Ctx**($\mathfrak{T}$), this quotient is included in the quotient by terms of the identity type. But for the fibration category Ctx($\mathfrak{T}$), where we omitted the latter quotient, we do still have to impose a quotient by judgmental equality – or, if our type theory has a terminating reduction algorithm (the technical term is "strongly normalizing"), use only the canonical forms to represent objects and morphisms.

The puzzling thing, of course, is how this equality ≡ is related to the equality *type* $x: A, y: A \vdash (x = y)$ type. Formally, the difference between these "two equalities" is analogous to the difference between the variables $x: A$ occurring in a context and the "metavariables" such as Γ that we use in describing the operations of the theory. Any inductive definition uses "metavariables" and can have an inductively defined equivalence relation; type theory is special because *internal* to the theory there are also notions of "variable" and "equality." The identity *type* is defined by a universal property, just like most other types; whereas judgmental equality, like the equivalence relation on words in a free group, is inductively defined as the smallest equivalence relation imposing the desired axioms (the computation rules from Section 2.5, which we denoted with ≡ for this very reason) and respected by all the other judgments. The latter condition means we have additional rules, such as

$$\frac{\Gamma \vdash a: A}{\Gamma \vdash a \equiv a}, \qquad \frac{\Gamma \vdash a \equiv b}{\Gamma \vdash b \equiv a}, \qquad \frac{\Gamma \vdash a \equiv b \quad \Gamma \vdash b \equiv c}{\Gamma \vdash a \equiv c},$$

$$\frac{\Gamma \vdash a: A \quad \Gamma \vdash A \equiv B}{\Gamma \vdash a: B}. \tag{4.10}$$

This formal description, however, does not really explain *why* we need two equalities, or what they mean intuitively. To start with, it cannot be emphasized strongly enough that *it is the identity type that represents mathematical equality*. Equality in mathematics is a proposition, and in particular something that can be *hypothesized* and *proven* or *disproven*. Judgmental equality cannot be hypothesized (added to a context), nor can it be proven (inhabited by a term) or disproven (we cannot even state internally a "negation" of judgmental equality). In its simplest form, judgmental equality is simply the algorithmic process of expanding definitions (hence the name "definitional equality"): for instance, the function $\lambda x.x^2$ is *by definition* the function that squares its argument, so $(\lambda x.x^2)(y+1)$ is *by definition* equal to $(y+1)^2$. But even the simplest equalities with mathematical content, such as the theorem that

$x + y = y + x$ for $x, y : \mathbb{N}$, are not a mere matter of expanding definitions but require proof.

What, then, *can* we do with judgmental equality? The main property it has that the identity type does not is (4.10): given $a: A$ and $A \equiv B$, the *same term a* is also an element of B (hence the name "substitutional equality"). In particular, if $a \equiv b$, then $(a = a) \equiv (a = b)$, so that $\text{refl}_a : a = b$; thus judgmental equality implies mathematical equality. By contrast, given $a : A$ and a mathematical equality $e: A = B$, it is possible to obtain a term of B, but that term is not syntactically equal to a; instead it is $e_*(a)$, involving the transport operation.

This need for explicit transports is somewhat annoying, so it is tempting to eliminate it by collapsing the two equalities with a *reflection rule*

$$\frac{\Gamma \vdash p: a = b}{\Gamma \vdash a \equiv b}.$$

Unfortunately, this makes it impossible to detect $\equiv$ using a reduction algorithm, since questions of mathematical equality cannot be decided algorithmically. This does not necessarily make such a type theory impractical,[39] but the reflection rule also turns out to imply UIP, which is a deal-breaker if we want to talk about $(\infty, 1)$-categories. The $(\infty, 1)$-categorical point of view also makes clear why we need to notate e in $e_*(a)$: since the type $A = B$ is (by univalence) the type of equivalences from A to B, it could have many *different* elements, so that $e_*(a)$ really does depend on the choice of e.

One might then be tempted to go to the other extreme and try to eliminate judgmental equality entirely. We could in principle express all the computation rules from Section 2.5 using elements of identity types rather than judgmental equalities. However, the resulting proliferation of transport operations would be so extreme as to render the theory essentially unusable. We need a happy medium, with a judgmental equality as strong as feasible but no stronger.

The intuitive meaning of judgmental equality is not entirely clear, although in some ways it is analogous to Frege's "equality of sense" (with mathematical equality analogous to "equality of reference"). Categorically, judgmental equality is analogous to the "point-set-level" or "strict" equality occurring in strict or semistrict models for higher categories, such as Quillen model categories or Gray categories. This finds a formal expression in

---

[39] Indeed, not infrequently, the name "extensional type theory" refers to type theory that includes the reflection rule, rather than assuming UIP as an axiom. Voevodsky and others have also proposed modified reflection rules that are compatible with $(\infty, 1)$-categories.

the fibration-category approach to semantics, where we need a "semistrictification" theorem presenting any $(\infty, 1)$-category by a fibration category satisfying all the judgmental equalities of our type theory strictly. Finding the right balance of strictness and weakness here is an active frontier of research.

5 Cohesive Homotopy Type Theory

5.1 Spaces versus ∞-Groupoids

Twice now we have encountered something called a "circle": in Section 3.4 we mentioned that $\mathbb{S}^1 = \{ (x, y) \colon \mathbb{R} \times \mathbb{R} \mid x^2 + y^2 = 1 \}$ has the correct topology, and in Section 4.5 we mentioned that $S^1 = \operatorname{coeq}(\mathbf{1} \rightrightarrows \mathbf{1})$ has the correct fundamental group. However, these two types $\mathbb{S}^1$ and S^1 are very different! The first $\mathbb{S}^1$ is a *set* in the sense of Section 4.2, whereas S^1 is definitely not, since its loop space is $\mathbb{Z}$. On the other hand, S^1 is *connected*, in the sense that its 0-truncation $\|S^1\|_0$ is contractible, whereas since $\mathbb{S}^1$ is a set, it is its own 0-truncation.

What is happening is that classical homotopy theory has led us to confuse two different things in our minds. On one hand, a *topological space* is a set with a notion of "cohesion" enabling us to define continuous functions and paths. The nearby points of a continuous path are "close" in some sense, but they are still distinct. On the other hand, an ∞-*groupoid* has a collection of "points" or "objects," plus for each pair of objects a collection $\hom(x, y)$ of equivalences or "ways in which x and y are the same," plus for each $f, g \in \hom(x, y)$ a collection $\hom_{\hom(x,y)}(f, g)$ of ways in which f and g are the same, and so on. When $\hom(x, y)$ is nonempty, x and y really *are* the same to ∞-groupoid theory, just as in plain category theory we do not distinguish between isomorphic objects.

The relation between topological spaces and ∞-groupoids is that from any space X, we can *construct* an ∞-groupoid $\int X$, called its *fundamental ∞-groupoid* or *shape*.[40] The objects of $\int X$ are the points of X, the objects of $\hom(x, y)$ are the continuous paths from x to y, the objects of $\hom_{\hom(x,y)}(f, g)$ are the continuous end point–preserving homotopies from f to g, and so on. The confusion arises because we can study $\int X$ without actually

---

[40] The symbol $\int$ is not an integral sign ($\int$) but an "esh," the IPA sign for a voiceless postalveolar fricative (English *sh*); in LaTeX it is available as with the package phonetic. An alternative notation is Π_∞, but the letter Π is overworked in type theory already. The term "shape" comes from "shape theory," which also studies generalizations of $\int$ for ill-behaved topological spaces.

constructing it (or even having a definition of "∞-groupoid"), by working with X itself and "doing everything up to homotopy"; and historically, people did this for a long time before they even thought of defining ∞-groupoids. Thus, algebraic topologists came to use the word "space" for objects that were actually being treated as ∞-groupoids.[41]

Homotopy type theory forcibly brings the distinction between topological spaces and ∞-groupoids front and center, since it allows us to talk about ∞-groupoids directly in a foundational system that is also strong enough to study topological spaces. In particular, we have the previously noted contrast between the types $\mathbb{S}^1$ and S^1. The relation between the two *ought* to be that $S^1 = \int \mathbb{S}^1$; but how are we to express this in type theory?

5.2 Combining Topology with Homotopy

The description of $\int X$ in Section 5.1 treats both topological spaces and ∞-groupoids as structures built out of sets. However, we have seen that in type theory, we can treat *both* of them synthetically, suggesting that $\int$ ought also to have a synthetic description. This requires combining the perspectives of Sections 3 and 4, obtaining a type theory in which topology and homotopy are synthetic *at the same time*. That is, we allow some types to have "intrinsic topology," like $\mathbb{S}^1$, and also some types to have "intrinsic homotopy," like S^1. It follows unavoidably that there must also be types with *both* nontrivial topology and nontrivial homotopy.

At this point the advantages of a synthetic treatment become especially apparent. Classically, to combine structures in this way, we have to define a new structure called a "topological ∞-groupoid" or a "topological ∞-stack": an ∞-groupoid equipped with a "topology" on its objects, another on its morphisms, and so on. If such a gadget has no nontrivial morphisms, it reduces to a topological space; while if all the topologies, are discrete it reduces to an ordinary ∞-groupoid. Formally, we might define these to be ∞-stacks on one of the sites $\{\mathbb{N}_\infty\}$ and $\{\mathbb{R}^n\}_{n\in\mathbb{N}}$ from Section 2.7, comprising ∞-toposes of *consequential ∞-groupoids* and *continuous ∞-groupoids* (or *smooth ∞-groupoids*). We would then need to develop a whole theory of such objects.

---

[41] Arguably, therefore, ∞-groupoids do not even belong in a book about notions of space. However, tradition is weighty, and moreover, ∞-groupoids do share some important attributes of notions of space, notably their ability to be present as "background structure" in the sense described in the introduction. It is to emphasize this aspect, but also their distinctness from other notions of space, that I sometimes call them *homotopy spaces*.

In type theory, however, we have seen that types "potentially" have *both* topological and homotopical structure, which we can draw out by asserting axioms such as Brouwer's theorem or Voevodsky's univalence axiom. Thus, to obtain a synthetic theory of "topological ∞-groupoids" is simplicity itself: we simply assert *both* groups of axioms at the same time. Of course, to model the theory in classical mathematics we still need to construct topological ∞-groupoids, but we do not need to bother about that when working *in* the theory.

Schreiber's chapter in the companion volume [2] argues that topological ∞-groupoids (or some enhancement thereof) are the correct context in which to formulate modern theories of physics. (For more general discussion of stacks, see Chapter 8.) The type theory modeled by $(\infty, 1)$-categories of this sort is an active field of current research called *cohesive homotopy type theory* [74, 96, 97, 105]. I will conclude by sketching some of its most appealing features.

5.3 Modalities and Cohesion

The synthetic description of $\int$ involves a different way to access the latent topological structure of types, based on Lawvere's ideas of *cohesion* [74]. Recall from Section 2.8 that most "topological" toposes come with a string of adjunctions

$$\text{topos of spaces}$$
$$\Delta \uparrow \ \dashv \ \downarrow \Gamma \ \dashv \ \uparrow \nabla$$
$$\text{topos of sets}$$

where Γ is the underlying-set functor, Δ constructs discrete spaces, and ∇ constructs indiscrete spaces, and Δ and ∇ are fully faithful. If we restrict our attention to the topos of spaces, then what is left of this adjoint triple is a monad $\sharp = \nabla \Gamma$ that reflects into the subcategory of indiscrete types, a comonad $\flat = \Delta \Gamma$ that coreflects into the category of discrete types, and an adjunction $\flat \dashv \sharp$ such that the induced transformations $\sharp \flat \to \sharp$ and $\flat \to \flat \sharp$ are isomorphisms.

We can incorporate $\sharp$ and $\flat$ in type theory as *higher modalities*. Traditional "modal logic" studies propositional modalities, most famously "it is necessary that P" (usually written $\Box P$) and "it is possible that P" (usually written $\Diamond P$), but also others, such as "so-and-so knows that P," "it will always be the case that P." Since these often have monad- or comonad-like properties (e.g., $\Box P \to P$ and $\Box P \to \Box\Box P$), and propositions are particular types (see Section 3.1), we may consider monads and comonads acting on all types as "higher-categorical modalities." I refer to type theory with ♭ and ♯ as *spatial type theory*, since it is designed for "topological" models such as consequential, continuous, and smooth sets or ∞-groupoids. We will not state its rules precisely here since they involve some technicalities, but the practical upshot is that ♭ and ♯ behave as described above; see [105] for a more extensive discussion.

As an example, these modalities allow us to state "discontinuous" versions of classicality axioms, such as LEM, that *do* hold in these intended models. The usual version of LEM is $\prod_{P:\Omega} \|P + \neg P\|$, which is false in the topological models, as discussed in Section 3.2, because a space is not generally the disjoint union of a subspace and its complement. But $\prod_{P:\Omega} \sharp\|P + \neg P\|$ and $\prod_{P:\flat\Omega} \|P + \neg P\|$ *are* true in these models: both equivalently express the true statement that any space is the smallest *subspace* of itself containing both any given subspace and its complement. They imply in particular that the (equivalent) subuniverses of discrete and indiscrete types satisfy ordinary LEM and thus are a place for classical reasoning inside synthetic-topological type theory. (Recall from Section 3.2 that the indiscrete spaces are also usually the ¬¬-sheaves. This often follows automatically in spatial type theory; see [105].)

Now, in many cases, the functor Δ also has a *left* adjoint, that is, the discrete spaces are *reflective* as well as coreflective. A map from a space A into a discrete set ΔB breaks A up as a coproduct of one disjoint piece for each element of B. Thus if A is a coproduct of "connected components," any map $A \to \Delta B$ is uniquely determined by where each connected component goes, that is, by a map $\pi_0(A) \to B$. Thus π_0 is left adjoint to Δ, or more precisely any left adjoint to Δ deserves the name π_0. Note that this "$\pi_0(A)$" is not the same as the 0-truncation $\|A\|_0$ discussed in Section 4; the latter treats types as ∞-groupoids, while this one treats them as topological spaces. In a moment we will see that $\pi_0(A) = \|\int A\|_0$.

Such a left adjoint π_0 exists for continuous sets and smooth sets, though not for consequential spaces (because the latter contain spaces, such as $\mathbb{N}_\infty$, that are not locally connected, hence not a coproduct of connected components). A topos with an adjoint string $\pi_0 \dashv \Delta \dashv \Gamma \dashv \nabla$ where Δ and ∇ are fully faithful and π_0 preserves finite products (and perhaps more; see [60, 74, 105]) is called *cohesive*.

Finally, this all works basically the same in the ∞-case: "cohesive ∞-toposes," such as continuous and smooth ∞-groupoids, are related to the ∞-topos of ∞-groupoids by a string of ∞-adjunctions, which can be represented by modalities in type theory.[42] For intuition, a "discrete" cohesive ∞-groupoid is one whose topologies are discrete at all levels, that is, neither its points nor its equalities between points, and so on, have any interesting topology. It could still have interesting ∞-groupoid structure; for instance, S^1 is discrete (but $\mathbb{S}^1$ is not!).

The magical thing is that for ∞-toposes, a left adjoint of Δ is no longer just π_0; instead, it deserves to be called the shape functor $\int$ discussed above! To prove this is technical (see [96, Proposition 4.3.32] or [26, Section 3]), but we can get a feel for it with examples.

First of all, by comparing universal properties, we see that (denoting a left ∞-adjoint of Δ by $\int$) the set $\|\int A\|_0$ is a reflection of A into discrete *sets* (i.e., homotopy 0-types). Thus, the 1-categorical argument above implies that $\|\int A\|_0$ deserves the name $\pi_0(A)$, which is what we expect for the shape of A.

Second, we have seen that $S^1 = \text{coeq}(\mathbf{1} \rightrightarrows \mathbf{1})$, and since the discrete types are closed under colimits (being coreflective), S^1 is also discrete. On the other hand, we have $\mathbb{S}^1 = \text{coeq}(\mathbb{R} \rightrightarrows \mathbb{R})$, where one map $\mathbb{R} \to \mathbb{R}$ is the identity and the other is "+1." Since left adjoints preserve colimits, we will have $\int \mathbb{S}^1 = S^1$ as long as $\int \mathbb{R} = \mathbf{1}$. This is true for continuous ∞-groupoids (an analogous fact about the smooth reals is true for smooth ∞-groupoids).

In fact, the discrete objects in continuous ∞-groupoids are essentially *defined* by the property that $\int \mathbb{R} = \mathbf{1}$. More specifically, a type A is discrete if and only if every map $\mathbb{R} \to A$ is constant, or more precisely, if the map const: $A \to (\mathbb{R} \to A)$ is an equivalence. This axiom is called *real-cohesion* [105]; it immediately implies that $\int \mathbb{R} = \mathbf{1}$. (The real-cohesion axiom also allows us to *construct* $\int$ as a higher inductive type by "localizing" in the sense of Section 4.4 at the map $\mathbb{R} \to \mathbf{1}$.)

We can make similar arguments in other examples. For instance, the topological 2-sphere $\mathbb{S}^2$ is the pushout of two open discs (each isomorphic to $\mathbb{R}^2$) under an open strip (isomorphic to $\mathbb{S}^1 \times \mathbb{R}$). Thus, as long as $\int$ preserves products, $\int \mathbb{S}^2$ is the pushout of two copies of $\mathbf{1}$ under S^1, that is, the homotopy-theoretic *suspension* of S^1, which is one definition of the homotopical 2-sphere S^2. Many familiar spaces can be presented as "open cell complexes" of this sort, thereby identifying their shapes with the expected discrete ∞-groupoids.

---

[42] Of course, the formal connection between cohesive ∞-toposes and cohesive type theory is at least as difficult as the ordinary case discussed in Section 4.6; indeed the cohesive case has not yet been studied formally at all. However, the cohesive type theory at least is fully rigorous as a formal system in its own right, with reference to ∞-toposes only for motivation.

We do have to avoid the more classical "closed cell complexes" that glue intervals and closed discs along boundaries. Gluing the end points of the unit interval [0, 1] in the topos of continuous sets does not produce S^1 but rather a circle with a "speed bump" at which any continuous path must stop for a finite amount of time before proceeding. This problem is avoided by consequential spaces, but as remarked previously, that topos fails to have $\int$. In fact, as discussed briefly in [58], it seems impossible to have both closed cell complexes and $\int$.

This description of $\int$ enables synthetic arguments that involve both topological spaces and homotopy spaces and their relationship. For instance, in [105], I used $\int S^1 = S^1$ to prove the Brouwer fixed point theorem synthetically. This is a theorem about the *topological* closed disc $\mathbb{D}^2$ (whose boundary is S^1), but its classical proof uses a *homotopical* argument, constructing a retraction $\mathbb{D}^2 \to S^1$ which is impossible since S^1 is not *homotopically* contractible. Synthetically, the proof can be done in almost exactly the same way, inserting $\int$ at the last step, and using the fact that $\Omega S^1 = \mathbb{Z}$ mentioned in Section 4.5 (which uses the univalence axiom) so that $\int S^1$ (being S^1) is not a retract of $\int \mathbb{D}^2$ (being **1**).

At a more advanced level, Schreiber's chapter in the companion volume [2] shows that smooth ∞-groupoids – and, by extension, cohesive type theory – are a natural setting for differential cohomology and gauge field theory, which involve the interaction between smooth spaces and homotopy spaces. The synthetic approach to $\int$ is thus not just a conceptual way to explain the difference between topological and homotopy spaces but a practical tool for combining them in applications.

6 Conclusion

What does the future hold for type theory and synthetic mathematics? Current research divides into two threads. One is "internal": developing mathematics in type theory. This includes both ordinary mathematics in constructive logic without LEM or AC, so as to be valid in all toposes (Section 3.2), and also more novel synthetic mathematics using nonclassical structure (Sections 3.4, 4, and 5).

The constructivization of ordinary mathematics has a long history, but plenty of fundamental questions remain unanswered, due in part to a tradition among some "constructivists" of neglecting propositional truncation and assuming countable choice. Synthetic mathematics is newer: synthetic differential geometry is several decades old but not well known outside topos theory, while synthetic homotopy theory is only a handful of years old [77, 112], and synthetic topology is in between. Thus, there are many open questions regarding which results of "analytic" mathematics can be reproduced synthetically.

The other thread of current research is "metatheoretic." As mentioned in Sections 4.6 and 4.7, there are many unsolved problems in the $(\infty, 1)$-categorical semantics of type theory. There are also purely syntactic open problems, such as reconciling the topological/homotopical point of view with the computational one from Section 3.3. For instance, can we make HITs and univalence "compute," i.e., maintain the property that all terms reduce to a "value"? At present the most successful approaches to this use "cubical" methods (e.g., [3, 4, 32, 34]).

Some problems involve both syntax and semantics. For instance, homotopy type theory is an excellent synthetic language for higher groupoids, but what about higher *categories*? Any classical definition of $(\infty, 1)$-category (such as quasi-categories) can be repeated inside the *sets* of type theory, but that would not be what we want: a good definition of $(\infty, 1)$-category in homotopy type theory should use the synthetic notion of ∞-groupoid provided by the types. The most promising approach is something like Rezk's "complete Segal spaces" [93]; this can be done for 1-categories [1], but for the ∞-case, it would require a notion of "coherent simplicial type," which so far has proven elusive.

This is a special case of another open question that I call the "problem of infinite objects," which also applies to other homotopy-theoretic notions like A_∞-spaces and structured ring spectra. Classically, such infinite coherence structures involve strict point-set-level equalities. For instance, A_∞-spaces are *strict* algebras for a topological operad; the weakness is in the operad. But homotopy type theory, in its most common form, severely restricts the use of strict equality: it can be accessed only using dependent types (e.g., terms belonging to a dependent type are strict sections of a fibration) and judgmental equality. This is good because it makes everything automatically homotopy invariant, but it means we lack a flexible way to assemble arbitrary higher coherence structures. (In particular, while synthetic homotopy theory can do a lot, further technical advances are needed before it could reproduce all of classical homotopy theory.) This problem might be solvable completely internally, but it might also require modifying the syntax, leading to a whole host of new metatheoretic problems.

Let me end with some remarks about the *philosophical* implications of synthetic mathematics. I have presented type theory in a way intended to seem useful and unobjectionable to a classical mathematician: as a syntax for reasoning about structured categories in a familiar language. Crucial to the usefulness of this syntax is the fact that it, like ZFC set theory, is general enough to encode all of mathematics, and therefore anything we can prove (constructively) in ordinary mathematics is automatically also "true internally" in any category.

This leads naturally to a slightly different question: can we *actually* use type theory as *the* foundation for mathematics? That is, must we consider the objects of mathematics to "really" be built out of sets, with "types" just a convenient fiction for talking about such structures? Or can we consider *types* to be the basic objects of mathematics, with everything else built out of *them*?

The answer is undoubtedly yes: the "sets" in type theory can encode mathematics just like the sets of ZFC can. Of course, there are subtleties. On one hand, if our type theory is constructive, we need to do our mathematics constructively. On other hand, type theory often suggests different ways to do things, using the synthetic spatial or homotopical structure of types instead of analytic topological spaces or ∞-groupoids.[43] Both of these involve their own open problems; but they are only potential *enhancements* or *refinements* of ordinary mathematics, so regardless of how they turn out, it is certainly *possible* to treat type theory as a foundation for all of mathematics.

The real question, therefore, is not "can we?" but "should we?" This is where things get more philosophical. Over the past century, mathematicians and philosophers have become accustomed to the fundamental objects of mathematics being discrete sets, with no spatial or homotopical structure. However, a priori there is no reason this has to be the case. Indeed, some of the early 20th-century constructivists, notably Brouwer, can (with a bit of hindsight) be read as arguing for the intrinsically spatial nature of mathematical objects.

But can spaces really be *fundamental* like sets are? A discrete set certainly seems simpler, and hence more fundamental, than a set equipped with spatial structure. But this argument merely begs the question, since if spaces are fundamental objects, then they are *not* just sets "equipped with spatial structure." In spatial type theory there is no obvious nontautological "structure" with which we can equip the discrete set of reals $\flat\mathbb{R}$ that determines the space of reals $\mathbb{R}$. Is $\flat\mathbb{R}$ "simpler" than $\mathbb{R}$? When we consider all the pathological nowhere-continuous functions supported by $\flat\mathbb{R}$ but not $\mathbb{R}$, it seems at least consistent to believe that $\mathbb{R}$ is the simpler. Moreover, discrete sets are just a

---

[43] In particular, for type theory to be an autonomous foundation for mathematics, it ought to suffice for its own metatheory, including the freeness of its own classifying $(\infty, 1)$-category; but we do not yet even know how to *define* $(\infty, 1)$-categories in homotopy type theory.

particular kind of space; so even if they are simpler, that does not necessarily argue that nondiscrete spaces cannot be fundamental. The empty set ∅ is probably simpler than $\aleph_\omega$, but in ZFC they are equally fundamental objects (i.e., sets).

Similar arguments apply to homotopy spaces, that is, ∞-groupoids. One of the central insights of category theory and homotopy theory is that no class of mathematical objects should be considered without the corresponding notion of isomorphism or equivalence: we study groups up to isomorphism, spaces up to homomorphism, categories up to equivalence, and so on. Thus, all mathematical collections naturally form groupoids, or more generally ∞-groupoids, when equipped with the relevant "notion of sameness." (See [106] for further philosophical discussion of this point.) The *set*[44] of all groups is much less tractable, and much less interesting, than the *category* of all groups; so even though the former is "simpler" in the sense of containing no nontrivial automorphisms, it is reasonable to regard the latter as being at least as fundamental.

One possible objection to treating spaces as fundamental is to ask how we should decide *which* rules our "spaces as fundamental" should satisfy. Indeed, we have already seen that there are different kinds of synthetic topology adapted for different purposes, modeled respectively by consequential, continuous, or smooth ∞-groupoids. Moreover, other kinds of synthetic mathematics, such as synthetic domain theory, synthetic differential geometry, and other fields waiting to be developed, will have their own toposes and their own type theories.

However, if we shift perspective a bit, we can see that this is a feature rather than a bug. Why must we insist on singling out some particular theory as "the" foundation of mathematics? The idea of a "foundation for mathematics" stems from the great discovery of 20th-century logic that we can encode mathematics into various formal systems and study those systems mathematically. But in the 21st century, we are sufficiently familiar with this process that we no longer need to tie ourselves to only *one* such system.[45] Even ZFC has a role from this point of view: it is a synthetic theory of well-founded membership structures!

Bell [18] makes an excellent analogy to Einstein's theory of relativity. In Newtonian physics, there is a special absolute "rest frame," relative to which all motion can be measured. There are moving observers, of course, but they are second-class citizens: the standard laws of physics do not always

---

[44] Or "proper class."

[45] In particular, it is meaningless to ask whether statements like the continuum hypothesis are "true"; they are simply true in some systems and false in others. This perspective is very natural to a category theorist but has recently made inroads in set theory as well [48].

apply to them. They feel "fictitious forces," like the centrifugal force and Coriolis force on a spinning merry-go-round or planet, that are not *really* forces but just manifestations of "truly" inertial motion in a noninertial reference frame.

By contrast, Einsteinian physics can be formulated equally well in *any* reference frame and obeys the same laws in each, with consistent rules for transforming between reference frames. Some frames, called "(locally) inertial," lead to a simpler formulation of the laws; but often this is outweighed by the relevance of some other frame to a particular problem (such as the noninertial reference frame of the Earth's surface). The centrifugal and Coriolis forces are exactly as real as any other force; in fact they are simply instances of gravitational force! To an observer on the Earth's surface, an inertial observer in a spaceship flying by is the one who is spinning (along with the rest of the universe), thereby feeling "fictitious" forces that cancel out these gravitational ones.

Similarly, in ZFC orthodoxy, there is an absolute notion of "set" out of which everything is constructed. Spaces exist, but they are second-class citizens, ultimately reducible to sets, and the basic axioms of set theory do not apply to them. But from a pluralistic viewpoint, mathematics can be developed relative to any topos, obeying the same general rules of type theory. We have consistent rules for translating between toposes along functors, and there are some toposes in which mathematics looks a bit simpler (those satisfying LEM or UIP). However, there is no justification for regarding any particular topos or type theory as the "one absolute universe of mathematics." An observer in a topos of classical mathematics can construct a topos of spaces in which all functions are continuous, thereby explaining its different behavior. But an observer inside a topos of spaces can also construct a topos of classical mathematics as the "discrete" or "indiscrete" objects, whose different behavior is explained by the triviality of their cohesion – and *both points of view are equally valid*. Just as modern physicists switch reference frames as needed, modern mathematicians should be free to switch foundational systems as appropriate.

This is particularly relevant for physicists and other scientists interested in *using* mathematics rather than debating its Platonic existence. If a particular synthetic theory is useful in some application domain (see, e.g., Schreiber's chapter in the companion volume [2]), we are free to take it seriously rather than demanding it be encoded in ZFC. Set theory and 20th-century logic were crucial stepping-stones bringing us to a point where we could survey the multitude of universes of mathematics; but once there, we see that there is nothing special about the route we took.

References

[1] B. Ahrens, K. Kapulkin, and M. Shulman, *Univalent categories and the Rezk completion*, Math. Struct. Compu. Sci. 25 (2015) 1010–39
[2] M. Anel and G. Catren, eds., *New Spaces in Physics: Formal and Conceptual Reflections*, Cambridge University Press (2021)
[3] C. Angiuli and R. Harper, *Computational higher type theory II: Dependent cubical realizability*, reprint, arXiv:1606.09638
[4] C. Angiuli, R. Harper, and T. Wilson, *Computational higher type theory I: Abstract cubical realizability*, reprint, arXiv:1604.08873
[5] P. Arndt and K. Kapulkin, *Homotopy-theoretic models of type theory*, in L. Ong, ed., *Typed Lambda Calculi and Applications*, Lecture Notes in Computer Science 6690, Springer (2011) 45–60
[6] *Algebraic set theory web site*, www.phil.cmu.edu/projects/ast/
[7] S. Awodey, *Type theory and homotopy*, in P. Dybjer, S. Lindström, E. Palmgren, and G. Sundholm, eds., *Epistemology versus Ontology: Essays on the Philosophy and Foundations of Mathematics in Honour of Per Martin-Löf*, Log. Epistemol. Unity Sci. 27, Springer (2012) 183–201
[8] S. Awodey, *Structuralism, invariance, and univalence*, Philos. Math. 22 (2014) 1–11
[9] S. Awodey, C. Butz, A. Simpson, and T. Streicher, *Relating first-order set theories, toposes and categories of classes*, Ann. Pure Appl. Logic 165 (2014) 428–502
[10] S. Awodey, Á. Pelayo, and M. A. Warren, *Voevodsky's univalence axiom in homotopy type theory*, Not. Amer. Math. Soc. 60 (2013) 1164–67
[11] S. Awodey and M. A. Warren, *Homotopy theoretic models of identity types*, Math. Proc. Camb. Phil. Soc. 146 (2009) 45–55
[12] S. Awodey and A. Bauer, *Propositions as [types]*, J. Logic Comput. 14 (2004) 447–71
[13] J. Baez, *The homotopy hypothesis* (2007), http://math.ucr.edu/home/baez/homotopy/
[14] J. C. Baez and A. E. Hoffnung, *Convenient categories of smooth spaces*, Trans. Amer. Math. Soc. 363 (2011) 5789–825
[15] J. C. Baez and M. Shulman, *Lectures on n-categories and cohomology*, in J. C. Baez and J. P. May, eds., *Towards Higher Categories*, The IMA Volumes in Mathematics and Its Applications 152, Springer (2009) 1–68
[16] A. Bauer, *Five stages of accepting constructive mathematics*, Bull. Amer. Math. Soc. (2016), https://doi.org/10.1090/bull/1556
[17] A. Bauer and D. Lešnik, *Metric spaces in synthetic topology*, Ann. Pure Appl. Logic 163 (2012) 87–100
[18] J. L. Bell, *Toposes and Local Set Theories*, Oxford Logic Guides 14, The Clarendon Press (1988)
[19] J. L. Bell, *A Primer of Infinitesimal Analysis*, Cambridge University Press (1998)
[20] E. Bishop, *Foundations of Constructive Analysis*, McGraw-Hill (1967)
[21] E. Bishop and D. Bridges, *Constructive Analysis*, Springer (1985)
[22] A. Boileau and A. Joyal, *La logique des topos*, J. Symbolic Logic 46 (1981) 6–16

[23] K. S. Brown, *Abstract homotopy theory and generalized sheaf cohomology*, Trans. Amer. Math. Soc. 186 (1974) 419–58
[24] G. Brunerie, *On the homotopy groups of spheres in homotopy type theory*, PhD thesis, Université de Nice (2016)
[25] V. Capretta, *General recursion via coinductive types*, Logical Methods Comput. Sci. 1 (2005) 1–28
[26] D. Carchedi, *On the homotopy type of higher orbifolds and haefliger classifying spaces*, Preprint, arXiv:1504.02394
[27] J. Cartmell, *Generalised algebraic theories and contextual categories*, Annals Pure Appl. Logic 32 (1986) 209–43
[28] G. Choquet, *Convergences*, Ann. Univ. Grenoble. Sect. Sci. Math. Phys. (N.S.) 23 (1948) 57–112
[29] A. Church, *A formulation of the simple theory of types*, J. Symbolic Logic 5 (1940) 56–68
[30] A. Church, *The Calculi of Lambda Conversation*, Princeton University Press (1941)
[31] P. Clairambault and P. Dybjer, *The biequivalence of locally cartesian closed categories and Martin-Löf type theories*, in *Proceedings of the 10th International Conference on Typed Lambda Calculi and Applications*, Springer (2011) 91–106
[32] C. Cohen, T. Coquand, S. Huber, and A. Mörtberg, *Cubical type theory: A constructive interpretation of the univalence axiom*, www.math.ias.edu/~amortberg/papers/cubicaltt.pdf
[33] R. L. Constable, S. F. Allen, H. M. Bromley, W. R. Cleaveland, J. F. Cremer, R. W. Harper, D. J. Howe, T. B. Knoblock, N. P. Mendler, P. Panangaden, J. T. Sasaki, and S. F. Smith, *Implementing Mathematics with the Nuprl Proof Development System*, Prentice Hall (1986)
[34] T. Coquand, *Cubical type theory*, www.cse.chalmers.se/~coquand/rules7.pdf
[35] D. Corfield, *Expressing "the structure of" in homotopy type theory*, https://ncatlab.org/davidcorfield/show/Expressing+'The+Structure+of'+in+Homotopy+Type+Theory
[36] P.-L. Curien, *Substitution up to isomorphism*, Fund. Inform. 19 (1993) 51–86
[37] H. B. Curry, *Functionality in combinatory logic*, Proc. Nat. Acad. Sci. 20 (1934) 584–90
[38] N. A. Danielsson, *Operational semantics using the partiality monad*, in *International Conference on Functional Programming 2012*, ACM Press (2012) 127–38
[39] P. Dybjer, *Internal type theory*, in *Selected Papers from the International Workshop on Types for Proofs and Programs*, Springer (1996) 120–34
[40] M. Escardó, *Synthetic topology of data types and classical spaces*, in *Electron. Notes Theor. Comput. Sci.*, Elsevier (2004), www.cs.bham.ac.uk/~mhe/papers/entcs87.pdf
[41] M. Escardó, *Topology via higher-order intuitionistic logic*, Unfinished draft, www.cs.bham.ac.uk/~mhe/papers/index.html
[42] M. Escardó and C. Xu, *The inconsistency of a Brouwerian continuity principle with the Curry–Howard interpretation*, Typed Lambda Calculi Appl. (2015) 153–64

[43] M. H. Escardo and T. Streicher, *The intrinsic topology of Martin-Löf universes*, www.cs.bham.ac.uk/~mhe/papers/universe-indiscrete.pdf
[44] K.-B. Hou (Favonia), E. Finster, D. Licata, and P. LeFanu Lumsdaine, *A mechanization of the Blakers–Massey connectivity theorem in homotopy type theory*, LICS (2016), arXiv:1605.03227
[45] M. Fourman, *Comparaison des réels d'un topos — structures lisses sur un topos elémentaire*, Cah. Topol. Géom. Différ. Catég. 16 (1975) 233
[46] D. Gepner and J. Kock, *Univalence in locally Cartesian closed ∞-categories*, Preprint, arXiv:1208.1749
[47] R. Goldblatt, *Topoi*, 2nd ed., Studies in Logic and the Foundations of Mathematics 98, North-Holland (1984)
[48] J. D. Hamkins, *The set-theoretic multiverse*, Rev. Symbolic Logic 5 (2012) 416–49
[49] R. Harper and D. R. Licata, *Mechanizing metatheory in a logical framework*, J. Funct. Program. 17 (2007) 613–73
[50] M. Hofmann, *On the interpretation of type theory in locally Cartesian closed categories*, in *Proceedings of Computer Science Logic*, Lecture Notes in Computer Science, Springer (1994)
[51] M. Hofmann, *Syntax and semantics of dependent types*, in *Semantics and Logics of Computation*, Publ. Newton Inst. 14, Cambridge University Press (1997) 79–130
[52] M. Hofmann and T. Streicher, *The groupoid interpretation of type theory*, in *Twenty-Five Years of Constructive Type Theory (Venice, 1995)*, Oxford Logic Guides 36, Oxford University Press (1998) 83–111
[53] W. A. Howard, *The formulae-as-types notion of construction*, in *To H. B. Curry: Essays on Combinatory Logic, Lambda Calculus, and Formalism*, Academic Press (1980) 479–91. Notes originally circulated privately in 1969.
[54] J. M. E. Hyland, *The effective topos*, in *The L.E.J. Brouwer Centenary Symposium (Noordwijkerhout, 1981)*, Stud. Logic Foundations Math. 110, North-Holland (1982) 165–216
[55] B. Jacobs, *Comprehension categories and the semantics of type dependency*, Theoret. Comput. Sci. 107 (1993) 169–207
[56] B. Jacobs, *Categorical Logic and Type Theory*, Studies in Logic and the Foundations of Mathematics 141, North-Holland (1999)
[57] P. T. Johnstone and I. Moerdijk, *Local maps of toposes*, Proc. of London Math. Soc. s3-58 (1989) 281–305
[58] P. T. Johnstone, *On a topological topos*, Proc. London Math. Soc. 38 (1979) 237–71
[59] P. T. Johnstone, *Sketches of an Elephant: A Topos Theory Compendium: Volumes 1 and 2*, Oxford Logic Guides 43, Oxford Science Publications (2002)
[60] P. T. Johnstone, *Remarks on punctual local connectedness*, Theory Appl. Categ. 25 (2011) 51–63
[61] A. Joyal and I. Moerdijk, *Algebraic Set Theory*, London Mathematical Society Lecture Note Series 220, Cambridge University Press (1995)
[62] A. Joyal, *Categorical homotopy type theory*, Slides from a talk given at MIT, https://ncatlab.org/homotopytypetheory/files/Joyal.pdf

[63] C. Kapulkin, *Locally Cartesian closed quasicategories from type theory*, Preprint, arXiv:1507.02648
[64] C. Kapulkin and P. LeFanu Lumsdaine, *The simplicial model of univalent foundations (after Voevodsky)*, Preprint, J. Eur. Math. Soc. (forthcoming)
[65] C. Kapulkin and P. LeFanu Lumsdaine, *The homotopy theory of type theories*, Preprint, arXiv:1610.00037
[66] A. Kock, *Synthetic Differential Geometry*, London Mathematical Society Lecture Note Series 51, Cambridge University Press (1981)
[67] N. Kraus and C. Sattler, *Higher homotopies in a hierarchy of univalent universes*, ACM Trans. Comput. Logic 16 (2015) 18:1–18:12
[68] J. Lambek and P. J. Scott, *Intuitionist type theory and the free topos*, J. Pure Appl. Algebra 19 (1980) 215–57
[69] J. Lambek and P. J. Scott, *Introduction to Higher-Order Categorical Logic*, Cambridge Studies in Advanced Mathematics 7, Cambridge University Press (1988)
[70] R. Lavendhomme, *Basic Concepts of Synthetic Differential Geometry*, Kluwer Texts in the Mathematical Sciences 13, Kluwer Academic (1996). Translated from the 1987 French original, revised by the author
[71] F. W. Lawvere, *Equality in hyperdoctrines and comprehension schema as an adjoint functor*, in *Applications of Categorical Algebra (Proc. Sympos. Pure Math., Vol. XVII, New York, 1968)*, American Mathematical Society (1970) 1–14
[72] F. W. Lawvere, *An elementary theory of the category of sets (long version) with commentary*, Repr. Theory Appl. Categ. 11 (2005) 1–35. Reprinted and expanded from Proc. Nat. Acad. Sci. U.S.A. 52 (1964), with comments by the author and Colin McLarty
[73] F. W. Lawvere, *Adjointness in foundations*, Repr. Theory Appl. Categ. (16) (2006) 1–16. Reprinted from Dialectica 23 (1969)
[74] F. W. Lawvere, *Axiomatic cohesion*, Theory Appl. Categ. 19 (2007) 41–49
[75] T. Leinster, *Rethinking set theory*, Amer. Math. Monthly 121 (2014) 403–15
[76] D. Licata and E. Finster, *Eilenberg–Mac Lane spaces in homotopy type theory*, LICS (2014), http://dlicata.web.wesleyan.edu/pubs/lf14em/lf14em.pdf
[77] D. R. Licata and M. Shulman, *Calculating the fundamental group of the circle in homotopy type theory*, LICS (2013)
[78] P. LeFanu Lumsdaine, *Weak omega-categories from intensional type theory*, Typed Lambda Calculi Appl. 6 (2010) 1–19
[79] P. Lefanu Lumsdaine and M. A. Warren, *The local universes model: An overlooked coherence construction for dependent type theories*, ACM Trans. Comput. Logic 16 (2015) 23:1–23:31
[80] J. Lurie, *Higher Topos Theory*, Annals of Mathematics Studies 170, Princeton University Press (2009)
[81] S. Mac Lane and I. Moerdijk, *Sheaves in Geometry and Logic: A First Introduction to Topos Theory*, corrected reprint, Springer (1994)
[82] M. E. Maietti, *The internal type theory of a Heyting pretopos*, in E. Giménez and Christine Paulin-Mohring, eds., *Types for Proofs and Programs*, Lecture Notes in Computer Science 1512, Springer (1998) 216–35

[83] G. Maltsiniotis, *Grothendieck ∞-groupoids, and still another definition of ∞-categories*, Preprint, arXiv:1009.2331
[84] P. Martin-Löf, *An intuitionistic theory of types: Predicative part*, in *Logic Colloquium*, North-Holland (1975)
[85] P. Martin-Löf, *Intuitionistic Type Theory*, Bibliopolis (1984)
[86] J. P. May, *Classifying spaces and fibrations*, Mem. Amer. Math. Soc. 1 (1975) xiii+98
[87] N. P. Mendler, *Quotient types via coequalizers in Martin-Löf type theory*, in G. Huet and G. Plotkin, eds., *Informal Proceedings of the First Workshop on Logical Frameworks, Antibes* (1990) 349–60
[88] W. Mitchell, *Boolean topoi and the theory of sets*, J. Pure Appl. Algebra 2 (1972) 261–74
[89] I. Moerdijk and G. E. Reyes, *Models for Smooth Infinitesimal Analysis* (1991)
[90] Á. Pelayo and M. A. Warren, *Homotopy type theory and Voevodsky's univalent foundations*, Bull. Amer. Math. Soc. (N.S.) 51 (2014) 597–648
[91] J. Penon, *De l'infinitésimal au local (thèse de doctorat d'État)*, Diagrammes (1985) S13:1–191, www.numdam.org/item?id=DIA_1985__S13__1_0
[92] C. Rezk, *Toposes and homotopy toposes*, www.math.uiuc.edu/~rezk/homotopy-topos-sketch.pdf
[93] C. Rezk, *A model for the homotopy theory of homotopy theory*, Trans. Amer. Math. Soc. 353 (2001) 973–1007
[94] C. Rezk, *Proof of the Blakers–Massey theorem*, www.math.uiuc.edu/~rezk/freudenthal-and-blakers-massey.pdf
[95] G. Rosolini, *Continuity and effectiveness in topoi*, PhD thesis, University of Oxford (1986)
[96] U. Schreiber, *Differential cohomology in a cohesive ∞-topos*, http://ncatlab.org/schreiber/show/differential+cohomology+in+a+cohesive+topos
[97] U. Schreiber and M. Shulman, *Quantum gauge field theory in cohesive homotopy type theory*, in *QPL'12* (2012), http://ncatlab.org/schreiber/files/QFTinCohesiveHoTT.pdf
[98] D. Scott, *Continuous lattices*, Lec. Not. Math. 274 (1972) 97–136
[99] D. Scott, *Data types as lattices*, SIAM J. Comput. 5 (1976) 522–87
[100] D. Scott, *A type-theoretical alternative to cuch, iswim, owhy*, Theoret. Comput. Sci. 121 (1993) 411–40. Reprint of a 1969 manuscript
[101] R. A. G. Seely, *Locally Cartesian closed categories and type theory*, Math. Proc. Cambridge Philos. Soc. 95 (1984) 33–48
[102] M. Shulman, *The univalence axiom for EI diagrams*, Reprint, arXiv:1508.02410
[103] M. Shulman, *The univalence axiom for elegant Reedy presheaves*, Homol. Homotopy Appl. 17 (2015) 81–106
[104] M. Shulman, *Univalence for inverse diagrams and homotopy canonicity*, Math. Struct. Comput. Sci. 25 (2015) 1203–77
[105] M. Shulman, *Brouwer's fixed-point theorem in real-cohesive homotopy type theory*. This paper has now appeared in Mathematical Structures in Computer Science 28 (2018), arXiv:1509.07584

[106] M. Shulman, *Homotopy type theory: A synthetic approach to higher equalities*, in E. Landry, ed., *Categories for the Working Philosopher*, Cambridge University Press (2017)

[107] W. P. Stekelenburg, *Realizability of univalence: Modest Kan complexes*, reprint, arXiv:1406.6579

[108] T. Streicher, *Semantics of Type Theory: Correctness, Completeness, and Independence Results*, Progress in Theoretical Computer Science, Birkhäuser (1991)

[109] K. Szumiło, *Two models for the homotopy theory of cocomplete homotopy theories*, Preprint, arXiv:1411.0303

[110] P. Taylor, *Practical Foundations of Mathematics*, Cambridge Studies in Advanced Mathematics 59, Cambridge University Press (1999)

[111] P. Taylor, *A lambda calculus for real analysis*, J. Logic Anal. 2 (2010) 1–115

[112] Univalent Foundations Program, *Homotopy Type Theory: Univalent Foundations of Mathematics*, 1st ed. (2013), http://homotopytypetheory.org/book/

[113] B. van den Berg and R. Garner, *Types are weak ω-groupoids*, Proc. London Math. Soc. 102 (2011) 370–94

[114] B. van den Berg and R. Garner, *Topological and simplicial models of identity types*, ACM Trans. Comput. Logic 13 (2012) 3:1–3:44

[115] J. van Oosten, *Realizability: An Introduction to Its Categorical Side*, Studies in Logic and the Foundations of Mathematics 152, Elsevier (2008)

[116] S. Vickers, *Topology via Logic*, Cambridge Tracts in Theoretical Computer Science 5, Cambridge University Press (1996)

[117] V. Voevodsky, *An experimental library of formalized mathematics based on the univalent foundations*, Math. Struct. Comput. Sci. 25 (2015) 1278–94

[118] V. Voevodsky, *Subsystems and regular quotients of C-systems*, in *Conference on Mathematics and Its Applications (Kuwait City, 2014)* (2015) 1–11

[119] V. Voevodsky, *C-system of a module over a Jf-relative monad*, Preprint, arXiv:1602.00352

[120] P. Wadler, *Propositions as types*, Commun. ACM (2015), http://homepages.inf.ed.ac.uk/wadler/papers/propositions-as-types/propositions-as-types.pdf

[121] M. A. Warren, *Homotopy theoretic aspects of constructive type theory*, PhD thesis, Carnegie Mellon University (2008)

[122] O. Wyler, *Are there topoi in topology?*, in *Categorical Topology (Proc. Conf., Mannheim, 1975)*, Lecture Notes in Mathematics 540, Springer (1976) 699–719

[123] O. Wyler, *Lecture Notes on Topoi and Quasitopoi*, World Scientific (1991)

Michael Shulman
Department of Mathematics, University of San Diego
shulman@sandiego.edu

PART III

Algebraic Geometry

7
Sheaves and Functors of Points

Michel Vaquié

Contents

Introduction		407
1	Sheaves on a Topological Space	410
2	Sheaves on a Category and Functor of Points	427
3	Relative Geometry	452
Conclusion		457
References		458

Introduction

To every type of geometrical space is associated a notion of regular functions on this space, which can be seen as the morphisms of the space in a field, generally $\mathbb{R}$ or $\mathbb{C}$. These functions play an essential role in the study of spaces, and a natural question is to wonder whether they make it possible to determine them.

We can deduce from the Gelfand–Naimark theorem [20] that there is an equivalence between a topological space and the algebra of functions on it. More precisely, let $Comp\mathcal{H}aus$ denote the category of compact Hausdorff spaces, where the morphisms are proper continuous mappings, and let $C^* - Alg$ denote the category of commutative unital C^*-algebras, where the morphisms are unit-preserving $*$-homomorphisms. Then the functor from the opposite category of $Comp\mathcal{H}aus$ to $C^* - Alg$

$$C: (Comp\mathcal{H}aus)^{op} \to C^* - Alg$$
$$X \mapsto C(X),$$

defined by

1. $C(X)$ is the commutative unital C^*-algebra of continous functions on X with values in $\mathbb{C}$,
2. $C(h) = h^*$ for all proper continuous mappings $h\colon X \to Y$,

is an equivalence of categories. (The role of the Gelfand–Naimark theorem in this result is to prove that C is an essentially surjective functor, i.e., every commutative unital C^*-algebra $\mathcal{A}$ is isometrically $*$-isomorphic to $C(X)$ for some compact Hausdorff space X, which is defined as the spectrum of $\mathcal{A}$.)

But if for topological spaces it is enough to know the set of the functions on X to characterize it, it is no more the case if we consider spaces with more structure, for instance, algebraic or analytic spaces. In these cases, the class of regular functions is formed by continuous functions verifying additional properties, to be analytic or algebraic. This specific class we have to consider could be very small, and the knowledge of the set of these functions is too poor to determine the space. For instance, if X is a projective algebraic variety, the only global regular functions on X are the constant functions.

It is thus necessary for us to consider also the functions defined partially, that is, defined on subsets of this space; the notion adapted to study these functions is the notion of *sheaf*. To any geometrical space X we can so associate not only the set of the regular functions defined on X but the sheaf of the regular functions on X.

A differential manifold is traditionally built as a topological space X obtained by gluing elementary pieces that are open subsets of $\mathbb{R}^n$ by means of open immersions. More precisely, we define an *atlas* of X as a collection of pairs $\{(U_\alpha, \varphi_\alpha)\}$ called charts, where the U_α are open sets that cover X, and for each α an homomorphism $\varphi_\alpha \colon U_\alpha \to \mathbb{R}^n$ of U_α onto an open subset of $\mathbb{R}^n$.

To generalize this approach in analytic geometry or in algebraic geometry, it is necessary to glue elementary pieces that are open of $\mathbb{C}^n$ by means of open analytic immersions, or affine spaces (defined as the spectrum of a commutative ring) along open immersion for a certain topology adapted to this context, the Zariski topology. But it is not sufficient to define the notion of analytic or algebraic space to say that space X is a topological space and that the maps φ_α of the charts are only homomorphisms. It is necessary that this space be more *structured*, more precisely, X has to be a *locally ringed space*, and the maps φ_α have to be morphisms of locally ringed spaces, and these definitions utilize in an essential way the concept of a sheaf of regular functions.

Another way to define a space X is to consider the applications from Y to X, when Y goes through the family of all the geometrical spaces; this is the point of view of the Yoneda embedding, a main result in category theory.

In this approach, we have first to define the category *Aff* of the objects that will play the role of the open subsets U_α of an atlas. The category of geometric spaces is obtained as a full subcategory of the category of *presheaves* on *Aff*, that is, the category of functors from $\textit{Aff}^{op}$ to the category *Set* of sets. To define the conditions that satisfy the geometric spaces, we have to define the notion of *Grothendieck topology* on the category *Aff*; this is a generalization of the usual notion of topology on a set, and we demand the spaces be sheaves for this topology. Moreover, they must satisfy another condition we can resolve by saying that they have to be locally representable; this is an analogue of the condition that there exists an atlas that covers the space.

In the first section of the article we recall the classical definition of sheaves on a topological space, and we study the elementary properties. Since the geometric spaces have been classically defined by gluing open subsets of $\mathbb{R}^n$, this notion of sheaf allows us to extend to these geometric spaces as well-known classical notions defined on these open subsets. Moreover, on these spaces, a sheaf plays a fundamental role, the sheaf of *regular functions*, and we will show how the use of this sheaf makes it possible to introduce a new object, a *locally ringed space*, which is a pair formed by a topological space and a sheaf of algebras on this space, and thus to define in a more intrinsic and general way the geometric spaces. In particular, it is thus that schemes, the fundamental object of algebraic geometry since the 1950s, have been defined.

In the second section, we will extend the notion of sheaves, and for this we must introduce the topologies of Grothendieck on a category, which are not topologies in the usual sense but a generalization of them. We give a brief overview of the notion of *topos*, which is fundamental in this theory. Then we show how these notions play an important role in algebraic geometry, in particular, to define good cohomology theory. This point of view also has the advantage of defining in a new way the spaces of the algebraic geometry, with the notion of *functor of points*, which gives a second definition of schemes as a sheaf on the category of affine schemes, which is by definition the opposite category of rings, endowed with a Grothendieck topology. Then we extend this method to define algebraic spaces, which are generalizations of the notion of scheme, which can in particular answer to the question of *representability* of the natural functor in algebraic geometry.

In the third section we will show how this construction, which has been introduced in algebraic geometry, can be extended, and we define the *geometric contexts* to get new geometric spaces. We show how we can use these ideas in a very general and abstract setting; we deal very rapidly with this aspect, which we call *relative geometry*, and give one definition of the field with one element.

1 Sheaves on a Topological Space

The usual structure on a space X to have a notion of object or property *defined locally* is the structure of *topological space*, that is, to endow the space X with a topology. We recall that a topology is defined by the datum of a set $O(X)$ of subsets of X, called *open subsets of X*, which is stable by finite intersections and any unions; in particular, the total space X and the empty set $\emptyset$ belong to $O(X)$.

The objects or structures defined locally on X are objects or structures defined on open subsets of X, for instance, we may define for any open subset U of X the set $C^0(U)$ of continuous functions on U, or if X is a manifold, the ring of differential forms $\Omega(U)$ or differential operators $\mathcal{D}(U)$, or the vector space $\mathcal{L}(U)$ of sections over U of a vector bundle L over X. The idea is to associate to any open subset U of X a set (or a ring, a module, or more generally an object in some fixed category) $\mathcal{F}(U)$ that gives some *local* information on X. Then we want to consider all these sets $\mathcal{F}(U)$ and the way they are related to obtain some global information on X. The theory of sheaves is a way to connect local and global behaviors.

1.1 Brief Historical Reminder

The origin of sheaf theory is algebraic topology; sheaves were introduced by Leray in the 1940s as local systems of coefficient groups, and he invented at the same time the notion of sheaf cohomology as well as spectral sequences for computing his sheaf cohomology [31, 32]. There are many articles or books about the history of algebraic topology and sheaves, and we refer to them ([8, 12, 28, 29, 39, 40, 60]).

Algebraic topology had its origins in the late 19th century with the work of Riemann, Betti on *homology numbers*, and Poincaré, who developed a more correct homology theory in *Analysis Situs*. The aim is to attach some numbers to topological spaces and to use them to study topological properties of these spaces, or of the maps between them, and in 1925, Emmy Noether pointed out that homology was an abelian group rather than just Betti numbers and torsion coefficients.

By the 1950s, there were several different methods of attaching (co)homology groups to a topological space; in each case, the inventors gave an ad hoc recipe for constructing a chain complex and defined their homology groups to be the homology of that chain complex. Eilenberg and Steenrod gave in [18] an axiomatic treatment of homology theory, rederiving the whole of homology theory for finite complexes from these axioms, and they also pointed out that singular homology and Čech homology satisfy the axioms. They defined a *cohomology theory* as a contravariant functor from the category of pairs of topological spaces (X, U), where X is a locally compact topological space and U is an open subset of X, and whose morphisms are the continuous maps of pair, to the category of *abelian groups*, satisfying the following axioms:

1. (exactness axiom) for any pair (X, U) we have the following exact sequence:
$$\ldots \to H^{r-1}(U) \to H^r(X, U) \to H^r(X) \to H^r(U) \to H^{r+1}(X, U) \to \ldots ;$$

2. (homotopy axiom) the map f^* depends only on the homotopy class of f;
3. (excision) if V is open in X and its closure is contained in U, then the inclusion map $(X \setminus V, U \setminus V) \to (X, U)$ induces an isomorphism $H^r(X, U) \to H^r(X \setminus V, U \setminus V)$;
4. (dimension axiom) if X consists of a single point, then $H^r(X) = 0$ for $r \neq 0$.

In the 1940s, Leray attempted to understand the relation between the cohomology groups of two spaces X and Y for which a continuous map $f: X \to Y$ is given. The idea to study the topology of a variety X by considering its projection on a variety of smaller dimension, $f: X \to Y$, and properties of the fiber F was first thought by Picard and Lefschetz.

The problem is that the cohomology of the fibers may vary, and it is necessary to study the function that associates to any closed subset $Z \subset Y$ the cohomology of the inverse image $H^q(f^{-1}(Z), k)$. This is the reason why Leray has introduced the notion of the sheaf as a local system of coefficient groups, to connect the cohomology of the different fibers. In his definition, a *faisceau*, a sheaf, $\mathcal{B}$, of modules on a topological space Y, is defined to be a function assigning to each closed set $F \subset Y$ a module $\mathcal{B}_F$ such that $\mathcal{B}_\emptyset = 0$, together with a homomorphism from $\mathcal{B}_F$ to $\mathcal{B}_{F'}$ whenever $F' \subset F$, satisfying a condition of transitivity for $F'' \subset F' \subset F$. The example of sheaf on Y is given by $H^r(-, A): F \mapsto H^r(F, A)$ for a ring A. But this definition is only appropriate for particular topological spaces, and Cartan has modified it to get the definition of a sheaf as a functor from the category of open subsets, as we will see in the next section.

1.2 Classical Point of View

To define locally an object, or a structure, on a topological space, we want to associate to any open subset U of X an object $\mathcal{F}(U)$ in some category $\mathcal{C}$, for instance, $\mathcal{C}$ may be the category of rings and $\mathcal{F}(U)$ the ring of continuous functions on U with values in $\mathbb{R}$. Moreover, for any inclusion of open subsets $V \subset U$, we want to have a map $r_{U,V} \colon \mathcal{F}(U) \to \mathcal{F}(V)$ in the category $\mathcal{C}$, which corresponds to the natural idea of *restriction* of objects of $\mathcal{F}(U)$ to V. In particular, we want to have transitivity; for any $W \subset V \subset U$, we have the relation $r_{U,W} = r_{V,W} \circ r_{U,V}$.

The partially ordered set by inclusion $O(X)$ of all open subsets of X may be considered as a category; indeed, any partially ordered set $(P, \leq)$ may be seen as a category such that for $x, y \in P$, we have $Hom_P(x, y) = \{*\}$ if $x \leq y$ and is empty otherwise. Then what we want to define is a functor

$$\begin{aligned} \mathcal{F} \colon O(X)^{op} &\longrightarrow \mathcal{C} \\ U &\mapsto \mathcal{F}(U), \end{aligned}$$

where, for any $j \colon V \to U$, there is the restriction map

$$\begin{aligned} \mathcal{F}(j) \colon \mathcal{F}(U) &\to \mathcal{F}(V) \\ s &\mapsto s_{|V}. \end{aligned}$$

We shall restrict for the moment to the case of a functor with values in the category $\mathcal{S}et$ of sets: for any open subset U of X, the object $\mathcal{F}(U)$ is a set; we forget any additional structure, and we call such a functor $\mathcal{F} \in Fun(O(X)^{op}, \mathcal{S}et)$ a *presheaf* on the topological space X. If we choose for the target category $\mathcal{C}$ the category of groups (resp. rings, resp. anything), we will say that we have a presheaf of groups (resp. rings, resp. anything).

Most of the objects we would like to consider are *determined locally*: if we have an open covering $U = \bigcup_{i \in I} U_i$ of U in $O(X)$, then we want a family (x_i) of elements in $\prod_{i \in I} \mathcal{F}(U_i)$ such that the restriction of x_i and x_j coincide, in $\mathcal{F}(U_i \cap U_j)$ for any i, j in I, determining a unique element x of $\mathcal{F}(U)$. Such a presheaf is called a sheaf, and we can give the following definition:

Definition 1.1 A sheaf $\mathcal{F}$ on a topological space X is a functor $\mathcal{F} \colon O(X)^{op} \to \mathcal{S}et$ such that, for any open subset U of X and for any open covering $U = \bigcup_{i \in I} U_i$ of U in $O(X)$, the diagram

$$\mathcal{F}(U) \longrightarrow \prod_i \mathcal{F}(U_i) \rightrightarrows \prod_{i,j} \mathcal{F}(U_i \cap U_j)$$

is an equalizer.

We can define the category $Pr(X)$ of presheaves on X as the category of functors from $O(X)^{op}$ to Set, the morphisms between two presheaves $\mathcal{F}$ and $\mathcal{G}$ are the natural transformations between functors, and the category of sheaves $Sh(X)$ is the full subcategory of $Pr(X)$ whose objects are the sheaves on X.

By definition of the category $Pr(X)$, a morphism $f \colon \mathcal{F} \to \mathcal{G}$ between two presheaves on X is a monomorphism, resp. an epimorphism, if and only if, for any open subset U of X, the map of sets $f_U \colon \mathcal{F}(U) \to \mathcal{G}(U)$ is a monomorphism, resp. an epimorphism. If $\mathcal{F}$ and $\mathcal{G}$ are sheaves, then if the morphism $f \colon \mathcal{F} \to \mathcal{G}$ is a monomorphism in $Sh(X)$, we still have a monomorphism $f_U \colon \mathcal{F}(U) \to \mathcal{G}(U)$ for any open subset U of X, but if f is an epimorphism, we cannot deduce anything about the applications f_U in general.

For any open subset U we can define a presheaf h_U by $h_U(V) = 1$, where 1 is the set $\{*\}$ with one element if $V \subset U$, and $h_U(V) = 0$, where 0 is the empty set otherwise, and it is easy to see that h_U is a sheaf. We call the sheaves of the form h_U for $U \in O(X)$ the *representable* sheaves on X. In particular, for $U = X$, we find that h_X is the constant sheaf 1, and there is a bijection between the ordered set of subsheaves of the constant sheaf 1 on X and the ordered set $O(X)$ of open subsets of X.

We can associate to any presheaf $\mathcal{P}$ a sheaf $a(\mathcal{P})$ with a morphism $\eta \colon \mathcal{P} \to a(\mathcal{P})$ satisfying the following universal property: for any sheaf $\mathcal{F}$ on X and any morphism $u \colon \mathcal{P} \to \mathcal{F}$ of presheaves, there exists a unique morphism $v \colon a(\mathcal{P}) \to \mathcal{F}$ such that the following diagram is commutative:

In other words, the functor of inclusion $i \colon Sh(X) \to Pr(X)$ has a left adjoint $a \colon Pr(X) \to Sh(X)$, called the *associated sheaf functor* or the *sheafification*. Moreover, the morphism $\eta \colon \mathcal{P} \to a(\mathcal{P})$ is an isomorphism if and only if $\mathcal{P}$ is a sheaf and the functor $a \colon Pr(X) \to Sh(X)$ is left exact, that is, it preserves finite limits.

Let $f \colon X \to Y$ be a continuous map between topological spaces; we want to define the direct and inverse images of a presheaf and of a sheaf by f. If $\mathcal{P}$ is a presheaf on X, the direct image of $\mathcal{P}$ by f, which is denoted by $f_*(\mathcal{P})$, is defined for any open subset V of Y by the following:

$$f_*(\mathcal{P})(V) := \mathcal{P}(f^{-1}(V)).$$

It is by definition a presheaf on Y, and it is very easy to see that if $\mathcal{P}$ is a sheaf, then the direct image $f_*(\mathcal{P})(V)$ is a sheaf.

It is more difficult to define the inverse image because the image $f(U)$ of an open subset U of X is not in general an open subset of Y. First we define the inverse image of a presheaf $\mathcal{G}$ on Y by f, which is denoted by $f^{-1}(\mathcal{G})$, by the following:

$$U \mapsto \varinjlim_{V \supseteq f(U)} \mathcal{G}(V),$$

where U is an open subset of X and the colimit runs over all open subsets V of Y containing $f(U)$.

Even if $\mathcal{G}$ is a sheaf on Y, in general, the presheaf $f^{-1}(\mathcal{G})$ does not satisfy the sheaf condition; then we define the inverse image of a sheaf $\mathcal{G}$ on Y, and we denote by $\tilde{f}^{-1}(\mathcal{G})$ the sheaf associated to this presheaf:

$$\tilde{f}^{-1}(\mathcal{G}) := a\big(f^{-1}(\mathcal{G})\big).$$

We have defined a pair of adjoint functors:

$$\tilde{f}^{-1} \colon Sh(Y) \rightleftarrows Sh(X) \colon f_*$$

We deduce that the functor $\tilde{f}^{-1}$, resp. f_*, respects colimits, resp. limits; in fact, we can prove that the functor $\tilde{f}^{-1}$ respects also all the finite limits.

For any presheaf $\mathcal{P}$ on a topological space X and for any point $x \in X$, we can define the *stalk* of $\mathcal{P}$ at the point X as the direct limit

$$\mathcal{P}_x = \varinjlim_{x \in U} \mathcal{P}(U),$$

indexed over all the open sets U containing x, with order relation induced by reverse inclusion. By definition (or universal property) of the direct limit, an element of the stalk is an equivalence class of elements $x_U \in \mathcal{P}(U)$, where two such sections x_U and x_V are considered equivalent if the restrictions of the two sections coincide on some neighborhood of x. For any open subset U of X and any point x in U, there exists a natural map $\mathcal{P}(U) \to \mathcal{P}_x$, and we denote by s_x the image of $s \in \mathcal{P}(U)$ and call it the *germ* at x of the section s.

A morphism $f \colon \mathcal{F} \to \mathcal{G}$ between two presheaves on a topological space X induces for any point x of X an application $f_x \colon \mathcal{F}_x \to \mathcal{G}_x$. Moreover, in the case of morphism between sheaves, we have the following result:

Proposition 1.2 *A morphism of sheaves $f \colon \mathcal{F} \to \mathcal{G}$ is a monomorphism, resp. an epimorphism, resp. an isomorphism, if and only if, for any point $x \in X$, the morphism $f_x \colon \mathcal{F}_x \to \mathcal{G}_x$ is a monomorphism, resp. an epimorphism, resp. an isomorphism.*

The associated sheaf $a(\mathcal{P})$ associated to a presheaf $\mathcal{P}$ on a topological space X is explicitly constructed in the following manner. For every open subset U of X, let $\mathcal{P}'(U)$ be the set of functions $s \colon U \to \coprod_{x \in U} \mathcal{P}_x$ such that for all $x \in U$, $s(x) \in \mathcal{P}_x$ and there exists an open neighborhood V of x, $V \subset U$, and $t \in \mathcal{P}(V)$ such as $t_y = s(y)$ for all $y \in V$. Then $\mathcal{P}'$ is the sheaf associated with $\mathcal{P}$, and for obvious reasons, it is also called the sheaf sections of $\mathcal{P}$.

We deduce from this description the following result:

Proposition 1.3 *The morphism $\eta \colon \mathcal{P} \to a(\mathcal{P})$ induces an isomorphism $\eta_x \colon \mathcal{P}_x \to a(\mathcal{P})_x$ for every point x of X.*

We will give now another way to describe the category of sheaves on a topological space. Let X be a topological space; a continuous map $p \colon Y \to X$ is called a *space over* X, and a morphism between two spaces $p \colon Y \to X$ and $q \colon Z \to X$ over X is a continuous map $f \colon Y \to Z$ such that $q \circ f = p$. The category of spaces over X is exactly the slice category **Top**$/X$, where **Top** is the category of topological spaces with continuous maps as morphisms.

A *cross section* of a space over X, $p \colon Y \to X$, is a continuous map $s \colon X \to Y$ such that we have $p \circ s = id_X$. And for any open subset $i \colon U \to X$, we define the set of cross sections of $p \colon Y \to X$ over U by

$$\Gamma_p(U) = \{s \colon U \to Y \mid p \circ s = i\}.$$

If $V \subset U$ are two open subsets of X, the restriction operation $s \mapsto s_{|V}$ defined a map from $\Gamma_p(U)$ to $\Gamma_p(V)$, and it is very easy to see that in fact we get a sheaf Γ_p on the space X. Moreover, if we have a map $f \colon Y \to Z$ of spaces over X, for any open subset U of X, the composition by f defines a map from $\Gamma_p(U)$ to $\Gamma_q(U)$. Then we have defined a functor

$$\begin{aligned} \Gamma_- \colon \quad &\mathbf{Top}/X &\longrightarrow \quad &Sh(X) \\ &p \colon Y \to X &\mapsto \quad &\Gamma_p. \end{aligned}$$

Now let $\mathcal{P}$ be a presheaf on a topological space X, and we recall we denote by $\mathcal{P}_x$ the stalk of $\mathcal{P}$ at x, that is, the set of germs s_x of sections of $\mathcal{P}$ at the point x. Then we define the set $\Lambda_\mathcal{P}$ of the germs of all the sections at all the points of X:

$$\Lambda_\mathcal{P} = \coprod_{x \in X} \mathcal{P}_x,$$

and a map $\pi \colon \Lambda_\mathcal{P} \to X$, sending each germ s_x to the point x where it is taken. By construction, the fiber of π over a point x is exactly the stalk $\mathcal{P}_x$ of $\mathcal{P}$ at x.

For each open subset U of X, for each element s of $\mathcal{P}(U)$ and each x in U, we get the germ s_x, which is a point of $\Lambda\mathcal{P}$. For any U and $s \in \mathcal{P}(U)$, the union of these points for all $x \in U$ is denoted by $s(U)$, and we take these sets $s(U)$ as a base of open sets in $\Lambda\mathcal{P}$. Notice that this topology induces the discrete topology on each stalk $\mathcal{P}_x$.

We call the topological space $\Lambda\mathcal{P}$ the *étale space* of the presheaf $\mathcal{P}$, and it is easy to see that any map of presheaves $\varphi\colon \mathcal{P} \to \mathcal{Q}$ induces a continous map from $\Lambda\mathcal{P}$ to $\Lambda\mathcal{Q}$ over X, that is, we have defined a functor

$$\begin{aligned} \Lambda_-\colon\ Pr(X) &\longrightarrow\ \mathbf{Top}/X \\ \mathcal{P} &\mapsto\ \pi\colon \Lambda\mathcal{P} \to X\,. \end{aligned}$$

Definition 1.4 A space over X, $p\colon Y \to X$ is said to be *étale* if p is a local homomorphism, that is, for any point $y \in Y$, there exists an open neighborhood V of y in Y such that the image $p(V)$ is open in X and the restriction $p_{|V}\colon V \to p(V)$ is a homomorphism.

The space over X associated to any presheaf $\mathcal{P}$ on X is étale, and if we consider the functor Γ_- as a functor with values in the category of presheaves over X, we have the following result:

Theorem 1.5 *For any topological space X, there is a pair of adjoint functors*

$$\Lambda\colon Pr(X) \rightleftarrows \mathbf{Top}/X\colon \Gamma\,.$$

The unit and counit of this adjunction define natural transformations

$$\eta_\mathcal{P}\colon \mathcal{P} \to \Gamma\Lambda\mathcal{P} \quad\text{and}\quad \epsilon_Y\colon \Lambda\Gamma Y \to Y\,,$$

and $\eta_\mathcal{P}$ is an isomorphism if and only if $\mathcal{P}$ is a sheaf and ϵ_Y is an isomorphism if and only if Y is étale over X.

If we denote by **Etale**$/X$ the full subcategory of **Top**$/X$ of étale spaces over X, we have the following:

Corollary 1.6 *The functors Λ and Γ restrict to an equivalence of categories*

$$\Lambda\colon Sh(X) \rightleftarrows \mathbf{Etale}/X\colon \Gamma\,.$$

Moreover, $Sh(X)$ is a reflective subcategory of $Pr(X)$, and **Etale**$/X$ *is a coreflective subcategory of* **Top**$/X$.

The notion of étale space was at the origin of the modern form of sheaf theory; in an exposé of the 1950–51 Cartan Seminar [13], dated April 8, 1951, Henri Cartan followed an idea of Michel Lazard and defined a sheaf as an *espace étalé*, a French term meaning roughly "spread out," with group structure, and he realized that the natural form of localization was to open sets rather than closed.

1.3 Homology and Cohomology

It is important to note that from the beginning, the definition of sheaves is about sheaves of modules and that it is related with the cohomology and the spectral sequences. When we work with sheaves with values in the category of rings, or more generally in a well-suited category, there is an abstract and very general way to define homology using *derived functors*.

Cartan and Eilenberg introduced and developped new foundations of algebraic homology and cohomology theories, and they revolutionized the subject in their textbook [7]. They limited themselves to functors defined on modules, but it was clear that there was more than a formal analogy with the cohomology of sheaves and that their methods worked in a more general setting. In his 1955 thesis, Buchsbaum defined the notion of an *abelian category* (the name "abelian category" is due to Grothendieck [22] and Heller) and extended the Cartan–Eilenberg theory of derived functors to such categories.

An abelian category is a category that is very similar to the category of modules on a ring; in particular, there exist good notions of *kernel* and *cokernel* for any map and a very useful notion of *exact sequence*. In general, a functor $F: \mathcal{A} \to \mathcal{B}$ between two abelian categories does not preserve exact sequences, and the *derived functors* have been defined to study the behavior of the exact sequences.

For instance, if the functor $F: \mathcal{A} \to \mathcal{B}$ preserves the monomorphisms, we say that F is *left exact*, but if it does not preserve the epimorphisms, the right-derived functors, $R^p F$ for $p \geq 0$ with $R^0 F = F$, if they exist, give a way to describe the situation. More precisely, if $u: x \to y$ is an epimorphism, we can define a short exact sequence $0 \to z \to x \to y \to 0$ in $\mathcal{A}$, where z is the kernel of u. Then by definition there exists a long exact sequence

$$\ldots \to R^{p-1} F(y) \to R^p F(z) \to R^p F(x) \to R^p F(y) \to R^{p+1} F(z) \to \ldots,$$

and the map $F(y) \to R^1 F(z)$ controls the obstruction to the surjectivity of the map $F(u): F(x) \to F(y)$.

To be sure that the right-derived functors exist, it is necessary to assume that the category $\mathcal{A}$ has *enough injectives*. We recall that an object a of $\mathcal{A}$ is said to be injective if, for every morphism $f: b \to a$ and every monomorphism $h: b \to c$, there exists a morphism $g: c \to a$ extending f, that is, such that

$g \circ h = f$. Then to get the right-derived functors $R^p F(x)$, we need an *injective resolution* of $x \in \mathcal{A}$, that is, a long exact sequence

$$0 \to x \to a^0 \to a^1 \to a^2 \to \ldots$$

such that all the a^i are injective. We have $R^p F(x) = H^p(F(a^\bullet))$, and this does not depend on the choice of the chosen injective resolution (in fact, it will be enough to choose the a^i's *acyclic* for the functor F).

In the same way, if we have a right exact functor F, that is, a functor that preserves epimorphisms, it is possible to define the *left-derived functors* $L^p F$ and to be sure to have the existence, the abelian category $\mathcal{A}$ must have *enough projectives*.

In what follows, we will consider the category of sheaves of abelian groups on a topological space X, and we denote this category by $\mathcal{A}b(X)$. Alexandre Grothendieck, in his famous 1957 Tohoku paper [22], showed that the category $\mathcal{A}b(X)$ is an abelian category with enough injectives, and so one can define the cohomology groups of the sheaves on a space X as the right-derived functors of the functor taking a sheaf to its abelian group of global sections.

According to what we saw in the previous paragraph, we can deduce that for any point x of X, the functor *fiber at the point x*, defined by

$$(-)_x : \quad \mathcal{A}b(X) \longrightarrow \mathcal{A}b$$
$$\mathcal{F} \mapsto \mathcal{F}_x,$$

is an exact functor. In the same way, for any open subset U of X, the functor *section over U* defined by

$$\Gamma(U, -) : \quad \mathcal{A}b(X) \longrightarrow \mathcal{A}b$$
$$\mathcal{F} \mapsto \mathcal{F}(U)$$

is left exact. In particular, for $U = X$, we get the functor *global section* $\Gamma : \mathcal{F} \mapsto \Gamma(X, \mathcal{F}) = \mathcal{F}(X)$. From the result of Grothendieck on the abelian category $\mathcal{A}b(X)$, we deduce that it has right-derived functors, which we call the *cohomology groups* of the sheaf $\mathcal{F}$ on X. We denote them by

$$R^p \Gamma(X, \mathcal{F}) = H^p(X, \mathcal{F}) \text{ for } p \geq 0.$$

The axioms for sheaf cohomology theory on a paracompact space X were introduced by H. Cartan in [5, Exposé 16]. R. Godement, in his book [21], summarized and refined all these developments, and he introduced the notion of *flabby* (*flasque*) sheaf as well as *fine* (*fin*) and *soft* (*mou*) sheaves, which play a very important role because these sheaves are acyclic for the functors *global section* Γ and *direct image* f_*.

An archetypal example is the sheaf of all set-theoretic (not necessarily continuous) sections of a bundle $E \to X$; regarding that every sheaf over a topological space is the sheaf of sections of the étale space $\pi : \Lambda_{\mathcal{F}} \to X$, every sheaf can be embedded into a flabby sheaf $\mathcal{C}^0(X, F)$ defined by $U \mapsto \prod_{x \in U} \mathcal{F}_x$, where $\mathcal{F}_x$ is the stalk of $\mathcal{F}$ at point x. Moreover, we can define a functorial flabby resolution of a sheaf $\mathcal{F}$ by iteration of the canonical embedding of $\mathcal{F}$ into $\mathcal{C}^0(X, \mathcal{F})$. This resolution was introduced by R. Godement in [21] and is called the *Godement resolution* of $\mathcal{F}$.

Let X be a topological space, and let $\mathcal{U}$ be an open covering of X. Let $N(\mathcal{U})$ denote the nerve of the covering. The idea of *Čech cohomology* is that, for an open covering $\mathcal{U}$ consisting of sufficiently small open sets, the resulting simplicial complex $N(\mathcal{U})$ should be a good combinatorial model for the space X. For such a covering, the Čech cohomology of X is defined to be the simplicial cohomology of the nerve.

Let X be a topological space, and let $\mathcal{F}$ be a sheaf of abelian groups on X. Let $\mathcal{U} = (U_i)_{i \in I}$ be an open covering of X such that the index set I is well ordered. The Čech cohomology of $\mathcal{U}$ with values in $\mathcal{F}$ is defined to be the cohomology of the cochain complex $(C^{\cdot}(\mathcal{U}, \mathcal{F}), \delta)$. More precisely, for every $J \subset I$ with $q + 1$ elements, $|J| = q + 1$, we denote by U_J the intersection $\bigcap_{i \in J} U_i$ and by $\mathcal{F}_{U_J}$ the sheaves on X defined by $\mathcal{F}_{U_J}(V) := \mathcal{F}(U_J \cap V)$ for any open subset $V \subset X$.

Consider the complex of sheaves on X

$$\mathcal{F}^{\bullet} := 0 \longrightarrow \mathcal{F} \longrightarrow \prod_{i \in I} \mathcal{F}_{U_i} \rightrightarrows \prod_{|J|=2} \mathcal{F}_{U_J} \Rrightarrow \ldots,$$

where the boundary maps are defined as follows. Denote by $\mathcal{F}^q$ the sheaf $\prod_{|J|=q+1} \mathcal{F}_{U_J}$, and let $\sigma = (\sigma_J)$, $(\sigma_J) \in \mathcal{F}(V \cap U_J)$ for some $J \subset I$ with $|J| = q + 1$. Then define $(\partial \sigma)_{\tilde{J}} \in \mathcal{F}(V \cap U_{\tilde{J}})$, where $\tilde{J} = \{j_0, \ldots, j_{q+1}\}$, $j_0 < \ldots < j_{q+1}$:

$$(\partial \sigma)_{\tilde{J}} = \sum_i (-1)^i \sigma_{\tilde{J} - \{j_i\} | V \cap U_{\tilde{J}}}.$$

It is not hard to see that the above sequence is a resolution of the sheaf $\mathcal{F}$ in the abelian category $\mathcal{A}b(X)$. It is called the Čech resolution with respect to the covering $\mathcal{U}$.

The groups

$$\check{H}^q_{\mathcal{U}}(X, \mathcal{F}) := H^q(\mathcal{F}^*(X))$$

are called the Čech cohomology groups relative to the open covering $\mathcal{U}$, and we have for each $q \geq 0$ a natural map, functorial in $\mathcal{F}$:

$$\check{H}^q_{\mathcal{U}}(X,\mathcal{F}) \to H^q(X,\mathcal{F}).$$

In the case that we choose a *good covering*, we have the following result:

Theorem 1.7 *If the space X is paracompact and if the open covering $\mathcal{U}$ has the property that $H^q(U_J,\mathcal{F}) = 0$ for every $q \geq 1$ and for every $U_J = \cap_{i \in J} U_i$, J finite, then*

$$\check{H}^q_{\mathcal{U}}(X,\mathcal{F}) = H^q(X,\mathcal{F}) \quad \forall q \geq 0.$$

1.4 Sheaves in Geometry

Classically, the spaces we consider in geometry are obtained by gluing elementary pieces, which are well known and on which there exist natural structures that allow us to make calculations, for example, systems of coordinates, differentiable functions and differential forms. In the most classic case, these elementary pieces are open subsets of the Euclidian space $\mathbb{R}^n$.

We will recall definitions about this construction of spaces in differential geometry. First, a *topological manifold* is a topological space (classically, we add the assumptions of second countable Hausdorff space) that is locally homomorphic to an open subset of $\mathbb{R}^n$, by a collection (called an *atlas*) of homomorphisms called *charts*. The composition of one chart with the inverse of another chart is a function called a *transition map* and defines a homomorphism of an open subset of $\mathbb{R}^n$ onto another open subset of $\mathbb{R}^n$. More precisely, a chart (U,φ) for a topological space M is a homomorphism φ from an open subset U of M to an open subset of $\mathbb{R}^n$, and an atlas for a topological space M is a collection $(U_\alpha,\varphi_\alpha)_{\alpha \in A}$ of charts on M such that $\bigcup_{\alpha \in A} U_\alpha = M$.

The transition map provides a way of comparing two charts of an atlas by considering the composition of one chart with the inverse of the other. To be more precise, suppose that $(U_\alpha,\varphi_\alpha)$ and (U_β,φ_β) are two charts for a manifold M such that $U_\alpha \cap U_\beta$ is nonempty; the transition map $\tau_{\alpha,\beta} \colon \varphi_\alpha(U_\alpha \cap U_\beta) \to \varphi_\beta(U_\alpha \cap U_\beta)$ is the map defined by $\tau_{\alpha,\beta} = \varphi_\beta \circ \varphi_\alpha^{-1}$. Note that since φ_α and φ_β are both homomorphisms, the transition map $\tau_{\alpha,\beta}$ is also a homomorphism.

Moreover, if we want to have additional structures on a manifold M, we consider an atlas such that the structure is defined on the open subset $\varphi(U)$ for every chart (U,φ) and such that all the transition maps respect this structure. For instance, we say that M is a differentiable manifold, or more generaly a manifold of class C^r, for $1 \leq r \leq \infty$, if all transition maps are C^∞-diffeomorphisms, resp. diffeomorphisms of class C^r. If we want to get a complex manifold, we have to demand that the charts have their image in $\mathbb{C}^n$ and that the transition maps are biholomorphic. We can say that a manifold is obtained by gluing local pieces, which are open subsets of $\mathbb{R}^n$ or of $\mathbb{C}^n$, and that the structure of the manifold M is controlled by the class of the transition maps.

We have the same construction in algebraic geometry, and we can define an algebraic variety over a field k as a space that locally looks like an *affine algebraic variety*, that is, a subset of the linear space k^n defined by algebraic equations. And more generally, we will define a scheme as a space that locally looks like the *spectrum* of a ring (see a precise definition of *scheme* later).

But this type of construction is not intrinsic, and in particular, we have to verify that the additional structure does not depend on the atlas as we have chosen to define it. There is a way to obtain globally these additional structures on a topological space X; for this we have to introduce the concept of *structured space*, which is by definition a topological space X provided with a sheaf of rings $\mathcal{O}_X$ that corresponds to the sheaf of *regular functions*. The general notion that is used in the usual geometric theories, as differential geometry, analytic geometry, or algebraic geometry, is the notion of *locally ringed space*.

We recall that a *local ring* is a ring A with a unique maximal ideal $\mathfrak{m}_A$ and that a morphism $\varphi \colon A \to B$ between two local rings is called *local* if the inverse image by φ of the maximal ideal of B is equal to the maximal ideal of A: $\varphi^{-1}(\mathfrak{m}_B) = \mathfrak{m}_A$.

Definition 1.8 A locally ringed space $(X, \mathcal{O}_X)$ is a topological space X endowed with a sheaf of rings $\mathcal{O}_X$, called the structure sheaf, such that the stalk $\mathcal{O}_{X,x}$ at any point x of X is a local ring. For any open subset U of X, the ring $\mathcal{O}_X(U)$ is the ring of *regular functions* on U, and for any point x of X, the ring $\mathcal{O}_{X,x}$ is the ring of *germs* of regular functions at the point x.

A morphism $(f, f^\sharp) \colon (X, \mathcal{O}_X) \to (Y, \mathcal{O}_Y)$ of locally ringed spaces is a continuous map $f \colon X \to Y$ with a morphism of sheaves of rings $f^\sharp \colon \mathcal{O}_Y \to f_* \mathcal{O}_X$ on Y, or, by adjunction, a morphism $\tilde{f}^{-1} \mathcal{O}_Y \to \mathcal{O}_X$ on X, such that for any point x in X, with image $y = f(x)$ in Y, the induced morphism of local rings $\mathcal{O}_{Y,y} \to \mathcal{O}_{X,x}$ is local.

In any locally ringed space $(X, \mathcal{O}_X)$ we have a notion of a sheaf of $\mathcal{O}_X$-modules as a sheaf $\mathcal{M}$ of abelian groups on X such that for any open subset U of X, the group $\mathcal{M}(U)$ is an $\mathcal{O}_X(U)$-module, and such that for each inclusion of open subsets $V \subset U$, the restriction morphism $\mathcal{M}(U) \to \mathcal{M}(V)$ is compatible with the structure of modules via the ring homomorphism $\mathcal{O}_X(U) \to \mathcal{O}_X(V)$.

Let $(f, f^\sharp) : (X, \mathcal{O}_X) \to (Y, \mathcal{O}_Y)$ be a morphism of locally ringed spaces; we want to consider the functor induced on the categories of sheaves of $\mathcal{O}_X$-modules and $\mathcal{O}_Y$-modules. By the map $f^\sharp : \mathcal{O}_Y \to f_*\mathcal{O}_X$, the direct image of a sheaf of $\mathcal{O}_X$-modules is a sheaf of $\mathcal{O}_Y$-modules. But the functor $\tilde{f}^{-1}$ is inappropriate, because in general, it does not even give sheaves of $\mathcal{O}_X$-modules. To remedy this, one defines in this situation for a sheaf of $\mathcal{O}_Y$-modules $\mathcal{G}$ its inverse image by

$$f^*\mathcal{G} := \tilde{f}^{-1}\mathcal{G} \otimes_{\tilde{f}^{-1}\mathcal{O}_Y} \mathcal{O}_X .$$

We have then a pair of adjoint functors

$$f^* : \mathcal{O}_Y - Mod \rightleftarrows \mathcal{O}_X - Mod : f_*$$

between the categories of $\mathcal{O}_Y$-modules and $\mathcal{O}_X$-modules.

The simplest case of structured space is the locally ringed space $(M, \mathcal{O}_M)$, where M is a topological manifold and $\mathcal{O}_M$ is the sheaf of continuous functions with values in $\mathbb{R}$; for any open subset U of M, the ring $\mathcal{O}_M(U)$ is the ring of continuous functions on U, and for any point $x \in M$, the stalk $\mathcal{O}_{M,x}$ is the local ring of *germs* of continuous functions at the point x, whose maximal ideal consists of the germs of functions that vanish at x. And if we consider a differentiable manifold M, we have to consider the locally ringed space $(M, \mathcal{O}_M)$ where now $\mathcal{O}_M$ is the sheaf of differentiable functions with values in $\mathbb{R}$.

In what follows, we will develop the case of algebraic geometry, and we will give the definition of a *scheme*, the basic concept of this geometry, as structured space; this is the definition given by Grothendieck in [17]. We can remark that it was in the context of algebraic geometry that the notion of locally ringed space was the most fruitful, but Grothendieck applied it in the context of analytic geometry at about the same time in the seminar Henri Cartan on foundations of analytic geometry [23].

We need a topology in the context of algebraic geometry that is well defined over any field without any reference to a topology on $\mathbb{R}$ or $\mathbb{C}$, even without any reference to a base field. This is the *Zariski topology*, which was introduced originally by Oscar Zariski when he considered the *local uniformization* problem [63]. The fundamental idea of this topology is to define the closed subset of an algebraic variety as the algebraic subsets.

First we define the notion of *spectrum* of any commutative ring A, which will play the role of elementary piece in the construction of these geometric spaces.

Definition 1.9 The *spectrum* of a commutative ring A is the topological space defined by the following: the underlying set X is the set of all the prime ideals of A, and the topology on X, the *Zariski topology*, is such that the closed subsets are exactly all the sets $V(\mathcal{I}) = \{\mathfrak{p} \in X / \mathcal{I} \subset \mathfrak{p}\}$, where $\mathcal{I}$ ranges the ideals of A.

We denote the spectrum of the ring A by $X = \operatorname{Spec} A$.

We have to verify that this defines a topology on the set X, and moreover, we can see that the closed subset $V(\mathcal{I})$ of $X = \operatorname{Spec} A$ is homomorphic to the spectrum of the quotient ring $A/\mathcal{I}$, that is, we have $V(\mathcal{I}) \simeq \operatorname{Spec} A/\mathcal{I}$.

By definition of the topology, a point $\mathfrak{p} \in X = \operatorname{Spec} A$ is a closed point if and only if the prime ideal $\mathfrak{p}$ is a maximal ideal of A. We can define the *maximal spectrum* of a ring A, and we denote by $X_{max} = \operatorname{Spec}_{max} A$ the set of all maximal ideals of A with a topology defined by the closed subsets $V_{max}(\mathcal{I}) = \{\mathfrak{m} \in X_{max} / \mathcal{I} \subset \mathfrak{m}\}$. It is then the set of closed points of $X = \operatorname{Spec} A$ with induced topology.

To have an idea of this topology, in the case where the ring A is the polynomial ring $k[x_1, \ldots, x_n]$, where k is an algebraically closed field, the maximal spectrum X is the affine space $\mathbb{A}^n_k$, and for any ideal $\mathcal{I}$ of $k[x_1, \ldots, x_n]$, the closed subset $V_{max}(\mathcal{I})$ is the set of points P of $\mathbb{A}^n_k$ satisfying $f(P) = 0$ for all $f \in \mathcal{I}$.

Let X be the spectrum of a ring A; then any $f \in A$ defines an open subset $D(f)$, which is the complement of the closed subset $V(f)$ associated to the ideal (f). It is easy to see that this open subset is homomorphic to the spectrum of the quotient ring A_f, and moreover, the open subsets $D(f)$ form a basis of the topology on X. We can then define a sheaf of rings $\mathcal{O}_X$ on X by

$$\mathcal{O}_X : \begin{array}{rcl} O(X)^{op} & \to & \mathcal{R}ing \\ D(f) & \mapsto & A_f . \end{array}$$

And the stalk of the structure sheaf $\mathcal{O}_X$ at a point x corresponding to the prime ideal $\mathfrak{p}$ of A is isomorphic to the localization of A at $\mathfrak{p}$, that is, $\mathcal{O}_{X,x} \simeq A_\mathfrak{p}$.

Definition 1.10 The *affine scheme* associated to a ring A is the locally ringed space $(X, \mathcal{O}_X)$, where X is the spectrum of the ring A and $\mathcal{O}_X$ is the sheaf of rings defined above.

Any morphism $A \to B$ of rings induces a morphism $(Y, \mathcal{O}_Y) \to (X, \mathcal{O}_X)$ of locally ringed spaces, where $X = \operatorname{Spec} A$ and $Y = \operatorname{Spec} B$, and by definition, the category *Aff* of affine schemes is the opposite category of rings, $A\!f\!f = \mathcal{R}ing^{op}$.

Definition 1.11 A geometric scheme is a locally ringed space $(X, \mathcal{O}_X)$ that is locally isomorphic to an affine scheme; the category of geometric schemes is the full subcategory of locally ringed spaces whose objects are geometric schemes. This category is denoted by *Sch*, and the category *Aff* of affine schemes is a full subcategory of *Sch*.

In what follows, we shall say simply "scheme"; the notion of *geometric scheme* is introduced to differentiate it from the notion of *categorical scheme*, which we shall introduce in Section 2.7. Moreover, if we consider schemes over a ring k, by considering k-algebras instead of rings, we obtain the categories Aff_k and Sch_k.

If $(X, \mathcal{O}_X)$ is an affine scheme associated to the ring A, any A-module M defines a sheaf of $\mathcal{O}_X$-module on X, which associates to any open subset $D(f)$ of X the A_f-module $M_f = M \otimes_A A_f$. We denote this sheaf by $\tilde{M}$, and the stalk of $\tilde{M}$ at a point x corresponding to the prime ideal $\mathfrak{p}$ is then isomorphic to the localized module $M_\mathfrak{p} = M \otimes_A A_\mathfrak{p}$.

Definition 1.12 A *quasi-coherent* sheaf $\mathcal{M}$ on a scheme $(X, \mathcal{O}_X)$ is a sheaf of $\mathcal{O}_X$-modules that is locally of the form $\tilde{M}$, i.e. such that there exists an open covering of X by affine subschemes $X_i = \operatorname{Spec} A_i$ and, for any i, an A_i module M_i with $\mathcal{M}_{|U_i} \simeq \tilde{M}_i$.

We denote by $QCoh(X)$ the full subcategory of $\mathcal{O}_X$-*Mod* of quasi-coherent sheaves on X.

These sheaves play a very important role in the study of schemes; in particular, the cohomology groups of quasi-coherent sheaves behave well. We notice first that for any affine scheme $X = \operatorname{Spec} A$ and for any quasi-coherent sheaf $\mathcal{M} = \tilde{M}$ on X, the higher cohomology groups $H^p(X, \mathcal{M})$, $p > 0$, vanish. In fact, it is a characterizaton of affine spaces; more precisely, for a Noetherian scheme X, the following conditions are equivalent [49]:

1. X is affine;
2. $H^q(X, \mathcal{F}) = 0$ for all quasi-coherent sheaf $\mathcal{F}$ and all $q > 0$;
3. $H^1(X, \mathcal{J}) = 0$ for all quasi-coherent sheaf of ideals $\mathcal{J}$.

We can calculate the cohomology groups $H^p(X, \mathcal{M})$ of a quasi-coherent sheaf $\mathcal{M}$ on a scheme X by using Čech cohomology relative to an open covering $\mathcal{U} = (U_i)$ of X such that any U_i is affine; such a covering is called an *affine covering* of X.

Theorem 1.13 *Let X be a separated scheme, let $\mathcal{U}$ be an affine covering of X, and let $\mathcal{F}$ be a quasi-coherent sheaf on X. Then for all $q \geq 0$, the natural maps*

$$\check{H}^q_\mathcal{U}(X, \mathcal{F}) \to H^q(X, \mathcal{F})$$

are isomorphisms.

Let $(f, f^\sharp): (X, \mathcal{O}_X) \to (Y, \mathcal{O}_Y)$ be a morphism of schemes; then the functors f_* and f^* respect the quasi-coherent sheaves on X and Y, and we have then a pair of adjoint functors

$$f^*: QCoh(Y) \rightleftarrows QCoh(X) : f_*.$$

Very often we will forget in notation the structured sheaf $\mathcal{O}_X$ of a scheme, and in the same way, we will forget the map $f^\sharp$. Then, by abuse of notation, we will say that X is a scheme and that $f: X \to Y$ is a morphism of schemes.

We recall that we can define a topological manifold as a locally ringed space $(X, \mathcal{O}_X)$, where X is a locally compact second countable Hausdorff space and where the sheaf of regular functions $\mathcal{O}_X$ on X is the sheaf of germs of continuous functions on X with values in $\mathbb{R}$ or $\mathbb{C}$. In this case the sheaf $\mathcal{O}_X$ is fine – this is a consequence of the classical Uryson theorem – and the higher cohomology groups $H^q(X, \mathcal{O}_X)$, $q > 0$, are zero. Moreover, we can deduce from the Gelfand–Naimark theorem that the group of global sections $\Gamma(X, \mathcal{O}_X) = H^0(X, \mathcal{O}_X)$ is a ring that completely determines the space X.

But this is no longer the case in the context of algebraic or analytic geometry. We remark that if A is a ring and $(Y, \mathcal{O}_Y)$ is a scheme, we can associate to any morphism $f: Y \to \operatorname{Spec} A$ a map on sheaves $f^\sharp: \mathcal{O}_{\operatorname{Spec} A} \to f_* \mathcal{O}_Y$, and taking global sections, we obtain a homomorphism $A \to \Gamma(Y, \mathcal{O}_Y)$. Thus there is a natural map $Hom_{Sch}(Y, \operatorname{Spec} A) \to Hom_{Ring}(A, \Gamma(Y, \mathcal{O}_Y))$, which is bijective. In other words, in the case of algebraic geometry, the functor global sections Γ from the category of schemes Sch to the opposite of the category of rings $(Ring)^{op}$ admits a right adjoint, which is exactly the functor Spec that associates to a ring A the affine scheme $X = \operatorname{Spec} A$:

$$\Gamma: Sch \rightleftarrows (Ring)^{op} : \operatorname{Spec}.$$

Moreover, for any ring A, the natural morphism from the ring of globlal sections of the sheaf $\mathcal{O}_X$ on $X = \operatorname{Spec} A$ to A is an isomorphism; then we recover that the category Aff of affine schemes, which is by definition the category $(Ring)^{op}$, is a full subcategory of the category of schemes by the functor Spec.

In this section, we have seen how the classical notion of sheaves on a topological space is useful for having a definition of geometric spaces and makes it possible to obtain a generalization of the notion of algebraic variety by that of schemes, and we have the same in the context of analytic geometry (cf. [6]). We want to finish this part by showing that the classical notion of a sheaf makes it possible to generalize the objects of differential manifold and we will present the notion of *abstract differential geometry* as it was introduced by Anastasios Mallios [37, 38].

Definition 1.14 Let X be a topological space; a *differential triad* over X is a triplet $\delta = (\mathcal{A}, \partial, \Omega)$, where $\mathcal{A}$ is a sheaf of algebras over X, Ω is an $\mathcal{A}$-module, and $\partial \colon \mathcal{A} \to \Omega$ is a $\mathbb{C}$-linear morphism of sheaves satisfying the Leibniz condition

$$\partial(x.y) = x.\partial(y) + y.\partial(x)$$

for any local sections $x, y \in \mathcal{A}(U)$ and for any open subset U of X.

The canonical example of differential triad is given by differential manifolds. More precisely, let X be a differential manifold, $\mathcal{A} := \mathcal{C}_X^\infty$ be the structure sheaf of germs of $\mathbb{C}$-value differentiable functions on X, and $\Omega := \Omega_X^1$ be the sheaf of germs of its smooth $\mathbb{C}$-value 1-forms on X; then the triplet $\delta = (\mathcal{C}_X^\infty, d_X, \Omega_X^1)$ is a differentiable triad, where d_X is the natural sheaf morphism $d_X \colon \mathcal{C}_X^\infty \to \Omega_X^1$, which sends a function to its differentiation.

There is a natural notion of morphism of triads:

Definition 1.15 Let X and Y be topological spaces, $\delta_X = (\mathcal{A}_X, \partial_X, \Omega_X)$, and $\delta_Y = (\mathcal{A}_Y, \partial_Y, \Omega_Y)$ be differential triads over X and Y, respectively.

A morphism $\tilde{f}$ between them is a triple $(f, f_{\mathcal{A}}, f_\Omega)$ subject to the following conditions:

1. $f \colon X \to Y$ is continuous;
2. $f_{\mathcal{A}} \colon \mathcal{A}_Y \to f_*(\mathcal{A}_X)$ is a morphism of sheaves of algebras;
3. $f_\Omega \colon \Omega_Y \to f_*(\Omega_X)$ is an $f_{\mathcal{A}}$-morphism;
4. the diagram

$$\begin{array}{ccc} \mathcal{A}_Y & \xrightarrow{f_{\mathcal{A}}} & f_*(\mathcal{A}_X) \\ {\scriptstyle \partial_Y} \downarrow & & \downarrow {\scriptstyle f_*(\partial_X)} \\ \Omega_Y & \xrightarrow[f_\Omega]{} & f_*(\Omega_X) \end{array}$$

is commutative.

Therefore, we can define the category $\mathcal{DT}$ of differential triads, and we have a natural functor that sends a differential manifold X to the triad $\delta = (\mathcal{C}_X^\infty, d_X, \Omega_X^1)$. This functor is fully faithful by the way we can see differential triads as generalizations of differential manifolds.

2 Sheaves on a Category and Functor of Points

2.1 Sheaves on a Category

The spaces we want to construct in geometry are more general than the usual spaces, or elementary spaces, that we know, for instance, an open subset of $\mathbb{R}^n$ in the case of differential geometry or the prime spectrum of a ring in the case of algebraic geometry. We have seen in the previous section that one way to construct these geometric spaces is to consider them as topological spaces with a structure that describes how they are locally of the form of these elementary spaces.

But there is a very intrinsic way of generalizing the objects of any category $\mathcal{C}$; the first step is to enlarge this category by dividing it into a bigger category that is defined canonically. For this we will introduce the category $Pr(\mathcal{C})$ of presheaves on a category $\mathcal{C}$.

Definition 2.1 The category of presheaves on $\mathcal{C}$ is the category $Fun(\mathcal{C}^{op}, Set) = Set^{\mathcal{C}^{op}}$ of functors from $\mathcal{C}^{op}$ to the category of sets. This category is usually denoted by $Pr(\mathcal{C})$ or $\widehat{\mathcal{C}}$.

Any object x of $\mathcal{C}$ defines a presheaf h_x by $h_x(y) := Hom_\mathcal{C}(y, x)$, and we obtain a functor

$$h_- : \mathcal{C} \longrightarrow Pr(\mathcal{C})$$
$$x \mapsto h_x,$$

called the Yoneda functor. We have the following fundamental result.

Proposition 2.2 (Yoneda lemma) *The functor h_- is fully faithful.*

We recall that a functor $F: \mathcal{C} \to \mathcal{D}$ is called *faithful*, resp. *full*, resp. *fully faithful*, if, for any x and y in $Ob(\mathcal{C})$, the natural application $Hom_\mathcal{C}(x, y) \to Hom_\mathcal{D}(F(x), F(y))$ between the sets of morphisms is injective, resp. surjective, resp. bijective. When the functor $F: \mathcal{C} \to \mathcal{D}$ is fully faithful, we can consider the category $\mathcal{C}$ as a subcategory of the category $\mathcal{D}$. The two categories are *equivalent* if, moreover, the functor F is *essentially surjective*, that is, if, for any u in $Ob(\mathcal{D})$, there exists x in $Ob(\mathcal{C})$ with $F(x)$ isomorphic to u in $\mathcal{D}$.

Then, by the Yoneda lemma, we can consider the category $\mathcal{C}$ as a subcategory of the category of presheaves $Pr(\mathcal{C})$, and we will identify an object x of $\mathcal{C}$ with its image h_x in $Pr(\mathcal{C})$. Moreover, for any $x \in \mathcal{C}$ and $P \in Pr(\mathcal{C})$, we have the following:

$$Hom_{Pr(\mathcal{C})}(h_x, P) = P(x).$$

For any presheaf $P \in Pr(\mathcal{C})$, we denote by $\mathcal{C}/P$ the category of elements of $\mathcal{C}$ over P; more precisely, the objects of $\mathcal{C}/P$ are the pairs (x, u) with $x \in \mathcal{C}$ and $u: h_x \to P$ in $Pr(\mathcal{C})$. There is an evident functor $\mathcal{C}/P \to Pr(\mathcal{C})$ defined by the projection $(x, u) \mapsto h_x$, and P is the colimit of this functor:

$$P = \varinjlim_{\mathcal{C}/P} h_x .$$

The category of presheaves behaves very well, as the category of sets; in particular, we have all the basic constructions as limits (product, pullback, etc.) and colimits (sum, pushforward, etc.). This category satisfies a universal property; more precisely, we have the following result:

Proposition 2.3 *The category of presheaves $Pr(\mathcal{C})$ is the completion by colimits of the category $\mathcal{C}$ in the following sense: if $\gamma: \mathcal{C} \to \mathcal{E}$ is a functor from the category $\mathcal{C}$ to a cocomplete category $\mathcal{E}$, then γ admits a unique colimit-preserving factorization*

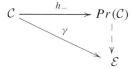

To be rigorous, we have to impose some condition on the *size* of the category $\mathcal{C}$, that is, we have to suppose that the category is *small*. In what folllows, at each time, it will be necessary to use this hypothesis.

2.2 Grothendieck Topologies

As we have seen, if we want to define spaces in some geometry, we have first to consider some elementary spaces, which we call in the following *affine spaces*, and let *Aff* be the category of these spaces. The presheaves $P \in Pr(Aff)$ can be considered as generalized objects of the category *Aff*, but they are too general to define geometric spaces, and we have to restrict ourselves to a smaller class to allow for the fact that we only allow gluing affine spaces along some class of applications.

By the Yoneda functor h_-, the category of presheaves $Pr(\mathcal{C})$ is a free co-completion of the category $\mathcal{C}$, but with this construction, we do not obtain the colimits we expect, for instance, we have already a coproduct in the category of topological manifolds $\mathcal{T}op$, namely, the disjoint union $M_1 \cup M_2$, and in the category of presheaves $Pr(\mathcal{T}op)$, the images h_{M_1} and h_{M_2} have a coproduct $h_{M_1} \coprod h_{M_2}$, which is different from the image $h_{M_1 \cup M_2}$.

So we want to introduce a new cocomplete category $Sh(\mathcal{C})$, a subcategory of $Pr(\mathcal{C})$, with a functor $\mathcal{C} \to Sh(\mathcal{C})$ such that the colimits in this new category are the ones we expect. For instance, if $(\phi_\alpha : U_\alpha \to M)$ is an atlas of a manifold M, M is built by gluing together all the U_αs along their intersections; in other words, the following is a coequalizer diagram:

$$\coprod_{\beta,\gamma} U_\beta \cap U_\gamma \rightrightarrows \coprod_\alpha U_\alpha \longrightarrow M .$$

We want the image of this diagram in this category $Sh(\mathcal{C})$ to be an equalizer.

For this we will introduce the notion of *Grothendieck topology* on a category $\mathcal{C}$, which is the datum, for any object x of C, of *covering families* that will play the role of coverings in a topological space. These families are collections of *cones* $(y_\alpha \to x)$ in $\mathcal{C}$ (where a cone is a diagram with a terminal vertex) that we want to become colimit cones in the category $Sh(\mathcal{C})$ (see the introductory article of D. Dugger for a very elegant exposition [16]).

If we want to define a topology in general, without any assumption about the category $\mathcal{C}$, we must use the concept of sieve. We recall that for an object x in $\mathcal{C}$, a *sieve* on x is a set $\mathcal{R}$ of maps $u : y \to x$ in $\mathcal{C}$ that is stable by right composition: for any $u : y \to x$ in $\mathcal{R}$ and any map $v : z \to y$, the composition $u \circ v : z \to x$ belongs to $\mathcal{R}$. We can also consider that the sieves on x are exactly the subpresheaves of h_x, where we associate to a sieve $\mathcal{R}$ on x the presheaf on $\mathcal{C}$, which we denote also by $\mathcal{R}$, defined by $\mathcal{R}(y) = \{u : y \to x | u \in \mathcal{R}\}$.

Definition 2.4 A topology τ on a category $\mathcal{C}$ is the datum for each object x of $\mathcal{C}$ of a set $J(x)$ of sieves on x satisfying some conditions[1] and is called the *covering sieves* of x for the topology τ.

A *site* is a pair $(\mathcal{C}; \tau)$ where $\mathcal{C}$ is a category and τ is a topology on $\mathcal{C}$; the category $\mathcal{C}$ is called the underlying category of the site. The sieves belonging to $J(x)$ are called the *covering sieves* of x for the topology τ.

In fact, it is not easy to define a topology by covering sieves, and it is more convenient to define a *pretopology*; for this we will assume for simplicity that the category $\mathcal{C}$ has fiber products $y \times_x x'$ for any diagram $y \to x \leftarrow x'$.

---

[1] We do not give here the precise and explicit conditions because it would be cumbersome and we do not need it in what follows.

Definition 2.5 A pretopology on a category $\mathcal{C}$ is the datum for every object x of $\mathcal{C}$ of a set $Cov(x)$ of families of maps with codomain x, called *covering families*, such that

1. (stability by base change) for any x, for any family $(u_\alpha: y_\alpha \to x)$ in $Cov(x)$ and any map $v: x' \to x$, the family $(u'_\alpha: y'_\alpha := y_\alpha \times_x x' \to x')$ is in $Cov(x')$;
2. (stability by composition) if $(u_\alpha: y_\alpha \to x)$ is a family in $Cov(x)$ and if, for each y_α, one has a family $(v_{\alpha,\beta}: z_{\alpha,\beta} \to y_\alpha)$ in $Cov(y_\alpha)$, then the family of composites $(u_\alpha \circ v_{\alpha,\beta}: z_{\alpha,\beta} \to x)$ is in $Cov(x)$;
3. the family $(id_x: x \to x)$ is in $Cov(x)$.

A pretopology on $\mathcal{C}$ generates a topology τ that is characterized by the following: a sieve $\mathcal{R}$ is a covering sieve for τ if and only if it contains a covering family. The pretopology is called a *basis* of the topology τ.

We can define an order on the set of topologies on a category $\mathcal{C}$. Let τ and τ' be two topologies on $\mathcal{C}$; the topology τ is *finer* or *bigger* than the topology τ', or the topology τ' is *coarser* or *smaller* than the topology τ, if, for any object x in $\mathcal{C}$, any covering sieve of x for τ is a covering sieve for τ'.

Let $(\tau_i)_{i \in I}$ be a family of topologies on $\mathcal{C}$; the topology τ defined by the intersection of these topologies, that is, for any object x in $\mathcal{C}$, the set of covering sieves for τ is the intersection of the covering sieves for the topologies τ_i, is the finest topology that is coarser than the τ_i. The topology τ is the greatest lower bound of the τ_i. There is also a topology that is the least upper bound of the topologies τ_i; this is the coarsest topology that is finer than the τ_i.

The topology on $\mathcal{C}$ in which the only covering sieve for an object x is the maximal sieve $\mathcal{T}_x$ is the coarsest topology on $\mathcal{C}$; it is called the *trivial* or *chaotic* topology on $\mathcal{C}$. The topology on $\mathcal{C}$ for which all the sieves are covering sieves is the finest topology on $\mathcal{C}$; it is called the *discrete* topology.

We can now define the category of *sheaves* on $\mathcal{C}$ for a topology τ.

Definition 2.6 Let $(\mathcal{C}, \tau)$ be a site; a *sheaf* on $(\mathcal{C}, \tau)$ or a *sheaf* on $\mathcal{C}$ for the topology τ is a presheaf $F \in Pr(\mathcal{C})$ such that, for each object x in $\mathcal{C}$ and any covering sieve $\mathcal{R}$ of x for τ, the natural map

$$Hom_{Pr(\mathcal{C})}(h_x, F) \longrightarrow Hom_{Pr(\mathcal{C})}(\mathcal{R}, F)$$

is a bijection.

The category of sheaves on $\mathcal{C}$ for the topology τ is the full subcategory of $Pr(\mathcal{C})$ generated by the sheaves on $(\mathcal{C}, \tau)$; we denote this category by $Sh_\tau(\mathcal{C})$.

Sheaves and Functors of Points 431

If the topology τ is generated by a pretopology, for any object x in $\mathcal{C}$ and any family $(u_\alpha : y_\alpha \to x)$ in $Cov(x)$, we get a cone

$$\coprod_{\beta,\gamma} y_\beta \times_x y_\gamma \rightrightarrows \coprod_\alpha y_\alpha \longrightarrow x,$$

and these are the cones, called the *distinguished cones*, associated to the topology τ, which we will want to become colimits in the category of sheaves. Then a presheaf F is a sheaf for τ if and only if, for any object x in $\mathcal{C}$ and any family $(u_\alpha : y_\alpha \to x)$ in $Cov(x)$, the diagram

$$F(x) \longrightarrow \prod_\alpha F(y_\alpha) \rightrightarrows \prod_{\alpha,\beta} F(y_\alpha \times_x y_\beta)$$

is an equalizer.

Example 2.7 Let X be a topological space and let $\mathcal{C} = O(X)$ be the category of open subsets of X; let τ the topology on $\mathcal{C}$ associated to the pretopology defined for any U in $O(X)$ by $Cov(U)$ be the set of coverings, that is, families $(U_i \to U)$ such that $U = \cup_i U_i$. Then the notions of sheaves on the topological space X and of sheaves on the category $O(X)$ coincide.

If τ and ρ are two topologies on a category $\mathcal{C}$, with τ finer than ρ, then we have the inclusion $Sh_\tau(\mathcal{C}) \subset Sh_\rho(\mathcal{C})$. For ρ equal to the trivial topology, we get $Sh_\rho(\mathcal{C}) = Pr(\mathcal{C})$.

Let τ be a topology on a category $\mathcal{C}$; then we can associate to any presheaf F a sheaf aF and a map $\eta \colon F \to aF$ in $Pr(\mathcal{C})$ universal from F to sheaves, that is, such that for any map $f \colon F \to S$ where S is a sheaf, there exists a unique morphism $g \colon aF \to S$ in $Sh_\tau(\mathcal{C})$ such that $g \circ \eta = f$:

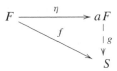

One has the more precise result.

Theorem 2.8 *The category of sheaves $Sh_\tau(\mathcal{C})$ is a reflexive left-exact subcategory of the category of presheaves $Pr(\mathcal{C})$, that is, the inclusion functor $i_\tau \colon Sh_\tau(\mathcal{C}) \to Pr(\mathcal{C})$ has a left adjoint*

$$a \colon Pr(\mathcal{C}) \longrightarrow Sh_\tau(\mathcal{C})$$

that commutes with finite limits.

The functor $a: Pr(\mathcal{C}) \to Sh_\tau(\mathcal{C})$ is called the *associated sheaf* functor. The Yoneda functor $h_-: \mathcal{C} \to Pr(\mathcal{C})$ composed with the functor a defines a functor

$$\epsilon_\tau : \mathcal{C} \longrightarrow Sh_\tau(\mathcal{C}),$$

which is called the *canonical functor* from $\mathcal{C}$ to $Sh_\tau(\mathcal{C})$.

The category of sheaves $Sh_\tau(\mathcal{C})$ is cocomplete because the left-adjoint functor a creates colimits. More precisely, if (F_α) is a diagram in $Sh_\tau(\mathcal{C})$, then we regard it as a diagram in $Pr(\mathcal{C})$; we take its colimit in that category and apply the functor a to it, and we obtain the colimit in $Sh_\tau(\mathcal{C})$. In the same way that the category of presheaves $Pr(\mathcal{C})$ is the completion of the category $\mathcal{C}$ for all colimits (see Proposition 2.3), the category of sheaves $Sh_\tau(\mathcal{C})$ is the completion for a class of colimits for the distinguished cones defined by the topology τ.

Proposition 2.9 *Let $(\mathcal{C}, \tau)$ be a site; then the canonical functor $\epsilon_\tau : \mathcal{C} \to Sh_\tau(\mathcal{C})$ takes the distinguished cones of $\mathcal{C}$ associated to τ to colimit cones in $Sh_\tau(\mathcal{C})$. Moreover, the category $Sh_\tau(\mathcal{C})$ has the following universal property: if $\mathcal{D}$ is a cocomplete category and $\gamma : \mathcal{C} \to \mathcal{D}$ a functor taking distinguished cones to colimits, then γ admits a unique colimit-preserving factorization*

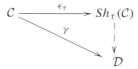

If the topology τ is *subcanonical*, that is, if all representable presheaves h_x are sheaves, the functor ϵ_τ is fully faithful and commutes with all small limits. In that case, we can consider the category of sheaves $Sh_\tau(\mathcal{C})$ as a generalization of $\mathcal{C}$, contained in the category of presheaves, and we have the following inclusions:

$$\mathit{Aff} \subset Sh_\tau(\mathit{Aff}) \subset Pr(\mathit{Aff}).$$

To have a good notion of geometric space, we must impose additional conditions on the sheaves that reflect the fact that a space is obtained by gluing affine spaces along a particular class of applications $\mathbb{P}$; in the classical case, they are open immersions. We will see later how we can define the category $\mathit{Geom\,Sp}$ of geometric spaces from the category Aff, a topology τ, and a class of maps $\mathbb{P}$ compatible with the topology τ.

2.3 Locale: Topology without Points

Classically, a topological space is defined as a set of points X endowed with a structure, a *topology* that is the ordered set $O(X)$ of its open subsets, and the definition of sheaves on the topological space X depends only on the ordered set $O(X)$. But we can take the notion of open set as primitive in the study of topological spaces, following the ideas of Hausdorff, Stone, Johnstone, and many other authors. The *pointless topology* is the theory of these generalized spaces obtained with this point of view.

We recall that the category $O(X)$ of open subsets of a topological space X is a lattice with finite meets (or inf) $\wedge$ equal to the intersection and all joints (or sup) $\vee$ equal to the union and that satisfies the distributive law: $a \wedge (\bigvee_i b_i) = \bigvee_i (a \wedge b_i)$.

We call such a lattice a *frame*, and we define a morphism of frames to be a morphism of an ordered set that preserves finite meets and all joins. We remark that in a frame, all the meets exist, but we do not have the distributivity law; in the same way, we do not demand a morphism of frames to respect them. By definition, the category of *locales* is the opposite of the category of frames: $Locale = (Frame)^{op}$.

The map that sends a topological space X to the set $O(X)$ of its open subsets defines a functor

$$Loc: \quad Space \longrightarrow Locale$$
$$X \quad \mapsto \quad O(X).$$

We recall that a point of a topological space X may be identified with a map from the point $\{*\}$ to X and that the point is the terminal object of the category of topological spaces. Then it is natural to define a point of a locale L as a map $\mathbf{1} \to L$, where $\mathbf{1} = \{0, 1\}$ is the terminal object in the category of locales; it is the locale associated to the frame $\mathbf{1} = \{0, 1\}$ with two elements: 0, which is the bottom element, and 1, which is the top element.

For a locale L, we denote by $pt(L)$ the set of points of L; this set is endowed with a topology for which the open subsets are the sets of the form

$$O(x) = \{\phi \in pt(L) \mid \phi(x) = 1\}$$

for any $x \in L$. This defines a functor $pt: Locale \to Space$, and we have the following result.

Theorem 2.10 *The functors Loc and pt define a pair of adjoint functors:*

$$Loc: Space \rightleftarrows Locale : pt$$

There is no strict equivalence between these two points of view, but this is the case if we add some hypothesis. A topological space X is said to be *sober* if every nonempty irreducible closed subset F of X is the closure of a unique point, and a locale L is said to be *spatial*, or to have *enough points*, if the elements of L can be distinguished by points of L, that is, for any two distinct elements x and y in L, there exists a point $\phi: L \to \mathbf{1}$ such that $\phi(x) \neq \phi(y)$.

Theorem 2.11 *The pair of adjoint functors (Loc, pt) induces an equivalence of categories:*

$$Loc: Sober \rightleftarrows Spatial: pt$$

where Sober is the full subcategory of Spaces of the topological spaces that are sober and Spatial is the full subcategory of Locale of the locales that have enough points.

The notion of Grothendieck topology was introduced as an extension of the usual notion of topology. To recover this notion, we have to take for underlying category $\mathcal{C}$ of the site the category $O(X)$ of open subsets of a topological space X.

We recall that a locale $\mathcal{C}$ as an ordered set is a category; for any objects x and y in $\mathcal{C}$, we have $Hom_{\mathcal{C}}(x, y) = \{*\}$ if $x \leq y$ and $Hom_{\mathcal{C}}(x, y) = \emptyset$ otherwise. This category is endowed with a natural Grothendieck topology τ defined by the following pretopology: for any $x \in \mathcal{C}$, a family $(u_i: y_i \to x)_{i \in I}$ is a covering family for τ if we have $\bigvee_I y_i = x$.

Then the two notions of sheaves are equivalent: for any topological space X, a functor

$$\begin{aligned} F: O(X)^{op} &\longrightarrow Set \\ U &\mapsto F(U) \end{aligned}$$

is a sheaf on the topological space X, in the usual sense, if and only if F is a sheaf on the site $(\mathcal{C}, \tau)$ associated to X. The category $Sh(X)$ of sheaves on the topological space X depends only on the locale $O(X)$ of open subsets of X, and this category is isomorphic to the category $Sh_\tau(\mathcal{C})$ of sheaves on the category $\mathcal{C} = O(X)$ endowed with the natural topology τ defined before.

2.4 Topos

One of the fundamental ideas underlying the previous approach is to say that the most natural, at least the most convenient, notion to approach a geometric theory and, first of all, to define the objects of this theory is that of sheaves, and even to distinguish among them a particular class of sheaves, those of regular functions.

The category of sheaves on a topological space, and more generally the category $Sh_\tau(\mathcal{C})$ of sheaves on a category $\mathcal{C}$ for a Grothendieck topology τ, is a category that looks like the category Set of sets, and many elementary and basic constructions are possible in it. We call such a category a *topos*; more precisely, there is an intrinsic definition of a topos as a category verifying some properties, and there is a fundamental theorem that claims that a topos is a category equivalent to the category of sheaves $Sh_\tau(\mathcal{C})$ for a Grothendieck topology on a small category. The simplest example of topos is the category Set of sets, which is the category of presheaves on the category $\underline{0}$ with one object and one morphism.

To have a natural definition of a morphism between topoi, we first consider the case of the topos defined as categories of sheaves on topological spaces. Let X and Y be two topological spaces, and let $\phi \colon X \to Y$ be a continuous function between them. Then we have defined a pair of adjoint functors

$$\phi^* \colon Sh(Y) \rightleftarrows Sh(X) \colon \phi_*$$

between the categories of sheaves on Y and X, respectively, and the functor ϕ^* is left exact, that is, it commutes with finite limits.

In the previous situation, the continuous function ϕ induces a functor $f = \phi^{-1} \colon O(Y) \to O(X)$ between the categories of open subsets, and we want to generalize this construction to any functor between categories $\mathcal{C}$ and $\mathcal{D}$ endowed with any Grothendieck topologies. Let $f \colon \mathcal{C} \to \mathcal{D}$ be a functor between two categories; then f induces two pairs of adjoint functors between the categories of presheaves

$$f_! \colon Pr(\mathcal{C}) \rightleftarrows Pr(\mathcal{D}) \colon f^* \qquad \text{and} \qquad f^* \colon Pr(\mathcal{D}) \rightleftarrows Pr(\mathcal{C}) \colon f_*$$

where f^* is defined by $f^*(G)(x) = G(f(x))$ for any presheaf G on $\mathcal{D}$ and any x in $\mathcal{C}$ and $f_!$ is defined on representable presheaves by $f_!(h_x) = h_{f(x)}$ for any x in $\mathcal{C}$.

Let $(\mathcal{C}, \tau)$ and $(\mathcal{D}, \rho)$ be two sites; a functor $f \colon \mathcal{C} \to \mathcal{D}$ is *continuous* if the functor $f^* \colon Pr(\mathcal{D}) \to Pr(\mathcal{C})$ sends the subcategory of sheaves $Sh_\rho(\mathcal{D})$ in the category of sheaves $Sh_\tau(\mathcal{C})$, that is, if one has the following commutative diagram:

$$\begin{array}{ccc} Sh_\rho(\mathcal{D}) & \xrightarrow{\tilde{f}^*} & Sh_\tau(\mathcal{C}) \\ \downarrow{i_\rho} & & \downarrow{i_\tau} \\ Pr(\mathcal{D}) & \xrightarrow{f^*} & Pr(\mathcal{C}) \end{array}$$

We define the functor $\tilde{f}_! \colon Sh_\tau(\mathcal{C}) \to Sh_\rho(\mathcal{D})$ to be the composition $\tilde{f}_! = a \circ f_! \circ i_\tau$. As the functor a and $f_!$ are left adjoint and i_τ is fully faithful, one gets an adjunction

$$\tilde{f}_! \colon Sh_\tau(\mathcal{C}) \rightleftarrows Sh_\rho(\mathcal{D}) \colon \tilde{f}^*.$$

In general there is no reason for the functor $\tilde{f}_!$ to be left exact; this is the case if the functor $f_!$ is left exact. When the functor $\tilde{f}_!$ is left exact, we say that $f \colon (\mathcal{C}, \tau) \to (\mathcal{D}, \rho)$ is a *morphism of sites*.

Definition 2.12 Let $\mathcal{E}$ and $\mathcal{F}$ be two toposes; a *geometric morphism* f between $\mathcal{E}$ and $\mathcal{F}$ is a pair of adjoint functors

$$f^* \colon \mathcal{F} \rightleftarrows \mathcal{E} \colon f_*$$

such that the left adjoint f^* is left exact. We call the functors f^* and f_*, respectively, the *inverse image* and the *direct image* parts of the geometric morphism f.

With this definition, we see that a morphism of site $f \colon (\mathcal{C}, \tau) \to (\mathcal{D}, \rho)$ induces a geometric morphism of toposes $\tilde{f} \colon Sh_\rho(\mathcal{D}) \to Sh_\tau(\mathcal{C})$. In particular, any continuous map $\phi \colon X \to Y$ induces a geometric morphism of toposes $\tilde{f} \colon Sh(X) \to Sh(Y)$.

If $f \colon \mathcal{E} \to \mathcal{F}$ is a geometric morphism defined by a pair of adjoint functors (f^*, f_*) such that the functor direct image f_* is fully faithful, which is equivalent to having the counit $\epsilon \colon f^* f_* \mathcal{E} \to \mathcal{E}$ be an isomorphism, we say that the geometric morphism f is an *embedding of toposes*.

We have seen before that if τ is a Grothendieck topology on a category $\mathcal{C}$, the functor $i_\tau \colon Sh_\tau(\mathcal{C}) \to Pr(\mathcal{C})$ has a left-exact left adjoint $a \colon Pr(\mathcal{C}) \longrightarrow Sh_\tau(\mathcal{C})$. Then we get an embedding of toposes $Sh_\tau(\mathcal{C}) \to Pr(\mathcal{C})$. We have the converse of this result; more precisely, we get the following:

Theorem 2.13 Let $\mathcal{C}$ be a category and let $\mathcal{G}$ be a category with a pair of adjoint functors

$$f^* \colon Pr(\mathcal{C}) \rightleftarrows \mathcal{G} \colon f_*$$

such that f^* is left exact and f_* is fully faithful; then there exists a Grothendieck topology τ on $\mathcal{C}$ such that $\mathcal{G}$ is equivalent to $Sh_\tau(\mathcal{C})$, and this adjunction defines the embedding of toposes $i_\tau \colon Sh_\tau(\mathcal{C}) \to Pr(\mathcal{C})$.

Then we see that a topos may be defined as a reflexive left-exact subcategory of a category of presheaves. More generally, any reflexive left-exact subcategory of a topos is again a topos.

As in the classical case of sheaves on a topological space, it is important to consider presheaves or sheaves on a category $\mathcal{C}$ with values in the category $\mathcal{A}b$ of abelian groups or of the category of rings. The more general notion introduced by A. Grothendieck and J.-L. Verdier is the notion of a *ringed topos* (*topos annelé*) [51, Exposé IV, §11].

As a topos $\mathcal{E}$ is a symmetric monoidal category for the product, we have the notion of a ring object in $\mathcal{E}$; in particular, if the topos $\mathcal{E}$ is the category of sheaves $Sh_\tau(\mathcal{C})$ on a site $(\mathcal{C}, \tau)$, a ring object corresponds to a sheaf with values in the category $\mathcal{R}ing$ of rings (see remarks in Section 3.1).

Definition 2.14 A ringed topos is a couple $(\mathcal{E}, A)$ where $\mathcal{E}$ is a topos and A is a ring object in $\mathcal{E}$, and a ringed site is a couple $(\mathcal{C}, A)$ where $\mathcal{C}$ is a site and A is a sheaf of rings on $\mathcal{C}$.

Let $(\mathcal{E}, A)$ be a ringed topos. Then we can define the category ${}_A\mathcal{E}$ of sheaves of A-modules; this is an abelian category. If we take $A = \mathbb{Z}$, the sheaf associated to constant presheaf $\mathbb{Z}$, the category ${}_\mathbb{Z}\mathcal{E}$ is the category of *sheaves of abelian groups*.

We can extend to the case of a ringed topos or ringed sites the concepts and results that we saw earlier for the classical sheaves of abelian groups on a topological space. If $(\mathcal{C}, A)$ is a ringed site, for any object $x \in \mathcal{C}$, we define the functor $\Gamma(x, -)$ of *sections over x*:

$$\Gamma(x, -): {}_\mathbb{Z}Sh_\tau(\mathcal{C}) \longrightarrow \mathcal{A}b$$
$$\mathcal{F} \mapsto \Gamma(x, \mathcal{F}).$$

Moreover, it is possible to show that the category ${}_\mathbb{Z}\mathcal{E}$ of abelian sheaves on $\mathcal{C}$ is an abelian category with enough injectives, the functor $\Gamma(x, -)$ is left exact, and we define its right-derived functors $R^p\Gamma(x, \mathcal{F}) = H^p(x, \mathcal{F})$.

In a similar way, if the topology τ is defined by a pretopology, for any covering family $\mathcal{U} = (u_\alpha: y_\alpha \to x)$, we can define the Čech cohomology groups $\check{H}^q_\mathcal{U}(x, \mathcal{F})$ of the sheaf $\mathcal{F}$ of A-modules relatively to the family $\mathcal{U}$ and the cohomology groups

$$\check{H}^q(x, \mathcal{F}) = \varinjlim_{\mathcal{U} \in Cov(x)} \check{H}^q_\mathcal{U}(x, \mathcal{F})$$

with canonical morphisms

$$\check{H}^q(x, \mathcal{F}) \to H^q(x, \mathcal{F})$$

but these morphisms are not isomorphisms in general.

2.5 Examples in Algebraic Geometry

The notions of topology on a category and, more generally, of site and topos were introduced by A. Grothendieck in algebraic geometry, and most of the results were demonstrated in this framework. In this section we shall show why it was necessary to introduce these notions and give the first examples of Grothendieck's topology. The principal motivation to define new topologies in algebraic geometry is the fact that the Zariski topology is not fine enough, for instance, it is too coarse, that is, has too few open subsets, to have good cohomology theories or to have local triviality in algebraic geometry. We will give two examples of pathologies that may appear.

1. For an algebraic variety X over the field of complex numbers $\mathbb{C}$, the set of points $X(\mathbb{C})$ acquires a topology from that on $\mathbb{C}$ (we will define more precisely in Section 2.6 how the set $X(R)$ of R-points of any algebraic variety or any scheme X can be defined), and so one can apply the machinery of algebraic topology to its study. For example, one can define the Betti numbers of X to be the dimensions of the vector spaces $H^r(X(\mathbb{C}); \mathbb{Q})$, and such theorems as the Lefschetz fixed point formula are available. But for a variety X over an arbitrary field k, we can only define the Zariski topology, which is too coarse for the methods of algebraic topology to be useful.

   A topological space X is said to be *irreducible* if any two nonempty open subsets of X have nonempty intersection, and there is a theorem of Grothendieck that says that if X is an irreducible topological space, then the cohomology groups $H^r(X; \mathcal{F})$ vanish for all constant sheaves $\mathcal{F}$ and all $r > 0$. Most varieties or schemes X considered in algebraic geometry are irreducible as a Zariski topological space, while the underlying usual topological space $X(\mathbb{C})$ in the case where X is defined over $\mathbb{C}$ is almost never. Then the groups $H^r(X; \Lambda) = 0$ for any ring Λ, computed using the Zariski topology, are zero for all $r > 0$ and so are very different from what we may expect.

2. A map $f: Y \to X$ is said to be a *vector bundle* if it is locally trivial in the sense that every point x of X has an open neighborhood U for which there is a commutative diagram

$$\begin{array}{ccc} f^{-1}(U) & \longrightarrow & U \times_k \mathbb{A}_k^n \\ {\scriptstyle f}\downarrow & & \downarrow{\scriptstyle pr_1} \\ U & =\!=\!= & U \end{array}$$

with the top arrow a k-linear isomorphism of algebraic varieties.

For a smooth variety over $\mathbb{C}$, a morphism $f: Y \to X$ as above is locally trivial for the Zariski topology if it is locally trivial for complex topology. Thus the Zariski topology is fine enough for the study of vector bundles (see [62]). However, it is not fine enough for the study of more exotic bundles, for example, for a regular map $f: Y \to X$ endowed with an action of an algebraic group G on Y over X, that is, a morphism $m: Y \times G \to Y$ satisfying the usual conditions for a group action and such that $f(yg) = f(y)$.

So as the Zariski topology is too coarse, Alexandre Grothendieck defined a new notion of topology in such a way that for any space, there are more covering families, but we have to allow covering families $(u_i: X_i \to X)$ of a space X where the maps u_i are no longer monomorphisms.

These topologies are defined on the category Sch of schemes, but by using *le lemme de comparaison* (cf. [51, Exposé III, Théorème 4.1 and Exposé VII, Proposition 3.1]), it is equivalent to consider the full subcategory Aff of affine schemes endowed with the topologies induced. More precisely, if we denote by τ the topology on Sch and the topology induced on Aff, the functor restriction $Sh_\tau(Sch) \to Sh_\tau(Aff)$ is an equivalence of categories. All these topologies are associated to pretopologies, and for each τ, we will give the set $Cov_\tau(X)$ of the covering families of an object X in Sch.

The first topology introduced by Grothendieck is the *étale topology*, $\tau = et$. A morphism $f: X \to Y$, in Sch is *étale* if f is locally of finite presentation, if it is flat, and if, for any $y \in Y$ the fiber $f^{-1}(y) = X_y$ is discrete and these local rings are finite separable extensions of the residue field $k(y)$. To give an intuitive idea a morphism between schemes defined on $\mathbb{C}$ is étale if it is locally an analytic isomorphism.

We define the *étale topology* on the category Sch to be the topology generated by the pretopology for which, for any X in Sch, the set $Cov_{et}(X)$ of covering families is the set of families $(u_\alpha: X_\alpha \to X)_{\alpha \in A}$ such that the u_α are étale maps and X is equal to the union $\bigcup_{\alpha \in A} u_\alpha(X_\alpha)$.

There is a variant of the étale topology that was suggested by Nisnevich [44], and which has proved useful in K-theory and in the study of the cohomology of group schemes, $\tau = Nis$. We take as its coverings the surjective families of étale maps $(U_\alpha \to U)_{\alpha \in A}$ with the following property: for each $u \in U$, there exists an $\alpha \in A$ and $u_\alpha \in U_\alpha$ such that map $\kappa(u) \to \kappa(u_\alpha)$ on the residue fields is an isomorphism. It is also called the *completely decomposed topology*.

Grothendieck also defined two *flat topologies* on Sch, the *fppf topology* and the *fpqc topology*; fppf stands for *fidèlement plate de présentation finie*, and fpqc stands for *fidèlement plate et quasi-compacte*. A family $(u_\alpha : U_\alpha \to X)_{\alpha \in A}$ is a covering family in $Cov_{fppf}(X)$ if any u_α is faithfully flat, of finite presentation, and quasi-finite and if $X = \bigcup_{\alpha \in A} u_\alpha(U_\alpha)$, and it is a covering family in $Cov_{fpqc}(X)$ if any u_α is faithfully flat and if, for each affine open $V \subset X$, there exists a finite subset B of A and affine opens $V_\beta \subset U_\beta$ such that $V = \bigcup_{\beta \in B} u_\beta(V_\beta)$.

One can compare these topologies, and one has the inclusions of categories of sheaves:

$$Sh_{fpqc}(Sch) \subset Sh_{et}(Sch) \subset Sh_{Nis}(Sch) \subset Sh_{Zar}(Sch).$$

And by the *faithfully flat descent theorem* [50, Exposé VIII], one has that any representable presheaf is a sheaf for the faithfully flat and quasi-compact topology, and then all these topologies are subcanonical.

Now we can define the étale cohomology of a scheme X, and for this we consider the *small étale site* on X. This site X_{et} has as underlying category Et/X, whose objects are the étale morphisms $U \to X$ and whose arrows are the X-morphisms $g : U \to V$, endowed with the restriction of the étale topology: covering families are the surjective families of étale morphisms $(U_\alpha \to U)$ in Et/X. The theory of étale cohomology was worked out by A. Grothendieck, M. Artin, and J.-L. Verdier [51]. The whole theory is closely modeled on the usual theory of sheaves and their derived functor cohomology on a topological space. For a variety X over $\mathbb{C}$, the étale cohomology groups $H^r(X_{et}, \Lambda)$ coincide with the complex groups $H^r(X(\mathbb{C}), \Lambda)$; when Λ is a finite ring, the ring of ℓ-adic integers $\mathbb{Z}_\ell$, or the field $\mathbb{Q}_\ell$ of ℓ-adic numbers (but not for $\Lambda = \mathbb{Z}$). When X is the spectrum of a field K, the étale cohomology theory for X coincides with the Galois cohomology theory of K. Thus étale cohomology bridges the gap between the first case, which is purely geometric, and the second case, which is purely arithmetic.

One of the main motivatons to introduce étale cohomology was the *Riemann hypothesis* and the *Weil conjectures*. For an algebraic variety V defined over a finite field $\mathbb{F}_q$, the aim of this conjecture is to determine the number N_n of points of V with coefficients in the finite field $\mathbb{F}_{q^n}$ (cf. Section 2.6 for the general notion of points in algebraic geometry) and to study the *zeta function* of V, which is by definition the formal power series with coefficients in $\mathbb{Q}$ given by

$$Z(V/\mathbb{F}_q, T) = \exp\left(\sum_{n \geq 0} N_n T^n / n\right).$$

Weil suggested that the conjectures would follow from the existence of a suitable *Weil cohomology theory* for varieties over finite fields, similar to the usual cohomology with rational coefficients for complex varieties, that is, a contravariant functor

$$X \mapsto H^\bullet(X) = \oplus H^i(X)$$

from the category of smooth, proper, irreducible algebraic varieties over a field k to the category of graded anticommutative algebras over a coefficient field K of characteristic zero, which satisfies some properties, which are the identities expected by analogy with the classical theory of singular homology for a differentiable manifold. His idea was that if ϕ is the Frobenius automorphism over the finite field, then the number of points of the variety V over the field $\mathbb{F}_{q^n}$ is the number of fixed points of ϕ^n (acting on all points of the variety V defined over the algebraic closure).

By considering the case of a *supersingular elliptic curve*, we can see that the coefficient field K of a Weil cohomology can be neither the field of rational numbers $\mathbb{Q}$, nor the field of real numbers $\mathbb{R}$, nor the field of p-adic numbers $\mathbb{Q}_p$, where p is the characteristic of k. However, it does not eliminate the possibility that the coefficient field is the field of ℓ-adic numbers $\mathbb{Q}_\ell$ for some prime $\ell \neq p$. By using étale topology on the category of schemes, A. Grothendieck and M. Artin managed to construct suitable cohomology theories over $\mathbb{Q}_\ell$, called ℓ-*adic cohomology*. The difficulty is that only cohomology with coefficients in torsion sheaves, with order prime to the characteristic of the field, will behave as expected. So they consider the groups $H^i_{\text{et}}(X, \mathbb{Z}/\ell^n\mathbb{Z})$, for $n \geq 1$, and they defined the theory with coefficients in the ring of ℓ-adic integers $\mathbb{Z}_\ell$ and with coefficients in the field of ℓ-adic numbers $\mathbb{Q}_\ell$ by the following:

$$H^i_{\text{et}}(X, \mathbb{Z}_\ell) = \varprojlim H^i_{\text{et}}(X, \mathbb{Z}/\ell^n\mathbb{Z}) \quad \text{and} \quad H^i_{\text{et}}(X, \mathbb{Q}_\ell) = H^i_{\text{et}}(X, \mathbb{Z}_\ell) \otimes_{\mathbb{Z}_\ell} \mathbb{Q}_\ell.$$

The ℓ-adic cohomology is a Weil cohomology for varieties over a field k of characteristics different from ℓ. Moreover, it satisfies the comparaison theorem: if V is smooth and proper over the field $k = \mathbb{C}$, we have

$$H^i_{\text{et}}(V, \mathbb{Q}_\ell) \otimes_{\mathbb{Q}_\ell} \mathbb{C} = H^i(V_\mathbb{C}, \mathbb{C}),$$

where $V_\mathbb{C}$ is the complex variety associated to V and $H^i(V_\mathbb{C}, \mathbb{C})$ is the group of Betti cohomology.

2.6 Points of an Algebraic Variety

Let V be an affine algebraic variety over a field k; by definition, V is a subset of the affine space $\mathbf{A}^n_k$ for some integer n, defined by a family $(f_1, \ldots, f_p)$ of polynomial in $k[X_1, \ldots, X_n]$.

We can consider that the variety V is the set of n-tuples $\underline{x} = (x_1, \ldots, x_n)$ in k^n that satisfy the equations $f_j(\underline{x}) = 0$ for $1 \leq j \leq p$, but in many situations, it is not enough to restrict to points $\underline{x}$ with coefficients in the field k. For instance, if we consider the variety V defined in $\mathbf{A}^2_{\mathbb{Q}}$ by the polynomial $f(X_1, X_2) = X_1^2 - 2X_2^2$, we have only the point $(0,0)$ if we consider points with coordinates in the field $\mathbb{Q}$. Then we may enlarge the definition of points of the variety V, and for any ring R, with R a k-algebra, we have the following definition:

Definition 2.15 The set of R-points of the variety V is the set of solutions of the equations $f_j(\underline{x}) = 0$, with coefficients in R. More precisely, if we denote this set $V(R)$, we have

$$V(R) = \{\underline{a} = (a_1, \ldots, a_n) \in R^n \mid f_j(\underline{a}) = 0 \text{ for } 1 \leq j \leq p\}.$$

We recall that we denote by Aff_k the category of *affine schemes* over k, the opposite category of the category of k-algebras, $\mathit{Aff}_k = k - \mathit{Alg}^{op}$, and that we denote by Spec A the affine scheme associated to any k-algebra A. The affine scheme associated to the previous variety V is equal to $V = \operatorname{Spec} A$, where A is the k-algebra $A = k[X_1, \ldots, X_n]/(f_1, \ldots, f_p)$, quotient of the ring of polynomials over k by the ideal generated by the f_j. And we remark that the set $V(R)$ of R-points of V, where R is any k-algebra, is exactly the set of morphisms of k-algebras from A to R:

$$V(R) = \operatorname{Hom}_{k-\mathit{Alg}}(A, R) = \operatorname{Hom}_{\mathit{Aff}_k}(\operatorname{Spec} R, V).$$

We can extend the definition of R-points to any algebraic variety or even to any scheme V over a ring k. As we have seen in Section 1.4, a scheme V over k, or a k-scheme, is defined as the union of a family of affine schemes V_α, with $V_\alpha = \operatorname{Spec} A_\alpha$ and where A_α are k-algebras (we need to add the structure of locally ringed space on V and to demand that the immersion of V_α in V is a map of locally ringed spaces where V_α has the structure associated to the ring A_α), and we have defined the category Sch_k of schemes over k and the full subcategory Aff_k of affine schemes.

Then we can define the set of R-points of V as the union of the sets $V_\alpha(R)$. But we have to be very careful when we do this union, because some points can belong to different affine charts V_α, and there is a more natural way to define an R-point of a scheme or a variety V by the following:

$$V(R) = \operatorname{Hom}_{\mathit{Sch}_k}(\operatorname{Spec} R, V).$$

If the variety V is defined over an integral domain k, we say that the k-points of V, or the F-points of V where F is the field of fractions of k, are the *rational points* of the variety. The study of rational points of varieties defined over $\mathbb{Z}$ is the subject of the *diophantime geometry*, and there are many important and very beautiful results, such as the Weil conjecture, the Mordell theorem, and so on.

In Section 1.4, we have defined a scheme V as a topological space, and we want to compare the notion of points of V as topological space, that is, the notion of elements of the underlying set, with this notion of R-points given previously.

If K is a field, any K-point p of $V = \text{Spec } A$ determines a morphism $\phi : A \to K$. The kernel of this morphism is a prime ideal $\mathfrak{p}$ of the ring A, and we have the factorization

$$A \longrightarrow A/\mathfrak{p} \hookrightarrow k(\mathfrak{p}) \hookrightarrow K,$$

where we denote by $k(\mathfrak{p})$ the residue field of the point $\mathfrak{p}$, which is equal to the quotient field of the domain $A/\mathfrak{p}$. Then any K-point, for K a field, determines a prime ideal $\mathfrak{p}$ of the ring A, that is, an element of the space $\text{Spec } A$, and we can associate to any point $\mathfrak{p}$ of $\text{Spec } A$ a K-point for a field K containing the residue field $k(\mathfrak{p})$. When K is an algebraically closed field, a K-point of V is called a *geometric* point of the variety; sometimes the notion of geometric point is used when K is a separably closed field.

By considering a scheme as a union of affine schemes, we see that in the same way, a K-point determines a point x of the scheme V and that we get all the points x of V if we choose a field K that is *big enough*, in the sense that for any x, there exists a nontrivial morphism of the residue field $k(x)$ in K. We can choose such a field Ω such that we have this property for every affine variety; this is the point of view of André Weil [61].

2.7 Representable Functor and Moduli Problem

The notion of R-points of a k-scheme V defines a functor from the category of k-algebras to the category of sets:

$$\begin{aligned} V: \quad k-Alg &\longrightarrow Set \\ R &\mapsto V(R). \end{aligned}$$

This functor, which is called the *functor of points* associated to the scheme V, is equivalent to a presheaf on the opposite category Aff_k, and it is in fact a sheaf for the Zariski topology.

We say that a morphism $j: Y \to X$ in $Sh_{Zar}(\text{Aff}_k)$ is an *open immersion* if, for any morphism $f: V \to X$ where V is an affine scheme, the product

$U := V \times_X Y$ is an affine scheme and the induced map $U \to V$ is a Zariski open immersion. Then we can give the definition of a *categorical scheme* by the following:

Definition 2.16 A categorical scheme X over k is a functor

$$X : \left(A\!f\!f_k\right)^{op} \to Set$$

such that

1. X is a sheaf for the Zariski topology;
2. there exists a family $(V_i)_{i \in I}$ of affine schemes over k and open immersions $u_i : V_i \to X$ such that the induced morphism $\coprod_I V_i \to X$ is an epimorphism in $Sh_{Zar}(A\!f\!f_k)$.

The category of categorical scheme over k is the full subcategory of $Sh_{Zar}(A\!f\!f_k)$ whose elements are schemes over k.

In Section 1.4, we have defined the category of geometric schemes as a full subcategory of locally ringed spaces, and we have the following deep result (cf. [11, Section 4.4]).

Theorem 2.17 *The functor that associates a scheme to its functor of points is an equivalence of categories between the category Sch_k of geometric schemes over k and the category of categorical schemes over k.*

The point of view of functor of points was introduced with the aim to define schemes with some properties that can be described by the notion of *representable functor*. We recall that a presheaf $F : C^{op} \to Set$ on a category C is called representable if there exists an object $x \in C$ such that F is isomorphic to $Hom_C(-, x)$, which is equivalent to saying that F is in the essential image of the Yoneda embedding $h_- : C \to Pr(C)$.

Then we can try to define some schemes by looking at some presheaves on the category Sch_k, or on the full subcategory category $A\!f\!f_k$, that is, a functor from $k - Alg$ to Set. There is no easy criterion to decide if a presheaf, or a functor from $k - Alg$ to Set, is representable by a scheme. As any scheme is obtained as some colimit of affine subschemes, it is necessary that this functor send these colimits in Sch_k to limits in Set. This property is exactly what we need when we demand this functor to be a sheaf for some Grothendieck topology on the category $A\!f\!f_k$.

Many classical problems in algebraic geometry can be described by a functor from the category Sch_k or Var_k to the category of sets; this is the class of *moduli problems* we are going to present now.

Let $\mathcal{B}$ be a *reasonable* class of objects, for instance, $\mathcal{B}$ could be all subvarieties of the projective space $\mathbb{P}^n$, all coherent sheaves on $\mathbb{P}^n$, all invertible sheaves, all smooth curves, or all projective varieties. The aim of the theory of moduli is to understand families of objects in $\mathcal{B}$ and to construct a *space* whose points are in one-to-one correspondence with the objects in $\mathcal{B}$. In general, it is not possible to find an algebraic variety that classifies the objects of $\mathcal{B}$; this is one of the principal reasons it is necessary to extend the notion of space and to introduce the notion of scheme, or of *algebraic space* or of *stack*.

If such a space exists, we call it the moduli space of $\mathcal{B}$; there are many classical examples of moduli problems as linear systems on a normal projective variety, effective cycles on a variety (the theory of Chow variety), closed subschemes of a given algebraic variety (the theory of Hilbert scheme), families of smooth projective curves of genus g, and so on.

More precisely, the moduli problem associated to a class $\mathcal{B}$ of some algebraic varieties is defined by a functor $M_\mathcal{B}$ in the following way. We fix a base field or ring k and let $\mathcal{V}_k$ be a reasonable full subcategory of the category Sch_k of schemes over k; it could be the category of varieties, that is, reduced irreducible schemes of finite type over k, or the category Aff_k^{pf} of affine schemes of finite type. Then the moduli problem is the functor

$$M_\mathcal{B}\colon (\mathcal{V}_k)^{op} \longrightarrow \mathit{Set}$$
$$T \mapsto M_\mathcal{B}(T),$$

where $M_\mathcal{B}(T)$ is the set of flat families $\lambda\colon B \to T$ such that every fiber is in the class $\mathcal{B}$. In fact, we do not consider all the families but families up to equivalence, for instance, isomorphism, and this is one of the most crucial difficulties in this problem.

We say that there exists a *fine moduli space* $\mathcal{M}_\mathcal{B}$ if this functor is *representable*, that is, if there exists a scheme $\mathcal{M}_\mathcal{B}$ in Sch_k and an isomorphism of functors between $M_\mathcal{B}(-)$ and $Hom_{Sch_k}(-, \mathcal{M}_\mathcal{B})$.

This point of view was developed by A. Grothendieck, and it is a very important way to define or construct abstract schemes. We will give some classical examples.

The affine line. Let the *forgetful functor*

$$F\colon \mathit{Aff}^{op} = Ring \longrightarrow \mathit{Set}$$
$$R \mapsto R$$

which sends a ring to the underlying set. This functor is representable by the affine scheme

$$\mathbb{A}^1 := \mathrm{Spec}\,\mathbb{Z}[T],$$

called the *affine line* or the *additive group*, and is also denoted by $\mathbb{G}_a$. For any commutative ring R, we have a functorial bijection between R and $Hom_{\mathit{Aff}}(\operatorname{Spec} R, \mathbb{A}^1) \simeq Hom_{Ring}(\mathbb{Z}[T], R)$. More precisely, to any element a in the ring R, we associate the unique morphism of rings $\mathbb{Z}[T] \to A$, which sends T on a.

The multiplicative group. Let the functor

$$G: \mathit{Aff}^{op} = Ring \longrightarrow \mathit{Set}$$
$$R \mapsto R^\times,$$

which sends a ring R to the set $R^\times$ of invertible elements of the ring. This functor is representable by the affine scheme

$$\mathbb{G}_m := \operatorname{Spec} \mathbb{Z}[T, T^{-1}],$$

called the *multiplicative group*.

The Grassmannian. Let S be a scheme, $\mathcal{E}$ a vector bundle on S, and r a positive integer less than the rank n of $\mathcal{E}$. Let

$$Gr_r(\mathcal{E}): \mathit{Sch}_S^{op} \longrightarrow \mathit{Set}$$
$$X \mapsto Gr_r(\mathcal{E})(X)$$

be the contravariant functor that associates to an S-scheme X the set of quotients of the $\mathcal{O}_X$-modules $\mathcal{E} \times_S X$, which are locally free of rank r.

The Hilbert scheme (cf. [19]). Let $X \to S$ be a morphism of schemes; we define the Hilbert functor of X/S as the functor $\mathcal{H}ilb_{X/S}$ that associates to a scheme $T \in \mathit{Sch}_S$ the set

$$\mathcal{H}ilb_{X/S}(T) = \left\{ \begin{array}{c} \text{closed subschemes } Z \subset T \times_S X \text{ which} \\ \text{are flat and proper over } T \end{array} \right\}.$$

The existence theorem of Hilbert schemes then says that, if X is quasi-projective over S, there is a scheme $Hilb_{X/S}$ that represents the functor $\mathcal{H}ilb_{X/S}$, that is, such that for any S-scheme T, we have

$$\mathcal{H}ilb_{X/S}(T) = Hom_{\mathit{Sch}_S}(T, Hilb_{X/S}).$$

Moreover, there exists a *universal family* $\pi: Univ_{X/S} \to Hilb_{X/S}$ such that the above isomorphism is given by pulling back the universal family. More precisely, the object $Z \subset T \times_S X$, flat and proper over T, associated to the morphism $u: T \to Hilb_{X/S}$ is defined by the following cartesian diagram:

$$\begin{array}{ccc} Z & \longrightarrow & Univ_{X/S} \\ \downarrow & & \downarrow \pi \\ T & \xrightarrow{u} & Hilb_{X/Z} \end{array}$$

This situation is in some sense the best we can demand, and we can generalize it. Let $\mathcal{Z}$ be a class of algebraic varieties, or schemes, or sheaves, or any reasonable geometric objects. We have to make some restrictions on the classes, for instance, we demand that for any field extension $k \subset K$, a k-variety X_k is in the class $\mathcal{Z}$ if and only if the K-variety $X_K := X_k \times_{\operatorname{Spec} k} \operatorname{Spec} K$ is in $\mathcal{Z}$.

For such a class $\mathcal{Z}$, we define the moduli functor

$$\begin{array}{rcl} F_{\mathcal{Z}}: Sch_S^{op} & \longrightarrow & \mathcal{S}et \\ T & \mapsto & F_{\mathcal{Z}}(T), \end{array}$$

where

$$F_{\mathcal{Z}}(T) = \left\{ \begin{array}{c} \text{flat families } f: Z \to T \text{ such that every fiber} \\ \text{is in } \mathcal{Z}, \text{ modulo isomorphisms over } T \end{array} \right\}.$$

Then a flat morphism $\pi: Univ_{\mathcal{Z}} \to Mod_{\mathcal{Z}}$ is a *fine moduli space* for the functor $F_{\mathcal{Z}}$ if the scheme $Mod_{\mathcal{Z}}$ represents the functor and if, for any $T \in Sch_S$, the isomorphism between $F_{\mathcal{Z}}(T)$ and $Hom_{Sch_S}(T, Mod_{\mathcal{Z}})$ is given by a pulling back of π. More precisely, the flat family $f: Z \to T$ in $F_{\mathcal{Z}}(T)$ that corresponds to the morphism $u: T \to Mod_{\mathcal{Z}}$ is given by the cartesian diagram

$$\begin{array}{ccc} Z & \longrightarrow & Univ_{\mathcal{Z}} \\ f \downarrow & & \downarrow \pi \\ T & \xrightarrow{u} & Mod_{\mathcal{Z}} \end{array}$$

Applying the definition to $T = \operatorname{Spec} K$, where K is a field, we see that the K-points of the fine moduli space $Mod_{\mathcal{Z}}$ are in one-to-one correspondence with the K-isomorphism classes of objects in $\mathcal{Z}$. The existence of a fine moduli space is the ideal possibility, but unfortunately, it is rarely achieved. In many cases, the obstruction of the existence of a fine moduli space is due to isomorphisms of objects in the class $\mathcal{Z}$. But it is sometimes possible to get a scheme whose K-points, for any algebraically closed field K, are still in one-to-one correspondence with the K-isomorphism classes of objects in $\mathcal{Z}$; this space is called a *coarse moduli space*.

2.8 New Geometric Spaces

Some of the more natural functors we study in algebraic geometry are not representable by an algebraic variety or even a scheme, but they are very close to being representable. To get it, we need to introduce new objects, new *geometric spaces* that will represent these functors, and we will use the functor to define these spaces.

The first example of natural functors that are not always representable by a scheme are the functors obtained as quotients of a scheme by a group. For instance, in the case of a quotient of a scheme X by a finite group G, a necessary and sufficient condition for the quotient scheme X/G to exist is that the orbit of every point of X be contained in an affine open subset of X [50, Exposé V, Proposition 1.8].

But there is a Hironaka's example of a proper nonprojective 3-fold with a $\mathbb{Z}/2$ action which is free (actually, there is one fixed point, but you can just throw it out) and whose quotient is not a scheme (see [27] or [26, Appendix B, Example 3.4.1]).

In his famous seminars [19], A. Grothendieck exposed the material on descent theory and existence theorems, including that for the Hilbert scheme for quasi-projective variety we have seen in the previous section. But he explained that the previous techniques do not allow us to solve the problems of existence in all generality; it is necessary to add hypotheses of projectivity to define, for instance, the Hilbert scheme.

One way to solve this problem is to extend the notion of *space* we authorize to represent a functor. The first condition is to suppose that this functor is a sheaf for a suitable topology, most of the time the étale or the fppf topology on Aff_k or on Sch/S. This condition is in general not difficult to verify, and if it is not the case, it is always possible to take the sheaf associated to the presheaf defined by the functor. But the most difficult is to get a condition of *algebricity* of this functor, in a way to be able to consider it as a *geometric object* with properties similar to those of an algebraic variety or a scheme.

To solve this problem, there are essentially two different approaches, but which give similar results:

1. a global method (following A. Grothendieck and D. Mumford [19, 42]) which consists in constructing a scheme and in taking the quotient by an algebraic group;

2. a local method (following M. Artin and M. Schlessinger [1–3, 47]) which consists in considering the theory of deformations of objects associated to the functor. It is possible to show that some moduli problem has a formal solution, that is, that some functor is representable by a pro-object (it is pro-representable) or an anlytic solution, and the problem of *algebrization of formal moduli* is to see if we can deduce from this formal or analytic solution an algebraic solution.

The notion of *algebraic space* was introduced first by M. Artin to solve the problem of *algebrization of formal moduli*. The first definition of an algebraic space is a sheaf for the étale topology that is locally in the étale topology representable [30]; more precisely, we have the following:

Definition 2.18 An algebraic space A over S is a functor

$$A \colon (Sch/S)^{op} \to Set$$

such that

1. A is a sheaf for the étale topology;
2. (local representability) there exists a scheme U over S and a map of sheaves $U \to A$ such that for all schemes V over S and all maps $V \to A$ over S the sheaf fiber product $U \times_A V$ is representable and the map $U \times_A V \to V$ is an étale map of schemes – we call $U \to A$ an *étale atlas* of A;
3. (quasi-separatedness) let $U \to A$ be as in (2), and then the map of schemes $U \times_A U \to U \times_S U$ is quasicompact.

We can also consider an algebraic space as a sheaf on the category Aff_k for the étale topology, and it may be convenient to modify a little the category Aff_k, for instance, to restrict it to a subcategory of affine schemes that satisfy some finiteness condition, for instance, to be locally of finite type over some excellent Noetherian base k.

The category $AlgSp/S$ of algebraic spaces over S is by definition the full subcategory of the category $Sh_{et}(Sch/S)$ of sheaves on Sch/S for the étale topology. It is easy to see that any scheme is an algebraic space, more precisely, given a scheme T over S, the representable functor h_T is an algebraic space. As the étale topology is subcanonical, the Yoneda functor is fully faithful, so then there is a fully faithful left-exact embedding of the category Sch/S in the category $AlgSp/S$.

Definition 2.19 Let U be a scheme over S. An étale equivalence relation on U over S is an equivalence relation $j: R \to U \times_S U$ such that the projections $s; t: R \to U$ are étale morphisms of schemes.

With these definitions, we can give the following characterization of an algebraic space.

Theorem 2.20 *Let S be a scheme. A sheaf $F \in Sh_{et}(Sch/S)$ is an algebraic space over S if and only if there exists a scheme U over S and an étale equivalence relation $(s; t): R \to U \times_S U$ such that F is isomorphic to the quotient U/R.*

We can see that the notion of algebraic space is a generalization of the notion of scheme, and algebraic spaces share many properties with schemes. More precisely, if $\mathcal{P}$ is a property of schemes that is stable in the étale topology, $\mathcal{P}$ extends to a property of algebraic spaces by taking, for any algebraic space X with an étale atlas $U \to X$, the algebraic space X has property $\mathcal{P}$ if and only if the scheme U has property $\mathcal{P}$. For instance, we can define an algebraic space to be separated, quasi-compact, Noetherian, normal, nonsingular, n-dimensional, and so on. In the same way, we can define properties of maps between algebraic spaces as open immersions, closed immersions, affine morphisms, and so on.

It is also possible to define the étale topology of algebraic spaces; moreover, for any algebraic space X, there is a structure sheaf $\mathcal{O}_X$, which is a sheaf of rings on the small étale site of X, and it is possible to define the notion of quasi-coherent sheaves on X.

Let X be an algebraic space; a point of X is a monomorphism $p: \text{Spec } k \to X$ of algebraic spaces, and the field k is called the residue field of p and is denoted $k(p)$. Two points $p_1: \text{Spec } k_1 \to X$ and $p_2: \text{Spec } k_2 \to X$ are equivalent if there is an isomorphism $e: \text{Spec } k_1 \to \text{Spec } k_2$ with $p_2 \circ e = p_1$.

Definition 2.21 Let X be an algebraic space; the underlying topological space of X, $|X|$, is defined as the set of points of X modulo equivalence of points. The set $|X|$ is given a topological structure by taking a subset $F \subset |X|$ to be closed if F is of the form $|Y|$ for some closed immersion $Y \to X$.

The topology on $|X|$ is called the *Zariski topology* on X.

We can think of algebraic spaces as being closer to analytic spaces than to schemes. In fact, every compact analytic space with enough meromorphic functions is an algebraic space. More precisely, we have the following definition:

Definition 2.22 Let X be a compact complex analytic space; X is called a *Moisezon space* if, for any irreducible component X_i of X, the transcendence degree of the field $K(X_i)$ of meromorphic functions on X_i is of degree $d_i := \dim X_i$ over $\mathbb{C}$.

By a simple extension of Serre's GAGA theorem, there exists a fully faithful functor that assigns to each algebraic space its underlying analytic space, and the image of any algebraic space proper over $\mathbb{C}$ is a Moisezon space. We define the full subcategory $AlgSp_{\mathbb{C}}^{prop}$ of algebraic spaces proper over $\mathbb{C}$ and the category $\mathcal{M}oisezon$ as the full subcategory of complex analytic spaces whose objects are Moisezon spaces, and then we get the following result:

Theorem 2.23 *The induced functor*

$$AlgSp_{\mathbb{C}}^{prop} \longrightarrow \mathcal{M}oisezon$$

is an equivalence of categories.

We want to finish this section by showing that some moduli problems are not representable by a scheme or even by an algebraic space and that it is necessary to introduce new notions that will be presented in Chapters 8 and 9 of this volume.

The first obstruction to get an algebraic space to represent a functor associated to a natural moduli problem is the existence of automorphisms. The most famous example is the case of the moduli space of smooth projective curves of fixed genus g. We want to study the moduli problem M_g given by the following:

$$\begin{aligned} M_g: \quad Sch^{op} &\longrightarrow \quad Set \\ X &\longmapsto \quad M_g(X), \end{aligned}$$

where $M_g(X)$ is the set of flat families $Z \to X$ such that every fiber is a smooth projective curve of genus g. Because of the existence of nontrivial automorphisms of a curve C of genus g, for $g \geq 2$, this functor is not representable, even by an algebraic space. To get a correct definition, we have to modify the *target category*, and in that case, we have to replace the category of sets by the category of *groupoids*, where a groupoid is a category such that all the morphisms are isomorphisms. Then we have to consider the new functor

$$\begin{aligned} \mathcal{M}_g: \quad Sch^{op} &\longrightarrow \quad \mathcal{G}roupoid \\ X &\longmapsto \quad \mathcal{M}_g(X), \end{aligned}$$

where $\mathcal{M}_g(X)$ is the groupoid of flat families $Z \to X$ such that every fiber is a smooth projective curve of genus g. This is a very general situation, and the notion of *stack* as functor whose values in the category $\mathcal{G}roupoid$ have been introduced to study these problems [13, 33]. In these cases, it is necessary to make a distinction between the *fine moduli problem*, which is represented by a stack, and the associated *coarse moduli problem*, a sheaf for the étale topology, which classifies the objects up to isomorphisms but does not represent the functor.

The second problem is related to the existence of the *tangent complex* associated to any good notion of geometric space, and the theory of deformation. With a naive definition of functors associated to some moduli problems, we cannot obtain a good *obstruction theory*; to get a correct definition, we have to modify the *source category* and replace the category of k-algebras by the category of *simplicial k-algebras*. The notion of *derived algebraic geometry* has been introduced by B. Toën, G. Vezzosi, and J. Lurie [35, 55, 59] to obtain the good setting to solve these problems.

We could find an example in [54, Section 1.2], where the necessity of these two extensions is presented. Let us consider the moduli functor defined by the finite-dimensional linear representation of a finitely presented group Γ. The naive functor associated to this moduli problem is

$$R(\Gamma) \colon \mathcal{C}omm \to \mathcal{S}et,$$

which sends a commutative ring A to the set of isomorphism classes of $A[\Gamma]$-modules whose underlying A-module is projective of finite type over A. But to get a more convenient and better-behaved object, we have to replace it by a functor

$$\tilde{R}(\Gamma) \colon s\mathcal{C}omm \to s\mathcal{S}et$$

from the category of simplicial commuative rings to the category of simplicial sets.

3 Relative Geometry

3.1 Geometric Context

We will see in this section a general notion of *geometric spaces* defined in a similar way to the one that we introduced in the previous section. There are two different approaches:

1. an approach inspired by the definition of the schemes as locally ringed spaces (following J. Lurie [36] and M. Anel and M. Vaquié [4]), which consists in constructing a topological space, or more generally a topos or an ∞-topos, with an additional structure;

2. an approach inspired by the definition of schemes as functors of points (following B. Toën, G. Vezzosi, and M. Vaquié [56, 57, 59]), which consists in constructing a site and in defining spaces as a class of sheaves that are locally representable.

For the first approach, we recall that in the first definition of scheme (cf. Section 1.4), we use the notion of ringed space, where a ringed space $(X, \mathcal{O}_X)$ is a pair where X is a topological space and $\mathcal{O}_X$ is a *ring object* in the category $Sh(X)$ of sheaves on X. By definition, for any category $\mathcal{C}$ with finite limits, a ring object x in $\mathcal{C}$ is an object x of $\mathcal{C}$ endowed with two composition laws that satisfy the same properties as the addition and the multiplication in a ring. In fact, for any *algebraic structure* $\mathbb{T}$, we do not give here the exact definition of an algebraic structure, and we refer to the fundamental articles of F. W. Lawvere [34]; we can define in the same way a $\mathbb{T}$-*structured object* in $\mathcal{C}$.

If we want to imitate the definition of schemes and define new sorts of geometric spaces, we have to choose the category associated to the algebraic structure $\mathbb{T}$ that will play the role of the category of rings. For instance, by this approach in the setting of derived geometry, it is possible to define *derived complex analytic geometry* and *derived non-Archimedean analytic geometry* [45, 46].

This idea was already used in some definitions of *synthetic differential geometry* (cf. A. Kock's article in Chapter 2); explicitly, the aim is to apply the techniques of algebraic geometry introduced by A. Grothendieck in the setting of differentiable geometry. The category of C^∞-*rings* is defined as a model of the algebraic theory that has as n-ary operations all the smooth functions from $\mathbb{R}^n$ into $\mathbb{R}$ and whose axioms are all the equations that hold between these functions. Then, in the same way as ringed space, we may define C^∞-*ringed space* $(X, \mathcal{O})$ as a pair where X is a topological space and $\mathcal{O}$ is a C^∞-ring object in the category of sheaves over X (cf. [14, 15]).

In the second approach, the idea is to define a geometric space as a sheaf on a category, and the objects of this category play the role of the elementary pieces, obtained by gluing representable sheaves along a class of morphisms. First we have to choose the category of the elementary pieces, which are the analogues of affine spaces, and to endow this category with a Grothendieck topology. Then we have to determine the class of morphisms along which we will glue the affine pieces. The notion we will introduce is the notion of *geometric context*.

Definition 3.1 A *geometric context* is a triplet $(C, \tau, \mathbb{P})$, where C is a category, τ a Grothendieck topology on C, and $\mathbb{P}$ a class of morphisms in C satisfying some compatibility conditions.

We do not specify the properties requested on the topology τ and the class $\mathbb{P}$ of morphisms; we can summarize these compatibility conditions by saying that we ask the topology to be subcanonical and the class to be local for the topology and to generate it. We refer to [56] for a precise definition of a geometric context and to [59] for a definition in *homological algebraic geometry*.

Let $(\mathcal{C}, \tau, \mathbb{P})$ be a geometric context; we will define the category of *geometric spaces* as a full subcategory of the category of sheaves $Sh_\tau(\mathcal{C})$. We recall that as the topology τ is subcanonical, we identify the category $\mathcal{C}$ to its essential image by the Yoneda embedding.

We recall that a sheaf $F \in Sh_\tau(\mathcal{C})$ is *representable* if it is isomorphic in $Sh_\tau(\mathcal{C})$ to an object h_X for $X \in \mathcal{C}$. In the same way, we say that a morphism $f: F \longrightarrow G$ of sheaves is *representable*, resp. *representable and in* $\mathbb{P}$, if, for any $X \in \mathcal{C}$ and any morphism $X \longrightarrow G$, the sheaf $F \times_G X$ is representable, resp. the sheaf $F \times_G X$ is representable and the induced morphism between representable sheaves $F \times_G X \longrightarrow X$ is in $\mathbb{P}$.

Definition 3.2 A sheaf F is *geometric* if there exists a family of objects $\{U_i\}$ in $\mathcal{C}$ and an epimorphism of sheaves

$$U := \coprod_i U_i \longrightarrow F$$

such that every morphism $U_i \longrightarrow F$ is representable and in $\mathbb{P}$.

The data of the U_i and of the morphism $\coprod_i U_i \longrightarrow F$ are called an *n-atlas for F*.

In fact this definition is too restrictive – the spaces defined in this way have an affine diagonal, which is a very strict *separateness condition*, and which it is possible to weaken, and even to suppress, to have the most general definition. We refer to the articles [56, 59] for the precise definitions that use an induction on the *level of geometricity* of a sheaf.

Let $(\mathcal{C}, \tau, \mathbb{P})$ be a geometric context; we will define the category of *geometric spaces* as a full subcategory of the category of sheaves $Sh_\tau(\mathcal{C})$. We recall that as the topology τ is subcanonical, we identify the category $\mathcal{C}$ to its essential image by the Yoneda embedding.

If we consider the geometric context $(\mathcal{C}, \tau, \mathbb{P})$, where $\mathcal{C}$ is the category Aff_k of affine schemes over k, $\tau := et$ the étale topology on $\mathcal{C}$ and $\mathbb{P} = et$ the class of étale morphisms in Aff_k, we recover a notion of algebraic spaces that is very similar to the one we have introduced in Section 2.8.

In the same way, by considering the geometric context $(D, \rho, \mathbb{Q})$, where D is the category Ste of Stein spaces, $\rho := top$ the usual topology on complex analytic spaces, and $\mathbb{Q} = et$ the class of étale morphisms between analytic spaces (i.e., local analytic homomorphisms), we define a notion of analytic spaces which is the analogue of the notion of algebraic spaces defined previously.

This notion of analytic spaces is more general than the classical analytic spaces; for example, if Γ is a discrete group that acts without fixed points but nonproperly on an analytic space X, the quotient sheaf X/Γ is an anlytic space with the previous definition. However, this quotient may not exist as an analytic space in the classical sense (e.g., in the definition of [24]). Moreover, even if there exists a quotient $X//\Gamma$ as an analytic space in the usual sense, the spaces X/Γ and $X//\Gamma$ may be different. For instance, for $X = \mathbb{C}^\times$ and $\Gamma = \mathbb{Z}$, which acts by $z \to q.z$ with $|q| = 1$, which is not a root of the unity, the usual quotient $X//\Gamma$ is reduced to a point, but the quotient X/Γ is not.

We end this section with the notion of change of geometrical contexts. For that purpose, let $(C, \tau, \mathbb{P})$ and $(D, \rho, \mathbb{Q})$ be two geometrical contexts, and let $f : C \longrightarrow D$ be a functor.

Proposition 3.3 *We assume that the functor f satisfies the following conditions:*

1. *The functor f commutes with finite limits and is continous for the topologies τ and ρ.*
2. *The image by f of a morphism in $\mathbb{P}$ is a morphism in $\mathbb{Q}$.*

Then the functor

$$\tilde{f}_! : Sh_\tau(C) \longrightarrow Sh_\rho(D)$$

defined in Section 2.4 preserves geometric sheaves and sends morphisms in $\mathbb{P}$ into morphisms in $\mathbb{Q}$.

This result enables us to compare different concepts of geometric spaces. For instance, the classical analytification functor (see [50, Exp. XII])

$$Aff_\mathbb{C}^{fp} \longrightarrow Ste$$

defined on the level of affine schemes satisfies the condition of the proposition, so then we can define the functor *analytification* that sends any algebraic space over $\mathbb{C}$ to an analytic space.

3.2 Relative Algebraic Geometry

In this last part, we want to present a special case of geometric context, which plays an important role in a possible definition of $\mathbb{F}_1$, the famous *field with one element*.

The idea of doing algebraic geometry over a closed symmetric monoidal category $(C, \otimes, \mathbf{1})$ has a long story (see [10, 25]) and was developped by many authors after the article of B. Toën and M. Vaquié [57]. We will give a short presentation of this approach; we refer to this last article to have more details.

Let $Comm(C)$ denote the category of commutative monoid objects in C. If $A \in Comm(C)$ is such a commutative monoid, we let $A - Mod$ denote the category of A-module objects in C. Then, it follows that $(A - Mod, \otimes_A, A)$ is also a closed symmetric monoidal category. Let $\mathit{Aff}_C := Comm(C)^{op}$ be the category of affine schemes over C. Given a commutative monoid object $A \in Comm(C)$, we let $Spec\ A$ denote the affine scheme corresponding to A.

Any morphism $p \colon Spec\ B \to Spec\ A$ in Aff_C, which corresponds to $u \colon A \to B$ in $Comm(C)$, induces a functor $p^* \colon A - Mod \to B - Mod$ defined by $- \otimes_A B$, which is the left adjoint of the forgetful functor $p_* \colon A - Mod \to B - Mod$. We say that the morphism p is flat, or that B is flat over A, if the functor p^* is left exact, that is, it commutes with finite limits.

Moreover, we say that the morphism p is a Zariski immersion if $u \colon A \to B$ is a flat epimorphism of finite presentation.

By considering the pseudo-functor

$$M \colon (\mathit{Aff}_C)^{op} \to Cat$$
$$Spec\ A \mapsto A - Mod,$$

we can define a Grothendieck topology on the category Aff_C. More precisely, we have the following definition:

Definition 3.4 A family of morphisms $\{p_i \colon X_i = Spec\ A_i \to X = Spec\ A\}_{i \in I}$ in Aff_C is a *Zariski cover* if it satisfies the two following conditions:

1. For any $i \in I$, the morphism $X_i \to X$ is Zariski immersion.
2. There exists a finite subset $J \subset I$ such that the family of functors $\{p_i^* \colon M(X) \to M(X_i)\}_{i \in J}$ is conservative (i.e., a morphism $u \colon M \to N$ of A-modules is an isomorphism if and only if, for any $i \in J$, the induced morphism $M \otimes_A A_i \to N \otimes_A A_i$ is an isomorphism).

The Zariski covers define a subcanonical pretopology on Aff_C; then we can define a geometric context by considering the category Aff_C, the topology Zar induced by the Zariski covers, and the class of Zariski immersions. The geometric spaces defined with this geometric context will be called *categorical schemes relative to* C, or *over* C, and we denote this category by $Sch(C)$.

It is important to notice that to define these spaces, we do not need any additional data on the symmetric monoidal category $(C, \otimes, \mathbf{1})$. If we take for the category C the category of abelian groups $C = \mathbb{Z} - Mod$, the category $Comm(C)$ is exactly the category of rings, and we recover the category $Sch(\mathbb{Z})$ of usual schemes over $\mathbb{Z}$.

If we take $(C, \otimes, \mathbf{1}) = (Set, \times, *)$ the category of sets endowed with the monoidal structure given by the direct product, we can define a category of schemes over Set, which we denote by $Sch(\mathbb{S})$, and we call these spaces *schemes over* $\mathbb{S}$.

We have a pair of adjoint functors

$$f : \mathbb{S} - Mod = Set \rightleftarrows \mathbb{Z} - Mod =: g$$

where g is the forgetful functor, which sends an abelian group to the underlying set, and f sends a set X to the free abelian group $X \otimes \mathbb{Z}$ generated by it. The functor f is symmetric monoidal and induces a base change functor

$$- \otimes_{\mathbb{S}} \mathbb{Z} : Sch(\mathbb{S}) \longrightarrow Sch(\mathbb{Z}).$$

The scheme $Spec\ \mathbb{Z}$ is the terminal object in the category of schemes; by the previous construction, we construct new algebraic spaces that are *under* $Spec\ \mathbb{Z}$.

In a strategy for proving the Riemann hypothesis, which follows E. Bombieri's proof of the Riemann hypothesis for function fields, it is important to find such a space under $Spec\ \mathbb{Z}$, which we will denote by $Spec\ \mathbb{F}_1$. If we succeed in defining such a suitable space, then the product $Spec\ (\mathbb{Z} \otimes_{\mathbb{F}_1} \mathbb{Z})$ should be studied as an analogue of the product $\overline{C} \times_{\overline{\mathbb{F}}_q} \overline{C}$ for a curve C defined over the finite field $\mathbb{F}_q$ (see [9, §2.3]). The relative algebraic geometry is an attempt to define such a space $Spec\ \mathbb{F}_1$, which will be associated to the *field with one element* $\mathbb{F}_1$. The space $Spec\ \mathbb{S}$ we have defined by considering the symmetric monoidal category $(Set, \times, *)$ seems not to suit, but there are other candidates that seem more adapted (see [9]).

Conclusion

As we have seen, the notion of sheaves was introduced to study the topology of some geometric space that was already defined, and it is now a very important and classical tool in the different sorts of geometry. After the fundamental work of A. Grothendieck, it appears that the notion of sheaves is also a way to construct new spaces, either the classical notion of sheaves on a preexisting topological space if we add a notion of structured sheaf or by extending the notion of topology and of sheaf on any category.

By generalizing these constructions, we can obtain new spaces that could be very far from the usual ones, and it is also the starting point of the generalizations as stacks or even derived higher stacks (cf. Chapters 8 and 9 in this volume).

But we can go further and consider that the object of the geometry is the category of sheaves. More precisely, we can associate to any algebraic variety, or to any scheme X, the derived category of quasi-coherent sheaves on X; we recall that this is the triangulated category obtained by considering complexes of sheaves of $\mathcal{O}_X$-modules with quasi-coherent cohomology, up to quasi-isomorphisms. In fact, this category, denoted by $D_{QCoh}(X)$, is a *dg-category*, that is, a category enriched over the symmetric monoidal category $C(k)$ of complexes of k-modules, if X is defined over a ring k. Or in the case of topological space M, we can consider the dg-category $D_{constr}(M)$ of complexes of sheaves on M whose cohomology sheaves are constructible with respect to a given stratification.

Let L_{perf} be the functor that associates to any variety over a ring k the dg-category of *perfect* complexes, which is the subcategory of compact objects in $D_{QCoh}(X)$. The dg-category $L_{perf}(X)$ may be seen as the *noncommutative space* associated to X, and it remembers many cohomological invariants of the variety X. Moreover, even if this functor does not give an embedding of categories, we can define an adjoint of L_{perf}, more precisely, we can define a functor $\mathcal{M}_-$ that associates to any dg-category T the higher derived stack $\mathcal{M}_T$ of moduli of objects in T (cf. [58]). Then, following M. Kontsevich, we can say that the idea of *noncommutative geometry* is to consider any dg-category as the category of quasi-coherent sheaves on a noncommutative space (cf. Chapter 10 in this volume).

References

[1] M. Artin, *Algebraic approximation of structures over complete local rings*, Publ. Math. IHES, 36 (1969) 23–58

[2] M. Artin, *Algebrization of formal moduli I*. In *Global Analysis (Papers in Honor of K. Kodaira)*, University of Tokyo Press (1969)

[3] M. Artin, *Versal deformations and algebraic stacks*, Invent. Math. 27 (1974) 165–89

[4] M. Anel and M. Vaquié, *Théorie spectrale en géométrie relative*, Preprint (2016)

[5] H. Cartan, *Séminaire Henri Cartan de topologie algébrique, 1950/1951, Cohomologie des groupes, suite spectrale, faisceaux*, Secrétariat mathématique (1955)
[6] H. Cartan, *Séminaire Henri Cartan*, Tome 13, *Famille d'espaces complexes et fondements de la géométrie analytique* (1960–61)
[7] H. Cartan and S. Eilenberg, *Homological Algebra*, Princeton University Press (1956)
[8] R. Chorlay, *L'émergence du couple local/global dans les théories géométriques, de Bernhard Riemann à la théorie des faisceaux (1851–1953)*, Thèse d'Histoire des mathématiques, Paris 7 (2007)
[9] A. Connes, *An essay on the Riemann hypothesis*, in *Open Problems in Mathematics*, Springer (2016) 225–57
[10] P. Deligne, *Catégories tannakiennes*, in *The Grothendieck Festschrift*, vol. II, Progr. Math. 87, Birkhäuser (1990) 111–95
[11] M. Demazure and P. Gabriel, *Groupes Algébriques, TOME I*, Masson & Cie (1970)
[12] J. Dieudonné, *A History of Algebraic and Differential Topology, 1900–1960*, Birkhäuser (1989)
[13] P. Deligne and D. Mumford, *The irreducibility of the space of curves of given genus*, Publ. Math. IHES 36 (1969) 75–109
[14] E. Dubuc, *Sur les modèles de la géométrie différentielle synthétique*, Cah. Top. Geom. Diff. 20 (1979) 231–79
[15] E. Dubuc, C^∞-*schemes*, Amer. J. Math. 103 (1981) 683–90
[16] D. Dugger, *Sheaves and homotopy theory*, http://math.uoregon.edu/ ddugger/cech.html
[17] A. Grothendieck and J. Dieudonné, *Eléments de géométrie algébrique*, Publ. Math. IHES 4 (1960) 8 (1961) 11 (1961) 17 (1963) 20 (1964) 24 (1965) 28 (1966) 32 (1967)
[18] S. Eilenberg and N. Steenrod, *Axiomatic approach to homology theory*, Proc. Nat. Acad. Sci. USA 31 (1945) 117–20
[19] A. Grothendieck, *Fondements de la géométrie algébrique*, Extraits du Séminaire Bourbaki, 1957–1962, Secrétariat mathématique (1962)
[20] I. M. Gelfand and M. A. Naimark, *On the embedding of normed rings into the ring of operators in Hilbert space*, Mat. Sbornik 12 (1943) 197–213
[21] R. Godement, *Topologie algébrique et théorie des faisceaux*, Hermann (1958)
[22] A. Grothendieck, *Sur quelques points d'algèbre homologique*, Tohoku Math. J. 9 (1957) 119–221
[23] A. Grothendieck, *Techniques de construction en géométrie analytique. II. Généralités sur les espaces annelés et les espaces analytique*, in *Séminaire Henri Cartan*, Tome 13, *Famille d'espaces complexes et fondements de la géométrie analytique* (1960–61)
[24] H. Grauert and R. Remmert, *Coherent analytic sheaves*, Grundlehren der MathematischenWissenschaften 265, Springer (1984)
[25] M. Hakim, *Topos annelés et schémas relatifs*, Springer (1972)
[26] R. Hartshorne, *Algebraic Geometry*, Springer (1977)
[27] H. Hironaka, *On the theory of birational blowing-up*, Thesis, Harvard University (1960)

[28] C. Houzel, *Histoire de la théorie des faisceaux*, in *Matériaux pour l'histoire des mathématiques au XXe siècle. Actes du colloque à la mémoire de Jean Dieudonné (Nice 1996)*, Séminaires et Congrès 3, SMF (1998)

[29] I. M. James, *History of Topology*, Elsevier (1999)

[30] D. Knutson, *Algebraic Spaces*, Lecture Notes in Mathematics 203, Springer (1970)

[31] J. Leray, *L'anneau d'homologie d'une représentation*, C. R. Acad. Sci. (Paris) 222 (1946) 1366–68

[32] J. Leray, *Stucture de l'anneau d'homologie d'une représentation*, C. R. Acad. Sci. (Paris) 222 (1946) 1419–22

[33] G. Laumon and L. Moret-Bailly, *Champs algébriques*, Ergebnisse der Mathematik und ihrer Grenzgebiete 39, Springer (2000)

[34] F. W. Lawvere, *Functorial semantics of algebraic theories and some algebraic problems in the context of functorial semantics of algebraic theories*, Reprints in Theory and Applications of Categories 5 (2004) 1–121

[35] J. Lurie, *Higher Algebra*, www.math.harvard.edu/lurie/ (2016)

[36] J. Lurie, *Derived Algebraic Geometry V: Structured Spaces*, www.math.harvard.edu/lurie/ (2011)

[37] A. Mallios, *Geometry of Vector Sheaves: An Axiomatic Approach to Differential Geometry*, Kluwer (1997)

[38] A. Mallios, *Abstract differential geometry, differential algebras of generalized functions, and de Rham cohomology*, Acta Appl. Math. 55 (1999) 231–50

[39] H. Miller, *Leray in Oflag XXVIIA: The origins of sheaf theory, sheaf cohomology, and spectral sequences*, Gazette Math. 84 suppl. (2000) 17–34

[40] J. Milne, *Lectures on étale cohomology*, version 2.21, www.jmilne.org/math/ (2013)

[41] S. Mac Lane and I. Moerdijk, *Sheaves in Geometry and Logic*, Universitext, Springer (1992)

[42] D. Mumford, J. Fogarty, and F. Kirwan, *Geometric Invariant Theory*, Ergebnisse der Mathematik und ihrer Grenzgebiete, Springer (1994)

[43] M. Nagata, *Imbedding of an abstract variety in a complete variety*, J. Math. Kyoto Univ. 2 (1962) 1–10

[44] Y. Nisnevich, *The completely decomposed topology on schemes and associated descent spectral sequences in algebraic K-theory*, in *Algebraic K-Theory: Connections with Geometry and Topology*, NATO Adv. Sci. Inst. Ser. C Math. Phys. Sci. 279, Lake Louise (1987) 241–342

[45] M. Porta, *Derived complex analytic geometry I: GAGA theorems*, Preprint, arXiv:1506.09042

[46] M. Porta and T. Y. Yu, *Derived non-Archimedean analytic spaces*, Preprint, arXiv:1601.00859

[47] M. Schlessinger, *Functors of Artin rings*, Trans. Amer. Math. Soc. 130 (1968) 208–22

[48] J.-P. Serre, *Faisceaux algébriques cohérents*, Ann. Math. 61 (1955) 197–278

[49] J.-P. Serre, *Sur la cohomologie des variétés algébriques*, J. Maths. Pures Appl. 36 (1957) 1–16

[50] A. Grothendieck, *Revêtements étales et groupe fondamental, 1960–1961*, Lecture Notes in Mathematics 224, Springer (1971)
[51] M. Artin, A. Grothendieck, and J.-L.Verdier, *Théorie des topos et cohomologie étale des schémas, 1963–1964*, Lecture Notes in Mathematics 269, 270, 305, Springer (1972–73)
[52] P. Deligne, *Cohomologie étale*, Lecture Notes in Mathematics 569, Springer (1977)
[53] A. Grothendieck, *Cohomologie l-adique et fonctions L, 1965–1966*, Lecture Notes in Mathematics 589, Springer (1977)
[54] B. Toën, *Derived algebraic geometry*, in I. Moerdijk and B. Töen (eds.), *Simplicial Methods for Operads and Algebraic Geometry*, Advanced Course in Mathematics, CRM Barcelona (2010)
[55] B. Toën, *Derived algebraic geometry*, EMS Surv. Math. Sci. 1 (2014) 153–245
[56] B. Toën and M. Vaquié, *Algébrisation des variétés analytiques complexes et catégories dérivées*, Math. Ann. 342 (2008) 789–831
[57] B. Toën and M. Vaquié, *Au-dessous de $spec(\mathbb{Z})$*, J. K-Theory 3 (2009) 437–500
[58] B. Toën and M. Vaquié, *Moduli of objects in dg-categories*, Ann. Sci. Ecole Norm. Sup. 40 (2007) 387–444
[59] B. Toën and G. Vezzosi, *Homotopical algebraic geometry II: Geometric stacks and applications*, Mem. Amer. Math. Soc. 193 (2008)
[60] C. A. Weibel, *History of homological algebra*, in I. M. James, ed., *History of Topology*, Elsevier (1999) 797–836
[61] A. Weil, *Foundations of Algebraic Geometry*, Colloquium Publications XXIX, AMS (1946)
[62] A. Weil, *Fibre spaces in algebraic geometry*, Notes of a course, University of Chicago (1952)
[63] O. Zariski, *The compactness of the Riemann manifold of an abstract field of algebraic functions*, Bull. Amer. Math. Soc. 50 (1944) 683–91

Michel Vaquié
Institut de Mathématiques de Toulouse, Université Paul Sabatier
vaquie@math.univ-toulouse.fr

8
Stacks

Nicole Mestrano and Carlos Simpson*

Contents

1	Introduction	462
2	Spaces and Coordinate Charts	463
3	Stacks in Algebraic Geometry	466
4	Gerbes	470
5	The Abstract Notion of Higher Stack	475
6	Artin Stacks	483
7	Non-abelian Cohomology	491
8	Prospects	501
References		503

1 Introduction

The theory of stacks is part of a more general trend in modern geometry to combine geometrical structures with structures from algebraic topology. The notion of stack allows us to make such a combination in novel ways. Homotopy types from algebraic topology are allowed to show up on a punctual scale with respect to the geometry.

---

* This research supported in part by ANR grant TOFIGROU (933R03/13ANR002SRAR). We would like to thank our numerous colleagues who have contributed over the years to the points of view presented here, particularly André Hirschowitz and Bertrand Toën. We thank Mathieu Anel and Gabriel Catren for their extensive comments and suggestions and Robert Ghrist for a significant conversation.

In this chapter, we describe some of the motivating examples and considerations leading toward the theory of higher Artin stacks. The guiding idea undergoes a transformation, changing from using algebraic topology to introduce fine structural information into algebraic geometry, to the idea of using algebraic geometry to attach new structure to homotopy types and thereby permit their utilization in algebraic contexts. At the end we shall see, with an example, how shape theory implied by the idea of non-abelian cohomology, permits to gain insight about the homotopical structure of non–simply connected spaces.

The discussion here is not uniform in level. It is not our intention to provide a mathematical introduction to the main definitions inherent in the theory. On the other hand, we would like to illustrate the genesis of stack theory by refering to the most basic examples. Therefore, some parts of the discussion will treat examples a student would learn very early in any study of the theory, whereas other parts of the discussion will float rather lightly over some of the most advanced concepts.

We would also like to say that we shall be dealing with the notion of stack as it occurs in algebraic geometry. Many closely related if not identical notions are current in nearby fields such as differential geometry. These are the subjects of other chapters in the book and our present emphasis on the vantage point of algebraic geometry is meant to be complementary to them.

We do not attack the question of philosophy head-on. Instead, we hope that our discussion will illustrate some of the basic philosophical ideas that arose as consequences of fundamental constructions and examples.

2 Spaces and Coordinate Charts

A basic element of the modern notion of "space" is the idea of a *coordinate chart*. This is something that allows us to give a precise measurement of the relationship between nearby points. One should think of a coordinate chart as expressing a view of the space, from a certain given viewpoint. For example, if we fix a point $p \in X$, then a neighborhood $p \in U \subset X$ corresponds to a collection of "points near to p," and a coordinate system $U \to k^n$ provides a way of viewing nearby points as being determined by their numerical coordinates $(x_1, \ldots, x_n) \in k^n$, where p corresponds to the origin $(0, \ldots, 0)$.

Seen from a different view point, with a different neighborhood $q \in V \subset X$, we would have a different system of coordinates $(y_1, \ldots, y_n)$. The global structure of the space is obtained from the collection of coordinate change expressions that say, for example, how the new coordinates are related to the old ones:

$$(y_1(x_1,\ldots,x_n), y_2(x_1,\ldots,x_n),\ldots,y_n(x_1,\ldots,x_n)).$$

Differentiating this expression gives the well-known Jacobian matrix, in turn providing gluing data to define the tangent bundle.

We have just described the most down-to-earth way that the structure of a space is built up out of its coordinate charts. One of the original objectives of the notion of stack was to give a fully general setting to this idea. Among other things, the notion of "point" itself can become secondary: rather than viewing coordinate charts as corresponding to neighborhoods of specific points, we just remember the maps $U \to X$. On the other side, the coordinate functions $(x_1,\ldots,x_n) : U \to k^n$ are abstracted to any kind of *concrete algebraic structure* that allows for a computational description of U.

A space is now viewed as something that is *covered* by a collection of maps $p_\alpha : U_\alpha \to X$, where the U_α themselves have a concrete algebraic structure. For the notion of scheme, the covering objects are *affine schemes* of the form $U_\alpha = \mathrm{Spec}(A_\alpha)$ with A_α a commutative ring, usually of finite presentation:

$$A_\alpha = \frac{k[z_1,\ldots,z_n]}{(f_1,\ldots,f_m)}.$$

It can be convenient to collect all of the charts together into a single object $U := \coprod_\alpha U_\alpha$. The coordinate change transformations correspond to the cartesian diagram

$$\begin{array}{ccc} R := U \times_X U & \longrightarrow & U \\ \downarrow & & \downarrow \\ U & \longrightarrow & X \end{array}$$

The object R together with its map $R \to U \times U$ is the equivalence relation gluing U to itself to obtain X. In other words, the above cartesian diagram is also cocartesian in an appropriate sense.

Defining the correct notion of pushout in order to make that work requires some kind of descent. Typically, this is where the notion of *Grothendieck site* comes in: we fix a convenient category with a Grothendieck topology, and our objects are considered as sheaves on the site. Then the gluing property says that X should be the coequalizer of our two maps $R \rightrightarrows U$, in the category of sheaves on the site. The passage to sheaves insures that the gluing notion is invariant with respect to the choice of covering: a different covering should give rise to the same object – that would not be the case if we took just quotient presheaves.

This level of generality leads, for example, to Artin's notion of *algebraic space*, which is something obtained by gluing together affine schemes by etale equivalence relations.

Once we have written things in this way, the motivation for introducing a notion of *stack* is easy to see: what happens if we try to construct a coequalizer

$$R \rightrightarrows U \to X,$$

but where the map $R \to U \times U$ is not injective? Geometrically, given two points r and s in R that map to the same pair (u, u'), it means that we would like to glue u to u' in two different ways. Now, one might like to keep track of which way was used to glue. Then, the composition of these two gluings will look like an *automorphism* of our point u.

The interpretation of noninjectivity of our relation in terms of groups is also seen when we consider the transitivity condition for an equivalence relation R. In the case $R \hookrightarrow U \times U$, transitivity may be viewed as a composition operation

$$R \times_U R \to R$$

relative to $U \times U$, using the first map of the first factor R and the second map of the second factor; the remaining two maps are absorbed in the fiber product expression.

To generalize to the case of a not-necessarily-injective map $R \to U \times U$, it is most natural to keep the composition map as a part of the data, then to impose associativity (which was automatic in the injective case). Existence of inverses corresponds to the symmetry property of a relation, and existence of identities corresponds to reflexivity. Altogether, U, R, and these maps now have a structure of *groupoid*. That is to say, they form an internal category with the invertibility property of a groupoid, in whatever category of spaces we are looking at.

Sometimes a stack X is given to us most naturally as a functor from the base category[1] to the category of groupoids, or more generally as a fibered category.[2] This is what happens for most moduli problems. In this case, we then choose in some way charts covering the functor. The union of charts gives us U, and we may put $R := U \times_X U$. If R is representable, and if the map $U \to \mathcal{F}$ has the right smoothness properties, then we get a groupoid $R \rightrightarrows U$,

---

[1] For us in algebraic geometry, the objects of the base category are usually some kind of schemes, and the category is provided with a Grothendieck topology making it into a site.
[2] See Section 5 for further discussion of the definitional aspects.

and the stack X is the "quotient" of this groupoid in the same way as discussed above.

What we have just described is the concrete way of looking at a stack: it is viewed as a quotient of a groupoid. The more abstract approach will be described in Section 5.3.

3 Stacks in Algebraic Geometry

Consider a very simple example of a stack in algebraic geometry. Suppose we are given the action of a finite group, say, $G = \mathbb{Z}/2$, on a smooth curve X. Suppose $p \in X$ is an isolated fixed point. The quotient curve $Y := X/G$ will also be smooth, but it has a point $y \in Y$ the image of p, which is special in some way because the map $X \to Y$ is ramified over y. The inverse image of y consists only of the point p, whereas the inverse image of a nearby point y' would contain two points x_1 and x_2 exchanged by the group action. We would then like to include the group-theoretical information of the stabilizer group $\mathbb{Z}/2$ of the inverse image point p, located at the point $y \in Y$. This information is included in the *quotient stack* $\mathcal{Y} := X /\!/ G$. We view the stabilizer group as corresponding to the homotopy type of a space $K(\mathbb{Z}/2, 1)$ located over $y \in Y$.

The groupoid presentation of this quotient stack is easy to describe; it is an example of the general notion of *action groupoid*. Let $R := X \times G$ with two maps $(x, g) \mapsto x$ and $(x, g) \mapsto gx$ to X. The groupoid structure map $R \times_X R \to R$ is given by $((x, g), (x', g'))_{gx=x'} \mapsto (x, g'g)$, and the resulting groupoid represents our quotient stack $\mathcal{Y}$.

This kind of example is known as an orbifold in differential or analytic geometry and as a *Deligne–Mumford stack* in algebraic geometry.

3.1 The Moduli Stack of Curves of Genus g

Deligne and Mumford introduced the theory in their paper [13], motivated by what is undoubtedly the first main example: their moduli stack $\mathcal{M}_g$ of smooth projective curves of genus $g \geq 2$, and its completion $\overline{\mathcal{M}}_g$ parameterizing stable nodal curves. To put this example in its natural setting, Deligne and Mumford looked at stacks in the etale topology with coordinate charts such that the transformations between different charts are etale maps.

Over the complex numbers, a smooth projective curve is the same thing as a compact Riemann surface. These quite naturally come in "families": a family of curves is a flat map $X \to S$ whose fibers are curves. The *moduli problem* consists in finding a scheme M_g such that families of curves $X \to S$

correspond to morphisms $S \to M_g$. The optimal way of solving such a moduli problem would be to have a *universal family* $X_g \to M_g$ such that for any X/S corresponding to a parameterization map $S \to M_g$, we have $X = S \times_{M_g} X_g$.

While there does exist a moduli space M_g in the category of schemes, it does not admit a universal family. This is mainly due to the existence of curves with automorphisms, and that is one of the main motivations to introduce the *moduli stack* $\mathcal{M}_g$. Working in the world of stacks, there is a universal family $\mathcal{X}_g \to \mathcal{M}_g$ solving the moduli problem in the best possible way.

To get back to a variety or scheme, we use the following definition:

Definition 3.1 Suppose X is a stack. We say that a morphism

$$X \to X^{\text{coarse}}$$

from X to a scheme X^{coarse} is *the coarse moduli space* if, for any other map to a scheme $X \to V$, there is a unique factorization $X \to X^{\text{coarse}} \to V$.

Deligne and Mumford construct both the moduli stack of curves and its coarse moduli space. While the coarse moduli space $M_g = (\mathcal{M}_g)^{\text{coarse}}$ is a scheme, it does not admit a universal family.

Another indication of the advantage of using $\mathcal{M}_g$ is the question of smoothness. The moduli stack and even its natural compactification $\overline{\mathcal{M}}_g$ are *smooth* Deligne–Mumford stacks of dimension $3g - 3$. However, at points corresponding to curves with automorphisms, the moduli scheme M_g is quite singular in general: it has finite quotient singularities, the local charts being indeed the quotients of the smooth charts in $\mathcal{M}_g$ by the finite automorphism groups.

The complex analytic stack associated to $\mathcal{M}_g$ is a smooth orbifold (except for $g = 2$, where it also involves a gerbe; see below). The local charts are open balls in $\mathbb{C}^{3g-3}$, as we shall understand further in the upcoming example. In the analytic or differential geometric setting, this moduli stack has a global quotient structure

$$(\mathcal{M}_g)^{\text{an}} = \mathcal{T}_g / \Gamma,$$

where $\mathcal{T}_g$ is the famous *Teichmüller space* [32] and Γ is the mapping class group of oriented automorphisms of the topological surface up to isotopy.

3.2 Natural Charts at Fully Degenerate Curves

The principle of natural local charts that are smooth is perfectly illustrated by looking at the degenerate boundary points in the compactification $\overline{\mathcal{M}}_g$. These boundary points are the stable curves which are unions of smooth curves meeting at nodes. Typical examples would be the "stick figures" in $\mathbb{P}^3$

composed of rational lines, if we assume that no more than two lines intersect at any given point.

Deligne and Mumford said that a curve is *stable* if any rational component has at least three nodes. A curve that is not rational can be deformed, and given a rational component with four or more nodes, it can also be deformed or "broken" into a union of rational curves with less nodes. Thus, the most degenerate points in the boundary, corresponding to zero-dimensional strata, are the ones where each component curve is rational, that is, $\mathbb{P}^1$, with exactly three nodes.

Recall that the universal deformation space of a nodal curve singularity is one-dimensional. The universal family may be written down very easily as just $xy = t$, with t being the deformation parameter. At $t = 0$, we have a nodal curve $xy = 0$, whereas for $t \neq 0$, the curve is smooth. We could cut out this deformation along a sphere centered at the origin and then glue it into a global picture.

Suppose $Y = \bigcup_{i=1}^{k} Y_i$ is a union of smooth rational curves meeting at nodes $z_1, \ldots, z_r$. Let Y^- be the curve obtained by cutting out small disks around each of the nodes, and glue the universal deformation spaces considered in the previous paragraph at each of the nodes. We get a family of curves parameterized by a smooth parameter space with r parameters $t_1, \ldots, t_r$, say, for example, parameterized by Δ^r, where Δ is the unit disk.

The genus of Y is given by

$$g(Y) = 1 + \text{\# nodes} - \text{\# components} = 1 + r - k.$$

In the fully degenerate case where each component has three nodes, a node is contained in two components, so $2r = 3k$. Thus $g(Y) = 1 + r - \frac{2}{3}r = 1 + \frac{1}{3}r$; in other words,

$$r = 3g(Y) - 3.$$

This tells us that our deformation space obtained by combining the universal deformations of each of the singularities has the same dimension as the moduli space M_g of genus g curves, and indeed one may calculate that the tangent map from $\mathbb{C}^r$ to the deformation space $\text{Def}(Y)$ is an isomorphism. Therefore, we would like to think of our family as providing a chart for the moduli space.

It is a particularly nice chart. The boundary divisor of singular curves appears transparently as a divisor with normal crossings, being given by the equation $t_1 t_2 \cdots t_{r-1} t_r = 0$.

In general, our nodal curve Y will have automorphisms. One can form the *dual graph* $\mathcal{D}$ with k vertices one for each component Y_i, and r edges joining two components when they intersect in a node. In our fully degenerate case,

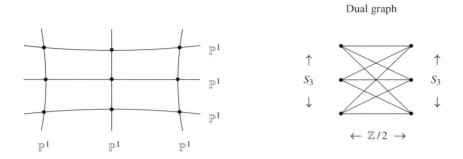

Figure 8.1 A curve Y of genus $g = 4$ with $9 = 3g - 3$ nodes and automorphism group $G = \mathrm{Aut}(Y) = (S_3 \times S_3) \rtimes \mathbb{Z}/2$.

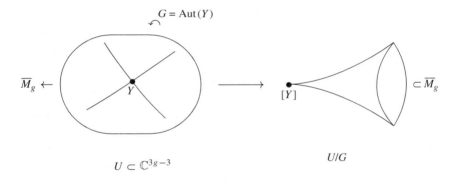

Figure 8.2 The chart over $\overline{\mathcal{M}}_g$ and its image in $\overline{M}_g$.

$\mathrm{Aut}(Y) = \mathrm{Aut}(\mathcal{D})$, and it is easy to write down graphs with automorphisms (Figure 8.1).

We now have a completely concrete situation showing the benefits of the notion of stack. In the moduli stack $\overline{\mathcal{M}}_g$, the family defined above is a coordinate chart, showing how $\overline{\mathcal{M}}_g$ has a structure of "smooth space" at the point Y. The automorphisms of Y induce automorphisms of the coordinate chart since they permute the nodes, but these maps are etale.

The only kind of variety we can get, the coarse moduli space $\overline{M}_g$, will have a quotient singularity at the point $[Y]$ (isomorphism class of Y), given by taking the quotient of Δ^r by the automorphism group of Y (see Figure 8.2).

It is clear that we would rather like to look at the smooth neighborhoods of Y with highly natural coordinate systems discussed above. On the smooth charts, the coordinate functions, defining the boundary divisor components, correspond to the nodes. The possibility of looking at such charts is afforded by the notion of the Deligne–Mumford stack.

This example was for us primordial. Explained by the first author to the second, it showed in a concrete way what the notion of stack was all about.

3.3 Orbifolds and DM-Stacks

Such considerations date back quite a long way. The classical Kuranishi theory provided versal deformation spaces, and the problem of automorphisms led quite early on to the definition of *V-manifold* by Satake. A *V*-manifold is essentially the same thing as a smooth Deligne–Mumford stack, but the generic stabilizer group is assumed trivial. In topology, the corresponding notion was called an *orbifold*.

The definition of Deligne and Mumford provides us with a few useful generalizations going beyond the notion of orbifold. For example, the stack does not need to be smooth or even reduced. This reflects the fact that we might very well, for some reason, have wanted to consider a compatible collection of singular subschemes of the coordinate charts.

Singular DM-stacks can fit as points in moduli spaces, giving new kinds of compactifications. Abramovich and Vistoli [1] introduced the notion of *twisted curve*, a nodal DM-curve with orbifold points at the normal crossings where the cyclic stabilizer groups of the orbifold points on the two branches are identified. Several authors [2, 12, 30] have constructed moduli stacks of twisted curves giving interesting compactifications of $\mathcal{M}_g$.

The other direction of generalization is quite interesting and leads to a more intricate topological structure. The local automorphism group in the stack, coming from the stabilizer group of a group action or the automorphism group of the objects whose moduli we are looking at, is not necessarily required to be trivial at the generic point. In other words, we can have "stacky" behavior over the whole space rather than just concentrated at some points lower down in a stratification. This is reflected in the notion of the *gerbe*.

4 Gerbes

The generalization to gerbes is quite natural and would in due course have necessarily been included in any development of the original notion of *V*-manifold. Indeed, consider some subvariety of a *V*-manifold over which the automorphism groups are nontrivial. Then this subspace has an induced stack structure, but the generic stabilizer is nontrivial.

When there is a nontrivial generic stabilizer group, but that group stays "the same" over all points of the stack, then we have a *gerbe*. A gerbe over a point is just a stack of the form BG, where G is a group. If G is a finite group (or a group scheme etale over the base field in positive characteristic), then BG is a Deligne–Mumford stack. More general cases include, for example, G is an algebraic group scheme, in which case, BG is an Artin stack.

Let us look at the easiest case when G is a finite group. Suppose X is a variety. Then a gerbe over X with group G is a stack $\mathcal{Y}$ together with a map $p : \mathcal{Y} \to X$, such that X has an etale covering $\{a_i : U_i \to X\}$ and $U_i \times_X \mathcal{Y} \cong U_i \times BG$. There will not exist in general a global trivialization, nor even a trivialization over a Zariski open covering. And there need not even exist local sections. The G-gerbes over X are classified by the 2-*stack* $B\operatorname{Aut}(BG)$. This is a connected 2-stack whose π_1 is $\operatorname{Out}(G) = \operatorname{Aut}(G)/G$ the group of outer automorphisms and whose π_2 is the center $Z(G)$. Thus, a G-gerbe over X corresponds first to a cohomology class $\alpha \in H^1(X, \operatorname{Out}(G))$, which is to say an $\operatorname{Out}(G)$-torsor over X, and second, if this is trivial, a class in $H^2(X, Z(G))$.

4.1 Stratification by Automorphism Group in $\overline{\mathcal{M}}_g$

Such structures may be found within the moduli stack $\overline{\mathcal{M}}_g$ of stable curves of genus g. We feel that this phenomenon is an interesting question for study. It furthermore illustrates how gerbes occur "in nature."

A first example is, of course, the whole moduli stack $\overline{\mathcal{M}}_2$ of curves of genus 2: any genus 2 curve is hyperelliptic, so it has an involution, and the generic stabilizer group on $\overline{\mathcal{M}}_2$ is $\mathbb{Z}/2\mathbb{Z}$.

More generally, there is a decomposition into locally closed subsets

$$\overline{\mathcal{M}}_g = \coprod_G \mathcal{S}_G,$$

where $\mathcal{S}_G$ is the locus of curves whose automorphism group is isomorphic to G. Now $\mathcal{S}_G$ is itself a stack, and it is a G-gerbe over its coarse moduli space S_G, where we have the corresponding decomposition $\overline{M}_G = \coprod_G S_G$ of the coarse moduli space.

It is an interesting question to try to understand the structure of these gerbes. A very combinatorial version is obtained by restricting to the dimension 1 pieces of the boundary stratification, corresponding to nodal unions of rational curves such that one component has four nodes and the rest have only three nodes.

These questions have been studied by many authors. Some references include, nonexhaustively, [14–16, 27, 44, 45].

4.2 Classification of Gerbes

The theory of stacks is a bridge from geometry and algebra to topology. A basic building block is the gerbe BG over a point. It is the stack classifying G-torsors, but we may also think of it as the unique (up to weak homotopy equivalence) connected pointed topological space with $\pi_1(BG, o) = G$. The correspondence between groupoids and 1-truncated homotopy types is the link between algebra and topology in this simplest case.

Theorem 4.1 (Giraud [17]) *Suppose G is a sheaf of groups over a site. There is a 2-stack $B\operatorname{Aut}(BG)$ over the site and a universal G-gerbe over it, such that for any object (or indeed stack) X, the 2-groupoid of G-gerbes over X is naturally equivalent to the 2-groupoid of maps $X \to B\operatorname{Aut}(BG)$. If G is a constant group, then this 2-stack has a fibration sequence*

$$K(Z(G), 2) \to B\operatorname{Aut}(BG) \to B\operatorname{Out}(G).$$

The obstruction classes in cohomology of X discussed previously are direct consequences.

Giraud gave the proof [17] in purely categorical terms. One can phrase the statement and give its proof in terms of cocycles and 2-cocycles. Breen discusses the generalization to 2-gerbes in this light [10].

To give a rather more simple view of the idea, we discuss the topological version of the statement, classifying fibrations with fiber BG, in Section 4.4. We consider the case of a constant discrete group G. The case of a sheaf of groups is more general, and indeed one should note that Giraud's theory treats the most general situation of gerbes under a *lien*, which basically means a sheaf of "groups up to inner automorphism" [17, 43]. This results in the complicated algebraic structures discussed in [17], which will not be covered by our topological discussion.

4.3 Structure Theorem for DM-Curves

The classification of gerbes works equally well if the base X itself is a stack. This is useful since we have the following structure theorem for one-dimensional Deligne–Mumford stacks:

Theorem 4.2 *Suppose $\mathcal{X}$ is a smooth one-dimensional Deligne–Mumford stack. Then there is a smooth one-dimensional orbifold X^{orb}, that is to say, a DM-stack with trivial generic stabilizer or equivalently a V-manifold, and* a map $p : \mathcal{X} \to X^{\text{orb}}$ such that p is a G-gerbe for G the generic stabilizer group of $\mathcal{X}$. Furthermore, X^{orb} is a root stack

$$X^{\text{orb}} = X[\frac{1}{n_1}D_1, \ldots, \frac{1}{n_k}D_k]$$

over the coarse moduli space X a smooth curve, for points $D_1, \ldots, D_k \in X$ with integer multiplicities $n_i \geq 1$.

The local structures are determined by the integer multiplicities, so the new topological information is contained in the gerbe p.

4.4 Classification of Fibrations

To gain some insight into what is going on in Theorem 4.1, let us consider a fixed discrete group G and look at how to classify fibrations $Y \to X$ such that the fiber is isomorphic to the space BG. Our discussion takes place in the world of topology.

Theorem 4.3 *Suppose F is a space. There is a space $B\operatorname{Aut}(F)$ classifying fibrations with fiber F. It has a universal fibration, and given a fibration Y/X with fiber F, the space of homotopy classes of pairs (f, ζ) where $f : X \to B\operatorname{Aut}(F)$ and ζ is a homotopy equivalence between Y/X and the pullback of the universal fibration is contractible. In the case when G is a discrete group and $F = BG$ then $B\operatorname{Aut}(BG)$, is a connected, 2-truncated space with*

$$\pi_1(B\operatorname{Aut}(BG)) = \operatorname{Out}(G) \text{ and } \pi_2(B\operatorname{Aut}(BG)) = Z(G).$$

This theorem comes from Segal's theory of classifying spaces [37]. The "group" of self-homotopy equivalences $\operatorname{Aut}(F)$ may be viewed as a grouplike Segal space in the following way. Define a simplicial set T. with $T_0 = *$, and $T_1 \subset \operatorname{Map}(F, F)$ equal to the union of connected components corresponding to maps which are weak homotopy equivalences. Let $T_n := T_1 \times \cdots \times T_1$ with the various face and degeneracy maps defined by the monoid structure of T_1. This tautologically satisfies the Segal conditions. Since the maps in T_1 are invertible up to homotopy, this simplicial space also satisfies Segal's grouplike condition. We may therefore look at the realization

$$B\operatorname{Aut}(F) := |T.|,$$

and Segal shows that it is a space such that the tautological map

$$\operatorname{Aut}(F) = T_1 \to \Omega_*|T.|$$

is a homotopy equivalence. One constructs the universal family and shows that $B \operatorname{Aut}(F)$ is the classifying space.[3]

In the case $F = BG$,

$$\pi_1(B\operatorname{Aut}(BG)) = \pi_0(\operatorname{Aut}(BG)) = \operatorname{Out}(G)$$

$$\pi_2(B\operatorname{Aut}(BG)) = \pi_1(\operatorname{Aut}(BG)) = Z(G).$$

The fibrations Y/X with fibers BG are essentially the same thing as G-gerbes. The 2-stack in Theorem 4.1, for the case of a constant group G, is just the constant 2-stack with value equal to the Poincaré 2-stack of the space $B\operatorname{Aut}(F)$ of the previous theorem.

In this topological framework we can easily understand how to get obstruction classes for gerbes. Suppose given a fibration Y/X with fiber BG. The classifying map $X \to B\operatorname{Aut}(BG)$ composes with the map from $B\operatorname{Aut}(BG)$ to $B\operatorname{Out}(G)$, to give a class in $H^1(X, \operatorname{Out}(G))$ or equivalently an $\operatorname{Out}(G)$-torsor over X. This torsor measures the way in which the fundamental groups of the fibers of Y/X change as we move around loops in the base. Since Y/X does not necessarily have a section, the fundamental groups of the fibers are not well defined, but one can choose sections locally and a change of section corresponds to an inner automorphism of G. The monodromy of the system of fundamental groups is therefore an element of the group of outer automorphisms of G.

Suppose this torsor is trivial. It means that the map $X \to B\operatorname{Aut}(BG)$ may be viewed as going into the fiber of the projection to $B\operatorname{Out}(G)$ (more precisely, a choice of trivialization of the torsor corresponds to a choice of homotopy to a map into the fiber). That fiber is the Eilenberg–Mac Lane space $B^2 S(G) = K(Z(G), 2)$ of degree 2. Thus, we get a cohomology class in $H^2(X, Z(G))$. This class is the obstruction to existence of a section. If this class is trivial too, then there exists a section and our fibration is trivial, $Y = X \times BG$.

More generally, given an $\operatorname{Out}(G)$ torsor α, we get a local system $Z(G)_\alpha$, and a gerbe with that $\operatorname{Out}(G)$ torsor gives a classifying element of $H^2(X, Z(G)_\alpha)$.

This very brief tour of obstruction theory and classification of fibrations is designed to correspond to Giraud's classification of gerbes within the theory of stacks [17]. Some kind of work needs to be done in order to obtain the

---

[3] Suppose we have a fibration with fiber F over the realization of a simplicial set $X = |X.|$. For $x \in X_0$, choose a weak equivalence $Y_x \cong BG$. For any $u_1 \in X_1$, we have a fibration over Δ^1, and choice of a trivialization compared with the previous choices at the end points gives a self-homotopy equivalence of BG, hence a point in T_1. For $u_2 \in X_2$, its faces are $u_2(01), u_2(12), u_2(02) \in T_1$. The 2-cell u_2 may be seen as a homotopy in T_1 between $u_2(12) \circ u_2(01)$ and $u_2(02)$. Continuing in this way, we can build a homotopy coherent map from X. to the simplicial space T., yielding the classifying map $X \to |T.|$ for the fibration.

analogous classification theory for stacks over a site, for example to treat the case when G is sheaf of groups or even a lien.

A modern treatment of the proof of Theorem 4.1 might appeal to the theory of simplicial presheaves in order to transpose somewhat more directly the topological classification theory presented here to the relative situation. We are not sure if a full treatment in this spirit has yet been proposed.

5 The Abstract Notion of Higher Stack

The discussion of the previous section concerned the classification of fibrations over a base topological space. In algebraic geometry, gerbes and more generally stacks exist over base algebraic varieties. For example, we have considered the natural gerbes over strata $S_G \subset \overline{M}_g$ in the moduli space of stable curves. In such a situation, the notion of fibration of topological spaces is replaced by the notion of stack itself, and stacks are viewed as families of spaces over algebraic varieties.

Such a dialogue between algebraic geometry and algebraic topology has proven fundamental to recent developments including the gradual move toward the idea of ∞-stacks:

Sets	*Groupoids*	*n-Groupoids*	*∞-Groupoids*
Discrete spaces	$K(\pi,1)$'s	n-truncated homotopy types	all homotopy types
Presheaves	prestacks	n-prestacks	∞-prestacks
Sheaves	stacks	n-stacks	∞-stacks

5.1 Gluing and Descent Data

The goal of this section is to discuss the more abstract definitional aspects of the notion of stack, those having been postponed in the earlier, more geometric sections.

We start by explaining with some pictures the basic idea for going from the presheaf/prestack line to the sheaf/stack line in the above table. Recall that the property characterizing *sheaves* among all presheaves is that they are required to satisfy a gluing condition. In this section, we will assume that we are talking about objects (sheaves, stacks, etc.) over a base topological space X.

A *presheaf* $\mathcal{F}$ over X is just a functor from the opposite of the category of opens of X to sets. Thus, it consists of a collection of sets $\mathcal{F}(U)$ for any open set $U \subset X$, together with restriction functors $a \mapsto a|_{U'}$ whenever $U' \subset U$, satisfying the natural transitivity condition.

Suppose we are given two intersecting open subsets:

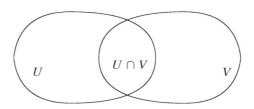

For a presheaf $\mathcal{F}$, the set of locally defined sections, or *descent data* with respect to the covering of $U \cup V$ by these two open sets, is the set of pairs (a, b), where $a \in \mathcal{F}(U)$ and $b \in \mathcal{F}(V)$ such that $a|_{U \cap V} = b|_{U \cap V}$ in $\mathcal{F}(U \cap V)$.

A presheaf $\mathcal{F}$ is a *sheaf* if $\mathcal{F}(U \cup V)$ is always isomorphic to the set of these locally defined sections. This is viewed as a *gluing condition*: given sections a and b over U and V, respectively, such that they "agree" on $U \cap V$, then they should "glue," that is, come from a (unique) section over $U \cup V$. In other words, the diagram

$$\begin{array}{ccc} \mathcal{F}(U \cup V) & \longrightarrow & \mathcal{F}(U) \\ \downarrow & & \downarrow \\ \mathcal{F}(V) & \longrightarrow & \mathcal{F}(U \cap V) \end{array} \quad (5.1)$$

is cartesian, the upper left corner being the limit of the lower right angle diagram.

Going rightward in the above table, we would like to generalize from presheaves that are families of sets $\mathcal{F}(U)$ indexed by opens $U \subset X$ to 1-, n-, or ∞-prestacks that are families of spaces. For now let us just think of one of these as being a functor from the opposite category of opens of X to spaces. The case of 1-stacks is, as indicated in the second column of the table, the case where these spaces are $K(\pi, 1)$s.

In case of two open sets intersecting as pictured above, we would like to define a *space* of locally defined sections in a way that is adapted to homotopy theory. So, instead of requiring an equality of points $a|_{U \cap V} = b|_{U \cap V}$ in the space $\mathcal{F}(U \cap V)$, it is better to ask for a path. Since the choice of path is not unique, it should be considered as part of the data. Thus, we arrive at the space of locally defined sections as being the space of triples (a, b, p) where $a \in \mathcal{F}(U)$ and $b \in \mathcal{F}(V)$ are points and $p : [0, 1] \to \mathcal{F}(U \cap V)$ is a path joining $a|_{U \cap V} = p(0)$ to $b|_{U \cap V} = p(1)$.

The condition for being a *stack* (relative to two-subset coverings, at least) is that the map from $\mathcal{F}(U \cup V)$ to the above space of locally defined sections should be a weak homotopy equivalence. This says that we can glue together sections a over U and b over V if a and b agree *up to homotopy* over $U \cap V$ and the gluing is uniquely defined up to homotopy once the path p has been specified. In other words, the diagram (5.1) should be *homotopy cartesian*, the upper left being the homotopy limit of the lower right angle.

In the case of finite coverings of a space, they can always be considered as gotten from a sequence of two-subset coverings as in the picture, and the above definition of the descent condition is sufficient.

Let us look, nonetheless, at what the natural notion of locally defined section should be for a more complicated covering, say, just one with three open subsets:

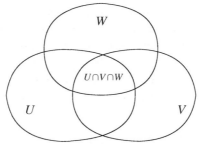

Our descent data will start with

$$a \in \mathcal{F}(U), \quad b \in \mathcal{F}(V), \quad c \in \mathcal{F}(W);$$

then it should also have

- a path p between $a|_{U \cap V}$ and $b|_{U \cap V}$ in $\mathcal{F}(U \cap V)$;
- a path q between $b|_{V \cap W}$ and $c|_{V \cap W}$ in $\mathcal{F}(V \cap W)$;
- a path r between $a|_{U \cap W}$ and $c|_{U \cap W}$ in $\mathcal{F}(U \cap W)$.

So far, if we project all of those things into the space $\mathcal{F}(U \cap V \cap W)$ by the appropriate restriction maps, denoting their images by $[a] := a|_{U \cap V \cap W}$, etc., we get the following picture:

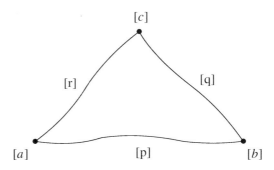

This suggests what the last piece of the descent data should be: a triangle mapping into $\mathcal{F}(U \cap V \cap W)$ whose boundary is $[p] \cup [q] \cup [r]$.

For 1-prestacks where the spaces are $K(\pi, 1)$s, existence of the triangle is a condition often called the *cocycle condition*. For the case of higher prestacks, this homotopy is itself part of the data.

More abstractly, the space of descent data as above may be expressed as a *homotopy limit (holim)*, namely, the homotopy limit of the diagram formed by the spaces $\mathcal{F}(U)$, $\mathcal{F}(V)$, $\mathcal{F}(W)$, $\mathcal{F}(U \cap V)$, $\mathcal{F}(V \cap W)$, $\mathcal{F}(U \cap W)$, and $\mathcal{F}(U \cap V \cap W)$ with the restriction maps between them. With more open sets, the reader may imagine the collection of higher-dimensional homotopy data called *higher coherencies* entering into the homotopy limit space of descent data.

The *descent condition*[4] for a prestack to be a stack is that for any covering, the map from the space assigned to the union of open sets to the homotopy limit of the corresponding diagram should be a weak homotopy equivalence.

5.2 Topology Relative to a Site

If the pictures in the previous section were drawn with a topological space as base, it is most often necessary to consider stacks over a more general kind of abstract structure designed precisely to allow the kinds of arguments that proceed from the intuition of gluing.

So, the first ingredient in the whole theory is the notion of *site*, a category $\mathcal{C}$ provided with a *Grothendieck topology* τ. It means that for any object $X \in \mathcal{C}$, we know when a family $U_i \to X$ is said to "cover" X. This information is supposed to obey appropriate axioms. When $\mathcal{C}$ admits disjoint unions, it suffices to consider one-object coverings $U \to X$ obtained from the

---

[4] The reader is encouraged to consult [40] for more technical and philosophical discussion about descent.

previous by setting $U := \coprod U_i$, and we shall generally assume this notational simplification. The topology allows us to say that something is happening "locally on X," meaning that it happens on some covering, in a coherent way.

A topological space relative to a site $(\mathcal{C}, \tau)$ should be viewed as a "presheaf of spaces" over the category $\mathcal{C}$. Such presheaves could come in many flavors. For instance, in many treatments of stack theory (the reader is referred to [41] in particular), the fundamental objects are fibered categories, whose fibers are groupoids. A fibered category is a kind of weak presheaf of groupoids: the assignment of a functor between fibers to every map in the base is not necessarily strictly compatible with composition but only compatible up to a natural equivalence satisfying further axioms. Nonetheless, a standard strictification process allows one to replace such weak presheaves by strict presheaves, so another valid viewpoint is to start from the beginning with strict presheaves of spaces over the category underlying our Grothendieck site.

The Grothendieck topology is taken into account via the condition of *descent*. One generalizes over a site the discussion we had in the previous section, where the base was the category of open subsets of a topological space.

For a presheaf of discrete or 0-truncated spaces, corresponding to a presheaf of sets $\mathcal{F}$, the descent condition is the condition of being a *sheaf*: for a covering $U \to X$, the diagram

$$\mathcal{F}(X) \to \mathcal{F}(U) \rightrightarrows \mathcal{F}(U \times_X U)$$

should be an equalizer diagram of sets. In other words, an element of $\mathcal{F}(X)$ is the same thing as $a \in \mathcal{F}(U)$ such that $p_1^*(a) = p_2^*(a)$ in $\mathcal{F}(U \times_X U)$. This latter condition is what we meant when we said "in a coherent way" three paragraphs ago.

For a 1-truncated presheaf of spaces, corresponding to a presheaf of groupoids $\mathcal{F}$, the descent condition is that of a *stack*. It means that the diagram

$$\mathcal{F}(X) \to \mathcal{F}(U) \rightrightarrows \mathcal{F}(U \times_X U) \stackrel{3}{\to} \mathcal{F}(U \times_X U \times_X U)$$

should express the groupoid $\mathcal{F}(X)$ as being equivalent to the "homotopy limit," or *groupoid of descent data* of the sequence on the right. An object of the homotopy limit consists of an object $a \in \mathcal{F}(U)$ and an isomorphism $\eta : p_1^*(a) \cong p_2^*(a)$ in $\mathcal{F}(U \times_X U)$ such that the natural diagram we can write in $\mathcal{F}(U \times_X U \times_X U)$ commutes (corresponding to the triangle in Section 5.1). Notice that now the "coherence" becomes a datum, that of the isomorphism η, which is itself required to satisfy a higher coherence condition.

The descent condition says that any such "descent datum" should be equivalent to an object coming from $\mathcal{F}(X)$. The morphisms in the groupoid of descent data are defined using the sheaf condition of the previous paragraph, and we also require that this set of morphisms be isomorphic to the set of morphisms in $\mathcal{F}(X)$.

The discussion of Section 4.4 brings out an important phenomenon: if 1-stacks came up when we looked for classifying spaces of regular objects having automorphisms, similarly when we try to classify 1-stacks, we end up talking about a 2-stack such as $B\operatorname{Aut}(BG)$. In general, the classifier for n-stacks is an $(n+1)$-stack. In this sense, we cannot avoid going upward in the ladder of degrees, and it becomes natural to consider the notion of n-stack for any n. Of course, in many practical situations, it suffices to stop after going up a level or two: for instance, when speaking of 1-stacks, it is occasionally useful to talk about 2-stacks, but we usually do not need to look at 3-stacks in the same context.

Another force pushing upward is the following question: what happens with a groupoid

$$\mathcal{R} \rightrightarrows U$$

where U is a scheme, but the morphism object $\mathcal{R}$ is itself a 1-stack? In this case the quotient is a 2-stack, as we shall see in the explicit example of Section 6.6 below. Continuing leads to n-stacks for any n, in a way closely related to the Artin condition, to be discussed in the next section.

The descent conditions generalize directly to the case of n-truncated[5] presheaves, saying that the space of sections over an object is equal to the homotopy limit of the space of sections over the nerve of a covering. An n-stack in n-groupoids is an n-truncated presheaf of spaces satisfying this descent condition. When we go to $n = \infty$ with arbitrary, nontruncated spaces, it turns out that there are two natural flavors of descent: hyperdescent and finite descent. This distinction was pointed out by Lurie, among others, and he closely investigated the difference [26]. The case of hyperdescent corresponds to the locality condition in the Jardine closed model structure on simplicial presheaves [21].

The idea of doing topology relative to a site connects naturally with the notion of *non-abelian cohomology*. Giraud's classification of gerbes was one

---

[5] Recall that a space Y is said to be n-truncated if $\pi_i(Y)$ is trivial for $i > n$. Allowing $n = \infty$, an ∞-groupoid is a space with no truncation condition. The theory of ∞-categories is not yet well developed, but for (∞, n)-categories, those ∞-categories in which morphisms are invertible in degrees $> n$, the various theories and their equivalence are fairly well understood. We leave it to the reader to investigate the numerous references on this aspect.

of the first instances. From the topological version of this discussion described above, we see that non-abelian cohomology really means looking at maps into some space. In algebraic geometry, this is happening relative to a site: spaces are replaced by presheaves of spaces required to satisfy a "descent condition." We will get back to the idea of non-abelian cohomology in this setting later.

5.3 Definitions of Stacks

We present, for comparison, some of the possible definitions following the discussion of the previous section. Suppose $\mathcal{C}$ is a category provided with a Grothendieck topology τ. We restrict here to the case of stacks of groupoids. There are now many different definitions of n-category, which have been shown to be equivalent by the axiomatic approach of Barwick and Schommer-Pries [5], but for groupoids, this can be understood easily in that an n-groupoid may be viewed as the same thing, up to homotopy, as an n-truncated space.

Definition 5.1 A 1-prestack over $\mathcal{C}$ is a functor of categories $\mathcal{F} \to \mathcal{C}$ with the property of being a fibered category such that the fibers $\mathcal{F}_x$ are 1-groupoids.

An n-prestack over $\mathcal{C}$ is a functor of n-categories $\mathcal{F} \to \mathcal{C}$ with the property of being a fibered category such that the fibers $\mathcal{F}_x$ are n-groupoids.

Definition 5.2 A 1-prestack over $\mathcal{C}$ is a functor $F : \mathcal{C}^{\mathrm{op}} \to 1\mathrm{Gpd}$, that is, a presheaf of groupoids.

An n-prestack over $\mathcal{C}$ is a functor $F : \mathcal{C}^{\mathrm{op}} \to n\mathrm{Gpd}$, that is, a presheaf of n-groupoids.

Definition 5.3 A 1-prestack over $\mathcal{C}$ is a presheaf of spaces $F : \mathcal{C}^{\mathrm{op}} \to \mathrm{Top}$ such that $F(x)$ is a 1-truncated space.

An n-prestack over $\mathcal{C}$ is a presheaf of spaces $F : \mathcal{C}^{\mathrm{op}} \to \mathrm{Top}$ such that $F(x)$ is an n-truncated space.

Here, Top may profitably be replaced by the category of simplicial sets, so 1- or n-prestacks may be viewed as simplicial presheaves over $\mathcal{C}$.

Given a prestack $\mathcal{F} \to \mathcal{C}$ in the sense of Definition 5.1, let $F(x)$ be the 1-groupoid of sections of $\mathcal{C}/x \to \mathcal{F}$. This has strict pullbacks corresponding to restriction of sections, so it is a presheaf of groupoids in the sense of Definition 5.2. In the other direction, given a presheaf of groupoids F we can let $\mathcal{F} := \int_{\mathcal{C}} F$ be the *Grothendieck integral*, also known as the category of elements of F. The objects of $\mathcal{F}$ are the pairs (x, u) where $x \in \mathrm{Ob}(\mathcal{C})$ and

$u \in F(x)$, and morphisms are defined in a natural way. These constructions are inverse up to equivalence. The same discussion relates Definitions 5.1 and 5.2 for n-prestacks.

To go between Definitions 5.2 and 5.3, recall that to a groupoid G, we can associate its classifying space $|G|$, which is a 1-truncated space, described, for example, as the realization of the simplicial set nerve of G. Given a 1-truncated space, its Poincaré groupoid is the corresponding groupoid. Again, these constructions extend to the equivalence between n-groupoids and n-truncated spaces, if one is not actually taking n-truncated spaces as the definition of n-groupoids. Being functorial, these constructions extend to the case of presheaves.

We may now use interchangeably any of the above definitions for prestacks. Given a prestack $\mathcal{F}/\mathcal{C}$, we obtain its higher groupoid of sections $\Gamma(\mathcal{C}, \mathcal{F})$, which should be calculated in a suitable way to give an answer invariant under homotopy equivalences relative to $\mathcal{C}$. The best way to do that is to use Quillen model category structures, although for sections of a 1-prestack, there is an algebraic definition predating model category theory. Taking the point of view that ∞-groupoids correspond to homotopy types, the higher groupoid of sections is the homotopy limit:

$$\Gamma(\mathcal{C}, \mathcal{F}) = \text{holim}_{\mathcal{C}} \mathcal{F}.$$

A prestack is called a *stack* if it satisfies the *descent condition*. There are many different viewpoints on this notion, and it is not our purpose to give an extensive discussion here. The idea is to encode the property we first saw in Section 5.1. One way to say it is as follows:

Definition 5.4 Suppose $x \in \mathcal{C}$, and suppose given a covering of x for the topology τ. The covering determines a *sieve* that is a subcategory $\mathcal{B} \subset \mathcal{C}/x$. The descent condition says that the map

$$\mathcal{F}(x) \sim \Gamma(\mathcal{C}/x, \mathcal{F}|_{\mathcal{C}/x}) \to \Gamma(\mathcal{B}, \mathcal{F}|_{\mathcal{B}})$$

should be an equivalence (of 1-groupoids or n-groupoids or spaces) for any x and any covering sieve $\mathcal{B}$. A prestack satisfying this condition is called a *stack*.

The homotopy theory of n-stacks is obtained by using the homotopy theory of n-prestacks, that is, diagrams of spaces, restricted to the subcategory of n-stacks. Jardine first constructed this homotopy theory by constructing a closed model category of simplicial presheaves, where the weak equivalences are "local" weak equivalences. An n-stack[6] is any n-truncated simplicial presheaf that is levelwise equivalent to a fibrant object in Jardine's model structure.

5.4 Homotopy Group Sheaves

A local weak equivalence is a map that induces weak equivalences on the "stalks" (defined, over sites with enough points, using a homotopy colimit in the same way as for sheaves). It can also be measured using the *homotopy group sheaves*. If $\mathcal{F}$ is a prestack, then the presheaf $x \mapsto \pi_0(\mathcal{F}(x))$ has an associated sheaf denoted $\pi_0(\mathcal{F})$. Similarly, if $u \in \Gamma(\mathcal{C}, \mathcal{F})$ is a section, then $\pi_i(\mathcal{F}, u)$ is the sheafification of the presheaf $x \mapsto \pi_i(\mathcal{F}(x), u(x))$. These notions are often most useful when restricted to slice categories $\mathcal{C}/x$. A map is a local weak equivalence if and only if it is an *Illusie weak equivalence*: it induces isomorphisms of homotopy group sheaves. The sheafification process used to define the homotopy group sheaves is the key aspect of this condition. In the same way as for sheaves, the stack associated to a prestack is the canonical model of its local weak equivalence class, the universal element to which the other ones map.

6 Artin Stacks

Once we have the theoretical framework to talk about stacks – and those could now mean ordinary 1-stacks, n-stacks, or similar generalizations – we can return to the original geometrical motivation. We wanted to represent objects presented as "quotients" of some kind of chart by a generalized equivalence relation. The quotient is viewed as an object in our category of stacks.[7] The

---

[6] If $n = \infty$, then there is a distinction [26] between descent for coverings and descent for hypercoverings. We presented here descent for a covering. The descent condition for a hypercovering can be expressed in the same way, but the sieve is replaced by the category of elements of the hypercovering, which has a functor $\mathcal{B} \to \mathcal{C}/x$. Without a truncation condition, in other words, for $n = \infty$, the fibrant objects in Jardine's model structure are the prestacks that satisfy hyperdescent; Lurie constructs a model category where the fibrant objects are the ones satisfying just descent for coverings [26].

[7] To define the quotient technically, suppose we are given a groupoid $R \rightrightarrows U$ in schemes. It represents a functor $\mathcal{C}^{op} \to 1\text{Gpd}$. From this 1-prestack, we take the associated stack. In terms of simplicial presheaves, take the Jardine fibrant replacement, but that can also be done by composing three times the functor of taking the groupoid of descent data. The result is a 1-stack, the universal 1-stack to which our prestack maps. It may also be viewed as the homotopy colimit of the groupoid (considered as a simplicial scheme) in the 2-category of stacks. Taking the associated stack makes it so that the resulting object stays the same even if we replace U by a different chart, say, a smooth cover $U' \to U$.

question can therefore be rephrased: what stacks $\mathcal{F}$ should be considered to admit a good presentation of the form

$$R \rightrightarrows U \to \mathcal{F},$$

as discussed in Section 2?

For 1-stacks, this condition was originally formulated by Artin [4]. He called such stacks *algebraic stacks*, but nowadays the terminology *Artin stacks* may be preferred, or – particularly in the higher context – *geometric stacks*.

As we have seen before, the dynamical relationship between the idea of a stack as a quotient of an equivalence relation and the idea of a stack as a functor with values in groupoids, n-groupoids, or ∞-groupoids (spaces or simplicial sets) can go in both directions. We might start with a natural chart and hence with a groupoid, or we might start with a functor and proceed later to look for a chart. In either case, we do eventually reach the situation of having a chart and should therefore ask what good conditions such a chart might satisfy in order to make it useful for studying the properties of the stack.

The basic idea is that the map $U \to X$ from the chart to the stack should be *smooth*. That is, as shall be seen from the example of BG below, the best possible condition we can hope for in general, allowing to study the infinitesimal structure of X. Artin's observation is that smoothness of $U \to X$ is reflected in smoothness of the two maps $R \to U$.

By looking at Artin's condition in the right way, it generalizes immediately to the case of n-stacks. This was pointed out to us by Charles Walter during discussions we were having with André Hirschowitz. It led to the preprint [38]. The discussion there is straightforward and self-contained; we will just sketch the basic outline, after the motivating example of BG in the next section.

6.1 A Chart for BG

Consider first one of the basic cases, the stack BG for G a group. If G is finite, BG is a Deligne–Mumford stack. It is natural to want to expand the collection of groups we can consider here, while staying within the algebro-geometric context. So let us look at BG for an algebraic group scheme G. To fix ideas for simplicity, say we work over a field k of characteristic zero. The most canonical map from a scheme to BG is just the base point

$$* \to BG.$$

This is a map whose fiber is G; indeed, suppose $f : X \to BG$ is any map. Then f corresponds to a G-torsor or principal G-bundle $F \to X$. The fiber product $X \times_{BG} *$ is the space representing trivializations of this principal bundle, so in fact

$$X \times_{BG} * = F$$

itself. We see that the map $X \times_{BG} * \to X$ has fiber isomorphic (non-canonically) to G, and this is what it means that $* \to BG$ is "a map with fiber G."

If G has positive dimension, this map is not etale, but since G is a group scheme and we are working in characteristic zero, it is smooth. Indeed, in the above formula $X \times_{BG} * = F$, the projection to X is the structural map of the torsor, which is smooth, and since we know it for any $X \to BG$, it means that the map $* \to BG$ is smooth.

The idea of Artin stacks is to allow "coordinate charts," which are smooth maps rather than just etale ones. An important further property of our map $* \to BG$, is that it is *representable*, meaning that for any map $X \to BG$ the fiber product as above is representable (it is best to require representability in the category of algebraic spaces). Now we can say that our stack BG is *locally a scheme*, that is to say, it has a smooth representable map from a scheme. Call this chart $U := * \to BG$.

The scheme of relations is

$$R := U \times_{BG} U = * \times_{BG} * = G.$$

Both maps $R \to U = *$ are of course the same; and the map $R \times_U R \to R$ is just the composition in the group G. We get the groupoid

$$(G \rightrightarrows *) \cong BG.$$

6.2 The Geometricity Condition

With the example of BG as motivation, we can now describe rather easily the general definition of an Artin stack. The point of view we shall present here was explained to us by Walter, and he suggested that it would generalize immediately to the case of higher stacks. The classical terminology used by Artin was "algebraic stack," but as Walter pointed out, it seems more intuitive to call the condition "geometricity."

Notice that if F is a 1-stack, and if X and Y are schemes, in particular, they are 0-stacks, then the fiber product $X \times_F Y$ is 0-truncated.[8]

---

[8] More generally, if F is n-truncated and X, Y are k-truncated for $k < n$, then $X \times_F Y$ is $n-1$-truncated. At the topological level, this may be seen by considering the exact sequences

We say that a map of 1-stacks $A \to B$ is *representable* if, for any map from a scheme $X \to B$, the fiber product $X \times_B A$ is representable by an algebraic space. We say that a representable map is *smooth* if, for any such X, the map $X \times_B A \to X$ is a smooth map of algebraic spaces. Now, the geometricity condition for a 1-stack F is that there exists a "smooth chart," that is to say, a surjective smooth representable map from a scheme $X \to F$.

If that exists, then we can give a presentation for F as a quotient of a nice equivalence relation: representability applies in particular to our chart itself, so the fiber product $R := X \times_F X$ is an algebraic space, and the two projection maps $R \to X$ are smooth. There is a natural composition $R \times_X R \to R$, and (X, R) gains a structure of groupoid in the category of algebraic spaces. Our stack F is the quotient $F = X /\!/ R$.

Following Walter's suggestion, the above definition generalizes immediately to the case of n-stacks. We define by induction on n simultaneously the notions of geometricity for an n-stack, for a morphism of n-stacks, and smoothness for a geometric morphism of n-stacks. Assuming these are known for $(n-1)$-stacks, the definitions for n are exactly as before: a morphism $A \to B$ from an $(n-1)$-stack to an n-stack is *geometric* (resp. smooth geometric) if, for any map from a scheme $X \to B$, the $(n-1)$-stack (see the previous footnote) $X \times_B A$ is geometric (resp. $X \times_B A$ is geometric and the map $X \times_B A \to X$ is a smooth map to a scheme); an n-stack F is *geometric* if it admits a geometric smooth surjective map $X \to F$ from a scheme of finite type, called a "chart"; and a map $F \to Y$ from a geometric n-stack to a scheme is smooth if, for any chart $X \to F$, the composed map $X \to Y$ is smooth. The reader may check that these definitions work together inductively to give the required notions, as is discussed in more detail in [38].

Notice that we start with a 0-geometric stack being an algebraic space. One can naturally go from just the notion of scheme to the notion of algebraic space by adding an additional step of the same form in the induction.

It has now become usual terminology to say that an *Artin n-stack* is an n-stack satisfying the above geometricity condition.[9]

for homotopy groups (cf. Section 5.4), first of the fibration X/F,

$$\pi_{n+1}(F) \to \pi_n(\text{fiber}(X/F)) \to \pi_n(X),$$

showing $\pi_n(\text{fiber}(X/F))$ trivial, then of the fibration $X \times_F Y/Y$, noting that the fiber is the same as before:

$$\pi_n(\text{fiber}(X/F)) = \pi_n(\text{fiber}(X \times_F Y/Y)) \to \pi_n(X \times_F Y) \to \pi_n(Y),$$

showing that $\pi_n(X \times_F Y)$ is trivial.

[9] **The Grothendieck–Pridham condition.** There is another way to look at the condition of being an Artin stack, much more directly related to the simplicial point of view. Pridham [36]

6.3 First Examples

The first examples of Artin 1-stacks are BG for G an algebraic group scheme. One may consider these as quotients $BG = */\!/G$ for the trivial action of G on a point. More generally, if G acts on an algebraic variety X, then the stack quotient $X/\!/G$ is an Artin 1-stack. The map $X \to X/\!/G$ is a smooth chart, and the relation is

$$R = X \times_{X/\!/G} X \cong G \times X,$$

with the two maps $G \times X \to X$ being the projection and the group action, respectively.

The quotient construction gives a huge array of examples, indeed, most moduli spaces are obtained by a quotient construction using Mumford's GIT. The corresponding moduli stacks are then quotient stacks of the above form.

One can play around with small examples. An interesting and useful one is the quotient

$$\mathcal{A} := \mathbb{A}^1/\mathbb{G}_m.$$

This stack has two points, corresponding to the two orbits [0] and [1]. The stabilizer group of [0] is $\mathbb{G}_m$, while the stabilizer group of [1] is trivial. This stack has many different uses, for example, families $V \to \mathcal{A}$ may be viewed as *filtered objects*, with $V_{[1]}$ being the underlying object and $V_{[0]}$ being the "associated graded" with its $\mathbb{G}_m$-action. This viewpoint is very useful for non-abelian filtrations occuring in non-abelian Hodge theory.

In a somewhat different direction, $\mathcal{A}$ is a classifier for Cartier divisors: given a scheme X, there is a one-to-one correspondence between Cartier divisors on X and morphisms $X \to \mathcal{A}$, with the divisor being the pullback of [0].

---

has shown the very nice characterization that an n-stack F is Artin if and only if it admits a presentation as a simplicial scheme X. (take the simplicial presheaf associated to the simplicial scheme and take the associated stack, or equivalently replace it by a fibrant object in the Jardine model structure) such that the pieces X_i are schemes of finite type and such that the attaching maps

$$X_i \to \lim_{i \to j < i} X_j = (\operatorname{csk}_{i-1}(X.))_i$$

are smooth. This condition was mentioned by Grothendieck in [18] as a possible definition of schematic homotopy type, so Pridham's theorem says that Grothendieck's intuition gives rise to the same very natural notion of Artin n-stack for which an inductive definition was discussed above. Many of the basic properties are easy to see from the geometricity definition but are not immediately clear from Grothendieck's definition. One could therefore view Pridham's theorem as a way of showing that Grothendieck's definition satisfies the properties one would want.

6.4 The Artin–Lurie Representability Theorem

Artin's approximation theorem is essential for translating charts constructed in an analytic or formal way, such as the Kuranishi deformation spaces, into algebro-geometric charts. This typical procedure for transforming a moduli problem, given as a functor, into an Artin stack was codified in the *Artin representability theorem* [3]. The statement combines the local infinitesimal considerations needed to get a formal chart with the algebraicity assumptions needed for the approximation theorem to get to maps from global schemes.

Lurie proves a vast generalization to higher stacks [25], and particularly to higher *derived stacks*. The introduction of derived structure is crucial for getting good control of the local infinitesimal theory. It would go outside of our scope to delve into the details. See, for example, Pridham's paper [35] for discussion and simplification.

6.5 The Higher Stack of Perfect Complexes

A motivating example for going to higher stacks and looking at Artin n-stacks is the stack **Perf** of perfect complexes. Fix a function $b : \mathbb{Z} \to \mathbb{N}$ which is zero outside of finitely many values, and let **Perf**$^{\leq b}$ be the higher stack whose value on a scheme X is the ∞-category of perfect complexes $C^{\cdot}$ over X such that at any point $x \in X$, we have $h^i(C_x^{\cdot}) \leq b(i)$. It is an n-stack, where n is the length of the interval on which b is nonzero. A main theorem says that **Perf**$^{\leq b}$ is an Artin n-stack of finite type. This was stated without proof in [20] and first proven by Toën and Vaquié [42]. They proved, in fact, a vast generalization concerning the moduli stack of perfect complexes relative to any projective scheme, and they considered its natural structure as a derived stack.

It is interesting to contemplate the structure of **Perf**$^{\leq b}$. For one thing, it has only finitely many points, much like in the example $\mathcal{A}$. We can enumerate the points of **Perf**$^{\leq b}$. Indeed, a perfect complex over an algebraically closed field of characteristic zero is determined up to equivalence by the dimensions of its cohomology groups. Thus, the points of **Perf**$^{\leq b}$ are in one-to-one correspondence with the functions $h : \mathbb{Z} \to \mathbb{N}$ such that $h^i \leq b(i)$.

On the other hand, the local coordinate charts of **Perf**$^{\leq b}$ are the very classical *Buchsbaum–Eisenbud schemes*. If we fix a point, corresponding to a function (h^i), then let V^i be the vector space of dimension h^i, and let $BE(h^{\cdot})$

be the scheme that parameterizes collections of differentials $d^i : V^i \to V^{i+1}$ such that $d^{i+1}d^i = 0$. Clearly $BE(h^{\cdot})$ parameterizes perfect complexes in the sense that there is a tautological perfect complex over it, and by semicontinuity, the cohomology dimensions are $\leq h^i \leq b(i)$, so we get a map

$$BE(h^{\cdot}) \to \mathbf{Perf}^{\leq b}.$$

These very natural charts were what we had in mind in [20], but they do not appear explicitly in the much more general situation of [42]. Benzeghli [6] proved that they do indeed provide smooth charts for the Artin n-stack and furthermore that these charts may be naturally completed to a simplicial scheme satisfying Pridham's smoothness criterion for giving an Artin n-stack.

Thus we may view the stack $\mathbf{Perf}^{\leq b}$ as being the natural quotient of the collection of Buchsbaum–Eisenbud schemes by the relation of weak equivalence. Notice that if one wanted to remain in the realm of 1-stacks, by taking the 1-truncation of the quotient, the geometricity property would no longer hold. We need to use the notion of n-stack in order to give this quotient its most natural nice property.

The theory of perverse coherent sheaves [3, 7] allows one to isolate a 1-truncated substack for which the previous comment does not apply, yielding an Artin 1-stack of moduli.

Kapranov and Pimenov have recently constructed derived schemes generalizing the Buchsbaum–Eisenbud schemes [22]. It is natural to expect that these should provide charts for the derived moduli stack of perfect complexes as it was constructed by Toën and Vaquié (applying [42] with X being a single point). We do not know if that has been done yet.

The Artin property of $\mathbf{Perf}^{\leq b}$ means that it can be considered as a geometric object of the same kind as a usual algebraic variety. One may propose the following:

Question 6.1 What does the cohomology $H^*(\mathbf{Perf}^{\leq b}, \mathbb{Q})$ look like? What is its mixed Hodge structure?

Cohomology classes $\eta \in H^i(\mathbf{Perf}^{\leq b}, \mathbb{Q})$ provide characteristic classes for perfect complexes whose cohomology is bounded by b. One should note that if we take the direct limit over bigger and bigger bounding functions b, then the structure of the cohomology becomes rather more simple, and the only characteristic classes are the usual Chern classes of the cohomology sheaves. Preliminary calculations suggest that if we maintain a fixed bound b, there can be more.

6.6 An Example: Perfect Complexes of Type $\leq (1,1)$

Let us close this section by examining explicitly the case of perfect complexes with Betti numbers $(1,1)$ and $(0,0)$. We will get something that looks a lot like the 1-stack $\mathcal{A}$ seen at the start of the section, but with an Artin 2-stack structure instead. This example, accessible without too much technical baggage, provides a useful window on the ideas behind higher Artin stacks.

Consider the simplest function defined by $b(0) = b(1) = 1$ but $b(i) = 0$ elsewhere. Use the superscript $\leq (1,1)$. The Euler characteristic is a locally constant function on **Perf**$^{\leq (1,1)}$, so we can fix the open and closed substack $\mathcal{A}'' \subset$ **Perf**$^{\leq (1,1)}$ of complexes with Euler characteristic zero.[10] This stack has two closed points corresponding to cohomology dimensions $(0,0)$ and $(1,1)$, respectively (the dimensions are required to be zero outside of degrees $i = 0, 1$). These points may be viewed as the complexes

$$e_1 \colon k \xrightarrow{1} k$$

and

$$e_0 \colon k \xrightarrow{0} k,$$

respectively. The complex e_1 is quasi-isomorphic to $0 \to 0$, that is, it is the zero complex; in particular, its space of automorphisms is reduced to a single point. It is the open point of $\mathcal{A}''$, very analogous to the point $1 \in \mathcal{A}$ considered at the beginning of the section.

The group of automorphisms of the complex e_0, on the other hand, is $\mathbb{G}_m \times \mathbb{G}_m$. This is different from the automorphism group $\mathbb{G}_m$ of the point $0 \in \mathcal{A}$. To understand this difference, we should note that $\mathcal{A}''$ is a 2-stack rather than a 1-stack, so we should consider not the group of all automorphisms of e_0 but rather the 1-stack of automorphisms modulo homotopy. The identity endomorphism $i \colon e_0 \to e_0$ has self-homotopies, and in fact the sheaf of homotopies is $\mathbb{G}_a$, corresponding to the functions $e_0^1 \to e_0^0$. The automorphism group $\mathbb{G}_m \times \mathbb{G}_m$ acts on the sheaf of homotopies via the product map. We therefore obtain a picture of the 1-stack of automorphisms of e_0 as having homotopy group sheaves (cf. Section 5.4) represented by $\pi_0 = \mathbb{G}_m \times \mathbb{G}_m$ and $\pi_1 = \mathbb{G}_a$. In terms of dimensions, everything works out, because this automorphism group stack still has dimension 1, so e_0 has dimension -1 and it is still a codimension 1 closed substack of $\mathcal{A}''$, as was the case for $0 \in \mathcal{A}$.

The Buchsbaum–Eisenbud scheme providing a neighborhood of the point e_0 is just the affine line $\mathbb{A}^1$ parameterizing complexes of the form

$$e_t \colon k \xrightarrow{t} k,$$

---

[10] The two other components corresponding to Betti numbers $(1,0)$ and $(0,1)$ are just $B\mathbb{G}_m$s.

so our Artin 2-stack $\mathcal{A}''$ may be thought of as a different way of taking a quotient of the affine line with the open subset of nonzero points as single open orbit.

Consider the relation $R \to \mathbb{A}^1 \times \mathbb{A}^1$ defining $\mathcal{A}''$. Itself a 1-stack, R is the quotient of a variety R_0 by an action of the group scheme $\mathbb{G}_a$, as follows. The variety R_0 is the space of quadruples (s, t, g, h), where (s, t) is a pair of elements in the parameterizing Buchsbaum–Eisenbud scheme $\mathbb{A}^1$ and (g, h) form a map of complexes

$$\begin{array}{ccc} k & \xrightarrow{s} & k \\ g \downarrow & & \downarrow h \\ k & \xrightarrow{t} & k \end{array}$$

meaning that $tg = hs$, such that, furthermore, (g, h) is a quasi-isomorphism. The quasi-isomorphism condition means that if either $s = 0$ or $t = 0$, then g and h are nonzero. Thus, R_0 is the open subset of the affine variety $tg = sh$ in $\mathbb{A}^4$, defined by removing the two planes $(s = 0, g = 0)$ and $(t = 0, h = 0)$. The action of $\mathbb{G}_a$ is by $a : (s, t, g, h) \mapsto (s, t, g + as, h + at)$. The quotient $R = R_0/\mathbb{G}_a$ is a 1-stack, and this 1-stack projects by (s, t) to $\mathbb{A}^1 \times \mathbb{A}^1$. It has a groupoid structure, and the quotient 2-stack is $\mathcal{A}''$.

We see again that the automorphism group of e_1, the fiber of R over $s = t = 1$, is given by equation $g = h$ modulo $g \mapsto g + a$ and $h \mapsto h + a$; this quotient of $\mathbb{A}^1$ by $\mathbb{G}_a$ is trivial. The automorphism group of e_0, the fiber of R over $s = t = 0$, is the "quotient" of the space of (g, h) with $g \neq 0$ and $h \neq 0$, by the trivial action of $\mathbb{G}_a$. Thus, the automorphism group of e_0 is isomorphic as a stack to $\mathbb{G}_m \times \mathbb{G}_m \times B\mathbb{G}_a$, although the group multiplication law is twisted.

This simplest example gives an idea of what the n-stack **Perf**$^{\leq b}$ looks like in general and of what it means to have a higher Artin structure. It illustrates the idea that we might have a groupoid where the relation object is itself a stack, leading to a 2-stack.

7 Non-abelian Cohomology

The introduction of higher algebraic stacks paves the way toward a new possibility of using them to investigate the structure of spaces, a technique that could go by the generic name of "shape theory."

Very roughly speaking, the shape of a space means the non-abelian cohomology functor it defines. In this section, we start by explaining the motivation, then we discuss an extended example that shows how shape theory using non-abelian cohomology leads to a new perspective on non–simply connected spaces, and then we get to a general discussion of the categories of coefficients that might be used.

7.1 From Topologizing Algebraic Geometry to Algebro-geometrizing Topology

Our discussion up until now has focused on the idea that new more complicated kinds of algebraic varieties, known as stacks, have local and even global topological structures attached to them, and indeed these topological structures are essential for the definition and theory of stacks. Here, topology is viewed as a tool needed to understand structures that arise naturally in algebraic geometry, and we can say that algebraic geometry becomes "topologized."

Developing this idea leads to the quite general point of view of presheaves of spaces (simplicial presheaves) over a Grothendieck site. But now, such a theory allows us to envision a new and different direction: algebro-geometrizing topology. By this, we mean envisioning new and more complicated kinds of "enriched spaces," where the additional structure is of an algebro-geometric nature. A first and typical example is the stack BGL_n. It is an Artin 1-stack classifying rank n vector bundles. Giving a morphism from an algebraic variety X to BGL_n is the same thing as giving a rank n vector bundle over X. Natural transformations between morphisms correspond to isomorphisms between bundles.[11]

One salient feature is that $\pi_1(BGL_n, o) = GL_n$ is a sheaf of groups, the sheaf represented by the group scheme GL_n. Thus, we can view BGL_n as a new kind of "space," whose homotopy groups are group schemes rather than just discrete groups. Generalizing this idea to higher stacks leads to the *schematization of homotopy*. The schematization operation was one of Grothendieck's overall aims in *Pursuing Stacks* [18]. Breen had an early paper on the schematic sphere [9], and one should note that his paper on cohomology calculations [8] provides in retrospect the foundation for schematization. More recently, Katzarkov Pantev and Toën [23, 24], Pridham [34], and others [28] defined the schematization in full generality and developed Hodge theory for it.

---

[11] Here is the construction of the stack BGL_n as a presheaf of spaces. For each affine scheme $\mathrm{Spec}(A)$, consider the group $GL_n(A)$ and form its classifying simplicial set $B(GL_n(A))$. The set of k-simplices here is $GL_n(A)^{k-1}$ with face maps given by projections to factors and multiplication. These simplicial sets are organized into a simplicial presheaf over our site. It is a 1-prestack because each simplicial set is 1-truncated. Then BGL_n is the associated stack, which may for example be described as the fibrant replacement of the prestack in the Jardine model structure. A more explicit description of the "stackification" is possible, generalizing the explicit description of the sheafification of a presheaf.

In the simply connected case, the schematization recovers the usual rational homotopy type. As indicated in the notation, if X is a simply connected homotopy type, then the schematization $X \otimes k$ corresponds to tensoring the rational homotopy type, considered for example as a dg commutative algebra, with the field k over $\mathbb{Q}$. Even in this case, additional insight is gained by thinking of $X \otimes k$ as a higher stack rather than just "the homotopy type whose dgca would be the tensor product one."

In the non–simply connected case, the schematization $X \otimes k$ provides a stack whose fundamental group is the *pro-algebraic completion* of $\pi_1(X,x)$, that is to say, the inverse limit of linear k-algebraic group schemes G over the index category of representations $\rho \colon \pi_1 \to G$. The map

$$\pi_1(X,x) \to \pi_1(X \otimes k, x)$$

is the universal representation of an affine group scheme over k. The higher homotopy of $X \otimes k$ records the relative Malcev completions of the homotopy of X, at all of the representations ρ.

When X is not simply connected, the schematization $X \otimes k$ does not provide a fully satisfactory answer to the problem of finding homotopical invariants "over k." It can be a very big object when $\pi_1(X)$ admits families of representations. In that case, $\pi_1(X \otimes k)$ contains a big direct product of all the target groups for the representations in the family, but the direct product does not reflect the continuous structure in the variation of these representations. If $k' \colon k$ is a transcendental field extension, then $X \otimes k'$ is not just the extension of scalars of the stack $X \otimes k$ from k to k', because we have a lot of new representations defined over k', so the algebraic fundamental group gets a lot of new factors.

7.2 Shape Theory

One possible answer to this question is to adopt a point of view of *shape theory* using non-abelian cohomology as the basic structure.

We let $\mathcal{C}$ be an appropriate site of schemes. For our purposes, we consider a field k of characteristic zero and look at the site $\mathcal{C} := \mathrm{Aff}_k^{\mathrm{ft}}$ of affine k-schemes of finite type, with the etale topology. Let Stack denote the ∞-category of stacks of ∞-groupoids over $\mathcal{C}$, with Stack_n denoting the subcategory of n-stacks.

Suppose $\mathscr{P}$ is an ∞-category with a functor ζ to Stack. If X is a space, let $\underline{X}$ denote the constant ∞-stack whose values are the Poincaré ∞-groupoid of X. In terms of simplicial presheaves, it means to take the (fibrant replacement of) the constant simplicial presheaf with values X. We define the $\mathscr{P}$-shape of X to be the functor

$$^{\mathscr{P}}\mathrm{Shape}_X : \mathscr{P} \to \mathbf{Stack}, \quad T \mapsto \mathbf{Hom}(\underline{X}, \zeta(T)).$$

We will discuss more about the choice of $\mathscr{P}$ below (and once chosen, it may usually be dropped from the notation).

Getting back to the simply connected case to begin, the k-rational homotopy type, or schematization, of a simply connected space may be viewed as an object representing the shape for an appropriate choice of category $\mathscr{P}$. For this, let $\mathscr{P}$ be the full subcategory of $\mathbf{Stack}_n$ consisting of the n-stacks that are 1-connected and whose higher homotopy group sheaves are of the form $\mathbb{G}_a^k$. Recall that $\mathbb{G}_a \cong \mathbb{A}^1$ is the group scheme represented by the affine line, which as a sheaf is just the structure sheaf $\mathcal{O}$. So we are asking that the higher homotopy group sheaves be of the form $\mathcal{O}^k$.

Then, for any simply connected finite CW-complex X, there is an n-stack $(X \otimes k)_{\leq n} \in \mathscr{P}$ together with a map

$$X_{\mathcal{C}} \to (X \otimes k)_{\leq n}$$

representing the functor $^{\mathscr{P}}\mathrm{Shape}_X$. For $i \leq n$, we have

$$\pi_i((X \otimes k)_{\leq n}) = \pi_i(X) \otimes_{\mathbb{Z}} k$$

in the sense that the left side is the sheaf of groups over $\mathcal{C}$ represented by the k-vector space on the right considered as an affine k-scheme.

This gives a functorial point of view to rational homotopy theory for simply connected spaces, although it does not lead to the introduction of any really new invariants.

7.3 An Example of Shape in the Non–Simply Connected Case

One of the main motivations for introducing non-abelian cohomology and for looking at the associated shape theory is the well-known problem that rational homotopy theory does not behave very well for non–simply connected spaces. Shape theory gives us new structures to use in this context, leading to a rather good improvement in some aspects.

This is an area where new ideas from the theory of stacks can shed light on phenomena in algebraic topology. Our goal in this section is to illustrate with an example.

We should clearly enlarge the category $\mathscr{P}$ of target stacks to include some non–simply connected ones. In the present section, we look at a first example.

Our relatively simple example will have an abelian fundamental group, so for simplicity, we consider target stacks whose fundamental group scheme is just $\mathbb{G}_m$.

Let $T_1 := K(\mathbb{G}_m, 1)$. Let us consider 3-stacks T with maps $T \to T_1$ inducing $\tau_{\leq 1}(T) \cong T_1$. We impose the condition that $\pi_2(T)$ and $\pi_3(T)$ be abelian group schemes of the form $\mathbb{G}_a^n$. Throughout this section, fix $\mathscr{P}$ to be the 4-category of stacks of this form.

Consider the space $X := S^1 \vee S^2$. We note that $\Gamma := \pi_1(X) = \mathbb{Z}$, and the universal cover $\tilde{X}$ is a wedge sum of one copy of S^2 for each element of Γ. Thus, $\pi_2(X) = H_2(\tilde{X}) = \mathbb{Z}[\Gamma]$. On the other hand, the higher homology groups vanish, so

$$\pi_3(X) \otimes \mathbb{Q} = \mathrm{Sym}^2(\pi_2(X) \otimes \mathbb{Q}) = \mathrm{Sym}^2(\mathbb{Q}[\Gamma]).$$

As a $\mathbb{Q}[\Gamma]$-module, it has infinite type.

Our goal in this paragraph is to illustrate how non-abelian cohomology with stack coefficients provides an invariant that has better finiteness properties yet still allows us to distinguish different classes in $\pi_3(X) \otimes \mathbb{Q}$.

The functor

$$\mathrm{Shape}_X : \mathscr{P} \to \mathrm{Stack}_3$$

is defined by $\mathrm{Shape}_X(T) := \mathbf{Hom}(\underline{X}, T)$, where $\underline{X}$ is the constant prestack with value X.

The image of this functor lies in the subcategory $\mathrm{Artin}_3 \subset \mathrm{Stack}_3$ of Artin 3-stacks of finite type (see Proposition 7.2 in Section 7.4).

The Artin finite-type property is what we mean by a better finiteness condition. The functor Shape_X relates categories of objects of finite type in this sense.

We would now like to see how the shape functor distinguishes elements of $\pi_3(X)$. The shape functor has a pointed version: fix a base point $x \in X$, and look at pointed stacks (T, t). Let $\mathrm{Shape}_X^*(T)$ be the space of pointed maps $(X, x) \to (T, t)$.

An element of $\pi_3(X, x)$ is a pointed map $\varphi \colon S^3 \to X$ inducing a natural transformation

$$\mathrm{Shape}_\varphi^*(T) \colon \mathrm{Shape}_X^*(T) \to \mathrm{Shape}_{S^3}^*(T).$$

That may also be extended to the case when $\varphi \in \pi_3(X, x) \otimes \mathbb{Q}$.

Proposition 7.1 *Different elements $\varphi \neq \psi$ in $\pi_3(X,x) \otimes \mathbb{Q}$ give different natural transformations: $\text{Shape}^*_\varphi \neq \text{Shape}^*_\psi$ from the X-shape to the S^3-shape.*[12]

[12] *Proof:* Consider pointed stacks $T^{a,b}$, which are as follows:

$$\begin{array}{ccc} F^{a,b} & \longrightarrow & T^{a,b} \\ & & \downarrow \\ & & B\mathbb{G}_m \end{array}$$

where, in turn,

$$\begin{array}{ccc} K(V^{a+b},3) & \longrightarrow & F^{a,b} \\ & & \downarrow \\ & & K(V^a \oplus V^b, 2) \end{array}$$

with V^a being the sheaf represented by an affine line, which as a $\mathbb{G}_m$-module of rank 1 has t acting by t^a. The structure of $F^{a,b}$ is the standard map

$$\mu \colon \text{Sym}^2(V^a \oplus V^b) \to V^{a+b}.$$

The base point t comes from the standard ones in the Eilenberg–Mac lane stacks.
A pointed map $(X,x) \to (T,t)$ is given by a pair of pointed maps $S^1 \to T$ and $S^2 \to T$, so up to homotopy, it is just given by a triple (t, v', v''), where $t \in \mathbb{G}_m$, $v' \in V^a$ and $v'' \in V^b$. Thus

$$\pi_0 \text{Shape}^*_X(T^{a,b}) = \mathbb{G}_m \times V^a \times V^b.$$

Similarly, a pointed map $S^3 \to T$ is given by $w \in V^{a+b}$:

$$\pi_0 \text{Shape}^*_{S^3}(T^{a,b}) = V^{a+b}.$$

Recall that we denote $\Gamma := \pi_1(X) \cong \mathbb{Z}$. An element of $\pi_3(X)$ may be written, rationally, as

$$\varphi = \sum_{i,j} p_{ij} \gamma^i \gamma^j \in \text{Sym}^2_\mathbb{Q}(\mathbb{Q}[\Gamma]),$$

with $p_{ij} = p_{ji}$, and where $\gamma^i \in \Gamma$ is the ith multiple (power) of the generator. We get an element of $\pi_3(X) \otimes \mathbb{Q}$ by letting p_{ij} be any rational coefficients.
The action of such a map φ on the shape is

$$\pi_0 \text{Shape}^*_\varphi(T^{a,b}) \colon \mathbb{G}_m \times V^a \times V^b \to V^{a+b}$$

$$(t, v', v'') \mapsto \sum_{i,j} p_{ij} \mu((t^i \cdot (v', v''))(t^j \cdot (v', v'')))$$

$$= \sum_{i,j} p_{ij}(t^{ai+bj} + t^{aj+bi}) v' v'' = 2 \sum_{i,j} p_{ij} t^{ai+bj}.$$

The last equation is due to the symmetry $p_{ij} = p_{ji}$.

In this example, the shape functor, with reasonable finiteness properties, is rich enough to distinguish the elements of the infinite-type group $\pi_3(X,x)\otimes \mathbb{Q}$.

It will be interesting to see how the result of the proposition generalizes to other fundamental groups and homotopy groups in higher degrees. The shape can certainly fail to see some parts of the fundamental group, such as the intersection of subgroups of finite index. Characterizing the homotopical information carried by the shape is undoubtedly a subtle problem.

7.4 Coefficients for Non-abelian Cohomology

Our new kinds of geometrized spaces allow us to envision a rich theory of non-abelian cohomology. As we can see from our discussion of the classification of gerbes, a non-abelian cohomology theory involves, in its simplest incarnation, fixing a target space T and looking at the functor

$$X \mapsto \mathrm{Hom}(X,T).$$

The abelian case is when $T = K(A,n)$ and $H^n(X,A) = \pi_0 \mathrm{Hom}(X,T)$. The first-degree non-abelian cohomology with coefficients in a group G is given by taking $T = BG = K(G,1)$, and $\mathrm{Hom}(X,T)$ is the 1-truncated space corresponding to the non-abelian cohomology groupoid $\mathcal{H}^1(X,G)$ of G-torsors over X.

In usual cohomology theory, we pretty quickly want to pass to cohomology with coefficients over, say, a field k. Thus, in the examples of the previous paragraph, A would be a k-vector space (such as k itself), and G would be

Suppose $\psi = \sum_{i,j} q_{ij} \gamma^i \gamma^j$ is a different element such that

$$\pi_0 \mathrm{Shape}^*_\varphi(T^{a,b}) = \pi_0 \mathrm{Shape}^*_\psi(T^{a,b})$$

for any a,b. It means that

$$\sum_{i,j} p_{ij} t^{ai+bj} = \sum_{i,j} q_{ij} t^{ai+bj}$$

for any $t \in \mathbb{G}_m$. Separating terms, we get that for any integer k,

$$\sum_{ai+bj=k} (p_{ij} - q_{ij}) = 0. \tag{7.1}$$

We claim this implies $p_{ij} = q_{ij}$. Indeed, let $\Sigma \subset \mathbb{Z} \times \mathbb{Z}$ be the set of pairs where $(p_{ij} - q_{ij}) \neq 0$. If it is nonempty, then we can choose a corner (i_0, j_0) of its convex hull and fix a rational slope of line that meets the convex hull only at that corner. That determines a,b,k such that the intersection of $ai + bj = k$ with Σ consists of just one point (i_0, j_0). The above relation (7.1) says $p_{i_0 j_0} - q_{i_0 j_0} = 0$, but that contradicts the choice of $(i_0, j_0) \in \Sigma$. Hence Σ is empty and $p_{ij} = q_{ij}$, so $\varphi = \psi$ in the rational homotopy group. This completes the proof of the proposition. ∎

an algebraic k-group scheme. In these cases, we can say explicitly how the cohomology retains a similar algebraic structure over k.

A general non-abelian cohomology situation will be when T is some kind of space with homotopy groups in various different degrees. What does it mean for T to be "algebraic over a field k"? This is where the idea of algebro-geometrizing topology comes in: rather than looking for just a space T, we look for a simplicial presheaf, or n-stack, over a conveniently chosen site, such as Aff $=$ Aff$_k^{\text{ft,et}}$, the affine schemes of finite type over k with the etale topology. This is the site to which we shall refer in the subsequent discussion.

If T is an n-stack over Aff and X is a space, then $\mathbf{Hom}(X, T)$ is again an n-stack over Aff by the formula

$$\mathbf{Hom}(X, T)(\operatorname{Spec} A) := \operatorname{Hom}(X, T(\operatorname{Spec} A)).$$

Thus, if T is an n-stack over k, so is the non-abelian cohomology stack $\mathbf{Hom}(X, T)$. This satisfies our requirement for having a theory of coefficients relative to k.

For Hodge theory with non-abelian coefficients, it is particularly necessary to have a theory of coefficients relative to the ground field of complex numbers $\mathbb{C}$; indeed, the natural structures of abelian Hodge theory exist only on cohomology with complex coefficients, and the same is true in the non-abelian case.

Consideration of some more or less pathological examples shows that we need to do more in order to obtain a nice theory of coefficients. We would like our cohomology stacks $\mathbf{Hom}(X, T)$ to have a geometrical structure. For example, the first non-abelian cohomology $\mathcal{H}^1(X, G) = \mathbf{Hom}(X, BG)$ with coefficients in a linear algebraic group G has a structure of an Artin algebraic stack whose coarse moduli stack is the classical *character variety* of $\pi_1(X)$ with coefficients in G.

So it is natural to look for an appropriate kind of geometrical structure to impose on the coefficients, in such a way that the cohomology stack maintains the same kind of structure. The notion of Artin n-stack provides such a structure, as is shown by the following proposition.

Proposition 7.2 *The category of Artin n-stacks of finite type is closed under finite products and fiber products. If X is a finite CW-complex and T is an n-stack, then $\mathbf{Hom}(X, T)$ is in the category of n-stacks generated by products and fiber products starting with T. Therefore, if T is an Artin n-stack of finite type, so is the non-abelian cohomology stack $\mathbf{Hom}(X, T)$.*

7.5 Non-abelian de Rham Cohomology

Hodge theory for a complex algebraic variety X is about the relationship between Betti cohomology, the cohomology of the usual topological space X^{top}, and other cohomologies defined using the algebraic structure of X, such as de Rham or Dolbeault cohomology.

Very briefly, we can define the non-abelian de Rham cohomology of X by introducing the sheaf X_{dR} on the site $\text{Aff}_{\mathbb{C}}^{\text{ft, et}}$ defined by

$$X_{\text{dR}}(Y) := X(Y^{\text{red}}).$$

When X is smooth, the de Rham stack X_{dR} is defined by a formal groupoid whose object is X and whose morphism object is the completion $\widehat{X \times X}$ of the diagonal in $X \times X$. The morphism object injects into $X \times X$, so it is really just a relation whose quotient is a sheaf: one should think of it as "gluing together infinitesimally near points."

If T is an n-stack on $\text{Aff}_{\mathbb{C}}^{\text{ft, et}}$, the non-abelian de Rham cohomology[13] of X with coefficients in T is $\mathbf{Hom}(X_{\text{dR}}, T)$. One may check that this recovers the usual notions of non-abelian de Rham cohomology. When $T = BG$ for a complex linear algebraic group, a map $X_{\text{dR}} \to BG$ consists, first, of a map $X \to BG$, that is, a principal G-bundle over X, and second, of gluing data over the relation $\widehat{X \times X}$ with a coherence condition, amounting to providing the principal bundle with a flat connection. Therefore, $\mathbf{Hom}(X_{\text{dR}}, BG)$ is the Artin 1-stack of principal G-bundles with flat connection.

When $T = K(\mathbb{G}_a, n)$, $\mathbf{Hom}(X_{\text{dR}}, T)$ yields Grothendieck's algebraic de Rham cohomology $H_{\text{dR}}^n(X, \mathbb{C})$.

One of the main tools used to link Betti and de Rham cohomology is Serre's GAGA theorem, saying that cohomology of a projective algebraic variety is the same as cohomology of the corresponding analytic variety. If we want to promote our abelian theory to one that involves non-abelian coefficients, then not only should the coefficients be "defined over $\mathbb{C}$" corresponding to n-stacks over $\text{Aff}_{\mathbb{C}}^{\text{ft, et}}$ but also there should hold an appropriate GAGA principle.

Consider a basic example: suppose $T = K(\mathbb{G}_m, n)$. This is an Artin n-stack. But for $n \geq 2$, $\text{Hom}(X_{\text{dR}}, K(\mathbb{G}_m, n))$ on the algebraic site is very different from the same space computed on the analytic site. The algebraic cohomology

---

[13] The natural extension to Dolbeault cohomology, and the Hodge filtration, are obtained by considering the deformation of this formal groupoid to its normal cone. The attachments between infinitesimally near points are deformed to loops located at each point, so X_{Dol} is the formal completion of the zero-section in the tangent bundle of X. Non-abelian Dolbeault cohomology is the moduli space of Higgs bundles in degree 1 and reflects the Dolbeault cohomology of Higgs bundles in higher degrees.

is torsion, whereas the analytic one is isomorphic to the Betti cohomology with $\mathbb{C}/\mathbb{Z}$ coefficients.

This example shows that we need to impose some condition on the higher homotopy group schemes of our coefficient stack T. A similar example in degree $n = 1$ shows that we cannot use an abelian variety A: again $\mathrm{Hom}(X_{\mathrm{dR}}, K(A, 1))$ on the algebraic site would be very different from the corresponding analytic cohomology group.

With these examples in mind, we make the following definition:

Definition 7.3 A *connected very presentable stack* T is an n-stack on $\mathrm{Aff}_{\mathbb{C}}^{\mathrm{ft, et}}$ such that $\pi_0(T) = *$, $\pi_1(T) = G$ is a linear algebraic group, and for $2 \leq i \leq n$, $\pi_i(T)$ is a direct sum of copies of $\mathbb{G}_a$.

For connected, very presentable coefficient stacks T, which are in particular Artin n-stacks, GAGA works in the usual way, as may be seen by induction on the Postnikov tower.

If X is a smooth projective complex algebraic variety, we obtain an algebraic de Rham cohomology stack $\mathbf{Hom}(X_{\mathrm{dR}}, T)$, which is an Artin n-stack whose analytification is the Betti cohomology:

$$\mathbf{Hom}(X_{\mathrm{dR}}, T)\mathrm{an} \cong \mathbf{Hom}(X^{\mathrm{top}}, T)^{\mathrm{an}}.$$

Recall that X^{top} is a finite CW complex, so the Betti cohomology is an Artin n-stack by Proposition 7.2.

If X is only quasi-projective, then $\mathbf{Hom}(X_{\mathrm{dR}}, T)$ will in general be considerably bigger, due to the presence of maps $X_{\mathrm{dR}} \to BG$ corresponding to G-bundles with irregular connections. For example, there are nontrivial such maps when $X = \mathbb{A}^1$. The de Rham shape, restricted to the 1-truncated stacks, recovers the information of the differential Galois group. We feel that the higher de Rham shape should lead to an interesting extension of differential Galois theory.

One might want to consider coefficient stacks T that are not necessarily "connected." Then $\pi_0(T)$ is a sheaf of sets on $\mathrm{Aff}_{\mathbb{C}}^{\mathrm{ft, et}}$. It is somewhat more subtle to write down the conditions analogous to Definition 7.3. One may in some cases still obtain a good structure on the de Rham cohomology, such as the case suggested by Bertrand Toën of coefficients in the stack $\mathbf{Perf}^{\leq b}$ that we explored above. The cohomology stack

$$\mathbf{Hom}(X_{\mathrm{dR}}, \mathbf{Perf}^{\leq b})$$

has a rich structure not yet completely elucidated [39]. For quasi-projective X, it should give a new differential Galois invariant.

8 Prospects

The theory of stacks has shown its usefulness in a wide variety of geometrical situations. Following current thinking that integrates a homotopical direction into geometry, higher stacks will find their place and contribute.

The theory of higher stacks involves a combination of working relative to a Grothendieck site and looking at functors with values in higher homotopical structures, such as ∞-categories. This allows us to perceive a rich and intricate interplay between homotopy, algebra, and geometry.

These inputs come at different scales. In the application of stacks to moduli problems in algebraic geometry, the homotopical information is often concentrated at the infinitesimally tiny scale of points or subvarieties, where the parameterized objects acquire extra automorphisms; but in the case of a gerbe, we may think of a small-scale homotopical phenomenon spread out over the variety and manifesting global cohomological behavior. In the other direction, the theory of stacks allows us to enrich homotopy theory by providing algebraic structure to homotopy types. Here, the component pieces of a homotopy type, for example, the homotopy groups, gain structure of algebraic group schemes.

The mixture of geometrical and homotopical directions leads to new ideas: higher stacks become the natural coefficient systems for higher non-abelian cohomology, leading to new geometrical structures spreading across the homotopical directions. The enrichment to an algebraic structure allows us to consider non-abelian de Rham cohomology. The shape theory implied by the schematization of homotopy types can allow us to get a new and useful viewpoint on the structure of non–simply connected homotopy types.

We have seen a number of specific questions that seem interesting for future research. These include, of course, the abstract questions about further development of the foundations and theory. Let us think instead about some more explicit geometrical questions.

One is the structure of substacks of the moduli space of stable curves. Substacks can be generically nontrivial gerbes, and little is understood about the classification of what possibilities can occur here. On the boundary, the same question takes on a distinctly combinatorial feel.

What does the moduli 2-stack of DM-curves look like?

Similar questions for moduli stacks of vector bundles and other sheaves move already from the realm of Deligne–Mumford stacks to Artin 1-stacks, where the stabilizer groups can be positive-dimensional. Understanding the local geometry here is already significantly more complicated; yet there still remain important questions about classification of gerbes that can occur.

The fine structure of Artin 1-stacks gives an internal approach to various aspects of geometric invariant theory [19].

As we move on to higher stacks, a first main question is to understand what kinds of Artin n-stacks can be produced by natural constructions, such as moduli of objects in dg-categories.

The moduli stacks of perfect complexes **Perf**$^{\leq b}$ are very natural first examples of n-stacks with only finitely many closed points. In this case, the stabilizer groups of the closed points are all of the form GL_r (more precisely, they are higher groups whose 0-truncations are GL_r). Are there other nice examples of n-stacks with finitely many closed points, say, with other groups appearing as stabilizers?

Once we have some basic examples, it opens up the question of whether more general Artin n-stacks can be mapped into our example stacks, such as **Perf**$^{\leq b}$, and if such maps can form an embedding, say, when combined with maps to ordinary projective space. For Deligne–Mumford stacks, this is the notion of *generating sheaf* [29, 31].

The classification question relates to non-abelian cohomology: we can ask what kinds of Artin n-stacks occur as non-abelian cohomology stacks, depending on the domain variety and the target coefficient stack. The notions of shifted symplectic and Poisson structures [11, 33] provide important constraints.

We can then ask how the position of non-abelian cohomology stacks in this classification relates to the geometry and homotopy theory of the domain variety. Indeed, this classification interacts with the construction problem for homotopy types of algebraic varieties. Information on the range of possibilities for the non-abelian cohomology stacks of algebraic varieties (satisfying various conditions, such as projective and smooth) should give information about the special properties of their homotopy theory.

This aspect is closely related to non-abelian Hodge theory: higher stacks, as natural coefficient systems for non-abelian cohomology, need to be given further data to generate non-abelian Hodge structures on the cohomology. The foundational structure of what this data should look like, and then the geometry of the resulting cohomology, are basic areas open for further study.

A fundamental question raised by the present project is, how do stacks and the various roles they play in the study of geometry interact with the new notions of space presented in the other chapters of this book?

References

[1] D. Abramovich and A. Vistoli, *Compactifying the space of stable maps*, J. Amer. Math. Soc. 15 (2002) 27–75

[2] D. Abramovich and T. Jarvis, *Moduli of twisted spin curves*, Proc. Amer. Math. Soc. 131 (2003) 685–99

[3] D. Arinkin and R. Bezrukavnikov, *Perverse coherent sheaves*, Mosc. Math. J. 10 (2010) 3–29

[4] M. Artin, *Versal deformations and algebraic stacks*, Invent. Math. 27 (1974) 165–89

[5] C. Barwick and C. Schommer-Pries, *On the unicity of the homotopy theory of higher categories*, Preprint, Arxiv:1112.0040

[6] B. Benzeghli, *Un schéma simplicial de Grothendieck-Pridham*, Preprint, Arxiv:1303.4941

[7] R. Bezrukavnikov, *Perverse coherent sheaves (after Deligne)*, Preprint, Arxiv:math/0005152

[8] L. Breen, *Extensions du groupe additif*, Publ. Math. I.H.E.S. 48 (1978) 39–125

[9] L. Breen and T. Ekedahl, *Construction et propriétés de la sphère schématique*, C. R. Acad. Sci. Paris Sér. I Math. 300 (1985) 665–68

[10] L. Breen, *On the classification of 2-gerbes and 2-stacks*, Astérisque 225, SMF (1994)

[11] D. Calaque, T. Pantev, B. Toën, M. Vaquié, and G. Vezzosi, *Shifted Poisson structures and deformation quantization*, J. Topol. 10 (2017) 483–584

[12] A. Chiodo, *Stable twisted curves and their r-spin structures*, Ann. Inst. Fourier 58 (2008) 1635–89

[13] P. Deligne and D. Mumford, *The irreducibility of the space of curves of given genus*, Publ. Math. I.H.E.S. 36 (1969) 75–109

[14] D. Edidin and D. Fulghesu, *Moduli stacks of curves with a fixed dual graph*, Arxiv:1007.2130

[15] D. Edidin and D. Fulghesu, *Normalization of the 1-stratum of the moduli space of stable curves*, Port. Math. 69 (2012) 167–92

[16] N. Giansiracusa, *The dual complex of $\overline{M}_{0,n}$ via phylogenetics*, Arch. Math. 106 (2016) 525–29

[17] J. Giraud, *Cohomologie non abélienne*, Grund. Math. Wissenschaften 179, Springer (1971)

[18] A. Grothendieck, *Pursuing stacks* (1983)

[19] D. Halpern-Leistner, *On the structure of instability in moduli theory*, Preprint, Arxiv:1411.0627

[20] A. Hirschowitz and C. Simpson, *Descente pour les n-champs*, Preprint Arxiv:math/9807049

[21] J. F. Jardine, *Simplicial presheaves*, J. Pure Appl. Algebra 47 (1987) 35–87

[22] M. Kapranov and S. Pimenov, *Derived varieties of complexes and Kostant's theorem for $\mathfrak{gl}(m|n)$*, in *Algebra, Geometry, and Physics in the 21st Century*, Birkhäuser (2017) 131–76

[23] L. Katzarkov, T. Pantev, and B. Toën, *Schematic homotopy types and non-Abelian Hodge theory*, Compos. Math. 144 (2008) 582–632

[24] L. Katzarkov, T. Pantev, and B. Toën, *Algebraic and topological aspects of the schematization functor*, Compos. Math. 145 (2009) 633–86
[25] J. Lurie, *Derived algebraic geometry*, PhD thesis, MIT (2004)
[26] J. Lurie, *Higher Topos Theory*, Annals of Mathematics Studies 170, Princeton University Press (2009)
[27] Binru Li and S. Weigl, *The locus of curves with D_n-symmetry inside $\mathcal{M}_g$*, Rendiconti Circolo Matematico Palermo 65 (2016) 33–45
[28] A. Mikhovich, *Proalgebraic crossed modules of quasirational presentations*, Extended Abstracts Spring 2015, Springer (2016) 109–14
[29] F. Nironi, *Moduli spaces of semistable sheaves on projective Deligne–Mumford stacks*, Preprint, Arxiv:0811.1949
[30] M. Olsson, *(Log) twisted curves*, Compositio Math. 143 (2007) 476–94
[31] M. Olsson and J. Starr, *Quot functors for Deligne–Mumford stacks*, Comm. Algebra 31 (2003) 4069–96
[32] A. Papadopoulos, ed., *Handbook of Teichmüller theory*, 4 vols., IRMA Lectures in Mathematics and Theoretical Physics, EMS (2007–14)
[33] T. Pantev, B. Toën, M. Vaquié, and G. Vezzosi, *Shifted symplectic structures*, Publ. Math. I.H.E.S. 117 (2013) 271–328
[34] J. Pridham, *Pro-algebraic homotopy types*, Proc. Lond. Math. Soc. 97 (2008) 273–338
[35] J. Pridham, *Representability of derived stacks*, J. K-Theory 10 (2012) 413–53
[36] J. Pridham, *Presenting higher stacks as simplicial schemes*, Adv. Math. 238 (2013) 184–245
[37] G. Segal, *Categories and cohomology theories*, Topology 13 (1974) 293–312
[38] C. Simpson, *Algebraic (geometric) n-stacks*, Preprint, Arxiv:alg-geom/9609014
[39] C. Simpson, *Geometricity of the Hodge filtration on the ∞-stack of perfect complexes over X_{DR}*, Moscow Math. J. 9 (2009) 665–721
[40] C. Simpson, *Descent*, in L. Schneps, ed., *Alexandre Grothendieck: A Mathematical Portrait*, International Press (2014) 83–141
[41] The Stacks Project Authors, *Stacks project*, http://stacks.math.columbia.edu
[42] B. Toën and M. Vaquié, *Moduli of objects in dg-categories*, Ann. Sci. E.N.S. 40 (2007) 387–444
[43] P. Venkataraman, *The 2-lien of a 2-gerbe*, PhD dissertation, University of Florida (2008)
[44] J. Zintl, *One-dimensional substacks of the moduli stack of Deligne–Mumford stable curves, Habilitationsschrift, Kaiserslautern*, Preprint, Arxiv:math/0612802
[45] J. Zintl, *The one-dimensional stratum in the boundary of the moduli stack of stable curves*, Nagoya Math. J. 196 (2009) 27–66

Nicole Mestrano
Laboratoire J.-A. Dieudonné, Université de Nice – Sophia Antipolis
nicole@math.unice.fr

Carlos Simpson
Laboratoire J.-A. Dieudonné, Université de Nice – Sophia Antipolis
carlos@math.unice.fr

9
The Geometry of Ambiguity: An Introduction to the Ideas of Derived Geometry

Mathieu Anel

Contents

1	Introduction	505
2	Tangent Complexes	511
3	Derived Spaces	520
4	Conclusion	545
	References	550

1 Introduction

Derived geometry is a theory of geometry whose aim is to provide better behavior of singular points than in algebraic, complex, or differentiable geometries. It is named after derived categories, derived functors, derived tensor products, and pretty much anything "derived," because it proposes a setting where the natural constructions of all the notions give directly the derived versions. Within derived geometry, nothing has to be derived anymore.

As often in the history of mathematics, most of the computing methods to deal with singular points were invented before their proper formalization within a theory organizing and justifying them (e.g., Koszul resolutions, Chevalley complexes, equivariant methods). Perhaps what is really new in derived geometry is not so much its methods as the new understanding it proposes. Derived geometry has successfully interpreted *in terms of geometry* these previously ad hoc constructions. We shall come back to this in our conclusion.

1.1 Tangent Complexes

The easiest way to introduce derived geometry is probably with the following analogy. Recall that homological algebra can be read as the enhancement of the theory of vector spaces into the theory of chain complexes; then derived geometry is to geometry (ordinary topological spaces, manifolds, schemes, etc.) what chain complexes are to vector spaces. This analogy is good enough because a number of features of chain complexes do have analogues in "derived spaces." For example, complexes in positive or negative degrees have corresponding derived spaces called, respectively, *stacks* and *derived schemes* (or *derived manifolds*, depending on the context). It is also possible to truncate derived spaces and extract analogues of Z_0 and H_0.

In fact, there exists a precise comparison between the two theories: it happens that the tangent spaces to derived spaces are naturally chain complexes and no longer vector spaces. The relationship between the different derived spaces and their tangent is summarized in Table 9.1, which is useful to keep in mind.

It turns out that these tangent complexes are not so difficult to compute in practice. Examples involve Hochschild complexes, tangent sheaf cohomology, and group cohomology. Actually, they are somehow so easy to compute that people stumbled upon them before realizing what they were. For example, it is a classical fact of deformation theory that the tangent space at a point of a moduli space[1] is given by a homology group of some complex, but the rest of these complexes were for a long time overlooked. It is only by realizing that they were more regular as a whole (they have a Lie algebra structure and are often perfect complexes) than the sole homology group of interest that

Table 9.1. *Tangent spaces*

Type of space	Structure of the tangent	
Scheme/manifold	vector space	T_0
Stacks	chain complex	$\ldots \to T_1 \to T_0$
Derived scheme/manifold	cochain complex	$T_0 \to T_{-1} \to \ldots$
General derived space	unbounded complex	$\ldots \to T_1 \to T_0 \to T_{-1} \to \ldots$

---

[1] A *moduli space* is a space classifying something: it could be the solutions to some equations but also a structure, such as the space of curves or the space of vector spaces. Inspired by the example of elliptic curves, *moduli* is the general name for the coordinates on those spaces.

people came up with the idea that moduli spaces would be more regular if they could be defined such that the whole complex would be the tangent space. This tangent Lie structure was used in deformation theory [23], emphasized by Drinfeld [15] and theorized by the "derived deformation theory" (DDT) [28, 29], and the "perfection" of tangent complexes can be understood as the meaning of the "hidden smoothness" of moduli spaces [12]. Altogether, these ideas sprung up mostly in the late 1980s to 1990s.

1.2 Singularities

This new geometry conserves the old one: classical manifolds or schemes embed faithfully in derived spaces. The new features concern in fact only singular points and thus spaces with singularities: smooth manifolds and smooth schemes behaved as they always did in this new geometry.

Singular points are classically defined as the points whose tangent space has a dimension bigger than expected, which is a way to say that not all tangent vectors can be integrated into a curve. They are essentially of two kinds: *quotient and intersection singularities*. For example, quotient singularities appear in a quotient by a group action when a point has a stabilizer under the action, and intersection singularities appear when the intersection is not transverse. We shall see in Section 2 how these two kinds of singularities create, respectively, a positive and a negative part in the tangent complexes. This is actually the first insight about tangent complexes: a point is singular if and only if its tangent complex has nonzero homology. This extra tangent structure measures the complexity of the singularity. Also, the whole formal neighborhood of a point can be reconstructed from the tangent complex equipped with its Lie structure (see 2.2.3).

1.3 Stacks

The daring idea of a new kind of space whose tangent spaces would be chain complexes, probably best summarized in the introduction of [12], could only be imagined because of a very mature ground. First of all, homological algebra had spread in algebraic geometry, where it had been rebirthed with *derived categories* and *total derived functors*. Since the 1960s, every geometer is accustomed to the "derived philosophy," which asserts that objects could have more regular properties if they were replaced by some "derived" version, enhanced with some "hidden" structure.

But most significant was the influence of *stack theory*, which had risen from an obscure topic for descent and nonabelian cohomology [22] to a proper geometrical theory after the works of Deligne, Mumford, and Artin [5, 13] in the late 1960s to 1970s. Stacks are a notion of highly unseparated space where points can form categories instead of just sets or posets, and this was perfectly suited to encoding the singular structures of moduli spaces.[2]

Stacks were for some time a kind of necessary evil in algebraic geometry, something efficient but whose geometrical nature was unclear. The proper context to understand them is in fact ∞-categories and an analogy with homotopy theory. Implicit in [22], this analogy was fully devised in the 1980s by Grothendieck in *Pursuing Stacks* [25], then successfully formalized in the 1990s by Hirschowitz and Simpson [30, 59] by using the ideas and methods of algebraic topology (simplicial presheaves and model categories).[3]

The importance of stacks for derived geometry was that they were already proposing a notion of space with tangent complexes. Within stack theory, a point can have a symmetry group, and the Lie algebra of this group is a part of the tangent structure at the point [43] (see also 2.1.2). In other words, the tangent complex of a stack at some point encodes not only first-order deformations of this point but also the symmetries of these deformations (which are in bijection with first-order deformations of the identity endomorphism). The tangent complexes of stacks, a notion that would eventually be better formalized by Simpson [59] in the context of higher stacks, are concentrated in (homological) nonnegative degrees and have been eventually recognized to explain half of the tangent complexes of the DDT program. In fact, stack theory has been recognized to be one-half of the pursued derivation of geometry, precisely the part regularizing the properties of quotient singularities [12, 28]. The other half, corresponding to intersection singularities and the negative part of the tangent complex, would demand new ideas.

1.4 The Good Formalism

The first formalisms of derived geometry were purely algebraic: since they were made to extract tangent chain complexes, they were using algebraic structures on chain complexes (Lie dg-algebras, commutative dg-algebras, and even dg-coalgebras) together with the natural extension of commutative

---

[2] It was folkloric at the time that the symmetries of points in moduli spaces were an obstruction to describe them locally by affine schemes or manifolds.

[3] These methods were actually available to Grothendieck, but he was not satisfied with them and refused to used them.

algebra to this setting.[4] This was efficient, but the spatial language was rather a psychological trick to justify the formal manipulations than yet a proper insight on a new geometry. Also, the formalisms were either for formal neighborhoods only [28, 40] or specialized to some examples [12, 37], and a better theory was called for.

The good idea was eventually found in the working philosophy of algebraic geometers, who, from the formalization of algebraic groups [14] to the formalization of moduli problems and algebraic stacks [13, 59], were constructing the objects of algebraic geometry in two steps: first by defining a notion of affine scheme (encoded faithfully by a commutative ring of functions), and then by completing the category of affine schemes by the quotient stacks of (étale or smooth) groupoids (we shall detail this in Section 3.1). The recipe for derived geometry was then to do the same starting with some "derived rings" (commutative dg-algebras). Hinich formalized this in the infinitesimal scale [28]. The attempt of Behrend [7], although correctly conceived on a conceptual level, was wronged by the technical limitations due to the use of 2-category theory instead of higher category theory. It was finally Toën and Vezzosi, both trained in homotopical algebra and ∞-categories, who were the first to formalize a proper setting by working out algebraic geometry within model categories [66–68].

Later, Lurie, with a better background in categorical logic, understood how to apply this approach to differentiable and complex geometries [34, 48, 55, 61]. He also improved the presentation of the theory by working out fully the higher categorical background. This eventually led him to re-prove and improve a number of major theorems of algebraic geometry in the derived setting and to develop a tremendous amount of higher categorical notions on the way [44–46].

Nowadays, derived geometry is out there, and its methods are spreading from algebraic geometry to other mathematical fields. The most important exportation of its ideas may yet be derived symplectic geometry, where it provides a definition of symplectic structures for singular spaces [4, Chapter 4], [9, 10, 54, 65].

1.5 Derived Rings

The nature of those "derived rings" with which to start derived geometry is actually an important degree of freedom of the theory, and this was the reason for the very general setting of Toën and Vezzosi in [68]. If the first motivation

---

[4] Perhaps the first occurence of this is to be found in physics; see, Section 3.5.

of derived geometry was to improve the tools of algebraic geometry to work on singular spaces, Toën and Vezzosi were also motivated by an original application in algebraic topology where a "brave new algebraic geometry" was emerging. The application was regarding elliptic cohomology [68, Chapter 2.4] and would eventually be worked out fully by Lurie [47].

The possibility of this choice of derived rings is also what authorizes the definition of derived differential geometry and derived complex geometry [34, 48, 55, 61]. We shall say more about this in Section 3.2.6.

1.6 Why Derivation?

Why stacks? Why derived rings? How can we have a geometric intuition of these objects? We shall try to answer these questions throughout the text. Let us only say for now that the deep reasons for the necessity for the derivation of geometry (and all derivations) are not to be looked for within geometry or algebra but within an insufficiency of set theory and even its extension category theory. We shall come back to this in our conclusion.

1.7 Other Texts

Previous texts have been written to explain the ideas of derived geometry – the main ones would be Toën's surveys [62, 64] and Lurie's introductions to his DAG series and to his book [46]. The present text has been written as a complement to these texts; I have tried to emphasize the conceptual guidelines of the theory rather than the applications (which are detailed in the aforementioned texts) and to give a more global view on the matter of derivation.

1.8 Notations

Through all the text, the word *space* shall be used in an informal way to refer to the general idea of space, independently of any mathematical formalization (topological space, topos, manifold, scheme, stack). A basic knowledge of category theory is assumed, in particular regarding limits and colimits.

1.9 For the Differential Geometer

We shall limit the presentation of derived *algebraic* geometry, but all considerations can be transposed into differential and complex geometries. For the reader uncomfortable with the notion of scheme, it is enough to know that the notion of a scheme differs from that of a manifold by the fact that it is allowed to have singularities. In the whole text, the word *scheme* can be always be understood as "manifold with possible singularities." The expression *affine scheme* means a scheme that can be defined as the zeros of some functions in some affine space $\mathbb{A}^n$, whereas *general schemes* are constructed by gluing affine schemes. Contrary to differential geometry, in which any manifold can be embedded in $\mathbb{R}^n$, not every scheme is affine, that is, a subspace of some $\mathbb{A}^n$ (the projective spaces, for example). This explains the double vocabulary of affine/non affine schemes in algebraic geometry, and in this text, it can be ignored by the differential geometer.

1.10 Acknowledgments

I was fortunate to learn derived geometry from Bertrand Toën while he was developing it. My mathematical training had left me frustrated by the absence of principles justifying what looked to me to be homologic and homotopic computational nonsense; my discovery with him of the ideas of higher category theory was illuminating.[5] May he find here my gratitude for proving to me that these maths are not just a bunch of incomprehensible techniques but actually do make sense!

Let this be also an opportunity to thank all the other people with whom I had discussions that helped me organize my views on derived geometry and other higher category matters: John Baez, Damien Calaque, Guy Casale, Gabriel Catren, Denis-Charles Cisinki, Eric Finster, Nicola Gambino, David Gepner, Clément Hyvrier, André Joyal, Joachim Kock, Damien Lejay, Jacob Lurie, Mauro Porta, Carlos Simpson, David Spivak, Joseph Tapia, Michel Vaquié, and Gabriele Vezzosi.

The research leading to these results has received funding from the European Research Council under the European Community's Seventh Framework Programme (FP7/2007-2013 grant agreement 263523).

2 Tangent Complexes

The easiest way to get familiar with derived algebraic geometry features is the computation of the so-called *(co)tangent complexes*. The tangent complex is an enhancement of the tangent vector space into a chain complex, it is defined

---

[5] In particular, the notion of homotopy colimit that I have chosen to put at the heart of this text (see Section 3.2.1).

at every point of a space and globally as a bundle. The cotangent complex is the corresponding algebraic object – it is to the tangent complex what the module of differential forms is to tangent spaces, and analogously, it is more suitable for computations.

The theory of cotangent complexes for affine schemes is just a more geometric name for André–Quillen cohomology of commutative rings. The motivation was the study and classification of extensions of rings by cohomological methods. The geometric interpretation is the theory of infinitesimal deformations where the cotangent complex helps to answer questions such as "can a tangent vector be integrated into a path?"[6]

Although the motivation was coming from geometry, the geometrical meaning of these complexes (as a whole) was not at all clear at the time (Grothendieck wonders about such an interpretation in the introduction of [24]). Derived algebraic geometry has given a clear answer to this question: within derived algebraic geometry, the natural notion of tangent *is* the tangent complex.

As we will show in examples, tangent complexes are easy enough to compute in practice. For smooth points, they are quasi-isomorphic to the tangent spaces, but they contain a lot more information at singular points. This latter fact is their whole interest. By focusing on the algebra of functions (rings) and not only on topology (algebra of open subsets), algebraic geometry had already established, with the notion of scheme, a good notion of space that could support singularities, but derived algebraic geometry goes further and provides a notion of space with an even better handling of singularities. For example, there is no longer any need for transversality lemmas to compute intersections in derived algebraic geometry (see Section 3.3.2).[7]

The most important (and the most intriguing, but see Section 2.2.2) property of tangent complexes, which has no counterpart in algebraic geometry, is that they have a natural Lie algebra structure. At smooth points, this structure is trivial (i.e., abelian), but otherwise it contains a lot of structure about the singularity. For example, the whole formal neighborhood of the singular point can be reconstructed from this structure (see Section 2.2.3). This is particularly useful

---

[6] The problem is obvious on a smooth manifold, but not if singular points are allowed, as in schemes. A point is singular if its tangent space has a dimension bigger than the dimension of the space. This means precisely that not all tangent vectors are tangent to paths. The structure of the tangent complex, in particular its Lie algebra structure, helps to describe the subset (a cone) of vectors that can be integrated. This is the so-called cohomological obstruction calculus.

[7] Difficulty in mathematics might be transformed but never really cancelled; the problem of having transversal intersections is replaced by that of finding a nice resolution (e.g., a Koszul resolution) of the rings at hand. However, we have transformed a geometrical problem into an algebraic problem, which is always more suited for computations.

to study moduli spaces. This Lie structure is computable in practice by means of an L_∞-structure[8] but often reduces to a simple Lie dg-algebra structure. The reason to be of this Lie structure will be explained in Section 2.2.2.

We shall not give a proper definition of the tangent complex (this would require the introduction of too many technical notions), but a few words will be said in 3.2.5 about the cotangent complex. We refer to [64] for an introduction and to references therein for precise definitions.

2.1 Examples

2.1.1 Subschemes and Intersections

In this section, we compute the tangent complex of an affine subscheme defined by some equations. If there is more than one equation, this case encompasses the intersection of subschemes.

Let $\mathbb{A}^n$ be the affine plane of dimension n; we consider the affine scheme Z of zeros of a polynomial function $f : \mathbb{A}^n \to \mathbb{A}^m$, defined by the fiber product

$$\begin{array}{ccc} Z & \longrightarrow & \mathbb{A}^n \\ \downarrow{\scriptstyle \Gamma} & & \downarrow{\scriptstyle f} \\ 1 & \xrightarrow{\;0\;} & \mathbb{A}^m. \end{array}$$

The function f is given by a ring map $f : \mathbb{C}[y_1, \ldots y_m] \to \mathbb{C}[x_1, \ldots x_n]$. The ring of functions on Z is the ring A, quotient of $\mathbb{C}[x_1, \ldots x_n]$ by the ideal generated by the elements $f(y_j)$.

Let x be a point of Z; it determines a point of $\mathbb{A}^n$ that we still call x, whose image by f is 0 in $\mathbb{A}^m$. The tangent space $T_x Z$ to x in Z is defined as the kernel of the differential $df_x : T_x \mathbb{A}^n \to T_0 \mathbb{A}^m$. The point x is called a *regular point* of Z if df_x is surjective. In this case, the dimension of $T_x Z$ is $n - m$, and it can be proved that so is the dimension of Z around x. If df_x is not surjective, x is called a *singular point*. If we are thinking Z as the intersection of the $Z_j = \{f_j = 0\}$, a point x is singular precisely when the intersection is not transverse.

---

[8] We shall not give the full definition of an L_∞-structure on a chain complex $\mathfrak{g}$ and we refer to [51] for details. It will be sufficient to know that L_∞-structures are essentially a Lie structure, but where all equations have been relaxed to hold only up to homotopy. Such a structure is given by higher-order brackets $[-, \ldots, -]_n : \mathfrak{g}^{\otimes n} \to \mathfrak{g}[2-n]$ having homological weight $2 - n$. The bracket of arity 1 is a differential, the bracket of arity 2 is the Lie bracket, the bracket of arity 3 is called the jacobiator and produces a chain homotopy ensuring Jacobi identity, and higher brackets are encoding higher coherence conditions. The structure is that of Lie dg-algebra if the brackets in arity > 2 are zero. Altogether, the brackets can be written as the homogeneous components of a single map $S(\mathfrak{g}[-1]) \to \mathfrak{g}$, where $S(\mathfrak{g}[-1])$ is the symmetric algebra on $\mathfrak{g}[-1]$.

The *tangent complex* of Z at x is defined as the chain complex

$$\mathbb{T}_x Z \;=\; T_x \mathbb{A}^n \;\longrightarrow\; T_0 \mathbb{A}^m,$$

where $T_x \mathbb{A}^n$ is in (homological) degree 0 and $T_0 \mathbb{A}^m$ in degree -1. The tangent space $T_x Z$ is the H_0 of this complex, and x is regular iff the homology of $\mathbb{T}_x Z$ is concentrated in degree 0. The H_{-1} measures the defect of submersivity of f at x and the excess of dimension of $T_x Z$, that is, the singularity of x. Also, the Euler characteristic of $\mathbb{T}_x Z$ is always the expected dimension of Z around x. In particular, it is locally constant as a function of x (and not only semicontinuous, as is the dimension of $T_x Z$).

As we mentioned, the main feature of $\mathbb{T}_x Z$ is that it is endowed with an L_∞-structure. This structure exists in fact on the shifted complex[9] $\mathbb{T}_x Z[1]$ concentrated in degrees -1 and -2 and is particularly simple to make explicit in this example. It turns out that an L_∞-structure on a complex concentrated in such degrees is given by a single map (satisfying no condition),

$$S(T_x \mathbb{A}^n) \;\longrightarrow\; T_0 \mathbb{A}^m,$$

where $S(T_x \mathbb{A}^n)$ is the symmetric algebra on $T_x \mathbb{A}^n$. Then the L_∞-structure is simply given by the Taylor series of f, and the brackets $[-, \ldots, -]_n$ of the L_∞-structure are given by the homogeneous components, that is, the higher differential $D^n f$ of f viewed as symmetric functions of n variables.[10]

2.1.2 Quotients

We shall now describe the tangent complex of the quotient of a smooth scheme by a group action.

Let G be a group with Lie algebra $\mathfrak{g}$, acting on a smooth scheme X. Let x be a point of X and $\bar{x}$ the corresponding point in the quotient X/G. The infinitesimal action induces a map $\mathfrak{g} \to T_x X$, where $T_x X$ is the tangent space at x. The image of this map is the tangent to the orbit of x, and the kernel is the Lie algebra of the stabilizer of x. Let us call x *regular* if $\mathfrak{g} \to T_x X$ is injective and *singular* if not.[11] If x is regular, the action of G is locally free[12] around x, and it is possible to find a local transversal section to the orbits. This section can

---

[9] The reason for this shift will be explained in Section 2.2.2.

[10] This example shows the versatility of L_∞-structures; they can be quite remote from actual Lie algebra structures. But this ability to interpolate between what could be called "formal neighborhood structures" and Lie algebra structures is somehow the main interest of L_∞-structures.

[11] Because of discrete groups, x can be regular in this sense but still be a singular point in the quotient (an orbifold point). However, this kind of singularity can be unfolded by an étale cover (a local diffeomorphism) and therefore does not count as a singularity from the point of view of infinitesimal calculus.

[12] By which we mean that a neighborhood of the identity of G is acting freely around x. In algebraic geometry, local Henselian rings need to be used.

be used to define the local structure around $\bar{x}$ in the orbit space. In particular, it is of dimension $d = \dim X - \dim G$, and we get that $T_{\bar{x}}(X/G) = (T_x X)/\mathfrak{g}$.

The *tangent complex* of X/G at $\bar{x}$ is defined as the chain complex

$$\mathbb{T}_{\bar{x}}(X/G) \quad = \quad \mathfrak{g} \quad \longrightarrow \quad T_x X,$$

where $T_x X$ is in (homological) degree 0 and $\mathfrak{g}$ in degree 1. As before, x is regular if and only if the homology of $\mathbb{T}_x(X/G)$ is concentrated in degree 0, in which case H_0 is the tangent space in the quotient. The H_1 of this complex is the Lie algebra of the stabilizer of x; it measures the singular nature of $\bar{x}$. The Euler characteristic of $\mathbb{T}_{\bar{x}}(X/G)$ is always the expected dimension of X; again, it is locally constant as a function of x (and not only semicontinuous as the dimension of $T_{\bar{x}}(X/G)$).

This tangent complex can be proven to be the tangent space of the *quotient stack* $X/\!/G$. In this setting, the points of $X/\!/G$ form a groupoid and not a set. The tangent space at any point is also a groupoid but with an extra linear structure, and such an object is the same thing as a chain complex in homological degrees 0 and 1 [1, 21]. In this tangent groupoid, the H_0 is the part encoding first-order deformation of $\bar{x}$ as an object in $X/\!/G$, and H_1 is the part encoding the symmetries of such deformations (which are equivalent to first-order deformations of the identity of $\bar{x}$ as a morphism in $X/\!/G$).

Again, the shifted complex $\mathbb{T}_x(X/G)[1]$, concentrated in degrees 0 and -1, is endowed with an L_∞-structure easy to describe explicitly. It turns out that an L_∞-structure on a complex concentrated in degrees 0 and -1 is given by two maps,

$$d' : S(T_x X) \otimes \mathfrak{g} \longrightarrow T_x X \quad \text{and} \quad d'' : S(T_x X) \otimes \Lambda^2 \mathfrak{g} \longrightarrow \mathfrak{g},$$

where $S(T_x X)$ is the symmetric algebra on $T_x X$. The brackets $[-, \ldots, -]_n$ of the L_∞-structure are the homogeneous components of these maps, viewed as symmetric functions of n-variables.

Recall that the action of the Lie algebra $\mathfrak{g}$ on X can be encoded by a Lie algebroid [52]. A Lie algebroid structure is characterized by two maps: the anchor and a Lie bracket. The maps d' and d'' are given, respectively, by the Taylor series of the anchor map and of the Lie bracket at the point x. We claim that the equations of the L_∞-structure give exactly the conditions on the anchor and the bracket of a Lie algebroid.

In particular, if X is a single point $*$, the quotient stack $*/\!/G$ is the stack BG (classifying G-torsors). Its tangent complex at $*$ reduces to $\mathfrak{g}$ in degree 1, and the L_∞-structure is simply given by the Lie algebra structure of $\mathfrak{g}$.[13]

---

[13] The proximity of L_∞-structures with Lie structures is clearer in this example than in the previous one, but once again, the L_∞-structure encompasses the extra structure of a formal neighborhood.

2.1.3 Fiber Products and Triangles

The previous computations are actually a particular case of a general formula for fiber products of derived stacks: a cartesian square of (pointed) derived stacks induces a cartesian square of tangent complexes [68, Lemma 1.4.1.16]:

$$
\begin{array}{ccc} F & \longrightarrow & X \\ \downarrow^{\ulcorner} & & \downarrow^{f} \\ * & \longrightarrow & Y \end{array} \quad \Longrightarrow \quad \begin{array}{ccc} \mathbb{T}_x F & \longrightarrow & \mathbb{T}_x X \\ \downarrow^{\ulcorner} & & \downarrow^{df} \\ 0 & \longrightarrow & \mathbb{T}_{f(x)} Y \end{array}
$$

In other words, $\mathbb{T}_x F \to \mathbb{T}_x X \to \mathbb{T}_{f(x)} Y$ is a distinguished triangle in the category of chain complexes, and the left (or right) object can be computed as the mapping cone (or the mapping cocone) of the other two.[14] We recover this way the computation of Section 2.1.1 as a cone. In the case of a quotient by a group action as in Section 2.1.2, we use the fact that the definition of the quotient as a stack $X /\!/ G$ always provides a cartesian square,

$$
\begin{array}{ccc} G & \longrightarrow & X \\ \downarrow^{\ulcorner} & & \downarrow^{q} \\ * & \xrightarrow{\overline{x}} & X /\!/ G \end{array}
$$

where the top map is the parameterization of the orbit of x.[15] Then the construction of the tangent complex follows from a mapping cocone construction. For the corresponding L_∞-structures, the reader can look in [51].

2.1.4 Tangent Complexes in Deformations

Perhaps the most common examples of tangent complexes are the chain complexes that appear in deformation theory.[16] We shall give some details about the case of principal bundles and only mention a few others.

Let $P \to X$ be a principal G-bundle on a fixed smooth scheme (or manifold) X. Such a bundle is classified by a 1-cocycle of X with values in G, and the first-order deformations of P can be shown to be classified by

---

[14] The reader not fluent in homological algebra can look at [3, 20] for the definitions of these notions.

[15] It is one of the nice features of quotient stacks $X /\!/ G$ that the fibers of the quotient map $X \to X /\!/ G$ are always isomorphic to G, even if the action is not free. Actually, the map $X \to X /\!/ G$ can be proved to be a G-torsor.

[16] Deformation theory deals with the problem of classifying infinitesimal deformations of a given object (a scheme, a bundle, a ring structure, a group representation, etc.). It is always possible to consider such an object as a point in the moduli space for the structure in question; then the infinitesimal deformations of the object correspond to the study of the infinitesimal neighborhood of the point. In particular, first-order deformations correspond to tangent vectors in the moduli space. Moduli spaces are difficult to construct, but first-order deformations are relatively easy since they consist of solving some linear equations.

1-cocycles of X in the adjoint bundle $ad(P) = P \times_G \mathfrak{g}$. This suggests introducing the cohomology complex $C^*(X, ad(P))$ or, better for our purpose, the chain complex $\mathbb{T}_P$ such that $(\mathbb{T}_P)_i = C^{1-i}(X, ad(P))$.[17] Remark that $\mathbb{T}_P[1]$ has a Lie dg-algebra structure inherited from the Lie structure of $ad(P)$.

The positive part of this bundle is easy to understand: $H_0(\mathbb{T}_P) = H^1(X, ad(P))$ is in bijection with the isomorphism classes of first-order deformations of P, and $H_1(\mathbb{T}_P) = H^0(X, ad(P))$ classifies the symmetries of such deformations. Let $Bun_G(X)$ be the stack classifying of G-bundles on X; then the truncated complex $(\mathbb{T}_P)_{\geq 0}$ can be proved to be the tangent complex of $Bun_G(X)$ at the point P. This is quite similar to the example of Section 2.1.2.

The interpretation of the negative part of $\mathbb{T}_P$ is more subtle. As in the example of Section 2.1.1, we can think of it as related to some nontransverse intersection feature.[18] It can also be explained in terms of obstruction theory or, better, in terms of deformations parameterized by dg-algebras, but this requires elements of derived geometry. We shall not explain this here (see [64, 70] or the introduction of [46] for details) – the only thing we need to know is that from a point P in $Bun_G(X)$, we have again produced a tangent complex with a Lie algebra structure.

We list a few other complexes related to deformation problems. If X is a scheme (or a manifold), the tangent at X to the moduli space of schemes is the H_0 of the tangent cohomology complex $(\mathbb{T}_X)_i = C^{i-1}(X, TX)$. Then $\mathbb{T}_X[1]$ has a structure of a Lie dg-algebra inherited from the Lie bracket of TX.

If Γ is a discrete group and M a linear representation of Γ, the tangent space to M in the space of representations (or character space) of Γ is the H_0 of the complex $(\mathbb{T}_M)_i = C^{i-1}(\Gamma, End(M))$ computing the cohomology of Γ in the adjoint representation $End(M)$ of M. Remark that the shifted complex $\mathbb{T}_M[1]$ has a structure of a Lie dg-algebra inherited from the Lie structure of $End(M)$. More generally, if X is a topological space (the previous case being $X = B\Gamma$) and M a local system on X, then $End(M)$ is again a local system, and the tangent space at M to the space of local systems on X can be enhanced into the complex $(\mathbb{T}_M)_i = C^{i-1}(X, End(M))$ of cohomology of X in M. And again, $\mathbb{T}_M[1]$ is a Lie dg-algebra.

If A is an associative algebra, the tangent space at A to the space of associative algebra is the H_0 of the (shifted) Hochschild complex $(\mathbb{T}_A)_i = C^{i-2}_{Ass}(A, A)$. An L_∞-structure can be proved to exist on $\mathbb{T}_A[1]$, but it is difficult to explicit. Similarly, if $\mathfrak{g}$ is a Lie algebra, the tangent space at $\mathfrak{g}$ to the space of

---

[17] The degrees are changed for convenience; see Section 2.2.1.
[18] This intersection has two sources: first the cutoff given by the cocycle condition and also the wild limit (indexed by the category of refinement of atlases) that has to be taken when bundles are described in terms of Čech cocycles.

Positive part	Zero part	Negative part
(quotient singularity)		(intersection singularity)
$\cdots \to T_2 \to T_1 \to$	T_0	$\to T_{-1} \to T_{-2} \to \cdots$
"stacky structure"	"real" tangent	"derived structure"
internal symmetries		outer intersection structure

Figure 9.1 Structure of the tangent complex.

Lie algebra is the H_0 of the (shifted) Chevalley complex $(\mathbb{T}_{\mathfrak{g}})_i = C_{Lie}^{i-2}(\mathfrak{g}, \mathfrak{g})$. Again, $\mathbb{T}_{\mathfrak{g}}[1]$ has a Lie algebra structure inherited from that of the coefficient $\mathfrak{g}$.

2.2 Geometry of Tangent Complexes

2.2.1 The Three Parts of the Tangent Complex

We have seen two notions of singularities: intersection and quotient singularities. We have seen in each case that the tangent structure could be encoded in a complex whose homology is concentrated in degree 0 if and only if the point is regular. The other homology groups are thus a reflection of the singular structure of the point. If there exist nontrivial positive homology groups, this means that some *bad quotient* was involved in the singular structure. Moreover, if the homology is nontrivial for some $n > 0$, this means at least n bad quotients had to be involved to create the singularity. If there exist nontrivial negative homology groups, this means that some *bad intersection* was involved in the singular structure. Moreover, if the homology is nontrivial for some $n < 0$, this means at least n nontransverse intersections had to be involved to create the singularity. Since a general space is constructed by taking both intersections and quotients (typically a symplectic reduction; see Section 3.5), its singularities will have both positive and negative tangent parts in general.

Altogether, our examples propose a picture of the tangent complex in three specific parts, as in Figure 9.1.

2.2.2 Lie Structure and Loop Stacks

We have seen also that tangent complexes $\mathbb{T}$ were always equipped with a kind of Lie structure on the shift $\mathbb{T}[1]$. It turns out that derived geometry provides a very nice and quite simple explanation for this fact.

First, recall that in homotopy theory, for a pointed space $x : * \to X$, the homotopy fiber product $* \times_X *$ is nothing but the loop space $\Omega_x X$ of X at x, which in particular is a group.[19] The interpretation is in fact the same in any ∞-category, so in particular in derived stacks where $\Omega_x X$ is the derived group stack of symmetries of x in X. Now, recall from Section 2.1.3 that a cartesian square of pointed derived stacks provides a triangle of tangent complexes. Applied to the square

$$\begin{array}{ccc} \Omega_x X & \longrightarrow & * \\ \downarrow {\scriptstyle \Gamma} & & \downarrow {\scriptstyle x} \\ * & \xrightarrow{x} & X \end{array}$$

it gives $\mathbb{T}_{id_x} \Omega_x X = cone(0 \to \mathbb{T}_x X) = \mathbb{T}_x X[1]$, that is, the shifted tangent complex at x is nothing but the tangent complex to the loop stack at the identity of x. Now, since the loop stack $\Omega_x X$ is a group, this explains the Lie algebra structure on the tangent.[20]

2.2.3 Lie Structure and Formal Neighborhood

Perhaps the most bizarre consequence of the existence of the Lie structure is the ability to reconstruct the whole formal neighborhood of a derived stack (called a *formal stack* or a *formal moduli problem*) from the tangent complex and its Lie structure. This idea has a long history [15, 23, 28, 29, 56] and was fully formalized within derived geometry in [50].

If a point x in a scheme X is regular, the formal neighborhood of x is the same as the formal neighborhood of a point in the tangent space $T_x X$[21] (thus encoded by the algebra of power series $k[\![T_x^* X]\!]$ generated by the cotangent module at x). When x is not smooth, $T_x X$ is too big, and the formal neighborhood is a subspace of $T_x X$. The equation of this subspace can be written from the Lie structure of the cotangent complex $\mathbb{T}_x X$ by means of the Maurer–Cartan equation. Example 2.1.1 shows clearly that the L_∞-structure can reconstruct the formal neighborhood; this should also be clear enough in the example of Section 2.1.2. We shall not say more about it – the matter is detailed in [64].

---

[19] The fiber product $A \times_C B$ of a diagram of sets $f : A \to C \leftarrow B : g$ is the set of pairs (a,b) such that $f(a) = f(b)$ in C. The homotopy fiber product $A \times_C B$ of a diagram of homotopy types $f : A \to C \leftarrow B : g$ is essentially the same thing but where the equality $f(a) = f(b)$ is taken up to homotopy, i.e., replaced by a path in C. More precisely, $A \times_C B$ is the homotopy type of the space of triplets (a, b, γ), where γ is a path in C from $f(a)$ to $g(b)$. When $f = g = x : * \to X$, this gives the loop space. See [16] for more details.

[20] Actually, this interpretation is not fully proven yet. Although Lie algebra structures have been proven to exist on tangent complexes [27, 36, 50], the above interpretation relies on a theory of Lie algebras of derived Lie groups that has not been developed yet.

[21] This is the algebraic analogue of the local homeomorphism between the tangent space and the neighborhood of a point in a manifold.

3 Derived Spaces

In this section, we explain how to build the spaces of derived geometry. This will use ideas from algebraic geometry crossed with ideas from higher category theory.

3.1 A View on Algebraic Geometry

As we mentioned in the introduction, it is convenient to split algebraic geometry into two parts: the theory of *affine schemes* and the theory of *nonaffine spaces* (general schemes, algebraic spaces). Affine schemes are those spaces that can be described faithfully by a commutative ring of functions. Nonaffine spaces are those spaces without enough functions and must be described by other means (functor of points, atlases).[22]

The theory of affine schemes consists of an almost perfect dictionary between geometric properties (points, open and closed subsets, étale and proper maps, connectedness, separatedness, bundles, dimensions, vector fields, etc.) and features of commutative rings (fields, localizations, quotients and ideals, separable and finite extensions, idempotents, valuations, modules, generating families and presentations, derivations, etc.). The tools involved in this first part are those of commutative algebra: proofs about affine schemes are ultimately proofs about modules over rings. Commutative algebra is greatly computational and improves the tools of the sole topology: the full power of this algebra is notably used in the definition of the infinitesimal calculus (Kähler differentials but also iterated powers of ideals); it is also used in the definition of closed subspaces (defined as quotients of rings) or intersections (defined as tensor products of rings). These ideas lead in particular to a nice formalization of intersection multiplicities, which topology alone is not able to get.[23] All this will generalize well to derived rings and derived affine schemes.

The second part of algebraic geometry consists in using affine schemes to build more sophisticated spaces. It happens that not every space of interest in algebraic geometry is an affine scheme: projective spaces, Hilbert schemes, and other moduli spaces do not in general have enough functions with values

---

[22] This general structure has vocation to be used in differential and complex geometries. Rings need simply to be replaced by the appropriate notions: C^∞-rings for differential geometry [34, 53, 61] and a similar notion for complex geometry [49, 55]. See also Section 3.2.6.

[23] This power of the algebra over geometry is one of the motivations for the extension of differential geometry with C^∞-rings methods [53], where there is an infinitesimal theory handier than in manifolds. In particular, such an extension allows one to apply the ideas and methods of synthetic differential geometry (see Chapter 2 of the current volume).

in the affine line to be described faithfully by a commutative ring.[24] However, practice proves that these spaces can often be constructed by pasting affine schemes, in the same way a manifold is a pasting of charts.[25] This is the theory of *(nonaffine) schemes* and of *algebraic spaces* and of general *sheaves of sets*[26] (Chapter 7). Contrary to affine schemes, which are defined individually by means of a ring, these new objects cannot be defined individually but only relatively to the previously defined affine objects. The tools here are those of category theory rather than commutative algebra (limits and colimits of diagrams, presheaves, universal properties, etc.). They are the tools to work on a collection of objects rather than on objects individually. These methods and notions will be derived into the theory of stacks.[27]

We shall present now a more conceptual understanding of these two parts. Recall that a commutative ring is always a quotient of a free ring by some system of equations. Geometrically, this says that affine schemes are constructed from affine spaces $\mathbb{A}^n$ (the affine schemes corresponding to free rings) as level sets of functions, that is, by fiber products, or by categorical limits. Then schemes and algebraic spaces are constructed from affine schemes by pasting, that is, by categorical colimits. The two steps of the construction of the spaces of algebraic geometry can therefore be read as the following procedure: start with affine spaces (free rings), then add some limits, then add some colimits.

Having this in mind, we can say that derived algebraic geometry is built with the same procedure but where we are going to change the way to compute level sets and quotients, i.e., limits and colimits. Changing the way to compute limits transforms the theory of affine schemes into that of *derived affine schemes* (technically, it is done by changing the way quotients are computed in rings; see Section 3.2.3), and changing the way to compute colimits is the replacement of sheaves by *stacks* (see Section 3.4). Both changes will require the introduction of higher categories.

---

[24] Methods of commutative algebra can be extended to projective spaces (through graded rings), but not all schemes are projective, so other methods are needed.

[25] Which is to say that they may not have enough globally defined functions but always have enough locally defined functions (coordinates).

[26] Let us recall that sheaves have two different uses in topology: the most common is to use sheaves of abelian groups as coefficients for cohomology, but sheaves of mere sets can be used as generalized spaces. In fact, from this point of view, sheaves (and stacks in a better way) are useful to solve the following conundrum: how to define a space that is not a manifold, i.e., that does not have an atlas (like orbifold) or, even worse, a space that is not a topological space, i.e., that does not have enough open subspaces (like an unseparated quotient)? Sheaves and stacks give a setting in which to define such spaces, provided we know what a map from a manifold (or an affine scheme, or a topological space, or any other "basic block") to this space is, i.e., provided we can define a functor of points. This is particularly suited to moduli problems.

[27] A good introduction to these ideas is Toën's course [63].

3.2 Homotopy Quotients and Derived Rings

Algebraic geometry has long enhanced commutative algebra by introducing homological or homotopical methods (chain complexes, dg-algebras, simplicial modules and algebras, etc.). Classical books on the matter justify these enhancements by the powerful computations they allow (essentially, the existence of long exact and spectral sequences) but not so much by principles.[28] The question of these underlying principles was actually a very difficult question for a long time; it has been solved only with the mutation of homotopical algebra into higher category theory in the 1990s. We shall explain only the part of the story that has to do with the operation of taking quotients.

3.2.1 Quotients of Sets

Quotients of sets are classically dealt with via equivalence relations. However, in practice, equivalence relations are often derived from other structures (graphs, group action, etc.) where two elements may have several ways to be identified (in a graph, there might be more than one edge between two vertices x and y; in a group action, there might be more than one element sending x to y if the action is not free). The associated equivalence relation remembers merely the *existence* of an identification between two elements and forgets about the potential *ambiguity* (in the sense of noncanonicity) of identifications.

It turns out that forgetting the multiplicity of identifications can generate irregularities. For example, in the case of a group action, it is not true that working equivariantly is the same thing as working over the quotient if the action is not free.

Another notion of quotient has been invented that takes into account the potentially multiple identifications of elements. However, this operation has values not in sets but in homotopy types.[29] In consequence, it is called the *homotopy quotient*. It has been developed in homotopy theory, where it is one of the most basic tool under the name of *homotopy colimit* [2, 10, 16, 19, 31]. Technically, the homotopy colimit is constructed as a simplicial set such that the classical quotient (the set of equivalence classes) can be identified with the set of connected components.[30] The construction procedure is fairly simple,

---

[28] Who never wonders about the necessity of homological/homotopical apparatus and how to make sense of all these constructions?

[29] Homotopy types are equivalence classes of topological spaces for the (weak) homotopy equivalence relation. They can be viewed as a generalization of the notion of groupoid and are sometimes called ∞-groupoids (see Chapter 5).

[30] For the reader unfamiliar with simplicial sets, it is sufficient to know that they look like triangulated topological spaces, but defined in a purely combinatorial way. In particular, they have a homotopy theory.

but the proper statement of its universal property requires some advanced homotopical algebra.[31]

The principle of the construction is to identify two elements in a set by putting an edge between them instead of equalizing them. Then three elements are identified by putting a triangle, four with a tetrahedron, and so on. It should be clear how this produces a simplicial set and that its set of connected components is indeed the classical quotient. The homotopy colimit of a diagram of sets is formally defined as the homotopy type of this simplicial set. We shall see some examples.

Let us consider first the case of the diagram describing a graph with two edges a and b between two vertices x and y,

$$\{a,b\} \xrightarrow[t]{s} \{x,y\},$$

where the two maps are the source and target maps from the set of edges with values in the set of vertices (here s sends a and b to x and t sends them to y). The colimit of this diagram is the quotient of the set $\{x, y\}$ by the relation $x \simeq y$, and the classical quotient is a singleton. But a and b provide two different identifications between x and y, and the homotopy quotient is simply the homotopy type of the graph, equivalent to the homotopy type of a circle:

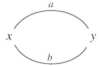

Its set of connected components is, as expected, in bijection with the classical quotient. If the set $\{a, b\}$ is replaced by a set with $n+1$ elements, the homotopy quotient can be computed to have the homotopy type of a wedge of n circles.[32]

Another important example is the homotopy quotient of the trivial action of a group G on a one-element set $*$. The recipe of the homotopy quotient gives the so-called *simplicial nerve* of the group, whose geometric realization is a classifying space for G [16, §4]. In consequence, the homotopy quotient is the

---

[31] This may explain why this construction is not more widely known. The best formulation of this universal property needs ∞-category theory: within this setting, the notion of homotopy colimit is simply the notion of colimit but computed in an ∞-category. We shall talk a bit more about this in Section 3.2.4.

[32] If both $\{a, b\}$ and $\{x, y\}$ are replaced by singletons, the homotopy colimit is a single vertex with a single edge, i.e., a circle. More generally, leaping from diagrams of sets to diagrams of homotopy types, if the set of edges $\{a, b\}$ is replaced by a homotopy type X of edges, the homotopy colimit is simply the unreduced suspension of X.

homotopy type of a classifying space for G. It is connected but has a nontrivial π_1 isomorphic to G.[33]

Finally, in the general case of a group G acting on a set E, the construction produces the nerve of the associated *action groupoid* $G \times E \rightrightarrows E$, and the homotopy quotient is the homotopy type of the classifying space $E/\!/G$ of this groupoid. If the action is free, this space is homotopy equivalent to the quotient set and the homotopy quotient coincides with the classical one, but if the action is not free, the groupoid remembers the stabilizer of a point as symmetries for this point, thus creating some π_1 in the quotient.[34]

Our examples have only nontrivial homotopy invariants in degree 0 and 1, but homotopy colimits of more complicated diagrams give rise to homotopy types with nontrivial homotopy invariants in any degree. In fact, any homotopy type can be described as the homotopy colimit of a diagram of sets.[35]

The main fact about the homotopy quotient is precisely that its higher homotopy invariants need not be trivial. In this sense, it encodes strictly more structure than the classical quotient. This extra structure is related to some kinds of syzygies:[36] the π_1 groups remember the ambiguity to identify elements (redundancy of relations), the π_2 groups remember the ambiguity to identify identifications between elements (redundancy of relations between relations), and so forth. Altogether, the homotopy quotient has the universal property to encode not only equivalence classes of elements but also the whole ambiguity about the identifications on those elements.[37]

---

[33] The quotient map is a map $* \to BG$. It is funny to remark that with this new notion of quotient, the point has many nontrivial quotients. This example shows that BG is the quotient of a point by the trivial action of the group G, but in fact, any connected homotopy type can be viewed as a quotient of a single point. This is an important point of view on homotopy types: they are in a sense a structure "below" sets. For example, if G is finite, the cardinality of BG is defined as $1/\#G$, which is indeed below $*$. The reader curious about cardinality of homotopy types can look at [18] and references therein.

[34] The notion of homotopy type can sometimes be replaced by the simpler notion of groupoid (taken up to equivalence of categories) to compute the homotopy quotients. This is the case when the homotopy quotients have trivial π_n for $n \geq 2$, for example, in a group action. Groupoids have a definition simpler than that of homotopy types; this explains why part of the literature focuses on them instead of full homotopy types. However, groupoids are limited by the fact that homotopy quotients of groupoids may have nontrivial π_2. So it is better, albeit more sophisticated, to work directly with full homotopy types.

[35] Any simplicial set is a diagram of sets, and the homotopy colimit of this diagram is the so-called geometric realization of the simplicial set. Any homotopy type can be described as a simplicial set (via the nerve of a contractible open covering; see [20]).

[36] This is an analogy with what will be told in Section 3.2.3.

[37] With this idea in mind, it is possible, and quite fruitful, to read the whole of homotopy theory as an enhancement of the theory of sets incorporating the "ambiguity of identifications." We shall come back to this idea in the conclusion. (This notion of ambiguity is compatible with the way the word is used in Galois theory. I have actually chosen the term *ambiguity* in reference to Galois theory.)

The main reason for considering homotopy quotients is their computational advantages over classical quotients. Let us mention a few. First, it is always true that working equivariantly is equivalent to working over the homotopy quotient (a very important property called *effectivity of groupoid quotients*[38]). Also, a fundamental computational tool is the long exact sequence of homotopy groups associated to any fibration sequence. And finally, they possess nice numerical invariants: when it is defined, the Euler characteristic provides a nice generalization of cardinality. For example, the Euler characteristic of $E//G$ is always (for *any* action of a finite group on a finite set) the rational number $\#E/\#G$, a formula that is false for classical quotients.[39]

The moral about homotopy quotients is that when no canonical identification exists between two elements to be identified, it is best to keep track of all identifications (which is the only canonical thing to do) by means of a homotopy type. This extra structure not only satisfies a stronger universal property but gives rise to an object with more regular properties. Projecting the identification data onto a mere equivalence relation truncates the structure and kills the nice computational regularities.

These facts actually make sense, not only for sets, but in a large variety of mathematical contexts, as we shall explain now for linear algebra, commutative algebra, and geometry.

3.2.2 Quotients of Vector Spaces

The issues with quotients of sets are inherited by quotients of any structure based on sets, although the technologies to take care of them may differ. For example, in linear algebra, chain complexes and quasi-isomorphisms turn out to be handier than simplicial objets and homotopy equivalences. In

---

[38] Let R be an equivalence relation on a set E with quotient $Q = E/R$ and quotient map $q: E \to Q$. The map q (or the relation R) is said to be *effective* if R can be reconstructed from q by as $R = E \times_Q E$. Let $G: G_1 \rightrightarrows G_0$ be a groupoid; it can be thought as a generalized equivalence relation on G_0 (where elements have several ways to be equivalent). There exists a notion of quotient for groupoids that we shall not define here. Let $q: G_0 \to Q$ be the quotient map; q (or G) is said to be *effective* if G_1 can be reconstructed from q as $G_1 = G_0 \times_Q G_0$. Since the whole set of arrows G_1 of the groupoid G can be reconstructed from the quotient, this means that the quotient contains the information about the multiplicity of identification between elements of G_0. In the category of sets, quotients of equivalence relations are always effective, but not quotients of groupoids. However, quotients of groupoids are effective when computed in homotopy types. The precise statement of the property of effectivity of groupoid quotients demands a definition of a groupoid object that we shall not give here (but see [44, 67]).

[39] This last point may seem anecdotic, but for objects more sophisticated than sets, this kind of invariant is related to the so-called virtual dimensions, virtual K-theory classes, etc.

consequence, homotopy colimits of diagrams of vector spaces are, rather, described as chain complexes.[40]

Let us consider the example of a map of vector spaces $d\colon E_1 \to E_0$. The classical quotient is the vector space $E_0/\operatorname{im}(d)$, which can be defined as the colimit of the diagram

$$E_1 \underset{0}{\overset{d}{\rightrightarrows}} E_0,$$

where 0 is the zero map. The homotopy quotient of d, which is the homotopy colimit of this diagram, is the construction called the *mapping cone* (see [3, 20, 71] for details). In our example, the mapping cone is simply $E_1 \to E_0$ viewed as a chain complex E_*, concentrated in (homological) degrees 1 and 0. In particular, $H_0(E_*)$ is the classical quotient, and $H_1(E_*) = \operatorname{Ker} d$ remembers the multiple identifications between elements of E_0. The space $H_1(E_*)$ plays the role of the π_1 of homotopy quotients of sets; it classifies nontrivial syzygies.[41]

In particular, we can understand now that the computation of the tangent complex of a quotient in 2.1.2 is nothing but the homotopy quotient of the map $T_e G \to T_x X$ between the tangent spaces. The nontrivial syzygies are essentially given by the stabilizer of the point. We shall see in Section 3.4.1 that the quotient stack $X/\!/G$ is nothing but the homotopy quotient of the group action, so space and tangent are computed the same way.

Homotopy colimits of more complicated diagrams give rise to chain complexes with larger homology amplitude. In fact, any chain complex in homological nonnegative degree can be obtained as the homotopy colimit of a diagram of vector spaces, and the homology groups H_i have exactly the same interpretation in terms of nontrivial syzygies as the homotopy groups π_i for homotopy quotients of sets.[42] In the same way that homotopy types can be understood as homotopy colimits of sets, chain complexes can be (and should be) understood as homotopy colimits of vector spaces.

Classically, the reason to consider homotopy quotients in the context of linear algebra is the computational power of long exact sequences and their

---

[40] It is possible to use simplicial methods in linear algebra, but the Dold–Puppe–Kan equivalence [35] proves that the two languages are in fact equivalent (provided we consider only complexes in nonnegative homological degrees).

[41] Again, this is an analogy with what will be said in Section 3.2.3.

[42] The chain complexes of nonpositive degree can be understood as generated by considering the dual notion of *homotopy limits*. The full category of unbounded chain complexes has a more subtle definition; it is generated by considering both homotopy colimits and limits of vector spaces, but with a constraint of commutation called *stability* imposing that finite homotopy limits and finite homotopy colimits should commute; see [45] and Footnote 87.

obstruction theory. Another reason is the effectivity of groupoid quotients, which, in this context, is reformulated into the property that the mapping cocone of a mapping cone is the identity.[43] We shall not expand on the homotopy theory of chain complexes. It is enough for this text to know that homotopy colimits of vector spaces are given by chain complexes up to quasi-isomorphism.

3.2.3 Quotients of Rings

Quotients of rings raise the same issues as with sets and vector spaces. The classical theory characterizes a quotient $A \to A'$ of a ring A by the data of an ideal $I \subset A$. In practice, though, quotients of A are not so much given this way but by systems of equations which are then interpreted as generators for an ideal and a quotient. The replacement of equations by an ideal is an operation of the same nature as truncating a graph or a group action into an equivalence relation and bears the same defects.

Let us consider the case of a system of equations $E = \{a_1 = 0, \ldots, a_n = 0\}$ for some a_i in A. The quotient of A by the system E can be presented as the colimit, in the category of rings, of the diagram

$$A[x_1, \ldots, x_n] \underset{x_i \mapsto 0}{\overset{x_i \mapsto a_i}{\rightrightarrows}} A.$$

It is classical that the quotient behaves well if the family of a_i is a *regular sequence* (neither element is a zero divisor relative to the previous ones). Let us recall why. An ideal I of a ring A that is generated by more than one element a_i is never free as an A-module, since there always exists the relation $a_i a_j - a_j a_i = 0$. Relations between generators of an A-module are called *syzygies*. Let us call *trivial syzygies* the relations that can be derived from the $a_i a_j - a_j a_i = 0$. The sequence of a_i is regular if and only if there are no nontrivial syzygies. Such syzygies exist when, for example, some a_{i_0} could be a zero-divisor, or equal to another a_{i_1} (repetition of equations), and this phenomenon can be understood as creating an ambiguity in the description of the elements of the ideal generated by E in terms of linear combinations of generators a_i.

For an element a in A, let $K(a) : A \to A$ be the chain complex concentrated in (homological) degree 0 and 1, where the differential is given by

---

[43] More precisely, for a map $f : E_1 \to E_0$ of (unbounded) chain complexes, let $F(f)$ be the mapping cone of f; there is a canonical map $g : E_0 \to F(f)$ whose mapping cocone $C(g)$ is quasi-isomorphic to E_1. In other words, homotopy cartesian squares are also cocartesian squares in chain complexes. The same property is true in the category of nonnegative chain complexes, but the map f needs to be surjective.

the multiplication by a. The complex $K(a)$ is a dg-algebra, called the *Koszul dg-algebra* of a (the multiplication of elements of degree 1 is nilpotent, and the other components are given by the multiplication of A; see [17, 20] for details). The group $H_0(K(a))$ is the classical quotient $A/(a)$, and the group $H_1(K(a))$ consists of all b in A such that $ba = 0$; it is nontrivial if and only if a is a zero-divisor. Hence, the dg-algebra $K(a)$, considered up to quasi-isomorphism, encodes the quotient but also keeps track of the regularity of the element a.[44]

More generally, the *Koszul dg-algebra* of a family of elements $E = \{a_i\}$ is defined to be the dg-algebra $K(E) = K(a_1) \otimes \cdots \otimes K(a_n)$. The full combinatorics of trivial relations between the a_i is encoded by the differential of $K(E)$: trivial syzygies (in a given degree) are defined as the image of this differential, and general syzygies as the kernel; the (classes of) *nontrivial syzygies* are defined as the homology of $K(E)$. The sequence E is regular if and only if $K(E)$ has only homology in degree 0, if and only if $K(E)$ is quasi-isomorphic to (or is a resolution of) the classical quotient $A/(a_1,\ldots,a_n)$.

Although this was not at all clear (or obvious) when it was invented, the construction $K(E)$ is now understood to be the homotopy quotient of A by the equations E. There exists a canonical map $A \to K(E)$ which image by H_0 is the classical quotient map $A \to H_0(K(E))$. The regularity of the sequence is the hypothesis under which the classical and the homotopy quotients agree. As for nonfree group actions, irregular sequences produce higher homotopical structure in the quotient, and the higher homology groups $H_i(K(E))$ have the same interpretation in terms of syzygies or ambiguity as the homotopy groups π_i for homotopy quotients of sets. Also, this notion of homotopy quotient in dg-algebras does satisfy the property of effectivity of groupoid quotients. We shall come back to this in Section 3.2.5.

The theory of dg-algebras provides also a good setting in which to define and compute cotangent complexes. Coming back to the example of Section 2.1.1, let $A = k[x_1,\ldots,x_n]$ and E be the system of m functions determined by f. Then, the derived fiber Z is encoded by $K(E)$, which is freely generated (as a graded algebra) by n generators x_i in degree 0 and m generators y_j in degree 1. Recall that the module of Kähler differential of a free ring is a free module on the same generators. The situation is the same for $K(E)$, and the cotangent complex at some point $x: K(E) \to k$ is a free module on the same

---

[44] Geometrically, a function a on a space X is a zero-divisor (or irregular) essentially when the subset $Z = \{a = 0\}$ has a nonempty interior. In such a case, the dimension of Z would not be $\dim X - 1$ as expected, hence the term *irregular*.

generators (renamed dx_i and dy_j); the only difference is a differential that can be computed to be the Jacobian matrix df of the function f, giving back the result of Section 2.1.1.

Finally, a word should be said about how homotopy colimits are related to better numerical invariants. The previous construction explained in terms of quotients can also be understood in terms of tensor products of rings (geometrically, this corresponds to describing a subspace as an intersection): $K(E)$ is also a model for the derived tensor product $A/a_1 \otimes_A^{\mathbb{L}} \cdots \otimes_A^{\mathbb{L}} A/a_n$, and $H_0(K(E))$ is the underived tensor product.[45] From this point of view, the higher homology of $K(E)$ is nothing but Tor modules. In the situation where A is local and of dimension n, the Serre intersection formula says that the correct multiplicity of the intersection is the Euler characteristic of the complex $K(E)$ and not the dimension of the crude $H_0(K(E))$.

The Serre formula is commonly used but totally ad hoc in classical algebraic geometry. It is only when one considers that the theory of algebras has to be enhanced into the homotopy theory of dg-algebras, which implies that colimits have to be replaced by homotopy colimits, that this formula finds its natural context.

3.2.4 Derivation and ∞-Categories

Let us organize what we have said so far. In the three contexts of sets, vector spaces, and rings, we have described a new operation of quotients: the homotopy colimit. It turns out that the proper setting to understand homotopy colimits (and in fact all homotopical phenomena) is higher category theory. Intuitively, if a category is thought as an enhancement of a set by allowing morphisms between elements, then an ∞-category is the enhancement of a category where morphisms (renamed 1-morphisms) have morphisms between them (named 2-morphisms), and 2-morphisms have 3-morphisms between them, ad infinitum. We shall not need much of the theory of ∞-categories; it will suffice to know a few things: (1) any category is an ∞-category; (2) the difference between the two notions is essentially that the arrows between two objects of an ∞-category do form a homotopy type rather than a mere set[46] and (3) the theory of ∞-categories is essentially the same as the theory

---

[45] The tensor product of commutative rings is a particular colimit (a pushout), and the derived tensor product is nothing but the corresponding notion of *homotopy* pushout of commutative dg-algebras.

[46] The easiest way to picture an ∞-category is to think of a category enriched in topological spaces, where these spaces are considered only up to homotopy equivalence. The 2-arrows are given by homotopies, 3-arrows by homotopies between homotopies, etc. All higher arrows are invertible; we shall not consider here higher categories with noninvertible higher arrows.

of categories: all the notions of adjoint functors, diagrams, colimits, and so on make sense in this new context and behave the same way.[47]

There exists an ∞-category S of homotopy types; this is essentially because morphisms between two homotopy types do form a homotopy type and not a set (the 2-arrows are given by homotopies of maps, the 3-arrows by homotopies between homotopies, etc.). Let *Set* be the category of sets; there exists an embedding *Set* $\to$ S by viewing sets as discrete homotopy types. With this in mind, we can explain that the homotopy colimit of a diagram of sets is nothing but the colimit of this diagram computed in the ∞-category S.

The same thing happens for vector spaces and rings. The homotopy theory of chain complexes and dg-algebras can be understood (and should) as the fact that they are naturally objects of ∞-categories rather than ordinary categories. Let *Vect* and *Ring* be the categories of vector spaces and rings, and let $Cplx_{\geq 0}$ and $dgAlg_{\geq 0}$ be the ∞-categories of chains complexes and dg-algebras in homological nonnegative degrees;[48] there exist embeddings

$$Vect \to Cplx_{\geq 0} \quad \text{and} \quad Ring \to dgAlg_{\geq 0}.$$

The operation of homotopy colimit in *Vect* and *Ring* is simply the natural notion of colimit in the ∞-categories $Cplx_{\geq 0}$ and $dgAlg_{\geq 0}$.

This kind of embedding of an ordinary category into an ∞-category, transforming automatically the way colimits (and also limits) are computed, can be taken as the formal meaning of the term *derivation* used in a homotopical/homological context. From this point of view, there is something arbitrary in the choice of the ∞-category used to derive a given category, and we could as well have considered the ∞-categories $Cplx$ and $dgAlg$ of unbounded chain complexes and dg-algebras. We could also have considered, as in [68], the notion of simplicial algebras instead of dg-algebras, or, as in [46], the notion of E_∞-ring spectra.

The purpose of derivation (as I see it) is to embed a category C into an ∞-category D having better behaved operations of colimit and limit. The choice of D depends on the kind of properties we are interested in. We already mentioned the property of effectivity of groupoid quotients, that is, the possibility to work on quotients of groupoids by equivariant methods. This property is also what allows the construction of classifying objects (e.g.,

---

[47] The references for the formal definitions of ∞-categories theory are [33, 44]. The introduction of [44] explains the relations of ∞-category theory and classical homotopical features. Another good introduction to these ideas is [6].

[48] If ∞-categories are pictured as categories enriched in spaces, the ∞-categories $Cplx_{\geq 0}$ and $dgAlg_{\geq 0}$ are classically constructed as Dwyer–Kan simplicial localizations of model categories.

for any kind of bundles in geometry or for extensions in ring theory; see Section 3.2.5).[49, 50]

Another nice property for $\mathcal{D}$ is to be generated by homotopy colimits from C. This explains the hypothesis of support for chain complexes: for example, $Cplx_{\geq 0}$ is generated by homotopy colimits by $Vect$ but not $Cplx$, and $dgAlg_{\geq 0}$ is also generated as such from $Ring$ but not $dgAlg$. Moreover, these examples of completion by colimits are free in a certain sense (see Footnote 55).[51]

3.2.5 Classifying Objects and Cotangent Complexes

In this section, we explain how the property of effectivity of groupoid quotients in the ∞-category of dg-algebras brings a nice feature to the theory of commutative rings.

Let us recall first that the category of homotopy types (and this would also be true in stacks) has classifying spaces and universal bundles. Let $p : P \to X$ be a principal G-bundle; then there exists a map $X \to BG$ (unique up to homotopy) and a (homotopy) cartesian square

where the map $* \to BG$ is the one of Footnote 33. The interpretation of this result is that $* \to BG$ is the universal G-bundle and BG is the classifying space for G-bundles.[52] The important remark for the sequel is that such a construction cannot exist in sets: even if G, X, and P are sets, that is discrete homotopy types, the classifying space BG is a nondiscrete homotopy type.

---

[49] The weaker property of *effectivity of equivalence relations* only (and not all groupoids; see Footnote 38) is one of the axioms of Grothendieck toposes, hence it has many instances within ordinary category theory. The stronger effectivity axiom becomes important once it is realized that most equivalence relations considered in practice are built by truncation of groupoids. Effectivity of groupoid quotients is one of the axioms of ∞-toposes. The strength of the effectivity axiom is precisely the difference between ordinary toposes and ∞-toposes, see [44, 57] and Chapter 4.

[50] In relation with homotopy type theory, the property of effectivity of groupoid quotients can be thought as a stronger form of the univalence axiom (it implies univalence since it implies the existence of classifying objects). However, since no good theory of quotients exists in HoTT, the effectivity property cannot be stated in this language.

[51] A natural question about derivation is the necessity of ∞-categories. Why is it necessary to go beyond the notion of ordinary category? We shall come back to this in our conclusion.

[52] Classically, this map is denoted $EG \to BG$ and built in topological spaces, with EG a contractible space, but since we are working with homotopy types and not spaces, there is no difference between EG and the contractible homotopy type $*$.

The possibility to construct such classifying objects is the main application of the property of effectivity of groupoid quotients.

Something similar exists in rings. Let A be a ring and M an A-module; then $A_M := A \oplus M$ has the structure of a ring where the product of two elements of M is zero. Such a ring is called a *trivial first order extension* of A. It has a surjective map $A_M \to A$ with kernel M. More generally, any quotient of commutative rings $A' \to A$ with kernel M is called a *first-order extension* of A by M if the product of two elements of M is zero.[53] It is classical that such extensions can be classified by cocycles of A with values in M.

The work of Quillen on cohomology of commutative rings brought a nice reformulation of these notions in more "geometrical" terms. He showed that trivial first-order extensions of A are exactly the bundles of abelian groups over A (defined in the category of rings) and that the first-order extensions of A by M are precisely the torsors over the group bundle $A_M \to A$.

Now, within dg-algebras with their effectivity of groupoid quotients, it is possible to build classifying objects and universal fibrations for these situations. If A and M are fixed, the classifying dg-algebra for A_M-torsors is the trivial first-order extension $A \oplus M[1]$, but where M is put in degree 1, and the universal extension by M is the canonical map $A \to A \oplus M[1]$. More precisely, a map of dg-algebras (up to quasi-isomorphisms) $C \to A \oplus M[1]$ classifies an extension of C by M (where M is viewed as a C-module through the induced map $C \to A$) given by the (homotopy) cartesian square

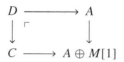

In other words, cocycles of rings can be represented by maps to some classifying object, but this object exists only in dg-algebras.[54]

This property, once interpreted geometrically, is the heart of the obstruction theory used in deformation theory, which was one of the motivations to define derived geometry. We shall not expand on this in this text; we refer to [64, 70] for a more detailed account.

The definition of first-order extensions of rings generalizes in the obvious way to dg-algebras. In particular, it is possible, even for a ring, to consider

---

[53] Recall that, geometrically, first-order and, more generally, any nilpotent extensions $B \to A$ correspond to infinitesimal thickenings.

[54] In comparison with the situation of homotopy types, the role of A is played by the contractible type $*$, and the fact that M is in degree 1 in $A \oplus M[1]$ is similar to BG, where G is the π_1.

extensions where M is a complex of A-modules. In particular, it is possible to prove the following result: any dg-algebra A can be built from $H_0(A)$ as the limit of a tower of first-order extensions (called the Postnikov tower of A)

$$A \to \ldots \to A_{\leq n} \to A_{\leq n-1} \to \ldots \to A_{\leq 1} \to H_0(A),$$

where the extension $A_{\leq n} \to A_{\leq n-1}$ is by the module $H_n(A)[n]$.

The work of Quillen actually went a bit further. If $A \oplus M$ is a trivial first-order extension, then a map of algebras $C \to A \oplus M$ is simply a derivation of $C \to M$ (where M is viewed as a C-module through the induced map $C \to A$). Classically, such maps are in bijection with maps of C-module $\Omega_C \to M$, where Ω_C is the module of differentials of C. In the context of dg-algebras where M can be a chain complex, Ω_C is not enough, and derivations $C \to M$ are now represented by a complex of C-module $\mathbb{L}_C$, called the *cotangent complex*, such that $H_0(\mathbb{L}_C) = \Omega_C$. This proves that the cotangent complex is nothing but the module of differentials but in the context of dg-algebras. In particular, the maps $C \to A \oplus M[1]$ classifying extensions by M are equivalent to maps of chain complexes $\mathbb{L}_C \to M[1]$, so $\mathbb{L}_C$ "controls" the theory of extensions.

Geometrically, this says something new: it says that infinitesimal thickenings of a derived affine scheme are controlled by its tangent bundle. We refer again to [64, 70] for more details.

3.2.6 Other Contexts of Derived Rings

We have presented the ∞-category of dg-algebras as a convenient context to enhance the theory of rings. However, this is not the only context that is used to derive the theory of commutative rings and the corresponding geometry. We briefly review a list of other contexts.

To derive algebraic geometry, simplicial rings can also be used. They should be thought as a commutative ring with an underlying homotopy type instead of a set. This is the original setting for derived algebraic geometry presented in [68]. The theory of simplicial rings is equivalent to that of dg-algebras in characteristic zero, but better suited in positive characteristic. One can also use *unbounded dg-algebras*, but the theory turns out to be a bit awkward (essentially the infinitesimal theory is more subtle); some details are given in [68].

Lurie is using yet another model, E_∞-ring spectra (i.e., multiplicative cohomologies theories), which can be thought as commutative rings with an underlying spectrum (in the sense of stable homotopy theory). These objects are what become commutative rings when the full structure of associativity and commutativity of both the product and the addition are relaxed homotopically (in comparison, simplicial rings are such that the addition and commutation of the product are kept strict). This context, which was an original motivation for the whole theory, is well suited to transform into geometry the results of stable homotopy theory. This compares to the way arithmetic properties have been translated into geometry with the theory of schemes (see the introduction of [46]).[55]

The fruitful (and elegant) algebraization of the geometry of polynomial and rational functions has always inspired attempts to turn the geometries of differentiable and holomorphic functions into algebra. An important problem to do so has been the extra-structure that the rings of those functions should have. The first idea to do so is probably to add some topology to these rings (in order to take care of convergence questions). But Lurie, inspired by synthetic differential geometry, has advertised another way to do so.

For the purposes of differential geometry, the theory of simplicial C^∞-rings seems well suited.[56,57] Some foundational work has been done in [34, 61], but the theory is not much developed yet. An obstruction is certainly the importance in the field of geometrical and analytical methods (in particular problems of convergence) over algebraic methods. Another might be the natural focus of the theory on smooth objects rather than singular ones, whereas scheme theory (derived or not) is intended for a study of singular objects. However, the recent development of derived symplectic geometry, and the potential application to Floer homology, could bring some motivations for the development of the theory.

---

[55] The plurality of these generalizations of commutative rings can be explained with the notion of algebraic theory (in the sense of Lawvere). The problem to define "homotopy rings" can be set into the problem of finding an algebraic theory such that the models in the category of sets are commutative rings; then the models in the category of homotopy types will be "homotopy rings." Dg-algebras, simplicial rings, and E_∞-ring spectra are different such theories. Moreover, if the algebraic theory is encoded by an ordinary category, the category of models in homotopy types is always a colimit completion of the category of models in sets.

[56] And probably the "dg-" version also, although I do not know if anybody has looked into this.

[57] The idea of a C^∞-ring is the following: in a classical ring A, the operations of sum and product can be combined to give more elaborated n-ary operations $A^n \to A$ indexed by any polynomial in $\mathbb{Z}[x_1, \ldots x_n]$. With this in mind, C^∞-rings are rings with n-ary operations given not only by polynomial functions but by any C^∞ function of n variables; for example, in a C^∞-ring, it makes sense to compute $\cos(e^x + y)$ for any elements x and y. An important difference with classical rings is that these operations cannot be generated from the sum and product only. The reader interested in learning more should look at [53] and learn about Lawvere theories. In contrast to topological rings, C^∞-rings, by enhancing the addition and product of the ring with operations indexed by C^∞ functions, provide a *purely algebraic* setting where methods from algebraic geometry can be adapted. In particular, the tensor product of C^∞-rings is the good one with no need for completion.

Finally, Lurie proposed also a similar theory for the purposes of complex geometry. The idea is the same – try to avoid topological rings and find an algebraic structure encoding holomorphic functions – but the definition is a bit more subtle than for C^∞-rings.[58] However, the theory seems to work nicely enough to have results such as a GAGA theorem [55].

3.3 Geometry of Derived Affine Schemes

We have explained how the theory of rings could benefit from its enlargement into the theory of dg-algebras. Although it should be clear enough that most results of commutative analogues have analogs for dg-algebras, it is not immediately clear that dg-algebras have also a geometrical side. The geometric objects corresponding to dg-algebras are named *derived affine schemes*, but what do they look like, and what are their new features with respect to ordinary affine schemes?

As we mentioned, classical commutative rings are related to geometry because it is possible to interpret the language and structure of geometry (points, open subsets, fibrations, etc., and their expected structural relations) in the opposite category of commutative rings. We shall sketch how it is possible to do the same for dg-algebras.

3.3.1 Comparison with Ordinary Schemes and Infinitesimal Structure

If A is dg-algebra (in nonnegative homological degrees), there always exists a quotient map $A \to H_0(A)$ (killing higher degree elements). This suggests that the derived affine scheme X corresponding to A is related to the classical affine scheme X_0 corresponding to $H_0(A)$ by a map $X_0 \to X$. X_0 is called the *classical part* or *truncation* of X.

It is reasonable to think that the points of this geometry will still be in correspondence with fields in dg-algebras. But, for reasons of degree, such fields have to be concentrated in degree 0, so there are no new points in the new geometry. Actually more is true: a map $A \to k$ from a dg-algebra A to a field k has to factor through $H_0(A)$. This says that the map $X_0 \to X$ induces a bijection on points.

It is also reasonable to expect that the correspondence between ring quotients and closed immersions will stay true in the new context. So $X_0 \to X$ would be a closed immersion inducing a bijection on points. Classically, this

---

[58] The trouble is essentially the fact that not every open subset is the complement of zeros of a holomorphic function. This forces us to encode within the ring of holomorphic functions the data of these extra open subsets. For this reason, the definition of a *holomorphic ring* is a bit more involved than the definitions of C^∞-rings and ordinary rings (see [49, 55]).

would say that X is an infinitesimal thickening of X_0. And this is the way it turns out to be: a derived affine scheme is an ordinary affine scheme endowed with an infinitesimal thickening, although this thickening can be of a new kind.[59]

3.3.2 Self-Intersections

One of the interests of schemes is that they give rise to spaces of infinitesimally closed points. For example, the ring $\mathbb{C}[x]/x^n$ is the ring of function on the subspace of the affine line $\mathbb{A}^1$ formed by $n+1$ points infinitesimally close. Derived geometry also gives rise to new spaces, for example, spaces of self-intersection.

Let us consider the simplest case of the self-intersection of a point in the line. The dg-algebra corresponding to this is the (derived) tensor product $\mathbb{C} \otimes_{\mathbb{C}[x]} \mathbb{C}$, which can also be written as a quotient of $\mathbb{C}[x]$ by the irregular system of equations $\{x = 0, x = 0\}$. With our previous notations, if $A = \mathbb{C}[x]$ and $E = \{x, x\}$, the computation of $K(E)$ gives the dg-algebra $\mathbb{C} \oplus \mathbb{C}[1]$ (with trivial differential). This algebra has to be understood as such: it is essentially $\mathbb{C}$, but with a nontrivial loop at any number (given the the generator of $\mathbb{C}[1]$).

Let X be the geometric object corresponding to $K(E)$; its classical part X_0 is a point x since $H_0(\mathbb{C} \oplus \mathbb{C}[1]) = \mathbb{C}$. What is the thickening of the point giving X? It is difficult to say: the structure is "purely derived" and not easy to describe geometrically, particularly in classical terms. However, a few things can be said. First, recall from Section 2.2.2 that, for the pointed scheme $(\mathbb{A}^1, 0)$, the homotopy fiber product $* \times_{\mathbb{A}^1} *$ is nothing but the loop space of $\mathbb{A}^1$ at 0. Since $* \times_{\mathbb{A}^1} *$ is also the self-intersection X, the nontriviality of X means that the affine line has a nontrivial loop space at 0!

This situation is not specific to $\mathbb{A}^1$: unless the point is isolated, the self-intersection $* \times_X *$ of a point x in any derived affine scheme X will be nontrivial. Then, since a loop space is a group, we can understand its Lie algebra and its actions. The Lie algebra of $* \times_X *$ can be proven to be $\mathbb{T}_x X[1]$, the tangent complex of X at x with degree shifted by -1 (see Section 2.2.2). As for the action of $* \times_X *$, let us say only that in the same way the loop group of a homotopy type acts on the fiber of any covering space, $* \times_X *$ acts on

---

[59] Formally, such a thickening is still encoded by nilpotent extensions, as mentioned in Section 3.2.5, but they are harder to interpret geometrically when the extension module has positive homology. With classical infinitesimal thickenings, the limit of two points converging to each other is understood as two points infinitesimally closed, i.e., a vector. But derived infinitesimal thickenings encompass also the limit of the intersection of two lines (say, in the plane) when the lines become the same. What then becomes of the intersection point at the limit? I do not know how to understand this in geometrical terms.

the fiber at x of any A-module (where A is the dg-algebra corresponding to X) [27, 36].

Finally, it can be proved that the classifying space (or better, the classifying scheme) of this group is the formal completion of X at x (see [64] for more on the matter).[60]

3.3.3 Intrinsic Geometry

The definition of geometric notions in derived algebraic geometry is done essentially the same way as in algebraic geometry. We refer to [45, 46, 68] for details.

Points correspond to fields (viewed as dg-algebras concentrated in degree zero), and the set of points of a derived scheme X is defined as in algebraic geometry by equivalence classes of maps from points. It is possible to define a notion of localization for dg-algebras and a corresponding notion of *Zariski open immersion*. It is also possible to define the notion of *smooth* and *étale* maps for dg-algebras (either by lifting properties of infinitesimal thickening or by using the cotangent complex). Moreover, all these notions satisfy the expected properties (stability by composition, base change, locality, etc.).

Once the idea that the extra structure of a derived affine scheme is infinitesimal is accepted, it is easy to reduce certain features of X/A to features of $X_0/H_0(A)$. For example, since open subsets contain the infinitesimal neighborhood of their points, the Zariski open subsets of X are in bijection with Zariski open subsets of its truncation X_0 in the classical sense. Also, using the infinitesimal lifting property, the category of étale maps over X can be shown to be equivalent to that of étale maps over X_0. As a consequence, a dg-algebra A is local (resp. Henselian) if and only if $H_0(A)$ is local (resp. Henselian). Another consequence is that the extra derived structure does not create more Zariski or étale topology, with the consequence that the Zariski and étale spectra of A and $H_0(A)$ will coincide.

There exists also a notion of *closed immersion*, corresponding to (homotopy) quotients of dg-algebras, and this is a notion with quite a different behavior than the classical one. Essentially, the main difference could be summarized by saying that closed immersions $Y \to X$ are not immersions.[61] If $Y \to X$ is a closed immersion, then it is not a monomorphism, i.e., the

---

[60] This relationship between the formal neighborhood of a point and its loop stack, called *infinitesimal descent* in [64], is still a mysterious feature of derived geometry for me.

[61] Algebraically, this is somehow related with the fact that the notion of ideal is no longer suited to describing homotopy quotients of dg-algebras: ideals only encode quotients by equivalence relations, but more is needed to describe all homotopy quotients.

canonical diagonal map $Y \to Y \times_X Y$ is not an isomorphism, as we saw in Section 3.3.2.

It is possible to define also notions of *maps of finite presentation, flat maps, separated maps*, and *proper maps* (although this notion is more interesting for nonaffine schemes). Altogether, the whole formal structure of algebraic geometry (in particular, all the aforementioned classes of maps) has been successfully transposed into derived algebraic geometry.

3.3.4 Geometrization via Spectra

Derived schemes are presented in [46] and [64] as particular ringed spaces. I have voluntarily chosen to delay this presentation because it is not intrinsic. Indeed, such a presentation using topological spaces is implicitly based on the notion of Zariski spectrum of a dg-algebra (generalizing the same notion for rings), but other notions of spectra exist (étale, Nisnevich, etc.), and there is no reason to prefer the Zariski one to the others (in fact, there are good reasons not to).

From the class of Zariski open immersions into a derived scheme X, it is possible to construct a topological space $Spec_{Zar}(X)$ called the *Zariski spectrum of* X (or of A if X is dual to A). Similarly, from the étale maps with target X, it is possible to construct a topos,[62] $Spec_{et}(X)$, called the *étale spectrum of* X. Both these spectra functors are not faithful (a lot of dg-algebras have the same spectrum), but it is possible to enhance them into faithful functors by adding a structure sheaf. Moreover, together with their structure sheaf, the spectra can be interpreted as moduli spaces such that the structure sheaf becomes the universal family: the Zariski spectrum of A is the moduli space of localizations of A and the étale spectrum of A is the moduli space of strict Henselizations of A.[63] This nice point of view, inherited from classical considerations on spectra (since at least [26]), is fully developed for derived geometry by Lurie in [46, 48].

Spectra are nice because they are genuine spaces and not the abstract objects of the opposite category of rings or dg-algebras. They are definitely useful for geometric intuition: for example, the Zariski spectrum of a ring can convince anybody immediately that rings indeed have something to do with geometry. However, the representation of the spatial nature of a ring via its Zariski spectrum (or any spectrum in fact) can also be misleading. The reason is that

---

[62] Toposes are what become topological spaces if the poset of specialization maps between points is allowed to be replaced by a category. They provide a notion of highly unseparated spaces (way below T_0-spaces) well suited to study spaces such as bad quotients of group actions and moduli spaces; see Chapter 4.

[63] This point of view is also suited for the Nisnevich spectrum, which is the space enlarging the étale spectrum by classifying all Henselizations of A and not only the strict ones.

some geometrical features can be well understood in terms of the underlying space of the spectrum, but others still hide within the structure sheaf. For example, a scheme can have many closed subschemes, but its Zariski spectrum can be a single point. Also, the underlying space of the Zariski spectrum does not commute with products (there are more points in $\mathbb{A}^2$ than pairs of points in $\mathbb{A}^1$). Hence, the Zariski topological space reflects quite poorly the intrinsic geometry of affine schemes (already in the underived setting). The étale spectrum is better[64] but also suffers the problems listed above. Another drawback of both Zariski and étale spectra, specific to derived geometry, is that they do not reflect the nontrivial nature of loop spaces at a point described in Section 3.3.2, and one is forced to hide this bit of geometry within the structure sheaf.[65]

In fact, each notion of spectrum should be taken as a reflection, or a projection, of the true, or intrinsic, geometric nature of affine schemes (derived or not); and these projections are preferred one to another, as Poincaré would say, for the sake of convenience (typically to define or to control cohomology theories). In my opinion, the most convincing fact that dg-algebras (or, already, rings) have a geometrical side is not so much their spectral theories as the successful interpretation of the whole language and expected structure of geometry they provide. To be able to talk about points, open subsets, étale maps, proper maps, fiber bundles, and other geometrical features is more important to doing geometry than to be able to faithfully describe a ring or a dg-algebra as a genuine space with a structure sheaf.[66]

3.4 Derived Stacks

3.4.1 Quotients of Spaces and Stacks

The construction of nonaffine spaces in algebraic geometry is done by taking affine schemes as elementary building blocks and by defining the new objects

---

[64] An important drawback of the Zariski spectrum is that it does not reflect étale maps of rings into étale maps of topological spaces, which forbids us from using it to define cohomology theories with étale descent. This problem was the motivation to introduce the étale spectrum in order to define étale cohomology, but the toll was to work with spaces even more unseparated than Zariski spectra. These spaces, which cannot be defined as topological spaces, were themselves the motivation for introducing toposes. The situation is the same in derived geometry, where ∞-toposes must be used.

[65] Is there a notion of spectrum that could reflect this "infinitesimal descent" (as Toën calls it in [64]) in terms of classical descent in toposes? See also Section 3.3.2 and Footnote 60.

[66] Such an interpretation of the language of geometry is precisely what is missing for noncommutative rings, essentially because there is no good notion of locality for noncommutative rings (which leads Kontsevich to joke about "noncommutative nongeometry"). In consequence, there exists no satisfying noncommutative geometry in the sense of building a spectral theory for noncommutative rings. There are, however, a number of features of geometrical objects that can still be defined for noncommutative objects (properness, smoothness, etc.) (Chapter 10).

to be formal pastings, that is, formal colimits, of these blocks. Let *Aff* be the category of affine schemes; the construction of all formal colimits from objects of *Aff* is the meaning of the category Pr(*Aff*) of presheaves of sets over *Aff*.[67] However, this construction is "too free" because it does not paste the pieces of a Zariski (or étale) atlas of an affine scheme X to X itself.[68] This leads to distinguishing certain pastings of affine schemes (Zariski, étale, or any other class of atlases) and to considering rather the completion Sh(*Aff*) of *Aff* that preserves these pastings. Technically, the distinguished atlases define a Grothendieck topology, and the completion is the category of sheaves of sets for this topology.[69]

Now, following the philosophy presented in Section 3.2.1, to have a better behavior of colimits, we should replace them by homotopy colimits in the construction of sheaves. This is precisely the definition of stacks: *stacks are to sheaves what homotopy types are to sets*. Or, in other terms, the category of stacks is the free homotopy colimit completion of the category of affine schemes (preserving atlases).[70] Technically, presheaves of sets are replaced by presheaves of homotopy types, called *prestacks*, and the definition of stacks copies that of sheaves by imposing that the homotopy colimits of atlases of

---

[67] It is a classical result of category theory that the Yoneda embedding $C \to \text{Pr}(C)$ of a category C into its presheaves of sets, sending an object to its so-called *functor of points*, is the free completion of C for the existence of colimits.

[68] Recall that an étale atlas for an affine scheme X is a family of étale maps $U_i \to X$ covering X. In the category *Aff*, X can be recovered as the quotient of an equivalence relation on the disjoint union $\coprod_i U_i$ (subcanonicity of étale topology). However, the quotient $|U_\bullet|$ of this equivalence relation in the category Pr(*Aff*) of presheaves (of sets) will not be (the image by Yoneda embedding of) X. Indeed, because colimits are "freely" added in Pr(*Aff*), the Yoneda embedding *Aff* $\to$ Pr(*Aff*) does not preserve the colimits existing on *Aff*. All we get is a canonical morphism $|U_\bullet| \to X$. The category of sheaves is then defined by imposing that the maps $|U_\bullet| \to X$, for all étale coverings of all affine schemes, be isomorphisms. Technically, such an operation is a localization of categories and corresponds to a quotient of Pr(*Aff*), but it can also be described as a full subcategory of $\mathcal{P}(Aff)$ (this is similar to the fact that a quotient can be described as a subset by imposing a gauge condition). Precisely, a presheaf F is called a *sheaf* (for the étale topology) if it "sees" all the maps $|U_\bullet| \to X$ as isomorphisms, i.e. if all the maps $\text{Hom}(X, F) \to \text{Hom}(|U_\bullet|, F)$ are isomorphisms. This condition is often called the *descent condition* [16, 60, 63].

[69] For the reader unfamiliar with sheaves, this completion is analogous to the more classical completion of $\mathbb{Q}$ into $\mathbb{R}$: a sheaf is very close to a Dedekind cut and sheaves are defined as "formal colimits" in the same way real numbers are defined as formal limits of sequences.

[70] Stacks were invented and used before this universal property was understood. This elegant description of the category of stacks (which only is true if stacks means higher stacks or ∞-stacks) is the result of a long interaction between algebraic geometry (where stacks were defined first), algebraic topology (where all the necessary homotopical techniques were invented) and the philosophy of higher category theory (where the notion of homotopy colimit completion was conceived).

X be X itself.[71] Intuitively, this means that we need to change our idea of a space with an underlying set of points to a notion of space with an underlying homotopy type (or ∞-groupoid) of points. The resulting morphisms between the points are of the same nature as the specialization morphisms in classical topology.

As with sets, complexes, and dg-algebras, a fundamental property of the category of stacks is the effectivity of groupoid quotients. This property is actually the essential reason to consider stacks and stacky quotients: if a group G acts on space X, then anything defined over the quotient stack $X/\!/G$ (function, bundle, sheaf, etc.) can equivalently be described by something over X that is equivariant for the action of G.

Finally, *derived stacks* are defined exactly the same way starting from the ∞-category $dAff$ of derived affine spaces instead of Aff, relative to one of the Grothendieck topologies existing on them (usually the étale topology) [46, 48, 68].

3.4.2 Geometric Stacks

Derived stacks are formal, or free, homotopy colimits of derived affine schemes in a special sense. The category of stacks is useful because of this completion property, but, as often with completions, it turns out that the general object of this category is quite wild and improper to the purposes of geometry. The problem is the local structure of these objects: for example, the tangent spaces may not have an addition.[72] Concretely, to have a tamed, or *geometric*, local structure (with tangent spaces, infinitesimal neighborhoods, and everything), we would like to describe the neighborhood of a point in a stack, as for affine schemes, by means of a local ring (maybe Henselian or more).[73, 74] However, this is not possible in general. Any stack Y can be described as

---

[71] This is done as in Footnote 68. Recall the inclusion $Set \to S$ of the category of sets into the ∞-category of homotopy types from 3.2.4. Let $\mathcal{P}(Aff)$ be the category of functors $Aff^{op} \to S$ (called *prestacks*). The inclusion $Set \to S$ induces an inclusion $Pr(Aff) \to \mathcal{P}(Aff)$ and a Yoneda embedding $Aff \to \mathcal{P}(Aff)$. The category of stacks (for the étale topology) is defined from $\mathcal{P}(Aff)$ by inverting the maps $|U_\bullet| \to X$, for all étale coverings of all affine schemes. It can be described as subcategory of prestacks: a prestack F is called a *stack* if the maps $\text{Hom}(X, F) \to \text{Hom}(|U_\bullet|, F)$ are all equivalences of homotopy types [16, 60, 63].

[72] It is only possible to define a tangent cone. The situation is the same with diffeologies (see Chapter 1). This is also related to the *microlinearity* condition in synthetic differential geometry.

[73] This problem is already present with sheaves; it has nothing to do with the homotopical nature of stacks.

[74] In the analogy of Footnote 69 between sheaves (or stacks) and real numbers, the wild nature of general sheaves and stacks compares well to that of general real numbers. To work with a given real number, it is better to give it more structure, like the property to be algebraic. Algebraic numbers compare well with "tamed," or "geometric," sheaves and stacks.

the homotopy colimit of some diagram of derived affine schemes X_i, and the local structure of Y can be tamed only if the neighborhoods of points of Y can be lifted to neighborhoods of points in some X_i for some diagram presenting Y, that is, if the quotient map $\coprod_i X_i \to Y$ has a lifting property for neighborhoods.

A map $f: X \to Y$ is said to have the *lifting property for étale neighborhoods* if, for any scheme Z, corresponding to a strict Henselian local ring A, with unique closed point named z (such a Z is an étale neighborhood of z) and for any commutative square as follows, there exists a diagonal lift

A sufficient condition for a map to admit a such lifting property is to be smooth (submersive); moreover, the lift is unique if the map is étale.[75] This leads to

Table 9.2. *Types of spaces*

	Basic objects	Objects with Zariski atlas	Objects with étale atlas	Objects with smooth atlas
Sheaves (ordinary AG)	affine schemes	schemes	algebraic spaces	
Stacks (AG + derived colimits)			Deligne–Mumford stacks	Artin stacks
Derived stacks (AG + derived limits and colimits)	derived affine scheme	derived schemes	derived D–M stacks	derived Artin stacks
Tangent complex	in d° ≤ 0	in d° ≤ 0	in d° ≤ 0	unbounded

[75] Several kinds of neighborhoods exist in (classical or derived) algebraic geometry. Let X be an affine scheme, A its ring of functions, k a field, and x a point of X given by a map $A \to k$. The most common neighborhoods of x in X are the Zariski, Nisnevich, étale (related to the corresponding Grothendieck topologies), and formal neighborhoods. They are defined, respectively, as the local ring of A at x, the Henselian local ring of A at x, the strict Henselian local ring of A at x, and the formal completion of A at x. Depending on the kind of local

considering those diagrams such that the quotient map $\coprod_i X_i \to Y$ is smooth or étale. This is done by restricting to specific diagrams, called smooth or étale *internal groupoid objects*, and by considering only colimits that are quotients of those groupoid objects. If X is the quotient of a groupoid G, then G is called an *atlas* for X. The choice of smooth or étale maps distinguishes two classes of geometric stacks called, respectively, *Artin stacks* and *Deligne–Mumford stacks*. For such stacks, it is possible to define local features such as tangent complexes, and we get back all computations of Section 2. We shall not say more here and refer to [62, 64] for more explanation and to [46, 56, 68] for the details. Table 9.2 summarized all the types of geometrical objects that we have mentioned.

3.4.3 Geometry of Derived Stacks

We have finally arrived at the notion of a derived geometric stack, which is the basic object of derived algebraic geometry.[76] It is convenient to compare derived geometric stacks with classical schemes or sheaves by means of a diagram (which I borrow from Toën and Vezzosi):

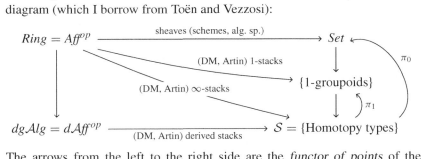

The arrows from the left to the right side are the *functor of points* of the differents objects of the theory. On the left side are the elementary blocks, or affine objects, and on the right are the categories of values for the *points* of the objects. It is convenient to read this as "the points of a scheme (or an algebraic space) form a set, the points of a 1-stack form a groupoid, and the points of a derived stack form a homotopy type (or ∞-groupoid)."

We mentioned in the introduction that derived stacks could be truncated like chain complexes; this is done by means of the previous diagram. If $X: d\mathit{Aff} \to \mathcal{S}$ is a derived stack, its *derived truncation* is the classical ∞-stack $Z_{\geq 0} X = \mathit{Aff} \to d\mathit{Aff} \to \mathcal{S}$, and its *coarse space* is the sheaf

---

structure in which we are interested, the lifting property for maps can be defined with respect to any of these notions. The most common choices are the *smooth* and *formally smooth* maps defined with the étale and formal neighborhoods, respectively. All this can be transcribed in analytic and differential geometry, where there are even more choices of neighborhood notions (e.g., Gevrey classes).

[76] Their definition can be transposed in differential or complex geometry with no trouble.

$\pi_0 Z_{\geq 0} X = \mathit{Aff} \to \mathit{Set}$. At the level of the tangent complex $\mathbb{T}_x X$, for some point x in X, these operations correspond to the positive truncation $Z_{\geq 0} \mathbb{T}_x X$ and to $H_0(\mathbb{T}_x X)$.

Finally, let us mention that all the notions developed in Section 3.3.3 for derived schemes (points, open immersion, étale, smooth and proper maps, etc.) can be generalized for derived stacks, so the language of geometry finds yet another model. In particular, it is possible to define Zariski and étale spectra of stacks. A very nice theorem of Lurie [48] states that Deligne–Mumford derived stacks can be described faithfully by their étale spectra (defined as an ∞-topos) together with the canonical structure sheaf. We refer to [46, 68] for the definitions of geometric features of derived stacks.

3.5 Derived Geometry and the BRST Construction

Applications of derived geometry are largely explained in [62, 64]. We shall say only a few things about it related to the connection with symplectic geometry and BRST construction.

We mentioned in the introduction that the first formalisms of derived geometry were algebraic, using dg-algebras (commutative or Lie) as opposed to geometric. Perhaps the origin of this is to be found in physics when the BRST construction was developed in the 1970s. We shall briefly tell how to interpret the algebraic construction of [41] in terms of derived geometry.

Recall first that a symplectic reduction is constructed in two steps: from a symplectic manifold, we extract a subspace (possibly with intersection singularities) and then take a quotient by a group action (possibly creating further quotient singularities). From everything that we have tried to explain in this text, it should be clear that derived geometry is the perfect setting in which to compute such a reduction. And indeed, such a reduction is one of the main operations of the new field of *derived symplectic geometry*, where it is proven that the resulting object, however singular, is still symplectic; see [4, Chapter 4], [8, 54] for more details.

Now, to make the connection with [41], we are going to construct a functor $\mathcal{O}: d\mathit{Stack} \to dg\mathcal{A}lg$ that sends a derived stack X to an unbounded dg-algebra $\mathcal{O}(X)$ playing the role of the algebra of (globally defined) functions in X. Let $d\mathit{Stack}$ be the ∞-category of derived stacks, and recall that the Yoneda embedding $d\mathit{Aff} \to d\mathit{Stack}$ is a homotopy colimit completion; then the construction of $\mathcal{O}$ is done using this universal property: if X is a derived affine scheme corresponding to a dg-algebra A, we put $\mathcal{O}(X) = A$, and for a general derived stack X, we write it as a colimit of derived affine schemes $X = \mathrm{colim}\, X_i$, and we put $\mathcal{O}(X) = \lim A_i$, where the (homotopy) limit is computed in the ∞-category $dg\mathcal{A}lg$ of unbounded dg-algebras. The resulting functor $\mathcal{O}\colon dStack \to dg\mathcal{A}lg$ is far from faithful, so it is not true that derived geometry can be replaced by a study of unbounded dg-algebras. This setting is more the poor man's approximation to derived geometry.

Coming back to [41], let us start with a symplectic algebraic variety[77] M with a group action $G \times M \to M$ and a moment map $\mu\colon M \to \mathfrak{g}^*$. The system of equations $\mu = 0$ (usually irregular) can be used to define the level set $\mu^{-1}(0)$ as a derived affine scheme X, encoded by a Koszul dg-algebra $K(\mu = 0)$ (see Section 3.2.3), which is $\wedge\mathfrak{g} \otimes \mathcal{O}(M)$ with $\mathfrak{g}$ in (homological) degree 1 and the Koszul differential d. Now recall that, by design, $\mathcal{O}$ sends colimits of derived stacks to limits of dg-algebras; in particular, if M_0 is the quotient of X by the canonical action of G, $\mathcal{O}(M_0)$ will be described as a limit construction involving $K(\mu = 0)$ and functions on G. This limit is the meaning of the Chevalley complex construction built from $\mathfrak{g}^*$ on $K(\mu = 0)$, which is $\wedge\mathfrak{g} \otimes \mathcal{O}(M) \otimes \wedge\mathfrak{g}^*$ (where $\mathfrak{g}^*$ is in degree -1) with total differential $d + \delta$, where the Koszul differential d is twisted by the Chevalley differential δ. This corresponds exactly to the two steps of the construction of [41] and interprets the resulting dg-algebra as the algebra of functions on the derived symplectic reduction M_0.

4 Conclusion

4.1 Success and Prospects

The background for the development of derived geometry was a number of techniques used in geometry (algebraic, complex, or differential) but not fully justified by geometrical reasons; they were ad hoc computational tools, efficient but somehow mysterious. Here is a short list:[78]

- homological features in commutative algebra (derived tensor products of rings, Serre formula, Koszul resolutions, cotangent complexes, virtual classes, etc.);
- equivariant techniques in geometry to work on bad quotients;
- cohomology with coefficients in bundles (or sheaves), derived categories and functors;
- the chain complexes with their Lie structure controlling deformation problems.

---

[77] The setting for symplectic manifolds has not yet been developed.
[78] To which one could add the use of geometric methods to understand some features of stable homotopy theory (the stack of formal groups for cobordisms, the stack of elliptic curves for topological modular forms), which were also ill justified by the theory. But I am going to forget these aspects and restrict to more classical geometry. See [46] for more on the matter.

We have tried to explain how derived geometry has successfully found *geometrical principles* behind these constructions:

- Once commutative rings (affines schemes) are enhanced into dg-algebras[79] (derived affine schemes), derived tensor products, Serre formulas, Koszul resolutions, cotangent complexes, and virtual classes become the proper notions of tensor product, intersection formula, quotient, Kähler differential, and fundamental classes (see Sections 3.2.3 and 3.2.5). Moreover, this enhancement is of a geometric nature (albeit surprising) because the full language and structure of geometry make sense in derived affine schemes (see Section 3.3.3).
- Equivariant techniques correspond indeed to work on the quotient if groupoid quotients are effective, which forces us to compute it in the higher categories of stacks (see Section 3.4.1). Moreover, as above, this extension of affine objects by stacks is of a geometric nature because the full language and structure of geometry extend to stacks (see Section 3.4.3).
- Once the setting of dg-algebras is accepted, the natural categories of modules are formed by chain complexes and not only modules, even for ordinary rings. From this point of view, the old abelian categories of modules are no longer regarded as fundamental objects, and the problem of derivation of functors that goes along with them somehow disappears. Derived functors are replaced by the natural functors between ∞-categories of chain complexes.[80] By extension, the categories of modules on a derived stack are also ∞-categories of chain complexes.[81]
- The chain complexes of deformation problems are simply the tangent spaces of derived geometry. The Lie structure is explained by the loop stacks (see Section 2.2.2), the obstruction theory by the structure theory of dg-algebras (see Section 3.2.5).

Altogether, this presents derived geometry as a new theory of geometry, not only encompassing the classical geometry, but having much better computational properties with respect to the whole homological/homotopical apparatus. The heart of this new geometry is to keep track of the inherent ambiguity to identify things when computing quotients (whether they be quotients of rings or quotients of spaces) by considering always homotopy

---

[79] Or simplicial algebras, or E_∞-ring spectra.

[80] This approach to the derivation of functors does even specify the role played by abelian categories where they become *hearts* of *t-structures*.

[81] Strictly speaking, when presented this way, only nonnegative chain complexes should be considered. The consideration of unbounded chain complexes is motivated because they have the nice extra property of *stability* (see Footnote 87).

quotients.[82] This is incredibly well suited to the study of moduli spaces, as can be checked by the list of examples in [62, 64].

Having said so, the above tools and techniques certainly do not exhaust the methods existing in geometry, and derived geometry is not the answer to all problems.[83] As we have tried to explain, it is only an enhancement taking better care of intersection and quotient singularities. But since singular spaces appear also in differential and complex geometries, derived geometry does have something to say in these contexts. For example, because of the more regular tangential structure at singularities given by the tangent complexes and not only the tangent spaces, it is possible to define on the whole of singular spaces (and not only on the smooth locus) a proper notion of differential forms and of symplectic or Poisson structures. These definitions have led to a huge extension of the notion of symplectic variety and to a very nice algebra of operations producing symplectic spaces from other ones (symplectic reductions of course, but also intersections of Lagrangian maps and mapping spaces) [4, Chapter 4], [9, 10, 54, 65]. So far this extension has only been done in the algebraic setting, but no doubt something similar could be done in differential and complex geometries .

4.2 Higher Categorical Mathematics

Some ideas subsuming derived geometry are not actually specific to geometry; they are ideas of higher category theory. The cross-breeding, in the 1990s to 2000s, of algebraic geometry, algebraic topology, and category theory, which gave birth at the same time to higher stack theory and higher category theory, may very well be one of the most fruitful of mathematics because it lays the ground work for a deep revolution that we would like to advertise.[84]

We left pending in Section 3.2.4 and Footnote 51 the question of the necessity of ∞-categories in the derivation process. It is indeed a natural question to ask why ∞-categories have become so important. The answer, which we can only sketch here, is simple and deep: ∞-categories provide computational properties inaccessible to ordinary categories. For example, we have underlined several times in this text the role played by the important property of *effectivity of groupoid quotients*, that is, the ability to work

---

[82] Instead of derived geometry, I think this should be more appropriately called the *geometry of ambiguity*, hence the title of this text.
[83] Notably, derived geometry has little to say about spaces of infinite dimension.
[84] Actually, derived geometry may very well be the first field born out of higher categorical ideas. Other fields could be stable category theory [45], homotopy type theory (see Chapter 6 or [32]), or derived symplectic geometry [8].

on quotients of groupoids by equivariant methods (see Section 3.2.4). This property of a category has the most remarkable property: if it is true in an ordinary category, then this category is the trivial category with one object.[85] This means than this property can only have nontrivial models in proper ∞-categories (and it does – we have seen a few)![86] Another very important such property, although outside the scope of this text, is the axiom of *stability*, which enhances and simplifies the theory of triangulated categories [45].[87, 88]

A second, quite natural question about ∞-categories is, why are homotopy types (or ∞-groupoids) so important? Here again there exists a simple and deep answer: because they can be seen as a notion more primitive than sets. This might seem silly, at first, because all mathematical objects, including homotopy types, are classically defined using sets as a primitive notion. But, assuming the notion of homotopy types, it is also possible to define sets as *discrete homotopy types*, that is, as homotopy types with a specific property.

Following this line of thought proposes to redefine the whole of mathematical structures with an underlying homotopy type instead of an underlying set. The first motivation for such a bold idea is the fact that examples of such notions of "structures with underlying homotopy type" exist (as we have tried to advertise in this text). But the main motivation is the fact that homotopy types provide a notion of identification for elements (through paths) more suited to talking about certain structures. We already mentioned how homotopy types could be seen as an enhancement of sets incorporating the ambiguity of identifications (see Section 3.2.1 and Footnote 37). Another situation is the following, of a more logical flavor. With respect to the manipulation of vector spaces, the set of all vector spaces is less natural than the groupoid of all vector spaces: since all constructions on vector spaces are expected to be invariant by isomorphisms, they should be defined with respect to the latter and not only the

---

[85] More precisely, $(n-1)$-groupoids can be effective in n-categories, but the effectivity of n-groupoids in an n-category forces it to be trivial. Only when $n = \infty$ can n-groupoids be effective in an n-category.

[86] This is in particular the case of all the so-called ∞-toposes and of all the ∞-categories of models of algebraic theories taken in an ∞-topos. For example, chain complexes or dg-algebras are models of algebraic theories in the topos S of homotopy types, hence their effectivity property.

[87] *Stability* is the property of a pointed category that any commutative square is a pushout if and only if it is a pullback, or equivalently that finite colimits commute with finite limits. Again, only the trivial ordinary category with one object is stable, but there are many nontrivial stable ∞-categories: ∞-categories of (unbounded) chain complexes are stable, and so are ∞-categories of spectra and parameterized spectra. See [45] for details and examples.

[88] From a logical point of view, one would say that there exist logical theories (or syntaxes) that have nontrivial semantics only in higher categories, for example, an object X such that $\Omega \Sigma X \simeq X$, where Ω and Σ are the loop space and suspension functors.

former. Developing a language based on homotopy types instead of sets would provide automatically that any construction be invariant by isomorphism.[89]

Since the 1960s, the best practical approach has been Quillen's model categories (and their variations), which underlie all approaches to higher categories. However, model categories are a way to reduce higher category theory to ordinary category theory, and the feeling is that there should be a proper theory for higher categories. Recently, homotopy type theory (see Chapter 6 or [32]) has proposed an interesting syntactic approach to homotopy types/∞-groupoids that could provide a foundational language for mathematical alternatives to set theory axioms. This is well suited to getting some aspects of homotopy types but largely insufficient for the purposes of the working mathematician. Another promising attempt is the development of tools to work directly in the quasi-category of quasi-categories [11].

4.3 Toward a New Axiomatization of Geometry?

The objects of derived geometry are more complex than ordinary manifolds or even schemes, but they enjoy better properties than their classical counterparts. So what should we prefer? This is actually an illustration of a tension that exists within mathematics about whether to define its objects individually by some intrinsic structure (the affine plane as $\mathbb{R}^2$) or in a family by an algebra of operations on the family (planar geometry with its figures and incidence rules). Algebraic geometry, particularly under the influence of Grothendieck, has rather favored the simplicity of properties over the simplicity of its objects: points at infinity (projective geometry), points with complex coordinates (complex geometry), and multiple points (schemes) were in particular motivated by the regularization of the intersection properties of the plane. Derived geometry, with its treatment of singularities, is but the latest step in the same direction. Its objects might be strange in their individual nature, and they may have new properties, but, as a whole, they behave more regularly, and remarkably, the language to talk about them is still the same.

We have insisted several times on the fact that the geometric nature of objects like rings, dg-algebras, sheaves, and stacks was in the successful interpretation of the language of geometry (together with some expected structural properties) such as points, étale and smooth maps, and proper maps (see Sections 3.3.3, 3.3.4, and 3.4.3). Actually, in front of the many settings where this language makes sense (topological spaces, toposes, manifolds, and

---

[89] This would be a strong version of Leibniz's principle of indiscernables. This strengthening is false with ZFC or axioms and the like. However, this issue is one of the motivations for Martin-Löf to have introduced its identity types [32]; within this syntax, all propositions about, say, groups are automatically stable by isomorphism.

schemes, and derived stacks being the latest), it is tempting to try to axiomatize an abstract setting for *geometry* as a category together with classes of maps corresponding to all the aforementioned classes. The notion of geometry of Lurie [48] is a first attempt in this direction, and so is Schreiber's setting for "differential cohesive homotopy theory" [58] and [4, Chapter 5]. Such an axiomatization of geometry, emphasizing the structure of the relations between the objects of the geometry rather than the structure of the objects themselves, would be the 21st-century version of Euclid's axioms.

References

[1] M. Anel, *Why deformations are cohomological*, http://mathieu.anel.free.fr/mat/doc/Anel%20-%20WhyDeformationAreCohomological.pdf

[2] M. Anel, *The homotopy quotient – When the notations mean more than what we make them say*, http://mathieu.anel.free.fr/mat/doc/Anel-Semiomaths-HomotopyColimit.pdf Notes available here.

[3] M. Anel, *Introduction aux catégories triangulées*, http://mathieu.anel.free.fr/mat/doc/Anel%20-%20CategoriesDerivees.pdf

[4] M. Anel and G. Catren, eds., *New Spaces in Physics: Formal and Conceptual Reflections*, Cambridge University Press (2021)

[5] M. Artin, *Versal deformations and algebraic stacks*, Invent. Math. 27 (1974) 165–89

[6] J. Baez and M. Shulman, *Lectures on n-categories and cohomology*, in *Towards Higher Categories*, IMA Vol. Math. Appl. 152, Springer (2010) 1–68

[7] K. Behrend, *Differential graded schemes II: The 2-category of differential graded schemes*, Preprint, arXiv:math/0212226

[8] D. Calaque, *Lagrangian structures on mapping stacks and semi-classical TFTs*, Preprint, arXiv:1306.3235

[9] D. Calaque, T. Pantev, B. Toën, M. Vaquié, and G. Vezzosi, *Shifted Poisson structures*, J. Topol. 10 (2017) 483–584

[10] D.-C. Cisinski, *Les préfaisceaux comme modèles pour les types d'homotopie*, Astérisque, no. 308 (2006)

[11] D.-C. Cisinski, *Higher categories and homotopical algebra*, Cambridge Studies in Advanced Mathematics 180, Cambridge University Press (2019)

[12] I. Ciocan-Fontanine and M. Kapranov, *Derived Quot schemes*, Ann. Sci. École Norm. Sup. 34 (2001) 403–40

[13] P. Deligne and D. Mumford, *The irreducibility of the space of curves of given genus*, Inst. Hautes Études Sci. Publ. Math. 36 (1969) 75–109

[14] M. Demazure and P. Gabriel, *Groupes algébriques. Tome I: Géométrie algébrique, généralités, groupes commutatifs*, Masson & Cie (1970)

[15] V. Drinfeld, letter to V. Schechtman, https://www.math.harvard.edu/?tdp%2Ftranslation_pdf

[16] D. Dugger, *A primer on homotopy colimits*, https://pages.uoregon.edu/ddugger/hocolim.pdf

[17] D. Eisenbud, *Commutative Algebra with a View toward Algebraic Geometry*, Graduate Texts in Mathematics 150, Springer (1995)

[18] I. Gálvez-Carrillo, J. Kock, and A. Tonks, *Homotopy linear algebra*, Preprint, arXiv:1602.05082

[19] N. Gambino, *Weighted limits in simplicial homotopy theory*, J. Pure Appl. Algebra 214 (2010) 1193–99

[20] S. I. Gelfand and Y. I. Manin, *Methods of Homological Algebra*, 2nd ed., Springer Monographs in Mathematics, Springer (2003)

[21] G. Ginot, *Introduction to differentiable stacks (and gerbes, moduli spaces...)*, https://webusers.imj-prg.fr/~gregory.ginot/papers/DiffStacksIGG2013.pdf

[22] J. Giraud, *Cohomologie non abélienne*, Die Grundlehren der mathematischen Wissenschaften 179, Springer (1971)

[23] M. Goldmann, *The deformation theory of representations of fundamental groups of compact Kähler manifolds*, Inst. Hautes Études Sci. Publ. Math. 67 (1988) 43–96

[24] A. Grothendieck, *Catégories cofibrées additives et complexe cotangent relatif*, Lecture Notes in Mathematics 79, Springer (1968)

[25] A. Grothendieck, *Pursuing stacks*, 1984, Archive Grothendieck cote 134, https://grothendieck.umontpellier.fr/archives-grothendieck/

[26] M. Hakim, *Topos annelés et schémas relatifs*, Ergebnisse der Mathematik und ihrer Grenzgebiete 64, Springer (1972)

[27] B. Hennion, *Tangent Lie algebra of derived Artin stacks*, J. die reine angewandte Math. 741 (2018) 1–45

[28] V. Hinich, *DG coalgebras as formal stacks*, J. Pure Appl. Algebra 162 (2001) 209–50

[29] V. Hinich and V. Schechtman, *Deformation theory and Lie algebra homology. II*, Algebra Colloq. 4 (1997) 213–40

[30] A. Hirschowitz and C. Simpson, *Descente pour les n-champs*, Preprint, arXiv:math/9807049

[31] P. S. Hirschhorn, *Model Categories and Their Localizations*, Math. Surv. Monogr. 99, AMS (2003)

[32] The Univalent Foundations Program, *Homotopy Type Theory: Univalent Foundations of Mathematics*, Institute for Advanced Study (2013), https://homotopytypetheory.org/book/

[33] A. Joyal, *Notes on quasi-categories*, http://www.math.uchicago.edu/~may/IMA/Joyal.pdf

[34] D. Joyce, *Algebraic geometry over C^∞-rings*, Mem. AMS (forthcoming)

[35] D. Kan, *Functors involving c.s.s. complexes*, Trans. Amer. Math. Soc. 87 (1958) 330–46

[36] M. Kapranov, *Rozansky–Witten invariants via Atiyah classes*, Comp. Math. 115 (1999) 71–113

[37] M. Kapranov, *Injective resolutions of BG and derived moduli spaces of local systems*, J. Pure Appl. Algebra 155 (2001) 167–79

[38] A. Kock, *Synthetic Differential Geometry*, 2nd ed., London Mathematical Society Lecture Note Series 333, Cambridge University Press (2006)
[39] M. Kontsevich, *Enumeration of rational curves via torus actions*, Progr. Math. 129 (1995) 335–68
[40] M. Kontsevich, *Deformation quantization of Poisson manifolds*, Lett. Math. Phys. 66 (2003) 157–216
[41] B. Kostant and S. Sternberg, *Symplectic reduction, BRS cohomology, and infinite-dimensional Clifford algebras*, Ann. Phys. 176 (1987) 49–113
[42] D. Knutson, *Algebraic Spaces*, Lecture Notes in Mathematics 203, Springer (1971)
[43] M.-B. Laumont, *Champs algébriques*, Ergebnisse der Mathematik und ihrer Grenzgebiete 3. Folge. A Series of Modern Surveys in Mathematics 39, Springer (2000)
[44] J. Lurie, *Higher Topos Theory*, Annals of Mathematics Studies 170, Princeton University Press (2009)
[45] J. Lurie, *Higher algebra*, http://people.math.harvard.edu/~lurie/papers/HA.pdf
[46] J. Lurie, *Spectral algebraic geometry*, http://people.math.harvard.edu/~lurie/papers/SAG-rootfile.pdf
[47] J. Lurie, *A survey of elliptic cohomology*, Preprint, http://people.math.harvard.edu/~lurie/papers/survey.pdf
[48] J. Lurie, *DAG V – Structured spaces*, Preprint, http://people.math.harvard.edu/~lurie/papers/DAG-V.pdf
[49] J. Lurie, *DAG IX – Closed immersions*, Preprint, http://people.math.harvard.edu/~lurie/papers/DAG-IX.pdf
[50] J. Lurie, *DAG X – Formal moduli problems*, Preprint, http://people.math.harvard.edu/~lurie/papers/DAG-X.pdf
[51] M. Manetti, *L-infinity structures on mapping cones*, Algebra Number Theory 1 (2007) 301–30
[52] J. Mrčun and I. Moerdijk, *Introduction to Foliations and Lie Groupoids*, Cambridge Studies in Advanced Mathematics 91, Cambridge University Press (2003)
[53] I. Moerdijk and G. Reyes, *Models for Smooth Infinitesimal Analysis*, Springer (1991)
[54] T. Pantev, B. Toën, M. Vaquié, and G. Vezzosi, *Shifted symplectic structures*, Publ. Math. Inst. Hautes Études Sci. 117 (2013) 271–328
[55] M. Porta and T. Y. Yu, *Higher analytic stacks and GAGA theorems*, Adv. Math. 302 (2016) 351–409
[56] J. P. Pridham, *Deformations of schemes and other bialgebraic structures*, Trans. Amer. Math. Soc. 360 (2008) 1601–29
[57] C. Rezk, *Toposes and homotopy toposes*, https://faculty.math.illinois.edu/~rezk/homotopy-topos-sketch.pdf
[58] U. Schreiber, *Differential cohomology in a cohesive ∞-topos*, https://arxiv.org/abs/1310.7930v1
[59] C. Simpson, *Algebraic (geometric) n-stacks*, Preprint, arXiv:alg-geom/9609014
[60] C. Simpson, *Descent, Alexandre Grothendieck: A Mathematical Portrait*, International Press (2014) 83–141
[61] D. Spivak, *Derived smooth manifolds*, Duke Math. J. 153 (2010) 55–128

[62] B. Toën, *Higher and derived stacks: A global overview*, Proc. Sympos. Pure Math. 80, Part 1 (2009) 435–87
[63] B. Toën, *Master course on stacks*, https://perso.math.univ-toulouse.fr/btoen/videos-lecture-notes-etc/
[64] B. Toën, *Derived algebraic geometry*, EMS Surv. Math. Sci. 1 (2014) 153–240
[65] B. Toën, *Derived algebraic geometry and deformation quantization*, Preprint, arXiv:1403.6995
[66] B. Toën and G. Vezzosi, *From HAG to DAG: Derived moduli stacks*, NATO Sci. Ser. II Math. Phys. Chem. 131 (2004) 173–216
[67] B. Toën and G. Vezzosi, *Homotopical algebraic geometry, I. Topos theory*, Adv. Math. 193 (2005) 257–72
[68] B. Toën and G. Vezzosi, *Homotopical algebraic geometry, II. Geometric stacks and applications*, Mem. Amer. Math. Soc. 193 (2008)
[69] J.-L. Verdier, *Des catégories dérivées des catégories abéliennes*. Astérisque, no 239 (1996)
[70] G. Vezzosi, *An overview of derived algebraic geometry*, Slides from a talk at RéGA (October 9, 2013), https://webusers.imj-prg.fr/~yue-tony.yu/rega/pdf/1314/09102013a.pdf
[71] C. Weibel, *An Introduction to Homological Algebra*, Cambridge Studies in Advanced Mathematics 38, Cambridge University Press (1994)

Mathieu Anel
Department of Philosophy, Carnegie Mellon University
mathieu.anel@gmail.com

10
Geometry in dg-Categories

Maxim Kontsevich

Contents

1	Introduction	554
2	Noncommutative Analogs of Usual Notions	561
3	Exotic Geometry in Noncommutative World	578
Appendix: Technical definitions		584
References		591

1 Introduction

1.1 Triangulated Categories and Spaces

Various spaces in algebraic geometry and topology naturally give rise to abelian categories of sheaves. In the case of schemes (algebraic varieties), we consider quasi-coherent sheaves, while for topological spaces, we consider local systems of abelian groups. More generally, for a given stratification of a topological space, we can consider constructible sheaves that are locally constant on open strata. As usual, it is much better to work with triangulated categories of complexes of quasi-coherent sheaves (resp. complexes of sheaves whose cohomology sheaves are constructible with respect to a given stratification). In both cases the triangulated categories of sheaves are *symmetric monoidal* categories with the monoidal structure given by the tensor product. The idea of **derived noncommutative geometry** is to interpret a general triangulated category[1] *without* the symmetric monoidal structure as (morally) "the category of quasi-coherent sheaves on a noncommutative space."

The goal of this text is to review noncommutative analogs of many notions from the commutative geometry[2] and also to show several genuinely new noncommutative phenomena. This is a part of the mental picture that I have tried to develop during the last two decades in collaboration with many colleagues. Homological mirror symmetry was the main source of inspiration (but not the only one). Definitely there are other viewpoints on what should be called "noncommutative geometry"; one can hope that a harmonious picture will emerge someday.

1.2 Examples of dg-Algebras and dg-Categories

1.2.1 Representation Theory: Quivers, Quantum Groups, and So On

For any associative algebra A over some ground field $\mathbf{k}$, we have an associated unbounded derived category $D(A\text{-mod})$ of A-modules. It has several dg-models (all equivalent in dg sense). The first one is by semi-free dg-modules (see Appendix A1), which is the dg-category of unbounded complexes $E^\bullet = \left((E^i)_{i\in\mathbb{Z}}, d\right)$ of free A-modules satisfying the following condition: there exists a bigraded vector space over $\mathbf{k}$

$$G = \bigoplus_{i\in\mathbb{Z},\ j\in\mathbb{Z}_{\geq 0}} G^i_j$$

such that $E^i = A \otimes_\mathbf{k} \oplus_{j\geq 0} G^i_j$ as A-module, and $dG^i_j \subset A \otimes_\mathbf{k} \left(\oplus_{j'<j} G^{i+1}_{j'}\right)$. Another version is by all dg-modules and canonical $RHom$-complexes (see Appendix A3), or by A_∞-modules over A (see Appendix A5; this description works for arbitrary A_∞-algebra A). Derived Morita-equivalence is an equivalence between dg-categories $A\text{-mod} \simeq A'\text{-mod}$ for two dg-algebras, and it is given (similarly to the usual Morita-equivalence) by a pair of dual bimodules perfect on each side (see Appendix A4).

Various branches of mathematics provide examples of associative algebras of quite different flavors. From the point of view of universal algebra, the most basic example is tensor algebra freely generated by vector space V:

$$Free(V) := \oplus_{n\geq 0} V^{\otimes n}, \quad V \in \mathbf{k}\text{-mod}.$$

---

[1] The most adequate formalism is that of stable $(\infty, 1)$-categories (see, e.g., [20]). Here we will use triangulated differential graded (dg for short) or A_∞-categories; see the appendix.

[2] It is quite astonishing that so many aspects of the commutative picture can be formulated in noncommutative terms, despite the fundamental lack of the notion of locality.

If we choose a basis $(v_i)_{i\in I}$ of V, then the basis of $Free(V)$ consists of finite words (including the empty one), in the alphabet I. A generalization of free algebra is the path algebra of a quiver (directed graph).

A traditional topic in the theory of rings is the study of finite-dimensional algebras and their representations. If the ground field **k** is algebraically closed, then any finite-dimensional algebra is Morita-equivalent to a finite-dimensional quotient of the path algebra of a finite quiver (the latter can be thought as dual to a generalized "noncommutative affine space").

Representation theory (related to Langlands program) of Lie groups and p-adic groups give rise to a different family of algebras. In the Archimedean case, one considers $U\mathfrak{g}$-modules where $\mathfrak{g}$ is a finite-dimensional Lie algebra over $\mathbb{C}$ (or Harish–Chandra modules). The universal enveloping algebra $U\mathfrak{g}$ is a deformation of the symmetric algebra $Sym(\mathfrak{g})$ and hence has polynomial growth (unlike the free algebra). A closely related class of algebras are algebras of polynomial differential operators on smooth affine algebraic varieties. Quantum groups give some other deformations of commutative algebras.

1.2.2 Algebraic Geometry

Let X be a scheme over field **k**. The unbounded derived category of the category of quasi-coherent sheaves admits (as a triangulated category) an essentially canonical dg-enrichment. Let $(U_i \subseteq X)_{i\in I}$ be a collection of open affine subschemes parameterized by a partially ordered set I such that

1. $i_1 \preceq i_2$ implies $U_{i_1} \subseteq U_{i_2}$;
2. $X = \cup_{i\in I} U_i$;
3. for any $i_1, i_2 \in I$, we have $U_{i_1} \cap U_{i_2} = \cup_{j\in I: j \preceq i_1, i_2} U_j$.

For example, the collection of *all* open affine subschemes in S suffices. Then consider the following small **k**-linear dg-category $\mathcal{D} = \mathcal{D}_{(U_i)_{i\in I}}$ (in fact, a category with morphisms only in degree 0): the set of objects is I, and for any $i, j \in I$, we define the space of morphisms $\mathrm{Hom}_{\mathcal{D}}(j,i)$ to be 0 if $i \not\preceq j$ and

$$\mathrm{Hom}_{\mathcal{D}}(j,i) := \mathcal{O}(U_i) \text{ if } i \preceq j.$$

Any $\mathcal{D}$-module gives by definition a collection $(\mathcal{E}_i)_{i\in I}$ of dg-modules over (usual) **k**-algebras $\mathcal{O}(U_i), i \in I$, related with each other by some auxiliary data. In particular, for any $i \preceq j$, we obtain a morphism of complexes

$$\mathcal{O}(U_i) \otimes^L_{\mathcal{O}(U_j)} \mathcal{E}_j \to \mathcal{E}_i,$$

where the tensor product is understood in the derived sense (see Appendix A3). The dg-category of semi-free dg-modules can be called the "dg-category of quasi-coherent presheaves" for the covering $(U_i \subset X)_{i \in I}$. It contains a naturally defined full subcategory of "quasi-coherent sheaves":

Theorem 1.1 *The full dg-subcategory of semi-free dg-modules consisting of $(\mathcal{E}_i)_{i \in I}$ such that for any $i \preceq j$, the morphism of complexes $\mathcal{O}(U_i) \otimes^L_{\mathcal{O}(U_j)} \mathcal{E}_j \to \mathcal{E}_i$ is a quasi-isomorphism ("sheaf property"), is a triangulated dg-category whose associated plain **k**-linear triangulated category is naturally equivalent to $D(QCoh(X))$.*

This result is slightly nonorthodox. The inclusion "sheaves" $\hookrightarrow$ "presheaves" has a *right* adjoint (not a left one, as for usual sheaves and presheaves). We will return to this description of $D(QCoh(X))$ at the end of Section 2.13.

1.2.3 Topology: Chains on Paths, Constructible Sheaves

Let X be a locally contractible Hausdorff topological space. Denote by P_X the *path category* of X, enriched over the monoidal category of topological spaces. The set of objects is X, while the spaces of morphisms are

$$Hom_{P_X}(x, y) := \{(t, \phi) \mid t \in \mathbb{R}_{\geq 0}, \phi \in X^{[0,t]}, \phi(0) = x, \phi(t) = y\},$$

with the composition given by the concatenation of paths. Applying to the spaces of morphisms the functor of simplicial chains

$Chains_\bullet$: Topological spaces $\to \mathbb{Z}_{\leq 0}$ — graded complexes of abelian groups

and the Eilenberg–Zilber product

$$Chains_\bullet(Y_1) \otimes Chains_\bullet(Y_2) \to Chains_\bullet(Y_1 \times Y_2),$$

we obtain a dg-category that we denote by C_X. The dg-category of semi-free C_X-modules is a dg-model of the category of complexes of sheaves with locally constant cohomology sheaves.

If X is homotopy equivalent to a simplicial complex, then one can make an explicit algebraic model of C_X. Let us assume for simplicity that X is the geometric realization of a subcomplex $X_\bullet$ of the standard simplex Δ^N. The set of objects is $X_0 \subset \{0, \ldots, N\}$. The dg-category $C_{X_\bullet}$ is freely generated by

$$a_{i_0, \ldots, i_k} \in \underline{Hom}^{1-k}_{C_{X_\bullet}}(i_k, i_0), \quad \forall (i_0 < i_1 < \cdots < i_k) \in X_k, \ k \geq 1,$$

with the condition that *all* arrows in degree 0 corresponding to nondegenerate 1-simplices are *invertible*. The differential is given by the formula

$$d(a_{i_0, \ldots, i_k}) = \sum_{j=1}^{k-1} (-1)^j \left(a_{i_0, \ldots, i_{j-1}, i_{j+1}, \ldots, i_k} + a_{i_0, \ldots, i_j} \cdot a_{i_j, \ldots, i_k} \right).$$

More generally, one can invert only a part of the arrows in degree 0. In this way one can model a modification of C_X: for a reasonable stratified space, consider spaces of paths staying in the closure of the stratum containing the starting point $\phi(0)$ of the path. Semi-free contravariant dg-modules form a dg-model for the derived category of complexes of sheaves constructible with respect to the given stratification.

Algebraic geometry should give interesting (especially in the case of positive characteristic) analogs of such categories, namely, derived categories of constructible l-adic sheaves on a given stratified scheme.

1.3 Examples of Noncommutative Equivalences

First, there are numerous examples of different algebras with equivalent derived categories of representations. For example, if Q is a quiver with a distinguished vertex v which is a source (i.e., there is no arrow with the head at v), then reversing directions of all arrows with the tail at v, we obtain a new quiver Q' such that path algebras $\mathbf{k}Q$ and $\mathbf{k}Q'$ are derived Morita-equivalent (Gelfand–Ponomarev mutation). Similarly, in algebraic geometry, there are numerous examples of different varieties with the same derived category of quasi-coherent sheaves (e.g., Fourier–Mukai transform for dual abelian varieties; see Section 3.1).

There are obvious "tautological" equivalences between triangulated dg-categories coming from algebra, algebraic geometry, and topology. Namely, if $X = \mathrm{Spec}\,(A)$ is an affine scheme where A is a commutative algebra over $\mathbf{k}$, then $D(QCoh(X)) = D(A\text{-mod})$. Also, if X is CW-complex, which is of homotopy type $K(\pi, 1)$ for certain (possibly noncommutative) group π), then the derived category of complexes of sheaves of $\mathbf{k}$-vector spaces on X whose cohomology sheaves are local systems is equivalent to the category $H^0(A\text{-mod}^{s.f})$, where $A := \mathbf{k}[\pi]$ is the group ring.

In particular, for $\pi = \mathbb{Z}$, we have three descriptions of the same category

$$D\left(QCoh(\mathbb{A}_\mathbf{k}^1 - \{0\})\right) \simeq D(\mathbf{k}[x^{\pm 1}]\text{-mod}^{s.f})$$
$$\simeq D_{constr}(S^1, \text{trivial stratification}).$$

If we replace the trivial stratification of S^1 by the one with two strata (zero- and one-dimensional cells), we obtain a new category that again has three descriptions (in the middle: representations of Kronecker quiver K_2):

$$D(QCoh(\mathbb{P}_\mathbf{k}^1)) \simeq D(\bullet \rightrightarrows \bullet) \simeq D_{constr}(S^1, \{\text{point}\} \sqcup \text{interval}).$$

1.4 Comparison with Other Geometries and Further Challenges

There are several versions of derived or noncommutative geometry where one has some of notion of spectrum. First of all, there is a very useful and well-developed **derived commutative algebraic geometry** by B. Toën and G. Vezzosi [27, 28], J. Lurie [21], and D. Gaitsgory and N. Rozenblyum [10]. This is a theory directed toward the construction of "right" moduli stacks and it is a direct derived generalization of the usual algebraic geometry, with Zariski topology on the nose. Roughly speaking, one uses dg-commutative rings instead of the usual commutative rings. The positively graded part is interpreted as a pro-nilpotent thickening of the degree 0 part (higher obstructions in deformation theory), while the negatively graded part is replaced by a stacky generalization (symmetries, symmetries of symmetries, etc.). Higher derived stacks in this theory give rise to tensor (i.e., symmetric monoidal) stable categories of complexes of quasi-coherent sheaves.

P. Balmer [1] proposed a notion of a **prime spectrum of a symmetric monoidal triangulated category** reproducing the usual spectrum in the commutative case. Besides derived stacks, one has many exotic symmetric monoidal triangulated dg-categories. For example, any dg PROP (i.e., a collection of multilinear operations with several inputs and outputs satisfying certain universal relations, such as the notion of a Hopf algebra) gives rise tautologically to a symmetric monoidal category that can be thought of as the category of sheaves on the "universal" moduli space of algebras over PROP. Thus any notion in mathematics that can be expressed in terms of linear algebra gives rise to a generalized derived commutative space.

In principle, there are weaker versions of symmetry for monoidal categories, for example, totally nonsymmetric or braided symmetry. I am not aware of any meaningful theory of spectra in such cases.

Another direction is to abandon completely the monoidal structure. Before going to categories, one can start just with noncommutative rings. There are some theories of spectra based on two-sided ideals, but they seem to be too rude, as many algebras of interest in representation theory are simple. There is a much more meaningful theory of left spectra (based on left ideals), where, for example, the spectrum of algebra of differential operators on smooth affine variety X looks roughly like the subset of the Zariski spectrum of T^*X consisting of generic points of coisotropic subvarieties in T^*X (with the induced topology). A. Rosenberg [24] defined a **spectrum of abelian category**, again producing the usual Zariski topology when applied to the abelian category of quasi-coherent sheaves, generalizing the earlier result by P. Gabriel for Noetherian schemes.

What I discuss here is an even more general framework, as I consider a general triangulated category *without* any tensor structure and any t-structure (hence no natural abelian category). This is a very dry environment with a restricted freedom to make meaningful general constructions. It seems that all the intuition related to locality breaks down. Natural analogs of open sets (see Section 2.7) *do not* form a lattice. The question is, why bother thinking about a "geometry" for triangulated or dg-categories at all? For me the initial impetus was remarkable discoveries in the early 1990s by A. Bondal and M. Kapranov of the first basic derived noncommutative notions and phenomena (saturated categories, semiorthogonal decompositions, Serre functors, the role of the Calabi–Yau condition). Later, with the advent of homological mirror symmetry, a new geometric intuition arose based on symplectic topology (I will mention it very briefly in Section 3.8). Some locality (sheaf property) is restored in the classical limit, when one gets a constructible cosheaf of dg-categories with very nice finiteness properties (finite type; see Section 2.5) on singular Lagrangian subsets. In this way, one obtains a new geometric description of many classical dg-categories, making various noncommutative phenomena manifestly clear.

In the other direction, it turned out that saturated categories (noncommutative generalizations of smooth proper algebraic varieties) are tailor-made objects for a theory of motives. Recently there was an explosive development concerning saturated (or, in other words, smooth and proper) dg-categories in positive characteristic leading to a noncommutative analog of crystalline cohomology and to the proof of degeneration of the Hodge–to–de Rham spectral sequence in zero characteristic by D. Kaledin [13]. It is obvious now that saturated dg-categories form a more natural and general environment for motives than usual commutative algebraic varieties, although it looks likely that the resulting category of motives (with rational coefficients) is essentially the same as the one coming from the commutative world.

There are several challenges, the most glaring of which is to define an analog of Betti cohomology (in a form of "topological K-theory") and a comparison isomorphism between Betti and de Rham cohomology (periodic cyclic homology). Here one can hope to make contact with A. Connes's version of noncommutative geometry, especially with his theory of integration in terms of Fredholm modules (see [6, Chapter IV, Section 1]). Another rich topic for speculation is a desirable extension of *Kähler geometry* to the derived noncommutative setting, in the framework of Fukaya categories and string theory (stable objects with "harmonic metrics" as algebraic encodings of D-branes).

2 Noncommutative Analogs of Usual Notions

2.1 Families

A. Grothendieck taught us that every notion in algebraic geometry should be (ideally) defined for families, over an arbitrary base scheme. In principle, one should aim for more general bases like higher stacks or stable cocomplete symmetric monoidal categories (e.g., super vector spaces) understood as categories of quasi-coherent sheaves on a generalized (commutative) space. We will not pursue this direction here; all notions will be defined over the base Spec **k**, where **k** is a field. In particular, we will define notions of *smooth* and *proper* **nc**-spaces, generalizing the notion of smooth (resp. proper) morphism to Spec **k** in the usual algebraic geometry. It is not clear whether there are reasonable generalizations of properties of smoothness or properness for general morphisms of **nc**-spaces defined below (see the end of Section 2.6 for a partially successful attempt).

Most of our definitions admit extensions to more general bases. For example, if one wants to define A_∞-algebras over an *arbitrary* commutative ring K, it is sufficient to consider underlying complexes that are semi-free dg K-modules. One can completely avoid the use of *injective* modules, which are ugly set-theoretic beasts in general.

2.2 Definition of nc-Spaces

Definition 2.1 An **nc**-space over base field **k** is a triangulated **k**-linear dg-category which is cocomplete (i.e., admits arbitrary sums; see Appendix A4).[3] Notation: for an **nc**-space X, the corresponding dg-category is denoted by $D(X)$.

Any usual (commutative) scheme (or stack) $Y/\mathbf{k}$ defines an **nc**-space Y_{nc} such that $D(Y_{nc})$ is $D(QCoh(Y))$ endowed with the dg-enrichment as in Section 1.2.2 (for the definiteness one takes I to be equal to the partially ordered set of all open affine subschemes in X).

Similarly, a dg-algebra A over **k** defines an **nc**-space denoted Spec A via

$$D(\text{Spec } A) := A\text{-mod}^{sf}.$$

Notice that unlike in the commutative case, the notation Spec A is completely formal, that is, we do not mean that we have some set of points endowed with

---

[3] There is a general notion (not discussed here) of homotopy diagrams and homotopy (co)limits in dg-categories. A triangulated dg-category has all homotopy colimits if and only if it admits arbitrary sums. This is intrinsically the main motivation for the definition.

a sheaf of rings. Still, the notation is compatible with the commutative one in the sense that for any commutative **k**-algebra A, we have

$$D(\operatorname{Spec} A_{nc}) \simeq D(\operatorname{Spec} A)_{nc},$$

where A_{nc} is A considered as dg-algebra over **k** *forgetting* commutativity. In other words, the dg-category of complexes of quasi-coherent sheaves on affine scheme $\operatorname{Spec}(A)$ is naturally equivalent to the dg-category of semi-free A_{nc}-modules.

From now on we will systematically abuse notation by skipping subscript nc. From our perspective, a *commutative* space (e.g., a scheme or a stack) over **k** is an **nc**-space *together* with the additional structure of a *symmetric monoidal category* (in higher homotopy sense) on $D(X)$.

Definition 2.2 For an **nc**-space X, an object $\mathcal{E} \in D(X)$ is called **perfect** if and only if it is a compact object in $D(X)$, that is, $\underline{Hom}(\mathcal{E}, \cdot) \colon D(X) \to D(\operatorname{Spec} \mathbf{k})$ commutes with direct sums. The full subcategory of $D(X)$ consisting of perfect (compact) objects is denoted by $\operatorname{Perf}(X)$.

The notation $\operatorname{Perf}(X)$ is compatible with the same notation both for schemes and for A-modules. One can also interpret perfect objects as *vector bundles* (or locally free coherent sheaves, generalizations of finitely generated projective modules for usual algebras). For any perfect object $\mathcal{E} \in \operatorname{Perf}(X)$, we have a dg-algebra $A_{\mathcal{E}} := \underline{Hom}_{D(X)}(\mathcal{E}, \mathcal{E})$ and a fully faithful embedding (see Proposition A.13 from Appendix A4)

$$A_{\mathcal{E}}^{op}\text{-mod}^{sf} \hookrightarrow D(X).$$

2.3 Being Affine

Definition 2.3 An **nc**-space X is called **affine** if there exists a perfect $\mathcal{E}$ that ind-generates $D(X)$ (in such case, we have an equivalence $X \simeq \operatorname{Spec}(A_{\mathcal{E}}^{op})$).

The following fundamental fact indicates that affine **nc**-spaces are sufficient for many purposes:

Theorem 2.4 *(A. Bondal, M. van den Bergh,[4]) For any separated scheme of finite type $X/\operatorname{Spec}(\mathbf{k})$ and any closed subset $Z \subset X$, the full subcategory $D_Z(X) \subset D(X)$ consisting of objects with support contained in Z is affine. Moreover, an object $\mathcal{E} \in D_Z(X) \subset D(X)$ is compact in $D_Z(X)$ if and only if it is compact in $D(X)$.*

An easy case is when X is smooth quasi-projective scheme and $Z = X$. One can show that both coherent sheaves

$$\mathcal{E}_1 := \mathcal{O}_X \oplus \mathcal{O}_X(1) \oplus \cdots \oplus \mathcal{O}_X(n)$$
$$\mathcal{E}_2 := (i_{X_0 \hookrightarrow X})_* \mathcal{O}_{X_0} \oplus \cdots \oplus (i_{X_n \hookrightarrow X})_* \mathcal{O}_{X_n}$$

are compact generators of $D(X) = D_X(X)$. Here $n = \dim X$, $\mathcal{O}_X(i)$ denotes the ith power of an ample line bundle on X, and $X_0 \subset X_1 \subset \cdots \subset X_n = X$ is a chain of closed subschemes such that X_0 is finite and each $X_i - X_{i-1}$ is smooth affine of dimension i. Also, if X is quasi-affine (an open subscheme of an affine scheme, e.g., $\mathbb{A}_\mathbf{k}^2 - \{0\}$), then $\mathcal{O}_X$ is an ind-generator of $D(X)$.

We will define below several "geometric" notions for affine **nc**-spaces in terms of the corresponding dg-algebras. All these notions will not depend on the choice of a compact generator, in other words, they will be invariant under the derived Morita-equivalence (see Definition A.16 in Appendix A4).

There are natural situations when there is no single perfect generator, but still a *set* of perfect objects that ind-generates $D(X)$, that is, when the category is equivalent to $\mathcal{C}$-mod^{sf} for some small dg-category $\mathcal{C}$ (see Appendix A1). All the considerations valid for the case of a single generator survive in this larger generality.

For example, let us assume that char($\mathbf{k}$) = 0, $\mathbf{k} = \overline{\mathbf{k}}$, and a reductive group $G/\mathbf{k}$ acts by strict automorphisms of dg-algebra A in such a way that the representation of G in the underlying space is a sum of algebraic finite-dimensional modules (an **nc**-analog of an Artin stack). In this case, the "correct" equivariant category consists of (countable infinite) collections of complexes (M_χ) labeled by the characters of the irreducible representations V_χ of G, together with a G-equivariant structure of dg A-module on $\oplus_\chi (M_\chi \otimes V_\chi)$. The small category $\mathcal{C}$ in this case has its set of objects naturally identified with the set of irreducible characters $\{\chi\}$.

2.4 Tensor Product, Opposite Spaces, and Duality

In what follows, we will use the notion of tensor product of dg-categories (see Appendix A1) and of split triangulated closure (see Proposition A.12 in Appendix A4).

Definition 2.5 For a finite collection of *nc*-spaces $(X_i)_{i \in I}$, define the dg-category of sheaves on their tensor product as the split triangulated closure of the naive tensor product $\otimes_{i \in I} D(X_i)$.

It follows from the definition that for affine spaces, the tensor product corresponds to the tensor product of dg-algebras: Spec $A \otimes$ Spec $B =$ Spec $A \otimes B$. Tensor product corresponds to the usual (cartesian) product for commutative schemes, but it is *not* equal to the categorical product for **nc**-spaces.

Definition 2.6 For any affine **nc**-space $X = \text{Spec}(A)$, define the **opposite** space X^{op} as $\text{Spec}(A^{op})$. For any $\mathcal{E} \in \text{Perf}(X)$, define the **dual** object $\mathcal{E}^{\vee}$ as $\underline{RHom}_{A\text{-mod}}(E, A_{diag}) \in A^{op}\text{-mod}$.

Duality does not depend on the choice of a compact generator and gives a canonical equivalence

$$\text{Perf}(X^{op}) \simeq (\text{Perf}(X))^{op}.$$

Properties of affine **nc**-spaces defined in the next section, such as properness and smoothness, and of finite type are preserved by $X \to X^{op}$ and by tensor product. The support of the dual perfect object $\mathcal{E}^{\vee}$ is opposite to $supp\,\mathcal{E}$.

For every usual scheme X, we have a canonical equivalence $X \simeq X^{op}$ of **nc**-spaces. Duality on perfect objects goes to the usual duality.

2.5 Finiteness Properties: Smooth, Proper, and So On

Definition 2.7 An affine **nc**-space $X = \text{Spec}(A)$ is called (homologically) **smooth** if the semifree resolution of the diagonal bimodule $A_{diag} = A \in A \otimes A^{op}$-mod is perfect.

A separated scheme of finite type $S/\text{Spec}(\mathbf{k})$ is smooth in the **nc**-sense if and only if it is smooth in the usual sense. Many interesting associative algebras are homologically smooth: path algebras of finite quivers (e.g., finitely generated free algebras), universal enveloping algebras of finite-dimensional Lie algebras, and algebras of differential operators on smooth affine varieties.

Definition 2.8 An affine **nc**-space $X = \text{Spec}(A)$ is called of **finite type** if and only if the A_∞-algebra A is a retract in the homotopy category of dg-algebras of the free finitely generated algebra $\mathbf{k}\langle x_1, \ldots, x_n\rangle$, $\deg(x_i) = k_i \in \mathbb{Z}$ endowed with a differential d such that, for any i, $d(x_i) \in \mathbf{k}\langle x_1, \ldots, x_{i-1}\rangle$.

Any **nc**-space of finite type is smooth. All examples of smooth **nc**-spaces from above are of finite type. For example, the affine plane $\mathbb{A}^2_{\mathbf{k}}$ is equivalent to $\text{Spec}\,A$, where A is the free noncommutative algebra in three generators:

$$k\langle x, y, \xi\rangle, \quad dx = dy = 0, \; d\xi = xy - yx, \quad \deg x = \deg y = 0, \deg \xi = -1.$$

An example of an **nc**-space that is smooth but not of finite type is the generic point of the affine line $\text{Spec}\,\mathbf{k}(t)$ (i.e., spectrum in the **nc**-sense of the field of rational functions in one variable considered as a **k**-algebra).

Definition 2.9 An affine **nc**-space $X = \text{Spec}(A)$ is called **proper** if and only if the underlying complex of A has total finite-dimensional cohomology:

$$\sum_{i \in \mathbb{Z}} \dim_{\mathbf{k}} H^i(A, d_A) < +\infty.$$

For any separated scheme X of finite type and *proper*, and any closed subset $Z \subset X$, the affine noncommutative space X_Z defined via

$$\text{Perf}(X_Z) := \text{Perf}_Z(X)$$

using Theorem 2.4 is proper in the **nc**-sense.

If A is a dg-algebra with total finite-dimensional cohomology, then it is A_∞-equivalent to an A_∞-algebra with $\mathbf{m}_1 = 0$ with *finite-dimensional* underlying $\mathbb{Z}$-graded space $V^\bullet := H^\bullet(A, \mathbf{m}_1)$. Choosing a graded basis of $V^\bullet$ and writing down explicitly the equation of higher associativity, we see that an A_∞-structure on V is given by a solution of a countable system of polynomial equations in countably many variables, that is, a $\mathbf{k}$-point in the countable projective limit of affine schemes of finite type over $\text{Spec } \mathbf{k}$.

Definition 2.10 An **nc**-space is called **saturated** if it is simultaneously smooth and proper. A small dg-category is called **saturated** if it is equivalent to $\text{Perf}(X)$ for saturated X.

Theorem 2.11 *Any saturated **nc**-space is of finite type. For any saturated category $\mathcal{D}$ and split-closed triangulated dg-category $\mathcal{C}$, the dg-category of functors $\mathcal{D} \to \mathcal{C}$ is equivalent to the split closure of $\mathcal{D}^{op} \otimes \mathcal{C}$. The opposite of a saturated space (or a category) is also saturated. For any two saturated spaces X, Y, the maps $X \to Y$ are the same as functors $\text{Perf}(Y) \to \text{Perf}(X)$ ($\Leftrightarrow$ objects in $\text{Perf}(Y^{op} \otimes X)$). All such functors have left and right adjoints.*

2.6 Morphisms

Definition 2.12 For two **nc**-spaces X, Y, define a **morphism** $f: X \to Y$ as a pair of adjoint[4] functors $f^* \dashv f_*$

$$f^*: D(Y) \to D(X), \quad f_*: D(X) \to D(Y),$$
$$\underline{Hom}_{D(X)}(f^* \cdot, \cdot) \simeq \underline{Hom}_{D(Y)}(\cdot, f_* \cdot)$$

such that f_* preserves arbitrary sums.

---

[4] See Definition A.15 in Appendix A4 for the notion of adjoint functors in a dg-setting.

The definition implies that the pullback f^* preserves compact objects:

$$f^*(\text{Perf}(Y)) \subset \text{Perf}(X).$$

In the case of affine **nc**-spaces $X = \text{Spec } A$, $Y = \text{Spec } B$ a morphism is the same as a dg-functor $\text{Perf}(Y) \to \text{Perf}(X)$ (the restriction of f^*), and it is given by a bimodule $M_f \in A \otimes B^{op}$-mod that is perfect as an A-module:

$$f^* = M_f \otimes_B \cdot, \quad f_* = (M_f)^{\vee_A} \otimes_A^L \cdot,$$
$$(M_f)^{\vee_A} := \underline{Hom}_{A\text{-mod}}(M_f, A) \in A^{op} \otimes B\text{-mod}.$$

Usual morphisms of schemes induce morphisms in the **nc**-sense. Also, any possibly *nonunital* morphism of dg-algebras $\rho \colon B \to A$ induces a morphism $\text{Spec } A \to \text{Spec } B$, with $M_f = A \cdot \rho(1_A)$. In fact, any morphism between affine **nc**-spaces comes from a nonunital morphism after replacing the initial dg-algebras by appropriate derived Morita-equivalent dg-algebras.

Recall that any functor between cocomplete categories that admits a right adjoint preserves arbitrary sums. Morally (modulo set-theoretic complications) the converse is true. Hence, to define a morphism, it is sufficient (and morally necessary) to have *three* adjoint functors $f^* \dashv f_* \dashv f^!$.

Notice that for given X, Y, morphisms $X \to Y$ are naturally objects of a (large) triangulated dg-category. The existence of the notion of isomorphism (and isomorphisms between isomorphisms) in dg-categories implies that we should consider **nc**-spaces together with their morphisms as an $(\infty, 1)$-category, that is, a category enriched in spaces in the sense of homotopy theory (Kan complexes).

In general, the $(\infty, 1)$-category of *small* split closed triangulated dg-categories is naturally realized as a full subcategory of $(\infty, 1)$-category of **nc**-spaces. The essential image consists of the **nc**-spaces X such that $D(X) = \mathcal{C}$-mod for some small dg-category $\mathcal{C}$. For such **nc**-spaces, one has obviously all homotopy limits and colimits. The full subcategory of affine spaces is closed under finite homotopy limits and colimits. The final and initial object is the empty space with zero category of quasi-coherent sheaves. Pullbacks of affine spaces translate to free cofibered products of dg-algebras, while pushouts translate to homotopy fibered products (see, e.g., beginning of Section 2.13).

One can try, like in the usual algebraic geometry, to introduce notions of smoothness and properness for morphisms. As a first naive attempt, one can call a morphism $f \colon X \to Y$ smooth (resp. proper) if and only if

$$\text{id}_X \in \langle f^* \circ f_* \rangle_{str}, \quad (\text{resp. } f_* \circ f^* \in \langle \text{id}_Y \rangle_{str}).$$

Here we work in the big triangulated categories of endofunctors of $D(X)$ (resp. of $D(Y)$) preserving arbitrary sums (i.e., continuous), and the notation $\langle \cdot \rangle_{str}$ denotes the split triangulated closure (see Proposition A.12 in Appendix A4). To avoid set-theoretic issues, one may assume that X, Y are affine and identify continuous functors with bimodules.

As a partial justification, observe that for an affine **nc**-space $X = \mathrm{Spec}\, A$, the morphism $f : X \to \mathrm{Spec}\, \mathbf{k}$ given by bimodule $M_f := A \in A\text{-mod} - \mathbf{k}$ is smooth (resp. proper) if and only if X is smooth (resp. proper). Also, smooth (resp. proper) morphisms as defined above are closed under compositions, and for the usual morphisms of separated schemes of finite type, the conventional smoothness (properness) implies the derived one. However, the notion of smoothness defined as above does not look satisfactory, since the direct sum of a smooth morphism with an arbitrary one is again smooth.

2.7 Recollement: Closed Subsets and Open Subspaces

Let X be a separated scheme of finite type and $Z \subset X$ be a closed subset for the Zariski topology. The complement $U := X \setminus Z$ is an open subscheme of X. Full subcategories $D_Z(X) \subset D(X)$ and $j^* D(U) \subset D(X)$ (here $j : U \hookrightarrow X$ is the obvious inclusion) form a **semiorthogonal decomposition** of $D(X)$, that is, $H^\bullet(\underline{Hom}_{D(X)}(\mathcal{E}, \mathcal{F})) = 0$ for any $\mathcal{E} \in D_Z(X), \mathcal{F} \in j_* D(U)$, and for any $\mathcal{G} \in D(X)$ there exists an (essentially unique) exact triangle

$$\mathcal{E} \to \mathcal{G} \to \mathcal{F}, \quad \mathcal{E} \in D_Z(X), \mathcal{F} \in j_* D(U).$$

One can mimic such a situation in the **nc** world.

Definition 2.13 For an **nc**-space X, a **closed subset** $Z \hookrightarrow X$ is given by a morphism $i: X \to X_Z$ of **nc**-spaces such that i^* is fully faithful embedding (equivalently, the adjunction morphism $\mathrm{id}_{X_Z} \to i_* \circ i^*$ is an equivalence). The category $i^* D(X_Z)$ is also denoted by $D_Z(X)$.

Any perfect object $\mathcal{E} \in D(X)$ defines a "closed subset" $\mathrm{supp}\,\mathcal{E} \hookrightarrow X$, which we can call the **support** of $\mathcal{E}$. Namely, we declare $D_{\mathrm{supp}\,\mathcal{E}}(X)$ to be the ind-completion of $\mathcal{E}$ in $D(X)$. The **nc**-space $X_{\mathrm{supp}\,\mathcal{E}}$ is affine, as $D_{\mathrm{supp}\,\mathcal{E}}(X) \simeq \mathrm{Spec}\,(B^{op})$, where $B := \underline{Hom}_{D(X)}(\mathcal{E}, \mathcal{E})$.

Definition 2.14 For an **nc**-space X, an **open subspace** in X is given by a morphism $j: U \to X$ of **nc**-spaces such that j_* is a fully faithful embedding (equivalently, the adjunction morphism $j^* \circ j_* \to \mathrm{id}_U$ is an equivalence).

Any closed subset defines an open subspace, and vice versa, for example, $D_Z(X) \hookrightarrow D(X) \hookleftarrow j_* D(U)$ form a semiorthogonal decomposition, or, equivalently, idempotent monad $j_* \circ j^*$ and coidempotent comonad $i^* \circ i_*$ in $D(X)$ are related by the exact triangle

$$i^* \circ i_* \to \mathrm{id}_X \to j_* \circ j^*,$$

where both arrows are the adjunction morphisms.

If $X = \mathrm{Spec}\, A$ is affine, then any open $U \xhookrightarrow{j} X$ is also affine, and the inclusion is given by a morphism of dg-algebras

$$A \to \underline{Hom}_{D(U)}(j^*A, j^*A)^{op}.$$

Conversely, let $\phi: A \to B$ be a morphism of dg-algebras. The corresponding morphism $f_\phi: \mathrm{Spec}\, B \to \mathrm{Spec}\, A$ is an open embedding if and only if the natural morphism (given by the multiplication) is a quasi-isomorphism:

$$B \otimes_A^L B \to B.$$

A morphism of dg-algebras inducing an open embedding will be called a **localization morphism**.

If X is smooth, then any open in X is also smooth. If X is of finite type, then the complement to $supp\, \mathcal{E}$ is also of finite type for any $\mathcal{E} \in \mathrm{Perf}(X)$.

A semiorthogonal decomposition as above is essentially the same as a recollement. Recall that for a morphism f between **nc**-spaces, one should have morally the third adjoint functor. Hence we should have altogether *six* functors:

$$D(U) \underset{j^!}{\overset{j^*}{\underset{\longleftarrow}{\overset{\longleftarrow}{\xrightarrow{j_*}}}}} D(X) \underset{i^!}{\overset{i^*}{\underset{\longleftarrow}{\overset{\longleftarrow}{\xrightarrow{i_*}}}}} D(X_Z).$$

Historically, the notion of recollement was introduced first for triangulated categories of constructible sheaves or of general sheaves (not quasi-coherent ones). In this case, both the notation and the geometric interpretation are different. For a closed embedding $i: Z \hookrightarrow X$ of *topological spaces* and the corresponding open embedding $j: U := X \setminus Z \hookrightarrow X$, one has the following recollement diagram:

$$Sh(Z) \underset{i^!}{\overset{i^*}{\underset{\longleftarrow}{\overset{\longleftarrow}{\xrightarrow{i_*}}}}} Sh(X) \underset{j_*}{\overset{j_!}{\underset{\longleftarrow}{\overset{\longleftarrow}{\xrightarrow{j^*}}}}} Sh(U).$$

2.8 Deformation Theory

In general, if a mathematical object can be described in terms of linear algebra over field **k** with char **k** $= 0$, then its deformation theory is controlled by a dg-Lie algebra over **k** (or, more generally, by a L_∞-algebra).

Any dg-Lie algebra $\mathfrak{g}$ defines a set-valued[5] functor $Def_\mathfrak{g}$ from the category of finite-dimensional commutative **k**-algebras R such that the semisimplification $R/\mathfrak{m}_R$ (here $\mathfrak{m}_R \subset R$ is the Jackobson radical) is equal to **k**. Define the set $Def_\mathfrak{g}(R)$ as

$$\{\gamma \in \mathfrak{g}^1 \otimes \mathfrak{m}_R \,|\, d\gamma + [\gamma,\gamma]/2 = 0\}/ \text{ action } \gamma \mapsto e^\phi d(e^{-\phi})$$
$$+ e^\phi \gamma e^{-\phi}, \; \phi \in \mathfrak{g}^0 \otimes \mathfrak{m}_R.$$

Any dg-algebra A carries a natural dg-Lie algebra structure, with Lie bracket given by the (super)commutant. For any object $\mathcal{E}$ of $D(X)$, the deformation theory is controlled by the dg-Lie algebra $(\underline{Hom}_{D(X)}(\mathcal{E},\mathcal{E}),[\cdot,\cdot],d)$.

If we want to deform not an object but a small dg-category, then we need the following dg-Lie algebra.

Definition 2.15 For a small dg-category $\mathcal{D}$, its **cohomological Hochschild complex** $C^\bullet(\mathcal{D})$ is defined as

$$\prod_{n \geq 0} \prod_{\mathcal{E}_0,\dots,\mathcal{E}_n \in \mathcal{D}} \underline{Hom}_{Comp_\mathbf{k}}\left(\otimes_{i=0}^{n-1} \underline{Hom}_\mathcal{D}(\mathcal{E}_{i+1},\mathcal{E}_i), \underline{Hom}_\mathcal{D}(\mathcal{E}_n,\mathcal{E}_0)\right)[-n],$$

with the differential equal to the sum of differentials coming from complexes $\underline{Hom}_\mathcal{D}(\cdot,\cdot)$ and an additional term δ increasing parameter n by 1:

$$(\delta\phi)(a_0,\dots,a_n) = \pm a_0 \cdot \phi(a_1,\dots,a_n) + \sum_{i=1}^{n-1} \pm \phi(\dots, a_i \cdot a_{i+1}, \dots)$$
$$\pm \phi(a_0,\dots,a_{n-1}) \cdot a_n.$$

The Lie bracket on $\mathfrak{g} := C^\bullet(\mathcal{D})[1]$ is given by

$$[\phi,\psi] = \phi \circ \psi - \pm \psi \circ \phi,$$
$$(\phi \circ \psi)(a_1,\dots,a_n)$$
$$:= \sum_{i<j} \pm \phi(\dots,\psi(a_i,\dots,a_j),\dots).$$

Remarkably, the complex $C^\bullet(\mathcal{D})$ is quasi-isomorphic to the derived endomorphism algebra

$$\underline{RHom}_{\mathcal{D}\otimes\mathcal{D}^{op}\text{-mod}}(\mathcal{D}_{diag},\mathcal{D}_{diag}).$$

One can show that the cohomological Hochschild complex is invariant (up to a quasi-isomorphism) by derived Morita-equivalences.

The Hochschild cohomology groups $HH^\bullet(X) := H^\bullet(C^\bullet(\texttt{Perf}(X)))$ of a smooth scheme X are equal to $R\Gamma(X, \oplus_i \wedge^i T_x[-i])$.

---

[5] One can extend this functor to take values in homotopy types (e.g., Kan complexes).

2.9 De Rham Cohomology

In this section, we assume that char $\mathbf{k} = 0$.

For an affine **nc**-space $X = \mathrm{Spec}\,(A)$, define its Hochschild chain complex as the derived tensor product:

$$C_\bullet(A) := A'_{diag} \otimes^L_{A \otimes A^{op}} A_{diag},$$
$$A'_{diag} := A \in A^{op} \otimes A\text{-mod} = (A \otimes A^{op})^{op}\text{-mod}.$$

A more convenient model for the Hochschild complex is

$$\oplus_{n \geq 0} A \otimes (A[1]/(\mathbf{k} \cdot 1_A)[1])^{\otimes n}$$

endowed with the differential traditionally denoted by b:

$$b(a_0 \otimes \cdots \otimes a_n) := \pm a_n a_0 \otimes \cdots + \sum_{i=1}^n \pm \cdots \otimes a_{i-1} a_i \otimes \cdots$$
$$+ \sum_{i=0}^n \pm \cdots \otimes da_i \otimes \cdots$$

If X is a smooth scheme, then cohomology of its Hochschild complex is equal to $R\Gamma(X, \oplus_{i \geq 0} \wedge^i T^*_{X/\mathbf{k}}[i])$. Notice that, unlike in the usual geometry, forms of degree $i \geq 0$ are placed in *negative* cohomological degree $-i$. The analog of the de Rham differential is *Connes differential B* of degree -1 satisfying $b^2 = bB + Bb = B^2 = 0$. In the convenient complex for dg-algebras, it is given by

$$B(a_0 \otimes \cdots \otimes a_n) := \sum_{i=0}^n \pm 1_A \otimes a_i \otimes a_{i+1} \otimes \cdots \otimes a_n \otimes a_0 \otimes \cdots \otimes a_{i-1}.$$

Finally, one defines **periodic cyclic homology** $HP_\bullet(X)$ as cohomology of the 2-periodic complex

$$C^{per}_\bullet(A) := C_\bullet(A)((u)), \quad \text{with differential } b + uB$$

of vector spaces over the field of Laurent series $\mathbf{k}((u))$ in a formal variable u of cohomological degree $+2$. Remarkably, for any separated scheme of finite type over $\mathbf{k}$ (no smoothness or properness assumed), the periodic cyclic cohomology coincides with crystalline cohomology (which is de Rham cohomology of the formal thickening of X in any ambient smooth scheme), tensored by $\mathbf{k}((u))$ over $\mathbf{k}$. In particular, $HP_\bullet(X)$ is a finite-dimensional space over $\mathbf{k}((u))$. There is no analog of such finiteness in dg-geometry, as the periodic cyclic homology of an **nc**-space of finite type could be infinite-dimensional.

In general, for any saturated space X, Hochschild homology $HH_\bullet(X)$ is a finite-dimensional $\mathbb{Z}$-graded space over $\mathbf{k}$, and also $HP_\bullet(X)$ is finite-dimensional $\mathbb{Z}/2$-graded space over $\mathbf{k}((u))$. One has always an inequality

$$\dim_{\mathbf{k}} HH_\bullet(X) \geq \dim_{\mathbf{k}((u))} HP_\bullet(X).$$

In the case char $\mathbf{k} = 0$, it is always an equality (the fundamental result of D. Kaledin [13] with simplified exposition in [22], a noncommutative version of the degeneration of Hodge–to–de Rham spectral sequence; see [8, 18] for origins and hypothetical generalizations).

2.10 Calculus Relating Hochschild Homology and Cohomology

The Hochschild homology and cohomology groups for a smooth affine scheme S are (respectively) differential forms and polyvector fields:

$$HH_\bullet(S) = \Omega^\bullet(S) := \Gamma(S, \wedge^\bullet T_S^*), \quad HH^\bullet(S) = T_{poly}^\bullet(S) := \Gamma(S, \wedge^\bullet T_S).$$

There are several basic universal operations involving such objects:

- cup-product for forms $\Omega^i(S) \otimes \Omega^j(S) \to \Omega^{i+j}(S)$;
- de Rham differential $\Omega^i(S) \to \Omega^{i+1}(S)$;
- cup-product for polyvector fields $T_{poly}^i(S) \wedge T_{poly}^j(S) \to T_{poly}^{i+j}(S)$;
- Schouten–Nijenhuis Lie bracket $T_{poly}^i(S) \wedge T_{poly}^j(S) \to T_{poly}^{i+j-1}(S)$;
- contraction $T_{poly}^i(S) \otimes \Omega^j(S) \to \Omega^{j-i}(S), \xi \otimes \omega \mapsto i_\xi(\omega)$;
- Lie derivative $T_{poly}^i(S) \wedge \Omega^j(S) \to \Omega^{j-i+1}(S); \xi \otimes \omega \mapsto Lie_\xi(\omega)$;
- dual contraction $\Omega^i(S) \otimes T_{poly}^j(S) \to T_{poly}^{j-i}(S)$.

These operations are related by a bunch of classical identities, the most prominent of which is the Cartan formula $Lie_\xi = [d, i_\xi]$. Remarkably, *all these operations, except the cup-product for forms and the dual contraction, generalize* (together with the relevant identities) to the noncommutative case. For example, for the case of a cohomological Hochschild complex, the Lie bracket and the cup-product define on $HH^\bullet(A)$ the structure of *Gerstenhaber algebra*, that is, the Lie bracket satisfies the graded Jacobi identity on $HH^\bullet(A)[1]$, the cup-product gives the structure of a unital associative graded commutative algebra on $HH^\bullet(A)$, and a graded version of the Leibniz rule holds:

$$[a, b \wedge c] = [a, b] \wedge c + (-1)^{(\deg(a)-1)\deg(b)} b \wedge [a, c].$$

One can check that for any $n \geq 0$, the space of all natural n-linear operations $G^{\otimes n} \to G$ on general Gerstenhaber algebra (obtained by compositions of cup-product and Lie bracket) has dimension $n!$ and is canonically isomorphic to the homology of the configuration space $\mathbb{C}^n - Diagonal$ of n distinct labeled points in $\mathbb{C} \simeq \mathbb{R}^2$. The latter space is homotopy equivalent to the space of collection of n disjoint disks in the standard unit disk in $\mathbb{R}^2$. These spaces (for all $n \geq 0$) form a topological operad called a little disks operad. The Deligne conjecture (proven in many ways by now) says that the associated dg-operad of simplicial chains acts naturally on the cohomological Hochschild complex $C^\bullet(A)$. The generalized Deligne conjecture (proven, e.g., in [18]) adds to this picture the homological Hochschild complex. Additional natural operations

$$\left(C^\bullet(A)\right)^{\otimes n} \otimes C_\bullet(A) \to C_\bullet(A)$$

form a complex quasi-isomorphic to the chain complex of the space of collections of n disjoint disks on the standard cylinder $S^1 \times [0, 1]$ together with a marked point on each boundary circle, modulo the free action of rotation group S^1.

The two-colored operad ("calculus") acting on $(C^\bullet(A), C_\bullet(A))$ is formal ($\Leftrightarrow$ quasi-isomorphic to its cohomology operad). Moreover, for a smooth affine scheme $S = \mathtt{Spec}\,(A)$, the algebra $(C^\bullet(A), C_\bullet(A))$ over the calculus operad is quasi-isomorphic to $\left(T_{poly}^\bullet(S), \Omega^\bullet(S)\right)$ endowed with the classical operations.

It can be deduced from the calculus that for a family of dg-algebras A_t depending on a point t in a base scheme T, the corresponding periodic cyclic homology carries (formally) a flat connection along T. This is the celebrated *noncommutative Gauss–Manin connection* originally introduced by E. Getzler [12].

2.11 Noncommutative Motives

By analogy with the Grothendieck approach to pure motives in terms of correspondences between smooth projective varieties, one can define noncommutative analogues. Namely, for any saturated **nc**-space X, one has a group $K_0(X) := K_0(\mathtt{Perf}(X))$, which is a counterpart of the direct sum of all Chow groups $\oplus_k CH_k(X)$ in algebraic geometry. The analog of intersection pairing is the (nonsymmetric) Euler pairing

$$\chi : K_0(X) \otimes K_0(X) \to \mathbb{Z}, \quad [E] \otimes [F] \mapsto \chi\left(H^\bullet\left(\underline{Hom}_{D(X)}(E, F)\right)\right).$$

The left kernel of this pairing coincides with the right kernel, and the quotient can be called the **nc** numerical Chow group. Then one defines numerical correspondences from X to Y as $CH_{num}(X^{op} \otimes Y) \otimes \mathbb{Q}$. Similarly, one defines an analog of a Voevodsky homotopy category. Namely, consider the

(∞, 1)-category enriched over spectra (in the sense of topology), whose objects are saturated **nc**-spaces, and morphism spaces are algebraic K-theory spectra of functor categories. Its triangulated envelope (obtaining by adding cones, shifts, and direct summands as in the dg case) is the stable (∞, 1)-category of mixed **nc** motives (see, e.g., [25]).

The only known Weil cohomology theories at present are Hochschild homology, periodic cyclic homology (if char $\mathbf{k} = 0$), and topological periodic Hochschild homology in positive characteristics; see [2]. As I mentioned in the introduction, the main challenge is to define Betti cohomology or étale cohomology.

2.12 Calabi–Yau Structures

Let $X = \text{Spec } A$ be a saturated **nc**-space. The identity functor for $D(X) = A\text{-mod}^{sf}$ is given by the tensor product with the diagonal bimodule:

$$\text{id}_{D(X)} \simeq A_{diag} \otimes_A^L \cdot$$

The semifree resolution (or replacement) $R(A_{diag})$, where $R := R_{A \otimes A^{op}}$, has two finiteness properties: (1) it is perfect as bimodule (follows from smoothness) and (2) it has finite-dimensional cohomology as a $\mathbf{k}$-module (follows from properness). Hence there are two dual (in two different senses) bimodules, both belonging to $\text{Perf}(X)$:

$$A^{\vee}_{diag} := R\left(\underline{Hom}_{A \otimes A^{op}\text{-mod}}(R(A_{diag}), A \otimes A^{op})\right),$$
$$A^{*}_{diag} := R\left(\underline{Hom}_{\mathbf{k}\text{-mod}}(A, \mathbf{k})\right).$$

Functors $A^{\vee}_{diag} \otimes \cdot$ and $A^{*}_{diag} \otimes \cdot$ are canonically inverse to each other. An intrinsic definition of the restriction of the **Serre functor** $S_X := A^{*}_{diag} \otimes \cdot$ to the full subcategory $\text{Perf}(X) \subset D(X)$ is given via the functorial quasi-isomorphism

$$\underline{Hom}_{D(X)}(\mathcal{E}, \mathcal{F})^* \simeq \underline{Hom}_{D(X)}(\mathcal{F}, S_X(\mathcal{E})), \quad \forall \mathcal{E}, \mathcal{F} \in \text{Perf}(X).$$

Moreover, $A^{\vee}_{diag} \otimes \cdot$ (resp. $A^{*}_{diag} \otimes \cdot$) considered as endofunctors of $D(X)$ (and not of $\text{Perf}(X)$) can be intrinsically defined for arbitrary smooth (resp. proper) affine **nc**-space and can be denoted by S_X^{-1} (resp. S_X). If X comes from a commutative smooth equidimensional proper scheme over $\mathbf{k}$, then

$$S_X = K_X[\dim_X] \otimes_{\mathcal{O}_X} \cdot = \wedge^{\dim_X}(T^*_{X/\mathbf{k}})[\dim_X] \otimes_{\mathcal{O}_X} \cdot$$

In particular, if X has trivial canonical class (Calabi–Yau variety in the algebro-geometric sense), then S_X is isomorphic to the shift functor $[\dim_X]$. This justifies the following:

Definition 2.16 Fix an integer $N \in \mathbb{Z}$. A smooth affine **nc**-space $X = \text{Spec } A$ is called **weak Calabi–Yau of dimension N** if and only if $A_{diag}^{\vee} \otimes_A \cdot \simeq [-N]$. A proper affine **nc**-space $\text{Spec } A$ is called **weak Calabi–Yau of dimension N** if and only if $A_{diag}^{*} \otimes_A \cdot \simeq [N]$.

The definitions are compatible for saturated X. An example of smooth weak Calabi–Yau is any usual smooth commutative scheme with trivial canonical class. If $Z \subset X$ is a proper closed subset in such a scheme, then X_Z is a proper Calabi–Yau **nc**-space.

In fact, there are notions of "correct" Calabi–Yau structures for smooth (resp. proper) affine **nc**-spaces rigidifying the isomorphisms in the above definition together with certain higher homotopy corrections. In the smooth case, a Calabi–Yau structure is the same as a cohomology class in **negative cyclic homology**

$$HC_\bullet^-(A) := H^\bullet(C_\bullet(A)[[u]], b + uB)$$

whose projection to Hochschild homology

$$HH_\bullet(A) \simeq H^\bullet(\underline{Hom}_{A\text{-mod-}A}(A_{diag}^{\vee}, A_{diag}))$$

induces an isomorphism. Similarly, for the proper case (see [18]), a Calabi–Yau structure is the same as a functional on **cyclic homology**

$$HC_\bullet := H^\bullet(C_\bullet(A)((u))/uC_\bullet(A)[[u]], b + uB)$$

whose image under the natural map

$$(HC_\bullet)^* \to (HH_\bullet)^* = H^\bullet(\underline{RHom}_{A\text{-mod-}A}(A_{diag}, A_{diag}^{*}))$$

induces an isomorphism.

2.13 Beyond Affineness

(The context of this section grew from conversations with the late A. Rosenberg about 10 years ago and, to my knowledge, is not yet in the literature.)

Let us return to the situation considered in Section 2.7 of a localization morphism

$$\phi : A \to B, \quad Cone(B \otimes_A^L B \to B) \simeq 0.$$

The **nc**-space Y corresponding to "sheaves with given support" ($Y = X_Z$ and $D(Y) = D_Z(X)$ in notation from Section 2.7) is in general *not* compactly generated. In fact, it is feasible that the category $D_Z(X)$ is nontrivial and at the same time has no nonzero compact objects. Still, it turns out that it is possible to work with $Y = X_Z$ as if it were an affine **nc**-space.

Definition 2.17 An nc-space Y is called **accessible** if and only if $D(Y)$ is equivalent to the full subcategory

$$\mathrm{Ker}(B \otimes_A \cdot) := \{\mathcal{E} \in A\text{-mod}^{sf} \mid H^\bullet(B \otimes_A \mathcal{E}) = 0\} \subset A\text{-mod}^{sf}$$

for some localization morphism $\phi \colon A \to B$. A choice of a localization morphism is called a **presentation**.

For accessible spaces, one can repeat essentially everything we have done with affine spaces or, more generally, generated by a small collection of compact objects, in terms of presentations. Set-theoretic difficulties evaporate for such spaces as they are concrete set-based structures. An intrinsic characterization of accessible spaces is as dualizable[6] objects in $(\infty, 1)$-category of cocomplete triangulated dg-categories and continuous (i.e., sum (or colimit) preserving) functors (N. Rozenblyum, private communication).

First, we introduce an auxiliary construction. Let

$$\begin{array}{ccc} C_2 & \xrightarrow{a_{24}} & C_4 \\ a_{12}\uparrow & & a_{34}\uparrow \\ C_1 & \xrightarrow{a_{13}} & C_3 \end{array}$$

be a commutative diagram of dg-algebras. Define[7] dg-algebra $C_2 \times^h_{C_4} C_3$ as

$$\left\{\text{matrices} \begin{pmatrix} C_2 & C_4[-1] \\ 0 & C_3 \end{pmatrix}\right\}, d_{C_2 \times^h_{C_4} C_3} := \left[\begin{pmatrix} 0 & 1_{C_4} \\ 0 & 0 \end{pmatrix}, \cdot\right] + \begin{pmatrix} d_{C_2} & d_{C_4}[-1] \\ 0 & d_{C_3} \end{pmatrix},$$

with the multiplication defined using homomorphisms a_{24}, a_{34}. There is a natural morphism of dg-algebras $C_1 \to C_2 \times^h_{C_4} C_3$ given by

$$t \in C_1 \mapsto \begin{pmatrix} a_{12}(t) & 0 \\ 0 & a_{13}(t) \end{pmatrix}.$$

Definition 2.18 For two morphisms $A_1 \to B_1$ and $A_2 \to B_2$ of dg-algebras, define their **tensor product** as the morphism

$$A_1 \otimes A_2 \longrightarrow (B_1 \otimes A_2) \times^h_{(B_1 \otimes B_2)} (A_1 \otimes B_2),$$

applying the above construction to the appropriate commutative diagram.

---

[6] Similarly, small categories $\mathrm{Perf}(X)$ for *saturated* X are characterized as dualizable objects in $(\infty, 1)$-category of small split closed triangulated dg-categories.

[7] This is just a specific model for the homotopy fibered product.

Proposition 2.19 *The tensor product of localization morphisms is again a localization morphism, and the corresponding category coincides with* $\mathrm{Ker}(B_1 \otimes_{A_1} \cdot) \cap \mathrm{Ker}(B_2 \otimes_{A_2} \cdot) \subset A_1 \otimes A_2\text{-mod}^{s.f}$.

Proposition 2.20 *If $\phi\colon A \to B$ is localization, then $\phi^{op}\colon A^{op} \to B^{op}$ is a localization.*

Theorem 2.21 *For two localizations $A \to B$, $A' \to B'$, a functor between corresponding dg-categories $\mathrm{Ker}(B \otimes_A \cdot) \to \mathrm{Ker}(B' \otimes_{A'} \cdot)$ preserving arbitrary sums is (up to a quasi-isomorphism) given by a tensor product: $M \otimes_A \cdot$, where*

$$M \in A' \otimes A^{op}\text{-mod}^{s.f}\colon\ B' \otimes_{A'} M = 0,$$
$$M \otimes_A B = 0 \Leftrightarrow M \in Ker(B' \otimes B^{op}) \otimes_{A' \otimes A^{op}}.$$

The composition of functors is given by the tensor product of bimodules (over the intermediate algebra). The identity functor for $Ker(B \otimes_A \cdot)$ is given by $R_{A \otimes A^{op}}(Cone(A \to B)[-1])$, where $R_{A \otimes A^{op}}$ is the semifree resolution functor.

The notion of a *morphism* for accessible spaces can be formulated also in terms of linear algebra (in a sense similar to the description of morphisms in the affine case in Section 2.6). To achieve this, one uses an explicit approach to adjunctions as in Appendix A4. A morphism f of **nc**-spaces is invertible if and only if the adjunction morphisms $id \to f_* \circ f^*, f^* \circ f_* \to id$ are quasi-isomorphisms at the level of bimodules.

Finally, Hochschild homology and cohomology can be defined for accessible spaces using a tensor product and $RHom$ for the "diagonal" bimodule responsible for the identity morphism. Independence of Hochschild (co)homology of the choice of a presentation can be proven using the description of invertible morphisms in terms of linear algebra data.

Here are three examples of the use of the formalism of accessible spaces:

1. Let $\mathcal{D}$ be a small dg-category. Denote by A_0 the nonunital dg-algebra $\oplus_{\mathcal{E}, \mathcal{F} \in \mathcal{D}} \underline{Hom}_{\mathcal{D}}(\mathcal{E}, \mathcal{F})$ and by $A := A_0 \oplus \mathbf{k} \cdot 1_A$ the corresponding unital algebra. Let $B := \mathbf{k}$ and $A \to A/A_0 = B$ be the augmentation morphism. This is a localization, and $Ker(B \otimes_A \cdot)$ is equivalent to $\mathcal{D}\text{-mod}^{s.f}$.

   In the whole discussion of accessible spaces, one can replace dg-algebras by small dg-categories. The above construction implies that this generalization is vacuous; we get the *same* class of accessible spaces. The reason is that the class of closed embeddings (i.e., morphisms as in Definition 2.13) is closed under composition.

2. Let A be a dg-algebra and $B \in A \otimes A^{op}$ be an A-coalgebra in the monoidal category of bimodules. Assume (for simplicity) that B is semifree as an A^{op}-module. We will define a dg-model for the "correct" $(\infty, 1)$-category of homotopy dg-comodules (this question appears when one studies descent). Let $\mathcal{D}$ be the following small dg-category. The set of objects of $\mathcal{D}$ is $\mathbb{Z}_{\geq 0} = \{\mathbf{0}, \mathbf{1}, \dots\}$. For $\mathbf{i}, \mathbf{j} \in \mathcal{D}$, we define the complex of morphisms as

$$\underline{Hom}_{\mathcal{D}}(\mathbf{i},\mathbf{j}) := \bigoplus_{0 \leq k \leq j} B^{\otimes_A k} \otimes \mathbf{k}^{\oplus S(j-k,i)},$$

where $S(a, b)$ is the set of monotonic maps $\{0, \dots, a\} \to \{0, \dots, b\}, a \mapsto b$. To define the composition, imagine the following. Let M be a B-comodule. Define for $n \geq 0$ complexes M_n as $(B^{\otimes_A} \cdot)^n(M)$. Associate with $(t \otimes f) \in \underline{Hom}_{\mathcal{D}}(\mathbf{i},\mathbf{j})$, $t \in B^{\otimes_A k}$, $f \in S(j-k, i)$, the linear operator from M_i to M_j given by the composition $M_i \xrightarrow{f^*} M_{j-k} \xrightarrow{t \otimes_A \cdot} M_j$, where f^* is naturally defined using the comodule structure.

It is easy to see that such operators are closed under composition. This gives a description of a dg-category structure on $\mathcal{D}$ ($\mathbf{i} \mapsto M_i$ should be a dg-functor). We define dg-category B-comod^h (homotopy comodules) as the full dg subcategory of $\mathcal{D}$-mod^{sf} consisting of such functors that the natural maps $B \otimes_A M_i \to M_{i+1}$ are quasi-isomorphisms. One can check that the fully faithful embedding i^*: B-comod$^h \hookrightarrow \mathcal{D}$-mod^{sf} (obviously preserving arbitrary sums) gives a coidempotent comonad in $\mathcal{D}$-mod^{sf}. This comonad is a given by certain explicit formulas involving only tensor products, hence also preserving arbitrary sums. By general formalism, i^* corresponds to a closed embedding.

3. Recall the description of $D(X)$ for a scheme X together with an affine covering, as in Section 1.2.2. The ambient dg-category of "sheaves" is a full triangulated subcategory of $\mathcal{D}_{(U_i)_{i \in I}}$-mod closed under arbitrary sums. If the covering is cubical, that is, I is the set of finite nonempty subsets in another set J (i.e., we cover X by open affine subsets such that all finite intersections are affine), one can show that the embedding functor is i^* for a closed subset in the sense of Definition 2.13 (i.e., the comonad in the category of "presheaves" preserves arbitrary sums). This gives a description as an accessible space of X_{nc} for quasi-separated X. To go to arbitrary X, notice that any quasi-affine scheme is affine in the derived sense. Hence, we embed arbitrary schemes (without the assumption of being quasi-compact and quasi-separated, as in the most general version of the Bondal–van den Bergh result [Theorem 2.4]) into an affine-like framework [14].

3 Exotic Geometry in Noncommutative World

3.1 Large Symmetry Groups, Fourier–Mukai Transform

If X is a saturated **nc**-space and $E \in \text{Perf}(X)$ is a **spherical object** (i.e., $H^\bullet(\underline{Hom}_{D(X)}(E,E)) \simeq H^\bullet(S^n)$ for some $n \geq 0$), then the endofunctor

$$F \mapsto Cone(\underline{Hom}_{D(X)}(E,F) \otimes_{\mathbf{k}} E \to F)$$

is invertible. In particular, for a smooth projective 3-dimensional Calabi–Yau variety X with $H^{1,0}(X) = 0$, we should have an enormous collection of such spherical reflections, as the typical vector bundle is expected to be simple and rigid, hence a spherical object of dimension 3.

If A is an abelian variety and $A^\vee := Pic_0(A)$ is the dual abelian variety, then categories $D(A)$ and $D(A^\vee)$ are naturally equivalent. The equivalence (Fourier–Mukai transform) is given by the functor

$$E \mapsto (pr_2)_*(\mathcal{L} \otimes_{\mathcal{O}_{A \times A^\vee}} (pr_1)^*(E)),$$

where $\mathcal{L} \in D(A \times A^\vee)$ is the universal line bundle trivialized over $(A \times \{0\}) \cup (\{0\} \times A^\vee)$ and pr_1, pr_2 are projections to A (resp. $A^\vee$).

If $A = E^n$, where E is an elliptic curve (hence $E \simeq E^\vee$), then combining Fourier–Mukai transforms along various factors, and the obvious action of $GL(n, \mathbb{Z})$, one obtains an action of the group $\widetilde{Sp(2n, \mathbb{Z})}$ on $D(A)$. This group is a central extension by $\mathbb{Z}$ (shift functors) of $Sp(2n, \mathbb{Z})$ and is defined as the fundamental group of the moduli stack of principally polarized abelian varieties endowed with the trivialization of the square of the canonical bundle.

A similar picture is expected for general Calabi–Yau varieties in the situation of mirror symmetry. Fundamental groups of moduli spaces act by automorphisms on derived categories of mirror dual Calabi–Yau varieties. Still, the mirror symmetry perspective gives only a very tiny portion of the whole automorphism group, as there are many autoequivalences without an obvious mirror counterpart, such as spherical reflections.

There exists a heuristic explaining why the Calabi–Yau property (at least in algebraic geometry) is linked to the property of having a large symmetry group. The reason is that the Serre functor is intrinsically defined and hence is preserved by any automorphism. If X is a smooth proper variety that is either of general type or Fano (i.e., the canonical class is ample or anti-ample), then X can be reconstructed from $D(X)$ using the algebra of natural transformations between various powers of the Serre functor.

3.2 Duality of Quotients

Monoidal dg-categories (when monoidal structure is understood in the homotopy sense similar to A_∞-algebras) can act on other dg-categories. In such a case, one can speak about *equivariant* objects, thus forming a new dg-category. Passing to **nc**-spaces, we can say that we form a *quotient* space. For example, if Γ is a group acting by automorphisms of dg-algebra A, then

$$\text{Spec}(A)/\Gamma = \text{Spec}(\mathbf{k}[\Gamma] \ltimes A).$$

Remarkably, in the **nc** world, the original space is a quotient of its quotient space! For, if Γ is finite abelian and acts on **nc**-space X, then the dual group scheme $\Gamma^\vee := Hom(\Gamma, \mathbb{G}_\mathbf{m})$ acts on $Y := X/\Gamma$, and we have $X = Y/\Gamma^\vee$.

This duality generalizes to the action of finite-dimensional Hopf algebras and also to infinite-dimensional examples. With appropriate definitions, one can achieve an equivalence between **nc**-spaces with the action of $\mathbb{Z}$ (i.e., an autoequivalence) and **nc**-spaces with the action of $\mathbb{G}_\mathbf{m}$. All this should be an instance of 2-categorical derived Morita-equivalence (as monoidal categories are 2-categories with a single object).

3.3 Fractional Calabi–Yau

Definition 3.1 A saturated **nc**-space X is called **fractional Calabi–Yau** if, for some $m \geq 1$, the corresponding power S^m of Serre functor is equivalent to a power $[n]$, $n \in \mathbb{Z}$, of the shift functor. Rational number $n/m \in \mathbb{Q}$ is called the **Calabi–Yau dimension** of X.

If X is a Deligne–Mumford stack which is the quotient of smooth projective Calabi–Yau variety by the action of a finite group (not necessarily preserving the algebraic volume element), then it is fractional Calabi–Yau of dimension equal to the usual geometric dimension.

If the Calabi–Yau dimension is an integer, then we obtain an action of the cyclic group $\mathbb{Z}/m\mathbb{Z}$ generated by $S[-n/m]$. By taking the quotient, we obtain a Calabi–Yau space $Y = X/(\mathbb{Z}/m\mathbb{Z})$. By previous considerations, we see that conversely, X is a quotient of Y by the group scheme of mth roots of 1.

Surprisingly, there are many fractional Calabi–Yau **nc**-spaces of noninteger dimension. For example, the path algebra of Dynkin quiver A_n (with arbitrary orientation of arrows) has dimension $(n-1)/(n+1)$. Many tensor products of such fractional **nc**-spaces happen to be cyclic quotients of Calabi–Yau varieties (if the total dimension happens to be an integer).

3.4 Completion at the Support, Koszul Duality

Let $X = \text{Spec } A$ be an affine **nc**-space and $\mathcal{E} \in D(X) = A\text{-mod}^{sf}$ be an object (not necessarily compact). Define a sequence of dg-algebras and modules in the following inductive manner:

$$A_1 := A, \quad \mathcal{E}_1 := \mathcal{E}, \quad A_{n+1} := \underline{\text{Hom}}_{A_n\text{-mod}}(\mathcal{E}_n, \mathcal{E}_n), \quad \mathcal{E}_{n+1} := R_{A_{n+1}}(\mathcal{E}_n),$$

where, in the last formula, $\mathcal{E}_n$ is interpreted as an A_{n+1}-module. We have natural homomorphisms of algebras $A_n \to A_{n+2}$ arising in the following way: $A_{n+1} = \underline{\text{Hom}}_{A_n\text{-mod}}(\mathcal{E}_n, \mathcal{E}_n)$ maps (using functoriality of $R_{A_{n+1}}$) to

$$\underline{\text{Hom}}_{A_n\text{-mod}}(R_{A_{n+1}}(\mathcal{E}_n), R_{A_{n+1}}(\mathcal{E}_n)) = \underline{\text{Hom}}_{A_n\text{-mod}}(\mathcal{E}_{n+1}, \mathcal{E}_{n+1}).$$

Hence $\mathcal{E}_{n+1}$ is an $A_n \otimes A_{n+1}$-module; therefore we have a map

$$A_n \to \underline{\text{Hom}}_{A_{n+1}\text{-mod}}(\mathcal{E}_{n+1}, \mathcal{E}_{n+1}) = A_{n+2}.$$

The situation is more transparent in the A_∞-world; in this case, algebras act on the same underlying complex and are iterated centralizers. One can show that for $n \geq 2$, the morphism $A_n \to A_{n+2}$ is a quasi-isomorphism; therefore, in our sequence, we have just three different (up to homotopy) algebras A_1, A_2, A_3.

Definition 3.2 Let A, B be two dg-algebras and $M \in (A \otimes B)\text{-mod}^{sf}$ be a bimodule. We say that M realizes a **generalized Koszul duality** between A and B if the natural morphisms

$$A \to \underline{\text{Hom}}_{B\text{-mod}}(M, M), \quad B \to \underline{\text{Hom}}_{A\text{-mod}}(M, M)$$

are quasi-isomorphisms.

This is a generalization of derived Morita-equivalence (see Definition A.16 and after). The difference is that now we do not assume that M is perfect as a one-sided module.

Theorem 3.3 *Generalized Koszul duality induces an isomorphism of cohomological Hochschild complexes (as algebras over little disks operad).*

For the above sequence of algebras, we see that $A_2 \simeq A_4 \simeq \ldots$ are generalized Koszul dual to $A_3 \simeq A_5 \simeq \ldots$

If $A = A_1 = \mathbf{k}[x_1, \ldots, x_n]$ and $M = M_1 = \mathbf{k} = A/\sum_i A \cdot x_i$, then

$$A_2 \simeq \wedge^\bullet(\mathbf{k}^n) = H^\bullet((S^1)^n, \mathbf{k}), \quad A_3 \simeq \mathbf{k}[[x_1, \ldots, x_n]].$$

In general, it seems that A_3 describes quasi-coherent sheaves in the formal completion of the support of M. This is literally true (see [7]) if $\text{Spec}(A)$ is equivalent to X_{nc} for a separated scheme X of finite type and M corresponding to a perfect generator of $D_Z(X)$ for a closed subset $Z \subset X$.

3.5 Smooth Spaces from Singular Schemes, Categories of Singularities, and Knörrer Periodicity

There is a counterintuitive, in a sense, class of **nc**-spaces of finite type (in particular, smooth) arising naturally in algebraic geometry.

Theorem 3.4 (*A. Efimov, V. Lunts, [9]*) *For every separated scheme of finite type $S/\text{Spec}\,\mathbf{k}$, the natural dg enhancement of triangulated category $D^b(Coh(S))$ has a generator $\mathcal{E}$, and the corresponding dg-algebra $\underline{Hom}_{D(S)}(\mathcal{E},\mathcal{E})^{op}$ is of finite type.*

Notice that the full subcategory $D^b_c(Coh(S)) \subset D^b(Coh(S))$ consisting of objects with compact support is naturally equivalent to the category of maps in **nc**-sense $\text{Spec}\,\mathbf{k} \to S_{nc}$.

Informally speaking, using the full embedding $\text{Perf}(S) \subset D^b(Coh(S))$, the picture is that the singular scheme is similar to **nc**-space of type X_Z corresponding to a closed subset Z in a smooth scheme X. The ind-completion of $D^b(Coh(S))$ is called the category of **ind-coherent sheaves**, and it differs for singular S form quasi-coherent sheaves.

The quotient category $D^b_{sing}(S) := D^b(Coh(S))/\text{Perf}(S)$ responsible for the structure of singularities is again of finite type. In the case when S is a zero locus of a function f on an ambient smooth variety X, the category $D^b_{sing}(S)$ is $2\mathbb{Z}$-periodic. It is equivalent (up to Karoubi completion) to the category of matrix factorizations $MF(X, f)$ prominent in mirror symmetry (see [23]), and hypothetically mirror dual to Fukaya categories of non-Calabi–Yau varieties. An alternative viewpoint on $MF(X, f)$ is as on a $\mathbb{Z}/2\mathbb{Z}$-graded deformation of $\text{Perf}(X)$ corresponding to the even class $[f] \in HH^0(X)$ in the deformation complex.

3.6 Semiorthogonal Decomposition and Action of Braid Group

If $\mathcal{C}$ is a saturated category and $\mathcal{D} \subset \mathcal{C} = \text{Perf}(X)$ is a full subcategory which is also saturated, then $\mathcal{D}$ has both left and right orthogonals in $\mathcal{C}$, and these orthogonals $^{\perp}\mathcal{D}, \mathcal{D}^{\perp}$ are also saturated (and equivalent to each other), and moreover, $\mathcal{D} = (^{\perp}\mathcal{D})^{\perp} = {}^{\perp}(\mathcal{D}^{\perp})$. One writes $\mathcal{C} = \langle \mathcal{D}_1, \mathcal{D}_2 \rangle$ when $\mathcal{C}$ is saturated and $\mathcal{D}_2 = {}^\perp \mathcal{D}_1, \mathcal{D}_1 = \mathcal{D}_2^\perp$ are two saturated subcategories. In this case, $\mathcal{C}$ has semiorthogonal decomposition into $\mathcal{D}_1$ and $\mathcal{D}_2$. The way $\mathcal{C}$ is built from $\mathcal{D}_1$ and $\mathcal{D}_2$ is uniquely determined by bimodule $\underline{Hom}_\mathcal{C}(\cdot,\cdot)\colon \mathcal{D}_1^{op}\otimes\mathcal{D}_2 \to Comp_{\mathbf{k}}$, which could be an *arbitrary* element of the split closure of $\mathcal{D}_1\otimes\mathcal{D}_2^{op}$. More generally, one can consider finite flags

$$0 = \mathcal{C}_0 \subset \mathcal{C}_1 \subset \cdots \subset \mathcal{C}_n = \mathcal{C}$$

of full saturated subcategories of a given saturated category $\mathcal{C}$. Taking subsequent right orthogonals, one obtains a more general semiorthogonal decomposition

$$\mathcal{C} = \langle \mathcal{D}_1, \ldots, \mathcal{D}_n \rangle, \quad Hom_{H^\bullet(\mathcal{C})}(\mathcal{E},\mathcal{F}) = 0 \text{ if } \mathcal{E}\in\mathcal{D}_i,\ \mathcal{F}\in\mathcal{D}_j,\ i>j.$$

Example: $D(\mathbb{P}^n) = \langle \mathcal{O}(0), \ldots, \mathcal{O}(n) \rangle$. Playing with left and right orthogonals, one can obtain an action of the braid group B_n on flags of length n (see [3]).

3.7 Calculus for Pre-Calabi–Yau Structures

At the end of Section 2.12, we described an **nc**-analog of Calabi–Yau varieties (both in smooth and proper cases). There exists a general notion of *pre-Calabi–Yau structure* on affine **nc**-space (no finiteness condition), which we will not describe here (see [19]), containing both smooth and proper Calabi–Yau **nc**-spaces as special cases. In algebraic geometry, pre-Calabi–Yau structure arises if we are given a section of anticanonical bundle on a smooth scheme. In topology, pre-Calabi–Yau structure arises when we consider local systems on an oriented manifold with boundary.

There is a natural source of pre-Calabi–Yau structures of dimension N associated with a relative Calabi–Yau structure (see [5]), describing in a sense smooth or proper spaces with $(N-1)$-dimensional Calabi–Yau boundary and relative fundamental class.

In general, for a pre-Calabi–Yau space $\mathtt{Spec}\,A$, there is a remarkable algebraic structure on homological Hochschild complex $C_\bullet(A)$, generalizing operations from Deligne conjecture. Namely, for any $n,m \geq 1$ and $g \geq 0$, there are natural maps (of degree depending on Calabi–Yau dimension)

$$C_\bullet(\mathcal{M}_{g,\vec{n}+\vec{m}}) \otimes C_\bullet(A)^{\otimes n} \to C_\bullet(A)^{\otimes m},$$

where $\mathcal{M}_{g,\vec{n}+\vec{m}}$ is the moduli stack of conformal structures on surfaces of genus g with $n+m$ boundary circles (first, n are oriented positively, and last, m

are oriented negatively), with a marked point on each circle. Gluing surfaces, we obtain a PROP acting on $C_\bullet(A)$ (see also [18] for proper Calabi–Yau).

3.8 About Fukaya Categories

Symplectic topology gives a new, fresh source of **nc**-spaces. This is a huge theme – I will just give some brief highlights. Roughly, objects of the Fukaya category are Lagrangian subvarieties in a symplectic manifold X. Spaces of morphism have a basis consisting of intersection points. The Fukaya category $\mathcal{F}(X)$ is naturally defined as A_∞-category, with higher compositions coming from counting holomorphic polygons weighted with the factor exp(−symplectic area). Convergence issues force us (at the moment) to consider Fukaya categories only over non-Archimedean fields. $\mathcal{F}(X)$ is $\mathbb{Z}$-graded if and only if X has trivial first Chern class (symplectic Calabi–Yau). Moreover, Fukaya categories belong in general to the realm of *rigid analytic* noncommutative dg-geometry and not algebraic noncommutative dg-geometry, as considered previously.

A Fukaya category itself is always Calabi–Yau (only $\mathbb{Z}/2\mathbb{Z}$-graded if $c_1(X) \neq 0$) if we consider only *compact* subvarieties. In general, one can define the so-called partially wrapped Fukaya categories allowing certain noncompact Lagrangian subvarieties with cylindrical behavior at infinity. There is an additional data controlling wrapping, consisting of complete in the positive direction Hamiltonian flow with Hamiltonian bounded below. Partially wrapped categories are in general not Calabi–Yau.

For a special class of wrappings, in the classical limit, when the contribution of holomorphic disks vanishes, one obtains a generalization of the categories of constructible sheaves with respect to a given stratification (a version of string topology). Namely, if $L \subset X$ is a closed, possibly noncompact singular (admitting Whitney stratification) Lagrangian subset, with a cylindrical behavior at infinity, then it is expected that L carries a cosheaf of dg-categories with stalks of finite type (see [16] for an original proposal with some examples and [11] for recent developments). The global section category $\mathcal{F}_L$ is a homotopy colimit of a finite diagram of dg-categories of finite type and hence is of finite type as well. The dg-categories from Section 1.2.3 appear as particular examples, and also one can get categories of coherent sheaves on many nice algebraic varieties like toric varieties or, cluster varieties. The smooth dg-category $\mathcal{F}_L$ carries a pre-Calabi–Yau structure of dimension dim $X/2$. Pseudo-holomorphic discs on X with boundary on L should produce a deformation of $\mathcal{F}_L$, and finite-dimensional modules over the deformed $\mathcal{F}_L$ should form a full subcategory of $\mathcal{F}(X)$.

Homological mirror symmetry (see [17]) predicts that in many cases, $\mathcal{F}(X)$ is equivalent to $D(X^\vee)$ for a mirror algebraic Calabi–Yau $X^\vee$ (or matrix factorization category if X is not Calabi–Yau). To have a commutative mirror (i.e., symmetric monoidal structure on $\mathcal{F}(X)$), it is sufficient to have a structure of a real integrable system on X. The main "usefulness" of Fukaya categories is that a large part of the automorphism group became transparent, as the group of symplectomorphisms up to Hamiltonian isotopy. The braid group action on flags of saturated subcategories is geometrically transparent for so-called Fukaya–Seidel categories, associated with symplectic manifolds endowed with a map to $\mathbb{R}^2$ and certain associated partial wrapping. Also, the existence of PROP structure on Hochschild homology related to surfaces, as in Section 3.7, is also geometrically transparent.

One expects further developments based on the symplectic viewpoint. In particular, it should be possible to generalize Fukaya categories by adding "perverse sheaves" of dg-categories as coefficients. Also, we expect that there exists a general machinery producing stability structures in the sense of Bridgeland on such categories, from stability structures on stalk categories.

Appendix: Technical Definitions

A1 dg-Categories and Semifree Modules

For simplicity, we will work over a field $\mathbf{k}$ (instead of a commutative ring or, more generally, a base scheme). The category $Comp_{\mathbf{k}}$ of $\mathbb{Z}$-graded complexes over $\mathbf{k}$ with differential of degree $+1$ is a symmetric monoidal $\mathbf{k}$-linear category, and the commutativity morphism is given by the Koszul rule of signs.

Definition A.1 A $\mathbf{k}$-linear dg category is a category enriched over $Comp_{\mathbf{k}}$. For such a category $\mathcal{C}$, and any two objects $\mathcal{E}, \mathcal{F} \in \mathcal{C}$ denote by $\underline{Hom}_{\mathcal{C}}(\mathcal{E}, \mathcal{F}) \in Comp_{\mathbf{k}}$ the corresponding complex of morphisms.

By general nonsense valid for any kind of enrichment, one can define a **tensor product** of arbitrary finite collection of dg-categories, as well as the **opposite** dg-category to a given one.

Any dg-category $\mathcal{C}$ gives a plain (not enriched) category with the same class of objects as $\mathcal{C}$ and with the set of morphisms $Hom_{\mathcal{C}}(\mathcal{E}, \mathcal{F})$ given by

$$Hom_{Comp_{\mathbf{k}}}\left(\mathbf{1}_{Comp_{\mathbf{k}}}, \underline{Hom}_{\mathcal{C}}(\mathcal{E}, \mathcal{F})\right)$$
$$= Ker\left(d : \underline{Hom}^0_{\mathcal{C}}(\mathcal{E}, \mathcal{F}) \to \underline{Hom}^1_{\mathcal{C}}(\mathcal{E}, \mathcal{F})\right).$$

We also define two other plain categories $H^0(\mathcal{C})$ and $H^\bullet(\mathcal{C})$ by applying functors $H^0, H^\bullet \colon Comp_{\mathbf{k}} \to \mathbf{k}$-mod to Hom-complexes in $\mathcal{C}$. Functor H^0 associates with any complex its zeroth cohomology space, and $H^\bullet := \oplus_{i \in \mathbb{Z}} H^i$.

Definition A.2 A dg-functor $F \colon \mathcal{C}_1 \to \mathcal{C}_2$ is called **fully faithful** if and only if the induced functor $H^\bullet(\mathcal{C}_1) \to H^\bullet(\mathcal{C}_2)$ is fully faithful. F is called an **equivalence** if and only if it is fully faithful and induces an equivalence $H^0(\mathcal{C}_1) \to H^0(\mathcal{C}_2)$.

Let A be a dg-algebra (the same as the dg-category $\mathcal{C} = \mathcal{C}_A$ with a single object $\mathcal{E}$ such that $\underline{Hom}_\mathcal{C}(\mathcal{E}, \mathcal{E}) = A$). The plain (not enriched) category A-mod of dg-modules over A is the same as the category of dg-functors $\mathcal{C}_A \to Comp_{\mathbf{k}}$. It has a natural dg enrichment defined via the adjunction

$$Hom_{Comp_{\mathbf{k}}}(G, \underline{Hom}_{A\text{-mod}}(M_1, M_2))$$
$$= Hom_{A\text{-mod}}(G \otimes M_1, M_2) \quad \forall G \in Comp_{\mathbf{k}}.$$

The dg-category A-mod defined above is not homotopically meaningful (at least on the nose), for example, a quasi-isomorphism of dg-algebras does not induce in general an equivalence of corresponding dg-categories. Still, there is a "good" part of A-mod that behaves well:

Definition A.3 A dg-module M over dg-algebra A is called **semifree** if and only if there exists a countable complete increasing filtration by sub-dg-modules

$$0 = F_0 M \subset F_1 M \subset \cdots \subset M = \cup_{n \geq 0} F_n M$$

such that for each $n \geq 1$, the quotient $F_n M / F_{n-1} M$ is isomorphic as a dg-module to $G_n \otimes A$ where $G_n \in Comp_{\mathbf{k}}$. The full dg-subcategory of A-mod consisting of semifree modules is denoted $A\text{-mod}^{sf}$.

Similar definitions can be made for arbitrary *small* dg-category $\mathcal{C}$ (instead of a category with a single object). We replace dg-category A-mod by the category $\mathcal{C}$-mod of dg-functors $\mathcal{C} \to Comp_{\mathbf{k}}$. Define the full dg-subcategory of semifree objects $\mathcal{C}\text{-mod}^{sf} \subset \mathcal{C}\text{-mod}$ as consisting of functors admitting a countable complete increasing filtration with the associated graded components of the form

$$\mathcal{F} \mapsto \oplus_{\mathcal{E} \in \mathcal{C}} G_{n,\mathcal{E}} \otimes \underline{Hom}_\mathcal{C}(\mathcal{E}, \mathcal{F})$$

for some collection of complexes $(G_{n,\mathcal{E}} \in Comp_{\mathbf{k}})_{n \geq 1, \mathcal{E} \in \mathcal{C}}$. We denote by $\mathcal{C}\text{-mod}^{fsf}$ the full dg-subcategory of $\mathcal{C}\text{-mod}^{sf}$ consisting of dg-functors such that there exists a filtration as above satisfying the *finiteness* condition $\sum_{n\geq 1, \mathcal{E}\in\mathcal{C}} \sum_{i\in\mathbb{Z}} \dim H^i(G_{n,\mathcal{E}}) < +\infty$. It is easy to see that

$$(\mathcal{C}\text{-mod}^{fsf})^{op} \simeq \mathcal{C}^{op}\text{-mod}^{fsf}.$$

Proposition A.4 *For any small dg-category $\mathcal{C}$, both categories $H^0(\mathcal{C}\text{-mod}^{sf})$ and $H^0(\mathcal{C}\text{-mod}^{fsf})$ carry natural triangulated structure. The Yoneda functors from $\mathcal{C}^{op}$ to $\mathcal{C}\text{-mod}^{sf}$ and to $\mathcal{C}\text{-mod}^{fsf}$ are fully faithful. An equivalence of small dg-categories $\mathcal{C}_1 \to \mathcal{C}_2$ induces an equivalence of the corresponding categories of large and small semifree modules.*

It is a long-standing tradition to define derived category as the quotient category. This is a sick construction set-theoretically, if applied to big (i.e., not small) categories (and requires consideration of universes, cardinals, etc.). If one ignores set-theoretic difficulties, then the following is true: for any small dg-category $\mathcal{D}$, the plain triangulated category $H^0(\mathcal{D}\text{-mod}^{sf})$ is equivalent to the quotient of $H^0(\mathcal{D}\text{-mod})$ by the multiplicative system of quasi-isomorphisms.

A2 Triangulated dg-Categories

Definition A.5 Small dg-category $\mathcal{C}$ is called **triangulated** if and only if the Yoneda embedding $Y: \mathcal{C} \to \mathcal{C}^{op}\text{-mod}^{fsf}$ is an equivalence.

Proposition A.4 implies that $H^0(\mathcal{C})$ is triangulated in the plain sense. Notice that in the dg world, the adjective *triangulated* is a *property* and not an additional structure on a category. The definition from above is concise but slightly misleading, because it is (at least apparently) "not local," as it uses Yoneda embedding.

One can reformulate the property of being triangulated in an elementary way, making sense *without* the assumption that $\mathcal{C}$ is small.

We will show (as an example) the criterion for the class of objects $H^0(\mathcal{C}) \subset H^0(\mathcal{C}\text{-mod}^{fsf})$ to be close (up to isomorphisms) under cones, assuming that it is already closed under shifts. Define dg-category $\mathcal{C}_3$ to have three objects **1, 2, 3** and to be freely generated as a $\mathbb{Z}$-graded category by 12 morphisms:

$$1 \leq i < j \leq 3: \gamma_{ij} \in \underline{Hom}_\mathcal{C}^{+1}(\mathbf{j},\mathbf{i}), \quad 1 \leq i,j \leq 3: \delta_{ij} \in \underline{Hom}_\mathcal{C}^{-1}(\mathbf{j},\mathbf{i})$$

The differential is defined on generators in the matrix form as

$$d(\gamma) = -\gamma \cdot \gamma, \quad d(\delta) = id_{3\times 3} - (\gamma \cdot \delta + \delta \cdot \gamma).$$

This category contains full dg-subcategory $\mathcal{C}_2$ consisting of two objects **1, 2** and one nontrivial morphism γ_{12}, $d(\gamma_{12}) = 0$.

Proposition A.6 *Let $\mathcal{C}$ be a small dg-category, $\mathcal{E}_1, \mathcal{E}_2 \in \mathcal{C}$ two objects, and $[\alpha]: Y(\mathcal{E}_1)[-1] \to Y(\mathcal{E}_2)$ a morphism in $H^0(\mathcal{C}^{op}\text{-mod}^{fsf})$. Let us choose a lift $\alpha \in \underline{\mathrm{Hom}}^1_{\mathcal{C}}(\mathcal{E}_1, \mathcal{E}_2)$, $d\alpha = 0$, defining a dg-functor*
$$C_2^{op} \to \mathcal{C}, \quad \mathbf{1} \mapsto \mathcal{E}_1, \quad \mathbf{2} \mapsto \mathcal{E}_2, \quad \gamma_{12} \mapsto \alpha.$$
Then there exists a lift $C_3^{op} \to \mathcal{C}$ of this functor with $\mathbf{3} \mapsto \mathcal{E}_3$ for some $\mathcal{E}_3 \in \mathcal{C}$ if and only if $Y(\mathcal{E}_3)$ is isomorphic to $\mathrm{Cone}(Y(\mathcal{E}_1)[-1] \xrightarrow{[\alpha]} Y(\mathcal{E}_2))[-1]$ in $H^0(\mathcal{C}\text{-mod}^{fsf})$.

We can use this criterion (and similar simpler criteria for the existence of shifts and finite sums) as an alternative definition of triangulatedness, making sense for arbitrary (not necessarily small) dg-categories. In this way, it becomes obvious that *any* dg-functor $\mathcal{C} \to \mathcal{C}'$ between triangulated dg-categories induces a triangulated (equivalently, exact) functor between conventional triangulate categories $H^0(\mathcal{C}) \to H^0(\mathcal{C}')$.

Proposition A.7 *For any triangulated dg-category $\mathcal{C}$ and arbitrary small full subcategory $\mathcal{D} \subset \mathcal{C}$, the smallest full subcategory of $\mathcal{C}$ containing $\mathcal{D}$ and closed under shifts, finite sums, and cones in $H^0(\mathcal{C})$ (called the **triangulated closure**) is naturally equivalent to $\mathcal{D}^{op}\text{-mod}^{fsf}$.*

A3 Resolutions and Tensor Products

Sometimes it is convenient to use arbitrary dg-modules, not necessarily semifree. For simplicity, we consider the case of modules over a dg-algebra A (and not over a small dg-category).

Any object $M \in A\text{-mod}$ admits a **canonical semifree resolution**: $R_A(M) := \oplus_{n \geq 1} A[1]^{\otimes n} \otimes M[-1]$ endowed with the differential equal to the sum of the standard differential on the tensor product of complexes and of the following correction term:

$$a_1 \otimes \cdots \otimes a_n \otimes m \mapsto \sum_{i=1}^{n-1} \pm a_1 \otimes \cdots \otimes a_i a_{i+1} \otimes \cdots \otimes m$$
$$\pm a_1 \otimes \cdots \otimes a_{n-1} \otimes a_n m.$$

The A-module structure is given by $a \cdot (a_1 \otimes \ldots) := a \otimes a_1 \otimes \ldots$. The natural transformation $R(M) \to M$ is a quasi-isomorphism (it induces an isomorphism of cohomology groups).

It is easy to see that the endofunctor $R := R_A$ is a comonad. This fact allows us to define a *new* structure of a dg-category on A-mod. Namely, for $\mathcal{E}, \mathcal{F} \in A\text{-mod}$, define the new complex of morphisms as

$$\underline{RHom}_{A\text{-mod}}(\mathcal{E}, \mathcal{F}) := \underline{Hom}_{A\text{-mod}}(R(\mathcal{E}), \mathcal{F}).$$

The underlying $\mathbb{Z}$-graded vector space of this complex is the same as for

$$\prod_{n \geq 0} \underline{Hom}_{Comp_{\mathbf{k}}}(A[1]^{\otimes n} \otimes \mathcal{E}, \mathcal{F}).$$

The identity morphism comes from the counit morphism $R \to Id$ and the composition from the coproduct morphism $R \to R \circ R$.

Proposition A.8 *The dg-category A-mod with the complexes of morphisms $\underline{RHom}_{A\text{-mod}}(\cdot, \cdot)$ is equivalent to $A\text{-mod}^{sf}$.*

Finally, the canonical semifree resolution is convenient for the definition of the derived tensor product.

Definition A.9 Let A, B, C be three dg-algebras and $\mathcal{E} \in A\text{-mod} - B$, $\mathcal{F} \in B\text{-mod} - C$ two bimodules (i.e., $\mathcal{E} \in A \otimes B^{op}\text{-mod}$, $\mathcal{F} \in B \otimes C^{op}\text{-mod}$). Define the **derived tensor product** $\mathcal{E} \otimes_B^L \mathcal{F}$ as the total complex of

$$\bigoplus_{n \geq 0} \mathcal{E} \otimes B[1]^{\otimes n} \otimes \mathcal{F} \in A\text{-mod} - C$$

with the differential equal to the sum of the standard differential on the tensor product of complexes and of the correction term

$$e \otimes \cdots \otimes f \mapsto (ea_1 \otimes \ldots) + \sum_{i=1}^{n-1} \pm (\cdots \otimes a_i a_{i+1} \otimes \ldots) \pm (\cdots \otimes a_n f).$$

This definition can be rewritten in two equivalent ways,

$$\mathcal{E} \otimes_B^L \mathcal{F} := R_{B^{op}}(\mathcal{E}) \otimes_B \mathcal{F} = \mathcal{E} \otimes_B R_B(\mathcal{F}).$$

where we use resolutions only for the B^{op} (resp. B) action and take the plain tensor product over B with *forgotten* differentials afterward. If one bimodule $\mathcal{E}$ or $\mathcal{F}$ is semifree as a right (resp. left) B-module, then the derived tensor product is quasi-isomorphic to the usual one, $\mathcal{E} \otimes_B \mathcal{F}$.

A4 Compact Objects

Definition A.10 A triangulated dg-category $\mathcal{C}$ admits **arbitrary sums** ($\Leftrightarrow$ co-complete) if and only if $H^0(\mathcal{C})$ admits arbitrary sums. An object $\mathcal{E} \in \mathcal{C}$ in such a category is **compact** if and only if $Hom_{H^0(\mathcal{C})}(\mathcal{E}, \cdot) : H^0(\mathcal{C}) \to \mathbf{k}\text{-mod}$ preserves sums.

For any triangulated dg-category $\mathcal{C}$ with arbitrary sums, the corresponding triangulated category $H^0(\mathcal{C})$ is **split closed** ($\Leftrightarrow$ Karoubi closed), that is, any projector comes from a direct sum decomposition. In fact, it suffices to assume that $\mathcal{C}$ is closed only under *countable* sums.

Proposition A.11 *For any small dg-category $\mathcal{D}$, category $\mathcal{D}$-mod^{s,f} has arbitrary sums. The full subcategory of compact objects in $\mathcal{D}$-mod^{s,f} consists of objects that are direct summands of objects from $\mathcal{D}$-mod^{f,s,f}.*

Proposition A.12 *For any split closed triangulated dg-category $\mathcal{C}$ and an arbitrary small full subcategory $\mathcal{D} \subset \mathcal{C}$, the smallest full subcategory $\langle \mathcal{D} \rangle_{str}$ of $\mathcal{C}$ containing $\mathcal{D}$ and closed under shifts, finite sums, cones, and summands in $H^0(\mathcal{C})$ (called the **split triangulated closure** of $\mathcal{D}$ in $\mathcal{C}$) is naturally equivalent to the category of compact objects in $\mathcal{D}^{op}$-mod^{s,f}.*

Proposition A.13 *For any triangulated dg-category $\mathcal{C}$ with arbitrary sums and a small full subcategory $\mathcal{D} \subset \mathcal{C}$ consisting of compact objects, there exists a canonical fully faithful embedding $i: \mathcal{D}^{op}$-mod$^{s,f} \hookrightarrow \mathcal{C}$. Moreover, for any object $\mathcal{E} \in \mathcal{C}$, there exists an (essentially unique) exact triangle in $H^0(\mathcal{C})$;*

$$\mathcal{E}_- \to \mathcal{E} \to \mathcal{E}_+, \mathcal{E}_- \in \widehat{\mathcal{D}}, Hom_{H^0(\mathcal{C})}(F, \mathcal{E}_+) = 0 \ \forall F \in \widehat{\mathcal{D}} := i(\mathcal{D}^{op}\text{-mod}^{s,f}).$$

The essential image of i as above is called **ind-completion** of $\mathcal{D} \subset \mathcal{C}$.

For a dg-algebra A, compact objects in A-mod^{s,f} are called **perfect** modules and form a full subcategory denoted by $\text{Perf}(A)$. They can be characterized as dualizable objects in an appropriate sense:

Proposition A.14 *Object $\mathcal{E} \in A$-mod^{s,f} is perfect if and only if there exist $\mathcal{E}^\vee \in \text{mod}^{s,f} - A$ and two morphisms*

$$A \otimes_\mathbf{k} A^{op} \xrightarrow{\phi} \mathcal{E} \otimes_\mathbf{k} \mathcal{E}^\vee \text{ in category } A\text{-mod} - A,$$

$$\mathcal{E}^\vee \otimes_A \mathcal{E} \xrightarrow{\psi} \mathbf{k} \text{ in category } Comp_\mathbf{k}$$

such that the induced morphisms

$$\mathcal{E} \xrightarrow{1_A \otimes \cdot} A \otimes_\mathbf{k} \mathcal{E} = A \otimes_\mathbf{k} A^{op} \otimes_A \mathcal{E} \xrightarrow{\phi \otimes \cdot} \mathcal{E} \otimes_\mathbf{k} \mathcal{E}^\vee \otimes_A \mathcal{E} \xrightarrow{\cdot \otimes \psi} \mathcal{E}$$

$$\mathcal{E}^\vee \xrightarrow{\cdot \otimes 1_{A^{op}}} \mathcal{E}^\vee \otimes_\mathbf{k} A^{op} = \mathcal{E}^\vee \otimes_A A \otimes_\mathbf{k} A^{op} \xrightarrow{\cdot \otimes \phi} \mathcal{E}^\vee \otimes_A \mathcal{E} \otimes_\mathbf{k} \mathcal{E}^\vee \xrightarrow{\psi \otimes \cdot} \mathcal{E}^\vee$$

induce identity morphisms in $H^0(A\text{-mod}^{s,f})$ (resp. $H^0(A^{op}\text{-mod}^{s,f})$). Moreover, $\mathcal{E}^\vee$ is essentially unique and perfect.

We will need a notion of adjunction for dg-functors; it is similar to the above criterion for perfectness:

Definition A.15 For two dg-categories $\mathcal{C}, \mathcal{C}'$ and a pair of dg-functors $F: \mathcal{C} \to \mathcal{C}'$, $G: \mathcal{C}' \to \mathcal{C}$, the (strict) **adjunction structure** is given by two natural transformations (called adjunction morphisms)

$$F \circ G \to id, \quad id \to G \circ F$$

such that both natural transformations

$$F \to F \circ G \circ F \to F, \quad G \to G \circ F \circ G \to G$$

induce identity morphisms after applying the functor $H^\bullet$ to the enrichment.

Definition A.16 Two dg-algebras A, B are called **derived Morita-equivalent** if and only if the dg-categories $A\text{-mod}^{sf} \simeq B\text{-mod}^{sf}$ ($\iff$ `Perf(A)` $\simeq$ `Perf(B)`).

It is easy to deduce from definitions that such an equivalence is given by a bimodule $M \in B \otimes A^{op}\text{-mod}^{sf}$ such that the natural morphisms of complexes

$$A^{op} \to \underline{Hom}_{B\text{-mod}}(M, M), \quad B \to \underline{Hom}_{A^{op}\text{-mod}}(M, M)$$

are quasi-isomorphisms, and M is *perfect* separately as A^{op} and as B-module.

A5 A_∞-Language

There exists another approach to (enhanced) triangulated categories, alternative to dg-categories. This approach (via A_∞-categories) is in a sense more natural, although slightly more heavy. At the end of the day, it is just another language describing the same reality. Here we review it briefly (see also [18]).

Definition A.17 A **nonunital A_∞-algebra** over **k** is given by a $\mathbb{Z}$-graded vector space A and a coderivation Q_A of degree $+1$ of the cofree nonunital coassociative coalgebra $\oplus_{n \geq 1} A[1]^{\otimes n}$ such that $Q_A \circ Q_A = 0$. For two nonunital A_∞-algebras A_1, A_2, a **morphism** $f: A_1 \to A_2$ is a morphism of corresponding dg-coalgebras.

Unwinding the definition, we see that Q is uniquely determined by its Taylor coefficients $\mathbf{m}_k: A^{\otimes k} \to A[2-k]$, $k \geq 1$, satisfying a tower of identities. The first identity says that $\mathbf{m}_1^2 = 0$, that is, that $(A, \mathbf{m}_1)$ is a complex. The second identity says that $\mathbf{m}_2: A \otimes A \to A$ is a morphism of complexes, the third identity means that it is associative up to homotopy given by $\mathbf{m}_3$, and so on. Similarly, a morphism of A_∞-algebras gives a morphism of complexes, preserving multiplication up to homotopy. A **nonunital module** over A_∞-algebra A is a complex $M \in Comp_\mathbf{k}$ and a morphism of

A_∞-algebras $A \to \underline{Hom}_{Comp_k}(M, M)$ (the latter is a dg-algebra, hence an A_∞-algebra).

Then one fixes the problem of (non)-unitality, either just by demanding the existence of units on the cohomology level or by introducing certain involved formalism, of higher homotopy units. Then one adds many objects (cf. dg-categories with a single object vs. small dg-categories). One can define a canonical A_∞-structure on A_∞-functors between two A_∞-categories. It is quite a hassle to define the *tensor product* for A_∞-algebras and categories (there are no good canonical formulas), hence it is often simpler to stay in the dg framework.

Formulas for complexes calculating A_∞-morphisms between two A_∞-modules and for the tensor product of left and right modules coincide (in the dg case) with the resolution formulas for $\underline{RHom}$ and $\otimes^L$ from Appendix A4.

References

[1] P. Balmer, *The spectrum of prime ideals in tensor triangulated categories*, J. Reine Angewandte Math. 588 (2005) 149–68

[2] A. Blumberg and M. Mandell, *The strong Künneth theorem for topological periodic cyclic homology*, Preprint, arXiv:1706.06846

[3] A. Bondal and M. Kapranov, *Representable functors, Serre functors, and mutations*, Math. USSR Izv. 35 (1990) 519–41

[4] A. Bondal and M. van den Bergh, *Generators and representability of functors in commutative and noncommutative geometry*, Mosc. Math. J. 3 (2003) 1–36, 258

[5] C. Brav and T. Dyckerhoff, *Relative Calabi–Yau structures*, Preprint, arXiv:1606.00619

[6] A. Connes, *Noncommutative Geometry*, Academic Press (1995)

[7] A. Efimov, *Formal completion of a category along a subcategory*, Preprint, arXiv:1006.4721

[8] A. Efimov, *Generalized non-commutative degeneration conjecture*, Preprint, arXiv:1506.00311

[9] A. Efimov, *Homotopy finiteness of some DG categories from algebraic geometry*, Preprint, arXiv:1308.0135

[10] D. Gaitsgory and N. Rozenblyum, *A Study in Derived Algebraic Geometry. Volume I. Correspondences and Duality*, AMS (2017)

[11] S. Ganatra, J. Pardon, and V. Shende, *Covariantly functorial Floer theory on Lioiville sectors*, Preprint, arXiv:1706.03152

[12] E. Getzler, *Cartan homotopy formulas and the Gauss–Manin connection in cyclic homology*, Israel Math. Conf. Proc. 7 (1993) 65–78

[13] D. Kaledin, *Non-commutative Hodge-to-de Rham degeneration via the method of Deligne–Illusie*, Pure Appl. Math. Q. (2008) 785–875

[14] L. Katzarkov, M. Kontsevich, and T. Pantev, **nc** *Descent*, in preparation
[15] B. Keller, *Deriving DG categories*, Ann. Sci. École Norm. Sup. 27 (1994) 63–102
[16] M. Kontsevich, *Symplectic geometry of homological algebra*, talk on Arbeitstagung 2009, http://www.ihes.fr/~maxim
[17] M. Kontsevich, *Homological algebra of mirror symmetry*, Proc. Int. Cong. Math. 1, 2 (1994) 120–39
[18] M. Kontsevich and Y. Soibelman, *Notes on A_∞-algebras, A_∞-categories and non-commutative geometry*, in K.-G. Schlesinger, M. Kreuzer, and A. Kapustin, eds., *Homological Mirror Symmetry*, Lecture Notes in Physics 757, Springer (2009) 153–219
[19] M. Kontsevich and Y. Vlassopoulos, *Pre-Calabi–Yau algebras*, in preparation.
[20] J. Lurie, *Higher algebra*, http://www.math.harvard.edu/~lurie
[21] J. Lurie, *DAG-...*, http://www.math.harvard.edu/~lurie
[22] A. Matthew, *Kaledin's degeneration theorem and topological Hochschild homology*, Preprint, arXiv:1710.09045
[23] D. Orlov, *Triangulated categories of singularities and D-branes in Landau–Ginzburg models*, Tr. Mat. Inst. Steklova 246 (2004), 240–62; transl. Proc. Steklov Inst. Math. 246 (2004) 227–48
[24] A. Rosenberg, *The spectrum of abelian categories and reconstruction of schemes*, in *Rings, Hopf Algebras, and Brauer groups*, Lecture Notes in Pure and Applied Mathematics 197, Marcel Dekker (1998) 257–74
[25] G. Tabuada, *Noncommutative motives*, AMS University Lecture Series 61 (2015)
[26] B. Toën and M. Vaquié, *Moduli of objects in dg-categories*, Ann. Sci. École Norm. Sup. 40 (2007) 387–444
[27] B. Toën and G. Vezzosi, *Homotopical algebraic geometry. I. Topos theory*, Advances in Mathematics (2005)
[28] B. Toën and G. Vezzosi, *Homotopical algebraic geometry. II. Geometric stacks and applications*, Mem. Amer. Math. Soc. (2008)

Maxim Kontsevich
Institut des Hautes Études Scientifiques, Bures-sur-Yvette
maxim@ihes.fr